T0142278

Advances in Intelligent Systems and Computing

Volume 982

Series Editor

Janusz Kacprzyk, Systems Research Institute, Polish Academy of Sciences,
Warsaw, Poland

Advisory Editors

Nikhil R. Pal, Indian Statistical Institute, Kolkata, India
Rafael Bello Perez, Faculty of Mathematics, Physics and Computing,
Universidad Central de Las Villas, Santa Clara, Cuba
Emilio S. Corchado, University of Salamanca, Salamanca, Spain
Hani Hagras, Electronic Engineering, University of Essex, Colchester, UK
László T. Kóczy, Department of Automation, Széchenyi István University,
Gyor, Hungary
Vladik Kreinovich, Department of Computer Science, University of Texas
at El Paso, El Paso, TX, USA
Chin-Teng Lin, Department of Electrical Engineering, National Chiao
Tung University, Hsinchu, Taiwan
Jie Lu, Faculty of Engineering and Information Technology,
University of Technology Sydney, Sydney, NSW, Australia
Patricia Melin, Graduate Program of Computer Science, Tijuana Institute
of Technology, Tijuana, Mexico
Nadia Nedjah, Department of Electronics Engineering, University of Rio de Janeiro,
Rio de Janeiro, Brazil
Ngoc Thanh Nguyen, Faculty of Computer Science and Management,
Wrocław University of Technology, Wrocław, Poland
Jun Wang, Department of Mechanical and Automation Engineering,
The Chinese University of Hong Kong, Shatin, Hong Kong

The series "Advances in Intelligent Systems and Computing" contains publications on theory, applications, and design methods of Intelligent Systems and Intelligent Computing. Virtually all disciplines such as engineering, natural sciences, computer and information science, ICT, economics, business, e-commerce, environment, healthcare, life science are covered. The list of topics spans all the areas of modern intelligent systems and computing such as: computational intelligence, soft computing including neural networks, fuzzy systems, evolutionary computing and the fusion of these paradigms, social intelligence, ambient intelligence, computational neuroscience, artificial life, virtual worlds and society, cognitive science and systems, Perception and Vision, DNA and immune based systems, self-organizing and adaptive systems, e-Learning and teaching, human-centered and human-centric computing, recommender systems, intelligent control, robotics and mechatronics including human-machine teaming, knowledge-based paradigms, learning paradigms, machine ethics, intelligent data analysis, knowledge management, intelligent agents, intelligent decision making and support, intelligent network security, trust management, interactive entertainment, Web intelligence and multimedia.

The publications within "Advances in Intelligent Systems and Computing" are primarily proceedings of important conferences, symposia and congresses. They cover significant recent developments in the field, both of a foundational and applicable character. An important characteristic feature of the series is the short publication time and world-wide distribution. This permits a rapid and broad dissemination of research results.

** Indexing: The books of this series are submitted to ISI Proceedings, EI-Compendex, DBLP, SCOPUS, Google Scholar and Springerlink **

More information about this series at http://www.springer.com/series/11156

Vera Murgul · Marco Pasetti
Editors

International Scientific Conference Energy Management of Municipal Facilities and Sustainable Energy Technologies EMMFT 2018

Volume 1

 Springer

Editors
Vera Murgul
Moscow State University of Civil
Engineering
Moscow, Russia

Marco Pasetti
Department of Information Engineering
Università degli Studi di Brescia
Brescia, Italy

ISSN 2194-5357   ISSN 2194-5365 (electronic)
Advances in Intelligent Systems and Computing
ISBN 978-3-030-19755-1   ISBN 978-3-030-19756-8 (eBook)
https://doi.org/10.1007/978-3-030-19756-8

© Springer Nature Switzerland AG 2020
This work is subject to copyright. All rights are reserved by the Publisher, whether the whole or part of the material is concerned, specifically the rights of translation, reprinting, reuse of illustrations, recitation, broadcasting, reproduction on microfilms or in any other physical way, and transmission or information storage and retrieval, electronic adaptation, computer software, or by similar or dissimilar methodology now known or hereafter developed.
The use of general descriptive names, registered names, trademarks, service marks, etc. in this publication does not imply, even in the absence of a specific statement, that such names are exempt from the relevant protective laws and regulations and therefore free for general use.
The publisher, the authors and the editors are safe to assume that the advice and information in this book are believed to be true and accurate at the date of publication. Neither the publisher nor the authors or the editors give a warranty, expressed or implied, with respect to the material contained herein or for any errors or omissions that may have been made. The publisher remains neutral with regard to jurisdictional claims in published maps and institutional affiliations.

This Springer imprint is published by the registered company Springer Nature Switzerland AG
The registered company address is: Gewerbestrasse 11, 6330 Cham, Switzerland

Preface

The XX annual international scientific conference Energy Management of Municipal Facilities and Sustainable Energy Technologies EMMFT 2018 took place in Voronezh on December 10–13, 2018.

The conference was hosted by Voronezh State Technical University, Russia.

Specialists from more than 10 counties participated in the EMMFT 2018 conference. This year, the authors submitted approximately 480 qualified papers.

The objective of the conference was the exchange of the latest scientific achievements, strengthening of academic relations with leading scientists of European Union, Russia, and the World, creating favorable conditions for collaborative researches and implementing collaborative projects in the fields of energy management and development of sustainable energy technologies. Experts invited to participate in the conference presented special lectures, demonstrated equipment and devices for HVAC systems, and shared the latest technologies of thermal protection of buildings. Special attention was paid to the development of renewable energy industry. The efforts of scientists, politicians, and heads of energy enterprises were united for developing specific research programs in the field of development of renewable energy sources.

During the conference, issues on the following topics were discussed within several workshops: building physics; heating, ventilation, and HVAC & R; renewable energy; energy management; energy efficiency in transport, modeling, and control in mechanical engineering.

The conference program also included seminars, round tables, and excursions to research laboratories and research and educational centers of the Voronezh State Technical University.

All papers passed a four-staged review. The first stage consisted in an examination for compliance with the subject of the conference. At the second stage, all papers were thoroughly checked for plagiarism. Acceptable minimum of originality was 90%. The third stage involved the review by a native speaker for acceptable English language. At the same time, papers were checked by a technical proofreader. The fourth stage involved a scientific review made by at least three reviewers, using double-blind review method. If opinions of the reviewers were radically different, additional reviewers were appointed.

The members of our organizing committee express their deep gratitude to the team of "Advances in Intelligent Systems and Computing" journal and to the editorial department of Springer Nature publishing house for publication of EMMFT 2018 conference proceedings.

Organization

Organizing Committee of the Conference

Sergei Kolodyazhni	Voronezh State Technical University, Russia
Marco Pasetti	Università degli Studi di Brescia (UNIBS), Italy
Vera Murgul	Moscow State University of Civil Engineering, Russia
Igor Surovtsev	Voronezh State Technical University, Russia
Svetlana Uvarova	Voronezh State Technical University, Russia
Norbert Harmathy	Budapest University of Technology and Economics, Hungary
Vadim Kankhva	Moscow State University of Civil Engineering, Russia

Scientific Committee of the Conference

Aleksander Szkarowski	Politechnika Koszalinska, Koszalin, Poland
Antony Wood	Illinois Institute of Technology, Chicago, USA
Iurii Tabunschikov	Corr. Member of RAASN, Honorary Member of the International Ecoenergy Academy of Azerbaijan, ASHRAE Fellow Member, REHVA Fellow Member, Corr. Member of VDI, Member of ISIAQ Academy, Winner of the 2008 Nobel Peace Prize as a Member of the Intergovernmental Panel on Climate Change
Viktor Pukhkal	Saint-Petersburg State University of Architecture and Civil Engineering, Russia

Sergey Anisimov	Wroclaw University of Science and Technology, Poland
Marianna M. Brodach	Moscow Architectural Institute (State Academy), Russia
Daniel Safarik	CTBUH Journal, Chicago, USA
Samuil G. Konnikov	Ioffe Physical-Technical Institute of the Russian Academy of Sciences, Russia
Alexander Solovyev	Research Laboratory of Renewable Energy Sources—Lomonosov Moscow State University, Russian Academy of Natural Sciences, Russia
Dietmar Wiegand	Technische Universität Wien TU, Wien
Luís Bragança	Building Physics and Technology Laboratory, Guimaraes, University of Minho, Portugal
Anatolijs Borodinecs	Institute of Heat, Gas and Water technology, Riga Technical University, Latvia
Alessandro Bianchini	University of Florence (UNIFI), Italy
Aleksandr Gorshkov	Peter the Great St. Petersburg Polytechnic University, Russia
Zdenka Popovic	Faculty of Civil Engineering, University of Belgrade, Serbia
Marco Pasetti	Università degli Studi di Brescia (UNIBS), Italy
Valerii Volshanik	Moscow State University of Civil Engineering, Russia
Mirjana Vukićević	University of Belgrade, Serbia
Sang Dae Kim	Korea University, Seoul, South Korea
Manfred Esser	GET Information Technology GmbH, Grevenbroich, Germany
Alenka Fikfak	University of Ljubljana, Slovenia
Milorad Jovanovski	Ss. Cyril and Methodius University in Skopje, Macedonia
Radek Škoda	Czech Technical University in Prague, Czech Republic
Nikolai Vatin	Peter the Great St. Petersburg Polytechnic University, Russia
Paulo Cachim	University of Aveiro, Portugal
Aires Camões	University of Minho, Portugal
Michael Tendler	Royal Institute of Technology, Stockholm—Kungliga Tekniska Högskolan (KTH), Sweden
Christoph Pfeifer	University of Natural Resources and Life Sciences, Vienna, Austria
Antonio Andreini	University of Florence (UNIFI), Italy
Pietro Zunino	DIME Universitá di Genova, Genoa, Italy

Olga Kalinina	Peter the Great St. Petersburg Polytechnic University, Russia
Tomas Hanak	Brno University of Technology, Czech Republic
Vera Murgul	Moscow State University of Civil Engineering, Russia
Darya Nemova	Peter the Great St. Petersburg Polytechnic University, Russia
Norbert Harmathy	Budapest University of Technology and Economics, Hungary
Igor Ilyin	Peter the Great St. Petersburg Polytechnic University, Russia

Contents

Contents

Building Physics, Building Energy Modeling, HVAC

Computational Analysis of the Influence of PCMs on Building Performance in Summer

Manuela Neri$^{(\boxtimes)}$, Paola Ferrari, Davide Luscietti, and Mariagrazia Pilotelli

University of Brescia, Via Branze 38, 25123 Brescia, Italy
manuela.neri@unibs.it

Abstract. The insulation of buildings in summer requires to exploit the heat capacity of materials in order to delay the heat transfer through the building structure. Phase change materials (PCM) installed in buildings can reduce the indoor temperature; however, given their high cost, their use must be evaluated carefully. This paper investigates the structures that can be coupled with PCM efficaciously, and it highlights some problems that could be caused by PCMs. The investigation has been performed numerically by means of the Energy-Plus software.

Keywords: PCM · Ambient comfort · Summer cooling

1 Introduction

Buildings thermal behavior in summertime and in wintertime is very different due to dissimilar boundary conditions. In wintertime the ambient temperature is constant and the contribution of solar radiation is negligible. In these conditions, indoor comfort can be easily achieved by designing structures with low thermal transmittance. At the contrary, in summertime the ambient temperature varies strongly and the effect of solar radiation is predominant. In this latter case, the use of materials with low thermal conductivity is not enough to guarantee the indoor thermal comfort, especially in areas where solar radiation is strong. In summertime, indoor ambient thermal insulation may be achieved by means of structures with high specific heat and a great mass: the coupling of these two properties allows increasing significantly the delay between the maximum temperature on the external surface of the wall, and the maximum temperature on the inner surface. In this way, it is possible to limit the storage of heat in indoor ambient and, consequently, the power necessary to remove it by means of energy systems. In the building sector, the need is growing for innovative materials which are capable of fulfilling the most diverse functions, like sustainability [1], noise protection [2] or electromagnetic shielding [3, 4]. A particularly topical issue is the possibility to achieve all-year thermal comfort while keeping energy consumption as low as possible.

The attention to environmental problems and to the reduction of energy consumption has led to the design of the so called Nearly Zero Energy Buildings (NZEB), that are, buildings that exploit energy systems in a limited manner [5–8] and exploit the

© Springer Nature Switzerland AG 2020
V. Murgul and M. Pasetti (Eds.): EMMFT 2018, AISC 982, pp. 3–15, 2020.
https://doi.org/10.1007/978-3-030-19756-8_1

energy stored. However, to do this, it is necessary to control energy load peaks by integrating buildings with Thermal Energy Storage (TES) systems, and by limiting the effect of thermal bridges [9, 10]. Many studies investigated how to exploit energy in buildings [11, 12], while other studies focused on solar radiation [13, 14]: it was investigated how much energy can be extracted from solar radiation in a given environment.

An example of TES is Phase Change Materials (PCMs) that store energy by changing their phase. Basically, they are characterized by three stages, that are, the charging stage, the conversion stage, and the discharging stage. As regards PCMs installed in buildings, the conversion stage brings the material from the solid state to the liquid state and vice versa. In the charging stage the heat (for example from solar radiation) is accumulated and the phase change process is activated. During the conversion stage heat is stored. When the phase change process is completely, the temperature of the material increases again. When the temperature decreases, the discharging stage takes place and the material becomes solid again.

A multiplicity of PCMs are nowadays available in the market, and they are basically identified by the melting point temperature [15–17]. Each type of PCM is characterized by pros and cons: in general, a good PCM is characterized by big heat capacity, long time duration and non-toxicity [15]. PCMs can be exploited in many applications such as the control of buildings indoor temperature [18–22] (this is the case analyzed in this paper), but also as fire retardants [23, 24]. Since to the authors knowledge, fire safety at chimney roof penetration [25–27] is still an open issue, a study could investigate whether they could prevent flammable materials overheating in this particular case. An interesting study was presented in [27], where the ANOVA technique was used to assess the influence of the factors on the observed phenomenon. In particular, it was investigated how the variables related to the roof and the chimney affect the flammable materials temperature in the vicinity of the chimney. The study required a preliminary investigation to assess the variables affecting the phenomenon and, then, a series of tests were performed to collect the data necessary to the statistical investigation. The result was a series of mathematical relationships by means of which it is possible to determine the roof temperature.

Since solar radiation and its energy content varies depending on the altitude and the latitude, the choice of PCMs to be installed in buildings walls must be done carefully by taking into account the ambient condition. For example, the phase change temperature of the PCM must be the one that increases the delay between the maximum temperature on the external surface, and the maximum temperature on the inner surface of the wall. A great delay has two effects. Firstly, it allows limiting the peak of energy necessary to cool the indoor ambient. Secondly, if the discharging phase occurs at night when the ambient temperature is lower than the temperature of the wall, the heat stored in the PCM layer is dissipated in both the external and in the indoor ambient; then, a lower quantity of energy must be removed by the air conditioning system.

Since PCMs are very expensive, their use must be justified by advantages: in other words, the money saving due to the lower use of energy must be greater than the cost of the PCM layer. This paper reports part of the results presented in [28] where the influence of different PCMs on the indoor temperature was investigated. In particular, the effect of PCMs on walls of different heat capacity is analyzed. Data here presented is the temperature-time curves obtained by means of numerical simulations performed with the Energy-Plus software. By comparing the temperature obtained in different conditions, it has been possible to assess when the use of PCMs in buildings is beneficial.

2 Methods and Results

The study investigates the influence of PCMs on indoor temperature for several buildings. In particular, it has been investigated whether a PCM layer may effectively reduce the indoor temperature, and it has been evaluated the influence of PCMs phase change temperature and thickness. The study analyzes the results of numerical simulations performed with the Energy-Plus software. Given a solar radiation distribution [29], the indoor temperature over a period of time of three months is shown and analyzed.

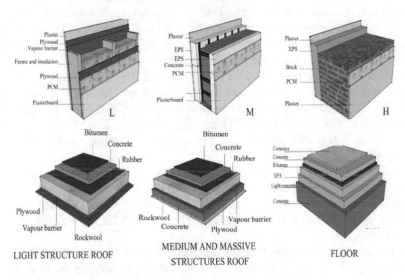

Fig. 1. Structures analyzed in the study: light (*L*), medium density (*M*) and massive (*H*).

A building situated in Brescia (Italy) has been considered: it is rectangular and its dimensions are 8 m × 10 m × 3 m with the long side facing south. The building has not windows. Thermal resistance and heat capacity of the walls has been varied: the three structures shown in Fig. 1 have been considered and they are classified as light density structure (L), medium density structure (M), and massive-heavy density structure (H). For each structure three different thermal transmittance values (U) have been chosen. As reported in Table 1, the total thickness of the walls has been maintained constant, but the thermal conductivity (λ) of the insulating layers has been varied.

The thermal transmittance U of a wall made of n layers is defined as

$$U = \frac{1}{R} = \frac{1}{\left[\frac{1}{he} + \frac{1}{hi} + \sum_{i=1}^{n}\left(\frac{si}{\lambda i}\right)\right]}$$

where R is the thermal resistance of the wall, s is the thickness of the layer, λ is the thermal conductivity, hi and he are the heat transfer coefficients (that take into account both convection and thermal radiation) on the internal and external surfaces. According to [30] for the Brescia district he and hi coefficients are 25 W/m^2 K and 7.7 W/m^2 K respectively.

In the following, each case is identified by an acronym where the first letter indicates the structure, and the second letter indicates the thermal transmittance: structures identified with letter A are characterized by U = 0.21 W/m^2 K, structures B are characterized by U = 0.32 W/m^2 K, structures C are characterized by U = 0.56 W/m^2 K. For example, structures LA, LB and LC are the three cases analyzed for the light structure. According to Fig. 1 and Table 2, it is supposed to install the PCM layer only on the vertical walls, while roofs and floor do not contain PCMs. One floor and two roofs have been considered: one roof is for the light structure (L), and one roof is for the medium (M) and the massive (H) structures.

Meteorological data from the 21st of June to the 21st of September for Brescia (Italy) district collected in [29] have been used, that are, the average ambient temperature T = 33 °C, the maximum daily temperature variation ΔT = 10.9 °C, the direct solar radiation Edr = 7.4 MJ/m^2, the diffuse solar radiation Edf = 17 MJ/m^2, the global solar radiation Egl = 24.4 MJ/m^2, and the incident radiation on horizontal plane Ih = 282 W/m^2.

In the analysis, four BioPCM [31] materials have been considered. Their entalpy-temperature curves are reported in Table 3. The Q21 material is characterized by a phase change temperature equal to 21 °C and by a latent heat of 55 kJ/kg, while these two values are 25 °C and 202 kJ/kg for Q25, 27 °C and 135 kJ/kg for Q27, 29 °C and 261 kJ/kg for Q29.

By comparing the temperature distribution obtained for the different materials reported in Figs. 3, 4 and 5, it has been possible to evaluate whether and how the phase change temperature affects the indoor temperature. To investigate the influence of the thickness of the PCM layer, numerical simulations have been performed on the light structure (L) and the thickness of the PCM layer has been set equal to 1.1 cm as in the simulation just described, and 2.1 cm respectively (indicated with * in the following), and temperatures are shown in Fig. 6. In Figs. 3, 4, 5 and 6, the area in yellow represents the indoor comfort zone, that is, the range of temperature (21 °C–25 °C) that does not require the use of cooling/heating systems.

Table 1. Properties of the layers for the different structures: s is the thickness of the layer, λ is the thermal conductivity, ρ the density and c the specific heat. R and U are the thermal resistance and the thermal transmittance of the wall respectively.

Light					LB		LC	
LA								
Layer	s [cm]	λ [W/mK]	ρ [Kg/m³]	c [J/kgK]	s [cm]	λ [W/mK]	s [cm]	λ [W/mK]
Plaster	1,5	0,65	1100	840	1,5	0,65	1,50	0,65
Plywood	1,3	0,41	460	1880	1,3	0,41	1,30	0,41
Vapour barrier	0,5	0,23	-	-	0,5	0,23	0,50	0,23
Frame and insulation	15,2	0,034	10	1470	15,2	0,06	15,2	0,11
Plywood	1,3	0,09	460	1880	1,3	0,09	1,30	0,09
BioPCM	1,1	2,3	235	1970	1,1	2,30	1,10	2,30
Plasterboard	1,3	3,3	640	1150	1,3	3,30	1,30	3,30
	Rp=					Rp=		Rp=
	R [m²K/W]=	4,87				R [m²K/W]= 3,13		R [m²K/W]= 1,77
	U [W/m²K]=	0,21				U [W/m²K]= 0,32		U [W/m²K]= 0,57
Medium								
MA					MB		MC	
Layer	s [cm]	λ [W/mK]	ρ [Kg/m³]	c [J/kgK]	s [cm]	λ [W/mK]	s [cm]	λ [W/mK]
Plaster	1	0,72	1860	840	1	0,72	1	0,72
EPS	9	0,034	25	1400	9,00	0,06	9,00	0,11
Concrete	20	2	2400	1000	20	2	20	2
EPS	6	0,034	25	1400	6	0,055	6	0,111
BioPCM	1,1	0,2	235	1970	1,1	0,2	1,1	0,2
Plasterboard	1,3	0,16	640	1150	1,3	0,16	1,3	0,16
	Rp=					Rp=		Rp=
	R [m²K/W]=	4,83				R [m²K/W]= 3,13		R [m²K/W]= 1,77
	U [W/m²K]=	0,21				U [W/m²K]= 0,32		U [W/m²K]= 0,57
Massive - Heavy								
HA					HB		HC	
Layer	s [cm]	λ [W/mK]	ρ [Kg/m³]	c [J/kgK]	s [cm]	λ [W/mK]	s [cm]	λ [W/mK]
Plaster	1,5	0,72	1860	840	1,5	0,72	1,5	0,72
XPS	12	0,03	35	1400	12	0,0536798	12	0,134
Brick	50	0,73	1910	840	50	0,73	50	0,73
BioPCM	1,1	0,2	235	1970	1,1	0,2	1,1	0,2
Plaster	1	0,72	1860	840	1	0,72	1	0,72
	Rp=					Rp=		Rp=
	R [m²K/W]=	4,94				R [m²K/W]= 3,18		R [m²K/W]= 1,84
	U [W/m²K]=	0,20				U [W/m²K]= 0,31		U [W/m²K]= 0,54

Table 2. Properties of the roof and the floor: s indicate the thickness, λ is the thermal conductivity, ρ the density and c the specific heat of the layer.

Floor

Layers	s [cm]	λ [W/mK]	ρ [Kg/m^3]	c [J/kgK]
Ceramics	1,5	1,2	2000	850
Concrete	6	0,42	1200	840
Bitumen	10	0,23	1100	1000
XPS	8	0,034	35	1400
Light concrete	1,3	0,09	460	1880
Concrete	25	2,3	2300	1000

Roof (light structure)

Layers	s [cm]	λ [W/mK]	ρ [Kg/m^3]	c [J/kgK]
Bitumen	2	0,43	1600	1000
Concrete	4	0,41	1200	840
Rubber	0,2	0,17	1500	1470
Roockwool	14	0,036	90	1030
Vapour barrier	1	-	-	-
Plywood	1,9	0,1	450	1880

Roof (medium and massive structures)

Layers	s [cm]	λ [W/mK]	ρ [Kg/m^3]	c [J/kgK]
Bitumen	2	0,43	1600	1000
Concrete	4	0,41	1200	840
Rubber	0,2	0,17	1500	1470
Roockwool	14	0,036	90	1030
Vapour barrier	1	-	-	-
Concrete	10	2,3	2300	1000
Plaster	1	0,72	1860	840

Table 3. Enthalpy-temperature curves for the PCMs considered in the analysis.

Q21		Q23		Q25		Q27	
T [°C]	h [J/kg]	T [°C]	h [J/kg]	T [°C]	h [J/kg]	T [°C]	h [J/kg]
0	12	0	12	0	8	0	5
10	25058	10	23058	10	19290	10	16458
15	34799	15	32580	15	27420	15	23562
20	38970	20	41280	20	26990	20	32561
21	55119	21,5	81820	23	42867	25	43078
21,5	80820	22	128509	24	56221	26	57014
22,5	128509	22,5	201879	24,5	83245	26,5	84146
23	201879	24	236860	25	133649	27	134578
24	225581	25	245462	25,5	201879	27,5	202864
25	231773	27	249194	26	236860	28	237015
30	233328	30	254503	28	247994	30	251278
35	246859	35	258813	35	257761	35	258320
45	254741	45	267178	45	266724	45	267324
100	289545	100	300420	100	322285	100	322093

3   Discussion

In this section, the temperature time-curves reported in Figs. 3, 4, 5 and 6 are discussed. The influence on the indoor temperature of the phase change temperature and of the thickness of the PCM layer is discussed.

Figures 3, 4 and 5 show the temperature distribution for different building structures and different wall thermal transmittance U. The yellow area represents the comfort zone: when indoor temperature is over this area, the support of an air conditioning system is necessary to guarantee the indoor thermal comfort.

Temperatures for the light structure L are shown in Fig. 3. Figure 3(a) shows that for the LA structure, the one with lower thermal transmittance, the indoor temperature exceeds the limit of 25 °C for about two weeks when the ambient temperature exceeds 30 °C. Figure 3(b) shows that for the LB structure the limit temperature is exceeded for a longer period than in the previous case (LA) because the limit temperature is exceeded for 21 days. Figure 3(c) shows that for the LC structure, the indoor temperature exceeds the limit temperature for a longer period and the values reaches 30 °C; in this case, the use of the air conditioning system is essential. The greater delay between the maximum external temperature and the maximum indoor temperature occurs with the $Q25$ for the LA structure (7.18 h instead of 6.52 h without PCM), with the $Q23$ for the LB structure (4.58 h instead of 4.03 h without PCM), and with $Q27$ for the LC structure (4.31 h instead of 3.16 without the PCM). However, the delay is in the order of one hour in all the three cases. Then, the delay between the maximum temperature on the internal and the external surfaces is function of both the thermal transmittance of the wall, and the phase change temperature of the PCM layer. As expected, the higher the walls thermal transmittance, the higher the indoor temperature but the phase change temperature does not affect the indoor temperature significantly. In almost all the cases, the same indoor temperature is detected with and without the PCM. Unexpectedly, in some cases the use of PCM is pejorative, for example when the wall thermal transmittance is high: this is due to the fact that the PCM layer delays the entering of energy but, at the same time, also the exiting of energy at night with a consequent storage of energy and increase in the indoor temperature. In this case, the use of PCM is not recommended because of the high cost of the material and the increase in the indoor temperature.

Temperatures for the medium structure M are shown in Fig. 4. In Fig. 4(a), the indoor temperature is always lower than 25 °C. The difference in temperature for the different PCMs is not appreciable. Then, the melting point seems not to be influential. In Fig. 4(b), the indoor temperature exceeds the limit temperature of 25 °C for 9 days but it is always lower than 26 °C. Then, the activation of an air conditioning system entails a limited effect on the peak load. In Fig. 4(c), the limit temperature is exceeded for a longer period and the maximum temperature reaches 27 °C for 9 days. However, the presence of the PCM layer does not entail a significant variation in temperature (Fig. 2).

The maximum estimated delay temperature is 11.08 h for structure MA with the Q25 (instead of 10.72 h without PCM), 11.4 for MB obtained with Q27 (instead of 9.80 h without PCM), and 11.59 h with $Q27$ (instead of 10.63 without PCM). The delay in temperature is comparable with that estimated for case L. It emerges that the

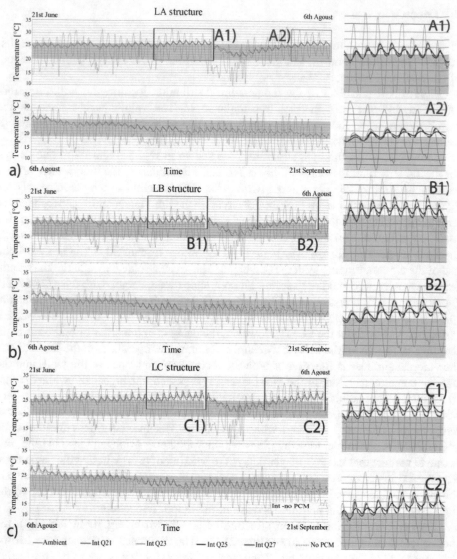

Fig. 2. Temperature-time distribution for the light structure: (a) U = 0.21 W/m²K, (b) U = 0.32 W/m²K, (c) U = 0.56 W/m²K.

most insulating the walls, the lower the indoor temperature but, given that the indoor temperature is the same with and without the PCM layer, the effect of this latter is negligible.

Temperatures for the heavy/massive structure *H* are shown in Fig. 5. As regards the *HA* and the *HB* structures in Fig. 5(a) and (b), it emerges that the limit temperature of 25 °C is never exceeded. For the *HC* structure, that is the one with higher thermal transmittance, the limit comfort temperature is exceeded for 17 days.

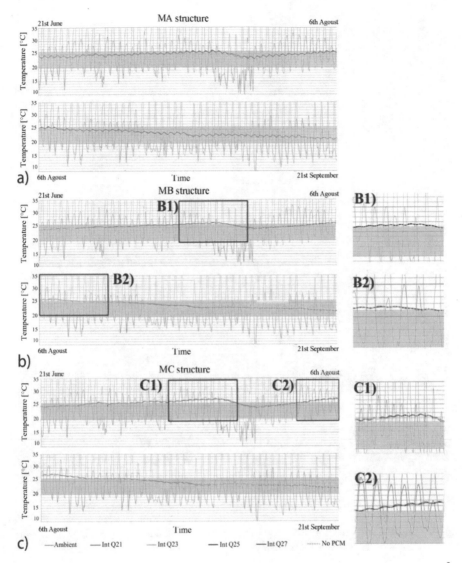

Fig. 3. Temperature distribution for the medium structure: (a) U = 0.21 W/m²K, (b) U = 0.32 W/m²K, (c) U = 0.56 W/m²K.

However, the indoor temperature is always lower or equal to 26 °C and, consequently, the load on the cooling system is limited. Since the estimated indoor temperature does not vary with and without the PCM layer, it can be stated that the presence of this latter is not influential. Since in this case the indoor temperature is more flatter, it can be stated that to limit the entering of energy in buildings it is necessary to design structures with a high thermal capacity, and delaying the peak in temperature on the internal surface is not enough.

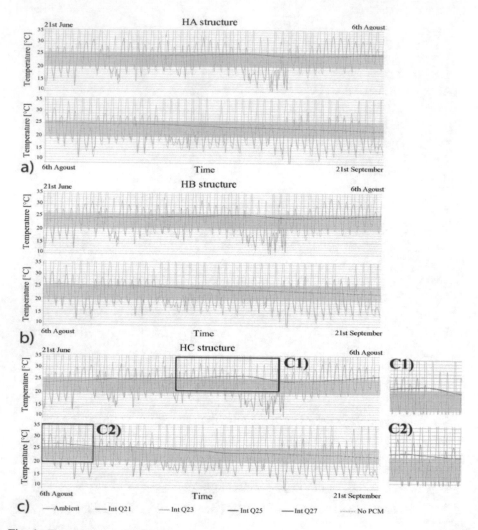

Fig. 4. Temperature distribution for the massive structure: (a) U = 0.21 W/m²K, (b) U = 0.32 W/m²K, (c) U = 0.56 W/m²K.

From Fig. 6(a) it emerges that for the *LA* structure lower indoor temperature has been estimated for high phase change temperature and thicker PCM layer (*M51-Q25*). The same occurs for the structures *M* and *H* in Figs. 6(b) and (c), but in these latter two cases the PCM layer is less influential because comparable temperatures have been obtained and they exceed the limit temperature of 25 °C for a long period. Therefore, the choice of the right thickness of the PCM layer is essential to guarantee thermal comfort in buildings. It can be stated that there is a correlation between the PCM layer thickness, the phase change temperature, and the structure thermal transmittance: more

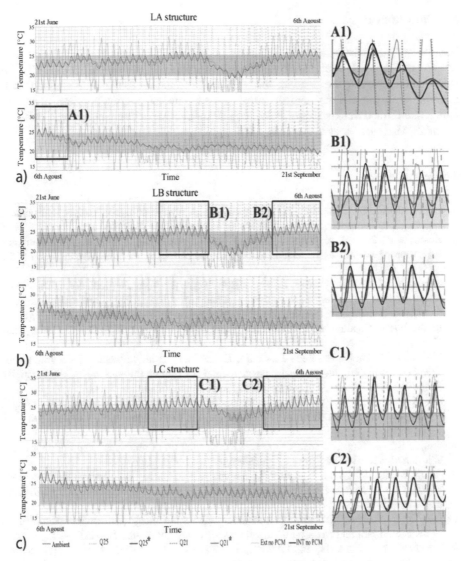

Fig. 5. Effect of the PCM thickness for the light structure: (a) U = 0.21 W/m²K, (b) U = 0.32 W/m²K, (c) U = 0.56 W/m²K.

cases could be analyzed statistically in order to determine the influence of each variable as was done in [33]. Then, to limit the indoor temperature the choice of the PCM layer must be done carefully. In particular the phase change temperature and the thickness of the layer must be evaluated case by case depending on the structure and the climate zone.

4   Conclusion

The study has investigated the influence of PCMs in structures of different heat capacity and thermal resistance. It has been shown that, in some cases, the use of PCMs is detrimental because it can reduce the heat flux from indoor to the ambient that occurs at night when the external temperature is lower than the indoor ambient. It is also shown that limited thicknesses of PCMs can be ineffective or even pejorative. Then, the use of PCMs must be evaluated carefully.

References

1. Asdrubali, F., D'Alessandro, F., Schiavoni, S.: A review of unconventional sustainable building insulation materials. Sustain. Mater. Technol. **4**, 1–17 (2015)
2. Scamoni, F., Piana, E.A., Scrosati, C.: Experimental evaluation of the sound absorption and insulation of an innovative coating through different testing methods. Build. Acoust. **24**, 173–191 (2017)
3. Donini, A., Spezie, R., Cortina, R., Piana, E.A., Turri, R.: Accurate prediction of the corona noise produced by overhead transmission lines. In: AEIT 2016 - International Annual Conference: Sustainable Development in the Mediterranean Area, Energy and ICT Networks of the Future. Institute of Electrical and Electronics Engineers Inc., Capri, Naples, Italy (2017)
4. Kolcunová, I., Pavlík, M., Beňa, L., Čonka, Z., Ilenin, S., et al.: Influence of electromagnetic shield on the high frequency electromagnetic field penetration through the building material. Acta Phys. Pol., A **131**, 1135–1137 (2017)
5. Svenfelt, A., Engström, R., Svane, O.: Decreasing energy use in buildings by 50% by 2050 —a backcasting study using stakeholder groups. Technol. Forecast. Soc. Chang. **78**, 785–796 (2011)
6. Sait, H.H.: Auditing and analysis of energy consumption of an educational building in hot and humid area. Energy Conversation Manage. **66**, 143–152 (2013)
7. Singh, M.K., Mahapatra, S., Teller, J.: An analysis on energy efficiency initiatives in the building stock of Liege Belgium. Energy Pol. **62**, 729–741 (2013). https://doi.org/10.1016/j.enpol.2013.07.138
8. Dall'O', G., Galante, A., Torri, M.: A methodology for the energy performance classification of residential building stock on an urban scale. Energy Build. **48**, 211–219 (2012)
9. Benedetti, M., Gervasio, P., Luscietti, D., Pilotelli, M., Lezzi, A.M.: Point thermal transmittance of rib intersections in concrete sandwich wall panels. Heat Transf. Eng. 1–10 (2018)
10. Luscietti, D., Gervasio, P., Lezzi, A.M.: Computation of linear transmittance of thermal bridges in precast concrete sandwich panels. J. Phys. Conf. Ser. **547**, 012014 (2014)
11. Dedé A., Della Giustina D., Massa G., Pasetti M., Rinaldi S.: A smart PV module with integrated electrical storage for smart grid applications. In: IEEE International Symposium on Power Electronics, Electrical Drives, Automation and Motion (SPEEDAM), pp. 895–900 (2016)
12. Marchi, B., Zanoni, S., Pasetti, M.: A techno-economic analysis of Li-ion battery energy storage systems in support of PV distributed generation. In: 21st Summer School Francesco Turco of Industrial Systems Engineering, pp. 45–149 (2016)

13. Neri, M., Luscietti, D., Pilotelli, M.: Computing the exergy of solar radiation from real radiation data. J. Energy Resour. Technol. **139**, 061201 (2017)
14. Pons, M.: Exergy analysis of solar collectors, from incident radiation to dissipation. Renew. Energy **47**, 194–202 (2012)
15. Elarga, H., Fantucci, S., Serra, V., Zecchin, R., Benini, E.: Experimental and numerical analyses on thermal performance of different typologies of PCMs integrated in the roof space, Energy Build. 150, 546–557 (2017)
16. Khudhair, A.M., Farid, M.M.: A review on energy conservation in building applications with thermal storage by latent heat using phase change materials. Energy Convers. Manage. **45**, 263–275 (2004)
17. Abhat, A.: Low temperature latent heat thermal energy storage. Heat storage materials. Sol. Energy **30**, 313–323 (1983)
18. Tyagi, V.V., Buddhi, D.: PCM thermal storage in buildings: a state of art. Renew. Sustain. Energy Rev. **11**, 1146–1166 (2007)
19. Farid, M.: A review on energy storage with phase changes. In: Proceedings of Chicago/Midwest Renewable Energy Workshop, Chicago, USA (2001)
20. Neeper, D.A.: Thermal dynamics of wallboard with latent heat storage. Sol. Energy **68**, 393–403 (2000)
21. Waqas, A., Din, Z.U.: Phase change material (PCM) storage for free cooling of buildings - a review. Renew. Sustain. Energy Rev. **18**, 607–625 (2013)
22. Stritih, U., Butala, V.: Experimental investigation of energy saving in buildings with PCM cold storage. Int. J. Refrig. **33**, 1676–1683 (2010)
23. Serrano, S., Barreneche, C., Navarro, A., Haurie, L., Fernandez, A.I., Cabeza, L.F.: Use of multi-layered PCM gypsums to improve fire response. Phys. Therm. Mech. Charact. Energy Build. **127**, 1–9 (2016)
24. Kontogeorgos, D.A., Semitelos, G.K., Mandilaras, I.D., Founti, M.A.: Experimental investigation of the fire resistance of multi-layer drywall systems incorporating Vacuum Insulation Panels and Phase Change Materials. Fire Saf. J. **81**, 8 16 (2016)
25. Neri, M., Pilotelli, M.: Data on temperature-time curves measured at chimney-roof penetration. Data Brief **20**, 306–315 (2018)
26. Neri, M., Luscietti, D., Bani, S., Fiorentino, A., Pilotelli, M.: Analysis of the temperatures measured in very thick and insulating roofs in the vicinity of a chimney. J. Phys: Conf. Ser. **655**, 012019 (2015). conference 1
27. Neri, M., Luscietti, D., Fiorentino, A., Pilotelli, M.: Statistical approach to estimate the temperature in chimney roof penetration. Fire Technol. **54**, 395–417 (2018)
28. Ferrari, P.: Influence of phase change materials (PCM) on thermal behavior of residential buildings, University of Brescia, Master Thesis (2014). (in Italian)
29. Fraunhofer ISE. https://www.enargus.de. Accessed 10 Dec 2018
30. UNI 6946: Building components and building elements - Thermal resistance and thermal transmittance - Calculation method (2007)
31. http://phasechange.com.au/. Accessed 20 Jan 2019

Precooling in Desiccant Cooling Systems with Application of Different Indirect Evaporative Coolers

Anna Pacak$^{(\boxtimes)}$ ⓘ, Demis Pandelidis ⓘ, and Sergey Anisimov ⓘ

Faculty of Environmental Engineering,
Department of Air Conditioning, Heating, Gas Engineering and Air Protection,
Wrocław University of Science and Technology,
Norwida st. 4/6, 50-373 Wrocław, Poland
anna.pacak@pwr.edu.pl

Abstract. In this study, four arrangements of the desiccant system with different Maisotsenko Cycle (M-Cycle) heat and mass exchangers (counter-flow and cross flow heat and mass exchangers) were selected for a comparative study. The proposed system is able to obtain high thermal COPs (up to 4.9) due to effective pre-cooling of the airflow with a highly effective evaporative heat and mass exchanger. The performance of the systems was analyzed numerically with ε–NTU models developed by the authors for moderate climatic conditions. To compare selected systems, different performance indicators were considered. These are the thermal COP (Coefficient of performance), the ERR (Energy Efficiency Ratio), and the humidity ratio decreases. The results show that the desiccant system with two counter-flow heat and mass exchangers (marked as System A) achieves highest humidity ratio decrease, as well as the highest thermal COP. In terms of the EER, factor for system with counter-flow and cross flow heat and mass exchanger (marked as System B) achieves highest effectiveness. There is a slight difference between COP values obtained by System A and System B. That is why it was concluded that System B is a solution which needs more analysis to be done to maximize this desiccant system potential.

Keywords: Solid desiccant wheel · Air conditioning · Moderate climate

1 Introduction

According to data presented in report published by International Energy Agency, a space cooling is the fastest growing use of electricity in buildings [1]. It is due to the economic and population growth and intensive urbanization with higher and higher indoor air-conditioning requirements. There are two main possibilities of electricity savings: increasing air-conditioning system efficiency or implementation an alternative environmentally friendly cooling systems [1]. The desiccant cooling systems utilize the heat energy to produce cooling energy. The solar collectors or thc hcat pump may be the source of heat, that is why they are environmentally friendly (typical vapor compression systems utilize great amount of electricity to produce the cool).

© Springer Nature Switzerland AG 2020
V. Murgul and M. Pasetti (Eds.): EMMFT 2018, AISC 982, pp. 16–25, 2020.
https://doi.org/10.1007/978-3-030-19756-8_2

Recent studies showed that there is a significant interest in desiccant systems. What is more, the researchers are looking for the most profitable solution to increase the efficiency of those system. For that reason, many complicated solutions have been proposed (such as two-stage, solar assisted heating system, multiple heat recovery exchangers, cooling towers etc. [2, 3]). There are some researchers who found out that precooling the desiccant wheel or the ambient air improves the desiccant system performance. Chen et al. [4] investigated the desiccant air conditioning system with the evaporator before the desiccant wheel, in order to provide the initial dehumidification and pre-cooling of the ambient air. The results show that the analyzed system outperforms the silica-gel based system by 130%. Moreover, it consumes less power than traditional dehumidification systems. Zhou et al. [5] studied the design, analyzed the performance of a novel tube-shell internally water-cooled desiccant wheel. The experiment was performed to validate the mathematical model. The results show that isothermal performance of the desiccant wheel can be achieved. Moreover, the improvement of dehumidification process is significant equaling 48% as compared with a conventional adiabatic desiccant wheel using a super-absorbent polymer as the desiccant material. Chen et al. [6] studied the precooling desiccant wheel in air conditioning system experimentally. They proposed the renewable low-grade heat source to perform the precooling of desiccant wheel. It allows to reduce the energy consumption of the system. Gadalla et al. [2] proposed three different types of two-stage desiccant air conditioning systems with Maisotsenko-Cycle units. To improve the two stage system performance, the precooling is considered. They examined five months of system operation for hot and humid climates. The obtained average COP of the system is 1.77 (between 12:00–14:00).

As shown above, the interest in improving desiccant systems performance by precooling the desiccant wheel or the ambient air before it enters the desiccant wheel is increasing. It follows from the fact that it significantly improves the system cooling capacity.

In this study, a novel system which is a simple way of improving the M-Cycle desiccant system performance is proposed. The new system uses precooling of the air stream before the dehumidification process. In this system, the ambient air is cooled two times, both before the dehumidification process and after. It allows lower supply air temperature and humidity ratios to be obtained. Moreover, the thermal COP of a proposed system (defined as obtained cooling capacity divided by required thermal energy) is significantly higher, reaching the level of 4.9 for specific operation conditions. The proposed system consists of a combination of two indirect evaporative air coolers combined with a desiccant wheel. The first M-Cycle unit is located before the desiccant wheel, to initially pre-cool the outdoor air, before it enters the dehumidification section. The second M-Cycle unit is located after the desiccant wheel post cooling the process air stream. The objective of this study is to establish which M-Cycle HMXs are the most profitable to use in a novel desiccant system. For that reason, the following most popular HMX devices are considered. These are: a counter-flow M-Cycle unit and a cross-flow M-Cycle unit. There are several important aspects of the HMX units' construction and performance which may affect the whole system effectiveness. That is why, systems analyzed in this study consists of different combinations of these devices.

It should be emphasized that adding the HMX units before and after the desiccant wheel does not significantly increase the cost of the system (M-Cycle air coolers are usually made out of plastic [7]). What is more, proper optimizing of the 1st HMX allows to decrease the additional pressure drop. The objective of the current study was to choose the most favorable application of desiccant system on the base on simulations results. Different key performance indicators considered will include the COP, EER, and humidity ratio decrease.

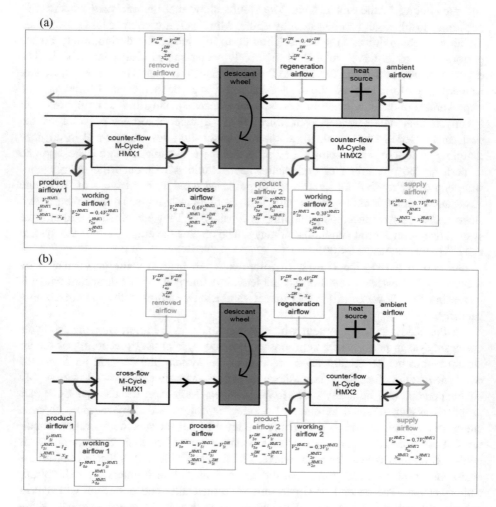

Fig. 1. Selected multi-stage systems with different evaporative heat and mass exchangers. (a) System A. (b) System B. (c) System C. (d) System D.

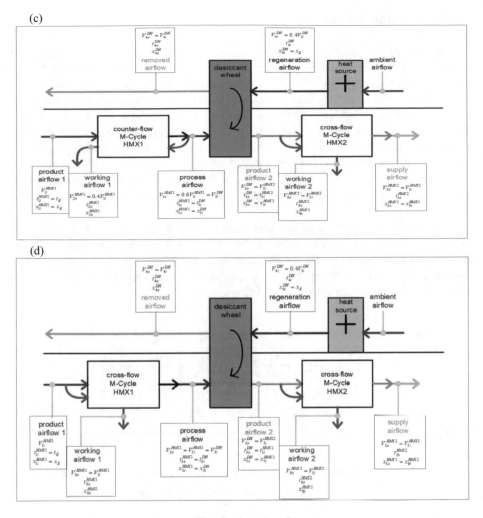

Fig. 1. (*continued*)

2   Methods

Both M-Cycle indirect evaporative air coolers and the desiccant wheel are simulated using an original ε-NTU heat and mass transfer model [8, 9]. These models are created with an assumption that the airflow is treated as gaseous fluid which has constant temperature, velocity and mass transfer potential. These parameters are equal to bulk average values in sections normal to the exchangers plate. All the basic assumptions for the method used in this study, including heat transfer characteristics, heat and mass transfer coefficients used for modeling and discretization process, were presented by authors [9–11]. The proposed models are implemented in a multi-module computer simulation program. A four-dimensional computational numerical code based on the modified Runge-Kutta method was implemented using the Wolfram Mathematica environment. Validation of the mathematical models has been presented in [8].

3 Results and Discussion

In this section, results of the simulation are presented and discussed. The proposed systems are shown in Fig. 1(b)–(e). Ambient air passes through each device before it enters the conditioning zone. Simulations were carried out for selected values of out-door and regeneration airflow inlet parameters, which were chosen as representative for moderate climates (ambient air temperature in range between 28 °C and 32 °C, ambient humidity ratio range between 10 g/kg and 14 g/kg). Due to the fact that low regeneration temperatures can be easily obtained by solar power, waste heat, district heating energy or heat pumps in moderate climates, low regeneration air temperatures were selected (treg = 40 °C, 50 °C, 60 °C). It is assumed that all systems examined use the same desiccant wheel, for which the regeneration to processed airflow ratio equals 0.4.

The specific rotor dehumidifier parameters are presented in Table 1. The four selected multi-stage systems consist of two M-Cycle heat exchangers in different arrangements. In this study, the counter-flow heat and mass exchanger and the cross-flow heat and mass exchanger are considered. The data for the heat and mass exchangers these data are presented in Table 2.

Table 1. The reference operating conditions for the analyzed desiccant wheel.

System	NTU_3/NTU_4	NTU_{3d}^*/NTU_{4d}^*	R	A_3/A_4	V_3/V_4	Ω
A	9.26/23.27	149.37/149.37	0.2	1	2.5	8
B	9.26/23.27	149.37/149.37	0.2	1	2.5	8
C	6.62/16.58	149.37/149.37	0.2	1	2.5	8
D	6.62/16.58	149.37/149.37	0.2	1	2.5	8

Table 2. The reference operating conditions for the analyzed heat exchangers.

		$NTU_{1/5}$	$NTU_{2/6}$	Working to primary air ratio
System A	HMX1	5.5	13.8	0.4
	HMX2	6.5	21.7	0.3
System B	HMX1	5.5	5.5	1.0
	HMX2	6.5	21.7	0.3
System C	HMX1	5.5	13.8	0.4
	HMX2	6.5	6.5	1.0
System D	HMX1	5.5	5.5	1.0
	HMX2	6.5	6.5	1.0

3.1 Humidity Ratio Decrease

In this section, the humidity ratio decrease obtained by selected systems is presented. In Fig. 2 there are graphs for humidity ratio decrease for different ambient air humidity ratio and for different regeneration air temperature. Higher inlet ambient humidity ratio results in less intensive evaporation process in the wet channels of first regenerative heat exchanger (HMX1). This results in higher air temperature delivered to the desiccant wheel, hence the dehumidification process occurs at higher temperatures which is less effective. That is why the secondary (second stage) HMX cools air with lower effectiveness. By increasing the regeneration air temperature to 60 °C, the air dehumidification increases by over a factor of two (see Fig. 2). At these conditions, System A performs most effective. This can be explained as follows: System A and System C are equipped with the counter-flow HMX before the desiccant wheel. This exchanger, as compared to the cross-flow HMX, is characterized by a high temperature effectiveness. Therefore, the airflow delivered to the desiccant wheel in Systems A and C have the same inlet temperature, which is lower than in case of cross-flow HMX in Systems B and D. When the dehumidification process occurs under lower process airflow temperatures, the core of desiccant wheel on the process side is characterized by lower temperatures. This results in lower partial water vapor pressure in the desiccant rotor on the process side of the heat exchanger, and, consequently, giving a higher difference of partial water vapor pressure between process air and matrix surface is observed. Therefore, the mass transfer from the process airflow to the desiccant matrix is characterized by higher effectiveness.

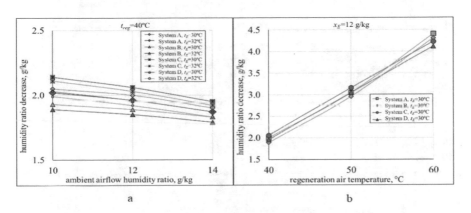

Fig. 2. Numerical simulation results of humidity ratio decrease obtained in selected systems (a) For different ambient air humidity ratio. (b) For different regeneration air temperature.

3.2 Coefficient of Performance

To compare systems, the coefficient of cooling performance (COP) is analyzed in this section. The COP is defined as a ratio, between the obtained system cooling capacity to energy required for regeneration:

$$\text{COP} = \frac{\Phi}{P} \qquad (1)$$

where: $\Phi = V_{1o}^{HMX2} \rho \left(i_E - i_{1o}^{HMX2} \right)$: obtained cooling capacity, kW

$P = V_4 \rho \, c_p (t_{reg} - t_E)$: required energy for regeneration purposes, kW

Figure 3 shows the results of COP calculations, obtained from the numerical simulations. It can be seen that Systems A and B are characterized by visibly higher COP values compared to Systems C and D (Fig. 3). This is due to the fact that Systems A and B are able to produce more cooling capacity than Systems C and D, with less amount of regeneration energy. This is caused by the specific of each HMX construction and airflow distribution through the channels. When the cross-flow HMX is used as a secondary heat and mass exchanger, the whole airflow (working and product) flows through the desiccant wheel. That is why a higher regeneration airflow is needed. In Systems A and B, a lower amount of the air is delivered to the desiccant wheel than in Systems C and D. That is why there is a significant difference in amount of regeneration airflow. The higher regeneration airflow, the lower the value of COP. Figure 3(a) shows that the higher inlet ambient airflow temperatures results in higher COPs because less energy is needed to heat the air from higher temperatures to the required regeneration temperatures.

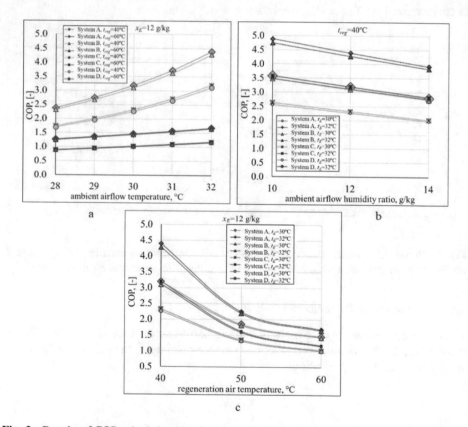

Fig. 3. Results of COP calculation based on numerical simulations. (a) For different ambient air temperature (b) For different ambient air humidity ratio. (c) For different regeneration air temperature.

It should be noted that the COPs for System A range between 1.27 and 4.39 for different ambient airflow temperatures (i.e., see Fig. 3(a)). Increasing the inlet ambient airflow humidity results in decreasing COP value (i.e., see Fig. 3(b)). The highest COP obtained by System A equals 4.88 and it is 1.40 more than that of System D for the same corresponding conditions (i.e., Fig. 3(b)). It has to be mentioned that COP values for each system are still higher than 1.0 (more cooling capacity is obtained than energy for regeneration is utilized). For System A, the COPs range between 2.8 and 4.88 for different ambient airflow humidity ratios (see Fig. 3(b)). The trends presented in Fig. 3 (c) show that increasing the regeneration airflow temperature results in a noticeable reduction in COP.

3.3   Energy Efficiency Ratio

The systems analyzed are equipped with two heat and mass exchangers and one desiccant wheel. To compare the electricity demand to cover the pressure losses, the EER indicator is calculated and presented. Where:

$$EER = \frac{\Phi}{N} \tag{2}$$

and $\Phi = V_{1o}^{HMX2} \rho \left(i_E - i_{1o}^{HMX2} \right)$ – System's cooling capacity, kW $N = \frac{\Delta p V_{1i}^{HMX1}}{\eta}$ – required fan power to pass the air through the system, kW In this study a fan efficiency of $\eta = 0.7$ was used kW.

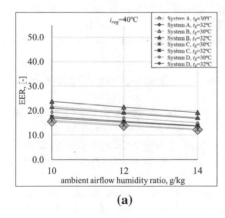
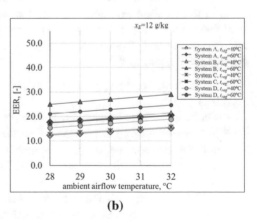

(a)   (b)

Fig. 4. Results of EER calculation based on numerical simulations (a) For different ambient air humidity ratio. (b) For different ambient air temperature.

The EER values depend on the HMX type used in each system. Each device has different pressure drop and flow distribution. For example, System B obtains the highest EER values for different ambient humidity ratios (see Fig. 4 (a)). On the other hand, System A represents the lowest EER. It is caused by the fact that a cross flow

HMX channels (dry and wet channel) are parallel to each other. In the regenerative HMX, the entire airflow needs to flow through the dry channel before it enters the wet channel. On the other hand, when cross-flow HMX is used as a secondary HMX, the whole airflow which has to be delivered to the dry and wet channels needs to flow through the desiccant wheel it causes a higher pressure drop in System D than in System B where regenerative HMX is used as a second step cooling. Nevertheless, the pressure drop of System B is the lowest while the COP and a values are also high. The discrepancies between COP value of System A and C are small. It can be concluded that from this analysis, System B is the most promising solution but without a doubt, more research needs to be done to maximize this M-Cycle desiccant system potential

4 Conclusions

The presented paper studies four different arrangements of new multi-stage desiccant system for moderate climates. The aim of this paper is to establish which system should be chosen for future research performed in moderate climate. Selected systems are a combination of Maisotsenko Cycle regenerative and cross-flow heat and mass exchangers cooperating with a desiccant wheel. The analysis is performed on the base of ε–NTU models validated against experimental data. In this study it was assumed that desiccant wheel is regenerated under low airflow temperatures which can be gained from solar collectors or a heat pump. The results of the analysis indicate that multi-stage desiccant system with cross-flow HMX and counter-flow HMX (System B) represents the one of highest thermal COP and the ability to meet the total cooling loads in the conditioned zones are the most favorable for this system. The calculated EER factor for System B is the greatest, hence it consumes least electricity. For specific climatic conditions, System A with two counter-flow HMX-es obtains the highest thermal COP equal to 4.9.

A – Rotor desiccant section area, m^2, cp – specific heat capacity of moist air, J/(kgK)
i – enthalpy, J/kg, N - required fan power to pass the air through the system, kW
P – required energy for regeneration purposes, kW, t – temperature, °C, V – volumetric airflow rate, m^3/h, x – humidity ratio, g/kg, Φ – cooling capacity, kW, Ω – rotation speed, 1/h, ρ – air density, kg/m^3, Δp – pressure drop, Pa, COP – coefficient of performance,

EER – energy efficiency ratio, NTU – Number of transfer units, NTU*d – number of heat transfer units: desiccant material, 1 – primary airflow in counter flow heat and mass exchanger, 2 – working airflow in counter flow heat and mass exchanger, 3 – product airflow in desiccant wheel, 4 – regeneration airflow in desiccant wheel, 5 – primary airflow in cross-flow heat and mass exchanger, 6 – working airflow in cross-flow heat and mass exchanger, E – ambient air parameters, i – inlet, o – outlet, DH – desiccant wheel, HMX – heat and mass exchanger, reg – referenced to regeneration airflow.

Acknowledgments. One of the co-authors, Demis Pandelidis, received financial support for his research from resources for scientific work for years 2016-2019 from Polish Ministry of Science and High Education (program "Iuventus Plus"), project number IP2015 058274.

References

1. International Energy Agency: The future of cooling. Report (2018)
2. Gadalla, M., Saghafifar, M.: Performance assessment and transient optimization of air precooling in multi-stage solid desiccant air conditioning systems. Energy Convers. Manag. **119**, 187–202 (2016)
3. Jani, D.B., Mishra, M., Sahoo, P.K.: Solid desiccant air conditioning – a state of the art review. Renew. Sustain. Energy Rev. **60**, 1451–1469 (2016)
4. Chen, C.-H., Hsu, C.-Y., Chen, C.-C., Chiang, Y.-C., Chen, S.-L.: Silica gel/polymer composite desiccant wheel combined with heat pump for air-conditioning systems. Energy **94**, 87–99 (2016)
5. Zhou, X., Goldsworthy, M., Sproul, A.: Performance investigation of an internally cooled desiccant wheel. Appl. Energy **224**, 382–397 (2018)
6. Chen, L., Chen, S.H., Liu, L., Zhang, B.: Experimental investigation of precooling desiccant-wheel air-conditioning system in a high-temperature and high-humidity environment. Int. J. Refrig **95**, 83–92 (2018)
7. Pandelidis, D.: Modelowanie procesów wymiany ciepła i masy w wymienniku z M-obiegiem pracującym w urządzeniach klimatyzacyjnych/Praca Doktorska/PhD Dissertation, Wrocław University of Technology (2015)
8. Anisimov, S., Pandelidis, D.: Theoretical study of the basic cycles for indirect evaporative air cooling. Int. J. Heat Mass Transf. **84**, 974–989 (2015)
9. Pandelidis, D., Anisimov, S., Worek, W.M., Drąg, P.: Comparison of desiccant air conditioning systems with different indirect evaporative air coolers. Energy Convers. Manag. **117**, 375–392 (2016)
10. Anisimov, S., Pandelidis, D.: Numerical study of the Maisotsenko cycle heat and mass exchanger. Int. J. Heat Mass Transf. **75**, 75–96 (2014)
11. Pandelidis, D., Anisimov, S.: Numerical study and optimization of the cross-flow Maisotsenko cycle indirect evaporative air cooler. Int. J. Heat Mass Transf. **103**, 1029–1041 (2016)

Rigidity of the Aluminum Window Profiles with Thermal Barrier

Aleksandr Konstantinov$^{(\boxtimes)}$, Egor Leontev, and Anastasia Remizova

Moscow State University of Civil Engineering,
Yaroslavskoe shosse 26, 129337 Moscow, Russia
apkonst@yandex.ru

Abstract. Mullions are the main elements of windows that provide their resistance to wind load (lack of blowing under the wind pressure influence) and, as a result, affect the heat lost through windows because of air infiltration (or exfiltration). Therefore, when designing windows, it is necessary to pay special attention to issues of the assignment of stiffness indicators for their mullions. The work is devoted to the research of the moments of inertia of the window mullions made of aluminum profiles with thermal barrier used in the glazing of civil buildings. Generally, these profiles should be considered as a composite in terms of their static operation. Under the action of the load in the aluminum profiles with a thermal barrier, the shift of their individual structural layers relative to each other is possible. Laboratory tests of aluminum profiles with thermal barrier were conducted. Samples of profile elements had a length of 3 m and were fixed at the edges of the hinge supports. During the tests, they were sequentially loaded by a concentrated load force acting on the center of the span of the elements. Comparison of the data obtained during laboratory tests with the results of theoretical calculations showed that at the deflections of profile elements that do not exceed the normalized indicators (no more than 1/200 of the span), the shift of their layers can be ignored. In this case, the calculation of the moment of inertia of the aluminum profiles with thermal barrier can be performed without the thermal barrier rigidity.

Keywords: Windows · Window unit · Aluminum profile ·
Aluminum windows · Aluminum profiles with thermal barrier

1 Introduction

Currently, modern types of translucent structures made of aluminum are widely used in the glazing of both unique objects and buildings by an individual design of civilian buildings, and objects of mass building.

Windows must fulfill a number of functions and requirements during their exploitation [1]. At the same time, one of the most important functions is providing of thermal protection of buildings [2–6]. Therefore, for the production of windows today, combined aluminum profiles with thermal barrier are used [7]. Combined aluminum profiles consist of internal and external elements of hollow aluminum profiles, which are interconnected by a thermal barrier, as a rule, of polymeric materials with low thermal conductivity. The applying of thermal barrier can significantly improve the

© Springer Nature Switzerland AG 2020
V. Murgul and M. Pasetti (Eds.): EMMFT 2018, AISC 982, pp. 26–34, 2020.
https://doi.org/10.1007/978-3-030-19756-8_3

thermal characteristics of the combined profiles as compared to hollow aluminum profiles. At the same time, more budget solutions of profiles (with a simple thermal barrier) are used at the mass building, and their more energy-efficient analogs find the application on unique and atypical projects (see Fig. 1).

A   B

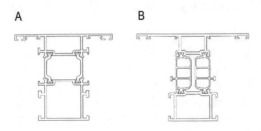

Fig. 1. Examples of combined aluminum profiles with thermal break. A – cross-section of a typical aluminum profiles with thermal break. B – cross-section of aluminum profiles with thermal break with high thermal performance

The windows should provide the perception of all the loads and actions that affect them during operation [8]. One of the key loads, whose action must be taken into account in the design and calculation of windows, is the wind load, because it affects their exploitation safety (for example, destruction/deformation of core elements and insulating glass units when exposed to wind pressure), and thermal protection and microclimate of premises (blowing off under the wind pressure influence and heat losses due to infiltration (or exfiltration) of air through windows) [9–11]. According to [8], the calculation of window and stained-glass structures for the action of wind load consists in selecting the necessary rigidity of their power profile elements (first of all, mullions). In this case, it is considered that the blowing of the structure will occur only when the limiting deflections of its mullions are exceeded.

As part of this work, researches were carried out on combined aluminum profiles with thermal barrier used for the installation of windows, primarily in residential buildings of mass development. A distinctive feature of the investigated type of the profile system is the use of profiles having a developed cross section from the street side. The widespread use of this type of profile systems in the glazing of mass building objects is caused by:

– their versatility and cost-effectiveness;
– the ability of installation on their basis of large-format translucent structures that are not externally different from ordinary window units (combination "sash + mullion + sash");
– the ability to perform the installation of structures included in the zone of external wall insulation without the use of additional installation systems, which can significantly improve the energy efficiency of translucent structures [12–14].

Currently, there is a method of calculating the actual moments of inertia of aluminum profiles with a thermal barrier in DIN EN 14024:2005-01. However, it requires additional initial data, which are often not available, as it requires additional laboratory tests of aluminum profiles. Therefore, the aim of this work was to develop a simplified method for calculating the actual moments of inertia of aluminum profile elements with a thermal barrier.

2   Methods of Laboratory Testing of Profile Elements

During the research, we studied 5 types of aluminum profiles used for the manufacture of windows, including 4 types of combined profiles with thermal barrier and one type of hollow aluminum profile. All aluminum profiles were manufactured in the same factory, using a single aluminum alloy and thermal barrier material (polyamide). Test combined aluminum profiles consisted of three parts. At the same time, the interior aluminum part and the middle part (thermal break 24 mm thick) were the same for the tested profiles, and the outer aluminum parts differed in height. Cross sections of the test profiles are presented in Fig. 2.

The length of the tested samples was 3.03 m.

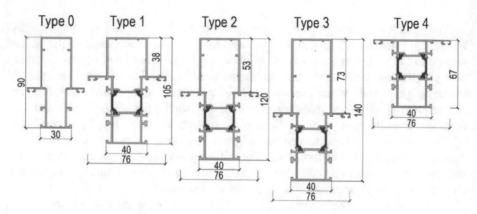

Fig. 2. Cross-sections of the test profiles

To conduct research, a special test bench was made. The test bench consisted of a metal power frame made of channels, on which hinged supports for the support of the test profile were installed on the lower side and a mechanical jack for the application of a load from the upper side. The distance between the supports was 3.0 m. The appearance of the test bench is shown in Fig. 3. The appearance of the hinged supports for the installation of profiles is shown in Fig. 4(a).

The load was measured using a universal electronic dynamometer and strain gauge sensor with the highest measurement limit of 2.0 kN and their accuracy of 0.0002 kN. The appearance of the strain gauge sensor with an installed node of load application is shown in Fig. 4(b). The profile deflection was measured using an electronic linear displacement sensor with a measurement range of 0-30 mm and a measurement accuracy of 0.01 mm.

To obtain reliable results, tests for each type of profiles were carried out on three identical samples.

Fig. 3. The general view of the test bench

(a)   (b)

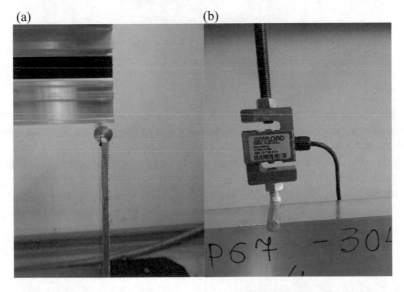

Fig. 4. (a) The appearance of the hinge supports the installation of profiles (b) Appearance of the strain gauge sensor with an installed node of load application

Laboratory researches of profiles were conducted in two stages. At the first stage, tests were carried out on the hollow aluminum profile with a known moment of inertia of its cross section. These tests were carried out to determine the actual modulus of elasticity of the aluminum alloy from which the profiles were made. The actual modulus of elasticity of the aluminum alloy was determined by formula 1. In this case, the beam on two hinged supports, loaded at the center of the span by a concentrated load and its own weight, was taken as the design scheme of the profiles (see Fig. 5). The concentrated load was applied to the sample under test in steps, with a step of 0.1 kN. The initial load was taken 0.3 kN (corresponds to the resultant force from the wind load of 0.1 kN/m for an mullion length of 3.0 m and load strip 1.0 m wide). The test of each sample was terminated after reaching a deflection of 1.5 cm (1/200 of the span). The test for large deflections was not performed due to the fact that the calculation of the profile elements of windows for deflections of more than 1/200 of the span is not done usually.

$$E_{Al} = \frac{1}{f_{exp}I_p}\left(\frac{PL^3}{48} + \frac{5qL^4}{384}\right),$$

(1)

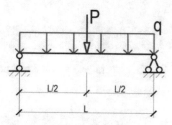

Fig. 5. The design scheme of the tested profiles

where P – the concentrated load on the test profile, kN;
q – uniformly distributed load of the own weight of the profile under test, kN/ mm;
L – the calculated span of the beam, mm^2;
Ip – the moment of inertia of the cross-section of the hollow profile, mm^4. The data obtained during the calculation of the geometric characteristics of the cross section of the basic tools of Autodesk AutoCAD based on the drawings of profiles;
f_{exp} – the actual deflection of the element obtained during laboratory tests, mm.

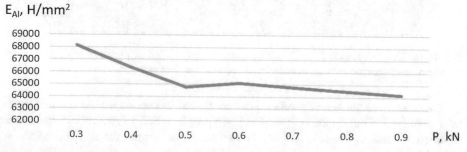

Fig. 6. Calculated values of the modulus of elasticity of aluminum alloy, obtained during laboratory tests.

The average account obtained during testing of hollow aluminum profiles (Type 0) was taken as the calculated elastic modulus of aluminum alloy (see Fig. 6). According to the test results, the actual modulus of elasticity of the aluminum alloy profiles was taken to be $E_{Al} = 64950$ H/mm^2.

At the second stage of laboratory studies, tests of combined aluminum profiles were carried out. These tests were conducted to determine the actual deflections of the combined aluminum profiles under the action of a concentrated load. The test technique was similar to the first stage of research. The results of laboratory tests of aluminum profiles are presented in Table 2.

3 Determination of the Theoretical Value of the Moment of Inertia of the Cross-Section of the Combined Profiles

Combined aluminum profiles from the point of view of their static work, in general, should be considered as a composite. Under the action of the load in the combined aluminum profiles, it is possible to shift their individual structural layers relative to each other. With minor deflections of the profile elements, the shift force of the aluminum layers relative to each other will be significantly lower than the normalized shift force (400 kg for samples with a length of 100 mm). Therefore, we first assume that with the deflection of the combined aluminum profiles up to 1/200 of the span, they can be calculated without taking into account the shift of the layers. In this case, the actual cross section of the combined aluminum profile can be replaced by a conventional cross section of aluminum alloy (the reduced section). In this reduced section, the geometrical characteristics of the thermal inset are introduced with the given characteristics, which are changed in proportion to the elastic module of the material of the thermal break and the aluminum alloy. As the material of the thermal break of the combined profiles, polyamide was used, which, according to the manufacturer, has a modulus of elasticity equal to Et = 2500 H/mm^2 (with an equilibrium moisture content in the polyamide). Thus, taking into account the previously determined actual moment of inertia of aluminum, the conversion coefficient will be $n = \frac{E_t}{E_{Al}} = \frac{2500}{65715} = 0,038$.

The calculated moment of inertia of the reduced section of the combined profiles can be calculated according to the formula (2) and the scheme in Fig. 7.

$$I_{pr} = I_1 + nI_2 + I_3, \qquad (2)$$

where I_1 – the moment of inertia of the external aluminum profile about the neutral axis, mm^4; I_2 – the moment of inertia of the profile of the thermal break relative to the neutral axis, mm^4; I_3 – the moment of inertia of the internal aluminum profile relative to the neutral axis, mm^4. The moments of inertia of each individual layer can be calculated according to the formula (3).

$$I_i = I_i^* + A_i y_i^2, \qquad (3)$$

where I_i^* – the moment of inertia of the i-th layer relative to its own center of gravity, mm^4; A_i – cross-sectional area of the i-th layer, mm^2; y_i – the distance from the center of gravity of the i-th layer to the neutral axis of the reduced section of the combined profile, mm.

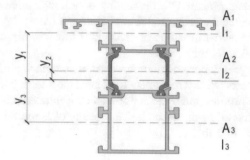

Fig. 7. Scheme for the determination of the calculated moments of inertia of the combined profiles

The results of the calculation of the theoretical values of the moments of inertia of the subjects of the combined profiles are presented in Table 1.

Table 1. Estimated moments of inertia of the tested combined profiles

Type of combined profile	Estimated moment of inertia of the combined profile, cm^4
Type 1	69.90
Type 2	98.15
Type 3	144.08
Type 4	30.09

4   Results. Data Analysis of Experimental and Theoretical Studies of Combined Profiles

The results of the comparison of experimental and theoretical data are presented in Table 2. At the same time, the theoretical accounts of the deflections of the tested samples were additionally calculated taking into account the effect of the concentrated load and the own weight of the profiles. Calculations of theoretical deflections were made according to formula 4.

$$f_{teor} = \frac{PL^3}{48E_{Al}I_{pr}} + \frac{5qL^4}{384E_{Al}I_{pr}} \tag{4}$$

The relative deviation between the data of the laboratory tests and experimental researches was within 9.5%, which indicates the validity of the assumptions made in theoretical researches. Calculations show that the rigidity of the thermal inset does not have a significant effect on the total moment of inertia of the combined section (for mullion profiles, the difference is no more than 1%).

Table 2. Comparison of the results of experimental and theoretical studies of combined profiles

P, kN	Type 1			Type 2			Type 3			Type 4		
	f_{exp}, mm	f_{teor}, mm	Δ, %	f_{exp}, mm	f_{teor}, mm	Δ, %	f_{exp}, mm	f_{teor}, mm	Δ, %	f_{exp}, mm	f_{teor}, mm	Δ, %
0.3	4.09	4.16	1.7	2.92	2.99	– 2.2	2.11	2.00	5.4	8.76	9.47	–8.1
0.4	5.4	5.40	0	3.82	3.87	– 1.3	2.72	2.58	5.1	11.78	12.35	–4.8
0.5	6.77	6.64	2.0	4.76	4.75	0.2	3.4	3.17	6.9	14.79	15.22	–2.9
0.6	8.07	7.88	2.4	5.7	5.63	1.2	4.08	3.75	8.1			
0.7	9.42	9.11	3.2	6.59	6.51	1.1	4.75	4.34	8.7			
0.8	10.77	10.35	3.9	7.55	7.40	2.0	5.39	4.92	8.7			
0.9	12.08	11.59	4.0	8.41	8.28	1.5	5.97	5.51	7.8			
1.1	14.69	14.07	4.2	10.26	10.04	2.1	7.35	6.68	9.2			
1.2	16.05	15.31	4.6	11.24	10.93	2.8	7.96	7.26	8.8			
1.3				12.11	11.81	2.5	8.64	7.84	9.2			
1.4				13.09	12.69	3.0	9.2	8.43	8.4			
1.5				13.94	13.57	2.6	9.88	9.01	8.8			
1.6				14.88	14.46	2.8	10.48	9.60	8.4			
1.7				15.78	15.34	2.8	11.15	10.218	8.7			
1.8							11.83	10.77	9.0			
1.9							12.55	11.35	9.5			
2.0							13.17	11.94	9.3			

Notice

Δ – relative deviation - the ratio of the difference between the data of experimental studies and theoretical calculations, to the data of experimental studies

5 Discussion

In the existing construction practice, in most cases, the calculation of the moments of inertia of the aluminum profiles with thermal break is made as for a solid section (that is, the material of the thermal break is conventionally replaced by aluminum). This approach greatly simplifies the process of designing translucent structures, however, it does not correspond to the actual scheme of operation of the profile elements of translucent structures during bending and cannot be applied to energy-efficient aluminum profile systems with a wide thermal break (see Fig. 1B), where the difference in the calculated moments of inertia reaches 10% or more. Therefore, it is more appropriate to use the approach proposed in this paper, in which the calculation of the combined profiles should be carried out without taking into account the rigidity of the thermal inset (to the reserve of rigidity).

To improve the exploitational reliability of modern types of translucent structures, it is also advisable to take into account in the calculations the possibility of deviations of the actual geometric dimensions of the profiles from their nominal accounts. For the researched profiles, according to the provisions of the current normative-technical documents, the thickness of the walls and shelves of the profiles may differ from the nominal account to ±0.25 mm. Calculations show that the moments of inertia of profiles are significantly reduced, especially for combined profiles with a developed cross-section (up to 10%). This question requires further research.

6 Conclusions

Researches show that when the combined aluminum profiles deflect no more than 1/200 of the span, the shift of their components can be ignored. Also, due to the significant difference between the elastic module of aluminum and polyamide (more than 20 times) when carrying out engineering calculations of windows on deflections on the effect of wind load, the rigidity of the thermal inset can be disregarded. Those, the calculated account of the moment of inertia of the cross-section of the combined aluminum profile can only be determined by taking into account the outer and inner aluminum layer.

References

1. Boriskina, I., Plotnikov, A., Zaharov, A.: Design of modern window systems for civil buildings (2008)
2. Grynning, S., Gustavsen, A., Time, B., Jelle, B.P.: Energy Build. **61**, 185–192 (2013)
3. Persson, M.L., Roos, A., Wall, M.: Energy Build. **38**(3), 181–188 (2006)
4. Pacheco, R., Ordónez, J., Martínez, G.: Renew. Sustain. Energy Rev. **16**(6), 3559–3573 (2012)
5. Verkhovsky, A.A., Zimin, A.N., Potapov, S.S.: Hous. Constr. **6**, 16–19 (2015)
6. Savin, V.K., Savin, N.V.: Hous. Constr. **10**, 47–50 (2015)
7. Boriskina, I.: Buildings and structures with translucent facades and roofs. Theoretical Bases of Designing of Glass Constructions (2012)
8. DIN EN 14351-1:2016-12 Windows and doors – Product standard, performance characteristics – Part 1: Windows and external pedestrian doorsets; German version EN 14351-1:2006 + A2:2016 (2016)
9. Konstantinov, A.P.: Sci. Prospects **1**(100), 26–31 (2018)
10. Konstantinov, A., Ratnayake, L.: E3S Web of Conferences, vol. 33, p. 02025 (2018)
11. Konstantinov, A., Motina, M.: IOP Conf. Ser. Mater. Sci. Eng. **463**, 032044 (2018)
12. Cappelletti, F., Andrea Gasparella, A., Romagnoni, P., Baggio, P.: Energy Build. **43**(6), 1435–1442 (2011)
13. Misiopecki, C., Bouquin, M., Gustavsen, A., Jelle, B.P.: Energy Build. **158**, 1079–1086 (2018)
14. Sierra, F., Gething, B., Bai, J., Maksoud, T.: Energy Build. **142**, 23–30 (2017)

Confirmed Method for Definition of Daylight Climate for Tropical Hanoi

Thi Khanh Phuong Nguyen[1,2]($\boxtimes$) [ID], Aleksei Solovyov[1] [ID],
Thi Hai Ha Pham[2] [ID], and Kim Hanh Dong[3] [ID]

[1] Moscow State University of Civil Engineering,
Yaroslavskoe Shosse 26, 129337 Moscow, Russia
[2] Faculty of Architecture and Planning,
National University of Civil Engineering (NUCE),
Giai Phong Road 55, Hanoi, Vietnam
phuongntk@nuce.edu.vn
[3] Thuyloi University, Tay Son Street 175, Dong Da District, Hanoi, Vietnam

Abstract. Daylight climate studies of the regions is the important process for the basics of theory and design practice of daylight prediction and assessment. As a result, they will give suggested codes, standards and guidelines to achieve the building energy efficiency. An important indicator of the daylight climate is cumulative diffuse horizontal illuminance, since this factor reflects the potential of the daylight climate in the regions and is the stabilized light source. The horizontal illuminance has a closer connection with the solar radiation and are determined by the luminous efficiency under the open sky from solar radiation data. In addition, horizontal illumination levels also heavily dependent on cloudiness statistics of the regions. The purpose of this study is confirmation of the model calculation of luminous efficacy to determine the daylight climate for Hanoi, Vietnam. To achieve this goal, a comparative analysis of luminous efficacy models based on calculated and measured solar radiation data and horizontal illuminance was conducted. Careful consideration of daylight climate data allows us to clarify the daylight-climatic factors and move to the territorial standards of daylight at the modern level. Based on the study, a general review of other climatic factors, such as statistics of solar radiation and cloud cover, was considered. The results of this study have shown that daylight climate in Hanoi characterized by a high level of horizontal illuminance, which causes a high thermal load into buildings. This problem should be developed in future studies of daylight assessment to achieve energy efficacy.

Keywords: Luminous efficacy · Daylight climate · Daylight assessment ·
Horizontal illuminance · Solar irradiance · Cloudiness statistic

1 Introduction

Vietnam is a tropical country located in Eastern South part of Asia with a population of 95 million people. Along with economic and social development, the construction industry has also constantly developed and opened up a period of rapid urbanization. According to a report from the World Bank in April 2012, Vietnam is the country with

© Springer Nature Switzerland AG 2020
V. Murgul and M. Pasetti (Eds.): EMMFT 2018, AISC 982, pp. 35–47, 2020.
https://doi.org/10.1007/978-3-030-19756-8_4

the highest level of urbanization in Southeast Asia with 813 urban networks. According to the Vietnam National Technical Regulation on Energy Efficient Buildings "Guiding the application on national regulation", by 2035, up to 50% of Vietnamese's population lives in urban areas. Afterwards, office buildings must be built at high speeds. At the same time, attention to reduce the environmental pollution, comfortable indoor climate, the impact on health, productivity of workers and energy efficiency should be considered. The energy used for air conditioning (cooling) and lighting is the highest proportion, respectively 34% and 18% for Hanoi (Fig. 1). Standards and practice codes of daylight prediction and assessment are in a state of renewal, since the use of an overcast sky in daylight calculations is not typical for the tropical conditions. It causes the problem of over-glazing buildings with excessive solar gain into the rooms.

This study analyzes the daylight climate in order to obtain the theory based on the definition of a real tropical type of sky for Hanoi instead of the overcast sky. Toward this task, the goal of this study is to validate the method for determining the daylight climate. Accordingly, the approach involves:

- Analysis of the characteristics of the daylight climate on the two most important indicators: solar radiation and cloudiness statistics.
- Comparison of average horizontal illuminance and luminous efficacy of modelled values and the experimental values. Validation of a suitable model for calculating the luminous efficacy for determining the daylight climate.

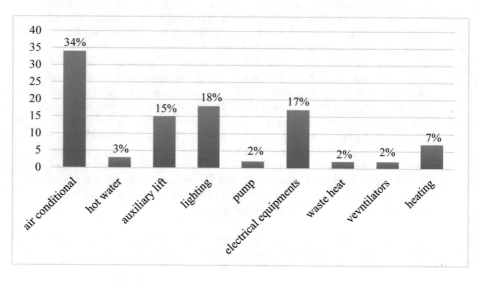

Fig. 1. Energy use distribution in office buildings.

2   Methodology

2.1   Characteristics of Tropical Climate of – An Overview

One of the most important factors influencing the daylight climatic conditions is the values of solar radiation. The climate of Vietnam characterized by a high level of diffuse and global irradiance. A study of solar radiation for Hanoi has been described in the works of Ha [1–3]. A set of data on solar radiation was collected over ten years from 1996 to 2005 and was calculated to obtain the distribution of diffuse and direct radiation on different facades of the building. For a typical period of summer days, the calculation was carried out for the average solar radiation of August 21, as shown in Table 1 and Fig. 2.

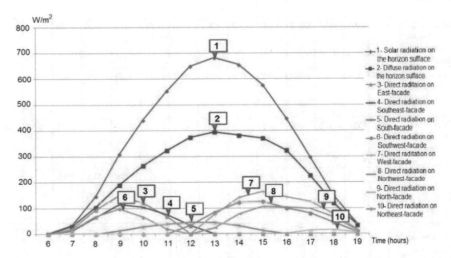

Fig. 2. Graph of direct, global and diffuse irradiance curves on August 21 [1–3] 1 - Direct solar radiation on the horizon surface; 2 - Diffuse radiation on the horizon surface; 3, 4, 5, 6, 7, 8, 9 and 10 - Direct radiation on facades different orientation, respectively E, S-E, S, S-W, W, N-W, N and N-E.

The relative values of diffuse to global solar radiation on facades of different orientations ID.ver/IG.ver from May to September were calculated and shown in Table 2, since this period characterize for tropical sky condition of Vietnam.

The lowest relative ratio $I_{D.ver}/I_{G.ver}$ on the western façade with values from 53.1% to 67.98% means a high distribution of direct solar radiation in this orientation. In the summer period, this ratio on the southern façade accepted almost 100% (in May, June and July), which means absolute diffuse solar radiation in this orientation. The values of direct radiation on west, north-west, south-west are 1.46 to 2.87 times higher than the values on the east, northeast, southeast facades. This characteristic of the distribution of solar radiation should be developed to obtain the diffuse orientation factors on daylight assessment.

Table 1. Solar radiation in August 21st [1–3].

Solar radiation on Surface/Façade (W/m^2)	Time (hours)													
	6	7	8	9	10	11	12	13	14	15	16	17	18	19
Direct solar radiation on horizon surface $I_{sun.hor}$	1.4	33.9	146.8	310.0	441.9	554.5	649.9	682.3	655.8	577.3	448.1	298.1	148.4	38.0
Diffuse radiation on horizon surface I_{Dhor}	1.4	29.7	104.1	191.0	264.7	324.9	375.1	395.1	382.7	371.2	324.9	226.5	118.4	34.6
Direct solar radiation on East facade $I_{sun.ver}$	0.0	19.9	94.1	150.1	121.2	67.3	0.0	0.0	0.0	0.0	0.0	0.0	0.0	0.0
Direct solar radiation on Southeast facade $I_{sun.ver}$	0.0	12.6	65.2	114.8	105.3	75.2	32.5	0.0	0.0	0.0	0.0	0.0	0.0	0.0
Direct solar radiation on South facade $I_{sun.ver}$	0.0	0.0	0.0	12.2	27.7	39.0	46.0	43.0	32.1	13.3	0.0	0.0	0.0	0.0
Direct solar radiation on Southwest facade $I_{sun.ver}$	0.0	0.0	0.0	0.0	0.0	0.0	32.5	83.0	122.1	125.3	99.3	79.5	43.8	7.8
Direct solar radiation on West facade $I_{sun.ver}$	0.0	0.0	0.0	0.0	0.0	0.0	0.0	74.3	140.5	163.8	143.2	125.9	77.1	15.8
Direct solar radiation on Northwest facade $I_{sun.ver}$	0.0	0.0	0.0	0.0	0.0	0.0	0.0	22.1	76.6	106.4	103.2	98.5	65.2	14.5
Direct solar radiation on North facade $I_{sun.ver}$	0.0	2.1	1.8	0.0	0.0	0.0	0.0	0.0	0.0	0.0	2.7	13.4	15.1	4.7
Direct solar radiation on Northeast façade $I_{sun.ver}$	0.0	15.6	67.8	97.5	66.1	20.1	0.0	0.0	0.0	0.0	0.0	0.0	0.0	0.0

Table 2. Relative values of diffuse to global solar radiation ID.ver/IG.ver (%).

Orientations	Eastern	Western	Southern	Northern	East-Southern	East-Northern	West-Southern	West-Northern
May	73.47	53.1	99.88	82.00	81.69	77.67	66.56	57.22
June	73.81	55.56	100.00	76.28	83.76	76.02	70.91	57.86
July	70.62	55.92	99.87	82.49	79.56	75.06	69.14	59.90
August	77.64	67.98	88.04	97.53	79.49	85.48	72.60	76.37
September	61.54	42.31	50.91	99.88	55.19	81.47	41.12	60.49

Other factors affecting the diffusion of horizontal illuminance are the cloud types and cloud amount, which are represented as cloudiness statistics. The study of Dang et al. about the climate in Vietnam [2–4], in accordance with the Vietnam Building Code Natural Physical & Climatic Data for Construction shows that cloudiness statistics have great influence on solar radiation, respectively, it affects the horizontal illuminance levels. The sky of Hanoi is characterized by intermediate sky conditions as a gap between sunny and cloudy sky conditions (Table 3).

Table 3. Cloudiness statistics covered and hours of sunlight exposure.

	January	February	March	April	May	June	July	August	September	October	November	December	Annual average
(1)	8.2	9.1	9.2	8.7	7.7	8.2	8.0	7.9	6.8	6.4	6.5	6.7	7.8
(2)	74	47	47	90	183	172	195	174	176	167	137	124	1585
	Typical with overcast and intermediate sky condition			Typical with sunny and intermediate sky condition									

(1): Cloudiness statistics.
(2): Hours of Sunlight exposure.

The high-altitude types of clouds Cirrus (Ci) and Stratus (St), Cumulus (Cu) from 4 to 12 km and middle-altitude types of clouds Cirrostratus (Cs), Stratocumulus (Sc) are marked for the sunny days. While the nasty days are characterized with the low-altitude types of clouds Cumulus (Cu) and Cumulonimbus (Cb), which are formed below 2 km. Until now, in Vietnam, measured data on external horizontal illuminance are not available, an approach is needed to estimate horizontal diffuse and global illuminance based on more widely available data on solar radiation using luminous equivalent, which is regulated by direct proportionality [4].

$$K = 697.33 \frac{\int_{\lambda_1=380}^{\lambda_2=780} \phi_e(\lambda)V(\lambda)d\lambda}{\int_{\lambda=0}^{\lambda=\infty} \phi_e(\lambda)d\lambda}, \frac{\text{Klux}}{\text{W} \cdot \text{m}^{-2}} \tag{1}$$

Where:

K - The luminous equivalent of radiation or the luminous efficacy;
697.33 - The maximum sensitivity for photopic vision, which occurs at 555 nm;
$\phi_e(\lambda)$ - Flux density at wavelength λ(nm);
$V(\lambda)$ - The value of the photopic spectral luminous efficiency function for that wavelength;

In fact, the luminous efficacy depends on various climatic factors:

- Sun position
- Transparency of the atmosphere and the atmospheric diffusion
- Cloudy statistic
- Ground reflection coefficient
- Other factor such as water vapor, ice particles at high altitude, dust and various gases, and other contaminants that enter the air because of human activity.

In recent studies of daylight climate in Vietnam [5–11], two calculated models of luminous efficacy diffuse KD and global KG, lm/W, which represented by authors Nguyen Sanh Dan et al. and Perez et al. were used. The results have shown for Hanoi, values KD and KG respectively 102 and 93 lm/W according to the model of Nguyen Sanh Dan et al. 122 and 113 lm/W according to the model of Perez et al. These theoretical models are described in more detail below:

The Model Nguyen Sanh Dan et al. In the published study [5–7], luminous efficacy determined with different heights of the Sun in the sky, which were calculated by the formulas (2) and (3).

$$h_0 = arcsin\{sin\delta \cdot sin\varphi + cos\delta \cdot cos\varphi \cdot cos[150(12 - T)]\}, (0) \tag{2}$$

$$\delta = 23.45 \sin\left[\frac{360}{365}(d - 81)\right], \quad \delta = 23.45 \sin\left[\frac{360}{365}(284 + d)\right], (0) \tag{3}$$

Where:

d - The serial number of the day in the year, counting from 1 January;
h_0 - The height of the sun (0);
δ - Declination of the sun on any day of the year (0);
T - Time in hours, for example, 16 h 15 m = 16.25 h;
φ - Latitude (South - minus sign) (0).

The luminous efficacy of solar radiation is calculated according to the formula (4), (5):

$$K_D = 0.1 \cdot h_0 + 67, (Klux/kcal \cdot cm^{-2} \cdot min^{-1}) \tag{4}$$

$$K_G = 0.1 \cdot h_0 + 62, (Klux/kcal \cdot cm^{-2} \cdot min^{-1}) \tag{5}$$

With: $1(lm/W) = 1.433(Klux/kcal \cdot cm^{-2} \cdot min^{-1})$.

The Perez Model et al. [5, 8] several meteorological variables are used to design luminous efficacy for all types of sky. This luminous efficacy model is based on parameters:

ε - Sky clearness;
Z - Solar zenith angle;
W - Atmospheric condensation of water vapor;
Δ - Sky brightness;
a_i, b_i, c_i, d_i - coefficients corresponding to the clearness of the sky (ε) and are provided by the authors for eight ranges of the sky, variable from clear sky to overcast sky [12].

$$K_{D,G} = a_i + b_iW + c_icosZ + d_iln(\Delta) \tag{6}$$

Calculations using a solar radiation data set from ASHRAE IWEC2 file – "Weather data for energy calculations", which were developed for ASHRAE by White Box Technologies, Ins. and based on the integrated hourly basis over the ISD surface for 3012 locations outside the US and Canada that have a minimum of 12 years of recording up to 25 years. When converting the raw integrated hourly surface database (ISD) into local time, the software fills or reduces data before hourly time steps and calculates solar radiation. Data on the weather of the daylight climate were obtained within twelve years, from 2005 to the end of 2017 and are typical data. The choice of these typical months is based not only on the average but also on statistical distributions of different climatic parameters by months, which were collected from the records of long-term observations [13].

2.2   Determination of Horizontal Illuminance and Luminous Efficacy from Experiments

To confirm the method of calculating the horizontal illuminance and the luminous efficacy K_D and K_G for Hanoi, experimental measurements irradiation and illuminance in the summer period from 15 to 21 August 2018 under the open sky were conducted. The measurements were divided into two stages:

- Measurement stage for obtaining cumulative diffuse horizontal illuminance: The illuminance data was measured under overcast sky condition (15, 16, 18th) and sunny sky condition (17, 19, 20th) from 8:00 to 17:00 in the actinometrical station. Mean values show the results of average hourly diffuse horizontal illuminance under intermediated sky condition.

Measurement stage for determining the global and diffuse luminous efficacy KG, KD,

$$K_D = \frac{E_D}{Q_D} \tag{7}$$

$$K_G = \frac{E_G}{Q_G} \tag{8}$$

which were calculated by formulas (7) and (8) [9, 10]:

The experiment was conducted on August 21. A set of measured illuminance and solar radiation at 5-min intervals were collected during the development, and then compared with modelled values. The solar radiation data was collected from a meteorological station, and horizontal illuminance data was measured using experimental equipment - a Light Meter Testo 545 (a product of Fotronic Corporation). During the measurement of diffuse horizontal illuminance, the Light Meter should be shaded from direct sunlight. The results then must be developed to get hourly average values.

3   Results and Discussion

In accordance with the methodology described above, Table 4 shows the hourly mean values of diffuse horizontal illuminance ED, which were obtained from the measured values under overcast sky and sunny sky condition (Figs. 3, 4). These values are considered as the average diffuse horizontal illuminances under the intermediate sky condition.

Fig. 3. Cloudy condition: (a) - Overcast sky condition and (b) - Sunny sky condition.

Fig. 4. Light Meter Testo 545.

Table 4. The results of the experimental measurement of horizontal illuminances by the hour.

Time (hour)	8:00	9:00	10:00	11:00	12:00	13:00	14:00	15:00	16:00	17:00
Diffuse illuminances on 15, 16, 18th August, lux under overcast sky condition (lux)										
Cloudy type: Cumulonimbus (Cb) and Cumulus (Cu), sun exposure: 0.4 (hours)										
E_D (lux)	23700	24421	29843	33430	38605	37960	35008	33010	26884	20758
Diffuse illuminances on 14, 17, 19th August, lux under sunny sky condition (lux)										
Cloudy type: Cirrus (Ci), Cirrostratus (Cs), and Stratus (St), sun exposure: 3.2 (hours)										
E_D (lux)	30530	40583	48500	57113	60225	60980	49685	44641.3	33583	25760
Mean values of diffuse illuminances (lux)										
	27115	32502	39171	45272	49415	49470	42347	38826	30234	23259

Table 5. Diffuse horizontal illuminances E_D in August according to the calculation models.

Time (hour)	8:00	9:00	10:00	11:00	12:00	13:00	14:00	15:00	16:00	17:00
Diffuse illuminances by Model Nguyen Sanh Dan et al. (lux)										
E_D (lux)		14500	23600	31200	38000	41100	39000	36600	31900	20900	12700
Diffuse illuminances by Model Perez et al. (lux)										
E_D (lux)		18149	28537	36206	42541	45202	43444	41799	36983	26280	17076

Table 6. Relative error of computational models in comparison with experimental results.

Models	Average relative error (%)	Relative error by hour (%)									
		8:00	9:00	10:00	11:00	12:00	13:00	14:00	15:00	16:00	17:00
N.S. Dan et al.	25.6	46.52	27.39	20.35	16.06	16.83	21.16	13.57	17.84	30.87	45.40
Perez et al.	12.5	33.07	12.20	7.57	6.03	8.53	12.18	1.29	4.75	13.08	26.58

Table 7. Determination of luminous efficacy K_D, K_G based on horizontal illuminance E_D, E_G and solar radiation Q_D, Q_G

Time (hour)	8:00	8:30	9:00	9:30	10:00	10:30	11:00	11:30	12:00	12:30	13:00	13:30	Avg.
Diffuse luminous efficacy definition													
E_D (lux)	17340	22915	27820	40034	43467	50069	47270	45750	44301	66898	68515	25309	
Q_D (W/m²)	173	210	217	231	249	280	270	265	271	411	431	156	
K_D (lm/W)	100	109	128	174	175	179	175	173	164	163	159	162	155.0
Global luminous efficacy definition													
E_G (lux)	45860	54123	75815	87460	92130	104484	107900	116428	122162	125859	120762	23983	
Q_G (lm/W)	459	496	625	721	760	869	897	968	978	1008	886	176	
K_G (lm/W)	100	109	121	121	121	120	120	120	125	125	136	137	121.4

When comparing the theoretical models for determining the diffuse horizontal illuminance and experiments, the results of studies Nguyen Sanh Dan et al. and Perez et al. should be considered. The comparison results of the calculation and measurement are shown in Fig. 5 and Tables 5 and 6.

To define the luminous efficacy by experiment, the measured set data of illuminance and solar radiation were used. Diffuse and global luminous efficacy by experiment were calculated using formulas (7) and (8). The results shown on Table 7.

The development of horizontal illuminance and solar radiation over time is illustrated in Figs. 6 and 7.

Comparing the diffuse and global luminous efficiency of modelled values with experimental values, accordingly, the model of Perez's et al. shown closer to the experiment (Fig. 8).

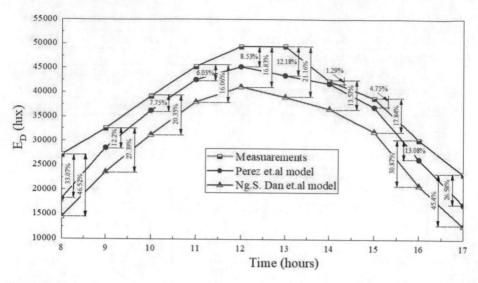

Fig. 5. Average values of horizontal diffuse illuminance of computational models and experiment.

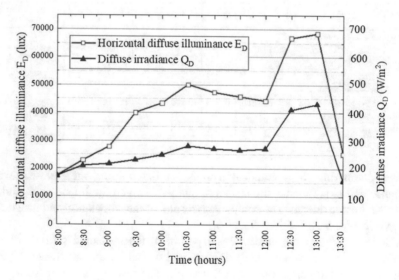

Fig. 6. Relativity of diffuse horizontal illuminance and diffuse solar radiation over time.

The comparison of horizontal illuminance and luminous efficiency between the two computational models with experiments showed:

- The average relative errors from horizontal illuminance comparison represented 12.5% and 25.6% with the better performance of Perez et al. model.

- The average relative errors from luminous efficiency comparison represented respectively K_D of 27% and 51.96%, K_G of 7.43% and 30.5% with the better performance of Perez et al. model.

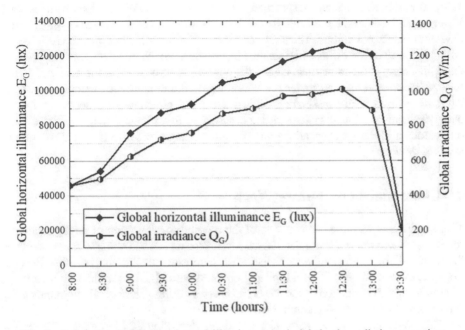

Fig. 7. Relativity of global horizontal illuminance and global solar radiation over time.

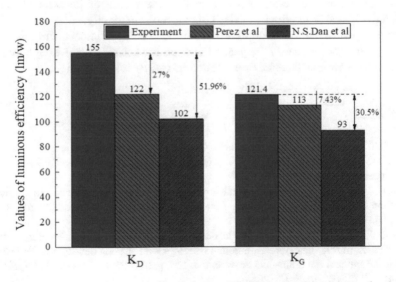

Fig. 8. Comparison of luminous efficacy by modelled values and experimental values.

In case study of Malaysia by authors Zain-Ahmed et al. [14–17], the Model of Perez et al. was used to obtain the luminous efficacy for Shubang. It is seen that the sky conditions in Malaysia is predominantly intermediate. The annual average values of diffuse and global luminous efficacies were found to be 120 and 112 lm/W. These value for Hanoi, Vietnam respectively 122 and 113 lm/W. It demonstrates that Vietnam and Malaysia are both in the humid- tropical zone and have the similar daylight climate conditions. The study of solar radiation and cloudiness statistic along with high values of diffuse horizontal illuminance, which confirm the potential of daylight usage, at the same time, causes the possibility of increasing the heat load in the rooms. This suggests that, under the condition of direct sunlight in tropical climate, daylight systems with sun shading devices are more effective. In order to assess the energy efficiency of such systems, it is necessary to investigate direct solar irradiation on facades at different orientations and to analyze the effect of different types of sun shading devices.

4 Conclusions and Future Work

This study deals with the confirmation of the method for definition of daylight climate for Hanoi, Vietnam. For more details, the influences of various climatic factors such as the solar radiation and cloudiness statistics were considered. The analysis describes the tropical sky conditions of Hanoi typically covered with high and middle-altitude types of clouds, which characterize for the high levels of diffuse horizontal illuminance for sunny and intermediate sky conditions.

This study confirms the suitable method for definition of illuminous efficacy for tropical skies condition by comparing the computed models and the experiment. Analysis of the average errors has shown that the model of Perez et al. performs significantly better in the tropical skies condition of Hanoi city.

The prospect of future research is needed to conduct experiments with other regions of Vietnam in order to clarify the general method for determining the daylight climate and move to territorial standards for predicting and assessing daylight.

References

1. Ha, P.T.H.: Passive architectural solution based on energy efficiency method of sunshading devices for high-rise apartment buildings in Hanoi. Ph.D. Thesis, National university of Civil Engineering, Hanoi (2018)
2. Ha, P.T.H.: A concept for energy-efficient high-rise buildings in Hanoi and a calculation method for building energy efficiency factor. Procedia Eng. **142**, 154–160 (2016). https://doi.org/10.1016/j.proeng.2016.02.026
3. Ha, P.T.H.: Energy efficiency façade design in high-rise apartment buildings using the calculation of solar heat transfer thorough windows with shading devise. In: IOP Conference Series: Earth and Environmental Science, vol. 143, p. 012055. CUTE (2018). https://doi.org/10.1088/1755-1315/143/1/012055
4. Dang, P.N., Ha, P.T.H.: Heat and Architecture Climatic. Construction Publisher, Hanoi (2002)

5. Phuong, N.T.K., Solovyov, A.K.: Potential daylight resources between tropical and temperate cities–a case study of Ho Chi Minh city and Moscow. In: Scientific Journal Matec Web of Conferences, vol. 193. EDP Science Publishing (2018). https://doi.org/10.1051/matecconf/201819304013

6. Fyong, N.T.H., Solov'yov, A.K.: Opredelenie svetovoj ehffektivnosti solnechnoj ra-diacii dlya V'etnama (in Russian). EHkonomika stroitel'stva i prirodopol'zovaniya **67**(2), 137–143 (2018)

7. Solov'yov, A.K., Fyong, N.T.H.: Metod rascheta svetovogo klimata po svetovoj ehffek-tivnosti solnechnoj radiacii: primer sravnitel'nogo analiza svetovogo klimata Ha-noya i Moskvy (in Russian). Svetotekhnika **5**, 21–24 (2018)

8. Kittler, R., Kocifaj, M., Darula, S.: Daylight Science and Daylight Technology. Springer, New York (2012). https://doi.org/10.1007/978-1-4419-8816-4

9. Fabian, M., Uetani, Y., Darula, S.: Monthly luminous efficacy models and illuminance prediction using ground. Sol. Energy **162** (2018). https://doi.org/10.1016/j.solener.2017.12.056

10. Dervishi, S., Mahdavi, A.: A comparison of luminous efficacy models for the diffuse component of solar radiation. In: Fourth German - Austrian IBPSA Conference Berlin University of Art, Berlin, pp. 117–120 (2012)

11. Seo, D.: Comparative analysis of all sky luminous efficacy models based on calculated and measured solar radiation data of four worldwide cities. Int. J. Photoenergy 9 p. (2018). https://doi.org/10.1155/2018/8180526, Article ID 8180526

12. Miroslav, F., Yoshiaki, U., Stanislav, D.: Monthly luminous efficacy models and illuminance prediction using ground. Sol. Energy (2018). https://doi.org/10.1016/j.solener.2017.12.056

13. Joe, H.: ASHRAE Research Project 1477-RP Development of 3,012 typical year weather files for international locations. White Box Technologics. Moraga, CA (2011)

14. Zain-Ahmed, A., Sopian, K., Zainol, A.Z., Othman, M.Y.H.: The availability of daylight from tropical skies-a case study of Malaysia. Renew. Energy **25**, 21–30 (2002)

15. Lim, Y.W., Mohd Hamdan, A., Ossen, D.R.: Empirical validation of daylight simulation tool with physical model measurement. Am. J. Appl. Sci. **7**(10), 1412–1429 (2010). https://doi.org/10.3844/ajassp.2010.1426.1431

16. Lim, Y.W., Mohd Zin, K., Mohd Hamdan, A., Ossen, D.R., Aminatuzuhariah, M.A.: Building facade design for daylighting quality in typical government office building. Build. Environ. **57**, 194–204 (2012). https://doi.org/10.1016/j.buildenv.2012.04.015

17. Lim, Y.W., Heng, C.Y.S.: Dynamic internal light shelf for tropical daylighting in high rise office buildings. Build. Environ. **106**, 155–166 (2016). https://doi.org/10.1016/j.buildenv.2016.06.030

Computer Modeling of the Creep Process in Stiffened Shells

Vladimir Karpov(iD) and Alexey Semenov$^{(\boxtimes)}$(iD)

Saint Petersburg State University of Architecture and Civil Engineering,
Vtoraya Krasnoarmeiskaya str. 4, 190005 Saint Petersburg, Russia
vvkarpov@lan.spbgasu.ru, sw.semenov@gmail.com

Abstract. A methodology for computer modeling of the buckling process for shell structures, taking into account creep of the material, is presented in this work. A mathematical model of shell deformation, accounting for transverse shears, geometric nonlinearity and presence of stiffeners, is provided. The process of creep is modeled according to a linear theory based on the Boltzmann–Volterra heredity theory. The model is written in the form of the functional of full potential deformation energy. The calculation algorithm is based on the Ritz method and iteration method. Calculation results for several options of shallow shells of double curvature, square of base, made of plexiglass, are presented. Contour fixation—fixed pin joints; uniformly distributed load is directed along the normal to the surface. "Deflection–time" relations are plotted for different load magnitudes. Values of critical time of buckling as a result of creep deformation are found. Curves of critical load decline over time as a result of creep development are obtained.

Keywords: Shells · Creep · Stress-strain state · Buckling · Rheology

1 Introduction

Complex rheological processes occur in a shell structure subjected to long-term loading [1–5], which may lead to irreversible changes in the material and loss of stability. Therefore, studying the stability of shell structures, it is necessary to take into consideration that creep deformations may occur [6–9]. Studies [10–13] should be attributed to recent works in this field. Effect of creep on the deformation of shells under long-term loading manifests in the slow change of equilibrium forms in the structure over time up to some critical point of time t_{cr}, when the equilibrium form changes rapidly (buckling of the shell occurs) [14].

We will use a linear theory of hereditary creep based on the Boltzmann–Volterra heredity theory as it reflects the deformation process of polymer materials reasonably well.

© Springer Nature Switzerland AG 2020
V. Murgul and M. Pasetti (Eds.): EMMFT 2018, AISC 982, pp. 48–58, 2020.
https://doi.org/10.1007/978-3-030-19756-8_5

2 Theory and Methods

In this work the mathematical model of shell structure deformation is based on the functional of full potential deformation energy, and also includes geometrical relationships (relating deformation and displacement), physical relationships (relating stress and deformation) and boundary conditions (selected depending on the method of contour fixation of the structure) [15].

2.1 Physical Relationships Taking into Account Creep Deformation

With account for creep in the shell's material, physical relationships take the following form:

$$\sigma_x = \frac{E}{1-\mu^2}\left[\varepsilon_x^z + \mu\varepsilon_y^z - \int_{t_0}^{t}\left(\varepsilon_x^z + \mu\varepsilon_y^z\right)R_1(t,\tau)d\tau\right],$$

$$\sigma_y = \frac{E}{1-\mu^2}\left[\varepsilon_y^z + \mu\varepsilon_x^z - \int_{t_0}^{t}\left(\varepsilon_y^z + \mu\varepsilon_x^z\right)R_1(t,\tau)d\tau\right],$$

$$\tau_{xy} = \frac{E}{2(1+\mu)}\left[\gamma_{xy}^z - \int_{t_0}^{t}\gamma_{xy}^z R_2(t,\tau)d\tau\right], \tag{1}$$

$$\tau_{xz} = \frac{E}{2(1+\mu)}\left[\gamma_{xz} - \int_{t_0}^{t}\gamma_{xz}R_2(t,\tau)d\tau\right], \quad \tau_{yz} = \frac{E}{2(1+\mu)}\left[\gamma_{yz} - \int_{t_0}^{t}\gamma_{yz}R_2(t,\tau)d\tau\right].$$

Here $R_1(t,\tau)$, $R_2(t,\tau)$—influence functions (relaxation cores) of the material under strain and shear.

2.2 Introduction of Reinforcing Stiffeners

We will consider shells reinforced by stiffeners (with thickness h and linear dimensions a, b along the x, y axes), when stiffeners (Fig. 1) are introduced discretely. In this case their contact with the shell occurs along the line, and their shear and torsional rigidities are taken into account. In the mathematical model, it is set by the following functions:

$$\bar{F} = \sum_{j=1}^{m}F^j\bar{\delta}(x - x_j) + \sum_{i=1}^{n}F^i\bar{\delta}(y - y_i) - \sum_{i=1}^{n}\sum_{j=1}^{m}F^{ij}\bar{\delta}(x - x_j)\bar{\delta}(y - y_i),$$

$$\bar{S} = \sum_{j=1}^{m}S^j\bar{\delta}(x - x_j) + \sum_{i=1}^{n}S^i\bar{\delta}(y - y_i) - \sum_{i=1}^{n}\sum_{j=1}^{m}S^{ij}\bar{\delta}(x - x_j)\bar{\delta}(y - y_i),$$

$$\bar{J} = \sum_{j=1}^{m}J^j\bar{\delta}(x - x_j) + \sum_{i=1}^{n}J^i\bar{\delta}(y - y_i) - \sum_{i=1}^{n}\sum_{j=1}^{m}J^{ij}\bar{\delta}(x - x_j)\bar{\delta}(y - y_i), \tag{2}$$

$$F^j = h^j, \quad S^j = h^j(h + h^j)/2, \quad J^j = 0.25h^2h^j + 0.5h(h^j)^2 + \tfrac{1}{3}(h^j)^3.$$

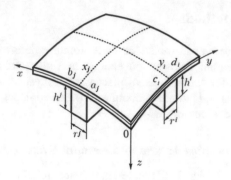

Fig. 1. A view of stiffeners.

Here $\bar{\delta}(x - x_j)$—a unit column function equal to one at $a_j \leq x \leq b_j$ $(a_j = x_j - r_j/$ $(2A))$, $b_j = x_j + r_j/(2A))$ and equal to zero at other x values; $\bar{\delta}(y - y_j)$—a unit column function equal to one at $c_i \leq y \leq d_i$ $(c_i = y_i - r_i/(2B), d_i = y_i + r_i/(2B))$ and equal to zero at other y values; h^j, r_j, m—height and width of stiffeners parallel to the y axis, and the number of stiffeners in this direction; h^i, r_i, n—similarly for stiffeners parallel to the x axis; $h^{ij} = \min\{h^i, h^j\}$; A, B – Lame parameters which are selected depending on the shell type.

2.3   Geometrical Relationships

Let us consider the shell deformation model of Timoshenko (Mindlin–Reissner) type, then three displacement functions $U = U(x, y)$, $V = V(x, y)$, $W = W(x, y)$ and $\Psi_x = \Psi_x(x, y)$, $\Psi_y = \Psi_y(x, y)$—unknown functions of normal angles of rotation in the xOz and yOz planes, respectively—will be the unknown functions. In this case, with account for geometrical nonlinearity and transverse shears, geometrical relationships in the middle surface will take the following form:

$$\varepsilon_x = \tfrac{1}{A}\tfrac{\partial U}{\partial x} + \tfrac{1}{AB}V\tfrac{\partial A}{\partial y} - k_x W + \tfrac{1}{2}\theta_1^2, \qquad \varepsilon_y = \tfrac{1}{B}\tfrac{\partial V}{\partial y} + \tfrac{1}{AB}U\tfrac{\partial B}{\partial x} - k_y W + \tfrac{1}{2}\theta_2^2,$$

$$\gamma_{xy} = \tfrac{1}{A}\tfrac{\partial V}{\partial x} + \tfrac{1}{B}\tfrac{\partial U}{\partial y} - \tfrac{1}{AB}U\tfrac{\partial A}{\partial y} - \tfrac{1}{AB}V\tfrac{\partial B}{\partial x} + \theta_1\theta_2,$$

$$\gamma_{xz} = k f(z)[\Psi_x - \theta_1], \qquad \gamma_{yz} = k f(z)[\Psi_y - \theta_2], \tag{3}$$

$$\chi_1 = \tfrac{1}{A}\tfrac{\partial \Psi_x}{\partial x}, \ \chi_2 = \tfrac{1}{B}\tfrac{\partial \Psi_y}{\partial y} + \tfrac{1}{AB}\tfrac{\partial B}{\partial x}\Psi_x, \ \chi_{12} = \tfrac{1}{2}\left(\tfrac{1}{A}\tfrac{\partial \Psi_y}{\partial x} + \tfrac{1}{B}\tfrac{\partial \Psi_x}{\partial y} - \tfrac{1}{AB}\tfrac{\partial B}{\partial x}\Psi_y\right),$$

where $\varepsilon_x, \varepsilon_y$—deformations of elongation along the x, y coordinates of the middle surface; $\gamma_{xy}, \gamma_{xz}, \gamma_{yz}$—shear deformations in the xOy, xOz, yOz planes, respectively; $\chi_1, \chi_2, \chi_{12}$—functions of curvature and torsional change; k_x, k_y—main curvatures of the shell along the x and y axes;

$$\theta_1 = -\left(\frac{1}{A}\frac{\partial W}{\partial x} + k_x U\right), \qquad \theta_2 = -\left(\frac{1}{B}\frac{\partial W}{\partial y} + k_y V\right);$$

$f(z)$—a function characterizing distribution of shear deformations γ_{xz}, γ_{yz} by shell thickness; k—a numerical coefficient:

$$f(z) = 6 \left(\frac{1}{4} - \frac{z^2}{h^2} \right), \quad k = \frac{5}{6}. \tag{4}$$

2.4 Functional of Full Potential Deformation Energy

For stiffened shells, with account for material creep under the linear theory of hereditary creep, the functional of full potential deformation energy can be written in the following form:

$$E_p = E_p^L - E_p^C, \tag{5}$$

where E_p^L—a part of the functional reflecting linear-elastic deformation; E_p^C—deformation associated with development of creep deformations. They are written in the following form:

$$
\begin{aligned}
E_p^L = \frac{E}{2(1-\mu^2)} \int_0^a \int_0^b & \left\{ (h + \overline{F}) \left[\varepsilon_x^2 + 2\mu\varepsilon_x\varepsilon_y + \varepsilon_y^2 + \tilde{\mu}\gamma_{xy}^2 + \tilde{\mu}k(\Psi_x - \theta_1)^2 + \tilde{\mu}k(\Psi_y - \theta_2)^2 \right] \right. \\
& + 2\overline{S}[\varepsilon_x\chi_1 + \mu\varepsilon_x\chi_2 + \varepsilon_y\chi_2 + \mu\varepsilon_y\chi_1 + 2\tilde{\mu}\gamma_{xy}\chi_{12}] \\
& \left. + \left(\frac{h^3}{12} + \overline{J} \right) \left(\chi_1^2 + 2\mu\chi_1\chi_2 + \chi_2^2 + 4\tilde{\mu}\chi_{12}^2 \right) - 2(1 - \mu^2)\frac{q}{E}W \right\} ABdxdy,
\end{aligned}
\tag{6}
$$

$$
\begin{aligned}
E_p^C = \frac{E}{2(1-\mu^2)} \int_0^a \int_0^b \int_{t_0}^t & \left\{ \left[(h + \overline{F}) \left(\varepsilon_x^2(\tau) + 2\mu\varepsilon_x(\tau)\varepsilon_y(\tau) + \varepsilon_y^2(\tau) \right) + 2\overline{S}(\chi_1(\tau)\varepsilon_x(\tau) \right. \right. \\
& + \chi_2(\tau)\varepsilon_y(\tau) + \mu\chi_2(\tau)\varepsilon_x(\tau) + \mu\chi_1(\tau)\varepsilon_y(\tau)) + \left(\frac{h^3}{12} + \overline{J} \right) \left(\chi_1^2(\tau) + \chi_2^2(\tau) \right. \\
& \left. + 2\mu\chi_1(\tau)\chi_2(\tau)) \right] R_1(t, \tau) + \left[(h + \overline{F}) \left(\tilde{\mu}\gamma_{xy}^2(\tau) + \tilde{\mu}k(\Psi_x(\tau) - \theta_1(\tau)) + \tilde{\mu}k(\Psi_y(\tau) - \theta_2(\tau)) \right) \right. \\
& \left. \left. + 4\overline{S}\tilde{\mu}\gamma_{xy}(\tau)\chi_{12}(\tau) + \left(\frac{h^3}{12} + \overline{J} \right) 4\tilde{\mu}\chi_{12}^2(\tau) \right] R_2(t, \tau) \right\} ABdxdyd\tau.
\end{aligned}
\tag{7}
$$

Here $\tilde{\mu} = (1 - \mu)/2$; q—external uniformly distributed transverse load directed along the normal to the surface.

If an iteration process on the coordinate t is used for the solution of the problem and the segment $[t_0, t_k]$ is divided into parts by points $t_1, t_2, \ldots, t_{k-1}$ with a step of $\Delta t = t_i - t_{i-1} = 1$ day, then, approximately, we can accept the following:

$$
\begin{aligned}
E_p^C(t_k) = \frac{E}{2(1-\mu^2)} \int_0^a \int_0^b 2 \Bigg\{ & \varepsilon_x(t_k) \sum_{i=1}^k \left[(h+\overline{F})\left(\varepsilon_x(t_{i-1}) + \mu\varepsilon_y(t_{i-1})\right) + \overline{S}\left(\chi_1(t_{i-1}) + \mu\chi_2(t_{i-1})\right) \right] \\
& \times R_1(t_k, t_{i-1})\Delta t + \chi_1(t_k) \sum_{i=1}^k \left[\overline{S}\left(\varepsilon_x(t_{i-1}) + \mu\varepsilon_y(t_{i-1})\right) + \left(\tfrac{h^3}{12} + \overline{J}\right)\left(\chi_1(t_{i-1}) + \mu\chi_2(t_{i-1})\right) \right] \\
& \times R_1(t_k, t_{i-1})\Delta t + \varepsilon_y(t_k) \sum_{i=1}^k \left[(h+\overline{F})\left(\varepsilon_y(t_{i-1}) + \mu\varepsilon_x(t_{i-1})\right) + S\left(\chi_2(t_{i-1}) + \mu\chi_1(t_{i-1})\right) \right] \\
& \times R_1(t_k, t_{i-1})\Delta t + \chi_2(t_k) \sum_{i=1}^k \left[\overline{S}\left(\varepsilon_y(t_{i-1}) + \mu\varepsilon_x(t_{i-1})\right) + \left(\tfrac{h^3}{12} + \overline{J}\right)\left(\chi_2(t_{i-1}) + \mu\chi_1(t_{i-1})\right) \right] \\
& \times R_1(t_k, t_{i-1})\Delta t + \tilde{\mu}\gamma_{xy}(t_k) \sum_{i=1}^k \left[(h+\overline{F})\gamma_{xy}(t_{i-1}) + 2\overline{S}\chi_{12}(t_{i-1}) \right] R_2(t_k, t_{i-1})\Delta t \\
& + 2\tilde{\mu}\chi_{12}(t_k) \sum_{i=1}^k \left[\overline{S}\gamma_{xy}(t_{i-1}) + 2\left(\tfrac{h^3}{12} + \overline{J}\right)\chi_{12}(t_{i-1}) \right] R_2(t_k, t_{i-1})\Delta t \\
& + k\tilde{\mu}\left(\Psi_x(t_k) - \theta_1(t_k)\right) \sum_{i=1}^k (h+\overline{F})\left(\Psi_x(t_{i-1}) - \theta_1(t_{i-1})\right) R_2(t_k, t_{i-1})\Delta t \\
& + k\tilde{\mu}\left(\Psi_y(t_k) - \theta_2(t_k)\right) \sum_{i=1}^k (h+\overline{F})\left(\Psi_y(t_{i-1}) - \theta_2(t_{i-1})\right) R_2(t_k, t_{i-1})\Delta t \Bigg\} ABdxdy.
\end{aligned}
$$

$$(8)$$

2.5   Ritz Method

In order to reduce the problem on minimum of energy functional to the solution of a system of nonlinear algebraic equations, we will use the Ritz method. According to this method, the unknown functions can be presented in the following form:

$$
U = U(x,y) = \sum_{k=1}^{\sqrt{N}} \sum_{l=1}^{\sqrt{N}} U_{kl} X1(k) Y1(l), \quad V = V(x,y) = \sum_{k=1}^{\sqrt{N}} \sum_{l=1}^{\sqrt{N}} V_{kl} X2(k) Y2(l),
$$

$$
W = W(x,y) = \sum_{k=1}^{\sqrt{N}} \sum_{l=1}^{\sqrt{N}} W_{kl} X3(k) Y3(l),
$$

$$
\Psi_x = \Psi_x(x,y) = \sum_{k=1}^{\sqrt{N}} \sum_{l=1}^{\sqrt{N}} PS_{kl} X4(k) Y4(l), \quad \Psi_y = \Psi_y(x,y) = \sum_{k=1}^{\sqrt{N}} \sum_{l=1}^{\sqrt{N}} PN_{kl} X5(k) Y5(l),
$$

$$(9)$$

where $U_{kl} - PN_{kl}$—unknown numerical parameters or variable t functions for the creep problems.

Substituting (9) into (6) and (8), we will find derivatives by numerical parameters $U_{kl} - PN_{kl}$. First, we will solve the linear-elastic problem, equating all unknown parameters depending on time to zero. As a result, we will obtain a system of nonlinear algebraic equations. Finding its solution at gradually increasing load values, we will choose several states, which will be used as initial conditions for the creep problem.

3 Numerical Results

Let us consider shallow shell structures of double curvature, square of base. In this case, Lame parameters and curvatures $A = B = 1$, $k_x = 1/R_1$, $k_y = 1/R_2$. Let us consider the middle shell surface as the coordinate surface. The x, y axes are oriented along the main shell curvatures lines, and the z axis—along the normal in the direction of the concavity (Fig. 2).

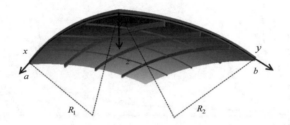

Fig. 2. Shallow shell of double curvature reinforced by stiffeners.

Geometrical parameters of shells are given in Table 1. Let us consider an orthogonal network of stiffeners with width $2h$ and height $3h$ as the structure support. During decomposition of the unknown functions into rows in the Ritz method, we will take the number of summands as $N = 9$.

Table 1. Geometrical parameters of shells.

No.	a, m	R_1, m	h, m
1	18	11.325	0.03
2	18	22.65	0.03
3	18	45.3	0.03

The values of critical loads decrease in the shell material under long-term loading as a result of creep growth. This fact also should be taken into account in the analysis of strength and stability of shells. Creep growth in metals can only occur at high temperatures, therefore, the study of the creep process will be conducted on shells made of plexiglass ($E = 0.33 \cdot 10^4$ MPa, $\mu = 0.354$) with a step along the time axis $\Delta t = 1$ day.

Influence functions for plexiglass will be as follows:

$$R_i(t, \tau) = A_i e^{-\beta_i(t-\tau)}(t - \tau)^{\alpha_i-1}, \quad i = 1, 2,$$
$$A_1 = 0.0269, \quad \beta_1 = 0.045 \cdot 10^{-3}, \quad \alpha_1 = 0.05, \tag{10}$$
$$A_2 = 0.01318, \quad \beta_2 = 0.833 \cdot 10^{-3}, \quad \alpha_2 = 0.20.$$

For further presentation of the results, let us introduce dimensionless deflection variable $\overline{W} = W/h$.

Relationships "$\overline{W} - t$" obtained for shell option 1 in the $(x = a/4, y = b/4)$ quarter of the structure are presented in Figs. 3, 4 and 5. In Fig. 3, they are presented for a shell without stiffeners. Curve 1 corresponds to the load of $q = 6.35 \cdot 10^{-2}$ MPa; 2—$q = 7.62 \cdot 10^{-2}$ MPa; 3—$q = 8.25 \cdot 10^{-2}$ MPa; 4—$q = 8.9 \cdot 10^{-2}$ MPa; 5—$q = 9.4 \cdot 10^{-2}$ MPa.

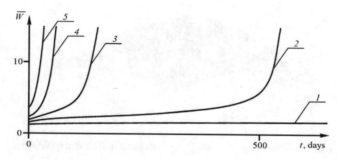

Fig. 3. Relationship "$\overline{W} - t$" for shell option 1 under different loading conditions as a result of creep deformation (shell is not reinforced by stiffeners).

The time of rapid deflection growth (5–10 times in comparison with the initial state at $t = 0$) is taken as critical time t_{cr}. Curves of deflection growth in time $\overline{W}$ for a shell reinforced by the orthogonal network of stiffeners 9×9 under corresponding loads are given in Fig. 4. Curve 1 corresponds to the load of $q = 8.9 \cdot 10^{-2}$ MPa; 2—$q = 9.5 \cdot 10^{-2}$ MPa; 3—$q = 10 \cdot 10^{-2}$ MPa; 4—$q = 11.43 \cdot 10^{-2}$ MPa; 5—$q = 11.9 \cdot 10^{-2}$ MPa.

Curves of deflection growth in time $\overline{W}$ for a shell reinforced by the orthogonal network of stiffeners 18×18 under corresponding loads are given in Fig. 5. Curve 1 corresponds to the load of $q = 0.1$ MPa; 2—$q = 0.109$ MPa; 3—$q = 0.119$ MPa; 4—$q = 0.127$ MPa; 5—$q = 0.136$ MPa.

In the analysis of shell creep, a quasistatic problem is solved when inertial members are neglected. With the increase in time, the process of deformation continues after loss of stability as a result of creep. It appears that a shell makes oscillatory movements in the post-critical phase. In Fig. 6 this process is shown for shell option 1 reinforced by the orthogonal network of stiffeners 18×18.

All obtained results (for three shell options) are given in Table 2.

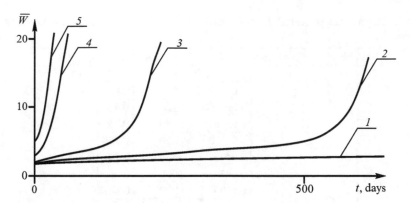

Fig. 4. Relationship "$\overline{W} - t$" for shell option 1 under different loading conditions as a result of creep deformation (number of stiffeners—9 × 9).

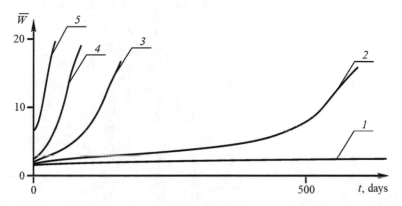

Fig. 5. Relationship "$\overline{W} - t$" for shell option 1 under different loading conditions as a result of creep deformation (number of stiffeners—18 × 18)

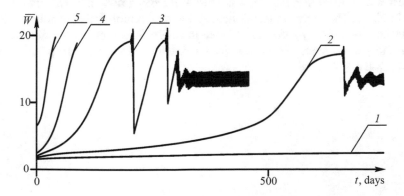

Fig. 6. Relationship "$\overline{W} - t$" for shell option 1 under different loading conditions as a result of creep deformation (number of stiffeners—18 × 18) taking into account stability loss.

Table 2. Dependence of critical time of creep initiation on the load for all shell options under consideration.

No.	Number of stiffeners	q, MPa	t_{cr}, days	q/q_{cr}	q_{cr}, MPa
1	0	$6.35 \cdot 10^{-2}$	850	0.67	$9.4 \cdot 10^{-2}$
		$7.62 \cdot 10^{-2}$	540	0.81	
		$8.25 \cdot 10^{-2}$	150	0.88	
		$8.9 \cdot 10^{-2}$	50	0.95	
		$9.4 \cdot 10^{-2}$	30	0.999	
	18	0.0890	800	0.74	0.1196
		0.0950	600	0.79	
		0.1000	230	0.84	
		0.1143	55	0.95	
		0.1190	35	0.99	
	36	0.100	950	0.73	0.136
		0.109	600	0.8	
		0.119	140	0.87	
		0.127	70	0.93	
		0.135	36	0.99	
2	0	0.0890	900	0.72	0.0125
		0.0102	430	0.82	
		0.0108	160	0.87	
		0.0114	70	0.92	
		0.0124	35	0.99	
3	0	$0.127 \cdot 10^{-2}$	900	0.67	$0.188 \cdot 10^{-2}$
		$0.152 \cdot 10^{-2}$	430	0.8	
		$0.165 \cdot 10^{-2}$	130	0.88	
		$0.178 \cdot 10^{-2}$	55	0.95	
		$0.187 \cdot 10^{-2}$	25	0.99	

Curves for critical load reduction as a result of creep growth in the shell material are given in Fig. 7. The curve number designates the shell option, index 0 indicates that the shell is not reinforced by stiffeners. As seen from Fig. 7, the greater the shell curvature is, the steeper becomes the critical load reduction curve, i.e. creep deformations develop faster.

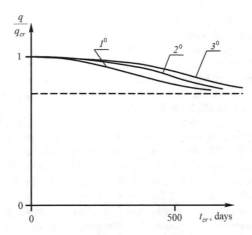

Fig. 7. Curves for critical load reduction as a result of the material creep for shells of different curvature.

4 Conclusions

The proposed computer modeling methodology can be used to study the creep process in shell structures. Based on the obtained results, we can conclude that accounting for possible development of creep deformation in the material of the structure leads to significant reduction of critical load of buckling in comparison with the calculation in the linear-elastic case.

Acknowledgements. The research was supported by RSF (project No. 18-19-00474).

References

1. Khokhlov, A.V.: Constitutive relation for rheological processes with known loading history. Creep and long-term strength curves. Mech. Solids **43**(2), 283–299 (2008). https://doi.org/10.3103/s0025654408020155
2. Betten, J.: Creep mechanics. Springer, Heidelberg (2008)
3. Pshenichnov, S.G.: Nonstationary dynamic problems of nonlinear viscoelasticity. Mech. Solids **48**(1), 68–78 (2013). https://doi.org/10.3103/S002565441301007X
4. Garrido, M., Correia, J.R., Keller, T., Cabral-Fonseca, S.: Creep of sandwich panels with longitudinal reinforcement ribs for civil engineering applications: experiments and composite creep modeling. J. Compos. Constr. **21**(1), 04016074 (2017). https://doi.org/10.1061/(ASCE)CC.1943-5614.0000735
5. Korobeynikov, S.N., Torshenov, N.G., Lyubashevskaya, I.V., Larichkin, A.Y., Chunikhina, E.V.: Creep buckling of axially compressed circular cylindrical shells of a zirconium alloy: experiment and computer simulation. J. Appl. Mech. Tech. Phys. **55**, 105–117 (2014). https://doi.org/10.1134/S0021894414010143

6. Zolochevsky, A., Galishin, A., Sklepus, S., Voyiadjis, G.Z.: Analysis of creep deformation and creep damage in thin-walled branched shells from materials with different behavior in tension and compression. Int. J. Solids Struct. **44**, 5075–5100 (2007). https://doi.org/10.1016/j.ijsolstr.2006.12.019

7. Hamed, E., Bradford, M.A., Ian Gilbert, R.: Nonlinear long-term behaviour of spherical shallow thin-walled concrete shells of revolution. Int. J. Solids Struct. **47**(2), 204–215 (2010). https://doi.org/10.1016/j.ijsolstr.2009.09.027

8. Kuznetsov, E.B., Leonov, S.S.: Technique for selecting the functions of the constitutive equations of creep and long-term strength with one scalar damage parameter. J. Appl. Mech. Tech. Phys. **57**(2), 369–377 (2016). https://doi.org/10.1134/S0021894416020218

9. Guryeva, Yu.A.: Eccentric compression of symmetrically reinforced concrete bar with due account for nonlinear creep of concrete. Bull. Civ. Eng. (6), 95–101 (2017). (in Russian). https://doi.org/10.23968/1999-5571-2017-14-6-95-101

10. Yankovskii, A.P.: Refined deformation model for metal-composite plates of regular layered structure in bending under conditions of steady-state creep. Mech. Compos. Mater. **52**(6), 715–732 (2017). https://doi.org/10.1007/s11029-017-9622-7

11. Chepurnenko, A., Neumerzhitskaya, N., Turko, M.: Finite element modeling of the creep of shells of revolution under axisymmetric loading. Advances in Intelligent Systems and Computing (2018). https://doi.org/10.1007/978-3-319-70987-1_86

12. Chepurnenko, A.S.: Stress-strain state of three-layered shallow shells under conditions of nonlinear creep. Mag. Civ. Eng. **8**, 156–168 (2017). https://doi.org/10.18720/mce.76.14

13. Siryus, V.: Minimization of the weight of ribbed cylindrical shells made of a viscoelastic composite. Mech. Compos. Mater. **46**(6), 593–598 (2011). https://doi.org/10.1007/s11029-011-9174-1

14. Miyazaki, N., Hagihara, S.: Creep buckling of shell structures. Mech. Eng. Rev. **2**(2), 14–00522 (2015). https://doi.org/10.1299/mer.14-00522

15. Karpov, V.V., Semenov, A.A.: Mathematical models and algorithms for studying strength and stability of shell structures. J. Appl. Ind. Math. **11**(1), 70–81 (2017). https://doi.org/10.1134/S1990478917010082

Energy-Saving Technologies in Design and Construction of Residential Buildings and Industrial Facilities in the Far North

Maria Berseneva$^{(\boxtimes)}$ ⓘ, Galina Vasilovskaya ⓘ,
Tamara Danchenko ⓘ, Ivan Inzhutov ⓘ, Sergei Amelchugov,
Alexandra Yakshina ⓘ, and Helena Danilovich ⓘ

Siberian Federal University, Svobodny Avenue 79, 660041 Krasnoyarsk, Russia
mari-leonm@yandex.ru

Abstract. The article discusses the main problems of energy saving and special features of design, construction and operation of residential buildings and industrial structures in extreme climatic conditions of the Northern territories. The impact of permafrost, seasonal thawing of soils and low bearing capacity of thawed soils on the choice of ways of arrangement of cities and pipelines in the Northern regions of Russia have been analyzed. There have been assessed the effect of climate change and the associated natural and man-made risks in the Far North on the development of a system of measures to ensure the reliability of energy facilities and options for emergency response. The solutions allowing to increase energy-saving capabilities of buildings and to reduce expenditures at their operation have been designated. Necessity of using non-traditional and renewable energy sources has been considered.

Keywords: Energy saving · Far North · Thermal insulation · Permafrost

1 Introduction

The development of the Far North is a strategic task to ensure the development of the national economy and security of Russia. Thanks to the discovery and development of natural resources in the North, large production and processing industrial complexes have been created, the maintenance of which and the construction of new ones with a developed social infrastructure will provide the basis for Russia's export potential.

The most acute problem in solving the development of the Northern territories of Russia is the issue of energy saving. The resource development of the Northern regions of Russia remains the economic base for the foreseeable future. A necessary condition for this development and the successful functioning of the industrial specialization is the accelerated development of the production base, the construction of industrial facilities and residential housing that provide complex development of the territory and the needs of the Far North population [1, 2]. Energy saving is a comprehensive solution of legal, organizational, scientific, industrial, technical and economic problems, the

© Springer Nature Switzerland AG 2020
V. Murgul and M. Pasetti (Eds.): EMMFT 2018, AISC 982, pp. 59–68, 2020.
https://doi.org/10.1007/978-3-030-19756-8_6

purpose of which is the rational use and consumption of heat and power resources that reduce energy loss. The shortage of basic energy resources, the increasing cost of their production, as well as the global environmental problems make energy saving the one of the priorities in the development of the Northern territories. The further economic development of our country and the standards of living for its' citizens depend on the results of that task. For transition from rotational ways of conducting activity and accommodation in the conditions of the Far North to the cities of permanent residence it is necessary to develop human settlements, the feasibility of which requires conducting a comprehensive study [3–5].

Construction in the areas of the Northern climatic zone is hampered by the characteristic extreme climatic conditions:

- negative annual average temperatures;
- permafrost soil conditions;
- polar nights;
- strong winds;
- snow drifts.

The duration of the heating season in the Northern territories can reach up to 11 months a year. For example, in the Norilsk industrial district the duration of the heating season is more than three hundred days [6].

Depending on the sum of the daily average negative air temperatures the Northern territories are divided into the following climatic zones (Fig. 1).

2   Materials and Methods

The tasks of energy saving in the North can be divided into several, the most important areas:

1. At the stage of design and planning of human settlements to make decisions to ensure the reducing of heat dissipation from buildings.
2. During construction, use effective insulation of walls, roofs, floors and heat networks.
3. Apply modern technologies and window designs that reduce heat loss.
4. In ventilation systems, use structures based on the recovery of heat losses due to air exchange in the premises [7].

From the very beginning of the development of the Northern territories, industrial, residential and administrative buildings were built on projects that do not take into account the subsequent costs arising from the maintenance of the built facilities, which led to unreasonably high material losses during the operation of these facilities. There was a need to create projects based on energy-saving technologies that reduce the amount of material investment in the construction and operation of facilities. Nowadays, further economic development of the Far North requires the use of new creative ideas and solutions at the design stage of industrial and civil facilities. Reducing material costs in the further operation of these facilities is one of the most important tasks that should be set in the development of project documentation for construction in the Far North [9].

Fig. 1. Map of zoning of the Northern territory by the sum of average daily negative air temperatures.

Over the years of development of the Northern territories in Russia there has been accumulated extensive experience in the design and planning of Northern cities. For example, in the Norilsk industrial district of the Krasnoyarsk region to combat the wind residential areas are built very compact with narrow gaps between the houses. Due to this layout in residential buildings significantly reduced wind speed. Smooth roofs of houses with simple facade profiles reduce snow drifts on houses. In order to minimize the infiltration of buildings in strong winds, when designing, the staircase, kitchens, etc. are brought to the windward side.

Together with extreme climatic conditions (Fig. 2) construction of industrial facilities and residential buildings is complicated with the existence of permafrost and requires the use of modern technologies, starting with the laying of the foundation.

During construction in conditions of permafrost in Russia to date, there are two options of building foundations: directly on the ground and on stilts when for ventilation of frozen surface there is created a gap between the ground and the base. Construction of houses on the ground, which is constantly changing its structure, is associated with great difficulties. During the operation of the building the frozen soil is heated and loses its solidity. This method of laying the Foundation requires the use of high-quality thermal insulation to prevent thawing [11].

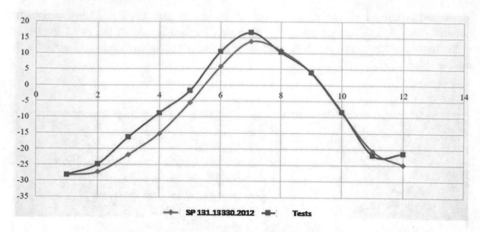

Fig. 2. Change in Norilsk air temperature.

The existence of permafrost, seasonal thawing of the soil, low bearing capacity of thawed soils and the instability of the permafrost, lead to the need to lay all pipelines on the surface of the soil – on supports or overpasses. As a result, pipelines laid in this way are exposed to a stronger influence of low temperatures than in underground laying. In winter, the heat loss from the surface of the pipes increases significantly, and there is a threat of freezing. To reduce heat losses and risk of freezing the pipelines it is necessary to apply heat insulation of pipes, to lay water pipelines with thermal satellites, to build intermediate boiler houses on water pipelines. The use and operation of these technologies require significant expenditures, do not provide a full guarantee of trouble-free operation of networks and the required level of energy saving. The problem of communications in permafrost conditions is an effective thermal insulation. Unlike small Northern towns, where communications are laid on top, in large cities, such as Norilsk, all the pipes laid underground at a depth of 6 m. In order to protect the frozen soil from heat, pipes are laid at a distance from homes, which reduces the risk of thawing the soil [14]. To reduce the heat loss of pipelines during installation, thermal insulation of various materials is used, depending on the level of their thermal conductivity and the conditions of their use in the form of special "shells", pipes with factory insulation are also used and only joints are isolated during installation (Table 1).

The use of these methods leads to an increase of energy saving in the operation of heating plants. Great opportunities for the successful solution of the problem of energy saving in the Far North consist in the reconstruction and replacement of equipment and pipelines that provide heat to the housing stock of cities and towns. At the moment, the heat supply of housing is carried out in most cases on a centralized basis. Heat power equipment, supplying heat and water to consumers, as well as the pipelines, have exceeded the standard service lifetime. According to the data of expertise carried out in the settlements of the Northern regions, the wear of this equipment reaches 60–70%. There are emergency situations that lead to the fact that at extremely low outer temperatures, residential buildings and even entire neighborhoods, with thousands of people living in them, remain without heat. In order to reduce the accident rate of equipment and reduce the amount of heat losses, it is necessary to intensify the modernization of obsolete equipment and to use modern high-performance thermal insulation when replacing the time-worn pipelines.

Table 1. Comparative characteristics of thermal insulation "shells" for different materials.

Material	Density, (kg/m^3)	Thermal conductivity, W/m^3	Material thickness, (mm)
Polyurethane foam	40–80	0.025	40
Foam polystyrene	15–50	0.038	40–150
Mineral wool	20–40	0.048	100

3 Results

On the territories of the Far North, energy saving should begin in the process of its generation. According to the up-to-date international parameters, the widespread thermal energy in Russia is considered to be low-efficient and, moreover, it often has a detrimental impact on the environment. Increasing the efficiency of the thermal power complex and reducing environmental pollution will ensure the transition from the consumption of energy produced by non-renewable sources (gas, oil, coal) to the widespread use of alternative energy (wind, ocean energy, etc.). Long-term observations and studies have shown the effectiveness of ventilated facade systems. The effectiveness of these technologies makes it possible to use them in new construction and for insulation of already built objects during their overhaul. A special advantage of this technology is the possibility of their installation regardless of the time of year and weather conditions, in contrast to the plaster systems, which can be used only at positive temperatures. Ventilated facade systems have a structural gap between the cladding and the bearing wall. The existence of a ventilated air layer shifts the condensation zone in the outer thermal insulation layer, increasing the heat storage capacity of the wall array (Fig. 3)

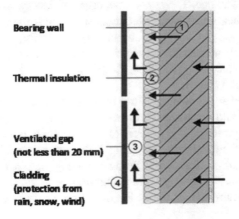

Bearing wall

Thermal insulation

Ventilated gap
(not less than 20 mm)

Cladding
(protection from
rain, snow, wind)

Fig. 3. Ventilated facade systems.

Meaning of the required resistance to vapor permeation from the condition of limiting moisture in the exterior wall panel for the period with negative air resistance temperatures is presented in Table 2.

Table 2. Meaning of the required resistance to vapour permeability R_{n2tp} with the condition of the moisture limitation in the building envelope during the period with negative monthly average temperatures, m^2 h·PA/mg.

	Dry moisture zone		Normal moisture zone		Wet moisture zone	
	req.	fact.	req.	fact.	req.	fact.
Mounted ventilated façade	3.47	5.34	2.62	5.34	2.08	5.36

For industrial facilities, an effective solution will be the use of sandwich panels in the construction of wall and roof. Sandwich panels have an optimal combination of indicators of the ratio of thermal characteristics with the cost of their production and quality of performance. The design and composition of sandwich panels allow to optimally balance decorative, heat-shielding, hydro-insulating and structural functions between different materials [15]. The use of stone wool as a heater in the production of sandwich panels has allowed to repeatedly increase the energy-saving quality of these building materials. Many years of experience in the use of these materials and technologies in the construction of settlements in the Far North, has shown that multi - layer panels with effective thermal insulation material- is the best method to increase the energy-saving effect in the construction of buildings. To ensure the required amount of energy saving, it is necessary in the manufacture of these building materials for walls, roofs and floors to use thermal insulation with a coefficient of thermal conductivity not higher than 0.04–0.07 $W/(m^2-K)$.

To further increasing of energy-saving qualities of these building materials in their manufacture, it is necessary to use modern thermal insulation materials with improved physical and mechanical properties, which will increase the service lifetime and reduce the cost of their maintenance. Therefore, now, compared to the previous period of time, requirements to the heat-insulating materials used in construction in the conditions of the Far North increase. in terms of thermal conductivity, fire safety, mechanical strength, etc. Main characteristics of physical and mechanical properties of modern thermal insulation materials (GOST 9573-2012) compared to the previously used materials are given in Table 3.

Operational and energy-saving qualities of buildings are defined not only by quality of finishing, physical and mechanical properties of heat-insulating materials, their sizes, designs, etc. The important factor is level of their protection against external adverse influences, such as temperature drops, long influence of negative temperatures and atmospheric precipitation. Therefore, during the construction of facilities in the Far North, special attention should be paid to the thermal properties of enclosing structures. High speed winds require special protection of building envelope from strong airflows.

Table 3. Compared characteristics of the main physical and mechanical properties of thermal insulation materials.

Physical and mechanical properties	Specifications	Units of measure	Advanced materials GOST 9573-2012	Earlier used materials GOST 9573-96
Density		kg/m^3	90–110	75–125
Combustibility		Degree	NG	NG (G 1)
Thermal conductivity	λ 10	W/(m K)	0.037	0.049
	λ 25		0.039	0.072
	λ A		0.044	-
	λ B		0.047	-
Compressibility, not more than		%	2	12
Tensile strength, not less than		kPa	8	-

It is necessary to calculate the temperature distribution over the thickness of the enclosing structures, especially on the inner surface, taking into account air permeability. To create a continuous thermal circuit of the building, it is necessary to choose the thickness of the insulation for all structures corresponding to this condition:

$$R_o^{pr} \geq R_o^{norm} \qquad (1)$$

This will prevent the emergence of "cold joints" that cause point cooling of surfaces, which may result in the formation of condensate.

A promising direction in increasing energy saving in the operation of buildings is the insulation of light-conducting structures. The relatively high cost of fuel and electricity in the North make it economically viable to use new technologies of active thermal buildings protection by using vacuum insulation and the possibility of obtaining additional energy from sunlight. These technologies reduce energy losses in buildings by up to 25%.

Heat-shielding characteristics of windows are also an important factor of energy-saving qualities of buildings and structures. The total heat loss through the translucent building envelope is comparable to the heat loss through the walls, despite the fact that the area of the windows in the overall structure of the building thermal circuit is much smaller than the area of the walls.

Heat transfer of window units must meet the following conditions:

$$R_o^{pr} \geq R_o^{norm} \qquad (2)$$

For the Northern territories, this requirement can be met by window structures with five or more profile cameras having low-emission glasses, the chambers of which are filled with argon. In addition, during construction to reduce heat loss in buildings and structures, it is advisable to install window blocks of smaller area. Since the length of daylight in the Northern regions in winter is very small, this will not affect the illumination of buildings, but the reduction of heat loss, according to experts, will lead to a reduction in expenditures up to one and a half times.

When eliminating the conditions leading to an undesirable level of infiltration, a sufficiently sealed thermal protective shell of the building is created, in which the functioning of natural ventilation, which occurs due to the pressure level difference inside and outside the room, is impossible. Therefore, in the design and construction of industrial and residential facilities, it is necessary to provide for the installation of mechanical supply and exhaust ventilation. One of the types of forced ventilation is a system with heat recovery of exhaust air, which can significantly reduce energy consumption for heating. Heat transfer occurs in the heat exchange cassette without mixing of the incoming flows and the supply air. (Figure 4)

Fig. 4. Principle of operation: ventilation with heat recovery.

4 Discussions

The analysis of the received information showed that the reliability of energy saving in extreme conditions of the Northern territories is one of the key issues, the solution of which will significantly improve the quality of life of the population and increase the viability of these regions. The main problems for the Northern territories are the wear of power equipment, a long period of the heating season, an isolated power supply system, low bearing capacity of soils, expensive fuel delivery.

A necessary condition for improving the reliability of power supply is to make decisions aimed at efficient energy saving at the design stage of facilities and the introduction of energy-saving materials, the use and operation of technologies to reduce heat losses and reduce operational risks. Provision of regions with qualified engineering and working personnel. Bringing to a qualitatively new technological level of energy production at the expense of modernization of power equipment. Increased use of local energy resources and renewable energy sources for efficient use of resources and reduction of environmental pollution.

5 Summary

Achieving effective energy saving in the development and development of the Far North is of great importance and is a long-term program, which is based on innovative projects of rational and efficient use of energy resources, the introduction of new technologies and materials in construction, modernization of obsolete equipment. In addition, the programme should identify the accompanying principles and mechanisms of action needed to achieve the goals and objectives, such as:

- ensuring rapid growth of production, through the introduction of science and new technologies;
- development and support of existing territorial production complexes and creation of new ones;
- development of new and maintenance of existing transport systems serving the main cargo transportation to the Northern territories;
- a differentiated approach to planning the development of the Northern territories, based on the characteristics of transport security and climatic conditions;
- the necessary social and economic development of the Northern territories.

It should also be noted that during the construction of buildings and structures in the Arctic zone of Eastern Siberia, the following most characteristic problems arise.

1. Research problems. Soil conditions of the Arctic zone have a widely heterogeneous composition and require detailed geo -, cryo-and hydrogeological studies. Changes in temperature, as well as wind, snow loads make significant adjustments to the forecast state of permafrost.
2. Design and construction problems. There is no reliable long-term information on the effectiveness of methods of freezing and/or maintaining frozen soils under buildings. Unsuitable for operation in the Arctic zone mixes and materials.
3. The problem of exploitation. Non-constant observance of the required operation is the basis for additional factors of changes in soil conditions within the boundaries of the building.

Combining existing solutions to heat efficiency problems separately for each specific site is the most appropriate way to carry out the construction and reconstruction of buildings and structures in the Far North.

Today, in the territories of the North, the living population has adapted to life and work in extreme conditions, the potential, the necessary qualifications of engineering and labor personnel. In order to prevent the outflow of population and attract new investments for the further development of areas rich in natural resources and energy resources, it is necessary to create comfortable living conditions.

At the same time, in order to avoid negative environmental consequences of the impact on the harsh and at the same time fragile nature of the Far North, careful preliminary studies should be carried out.

References

1. Ovsyannikov, S.I., Rodionov, A.S.: Justification of effective structures for the Far North. Bull. Sci. Educ. North-West Russ. **1**, 107–114 (2017)
2. Kornilov, T.A., Gerasimov, G.N.: On some errors in the design and construction of low-rise houses of light steel thin-walled structures in the far North. Ind. Civ. Constr. **3**, 41–45 (2015)
3. Ivanov, V.: Study of the aerodynamic region of the cooling system of the foundations of buildings on the filling soil in the conditions of the Far North. In: MATEC Web of Conferences, vol. 245, p. 10005 (2018)
4. Podkovyrkina, K.A., Podkovyrkin, V.S., Nazirov, R.A.: Design features of buildings and structures in the Northern latitudes from the point of view of building physics. Urban Stud. **4**, 78–85 (2017)
5. Ignatkin, I.Y.: Energy savings for heating in the far North. Vestnik NGIEI **1**(68), 52–58 (2017). (in Russian)
6. Kornilov, T.A., Nikiforov, A.A., Mordovskoy, S.B., Danilov, N.: On the experience of constructing a vented under-floor space with heat-insulated fences under the buildings of the lightweight steel-framed constructions on permafrost soils. In: IOP Conference Series: Materials Science and Engineering, vol. 463, no. 3, p. 032017 (2018). 31
7. Arenson, L.U., Phillips, M.: Living and building in permafrost regions. Geographische Rundschau **70**(11), 16–21 (2018)
8. Vaganova, N.A.: Simulation of thermal stabilization of bases under engineering structures in permafrost zone. In: AIP Conference Proceedings, vol. 2048, p. 030010 (2018)
9. Wang, T., Zhou, G., Chao, D., Yin, L.: Influence of hydration heat on stochastic thermal regime of frozen soil foundation considering spatial variability of thermal parameters. Appl. Therm. Eng. **142**, 1–9 (2018)
10. Kudryavtsev, S., Borisova, A.: The research of the freezing and thawing process of the foundations with the use of season and cold-producing devices. In: MATEC Web of Conferences, vol. 193, p. 03040 (2018)
11. Markov, E.V., Pulnikov, S.A., Yu, S.S.: Methodology for calculating the parameters of the thermal interaction between the pipeline and the soil. Int. J. Civ. Eng. Technol. **9**(6), 1397–1403 (2018)
12. Baryshnikov, A.A., Mustafin, N.Sh., Shadrina, A.A.: Application of composite building materials in the North. Reg. Dev. **8**, 6 (2015)
13. Podkovyrina, K.A.: Optimization of external enclosing structures with regard to energy efficiency and economic feasibility. PhD Thesis (2017)
14. Sychev, S.A., Shevtsov, D.S.: Prefabricated high-rise buildings from modular transformable building systems of high factory readiness in the far North. Bull. Civ. Eng. **1**(60), 153–160 (2017)
15. Blum, A.D., Luchkova, V.I.: Features of the development of low-rise construction in the far North of Russia and equivalent localities. In: New ideas of the new century: proceedings of the international scientific conference FAD PNU, vol.1, pp. 15–19 (2012)

Probabilistic-Statistical Model of Climate in Estimation of Energy Consumption by Air Conditioning Systems

Elena Malyavina⬛, Olga Malikova$^{(\boxtimes)}$⬛, and Van Luong Pham⬛

Moscow State University of Civil Engineering,
Yaroslavskoe Shosse 26, 129337 Moscow, Russia
KryuchkovaOU@mgsu.ru

Abstract. For a reliable assessment of energy consumption by the air conditioning system, data on the frequency of occurrence of combinations of temperature and enthalpy of outside air in the construction area is needed. The purpose of this article is to consider the influence of the detail of the probabilistic-statistical climate model on the results of calculating the consumption of heat, cold, and electricity by a direct-flow air conditioning system with a controlled cooling process serving office premises in Hanoi (Vietnam). To achieve this goal, the results of energy consumption calculations are considered in two versions: details of the climate model: the repeatability of combinations of climate parameters in option 1 is given for cells with a step of 2 °C, relative humidity of 5%, and in option 2 for cells with a step of 1 °C, relative humidity of 2.5%. The calculation results indicate a significant increase in the accuracy of calculations with a smaller step of the climate model.

Keywords: Climate model · Temperature · Relative humidity · Air conditioning · Consumption

1 Introduction

A correct assessment of the supply air treatment energy consumption plays a significant role in the building design works in different countries [1–3]. In Russia, this issue is also given a great attention [4–9]. This assessment is important for Hanoi [10–12] as well. The air conditioning systems of Hanoi are operating in specific climatic conditions of the Northern Vietnam central part. Hanoi is located at 21° of the northern latitude and 106° of the east longitude.

The climate of Hanoi belongs to the tropical monsoon climate. The average temperature in winter is 16.5 °C (with decreases to 2.7 °C), an average summer temperature is 29.5 °C (with increases to 43.7 °C). The average annual temperature is 23.2 °C, the average annual rainfall is 1,800 mm. The hot season begins in mid-April and lasts until mid-September, it is cool and dry from late September to November in Hanoi. The cold season begins in early November and lasts until the end of March next year. From the end of November to January the weather is cold and dry, from February to the end of March the weather is cold with long hoarfrost. However, due to the strong influence of

© Springer Nature Switzerland AG 2020
V. Murgul and M. Pasetti (Eds.): EMMFT 2018, AISC 982, pp. 69–77, 2020.
https://doi.org/10.1007/978-3-030-19756-8_7

the monsoon, the start and end times of each season often vary from year to year, so the division into seasons by month is quite relative.

2   Method of Calculation

To calculate the energy consumption of the air conditioning utilities, the construction area climatic model is taken as the climate basis being a subject of research as well [13, 14]. In this paper, a probabilistic-statistical model [15] of the Hanoi climate is adopted. It is represented as a Table with cells that indicate the repeatability of the temperature and relative humidity combinations given in each cell. Since the aim of the work is to clarify the role of the model detailing influence on the accuracy of the calculation results of the air conditioning system energy consumption, the climate model is considered in two versions: the first option with cells of a 2 °C step by temperature and a 5% by relative humidity, as well as the second option with cells of a 1 °C step by temperature and 2.5% by relative humidity.

The calculation of energy consumption is made for a direct-flow air conditioning system with a controlled cooling process. Due to the almost constant high relative humidity of the outdoor air during the year, its humidification is not provided in the air conditioning unit. The operation period is accepted in two versions: from 8 am to 17 pm and the day as a whole. The duration of the system operation makes 2223 h per year for operation hours from 8 am to 17 pm and 5928 h per year with 24 working hours. The outdoor air flow rate for all the utilities is accepted the same and equal to 10 000 kg/h.

Taking into account the possibilities of the air treatment by the mentioned air conditioning unit, the area of all combinations of temperature and relative humidity of Hanoi can be divided into weather zones with the outdoor air parameters within which the system operates in the same mode. For an illustration of such a division of one of the calculation variants, see the Fig. 1. With the parameters of the outdoor air, characterized by the center of each cell, the instantaneous consumption of heat, cold, electricity and water may be determined. Divisions into the weather zones are known and published in the literature, for example, [16]. The energy consumption calculation has been performed for each combination of temperature and relative humidity that took place during the year, taking into account its repeatability. In Fig. 1, the numbering of the weather zones is consistent with the accepted numbering in the Russian literature on air conditioning.

The air is heated in the zone 3. At the same time, heat and electricity are consumed for the operation of the pump transferring the heating agent. The zone 4 corresponds to the outdoor air parameter combinations that do not require its treatment. In the weather zone 5 the cooling is carried out without the air drying (while the outdoor air moisture content is less than this one of the air with the Ko point parameters), i.e. the zone 5a or with its drying – the zone 5b. The Ko point characterizes the limiting (with the maximum possible relative humidity $\varphi = 100\%$) condition of the supply air at the outlet of the air cooler. In this case, the temperature of the Ko point is assumed [16] to be 3 °C higher than the temperature of the cold water or the refrigerant (for a direct cooling air cooler). In zone 5, the electrical power is used to drive the compressor.

Combinations of the outdoor air parameters in the zone 7 require the air preheating to the air treatment beam from the Ko point through the point of maximum values of the temperature and relative humidity of the required supply air (the upper right point of the zone 4).

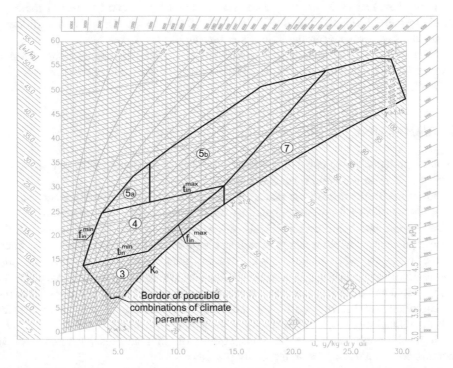

Fig. 1. Division of the outdoor climate area in the Hanoi city into the weather zones, where operate the utilities of the direct-flow air conditioning system without the supply air humidification under a controlled cooling, which supports the room indoor air parameters from 8 a.m. to 17 p.m. within 18–28 °C, 30–60% with a heat-humidity ratio of the room air parameter changes of 20 000 kJ/kg. Symbols: t_{in}^{max}, t_{in}^{min} – the maximum and minimum values of the required supply air temperature; f_{in}^{max}, f_{in}^{min} – maximum and minimum values of the required supply air relative humidity.

It should be borne in mind that the configuration of the zone 4, the upper boundary of the zone 3, and the lower boundary of the zone 5 are completely determined by combinations of the required indoor air parameters, the working temperature difference between the indoor and the supply air, as well as the heat - humidity ratio of the heat and moisture income to the room. The size of the zone 7 depends both on the required combination of maximum values of the supply air temperature and relative humidity, and on the temperature of water or the refrigerant in the air cooler.

The calculation has been carried out to maintain the following options of the indoor air parameters in the office building rooms during the year: a working room – the temperature within 18–28 °C, the relative humidity 30–60%; the head's office – the temperature within 20–27 °C, the relative humidity 35–60%; the meeting room, the conference room – the temperature within 16–27 °C, the relative humidity 30–60%.

It cannot be said that only the above mentioned premises have been subject to an analysis, since it has been believed that each of them may have a different ratio of the incoming heat and moisture to be taken into account by different values of the heat-humidity ratio: 5000, 10 000, 20 000, 30 000, 60 000 kJ/kg, when the working temperature difference between the indoor and the supply air makes 5 °C.

When the heat-humidity ratio is 0 kJ/kg, the working temperature difference is assumed to be 0 °C. The temperature of cold water in the air cooler was considered to be 7 °C.

3 Calculation Results

Table 1 presents the calculation results of the energy consumption by an air conditioning system in the options specified above.

From the Table 1 it follows that a different detailing of the climatic model did not affect the trends in energy consumption by the air conditioning system with increasing values of the heat-humidity ratio. Thus, the heat consumption by the heater and the electric power by the heater pump, as well as the number of the heat consumption hours by the air heater, when maintaining a constant operating temperature difference and achieving the same internal air parameters have been remaining permanent, while the costs of cold and electricity for the compressor drive, as well as the number of the cold consumption hours have been falling under the same conditions. The number of the unit operation hours without the supply air processing increases slightly.

It is absolutely clear that the consumption of all resources increases with an increase in the number of working hours per day.

It is clear that increasing the lower limit of the internal air temperature, to which the air conditioning system heats the room in the cold season, increases the need for heat. Similarly, it seems logical that in the warm period of the year to cool the room to the same temperature of 27 °C less cold is required in the variant, where the humidity of the internal air is allowed to be higher (65% > 60%). Also, if at the same humidity of 60% it is allowed to maintain a higher room temperature indoors (28 °C > 27 °C), the cold consumption will be reduced.

Table 1 also shows that in Hanoi, the need for heat in the zone 3 is significantly (by 2–4 orders of magnitude) less than the need for cold. At the same time, in the zone 7 it is necessary to provide the incoming air preheating before cooling it with drying. Moreover, the heat consumption, although it does not reach the value of the need for cold, is 2.3 to 4.6 times less than it. For most of the year, the air conditioning unit is set to cooling, although the unit operation time without processing the supply air is also significant and far exceeds the heating time of the supply air in the zone 3.

The estimation of the accuracy of the obtained results is given in the Table 2, which shows the differences between the results of the two variants of the probabilistic-statistical model in percentage terms as to the results of the model 2 (more detailed), that is why they should be considered more accurate.

Table 1. Heat and electrical power consumption and the duration of the year consumption of these resources, obtained by different climate model detailing.

Heat-moisture ratio, kJ/kg	Heat consumption by the air heater in the zone 3, kW·h	Heat consumption by the air heater in the zone 7, kW·h	Electrical power consumption by the air heater pump, kW·h	Number of hours of heat consumption, h	Cold consumption, kW·h	Number of hours of cold consumption, h	Number of operation hours without air treatment, h
Model No. 1: graduation 2 °C for temperature and 5% for relative humidity							
Operation time is from 8 am to 17 pm							
Indoor air parameters 16–27 °C, 30–65%							
0	2571	41250	244	1482	77747	1195	707
5000	153	54957	250	1515	132657	1558	583
10000	153	22771	206	1247	83929	1442	700
20000	153	17318	195	1179	72570	1442	700
30000	153	16030	190	1149	69602	1442	700
60000	153	14605	178	1081	67176	1392	749
Indoor air parameters 20–27 °C, 35–60%							
0	7975	64564	326	1977	109005	1361	231
5000	1661	97351	316	1917	186162	1675	227
10000	1661	42665	284	1722	114755	1558	344
20000	1661	33164	269	1630	101050	1499	403
30000	1661	31014	266	1610	97185	1499	403
60000	1661	31014	266	1610	97185	1499	403
Indoor air parameters 18–28 °C, 30–60%							
0	4831	53785	286	1733	88327	1274	473
5000	614	75946	279	1694	152383	1559	479
10000	614	34972	246	1494	95377	1436	602
20000	614	27188	228	1385	83153	1365	673
30000	614	25491	225	1366	80186	1365	673
60000	614	24014	223	1350	77675	1365	673
24-hour operation							
Indoor air parameters 16–27 °C, 30–65%							
0	7135	120171	675	4090	217215	3269	1792
5000	473	159770	701	4249	369249	4194	1498
10000	473	68630	588	3566	232958	3838	1854
20000	473	52880	565	3423	202356	3838	1854
30000	473	49078	564	3420	195471	3838	1854
60000	473	44848	530	3212	188235	3688	2004
Indoor air parameters 20–27 °C, 35–60%							
0	21751	186603	900	5457	304643	3786	451
5000	4676	278889	880	5331	518536	4570	491
10000	4676	125286	797	4832	319631	4194	867
20000	4676	98394	756	4580	281061	4018	1044
30000	4676	92317	751	4552	270876	4018	1044
60000	4676	87044	746	4522	262013	4018	1044
Indoor air parameters 18–28 °C, 30–60%							
0	13268	155982	793	4805	247060	3532	1102
5000	1809	220211	790	4788	424930	4347	1075
10000	1809	103506	696	4219	265620	3898	1524
20000	1809	81221	649	3932	231857	3669	1753
30000	1809	76391	643	3896	223552	3669	1753
60000	1809	72175	635	3850	216360	3669	1753

(*continued*)

Table 1. (*continued*)

Heat-moisture ratio, kJ/kg	Heat consumption by the air heater in the zone 3, kW·h	Heat consumption by the air heater in the zone 7, kW·h	Electrical power consumption by the air heater pump, kW·h	Number of hours of heat consumption, h	Cold consumption, kW·h	Number of hours of cold consumption, h	Number of operation hours without air treatment, h
Model No. 2: graduation 1 °C for temperature and 2.5% for relative humidity							
Operation time is from 8 am to 17 pm							
Indoor air parameters 16–27 °C, 30–65%							
0	1710	50537	256	1553	98856	1354	617
5000	73	64810	256	1551	157731	1658	541
10000	73	28268	216	1311	103616	1553	647
20000	73	21710	202	1223	91482	1532	667
30000	73	20076	196	1186	88095	1517	682
60000	73	18749	192	1166	85373	1517	682
Indoor air parameters 20–27 °C, 35–60%							
0	6241	75340	324	1964	133657	1439	230
5000	1092	112116	304	1845	217189	1738	300
10000	1092	50518	270	1635	137840	1621	417
20000	1092	39986	259	1571	122440	1587	450
30000	1092	37493	256	1552	118502	1584	453
60000	1092	37493	256	1552	118502	1584	453
Indoor air parameters 18–28 °C, 30–60%							
0	3540	63792	287	1737	110838	1365	460
5000	353	88477	274	1663	180287	1654	488
10000	353	41767	238	1441	116598	1526	615
20000	353	33283	228	1383	103474	1494	648
30000	353	31265	225	1363	100033	1490	651
60000	353	29546	223	1349	97048	1490	651
24-hour operation							
Indoor air parameters 16–27 °C, 30–65%							
0	4792	146979	713	4321	275737	3713	1538
5000	233	187731	720	4365	438887	4479	1377
10000	233	84466	621	3762	288001	4155	1701
20000	233	65406	585	3548	254955	4060	1797
30000	233	60883	573	3470	246154	4041	1816
60000	233	57020	566	3428	238648	4041	1816
Indoor air parameters 20–27 °C, 35–60%							
0	17042	216891	895	5427	372869	3977	465
5000	3115	320238	848	5141	604457	4737	685
10000	3115	147475	761	4611	383544	4358	1064
20000	3115	117759	734	4451	340511	4260	1162
30000	3115	110666	727	4406	329628	4248	1174
60000	3115	104377	720	4364	320210	4222	1200
Indoor air parameters 18–28 °C, 30–60%							
0	9731	183397	789	4780	309156	3743	1115
5000	1059	254055	767	4647	501997	4512	1180
10000	1059	122786	677	4105	324558	4119	1573
20000	1059	98729	651	3944	287631	4020	1672
30000	1059	92906	642	3888	278187	3993	1699
60000	1059	88024	639	3870	270012	3993	1699

Differences in the assessment of individual components of the energy resource consumption with a greater climatic model detailing are explained by the fact that such a model enables a more accurate setting of the boundaries of possible climate combinations in the construction area, as well as the boundaries between the weather zones. Therefore, a more detailed consideration of the operation of each device in different weather zones has a significant impact on the obtained results.

The minus sign refers to the option estimations. It is noteworthy that large relative deviations are associated with small absolute values of heat and electricity consumption in the cold period of the year, as well as with short durations of heat consumption. However, the clarification of the cold consumption value of 15%–21.5% of the significant values of this need during the year should also be attributed to a very significant one. In addition, the heat consumption depends only on the outdoor air temperature, while the cold one depends on the combination of the temperature and humidity, the accuracy of which affects the assessment of the energy consumption more noticeably than the accuracy of one parameter.

It is interesting to compare the results presented here for Hanoi with those obtained earlier [17] for Moscow. The fact, that in the Moscow conditions the discrepancy between the results of the heat consumption calculations makes within 14%–17.3% and of the cold within 29.8%–38.9%, is explained by a longer supply air heating period than a cooling one, as well as the fact that the annual consumption of heat is bigger than of cold. The absolute values of the cold consumption in Moscow are much greater than the value of the heat consumption in Hanoi, so the relative errors in the consumption of cold in Moscow against their absolute values are substantially less than the relative errors of the heat consumption in Hanoi against their small absolute values.

Table 2. Comparison of the calculation results of annual energy consumptionby air conditioning devices at different climate model detailing.

Heat-moisture ratio, kJ/kg	Increased accuracy, %, when calculated according to the more detailed model 2						
	Heat consumption by the air heater in the zone 3, kW·h	Heat consumption by the air heater in the zone 7, kW·h	Electrical power consumption by the air heater pump, kW·h	Number of hours of heat consumption, h	Cold consumption, kW·h	Number of hours of cold consumption, h	Number of operation hours without air treatment, h
Operation time is from 8 am to 17 pm							
Indoor air parameters 16–27 °C, 30–65%							
0	−50	18	5	5	21	12	−15
5000	−110	15	2	2	16	6	−8
10000	−110	19	5	5	19	7	−8
20000	−110	20	3	4	21	6	−5
30000	−110	20	3	3	21	5	−3
60000	−110	22	7	7	21	8	−10
Indoor air parameters 20–27 °C, 35–60%							
0	−28	14	−1	−1	18	5	0
5000	−52	13	−4	−4	14	4	24
10000	−52	16	−5	−5	17	4	18

(continued)

Table 2. (*continued*)

Heat-moisture ratio, kJ/kg	Increased accuracy, %, when calculated according to the more detailed model 2						
	Heat consumption by the air heater in the zone 3, kW·h	Heat consumption by the air heater in the zone 7, kW·h	Electrical power consumption by the air heater pump, kW·h	Number of hours of heat consumption, h	Cold consumption, kW·h	Number of hours of cold consumption, h	Number of operation hours without air treatment, h
20000	−52	17	−4	−4	17	6	10
30000	−52	17	−4	−4	18	5	11
60000	−52	17	−4	−4	18	5	11
Indoor air parameters 18–28 °C, 30–60%							
0	−36	16	0	0	20	7	−3
5000	−74	14	−2	−2	15	6	2
10000	−74	16	−3	−4	18	6	2
20000	−74	18	0	0	20	9	−4
30000	−74	18	0	0	20	8	−3
60000	−74	19	0	0	20	8	−3
24-hour operation							
Indoor air parameters 16–27 °C, 30–65%							
0	−49	18	5	5	21	12	−17
5000	−103	15	3	3	16	6	−9
10000	−103	19	5	5	19	8	−9
20000	−103	19	3	4	21	5	−3
30000	−103	19	2	1	21	5	−2
60000	−103	21	6	6	21	9	−10
Indoor air parameters 20–27 °C, 35–60%							
0	−28	14	−1	−1	18	5	3
5000	−50	13	−4	−4	14	4	28
10000	−50	15	−5	−5	17	4	19
20000	−50	16	−3	−3	17	6	10
30000	−50	17	−3	−3	18	5	11
60000	−50	17	−4	−4	18	5	13
Indoor air parameters 18–28 °C, 30–60%							
0	−36	15	−1	−1	20	6	1
5000	−71	13	−3	−3	15	4	9
10000	−71	16	−3	−3	18	5	3
20000	−71	18	0	0	19	9	−5
30000	−71	18	0	0	20	8	−3
60000	−71	18	1	1	20	8	−3

4   Conclusion

Comparison of the energy consumption by direct-flow air conditioning systems without humidification with controlled cooling process in the conditions of Hanoi (SR of Vietnam) shows that the initial climatic information in a more detailed form enables better accuracy results of the energy consumption for a building cooling by 15%–21.5%. The energy consumption for the air heating in Hanoi is too small to focus on the accuracy of its calculation. Therefore, it is reasonable to provide a detailed development of the probabilistic-statistical model, i.e. with a step of 1 °C in temperature and 2.5% in relative humidity.

References

1. Chua, K.J., Chou, S.K., Yang, W.M., Yan, J.: Achieving better energy-efficient air conditioning – a review of technologies and strategies. Appl. Energy **104**, 87–104 (2013)
2. Oldewurtel, F., Sturzenegger, D., Morari, M.: Importance of occupancy information for building climate control. Appl. Energy **101**, 521–532 (2012)
3. Afram, A., Janabi-Sharifi, F.: Review of modeling methods for HVAC systems. Appl. Therm. Eng. **67**(1–2), 507–519 (2014)
4. Samarin, O.D., Lushin, K.I., Kirushok, D.A.: Energy-saving air handling circuit with indirect evaporative cooling in plate heat exchangers. Hous. Build. **1–2**, 43–45 (2018). (in Russian)
5. Malyavina, E.G., Kryuchkova, O.Y.: Analysis of annual power consumption by central air conditioning systems using the climatic data stochastic statistics model. In: Environmental engineering. 8-th International Conference, Vilnius, Lithuania, vol. 19–20, pp. 776–780 (2011)
6. Pshenichnikov, V.M.: Building heat and energy consumption management based on mathematical modeling. HVAC **8**, 48–53 (2018). (in Russian)
7. Kostin, V.I., Rakova, E.A.: Constructive schemes of control systems of a microclimate of premises with the constant temperature of internal air. News High. Educ. Inst. Constr. **4**, 59–66 (2018). (in Russian)
8. Kostin, V.I., Russkikh, E.Yu.: Calculation of expenses of cold on air conditioning systems of industrial buildings. HVAC **5**, 18–21 (2012). (in Russian)
9. Arbatsky, A.A., Afonina, G.N.. Calculation of annual energy use by refrigeration centers. HVAC **6**, 10–13 (2017). (in Russian)
10. TCVN 5687.2010: Ventilation, air conditioning - Design Standards Vietnam, Hanoi (2010) (in Vietnamese)
11. Chẩn, T.N.: Air Conditioning. Building Publishing, Hanoi (2002). (in Vietnamese)
12. Van Lương, P.: Energy saving solution for a large air conditioning system. Vietnamese Association of Civil Environment, Hanoi (2015). (in Vietnamese)
13. Afram, A., Janabi-Sharifi, F.: Theory and applications of HVAC control systems – a review of model predictive control (MPC). Build. Environ. **72**, 343–355 (2014)
14. Oldewurtel, F., Jones, C.N., Parisio, A., Morari, M.: Stochastic model predictive control for building climate control. IEEE Trans. Control Syst. Technol. **22**(3), 1198–1205 (2013)
15. Malyavina, E.G., Kryuchkova, O.Y., Kozlov, V.V.: Comparison of climate models for calculating energy consumption central air conditioning systems. Hous. Build. **6**, 24–26 (2014). (in Russian)
16. Belova, E.M.: Central air conditioning systems in buildings. Euroclimate, Moscow (2006). (in Russian)
17. Malyavina, E.G., Malikova, O.Y. Comparison of the completeness of the climate probability-statistic model and the reference year model. In: IOP Conference Series: Materials Science and Engineering, Moscow, vol. 365 (2018). https://doi.org/10.1088/1757-899X/365/2/022009

Local Air Humidifiers in Museums

Darya Abramkina$^{(\boxtimes)}$ (iD) and Angelina Ivanova (iD)

Moscow State University of Civil Engineering,
Yaroslavskoye Shosse 26, 129337 Moscow, Russia
dabramkina@ya.ru

Abstract. The main purpose of the study is to evaluate the performance of local air humidifiers in museum premises with natural ventilation systems. This article discusses the possibility of using local air humidifiers not only to ensure required relative air humidity, but also as an additional way to reduce the concentration of fine suspended particles in principal premises of museums. The relative air humidity was measured in art gallery using Testo data logger. Also, counting concentration of suspended particles was measured in the internal air with working and disconnected air humidifier. The greatest decline of concentration is observed for suspended particles with diameter size larger than 5 microns, which is probably due to a greater sedimentation rate. Installing local humidifiers is extremely important in buildings with natural ventilation systems, because it can provide not only required relative air humidity but also reduce the concentration of fine suspended particles, which helps exhibits to retain its initial outlook.

Keywords: Local air humidifier · Fine suspended particles · Museums

1 Introduction

1.1 Requirements for Relative Air Humidity in Museums

Maintenance of required relative air humidity in principal premises of museums in cold period of the year is one of the most necessary problems of historic exhibits preservation. The principal premises of museum are exposition halls, storages of museum collections, restoration studios and temporary storage areas for exhibits. Required parameters of internal microclimate during the storage of different objects are in Table 1 [1].

Especially valuable exhibits, which should be located in places with special temperature and relative air humidity, are accommodated in glass cases, which are called "closed storage" (see **Ошибка! Источник ссылки не найден.**). In addition, it is necessary to take into account the history of environmental characteristics changes that means that we should consider adaptation of exhibits during their exposure and storage.

For technical and esthetic reasons, sometimes it is not possible to use air conditioning systems in heritage buildings [2, 3]. In such cases, local air humidifiers are used to provide required air humidity in exhibition halls and other principal premises (see Fig. 2), most importantly, when exhibits are installed near external walls and heating devices (Fig. 1).

© Springer Nature Switzerland AG 2020
V. Murgul and M. Pasetti (Eds.): EMMFT 2018, AISC 982, pp. 78–83, 2020.
https://doi.org/10.1007/978-3-030-19756-8_8

Table 1. Required parameters of internal microclimate in museums.

Exhibits	Temperature, °C	Relative humidity, %
Archeological artifacts	18–22	20–30
Leather artifacts, parchment	18–22	25–40
Textile, clothing, carpets	18–22	30–50
Painting on canvas	18–22	45–55
Painting on wood	18–22	40–60

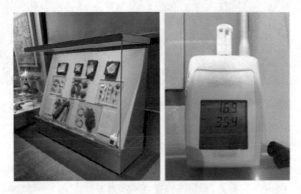

Fig. 1. The Pushkin State Museum of Fine Arts. Left – photo of closed storage, right – photo of temperature and relative air humidity in glass case.

The most common design of local humidifiers used in the exhibition halls, are devices that operating as cold evaporation, which completely excludes air overmoistcning and white coating formation on the surfaces. With different humidity parameters, it is possible to observe the processes occurring with the exhibits materials. For example, rapid corrosion of metal objects occurs during the high humidity, pests and mold growth more frequently, but shrinkage and cracking of organic materials occur more often at low relative air humidity.

Modern air humidifiers can be equipped by automatic water supply, additional membrane filters for air purification, and sensors of air relative humidity, touch-sensitive temperature display, work timers, remote control and monitoring.

A lot of museum exhibits absorb and excrete water. These materials and indoor environment can slow down relative air humidity changes, which can minimize the damage. The moisture sources in indoor environment are: human breathing, perspiration, the level of external humidity, walls moisture, floor washing, condensation and evaporation processes.

Fig. 2. The Pushkin State Museum of Fine Arts. Layout of engineering equipment.

1.2   Gas Composition of Air in Museums

Gas composition of air has a major impact on the preservation of museum exhibits. The air of modern cities is usually polluted by Sulphur dioxide, hydrogen sulfide, ozone, nitrogen dioxide, smoke, dust and other harmful components [4]. The most dangerous pollutant of air is fine suspended particles (dust), because they are not only easily gets into the building envelope, but can also stay suspended for a long time in the internal air [5]. Fine suspended particles can penetrate through the insignificant slots of closed storage. The main sources of indoor dust pollution are interior objects, visitors and finishing materials [6]. The composition of dust is different depending on its origin. For example, fine suspended particles that enter the room along with external air contain a high percentage of soot and tar, and indoor dust contains fibers of various origins, pieces of tissue, rubber and leather. Fine suspended particles enter the smallest cracks of paintings and can lead the destruction of its structure, causes blackout colors. Cement dust is also harmful to textile fibers and paintings, because it has abrasive and alkaline properties. A lot of fine suspended particles have acidic properties contribute to metal corrosion, textile aging, and blackening of stones, because they are caustic.

Besides dust, different chemical compounds such as acetic acid, hydrochloric formic acid, formaldehyde, and various organic radicals resulting from the oxygen reaction with exhaust gases have an aggressive effect on museum exhibits [7]. Formic and acetic acids contribute to metal corrosion, while formaldehyde promotes aging of various materials [8, 9]. Nitrogen dioxide action causes paint bleaching and destruction of unstable lacquers, hydrolysis of cellulose, corrosion of metals, rapid aging of calcium-containing minerals and murals. Metals, marble, and limestone are more vulnerable to such effects. The presence of oxygen in the air causes constant inevitable aging of organic materials, intense fading of fabrics, dyes, oxidation of oils, corrosion of metals even in a dry atmosphere. Especially strong destructive effect for organic materials has oxygen and light: fading of painted wooden exhibits, fabrics and oils.

This article discusses the possibility of using local air humidifiers not only to ensure required relative air humidity, but also as an additional way to reduce the concentration of fine suspended particles in principal premises of museums.

2 Materials and Methods

The first step of the study was surveillance of relative air humidity in the exhibition hall of private art gallery. The exhibition hall is located in heritage building equipped with a natural ventilation system. Measurements were conducted using Testo 174H logger during the cold period of a year.

Exhibition hall has been equipped by air humidifier, which operates according to the following principle: the air from the room enters the internal part of humidifier through the ventilation grids, and then passes through the cellular cylinder, which slowly rotates in the water (see Fig. 3). Water is able to detain the smallest dust particles, thereby cleansing the supplied air.

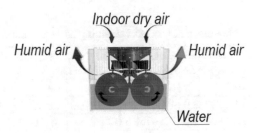

Fig. 3. Local humidifier with cellular cylinders.

To determine the average daily counting concentration of suspended particles in the indoor air, a Fluke 985 counter was used. Registration of aerosol particles is based on the photoelectric method, using the dependence of scattered light intensity on the particle size.

3 Results

According to the results of the study, graphs of relative air humidity daily variations were presented in Fig. 4. Horizontal lines show the range of required relative air humidity for storing paintings on canvas and wood. When humidifier was turned off, relative air humidity remains almost constant, below normalized values. This problem is precipitated by dehumidification of air due to the operation of heating system. When the device is in operation, the required relative air humidity for storing paintings on canvas is achieved 20 min after switching on. At the same time, humidifier causes fluctuation of parameters, however they are still remain optimal values, without causing a destructive impact on the exhibits. Air mobility control was carried out by measuring velocity with a hot-wire anemometer.

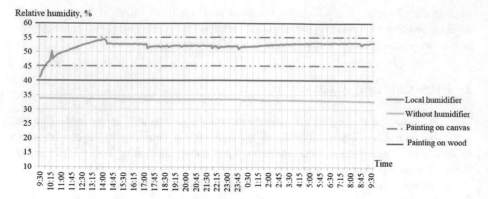

Fig. 4. Relative air humidity measurement results.

The Table 2 presents the results of counting concentration of suspended particles measuring in the air of art gallery.

Table 2. Measurement results of counting particles concentration

Operation mode	Concentrations of particles N, particles/m^3, with dimensions equal to or greater than the following values:					
	0.3	0.5	1	2.5	5	10
Without humidifier	41879768.0	4466799.0	762361.2	205865.0	172216.6	6287.1
With humidifier	37893432.0	4296769.5	409325.6	64164.3	4357.7	1395.4

4 Discussion

Air mobility in art gallery did not exceed 0.15 m/s. In still air fine suspended particles with diameter size smaller than 2.5 microns are practically not deposited and staying in permanent Brownian motion [10]. The greatest decline of concentration is observed for suspended particles with diameter size larger than 5 microns, which is probably due to a greater sedimentation rate [11]. In Russian hygienic standards, maximum permissible concentrations of dust with various sizes ($PM_{2.5}$ and PM_{10}) are given only in assessing the quality of atmospheric air. For internal air, particle size is not taken into account, which is a shortcoming, especially in assessing of indoor air quality in museums. The values of countable concentration of suspended particles are rationed only for clean rooms. Studies [12] show that the air dustiness in museums can cause not only the destruction of exhibition objects, but also affect occupant and visitors health. It is necessary to carry out more studies of dangerous effects of suspended fine particles ($PM_{2.5}$ and PM_{10}) towards different types of exhibition objects, including the effects of seasonal variations. In museums, equipped natural ventilation systems, with insufficient air exchange during warm and transitional periods of the year, the concentration of suspended particles may exceed the concentration in atmospheric air due to the influence of internal dust sources. Some researchers also believe that there is a need for investigation more accurate terminology relating to suspended particles in museums, allowing identifying different types of impacts on cultural objects [13].

5 Conclusion

Without a humidifier, the concentration of particles increases, which negatively affects the preservation of exhibits. Installing local humidifiers is extremely important, especially in buildings with natural ventilation systems, because it allows not only maintaining the required relative air humidity for various materials, but also reducing the concentration of fine suspended particles, which helps exhibits to retain its initial outlook.

References

1. Tabunshchikov, Yu.A., Bolotov, E.N., Brodach, M.M.: Museums. Heating, ventilation, air conditioning. AVOK, Moscow (2018)
2. Murgul, V., Vuksanovic, D., et al.: Development of the ventilation system in historical buildings of St. Petersburg. Appl. Mech. Mater. **633–634**, 977–981 (2014)
3. Mohd Dzulkifli, S.N., Abdullah, A.H., Leman, A.M.: Design and material in museum: does it affect the ventilation in indoor air quality? ARPN J. Eng. Appl. Sci. **11**(11), 7341–7348 (2016)
4. Sharma, S.B., Jain, S., Khirwadkar, P., Kulkarni, S.: The effects of air pollution on the environment and human health. Indian J. Research in Pharm. Biotechnol. **1**(3), 391–396 (2013)
5. Abramkina, D.V., Agakhanova, K.M.: Influence of natural air exchange on the concentration of suspended particles. Russ. Automobile Highw. Ind. J. **15**(6), 912–921 (2018)
6. Katiyar, V., Khare, M.: Indoor Dust. Chery Publication, Mumbai (2011)
7. Ligterink, F., Di Pietro, G.: The limited impact of acetic acid in archives and libraries. Heritage Sci. **6**(1), 59 (2018)
8. Gibson, L.T., Watt, C.M.: Acetic and formic acids emitted from wood samples and their effect on selected materials in museum environments. Corros. Sci. **52**(1), 172–178 (2010)
9. Englund, F., Fjaestad, M., Ferm, M.: Corrosivity of the air and the influence of building and furnishing materials in museums. In: 12th International Conference on Indoor Air Quality and Climate 2011, pp. 2234–2239. Curran Associates, New York (2011)
10. Wu, Y., Liu, J., Zhai, J., et al.: Comparison of dry and wet deposition of particulate matter in near-surface waters during summer. PLoS ONE **13**(6), e0199241 (2018)
11. Hanapia, N., Mohd Dinb, S.A.: A study on the airborne particulates matter in selected museums of Peninsular Malaysia. Procedia Soc. Behav. Sci. **50**, 602–613 (2012)
12. Lazaridis, M., Katsivela, E., et al.: Indoor/outdoor particulate matter concentrations and microbial load in cultural heritage collections. Heritage science **3**, 34 (2015)
13. Grau-Bove, J., Strlic, M.: Fine particulate matter in indoor cultural heritage: a literature review. Heritage Sci. **1**, 8 (2013)

Numerical Simulation of a Stable Microclimate in a Historic Building

Inna Sukhanova[(⊠)] and Kirill Sukhanov

Saint-Petersburg State University of Architecture and Civil Engineering,
2-nd Krasnoarmeiskaya st. 4, 190005 St. Petersburg, Russia
inna.suhanova@mail.ru

Abstract. A numerical model of the microclimate and air quality in the stables is built in the STAR CCM+ software. A fragment of a historic building in St. Petersburg has been taken for modeling. The model takes into account features of architecture, ventilation equipment, the number of horses. To compensate for the loss of heat, air heating is provided, combined with a ventilation system. Two options for supplying fresh air to the service area are considered: from the top downwards from the center of the corridor along the arches to each stall by horizontal jets and into the lower zone of the room. Air removal is carried out from the upper zone of the corridor. The distribution of velocity and temperature fields, relative humidity, and CO_2 concentrations in the stalls is obtained. The conclusion is made about the provision of normalized microclimate parameters in stables with the proposed solutions for heating and ventilation systems.

Keywords: Microclimate of stables · Internal air quality ·
Numerical simulation

1 Introduction

Equipping historic buildings with modern engineering systems to ensure the micro-climate is an essential aspect of all restoration programs, especially if the building is an architectural monument and at the same time is used as an operating stable. Heating and ventilation systems should ensure the integrity of the building and comfortable conditions for keeping the horses, as well as for the work of the staff. Making any changes to the interior of the building is not allowed in this case [1–4]. In the case of the object under study, there are animals in the room, which are a source of heat, moisture, carbon dioxide (CO_2) release, and ammonia is released due to their vital activity. An air environment with a high relative humidity and temperature is formed in a stable. In the cold period of the year, a significant amount of water vapor moves through the outer walling, saturating them with moisture and destroying building materials [5]. That is why numerical simulation methods are used at the design stage to select a variant of ventilation systems, assess the possibility of providing microclimate and air quality parameters [6–11]. The temperature of the internal air in the cold period of the year in the stall should be 8-13 °C. The optimum value of relative humidity is

© Springer Nature Switzerland AG 2020
V. Murgul and M. Pasetti (Eds.): EMMFT 2018, AISC 982, pp. 84–90, 2020.
https://doi.org/10.1007/978-3-030-19756-8_9

70%, permissible fluctuations from 60 to 85%. A very important indicator is the air velocity in the room: in winter - 0.3 m/s, in spring and autumn - 0.5 m/s, in summer - 1.0 m/s. Air exchange per animal is also different in seasons: in winter - 50 m^3/h, in spring and autumn - 70 m^3/h, in summer - 100 m^3/h. The maximum permissible concentration of ammonia is 20 mg/m^3.

2 Materials and Methods

The task is to study the microclimate parameters and the quality of the internal air in the stable building in St. Petersburg, and obtain the distribution of temperature and velocity fields, relative humidity, and CO_2 concentrations in the room. The research tool in this work is the STAR-CCM+ hydrodynamic computing complex [6–8, 11]. The calculations use the quadratic form of the k-ε turbulence model, which, in contrast to standard linear models, takes into account the effects of turbulence anisotropy.

This program allows obtaining the distribution of temperature and velocity fields in the rooms to assess the microclimate and air quality in the service area. The historic building built in 1855 (architectural monument) in St. Petersburg has been taken as a model. Only part of the building, which includes stalls for 8 horses, is considered in detail. The length of the simulated fragment is 16 m, width - 10 m, height - 8.5 m.

The model reflects the geometry of the room itself, the walls of the stalls, the horses, and also the air ducts and grids of the supply and exhaust ventilation (Fig. 1).

Heating and ventilation systems are designed in accordance with the requirements.[1]

The design outdoor temperature - minus 24 °C. The design temperature of the internal air - plus 10 °C. Heat losses of stalls are 22.4 kW.

To compensate for the loss of heat in the room, an air heating system is provided, combined with a ventilation system. The design temperature of the air supplied to the room is 23 °C. The design air exchange is determined from the conditions for ensuring the design parameters and the quality of the internal air. Standards for the allocation of hazards from sports horses are given in Table 1.

Table 1. Standards for the allocation of hazards from sports horses.

Groups of horses	Live weight, kg	Standards for the allocation for 1 horse per hour			
		Warmth, KJ		Carbon dioxide, l	Water vapor, g
		Total	Free		
Sports horses	400	3188.6	2295.3	114	357
	600	4399.5	3167.5	158	526
	800	5363.2	3861.5	192	300

(NTP APK (Russian technological design standards) 1.10.04.003-03 Norms of technologi-cal design of equestrian complexes; NTP-APK (Russian technological design standards) 1.10.04.001-00 Norms of technologi-cal design of horse-breeding enterprises).

The flow of fresh air is determined on the basis of heat loss compensation, but not less than 100 m^3/h per horse. Two variants of the supply ventilation system are considered:

- mixing, the air is supplied from the top downwards from the center of the corridor along the arches to each stall by horizontal jets;
- displacing, the air is supplied in the lower zone of the room.

Air removal is carried out from the upper zone of the corridor.

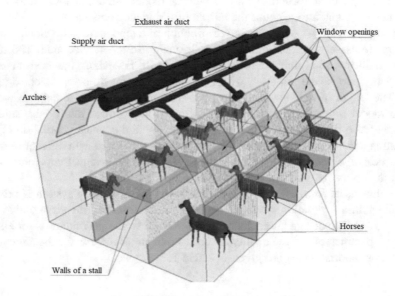

Fig. 1. Geometric model of a stable.

3 Results

The results of calculations of temperature and velocity fields, relative humidity, and CO_2 concentration in the stalls are shown in Figs. 2, 3, 4 and 5.

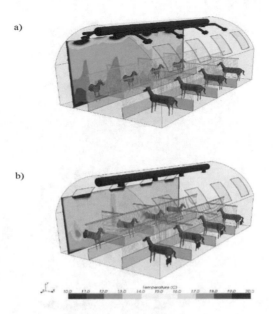

Fig. 2. Temperature field in a vertical section of a stable: a - air supply to the upper zone; b - air supply to the lower zone.

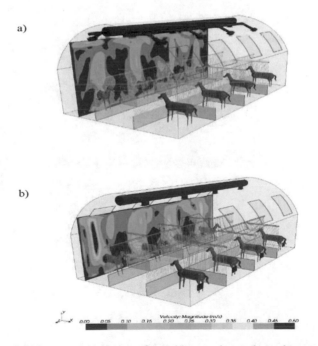

Fig. 3. Velocity field in a vertical section of a stable: a - air supply to the upper zone; b - air supply to the lower zone.

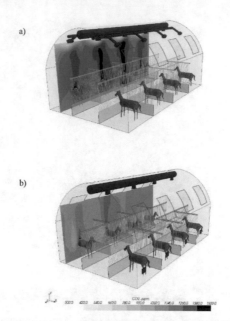

Fig. 4. CO_2 concentration field in a vertical section of a stable: a - air supply to the upper zone; b - air supply to the lower zone.

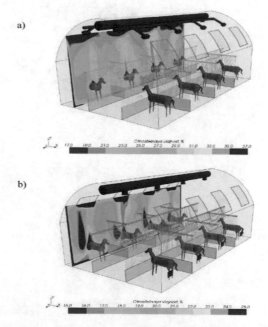

Fig. 5. Relative humidity field in a vertical section of a stable: a - air supply to the upper zone; b - air supply to the lower zone.

As a result of the study of the model, the values of temperatures and CO_2 concentrations in the service area were obtained. For the rooms under the study, the space in the room at a height of up to 2 m above the floor level is taken as the service area.

The results of the calculation of concentrations take into account the value of the CO_2 concentration of the supply air (400 ppm).

4 Discussion

According to the results of the study, it can be said that in the area where horses are located, when the air is supplied to the upper zone of the room, a temperature of 13.1 °C is created, and when it is supplied to the lower zone - 14.6 °C. That is, at the same supply air temperature in the second variant, the air temperature in the service area is higher. Regulation of air heating is quite flexible. Therefore, the temperature of the air supply can be reduced to achieve normalized parameters in the working area, which will reduce energy consumption. As for the air velocity, when air is supplied to the lower zone, the air velocity in the area where the horses are placed reaches 0.45 m/s, which exceeds the normalized parameters (in the cold period - 0.3 m/s). However, the standard indicators imply the supply of cold air and not the use of air heating. Accordingly, overspeed is permissible. By the air flow lines, it can be said that warm air quickly rises to the upper zone of the room along the arches. In this case, for more efficient air distribution, it is possible to reduce the air flow rate by increasing the cross-section of the supply grids.

The supply system with air supply to the lower zone of the room provides greater air mobility in the summer season, reduces energy consumption in the winter season, and provides higher air quality. However, with its theoretical advantages, this system is difficult to implement in the realities of the object under study due to the fact that the placement of air ducts in the lower part of the stall does not guarantee their safety. All space around the horses must be shockproof and have high strength. In addition, there are no ventilation channels in the walls of this building, and the status of an architectural monument does not allow for their installation. The historical value of reconstructed interiors often competes with engineering decisions, and lowering ducts to the lower zone can create architectural and design problems.

5 Conclusions

1. The mixing (air supply from top downwards from the center of the corridor along the arches to each stall by horizontal jets) and displacing ventilation schemes (air supply to the lower zone of the room) provide the required microclimate parameters in the area of horse location.
2. The concentration of CO_2 in the zone where the animals are located is permissible.
3. The relative humidity in the stalls is below 50%; this protects the enclosing structures of a historic building from premature destruction.
4. The supply system with air supply to the lower zone of the room allows reducing energy consumption in the winter season. However, its organization is difficult due to the lack of wall channels in this building. Despite this, the above study is applicable to the design of air heating at other equestrian complexes.

References

1. Dorokhov, V.B., Fomin, I.V.: Ways and Possibilities of Climatological Certification of Museum Buildings and Architectural Monuments (in Russian). INDRIK, Moscow (2008)
2. Dmitrieva, O.A.: Imperial ыtables. Protected State **2**, 29–38 (2014)
3. Razumov, V.: Stud farms in Russian estates. Protected State **6**, 22–31 (2017)
4. Zybina, D.D.: The impact of technological requirements on the architecture of equestrian complexes. Archit. Constr. **3**, 29–35 (2014)
5. Devina, R.A., Dorokhov, V.B.: The Microclimate of Church Buildings (The Basis for the Normalization of Temperature and Humidity of Monuments of Religious Architecture). GosNIIR, Moscow (2000). (in Russian)
6. Denisikhina, D.M., Ivanova, Yu.V., Mokrov, V.V.: Numerical simulation of outflows from modern air distribution devices. Eng. J. Don **2** (2018)
7. Puhkal, V.A., Sukhanov, K.O., Grimitlin, A.M.: Providing thermal comfort in rooms with a plinth water heating system. Bull. Civil Eng. **6**(59), 156–162 (2016)
8. Taurit, V.R., Korableva, N.A.: The choice of parameters for the calculation of the displacing ventilation of a new generation with high air quality in the area where people stay. Bull. Civil Eng. **3**(62), 166–170 (2017)
9. Omori, T., Tanabe, S.I., Akimoto, T.: Evaluation of thermal comfort and energy consumption in a room with different heating systems. In: IAQVEC 2007 Proceedings - 6th International Conference on Indoor Air Quality, Ventilation and Energy Conservation in Buildings: Sustainable Built Environment, vol. 2, pp. 51–58 (2007)
10. Ploskic, A., Holmberg, S.: Heat emission from thermal skirting boards. Build. Environ. **45**(5), 1123–1133 (2010)
11. Sukhanov, K., Smirnov, A.: Employment of skirting board heating water system in accommodation. Adv. Intell. Syst. Comput. **692**, 592–597 (2017)

A Testing Facility for the Thermal Characterization of Building Envelopes in Outdoor Operating Conditions

Alessandra Mesa[1]([✉]) [iD], Alberto Arenghi[1] [iD], and Marco Pasetti[2] [iD]

[1] Department of Civil, Environmental,
Architectural Engineering, and Mathematics, University of Brescia, Brescia, Italy
alessandra.mesa@unibs.it
[2] Department of Information Engineering, University of Brescia, Brescia, Italy

Abstract. The experimental assessment of building components represents a complex task, which involves the measurement and control of a wide number of physical phenomena. The use of the dynamic approach in outdoor real scale facilities provides a good representation of real operating systems, thanks to the inclusion of rather complex parameters, such as the occupants' behaviour. However, the adoption of outdoor test solutions is usually characterized by a high uncertainty of the results, due to the complexity of the physical model and to the large variability of the input parameters. On the other hand, the use of indoor tests under controlled conditions has proved to be able to provide reliable results, thanks to the strict control of boundary conditions and of input parameters. The main drawback of this approach is represented by the scarce significance of the results, due to the rather simple modelling of the real behaviour of building physics. The aim of this paper is thus to present a test facility which represents a compromise between the dynamic control approach in real scale systems and the use of indoor tests under controlled conditions: the Building Envelopes ouTdoor Thermal Test (BEsT3) facility of the University of Brescia. Thanks to application of the dynamic behaviour of real environmental conditions to outdoor test cells with controlled indoor thermal parameters, the proposed system has proved to be able to provide reliable results, while also satisfactorily reproducing the conditions of real operating systems. Experimental studies have been conducted to assess three different window solutions under real dynamic conditions. Measured data have been used to create a correspondent numerical model designed in Energy Plus. The model has been validated with different dynamic simulations, in which the complexity of the parameters has been increased step by step. The numerical results provided by the model have shown a good correspondence with the real behaviour of the outdoor test cells.

Keywords: Test cells · Building envelope · Thermal test · Outdoor testing facility

© Springer Nature Switzerland AG 2020
V. Murgul and M. Pasetti (Eds.): EMMFT 2018, AISC 982, pp. 91–104, 2020.
https://doi.org/10.1007/978-3-030-19756-8_10

1 Introduction

The Buildings have a huge impact on our ordinary day both in terms of the quality of life and in terms of the environmental impact. The assessment of the various physical phenomena through the building's envelope is a crucial point to improve their performance and our comfort. Nevertheless, the characterization of buildings components represents a complex task, due to the wide number of involved processes.

The study of the effectiveness of buildings elements is an interesting topic both for designers and final users. Nowadays, neither researchers nor standardization body have found a proper procedure to follow, which combine simplicity and affordable costs. The main assessment methods are based on three different experimental setups, which differ for the boundary conditions [1]. The first approach is performed by the use of outdoor real-scale facility, described in the ISO 9869-1:2014 [2]. This method is based on in-situ measurement, considering dynamic boundary conditions both for inside and outside. The weather is, indeed, monitored from the real conditions, collecting solar radiation and wind data, while inside parameters are highly influenced by the occupants' behaviour. The strength of this approach is the opportunity to involve real dynamic parameters, using an experimental set up that faithfully represents reality. It is worth to note that this approach has been also proved to be particularly effective for the field testing of energy efficiency measures and of related control architectures, whenever the variation of environmental conditions or the participation of active users could strongly affect the response of the supervised system. Examples are the characterization of energy storage systems in support of renewable generators [3], or the integration of electric vehicle charging systems in smart grids [4].

However, together with this great added value, it is necessary to discuss about the problematic issues that this approach contains too. In-field measurements are indeed influenced by numerous simultaneous factors, such as users' behaviour, or heating/cooling system, that make the isolation of a single parameter a complex task to reach. Furthermore, the unique nature of every single test, characterizing this kind of approach, doesn't allow comparing the obtained measurements with other dataset. This approach is highly recommended to be adopted when the goal is to understand the real behaviour of a particular building component, while it is not appropriate to its general characterization. To overcome these problems, laboratory indoor facilities may be chosen. This approach is based on the use of controlled boundary conditions, both for internal and external environment. This laboratory set-up allows analyzing specific parameters one at the time, understanding their influence on the building component. The use of fixed boundary conditions is also necessary to compare and replicate the test in other scenarios. The greatest limit of this method is the big difference in comparison to the real behavior of the element. Some essential parameters, such as dynamic weather conditions or the presence of occupants, are indeed omitted, producing misleading results. A good compromise between these two methods can be provided by outdoor test cells. These facilities allow assessing building components with controlled interior environment and maintaining real climate data for the outside boundary condition. This method obtains results that better suit the real behavior of the building element and allows also assessing the influence of each single parameter involved in the physical phenomena. The main obstacle remains the peculiar features of every single trial, that makes the test impossible to be replicated and compared to other similar studies. In Table 1, the main characteristics of three different approaches are summarised, highlighting pros and cons of each and referencing some significant studies as examples.

Table 1. Review of existing test facilities.

Test facilities	Boundary conditions	Pros & Cons	Examples
Real scale facilities	Dynamic for indoor and outdoor	Pros-Faithfull to real behavior Cons-Difficult to compare and to analyze single variable	[5–7]
Laboratory indoor test facilities	Steady state for indoor and outdoor	Pros-Assessment of single parameter, comparable to other tests Cons-Far from real behavior	[8, 9]
Outdoor test cells	Steady state for indoor, dynamic for outdoor conditions	Pros-Similar to real behavior, assessment of single parameter Cons-Difficult to compare	[10–14]

2 The Building Envelopes Outdoor Thermal Test (BEST3) Facility

According to above mentioned, outdoor test cells seems to be a proper compromise to assess a building component. The University of Brescia has developed a laboratory set-up in order to characterize building elements and up to now two facilities, which allow this kind of tests, are hosted. These facilities are located inside the campus and differ in terms of methods and goals. The first laboratory equipment is a climatic chamber, useful to generally characterize hygrothermal properties of materials. This facility allows testing walls under controlled boundary conditions, following a steady state calculation method. Tests under dynamic boundary conditions are conducted by means of the Building Envelops ouTdoor Thermal Test (BEsT3) facility. This facility, which is part of the energy living laboratory eLUX (energy Laboratory as University eXpo [15]), consists in three outdoor test cells, that allow testing different stratigraphy setting a controlled inside environment and considering the real weather as the outdoor boundary condition. These kinds of tests are based on dynamic calculations and allow consideration of many parameters usually involved in real life. The following chapter will focus on the outdoor cells, showing their construction and their instruments.

2.1 Outdoor Test Cells Envelope

The outdoor test cells are based on a square plan and have a volume of 34,08 m^3. They are composed by a load-bearing bolted structure made in steel L shape profile, with overall dimensions of 150 × 150 × 10 mm. Each cell is covered with a galvanized corrugated steel sheet roof, HI-BOND type A75/P760, properly sloped and projecting to protect the structure from rainfall. A similar solution is proposed for the cells floor, which are made by steel slab raised from the ground in order to prevent settlement. The roof and the floor are both stiffened with a beam respectively parallel and perpendicular to the testing wall. The whole structure was moreover finished with a zinc-coat to avoid rust. The raw cell structure is depicted in Fig. 1.

Fig. 1. The structure of the outdoor test cells.

The goal of the tests conducted in the outdoor cells is to make a quantitative comparison among the three different testing samples. All the cells must be then identical in terms of structure and surfaces, excepting the wall to test. Five permanent side behaves as close as possible to an adiabatic component, in order to not interfere to experiment walls. The "adiabatic" parts of the envelope are mainly composed by wide layers of insulating materials, as EPS for the walls and XPS for the roof and floor. The thermal properties, computed according to EN ISO 6946:2017 [16] (Table 2).

Table 2. Thermal properties of the permanent element of the cell

Building element	U-value (W/m^{2}K)	R-value (m^{2}K/W)
Walls	0.0740	13.56
Roof	0.0853	11.72
Floor	0.1096	9.12

The cells are positioned very close to each other, with a distance of nearly 1.50 m. To maintain homogeneity between them, two plastic panel were positioned to create an artificial shadow (Fig. 2). This structure allows comparing the peripheral cells, more exposed to the solar radiation and to the wind, to the central one.

Fig. 2. Plastic panel structure

2.2 Outdoor Test Cells Central Heating System

The singularity of outdoor test cells is to assess different samples considering real weather conditions for the outside and setting fixed value for the inside. The indoor environment can then vary on the basis of specific experimental requirements. The central heating is designed to have a good flexibility, in order to quickly respond to temperature variations inside each cell. Another essential property of the central heating is the low response time to the stress caused by the external weather. The system should, indeed, quickly react when an internal temperatures fluctuation is monitored. To satisfy these needs, cooling and heating system are split to keep them independent.

The system is composed by an air-water heat pump connected to two tanks, that refer to the heating and the cooling demand. Linked to this primary circuit there's a secondary system, that brings directly to the cells. Each cell has its own system, and this independency give lot of experimental freedom to the tests conducted with this method. The secondary circuit terminal is a fan coil, which allows achieving indoor temperatures set for the studies. In order to maintain the chosen internal boundary condition and to avoid overheating problems, a temperatures controller is connected to the fan coil so that it can be switched off as soon as experimental test conditions are reached.

2.3 Outdoor Test Cells Instruments

Dynamic testing of building components requires a well-controlled set of sensors with a correct measuring and control system. To assess energy and thermal performances of a building element is necessary to have an accurate and continuous monitoring of the physical phenomena involved in the heat exchange. The outdoor test cells are equipped with tools capable of guarantee precise measurements of internal humidity and temperature, surface temperatures, electrical absorbance and energy calculation.

The data collection system of the three cells is simple. Sensor measurements are monitored with a one-minute time step and collected by a data logger. The overall data are temporary collected, until arithmetic average, maximum and minimum values are calculated with a fifteen-minutes time step. All the data undergo to a pre-filtering process, which allows discarding false measurements due to tools accuracy problems, errors in data transmission or other general issues. The process is based on a simple method: a range of reasonable values is set for every parameter, if the recorded measurement is out of this interval, it is automatically left out from the collected data. All the information is stored by the server in two independent databases for raw data and filtered data, respectively. A schematic layout of the measurement data flow is depicted in Fig. 3.

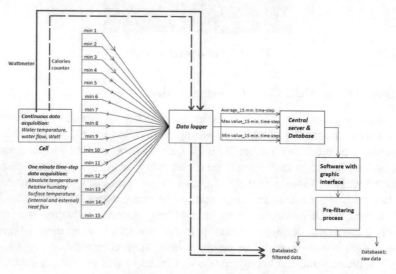

Fig. 3. Data collection system scheme.

Each cell is equipped with various instruments, as specified in Table 3:

Table 3. Cells equipment.

Instrument	Quantity per cell	Measured phenomena	Accuracy
Thermo-hygrometer	3	Temperature and relative humidity	±0.1 °C & ±2%(f.s.)
Heat flux plate	1	Heat flux	±5%
Thermocouples	6	Surface temperature	±0.1 °C
Wattmeter	1	Electrical power consumption	±2%
Flowmeter	2	Water flow	±5%
Immersion thermometer	6	Water temperature	±0.1 °C

Some additional instruments were then installed to monitor the external ambient surrounding the cells. A solarimeter was provided to assess solar radiation, and sensors were positioned to characterize the outdoor boundary condition in terms of temperatures and humidity.

3 Method

The three outdoor test cells were used to conduct experimental assessment on walls with openings. The three stratigraphy were tested under different periods and the thermal properties were assessed to characterize the windows. The goal of the tests was to design a numerical model with Energy Plus that faithfully replicates outdoor test cells, to conduct predictive simulations on different building components.

3.1 Outdoor Test Cells: Experimental Site, Testing Elements and Laboratory Set-up

The three cells were located in a wide space, with no particular shadowing element to notice, as shown in Fig. 4.

Fig. 4. In situ outdoor test cells.

The experimental site had the following geographical characteristic (Table 4).

Table 4. Experimental site had the following geographical characteristic

Address	Latitude	Longitude	Altitude
Via Branze 43, 25123 Brescia, Italy	45°33′	10°13′	167 m a.s.l.

Weather data were measured with an onsite weather station, located next to the cells. The monitored variables were: dry bulb temperature, relative humidity, atmospheric pressure, global horizontal solar radiation, wind speed and direction.

The goal of the tests was to assess the performance of three different windows that differed for the connection with the envelope. The wall was identical for all the case of study, a typical external wall of a residential building made by a layer of hollow bricks (25 cm thk) and an EPS insulating layer (8 cm thk). The thermal values of the wall under study were: U-value of 0.27 W/m²K and an equivalent R-value of 3.70 m²K/W.

The windows provided for the test were composed by a three-chamber aluminium frame and a low-energy double glass. The windows had overall dimensions of 1.82 m2, 1230 mm long and 1480 mm high, in accordance with the dimensions suggested in ISO 12567-1:2010 [17], ISO 10140-1:2016 [18] and ISO 10140-3:2010 [19].

The aim of the tests was to assess thermal properties of the three different solutions, comparing the energy consumption of the cells. The performance of the testing samples was indirectly derived from the energy consumption of the whole cells, assuming the contribution of the adiabatic walls as negligible. Internal temperature was set as a constant value of 23 °C, so that the energy required to maintain this condition was measured and compared. In order to maintain the temperature values under control and consequently limit measurement errors of the energy consumption, the reduction of the transitional period was necessary. To achieve this goal, the temperature controller was set up so that the deviation from the fixed 23 °C should not exceed ±1 °C. To maintain comparable thermal load between the test cells, the fan coil was design to be always on and to work in a steady state regime. For the same reason, flow and temperature of the heat transfer fluid were monitored and designed to maintained fixed values thanks to relaunch pumps.

3.2 Numerical Model: Assumptions and Design

In order to run simulations as close as possible to real behaviour, Energy Plus (developed by U.S. Department of Energy) was chosen to design a numerical model. To provide an accurate energy performance assessment, every detail, which has a significant thermal impact, must be studied and involved in the model. Energy Plus analyses simple opaque surfaces, for that reason thermal bridges were calculated in Therm (developed at Lawrence Berkeley National Laboratory (LBNL)) software that provides a two-dimensional building heat transfer assessment, and the results were then included in the model as numerical values. The external boundary condition was modelled according to the parameters monitored by the onsite weather station, to faithfully represent the environment surrounding the outdoor cells. Surfaces and shading elements were modelled in Energy Plus, excepting for windows, which have been designed by a software WINDOW (developed by Berkeley Lab WINDOW), that accurately describes glass surfaces.

The central heating system was modelled in Energy Plus separating heating and cooling circuit. This solution represents the real plants provided for the test, and it also allows better control of the three different cells, designed as independent thermal zones. Moreover, to represent the thermostat inside each cell, Energy Management System (EMS) programmes were used in the model.

4 Results

The numerical model was validated in different steps. The first approach was to collect all the data monitored during the experimental tests. The second step consisted in the reprocess of values recorded by the sensors. This work had the main target to create input files, as weather file, necessary to run the simulations. The last step was to create the model and compare the results with the values obtained in tests under real conditions. The validation of the model was subdivided in three different phases, slowly increasing the complexity of the parameters involved in the simulations.

4.1 Validation 1: Model with No Openings and No Central Heating

The first step was to assess the model considering only energy performance of the building envelope. The holes prepared for windows were covered with a multilayer spruce panel and the central heating was ignored. The simulation represented a 5-days trial conducted during summer period.

The results carried out from the three cells showed similar trend and led to the same observation, only the case of cell 2 will be then presented as example.

The first period of the trial was characterized by a clear deviation, nearly 2 °C, between the results, as it can be seen in Fig. 5. This gap is probably due to the different initial conditions of the real test and the dynamic simulation. This first transitional period was then followed by a second phase in which the results had the same trend. The deviation between the curves was low and the daily percentage error, calculated as function of the temperatures difference between real and simulated conditions, was lower of 3%. Another important validation of the numerical model was the comparison between the mean internal temperature values, as shown in Table 5.

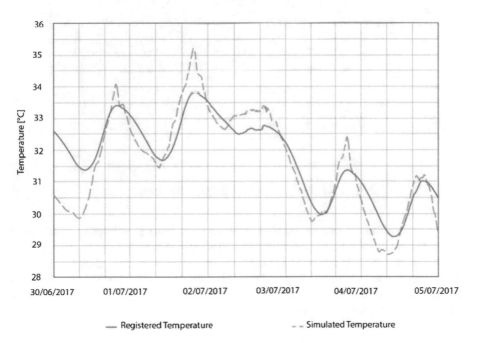

Fig. 5. Validation 1: internal temperatures trend, cell 2

Table 5. Mean internal temperature values, cell 2

Ti outdoor test cell [°C]	TiEnergy plus [°C]	ΔT [°C]	Percentage error [%]
31.8	31.6	0.2	0.8

4.2   Validation 2: Model with Openings and No Central Heating

The assessment of the building envelope founded a good correspondence between real tests and numerical model. Simulations complexity was then increased to study windows energy impact and their influence on the model. Dynamic simulations were run during summer period and had a duration of 3 days.

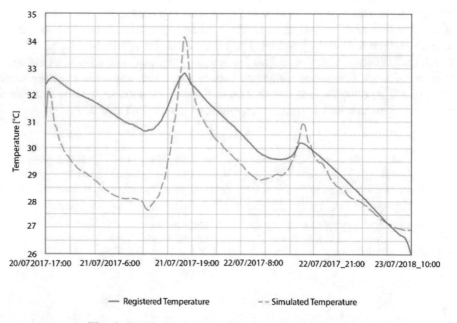

— Registered Temperature   – – Simulated Temperature

Fig. 6. Validation 2: internal temperatures trend, cell 2.

Figure 6 shows internal temperature trend of cells 2. As in validation 1, also these simulations were characterized by a transitional period, in which a gap between the two results can be noticed, followed by a stable period in which measured values and simulations values converge. Despite the short duration of the trials, the assessment of the mean internal temperatures in Table 6 showed similar results between the values, making the numerical model reliable.

Table 6. Mean internal temperature values, cell 2

Ti outdoor test cell [°C]	TiEnergy plus [°C]	ΔT [°C]	Percentage error [%]
30.33	29.18	1.15	3.79

4.3   Validation 3: Model with Openings and Central Heating

The last simulation run to validate the model included the heat gain from the central heating. Also, these tests were conducted during summer period, with a duration of 4 days. The goal of the simulations was to assess the energy needed to maintain the fixed internal temperature of 23 °C. The comparison between the results was made analyzing the internal temperatures and the energy provided.

Figure 7 shows the internal temperature comparison between outdoor tests measured values and simulations results. Synchronization of the two thermostats, the one of the models and the one inside the cells, was impossible to achieve, the results are shown, then, considering the daily mean value instead the hourly mean value.

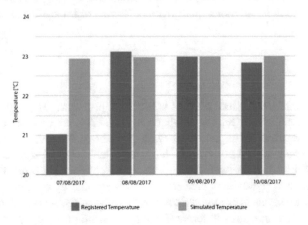

Fig. 7. Validation 3: internal temperature trend, cell 1

This simulation pointed out, once again, the difference concerning the initial period of the two results. Internal temperatures worked out, indeed, didn't need a transient period to reach the fixed temperature inside the cell. The measured data showed, instead, the requirement of a first transitional period, in order to achieve the set-up internal boundary condition. Regarding the whole trial, it can be noticed that internal temperature of the model remained stable during the entire simulation, while the values monitored by the thermostat slightly swung near the fixed temperature. Another comparison was made regarding the measured energy provided. This analysis, Fig. 8, showed results that confirm what was observed in the other simulations.

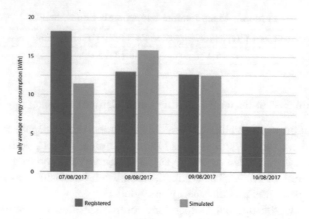

Fig. 8. Validation 3: energy consumption trend, cell 2

The first phase of the trial pointed out a transient period which mostly influence the data from the outdoor tests. This phase was then followed by a stable condition, in which results from simulations and from outdoor test converged, validating the numerical model.

5 Discussion

Outdoor tests were conducted without significant problems to point out. The only aspect to remark is the role that the adiabatic walls had on the energy need assessment. The conductance values of the adiabatic walls have been, indeed, a cause of concerns, and they had to be assessed in detail. Outdoor cells have been designed two years ago, the U-value of the adiabatic walls cannot be then kept pace with all the new technological solutions and regulations concerning buildings energy efficiency. The U-value of the adiabatic elements is comparable to the U-value of the wall under study and this may lead to misleading results in terms of energy consumption and performance of the element to assess. To limit or avoid this issue a detailed thermal analysis of the whole samples under study was then necessary to understand the phenomena involved in the heat transfer process and to recreate a reliable numerical model, which take this aspect into account.

Beside this issue, the outdoor test cells measured data were reliable enough to design a numerical model correspondent to reality. The simulations, which have been run for the validation of the model, have shown a good affinity between the measured and obtained data. The parameters under investigation had a similar trend, excepting for the first period of trials. This divergence may be related to the inevitable difference between real cells and numerical model and to the different complexity of the variables involved to assess parameters. Real behaviour is indeed characterized by a first transitional period, which can be limited but no completely avoided. Instead, numerical model behaviour doesn't need time to reach fixed set-up values, and this it can be easily pointed out in every monitored parameter. The first transitional period has always been followed by a correspondence of the results trend. Furthermore, the analyses of the mean temperatures' values have confirmed the reliability of the numerical model concerning real cells. For this reason, the numerical model has been validated, considering its great reproduction of real scale model. The creation of a numerical model correspondent to reality, is a useful method to conduct predictive analyses on buildings. An interesting development of these research can be the used of the validated model to assess building elements under different internal or external conditions.

6 Conclusions

BEsT3 turns to be a great facility to assess thermal properties and energy performance of building components. Cells are indeed very adaptable, both in terms of internal boundary condition and in terms of element to be investigated. The experimental set-up can be re-designed every time from new in accordance to the needs required to achieve the purpose of the research. The use of the outdoor cells is also related to relevant problems that must be taken into account in order to conduct reliable assessment. Outdoor test cells, as all the other facilities under dynamic conditions, are indeed characterize by high uncertainty of data measurement. For this reason, the cells have to be furnished with specific and accurate equipment to conduct the trial.

The tests conducted with BEsT3 facility allowed assessing three different window solutions under real dynamic conditions, characterizing their thermal behaviour and permitting to compare it with dynamic simulations.

The numerical model designed with Energy Plus have been analysed in different steps, increasing the complexity of the simulations, and a correspondence to the real test cells have been remarked. The validation of a numerical model allows conducting various predictive studies. These assessments are a good method to outline thermal performances of components during the design phase, so that the behaviour of buildings during their lifecycle can be predict. The greatest added value of this approach is the flexibility and the wide variety of simulations that can be conducted.

Acknowledgments. The authors would like to thank AIB (Associazione Industriali Bresciani), who granted the construction of BEsT3 and the eLUX laboratory (energy Laboratory as University eXpo) of the University of Brescia, which developed the data acquisition system.

References

1. Cattarin, G., Causone, F., Kindinis, A., Pagliano, L.: Outdoor cells for building envelope experimental characterization – a literature review. Renew. Sustain. Energy Rev. **54**, 606–625 (2016)
2. ISO 9869-1:2014: Thermal insulation – Building elements – In-situ measurement of thermal resistance and thermal transmittance Heat flow meter method, 29 July 2014
3. Rinaldi, S., Pasetti, M., Flammini, A., De Simone F.: Characterization of energy storage systems for renewable generators: an experimental testbed. In: Proceedings of the 2018 IEEE International Workshop on Applied Measurements for Power Systems (AMPS), Bologna, Italy, 26–28 September 2018 (2018)

4. Ferrari, P., Flammini, A., Pasetti, M., Rinaldi, S., Simoncini, F., Sisinni, E.: Testing facility for the characterization of the integration of E-Vehicles into smart grid in presence of renewable energy. In: Proceedings of the 2018 IEEE International Conference on Applications in Electronics Pervading Industry, Environment and Society (ApplePies), Pisa, Italy, 26–27 September 2018 (2018)
5. EnergyFlexHouse-Technology to the global challenge. https://www.dti.dk/labs/energy flexhouse/technology-to-the-global-challenge/25348. Accessed 12 Oct 2018
6. Swinton, M.: NRC-IRC full-scale facilities for hygrothermal and whole house performance assessment. In: International DYNASTEE Workshop, Bruxelles, Belgium, 30–31 March 2011 (2011)
7. Asdrubali, F., D'Alessandro, F., Baldinelli, G., Bianchi, F.: Evaluating in situ thermal transmittance of green buildings masonries—a case study. Constr. Mater. 1(2014), 53–59 (2014)
8. Ferrari, S., Zanotto, V.: The thermal performance of walls under actual service conditions: evaluating the results of climatic chamber tests. Constr. Build. Mater. 43(2013), 309–316 (2013)
9. Baldinelli, G., Bianchi, F., Lechowska, A.A., Schnotale, J.A.: Dynamic thermal properties of building components: hot box experimental assessment under different solicitations. Energy Build. 168(2018), 1–8 (2018)
10. Serra, V., Zanghirella, F., Perino, M.: Experimental evaluation of a climate façade: energy efficiency and thermal comfort performance. Energy Build. 42(2010), 50–62 (2010)
11. Corgati, S.F., Perino, M., Serra, V.: Experimental assessment of the performance of an active transparent façade during actual operation conditions. Sol. Energy 81(2007), 933–1013 (2007)
12. Dimoudi, A., Androutsopoulos, A., Lykoudis, S.: Experimental work on a linked, dynamic and ventilated. Wall Compon. 36(2004), 443–453 (2014)
13. Zanghirella, F., Perino, M., Serra, V.: A numerical model to evaluate the thermal behaviour of active transparent façades. Energy Build. 43(2011), 1123–1138 (2011)
14. Iribar-Solaberrieta, E., Escudero-Revilla, C., Odrizola-Maritorena, M., Campos-Celador, A., Garcìa-Gàfaro, C.: Energy performance of the opaque ventilated façade. In: 6th International Building Physics Conference, IBPC 2015 (2015)
15. Flammini, A., Pasetti, M., Rinaldi, S., Bellagente, P., Ciribini, A.C., Tagliabue, L.C., Zavanella, L.E., Zanoni, S., Oggioni, G., Pedrazzi, G.: A living lab and testing infrastructure for the development of innovative smart energy solutions: the eLUX Laboratory of the University of Brescia. In: Proceedings of the 2018 AEIT International Annual Conference, Bari, Italy, 3–5 October 2018 (2018)
16. EN ISO 6946:2017, Building components and building elements - Thermal resistance and thermal transmittance - Calculation method, 21 June 2017
17. ISO 12567-1:2010, Thermal performance of windows and doors - Determination of thermal transmittance by the hot-box method Complete windows and doors, 14 June 2010
18. ISO 10140-1:2016, Acoustics - Laboratory measurement of sound insulation of building elements Application rules for specific products, 05 August 2016
19. ISO 10140-3:2010, Acoustics - Laboratory measurement of sound insulation of building elements Measurement of impact sound insulation, 17 August 2010

Green Energy Technologies of Tall Buildings for Air Pollution Abatement in Metropolises

Natalia Potienko[✉][iD], Anna Kuznetsova[iD], and Darya Soya[iD]

Samara State Technical University,
Molodogvardeyskaya 244, 443100 Samara, Russia
natalia.potienko@mail.ru

Abstract. The paper considers the natural and anthropogenic causes of environmental pollution in the context of the modern urban development. There are analyzed processes of formation of urban agglomerations and the main tendencies of air pollution at increasing density of population and the amount of motor transport on their territory. The classification of modern types of urban smog and methods of dealing with it are given. Analyzed modern scientific research aimed at improving the ecological climate of cities. The main directions of ecologization of the over-urbanized environment in the context of the development of high-rise construction are considered on the basis of modern environmental techniques. These are such systems in the structure of buildings as: aeration, landscape gardening, the use of vertical gardens, modern facade technologies, air cleaning elements. The paper presents an overview of modern high-rise construction projects in which green technologies have been successfully implemented. The analyzed objects are located in the structure of large megacities with a high level of pollution of their atmosphere. It is proposed to introduce competitive architectural designing of futuristic projects while searching the new ways of ecologization of megalopolises.

Keywords: High-rise buildings · Air pollution · Green energy

1 Introduction

One of the most serious global problems faced by humanity in the XXI century was the problem of air pollution. By itself, this problem is not new, as it appeared in the era of the emergence of industry and transport, which worked first on coal and then on oil. For two centuries, air pollution was local in nature, since air masses could cope with small amounts of emissions on their own. All sources of air pollution are conventionally divided into natural and artificial, i.e. anthropogenic. The first group includes pollutants generated by nature itself: volcanoes, peat or forest fires, dust storms. The main artificial sources causing air pollution are industry, transport, thermal power plants, household waste, agricultural production. Atmosphere of cities contains a large amount of harmful substances, which include formaldehyde, hydrogen fluoride, lead compounds, ammonia, phenol, benzene, carbon disulphide, etc. However, the main air pollutants are sulphur dioxide (SO_2), nitrogen oxides (NOX), carbon monoxide (co) and particulate matter. In total emissions of harmful substances, they make about 98%. It is their concentration that most often exceeds acceptable levels in many cities of the world.

© Springer Nature Switzerland AG 2020
V. Murgul and M. Pasetti (Eds.): EMMFT 2018, AISC 982, pp. 105–115, 2020.
https://doi.org/10.1007/978-3-030-19756-8_11

Human activity has led to an increase in the concentration of carbon dioxide in the air by almost 30% over the past 200 years. The new air quality tracking model developed by the World Health Organization (WHO) confirms that 92% of the world's population lives in areas where air quality indicators exceed WHO limits. According to the World Bank, Russia today has one of the lowest levels of air pollution among developed countries, ahead of both traditional outsiders – India and China, and also Japan, the United States and European countries. From the 20th century, parallel to the growth of the population of the planet, the process of urbanization began, the main characteristics of which were population growth and concentration of the economy in cities. Today, the share of the urban population in developed countries already exceeds 70%; in Russia, this figure is 73%, and in the UK, it is 91%. More than 400 large cities of this day have the status of million-plus cities, about 20 very large cities have a population of more than 10 million people: Tokyo, New York, Shanghai, Bu Enos Aires, São Paulo, Delhi, Mumbai, Kolkata, Mexico City other. In addition, there is a rapid growth of the so-called urban agglomerations or merged cities, for example, Washington - Boston and Los Angeles - San Francisco in the United States. Large cities and neighboring small cities, such as Moscow and St. Petersburg, often merge with the surrounding cities. The growth of population, transport and industrial enterprises in small areas make cities dangerous for living. Thus, motor transport gives 60% to 70% of gas pollution [1–6]. Traffic jams, which contribute to a significant increase in air pollution by exhaust gases, arise not only from a large number of road transport, but also from natural and man-made barriers in the transport network, as well as the type of relief, the nature and location of green spaces and the accepted transport planning city structures [7].

Air pollution is determined by a measure called PM10, which reports the number of small particles found in the air [8]. Kanpur is the second most polluted city in India, with an index of 209 $\mu g/m^3$. The Iranian cities of Yasuj, Kermanshah, Sanandaj and Ahvaz have indicators of 215 $\mu g/m^3$, 229 $\mu g/m^3$, 254 $\mu g/m^3$ and 372 $\mu g/m^3$, in the latter the air pollution is aggravated by dust storms. The city of Gaborone, the capital of Botswana has a figure of 216 $\mu g/m^3$. In Pakistan, the cities of Peshawar and Quetta with indicators of 219 $\mu g/m^3$ and 251 $\mu g/m^3$ are the dirtiest in the country. In Russia, 151 cities have exceeded the maximum permissible concentration (MPC) of air pollution by 5 times, 87 cities - 10 times [2].

Air pollution reduces the transparency of the atmosphere, increases the occurrence of fog by 50%, gives more precipitation by 10% (among which the greatest danger is acid precipitation), reduces solar radiation and wind speed by 30%. In addition, under certain weather conditions, temperature inversions occur, leading to smog. It is formed in the lower layers of the troposphere, with the accumulation of a large amount of pollutants and persists for a long time due to windless weather. Existing types are wet (London type), ice (Alaskan type) and dry or photochemical (Los Angeles type) smog [9]. The latter type of smog is a complex multicomponent mixture of gases and aerosol particles of different origin. The main components of smog are: ozone, nitrogen and sulfur oxides, photooxidants. The cities whose inhabitants suffer from smog are such famous metropolises as London, Mexico City, Beijing, Hong Kong, Athens, Los Angeles, Moscow and many other megalopolises. According to the World Health Organization, according to data for 2014, about 4 million people died due to air pollution in the world. Nearly 7 million people died from exposure of polluted air to the organism [9]. The scale of this environmental problem has almost reached its apogee, as a result, the search for solutions and ways to eliminate it is carried out in all scientific fields. The architecture, being an inherently important mechanism of the huge high-tech world, also takes an active part in questions of solving ecological catastrophe.

2 Materials and Methods

Air protection from pollution must be maintained at the state level, and in many countries, there are already laws to protect it. In addition, laws are being passed at the government level, for example, regulating deforestation and the use of natural landscapes by the population, since the preservation of forests is a key factor in the purification of atmospheric air [10]. Many countries are tightening the requirements for exhaust gases of cars, so in Europe over the past 20 years, the standards for carbon oxides were reduced by 20 times, and for hydrocarbons and nitrogen oxides - by 17 times. Work is underway to reduce the amount of fuel burned in a car's engines on the one hand and to use alternative energy sources, such as solar panels or electricity, on the other. In many countries, there are architectural and planning measures governing the location of enterprises, planning for urban development based on environmental factors, using various techniques to improve the air quality in the city - organizing sanitary protection zones, greening cities, and, if necessary, preserving wetlands and coastal areas, and farmland inside the city.

Technological and sanitary measures most widely used in the world include: improvement of the processes of fuel combustion and waste disposal or their processing; conversion of boiler and thermal power plants to gas fuel or alternative fuels, such as water, wind, sun and others; improving the tightness of factory equipment; installation of high pipes; global use of treatment facilities and filtering devices from dust, aerosols and gases (the most efficient technological schemes developed for air purification: mechanical, plasma, adsorption or coal, photocatalytic), etc.

In the search for solutions to the ecological disadvantage of cities, a significant part of the effort, undoubtedly, belongs to scientists. Analyzing the situation, they offer the most rational and effective options for improving the ecological climate of cities, including reducing the negative impact of emissions. One such researcher is Italian architect Stefano Boeri. In his study, he reflected the value of green gardens and suggested that the greenery on the facades serves not only for aesthetic perception, but also is urgently needed for human life.

3 Results

In a modern super-urbanized environment, the construction and design of high-rise buildings occupies one of the key places. But height is not the main trend in high-rise construction. Today, there are two main promising directions: architectural unusualness and environmental friendliness of projects. As a part of considering the problem of urban air pollution, the second direction is of the greatest interest.

One of the most effective ways to improve the atmosphere of the city is the use of aeration techniques. Aeration system is formed by planning, engineering, technical and structural elements of the building. It is important to note the architectural object affects the air flow, distorts its direction and changes the speed. The building is able to cause airflows also called artificial breezes. This technique is used in the project of twin towers "Duo towers" [11, 12]. This multifunctional high-rise complex, including office space, a hotel, premium housing, a network of retail stores and fast food outlets, is located in Singapore on the border of the business district and the Kampong Glam quarter (see Fig. 1).

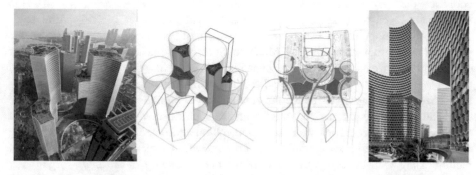

Fig. 1. Multifunctional high-rise complex "Duo towers", 2013–2017, Singapore (architectural bureau Ole Scheeren) [11, 12].

The project proposed to divide the site into two separate parts, leaving a lot of free space in the interval between both towers and between neighboring, smaller buildings of the district. The spatial orientation of the buildings is optimized with respect to the prevailing wind direction and the movement of the sun, while the concave facades of the towers capture the air flow, contributing to better ventilation not only of the interior, but also of the entire site, and contributing to the creation of a cool microclimate inside the complex. Concave semicircle of the facades contributes to a better aeration of the territory. The ecological situation is also improved by the gardens and green alleys with recreation areas located around the high-rise buildings. The total volume of landscape gardening at the level of the earth, green terraces and high-rise gardens is equal to the entire area of the site. Along with this, the complex meets all the requirements of passive and active energy-efficient design, natural ventilation of premises and territory. Today the search for the ideal material for facade finishing does not end up with the choice of an economical and aesthetic sample. An environmentally important feature of the 25-storey condominium "570 Broome" in Manhattan, is that its facade clears the air [13]. The building, clad in 2000 m^2 of Neolithic, is sheathed with panels, the material of which consists of natural raw materials subjected to high temperature and pressure in order to imitate the natural stone, which, in turn, is coated with Pureti nanoparticles titanium dioxide (see Fig. 2). Architect Tahir Demircioglu - director of Buildd and the author of the project, explained that the team initially chose the exterior material of the building without knowing its environmental impact.

Fig. 2. Condominium "570 Broome", 2016–2018, New York (USA) [14].

Travis Conrad, architectural consultant at Neolith, explains that the treated panels self-clean and actively change the chemical composition of the ambient air. The panels do not allow to leave stains, they repel water and dirt from the building. In addition, it is a photocatalytic material, that is, it reacts to sunlight and moisture contained in the air, destroying volatile compounds in the air, just as wood destroys greenhouse gases. In the construction this material was used for the first time and on a large scale, although the chemicals in its composition were discovered about 40 years ago. It was necessary to obtain as much of the mineral as possible on a flat surface, because the material must interact with moisture in the air and sunlight. Thus, spraying the solution onto a large thin sheet of Neolithic where all particles can live on the surface turned out to be the right solution. On the one hand, it had a positive impact on the environment, and on the other, it helped to reduce maintenance costs, because the used material limits the amount of chemicals in order to clean the facade. Special attention should be paid to the project "Bosco Verticale", built in Milan, whose air is very polluted [14, 15]. Geographically, the obstacles to the winds on the way to the metropolis from the north are mountains that interfere with the circulation of air masses and create a cloud of smog over the city. Architects Stefano Boeri, Gianandrea Barrek and Giovanni La Varra developed the concept of symbiosis of a residential high-rise building and green spaces. The lack of free land space in the city center, the desire to contribute to improving the ecological background of their homeland and ennoble the city led the architects to the idea of creating a vertical forest (see Fig. 3). To realize this concept, trees of various heights and life cycles were used. This allowed to create a closed ecosystem of more than 800 trees, 5 thousand different shrubs, more than 10 thousand perennial green areas and a huge number of flowers and herbs. The dynamics of the facade was, therefore, created by natural processes - spring flowering, the appearance of summer greenery and autumn wilting.

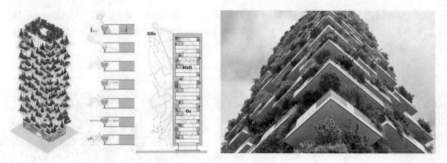

Fig. 3. Residential high-rise building "Bosco Verticale", Milan (Italy), 2014 [15, 16].

For the maximum possible organization of the space for landscaping, vertical communications and building entrances were brought to one of the 4 sides of the house. Careful study in the field of landscape design has led to the creation of comfortable conditions for plant growth by designers. Such parameters as air humidity, illumination, certain wind strength were taken into account. Gradation of the number of floors is selected taking into account the growth of plantings. Thus, the 27-storey skyscraper was able to replace the park area of 50 thousand m^2. The green massif, together with the aesthetic ones, also received environmental benefits, as it became a producer of oxygen and a city air purifier from dirt and dust. China is building the first "Vertical forest" in Asia – "Nanjing Green Towers", designed by architect Stefano Boeri as a continuation of the idea of greening high-rise buildings [16]. The complex consists of two towers of a height of 200 and 108 m with facades of natural greenery (see Fig. 4a).

Fig. 4. Vertical landscaping of high-rise buildings: a - Multifunctional high-rise complex "Nanjing Green Towers", Nanjing (China), 2018 [30]; b - The Forest City project in Liuzhou (China) [17].

The highlight of the main high-rise of the project is a dome of dense green vegetation, under which the club will be located. The floors from the eighth to the thirty-fifth will be occupied by offices, a Museum and a school of "green architecture". The second tower will be home to a new Hyatt hotel, commercial premises, entertainment and training facilities. The towers of the Chinese "Vertical forest" will be planted with a total of 1110 trees of local species and 2500 shrubs. This will increase humidity, lead to the absorption of carbon dioxide and the production of acid, and in addition, protect people and buildings from excessive solar radiation and noise. The green towers will help restore local biodiversity and absorb 35 tons of CO_2 annually, producing 60 kg of pure oxygen per day. Since a pair of "green skyscrapers" is not enough to overcome the Chinese environmental crisis, Boeri proposes to create a network of green mini-cities consisting of high-rise buildings with vertical gardens (see Fig. 4b), which are planned to be built in Liuzhou by 2021.

The use of about 40 thousand trees and about 1 million plants of more than 100 different species in buildings will contribute to the production of oxygen, protecting the area from noise and cleaning the air from dust. According to calculations, the proposed variety of greenery is capable of producing up to 900 tons of oxygen per year, absorbing 10 thousand tons of carbon dioxide and 57 tons of pollutants. It is planned for the irrigation of plants to apply the latest progressive developments, to use solar energy, to combine vertical gardening with horizontal to the maximum, thereby integrating man into nature to the greatest extent. In addition, the architect suggested using the same idea in the project for Shijiazhuang, the most polluted city in China. The conceptual design of the city implies maximum implementation of landscaping at all levels: from streets and bridges to buildings (see Fig. 5). This project should be, according to Boeri, the prototype of a new generation of small compact cities that can absorb the maximum amount of carbon dioxide. It is assumed that the green towers of this city will be able to absorb up to 1,750 kg of CO_2 per year.

Fig. 5. The design of the "forest city" in Shijiazhuang "Forest City Shijiazhuang" (China) [18].

Today, in addition to landscaping buildings, air purifying elements integrated into the structure of the building are used to solve the problem of atmospheric pollution in large cities. The first example was the project of the Dutch artist Daana Roosegaard, who in October 2016 presented the project of a 7-m tower capable of purifying the air in Beijing. The principle of operation of the tower was to blow in polluted air, its subsequent cleaning and "exhaling" back. Built in 2017 in the province of Shaanxi in Central China according to this principle, the 100-m tower (see Fig. 6) has proven to be effective in fighting smog. Tests conducted by scientists of the Academy of Sciences of China, showed that the tower significantly improved the air quality in the vicinity of the city of Xi'an, providing 10 million cubic meters of clean air per day to an area of 10 sq. km. The experts noted a reduction in air pollution in the area by 15%. Scientists hope in the near future to build a 500-m tower, which will be able to clean the surrounding air on an area of 30 km².

Fig. 6. The 100-m tower in the province of Shaanxi (Central China), 2017 [19].

Dubai architecture firm Znera has developed a conceptual project for air purification in Delhi, one of the most polluted cities in the world. The level of pollution is aggravated by burning the crop in the areas adjacent to the city. Burning acts as the main cause of toxic smog, covering the city and its neighboring areas. The network of free towers offered by the architect firm will help the city to clean the air and make it breathable (see Fig. 7).

Fig. 7. Conceptual design of "Smog project" for air purification in Delhi [20].

The project is a hexagonal network that uses solar energy as a renewable source. Each 100-m tower is carefully located in the center of the key city and produces clean air within two kilometers. The vertical air cleaning tower above the ground is connected by guiding elements that produce hydrogen and power the towers. Construction will begin with a single area to assess the degree of efficiency before a larger network is built. Specifically designed to remove smog and contaminating elements from the air, the project uses a filtering beam on the base propellers at the top. The tributaries at the base of the tower blow air in and out through five stages of filtration—including coal, negative Ion generators, and electrostatic plasma in order to absorb air particles. The air rises up where it passes through the photocatalyst filter to be sterilized against bacteria and viruses, before being released into the atmosphere. The towers will produce 3.2 million cubic meters of clean air per day.

One of the ways of finding a solution to the air pollution problem of a city is the competitive design of futuristic projects.

A key aspect of the multifunctional high-rise complex designed, within the framework of the course design for the discipline "Architectural Design" based on the architectural faculty of Samara State Technical University, is solving the problem of air pollution in Milan within the framework of the corresponding building typology. The number of especially super-heavy suspended particles that are especially harmful to health in the city exceeds the permissible standard according to the standard Pm 2.5, and the air pollution index is 107 mg/m^3, which also greatly exceeds the permissible values. The project proposes to create a complex in the developing area "City Life" with the inclusion of an air cleaning tower in its structure. The vertical pipe will take the polluted air at the level of the highest concentration of smog (4–15 m from ground level). Passing through several layers of cleaning filters, the purified air will enter the atmosphere, as well as inside the building through the air conditioning system. Along with this problem, the complex will also solve the main urban problems, such as traffic congestion and the shortage of affordable housing. As part of a transit-oriented development, the designed vertical area includes underground parking, vertical park, shopping center, office premises, a hotel, apartments, as well as restaurants, a viewing platform and a radio broadcast tower on the top floors (see Fig. 8a). The image and shape of the building is inspired by the "living" architectural sculptures of Philip Beasley (see Fig. 8b). The created network, which also forms the bearing frame of the building, is the personification of the air flow that rose from the ground level and dissolved at a height of several hundred meters. The structure of the network created from the columns includes transparent tubular elements with microalgae, which will also absorb carbon dioxide emissions.

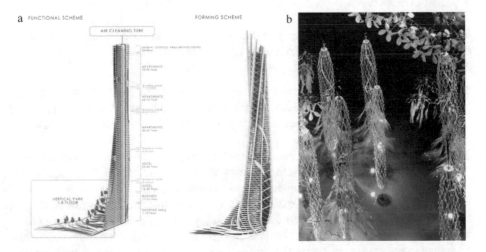

Fig. 8. Conceptual design in the framework of course design: a - functional diagram and volumetric solution of the project of a multifunctional high-rise complex; b - compositional prototype - architectural sculptures by Philip Beasley [21].

4 Discussion

Considering that atmospheric air pollution is one of the serious environmental problems of the modern city, causing significant damage to human health, the material and technical base of the city and the environment, we can talk about the need for maximum ecologization of the urban environment. This is most difficult to accomplish in large metropolitan areas and urban areas, the upward trend of which has been seen in the last decade. In addition to new environmental protection and restoration solutions, new environmental and economical technologies are also required for their introduction into the structure of buildings. Development of such technological solutions is made today in conceptual competitive projects around the world. Speaking about the real design of high-rise buildings, it can be said that there is a steady practice of synthesizing high-rise architecture and architectural planning techniques, technological and sanitary facilities to address the problem of air pollution in cities. Recently, quite a few projects have been implemented to integrate the corresponding methods and facilities into the structure of high-rise buildings. Often, they become elements of both the material and structural part of the building and the elements of architectural formation. The choice of methods and means to improve the state of atmospheric air depends largely on environmental conditions (degree of air pollution, climatic conditions), operational data, type of high-rise building (residential, multi-functional, technical). The only limitation for using such an approach as vertical gardening of buildings can be a harsh cold climate. The integration of methods and means of cleaning the polluted atmospheric air of cities into the structure of a high-rise building leads to the fact that they begin to act as architectural elements, and this is a rich ground for a creative activity of an architect.

5 Conclusions

In the examples analyzed above, you can see an attempt to combat an increasing environmental catastrophe. The architects try to solve the problem in different ways, each of which has its own advantages:

1. Creation of artificial breezes and necessary airing of the territory by competent solutions on the object aeration. Formed for this purpose, the space-planning solution and spatial organization make it possible to reveal the particular shape and aesthetic appearance of the building.
2. Synthesis of existing natural materials serves the purpose of developing an innovative and environmentally useful building envelope, and materials of façade systems, which increases the uniqueness and manufacturability of the object.
3. The maximum use of green space in high-rise construction compensates for the minimum amount of vegetation that is being replaced by rapidly expanding cities.
4. Development of air purifying elements directly in the building structure will allow reducing the content of harmful particles in the atmosphere on the largest scale.

Combining the methods already used, as well as creating and searching for new ones, will help you find the most effective formula for solving problems that are so important for humanity and the planet. When developing these or other measures it is recommended to take into account the economic component. Pollution control methods should be as efficient as possible and minimally costly. Currently, scientists propose to combine the main ways of solving the air pollution problem and use complex measures.

References

1. Potienko, N., Kalinkina, N., Bannikova, A.: Low-grade energy of the ground for civil engineering. MATEC Web Conf. **73**, 06026 (2017)
2. Potienko, N.D., Kuznetsova, A.A., Solyakova, D.N., Klyueva, Y.E.: The global experience of deployment of energy-efficient technologies in high-rise construction. E3S Web Conf. **33**, 01017 (2018)
3. Zhogoleva, A., Teryagova, A.: On methods of sustainable architectural design of bio-positive buildings in the low-rise residential development structure. In: MATEC, vol. 105, p. 01039 (2017)
4. Generalov, V., Generalova, E.: Urban Constr. Archit. **4**, 23–29 (2015)
5. Ivanov, V., Bakhtina, I., Ivanova, T., Ilinykh, S.: Urban Constr. Archit. **2**, 88–93 (2015)
6. Generalov, V.: High-rise residential buildings and complexes. Singapore. Experience in the design and construction of high-rise housing. LLC "Book", Samara (2013)
7. Wong, L.T., Mui, K.W., Law, L.Y.: An energy consumption benchmarking system for residential buildings in Hong Kong. Building Serv. Eng. Res. Technol. **2**(30), 135–142 (2009)
8. Krygina, A.: Resource, energy saving and environmental friendliness of construction as the basis for innovative sustainable development of residential real estate. Hous. Constr. **6**, 57–59 (2015)
9. Paiho, S., Seppa, I.P., Jimenez, C.: An energetic analysis of a multifunctional façade system for energy efficient retrofitting of residential buildings in cold climates of Finland and Russia. Sustain. Cities Soc. **15**, 75–85 (2015)
10. Windsweepers. High rise buildings, pp. 44–49, Moscow (2013)
11. https://www.admagazinc.ru/architecture/neboskreb-iz-medovyh-sot-v-singapure-ot-byuro-ole-shirena. Accessed 21 Nov 2018
12. https://www.architecturaldigest.com/story/new-york-city-building-facade-cleans-air. Accessed 21 Dec 2018
13. https://www.archdaily.com/777498/bosco-verticale-stefano-boeri-architetti. Accessed 21 Nov 2018
14. Vertical forest: High-tech buildings 1, pp. 92–126 (2013)
15. https://realt.onliner.by/2017/02/10/bosco-2-comments. Accessed 31 Nov 2018
16. http://www.stoletie.ru/zarubejie/tam_budet_gorod-sad_916.htm. Accessed 21 Nov 2018
17. https://robo-hunter.com/news/zelenie-doma-budushego-novii-put-spaseniya-kitaya7074. Accessed 21 Nov 2018
18. http://skuky.net/140255. Accessed 21 Nov 2018
19. https://www.archdaily.com/902403/znera-proposes-a-network-of-smog-filtering-towers-across-delhi. Accessed 21 Nov 2018
20. Marchenko, M., Lapchenko, A.: Futuristic solutions to the problem of air pollution in industrial design. Young Scientist **7**, 1081–1086 (2016)
21. https://pragmatika.media/zhivaja-arhitektura-oni-nabljudajut-za-nami/. Accessed 21 Nov 2018

Computational Study of a Natural Exhaust Ventilation System During the Heating Period

Elena Malyavina[(✉)] and Kaminat Agakhanova

Moscow State University of Civil Engineering,
Yaroslavskoe shosse 26, 129337 Moscow, Russia
emal@list.ru

Abstract. The article presents the data obtained according to the calculation results on the change of the floor exhaust air consumption at three combinations of the outdoor air temperature and the wind speed. These combinations of the temperature and wind correspond to the design ones of the Moscow climatic conditions calculated for designing natural ventilation, the coldest design conditions for heating and close to the average season ones for the heating period. A feature of the calculated studies is the presence of three isolated natural ventilation systems in each apartment located one above the other from the 2nd to the 19th floor. Each apartment is traditionally considered as one volume. The calculations confirmed that to achieve the specified exhaust air flow rates in each of the ventilation systems with floor branches (satellites) and one collecting channel (trunk), it is necessary to reduce the aerodynamic resistance of the trunk and increase the aerodynamic resistance of the satellites. In addition, it was found that the opening of the window rotary shutter for the inlet of the supply air from $3°$ (slit ventilation) to $30°$ does not lead to a significant increase in the supply air consumption under the same outdoor weather conditions. Much more important are the outside temperature and the wind rate.

Keywords: The system of equations of the air balance of the premises ·
Air-permeable element · Full excess pressure · Air consumption

1 Description of the Building Air Mode

The goal of this article is to investigate the change of the exhaust air consumption through exhaust grids of a natural ventilation system during the heating period.

Calculation of the air exchange in premises of a multi-storey building can be performed only taking into account the topology of the building, because the premises are connected by the air permeability of the structures between them, as well as the ventilation systems or stairway and elevation lobbies [1–4]. The building is a complex aerodynamic pipeline. The air exchange of the building premises occurs under the influence of the difference of the total overpressure formed on each air-permeable element (hole) between the premises, including the ventilation system elements.

The calculation of the total excess pressures formed under certain weather conditions in each room is generally a solution of a system of air balance equations for each room and each unit of ventilation systems. The number of equations of the system under

© Springer Nature Switzerland AG 2020
V. Murgul and M. Pasetti (Eds.): EMMFT 2018, AISC 982, pp. 116–124, 2020.
https://doi.org/10.1007/978-3-030-19756-8_12

investigation is equal to the number of rooms in the building summarized with the number of the ventilation system nodes. Moreover, when describing the air balance of each room or the ventilation system node unit, two types of equations are used [5–11]. One of them is the first Kirchhoff's law, which states that the amount of air flow through all the breathable elements of the room or node under consideration should make zero. The second is the Bernoulli equation, which is a form of the energy conservation law, describing the dependence of pressure losses on the consumption of the air, which passes through a breathable element.

As the boundary conditions, provision has been made of the outside air pressures at each air permeable element connecting any room with the external environment: windows, entrance doors, etc. The outdoor pressure is composed of the gravitational pressure decreasing with the higher location of the hole above the ground and the wind pressure depending on the wind speed increasing with height, as well as the wind direction forming the windward and leeward sides and the lateral facades of the building [1, 2, 5]. In addition, at the suggestion of V. Baturin for the buildings with approximately the same temperature in all rooms, a variable gravitational component of the indoor pressure is applied to the exterior pressure with a "minus" sign.

With all this, the exterior pressure P_{ext}, Pa, becomes:

$$P_{ext} = g(H - h)\rho_{ext} + (C - C_1)\rho_{ext}\, kV^2/2 \qquad (1)$$

where: g – acceleration of gravity, m^2/s; H – the height of the building from the ground to the highest point, m; h – height above ground of the center of the air permeable element, m; ρ_{ext} – density of exterior air (taking into account its temperature at the considered time), kg/m^3; C, Cl – aerodynamic coefficients of the calculated air permeable element and on the leeward side; k – the coefficient of account of the wind pressure change depending on the height over the ground and the landscape type; V – wind speed, m/s, according to the weather station informative data.

As a result of this methodological technique, i.e. removal of the variable part of the indoor pressure to the external one, the indoor pressure of each room P_{int}, Pa, has a constant value by the height, which facilitates the solution of the system of equations. Typically, a point model of the building rooms is considered, in which the total overpressure of each room is applied in the center of the room.

In addition to meteorological factors in the Bernoulli equation, provision shall be made of setting the air permeability of the elements, which is described by the resistance characteristics S, Pa/(kg/h)^2 of the elements. The S values remain permanent for most of the holes and depend on the air consumption only in t-pieces of ventilation systems.

2 Object of Investigations

As the object of investigation provision has been made of a 18-storey residential part of the building (from the 2nd to the 19th floors). For a fragment of the building plan see the Fig. 1.

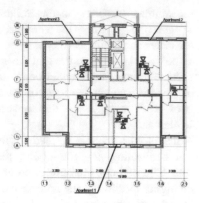

Fig. 1. Fragment of the section plan of a typical floor of a multistory residential building.

On each residential floor (from the 2nd to the 19th floors), there are three apartments: two 2-room apartments: one is with the single-side orientation (apartment 1) and the other is a double – side oriented apartment (apartment 3), as well as one 3-room apartment with the windows on two opposite facades (apartment 3). The building section has been designed with an elevator lobby having passenger lifts. On each floor the lobby has an exit to the balcony for passage to the isolated evacuation smoke-free stairs and to the inter-apartment corridor. The doors of all three apartments and the staircase-lift lobby on each floor face the mentioned one common corridor.

In residential areas, a natural system of supply and exhaust ventilation is provided, which is made in the form of three independent systems that remove the exhaust air from kitchens, toilets and bathrooms. Each of the three systems is designed as a combined channel (trunk) and floor branches (satellites). The air flow is supplied through the swivel-flap window sashes. The system of exhaust air ducts is adopted with 2 m long satellites, connected to the precast vertical channel under the ceiling of the above floor at an angle of 90°.

The house has an unheated attic, 3 m high, through which insulated exhaust shafts pass en route increasing the available pressure for all floors of the building, including the upper stories. Moreover, the upper two floors have floor-by-floor channels, brought into the exhaust shaft independently without joining the common trunk.

Steel air ducts of the exhaust ventilation have rectangular sections, gradually increasing from the lower floors from 200×250 mm to 350×400 mm on the floors from the 9th to the 17th for the trunk serving the kitchens. Sections of the trunks serving toilets and bathrooms are increasing from 200×200 mm to 200×350 mm. The floor-to-floor branches of all systems have the same sections on all floors being of 150×150 mm with adjustable exhaust grilles. Construction shafts, which locate vertical precast channels, are higher than the level of the roof and end with an "umbrella" hood. The design project provides the air removal with the following rates:

- from the kitchen - 60 m³/h;
- from the bathroom - 25 m³/h;
- from the toilet - 25 m³/h.

3   Calculation Methods

The calculation of the air regime of a multi-storey residential building begins with the setting of initial pressures in all rooms of the building [5–11]. These pressures have values between the maximum and minimum pressures of the outdoor air on each floor. The elevator lobbies are considered a single volume, as they are on all floors combined by elevator shafts, and there is nearly no sealing between the elevator cabin and the elevator shaft. For example, as the internal pressure in the elevator lobby provision is made of an averaged outdoor air pressure value according to the aerodynamic resistances of windows and entrance doors. Taking into account the indoor pressure in each room, we have checked whether the material balance has been respected within them. In case of non-compliance with the material balance in any room, in the iterative process it is necessary to change the value of the indoor pressure with a small step (0.0001 Pa in the program) to set the air balance to the value of 0.5 kg/h. The sequence of the air balance checking is as follows: first, the balance of the elevator lobbies is checked, then – this one of each apartment, at the end, the air balances of the floor corridors are checked. When considering the air balances of the apartments, provision is made of the air consumption not only through the windows and the doors, but also through all three exhaust grilles available in each apartment. The calculation algorithm for each apartment is the same. The number of air-permeable holes, taking into account the exhaust systems in each apartment, is individual.

In the scheme with a prefabricated vertical duct, it is necessary to recalculate the aerodynamic resistance characteristics of ventilation systems, since the values of the local resistance coefficients of tees or cross-parts depend on the changing ratios of air consumption through individual branches of a tee or a cross-piece. After each iteration, the aerodynamic drag characteristics of the ventilation system sections shall be recalculated depending on the air flow rates obtained in this iteration. The resulting new characteristic values are applied in the next iteration. Thus, when calculating the air consumption through ventilation systems with a precast vertical channel, not only the pressures in the rooms applied to the exhaust grilles are to be determined, but also the pressures at the flow confluence points in the ventilation system. In this case, the pressures in the ventilation system node units are equal to the difference of the pressure in the next air flow point of confluence of the flows and pressure losses between the two node points. The calculation starts from the last floor. After determining the pressure in all apartments and in the elevator lobby, the calculation of the material balance of each floor corridor shall be provided. The calculation of the material balance of the elevator lobby and the apartments is repeated at the newly obtained pressures in the corridors. The room pressures are taken to be the values obtained in the previous iteration. The iterative repetition is considered to be finished when the values of internal pressures in the elevator lobby, apartments and corridors, and, accordingly, the air flow through ventilation systems and air-permeable holes, are the same in the last two iterations. The discrepancy in the air flow rate is not more than 1 kg/h for each room.

4 Calculation Results

The calculation determined the internal pressure values in each room and in each node unit of the exhaust ventilation systems, and most importantly, the exhaust air consumption of each exhaust grille in each apartment at three combinations of outdoor air temperature and the wind rate. First, the calculated external conditions for the calculation of natural ventilation systems were taken: temperature +5 °C and absence of wind (V = 0 m/s). These conditions are accepted regardless of the local climate in any area of the Russian Federation. Under these conditions, the required exhaust air flow consumption shall be maintained. The supply air consumption shall compensate the exhaust air one in case of open windows in at least one room or kitchen of the apartment. These are the most severe weather conditions for the formation of a pressure difference, under the influence of which the removed air flows. Therefore, just for these conditions, provision is made of selection of the ventilation system design: the diameters of the air ducts, the characteristics of aerodynamic resistances on the floor grids of each system. These characteristics of the systems remain unchanged throughout the heating period.

The Fig. 2 shows graphs of changes in the characteristics of floor-by-floor aerodynamic resistances of the ventilation system satellites.

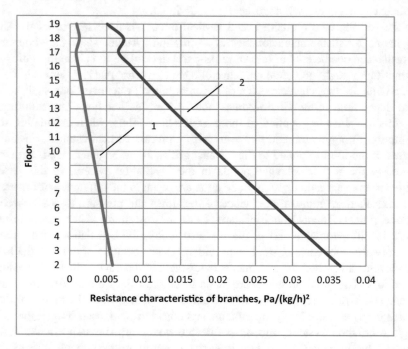

Fig. 2. Resistance characteristics of floor-to-floor branching of exhaust ventilation systems 1 – from kitchens, 2 – from toilets and bathrooms.

Table 1. Main calculation results

Variant	Temperature °C	Wind rate, m/s	Parameter	Opening angle of the window swing-flap sash		
				$\alpha = 3°$	$\alpha = 15°$	$\alpha = 30°$
All	Any	Any	Sin α	0,05	0,26	0,5
			Consumption coefficient, μ	0,032	0,1594	0,31
7th floor						
1	+5	0	Indoor pressure, Pa	27,72	28,58	28,61
			Outdoor pressure, Pa	28,62	28,62	28,62
			Water consumption through the window sash, kg/h	108,78	111,26	111,18
2	−5	1	Indoor pressure, Pa	47,21	48,62	48,69
			Outdoor pressure, Pa	48,68	48,68	48,68
			Air consumption through the window sash, kg/h	141,95	145,31	145,49
3	−25	2	Indoor pressure	91,50	94,04	94,12
			Outdoor pressure	94,15	94,15	94,15
			Air consumption through the window sash	197,66	202,20	202,32
17th floor						
1	+5	0	Indoor pressure, Pa	7,81	8,55	8,58
			Outdoor pressure, Pa	8,59	8,59	8,59
			Air consumption through the window sash, kg/h	101,43	109,02	109,36
2	−5	1	Indoor pressure	13,84	15,15	15,20
			Outdoor pressure, Pa	15,22	15,22	15,22
			Air consumption through the window sash, kg/h	136,98	147,26	147,69
3	−25	2	Indoor pressure	28,18	30,73	30,83
			Outdoor pressure	30,86	30,86	30,86
			Air consumption through the window sash	198,71	213,00	213, 48

In accordance with the inevitable increase in the available pressure on the lower floors of the building, the resistance of the floor branches increases from the upper floors to the lower ones. The adjustment shall be made by regulating the aerodynamic resistance of the exhaust grilles. Secondly, the weather conditions have been accepted for Moscow close to the average season ones for the heating period: the temperature −5 °C and the wind rate V = 1 m/s. Third, we considered the Moscow coldest design period – the design conditions for selection of the heating system thermal power: −25 °C temperature and the wind rate V = 2 m/s. Thus, in each apartment for the admission of supply air one window in three positions of a rotary-folding shutter was considered open: on $\alpha = 3°$, on $\alpha = 15°$ and $\alpha = 30°$. For the apartment 1, which is located on the windward facade, the main calculation results are shown in the Table 1. It follows that the air flow rates, which are proportional to the coefficient of consumption of the slit at the opening of the swing-flap window sash are little dependent on the sash opening angle, but are highly dependent on weather conditions. From the Table 1 it also follows that having selected the elements of the system when it is set with sufficiently high aerodynamic resistances, it can be achieved that even with a decrease in the outside air temperature to the calculated heating values, the exhaust air consumption differ little from the consumption under the design conditions for ventilation in case of the same angle of the window sash opening.

5 Discussion of the Results

The Fig. 3 shows the change in air consumption flowing through the exhaust grilles of different floors of the ventilation systems.

The figure shows that in case of well-adjusted systems from the 2nd to the 10th floors, the exhaust air rate does not deviate from the specified value even by 0.5 m³/h. On the upper floors at high outdoor temperature and windless, even with good setting of the systems, there are observed the consumption reductions of some m³/h of the required values in case of a slotted ventilation. It is clear that under the specified weather conditions for airing of the apartment, it is necessary to open the window swivel-flap sash on 15 or even on 30°. At first glance, the obtained results are incredibly good. However, if we put aside the difficulty of the fine adjusting of ventilation systems and consider the reasons for the got results, we can see that these results have the right to life.

First, consider the evident fact of an extra height of the exhaust shaft, which passes in the attic with a height of 3 m. Second, we shall agree, that when you open the window sash, the pressure in the room gets much closer to the pressure in the outer air behind this sash. Due to this, when opening the flap, the pressure difference between the outside and the inside air is reduced, the air consumption through the open sash would also be reduced, if not for the increased area of the open slit, as evidenced by the growing flow coefficient μ. As a result, the air consumption may be virtually considered as the unchangeable one.

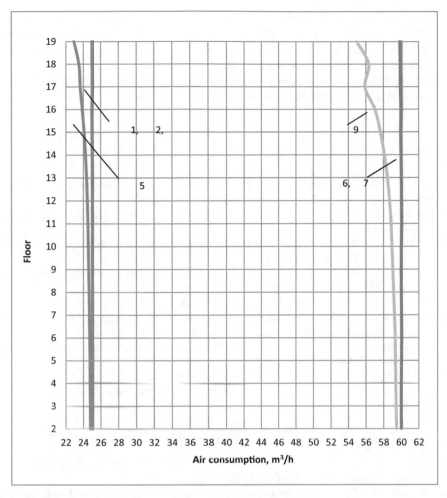

Fig. 3. Change in the exhaust air consumption by the floors at the outdoor temperature +5 °C and windless weather and the different sash opening angles of a swivel-flap window: 1, 2, 3, 4 – from the bathroom and the toilet at the sash opening angles of 30° and 15°; 5 – from the bathroom and the toilet at the sash opening angles of 3°; 6, 7, 8 – from the kitchen at the sash opening angles of 30° and 15°; 9 – from the kitchen at the sash opening angles of 3°.

6   Conclusions

1. The calculations have shown that with careful adjusting of natural exhaust ventilation systems it is possible to achieve a satisfactory operation of the systems in the climatic conditions of Moscow during the entire heating period. In this case, the division of the ventilation system trunk in height in an 18-storey building is not required.

2. Providing high aerodynamic resistances at the exhaust air grilles the opening of a window swivel-flap sash for the inlet of the fresh air is markedly seen with increasing of the opening angle up to $15°$. With a larger opening of the sash, there is practically no change in the air consumption rate.
3. The amount of the air exchange in a room is greatly affected by the weather conditions: the outdoor air temperature and the wind rate.

References

1. Malakhov, M.A., Savenkov, A.E.: Experience in designing natural mechanical ventilation of residential buildings with warm attics. AVOK **6**, 20–32 (2008)
2. Livchak, I.F., Naumov, A.L: Ventilation of Multi-Storey Residential Buildings. AVOK-PRESS, Moscow (2005)
3. Krivoshein, A.D.: Provision of regulated air flow in residential buildings: problems and solutions. AVOK **4**, 32–38 (2018)
4. Krivoshein, A.D.: Provision of regulated air flow in residential buildings: problems and solutions. AVOK **5**, 40–44 (2018)
5. Tertichnik, E.I., Agakhanova, K.M.: To the question of the decision of problems of air filtration through leaks in the enclosure and non-organized exchange in the building rooms under the action of natural forces. Sci. Rev. **8**, 62–66 (2015)
6. Malyavina, E.G., Biryukov, S.V.: Calculation of the air regime of multi-storey buildings with different indoor air temperatures. AVOK **2**, 40–44 (2008)
7. Malyavina, E.G., Biryukov, S.V., Dianov, S.N.: Air mode of a high-rise residential building during the year. Part 2. Air mode with mechanical exhaust ventilation. AVOK **1**, 26–33 (2005)
8. Agakhanova, K.M.: Calculation of air regime of a residential building with independent air. IOP Conf. Ser. Mater. Sci. Eng. **365** (2018). 022036. https://doi.org/10.1088/1757-899x/365/2/022036
9. Krivoshein, M.M.: To the question of mathematical modeling of air distribution in ventilation systems. Omsk Sci. Bull. **5**, 98–103 (2017)
10. Datsyuk, T.A.: Evaluation of the effectiveness of natural ventilation of residential buildings. Plumbing Heating Air conditioning **1**, 112–115 (2014)
11. Samarin, O.D.: On the methods of calculating the air regime of buildings. Plumbing Heating Air Conditioning **3**, 78–79 (2011)

Feasibility Study of Energy Efficient Repair of Residential Buildings of the First Mass Series

Olga Popova[1](✉) ⓘ, Nina Brenchukova[1] ⓘ, Leonid Yablonskii[2] ⓘ,
and Varvara Dikareva[3] ⓘ

[1] Northern (Arctic) Federal University named after M. V. Lomonosov,
Severnaya Dvina Emb. 17, 163002 Arkhangelsk, Russia
oly-popova@yandex.ru
[2] Saint Petersburg State University of Architecture and Civil Engineering,
Vtoraya Krasnoarmeiskaya Street 4, 190005 St. Petersburg, Russia
[3] Moscow State University of Civil Engineering,
Yaroslavskoye shosse 26, 129337 Moscow, Russia

Abstract. The goal of the study is to estimate the technical and economic efficiency of energy-efficient repairs of the first mass series buildings. Within the study, objects representing the apartment buildings of the first mass series, which are most common in the city of Arkhangelsk and constitute together more than 70% of the housing stock, were selected. The study of the technical characteristics and the state of the enclosing structures of the objects representing the housing stock has shown the need for carrying out the repair and restoration measures—planned major repair. The calculation of the current heat losses of the buildings has shown that buildings do not meet the modern standards of thermal protection. As a result, the assumption was made that, during the major repair of apartment buildings, works carried out, as a part of such repair should meet the requirements of the energy efficiency. As the measures of energy-efficient repair, we have chosen the options for wall insulation, coatings, attic and basement ceilings, as well as filling the openings of external walls. The calculation of the economic efficiency of individual measures of energy-efficient repair and their combination was made. The study has shown that the integrated energy-efficient repair brings to the more than 2 times reduction in heating costs. The greatest thermal effect is achieved from the installation of the individual heating point, wall insulation with sandwich panels, and a basement slab with fiberglass slabs. The term of completion of the complex energy-efficient repair of the first mass series buildings is 9–10 years. Thus, projects are economically effective and can be recommended for implementation.

Keywords: Energy-efficient · Building envelop ·
Buildings heat insulation · Complex renovation · Planned major repair

1 Introduction

Relevance to save the energy at the level of the separate states is related to the need to improve the environment (to reduce emissions of pollutants) and to ensure energy safety and competitiveness of national economies. Russia ranks third in the world according to

© Springer Nature Switzerland AG 2020
V. Murgul and M. Pasetti (Eds.): EMMFT 2018, AISC 982, pp. 125–136, 2020.
https://doi.org/10.1007/978-3-030-19756-8_13

the absolute indicator of total energy consumption (after the USA and China), and among the ten largest energy consumers, it has the highest energy intensity: it spends more energy per unit of gross domestic product (GDP) than any of these countries. High-energy intensity indicates the inefficient use of the energy resources, has a negative impact on the Russian economics, its energy security and its environment [1].

According to the opinion of the experts of the energy safety area, the housing sector in Russia takes second place after the manufacturing industry by the value of final energy consumption: more than ¼ of the total energy consumption [2].

The housing stock of the most cities in Russia for more than 70% consists of buildings built between 1950 and 1999. These are the buildings of the first mass series ("Stalin Project", "Khrushchev Project", "Ulyanov Project", "Brezhnev Project" and other serial houses), the construction of which was a salvation for millions of people who received separate apartments in houses with the necessary engineering systems. However, they were designed according to old building codes and do not meet modern requirements for thermal protection of buildings (heat loss through building envelope - up to 40%). Therefore, today the multi-apartment buildings built before 1995 are one of the main reasons for the low energy efficiency of the housing stock in Russia.

Today, according to the current standards, measures to increase energy efficiency are required not only for a new construction, but also for the reconstruction and major repair of residential buildings. According to the state program "Energy Strategy of Russia for the Period until 2030", one of the areas of energy-saving is the reconstruction and new construction of the buildings using heat-resistant structures, thermal automatics, and energy-efficient equipment.

According to the Housing and Utilities Reform Facilitation Fund, complex major repair can save up to 49% of energy resources. That is, during the major repair of apartment buildings, works carried out, as part of such repairs must comply with the requirements of energy efficiency [3].

Thus, the current direction of modern construction production is the rational use of energy not only new, but also buildings built in the second half of the 20th century.

Various aspects of this problem have been considered in many scientific papers, including the following areas of researches: methods of reconstruction of the first mass series buildings [4], the technological features of buildings [5–8], the calculation of the cost of facades insulation and the economic justification for the redesign [8, 9], the prospects for solving the problem of energy efficiency of buildings [11].

The purpose of this study is to identify the most optimal measures for increasing the climatic conditions of the north to increase the thermal protection of residential buildings of the first mass series, taking into account their structural features, as well as to control the heat transfer agent consumption.

2 Materials and Methods

The objects of study are the most common in the city of Arkhangelsk series of residential apartment buildings: "Stalin Project", "Khrushchev Project", "Ulyanov Project", "Brezhnev Project", the buildings of the 93-rd series. Their informal name indicates the period of their construction.

The subject of study is a complex energy-efficient renovation of the first mass series buildings in the climatic conditions of the north on the example of Arkhangelsk. Under the complex energy-efficient repair is understood a set of works carried out within the such repair and aimed at achieving the energy-saving effect (as a rule, by the insulation of building envelopes and installation of collective (general house) metering devices for resource consumption and control units).

The sequence of the study derives from the totality of the tasks to be solved, including the following:

1. To analyze the technical characteristics and condition of residential buildings of the first mass series on the example of the housing stock of Arkhangelsk.
2. To select the optimal energy-saving measures basing on the analysis of the design features of the objects of study and the climatic conditions of the city of Arkhangelsk.
3. To calculate heat losses through the walls, windows, doors, coatings, socle and attic floors of the first mass series buildings before and after the energy-saving measures.
4. To determine the savings in cash when paying for utilities (for heat supply) according to the applicable tariffs for each object of study. To choose the energy-saving measures with the highest thermal efficiency.
5. To implement a calculation of the economic efficiency of major energy-efficient repairs for the studied objects. To choose the most efficient energy saving measures.

The study uses the following assumptions and limitations:

1. Only the most common building series in Arkhangelsk are considered;
2. Only calculative methods are applied, without using actual data of instrumental heat loss control.

Research methods:

- The technique of thermal technical calculation of envelope structures is described in SP 50.13330.2012 "Thermal protection of buildings", SP 23-101-2004 "Design of thermal protection of buildings" and GOST R 54851-2011 "Heterogeneous building envelope structures".
- Calculation of the economic efficiency of reconstruction projects was carried out in accordance with the "Methodological recommendations for evaluating the effectiveness of investment projects and their selection for financing".

Objects representing the apartment buildings (MCD) of the first mass series, which are most common in the city of Arkhangelsk and constitute in total more than 70% of the housing stock, were chosen to be the objects of study. Table 1 presents the main technical characteristics of the studied residential buildings.

The study of the technical condition of the envelope structures of housing facilities by determining physical wearout showed (Fig. 1) that in order to eliminate their physical wearout, it is necessary to take mandatory repair and restoration measures. These works are planned in advance as a part of the mandatory major repair in accordance with the major repair program. Funds for these types of works are accumulated on the accounts of residential houses by paying monthly installments.

3   Results

3.1   Analysis of the Technical Characteristics and Condition of Residential Buildings of the First Mass Series on the Example of the Housing Stock of Arkhangelsk

At the next stages of the study, an assessment of the effectiveness of additional capital investments in repair and construction activities in order to implement the energy-efficient measures for complex repair of buildings was made.

Table 1. Studied residential buildings and their characteristics.

Group of the buildings	«Stalin project»	«Khrushchev project »		«Ulyanov project»	«Brezhnev project»	93-series
Series		1-447	1-464	1-335-AK-11	114-85	
Years of construction	1954-1959	1958-1964		1972 – 1985	1974-2005	1979-1990
Structural scheme	Crossing walls					Complete frame
Walls	Ceramic brick (640 mm)	Silicate brick (770 mm)	Two-layer panels made of haydite (foamed carbon aggregate mineral found at Shunga, Karelia) (350 mm)	Two-layer panels made of haydite (foamed carbon aggregate mineral found at Shunga, Karelia) (350 mm)	Silicate brick (770 mm)	Two-layer panels made of haydite (foamed carbon aggregate mineral found at Shunga, Karelia) (350 mm)
Floors	Concrete round hollow core slabs (220mm)		Solid reinforced concrete slabs (100 mm)	Concrete round hollow core slabs (220mm)		
Roof	pitched with technical floor, covering - asbestos-corrugated sheet		flat combined, coating – hydrobarrier rolled-strip roofing	flat with technical floor, coating – hydrobarrier rolled-strip roofing		
Basement/ technical floor	warm	cold				
Windows	wooden with the double-glassing in the separate binding					
External doors	metallic					
Number of porches	2	3	4	5	4	5
Number of floors	5	5	5	9	9	10
Floor height, m	3	2,8	2,65	2,95	2,8	2,95

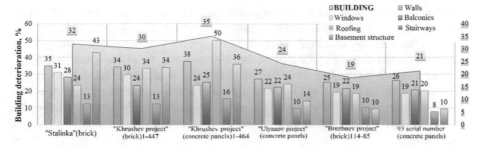

Fig. 1. Physical wearout of the first mass series buildings and their structures.

3.2 Selection of the Optimal Energy Saving Measures

Options of wall insulation, coverings, attic and basement floors, as well as replacing the old wooden windows by PVC two-chamber double-glazed windows and installing an individual heat supply point with a heat-metering unit (Table 2) were selected as energy saving measures. Materials for insulation were selected in accordance with the requirements of regulatory documents and climatic features of the city of Arkhangelsk.

In addition, Table 2 shows the planned repair works without taking into account energy-saving measures, which must necessarily be carried out within the programs for the residential buildings repair. Scope of such obligatory repairs is determined, among other things, by the amounts of cash savings planned for the major repairs under the major repair programs.

Table 2. Works for the complex energy efficient and planned major repair of the residential buildings.

Works for the complex energy-effective major repair	Works for the planned major repair	«Stalin project»	«Khrushchev project» (brick) 1-447	«Khrushchev project» (concrete panel) 1-464	«Ulyanov project» 1-335-AK-11	«Brezhnev project» 114-85	93-series
Installation of the heating point	–	+	+	+	+	+	+
Insulation of the basement floor	–		+	+	+	+	+
Replacing of the common property windows	Restoration of the common property windows	+	+	+	+	+	+
Thermal insulation of the facade with sandwich panels	Restoration of the facade	+	+	+	+	+	+
Complete replacement of the roofing structure	Restoration of certain places of the roofing structure of the combined roof			+			
Replacement of private property windows	–	+	+	+	+	+	+

Complex energy efficient repairs require additional capital investments. But it is the additional capital investments that will make it possible to increase the energy efficiency of the building, i.e. to reduce the amount of the consumed thermal energy, and, consequently, the scope of utility payments by the residents of the house. However, from the schedule on the Fig. 2 one can see that the cost of energy efficient repairs for all series of residential buildings significantly exceeds the cost of the obligatory works. Therefore, the apartment owners need a clear understanding of the economic effect that can be obtained before making a decision on the advisability of additional capital investments.

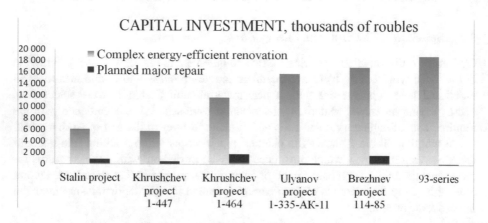

Fig. 2. Capital investments into the repairs

3.3 Calculation of Heat Losses of the Buildings Before and After the Energy-Efficient Repair

Thermal engineering calculation of envelope structures and elements of residential buildings was implemented by calculation according to the following categories: SP 50.13330.2012 "Thermal protection of buildings", SP 23-101-2004 "Design of the buildings thermal protection" using GOST R 54851-2011 "Inhomogeneous fencing building structures", STO 00044807-001-2006 "Thermal protection properties of envelope structures", GOST-30494-2011 "Microclimate parameters in the premises", "Construction climatology" and SP 230.1325800.2015 "Building envelope structures. Characteristics of heat engineering heterogeneities".

The results for all objects of study are presented in Fig. 3 and in Table 3.

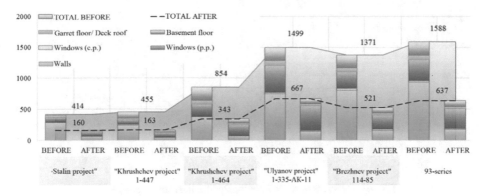

Fig. 3. Heat losses before and after reconstruction, Gcal.

The histogram (Fig. 3) shows the terms of total heat losses. One can see that the losses are reduced by more than 2 times, and the greatest heat losses occur on the walls, windows and basement floor.

Table 3. Heat losses of buildings before and after energy efficient repair.

Constructions and elements	Heat loss BEFORE and AFTER reconstruction, Q, Gcal											
	«Stalin project»		«Khrushchev project» (brick) 1-447		«Khrushchev project» (concrete panel) 1–464		«Ulyanov project» 1-335-AK-11		«Brezhnev project» 114-85		93-series	
	Before	After	Before	After	Before	After	Before	After	Before	After	Before	After
Walls	286	65	242	44	377	71	751	142	805	155	939	178
Windows (common)	8	5	12	8	5	4	19	13	31	21	19	13
Windows (personal)	71	61	68	61	199	169	436	395	266	228	339	315
Attic floors/combined roof	36	18	30	15	58	27	74	32	67	29	70	31
Warm basement/basement floor	13	11	102	34	214	73	219	85	202	89	221	100
In total	414	160	455	163	854	343	1499	667	1371	521	1588	637

3.4   Calculation of Overpayment for Heat Supply

After calculation of the heat losses, overpayments for heating for the heating season were calculated for each object of study as a whole and for its individual structures. The largest overpayment happens due to the walls, windows and basement floor.

Overpayments for heating due to losses through the building envelope of the considered series are presented in Table 4 and in Fig. 4.

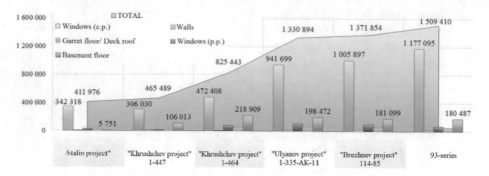

Fig. 4. Overpayment for heating for the heating season, RUR.

Table 4. Overpayment for heating for the heating season.

Constructions and elements	Overpayment for heating for the heating season, RUR					
	«Stalin project»	«Khrushchev project» (brick) 1-447	«Khrushchev project» (brick) 1-447	«Ulyanov project» 1-335-AK-11	«Brezhnev project» 114-85	93-series
Walls	342 318	306 030	472 408	941 699	1 005 897	1 177 095
Windows (common)	7 775	10 487	4 639	16 368	26 189	18 686
Windows (private)	28 359	19 999	80 971	109 721	100 791	72 719
Attic floors/combined roof	27 773	22 960	48 516	64 633	57 877	60 423
Warm basement/basement floor	5 751	106 013	218 909	198 472	181 099	180 487
In total	411 976	465 489	825 443	1 330 894	1 371 854	1 509 410

From Table 3 and the diagram in Fig. 4 one can see that the greatest decrease in heat losses is observed after the walls and basement floors insulation.

3.5 Calculation of the Economic Efficiency of Complex Energy-Efficient Repair

Calculation of the effect from the additional capital investments in the energy-efficient repairs is based on the payment saving for the heat supply agent according to the tariffs for thermal energy and the dynamics of their growth. Based on the tariffs of previous years (2014–2017), an average value of the relative tariff growth was found - 4.5%. Accounting of changes in the value of money over time is based on the risk-free rate of possible reinvestment - a deposit at a rate of 6% per annum.

Key project performance indicators and net present value schedules for each building are shown in Table 5.

Table 5. Performance Indicators for Energy Efficient Reconstruction Projects.

Series	Constructions, elements, systems	PI[1]	DPP[2], years	Payback schedule, years
1	2	6	7	8
«Stalin project»	individual heat points (ITP)	11,69	2	
	common property windows	2,87	8	
	walls	1,47	16	
	TOTAL:	2,21	10	
	private property windows	3,21	7	
«Khrushchev project» (brick) 1-447	individual heat points (ITP)	10,23	2	
	basement floor	1,82	12	
	common property windows	2,45	9	
	walls	1,75	13	
	TOTAL:	2,37	9	
	private property windows	2,66	8	
«Khrushchev project» (concrete panel) 1-464	individual heat points (ITP)	19,43	1	
	basement floor	1,98	11	
	common property windows	2,51	9	
	walls	1,44	16	
	roofing	2,67	8	
	TOTAL:	2,30	10	

Series	Constructions, elements, systems	PI[1]	DPP[2], years	Payback schedule, years
	private property windows	2,80	8	
«Ulyanov project» 1-335-AK-11	individual heat points (ITP)	20,80	1	
	basement floor	1,75	13	
	common property windows	2,51	9	
	walls	1,56	15	
	TOTAL:	2,42	9	
	private property windows	2,81	8	
«Brezhnev project» 114-85	individual heat points (ITP)	18,98	1	
	basement floor	1,80	13	
	common property windows	2,51	9	
	walls	1,69	13	
	TOTAL:	2,47	9	
	private property windows	2,81	8	
93-series	individual heat points (ITP)	19,81	1	
	basement floor	1,74	13	
	common property windows	2,87	8	
	walls	1,54	15	
	TOTAL:	2,30	10	
	private property windows	3,22	7	

[1] PI, payback index.
[2] DPP, discounted payback period.

For a more detailed analysis, the payback is calculated both for each project and for individual types of work.

Please notice, that replacement of private property windows is not included in the list of scheduled major repair works and was calculated separately from the main project. The owners calculate the payback of windows replaced in apartments in order to show what the effect will be of replacing of all wooden windows in the premises of the apartments with plastic ones with a two-chamber glass unit by the owners.

As one can see from Table 5, the installation of a heating point with a weather-dependent temperature control system for the heat supply agent (DPP = 1 ... 2 years) gives the greatest economic effect. This event is quite enough for the payment of utility bills to be reduced by 20–30%. And, in relation to the relatively low cost of this measure compared to the other types of work, the payback will be maximum.

It should be noticed that the effect of works to increase the thermal protection of envelope structures will also have a maximum value only if the optimum (normative) temperature inside the residential building is observed. Thus, it can be argued that the heating point is a priority measure for the energy-efficient repair.

4 Conclusions

In conclusion, of this study the following conclusions can be made:

1. the buildings of the first mass series do not meet modern requirements for thermal protection and are in a satisfactory condition when major repair or reconstruction are more appropriate;
2. the study showed that complex energy-efficient repair lead to a reduction in heating costs by more than 2 times;
3. the following energy-saving measures are the most optimal for Arkhangelsk: installation of an individual heating point with a weather-dependent regulation system; insulation of walls, combined roof and basement floor; installation of plastic windows;
4. the greatest thermal effect is achieved from the installation of an individual heating point, wall insulation with sandwich panels and a basement slab with fiberglass slabs;
5. the estimated payback period of the complex energy-efficient repair of buildings of the first mass series was 9–10 years, thus the projects are cost-effective and can be recommended for implementation. This is quite a convincing fact for the owners when making a decision to raise funds for the implementation of an aggregate or a part of the proposed activities and works.

References

1. Genzler, V., Petrova, E.F., Sivaev, S.B., Lykova, T.B.: Energy Saving in a Multi-apartment Building: Information and Methodological Manual. Scientific Book, Tver (2009)
2. The World Bank Group in collaboration with the Center for Energy Efficiency (CENEf): Energy Efficiency in Russia: Hidden Reserve (2007). http://www.cenef.ru/file/FINAL_EE_report_rus.pdf. Accessed 07 Feb 2019

3. Foundation for assistance to the reform of housing and public utilities homepage. http://energodoma.ru/kapitalnyj-remont-mkd. Accessed 07 Feb 2019
4. Grigorenko, K.A.: Reconstruction of the houses of the first mass series as a way to increase the usable area. Bull. PNRPU **7**(1), 47–55 (2016). Construction and architecture T
5. Bulgakov, S.N.: Payback Reconstruction of Residential Buildings of the First Mass Series. DIA Publishing House, Moscow (2016)
6. Afanasyev, A.A.: Reconstruction of residential buildings. Part I. Technologies for restoring the operational reliability of residential buildings, Moscow (2008)
7. Zhadanovsky, B.V.: Organizational and technological preparation for the reconstruction of civil and industrial buildings and structures. Ind. Civ. Eng. **10**, 59–60 (2009)
8. Tseytin, D.: A feasibility study of facade insulation during the renovation of the first mass residential buildings. Constr. Unique Build. Struct. **1**(40), 30–31 (2016)
9. Spirin, A.V.: Heat insulation of external walls in the reconstruction of brick buildings. Perm National Research Polytechnic University, Perm. http://docplayer.ru/28912404-Uteplenie-naruzhnyh-sten-pri-rekonstrukcii-kirpichnyh-zdaniy-insulation-of-outside-walls-during-reconstruction-of-brick-buildings.html. Accessed 07 Feb 2019
10. Gorshkov, A.S.: Method of calculating the payback on investment for the renovation of facades of existing buildings. Constr. Unique Build. Struct. **2**, 82–106 (2014)
11. Ovsyannikov, S.N.: Prospects for the reconstruction of residential buildings of the first mass series in the city of Tomsk. Bull. TGASU **2**, 201–212 (2010)

Green Roofs as a Response to a Number of Modern City Problems

Elena Sysoeva(✉)📧 and Ivan Aksenov📧

Moscow State University of Civil Engineering, Yaroslavskoye Shosse 26,
129337 Moscow, Russia
SysoevaEV@mgsu.ru

Abstract. The urban environment becomes a place of extremely high population density, which has such environmental defects as high risk of infectious
and chronic diseases, air pollution, vulnerability to climate change, etc. On
account of that, the problem of energy-efficient and safe construction methods
development that would mitigate the effects of air pollution (to create areas for
health recovery, increase the length of working period of human life) becomes
especially important. As a tool for solving these problems, the technology
of «green roof» is proposed. To confirm the theoretical studies on the energy
efficiency of roofing with natural landscaping, a test stand was created, and an
experiment to determine the thermal properties of a «green roof» was conducted
in a laboratory of NRU MGSU (Moscow, Russia). The results of the experiment
are presented. The energy efficiency of a public building roof was tested on three
models of «green roof» with different composition and thickness. The results
were compared with theoretical calculations.

Keywords: Urban environment · Energy-efficiency · Green roofs ·
Air pollution · Public building roof

1 Introduction

Nowadays, problems of urban environment health are becoming extremely important.
In 2008, the urban population of the planet for the first time exceeded the rural one and
it is still growing [1]. According to the United Nations organization (UNO) forecast, it
will almost be doubled by 2050 [2, p. 2]. The experts of World Health Organization
(WHO) stated that the contribution of biosphere quality to people health formation can
be estimated in the range from 20 to 40% [3]. A city becomes such form of human
existence that not only provides comfort and facilitates in everyday life but also significantly transforms the natural environment, leading to a violation of natural processes that negatively affects the main value of humanity such as the health of living
and future generations. In consequence of this situation, cities development are given
much attention at international level. At the United Nations Conference on Housing
and Sustainable Urban Development (Habitat III), which was held in October 2016 in
Quito (Ecuador), a program «New Urban Agenda» was adopted. This paper determines the main problems and challenges in a field of urbanization as well as general

© Springer Nature Switzerland AG 2020
V. Murgul and M. Pasetti (Eds.): EMMFT 2018, AISC 982, pp. 137–145, 2020.
https://doi.org/10.1007/978-3-030-19756-8_14

principles of their solution. In the strategy approved by this program, the following promising ways of buildings design and construction development are marked out:

1. Development of energy-efficient buildings and construction methods [2, p. 75];
2. Promotion of environment cleanness with taking into consideration air quality guidelines [2, p. 55];
3. Improving cities resilience to climate change [2, p. 67];

Wide application of «green roof» technology in urban planning practice can become a solution to this range of problems.

2 «Green Roofs» Impact on Building Energy Efficiency

For many regions of Russia, the problem of minimizing a building heat loss in the cold season is the most important issue in terms of energy efficiency. According to thermophysical indicators, the most vulnerable element of any building is a roof. Heat loss through it can reach 40% of the total one. In view of that, reducing the heat migration through a roof can lead to a significant increase of building energy efficiency. It is known that greening is able to improve roof thermal properties by adding new layers (drainage, substrate and vegetation). However, the vast majority of studies on this issue are conducted outside of Russia. They don't take into account the specifics of the Russian market of materials used for landscaping the roof and they are mostly devoted to a cooling capacity of «green roofs» during the summer [4–7]. That served as a pretext for a pilot study.

2.1 Preparing for the Experiment

In Scientific Research Center «Translucent Structures» in National Research Moscow State University of Civil Engineering (NRU MGSU) (Russia), an experiment was conducted, which purpose was to determine an actual thermal resistance of the upper part of a «green roof» multilayer structure. Three models of «green roofs» were studied (Table 1). As a result, it was found out how much a traditional roof total thermal resistance will be increased due to its landscaping, so to adding the layers studied in the experiment.

During the experiment, the following tasks were solved:

1. A temperature difference between opposite surfaces of the models was created. For this purpose, the test bench KS 3025/650 (Fig. 1) was used. It is a niche enclosed with adjustable partitions to create a closed volume. There is a cooling grid on the left side of the niche, which reduces temperature in the closed volume to a necessary value.
2. A heat flow through the models was organized thus that the total amount of heat passing through a cross section (a section parallel to the top and the bottom faces of a model) does not depend on a location of the section. For that, heat losses through the lateral models' surfaces were minimized by significant thermal insulation.

Table 1. The test specimens' characteristics.

	No	Layer	δ, m	ρ_V, kg/m^3	ρ_S, kg/m^2
Model 1	1	Substrate (growth media)	0.1	600	46.5
	2	Floraset 75 (drainage element)	0.075	–	1.7
	3	Protection mat	0.005	–	0.65
	4	Root-resistant waterproofing	–	–	–
	In total		0.135		48.85
Model 2	1	Substrate (growth media)	0.1	600	46.5
	2	Floraset 75 (drainage element)	0.075	–	1
	3	Floraset 75 (drainage element)	0.075	–	1
	4	Protection mat	0.005	–	0.65
	5	Root-resistant waterproofing	–	–	–
	In total		0.21		49.15
Model 3	1	Substrate (growth media)	0.125	600	75
	2	Filter sheet	–	–	–
	3	Floradrain 40 (drainage element)	0.04	–	2
	4	Protection mat	0.005	–	0.65
	5	Root-resistant waterproofing	–	–	–
	In total		0.17		77.65

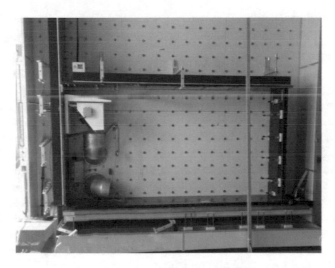

Fig. 1. The test bench KS 3025/650.

3. Temperature values on opposite model sides were measured, and there vertical distribution in the thickness of the models was fixed (for more detailed data). For this purpose, contact temperature sensors DS18B20 with digital thermometer "RODOS-6" were used. The layout of the sensors and their marking are shown in Fig. 2.

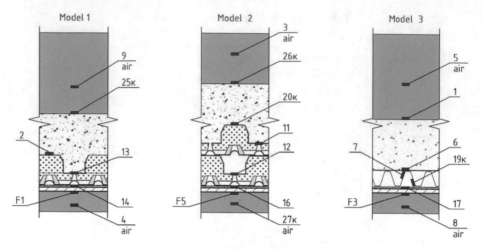

Fig. 2. The sensors layout and marking.

4. The stationary heat flux passing through the models was measured using a heat flux density measuring device ИТП-МГ4.03/Х(Y) «Potok» . Density heat flux sensors are shown in Fig. 2 (with names F1, F3 and F5).

2.2   Conduct and Results

The experiment lasted 6 days. The refrigeration unit upheld above the models the temperature of −20 °C, that is closed to the rated winter temperature for Moscow (−25 °C) [8]. During the first three days, a gradual freezing of the models was observed. By the 6th day of cooling system operation, the heat flow passing through the models became stationary. This day, a number of heat flux density measurements were made. Figure 3 shows how the temperature difference between the upper (cold) and lower (warm) surfaces of the models was changing during the experiment. According to the obtained data, the thermal resistance of the models was determined (Figs. 3, 4).

These results are significant. For example, for public buildings in Moscow the basic value of the required roof total thermal resistance is 3.42 $m^{2.0}C/W$ [9], the thermal resistance of the models amounts to 27–81% of this value. This result is similar to other studies of this kind. Thus, it was found that for the winter conditions of the La Rochelle (France), the use of intensive green roof helps to reduce a heat migration by 39% [4]. Therefore, it can be seen how the greening contributes to a roof thermal insulation property, and consequently to building energy efficiency.

The experimental data also enables us to determine the heat emission coefficient of the substrate surface. Since this surface is represented in all models by the same dispersed structure being in the same condition, the heat emission coefficient for all models should be approximately the same, which is confirmed by the calculation (Fig. 5).

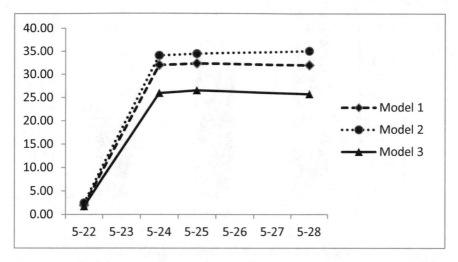

Fig. 3. The temperature difference between the cold and the warm surface of the models, °C.

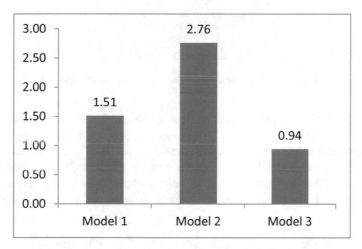

Fig. 4. The thermal resistance of the models, $m^{2.0}C/W$.

2.3   Conclusions from the Experimental Results

Let's compare model 1 and 2. Their layers composition is the same except that the model 2 has two drainage layers Floraset 75 instead of one. Obviously, it is this difference that leads to a significant increase in the thermal resistance of the model 2 (by 1.61 $m^{2.0}C/W$ in comparison with the model 1). Drainage element Floraset 75 is made from extruded polystyrene, has thick walls (due to a small strength of the material it is impossible to make it thin-walled) and has a pronounced corrugated geometry (Fig. 6), due to which it forms large air voids. That explains its significant impact on the overall thermal resistance of the structure.

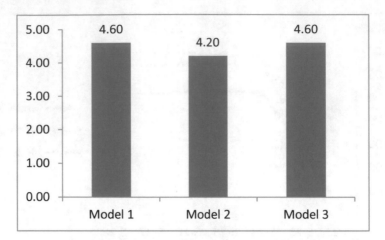

Fig. 5. The heat emission coefficient, $W/(m^{2 \cdot 0}C)$.

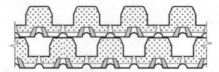

Fig. 6. Two Floraset 75 lairs form air voids in the model 2.

Model 3 has the lowest thermal resistance, although the substrate in it is a bit thicker than it is in the other models. The reason for that is a different type of drainage element (Floradrain 40) which was laid in the 3d model (Fig. 7). Floradrain 40 is made of polyethylene, has a small wall thickness and the total height in comparison with Floraset 75 and therefore forms smaller air voids and promotes intensive heat exchange (through thin walls) between the air voids located above and below the element.

Fig. 7. The drainage element Floradrain 40.

As it is seen, a type of drainage element laid in a «green roof» structure has a large influence on its total thermal resistance in winter conditions, especially if a substrate lair is thin. Substrate unavoidably contains moisture that turns into ice during the freezing. As known, ice conducts heat well enough but it also becomes an obstacle for air exchange between voids formed by a drainage element and external space. Owing to that inside air emptiness acquires a significant role in thermal insulation.

3 Promotion of Environment Cleanness

One of the main health risks concerned with an environment is air pollution. The higher it is, the greater likelihood of chronic lung illnesses and cardiovascular diseases is, both in the short and long terms. The main hazard for health is small dust particles suspended in the air [10]. According to J. Johnson and J. Newton studies, 2000 m^2 of uncut lawn can absorb 4000 kg of solid dust particles [11]. Consequently, the building envelope covered with vegetation can greatly contribute to the purification of the urban atmosphere, which can have a positive impact on population health [12–14].

The air quality guidelines developed by HWO experts [15] speak eloquently about the consequences of the decrease in particulate matter (PM) concentration (Table 2).

Table 2. Average annual particulate matter concentrations: intermediate and recommended values.

	PM_{10}, $\mu g/m^3$	$PM_{2.5}$, $\mu g/m^3$	Consequences
The first interim level	70	35	These concentrations lead to about a 15% higher long-term mortality risk relative to recommended concentrations
The second interim level	50	25	The decrease by these levels makes it possible to lower the risk of premature mortality by 2–11% in comparison with the previous level
The third interim level	30	15	The decrease by these levels makes it possible to lower the risk of premature mortality by 2–11% in comparison with the previous level
The values recommended by WHO	20	10	These are the lowest levels, the excess of that leads to increase of cardiopulmonary and lung cancer mortality with more than 95% confidence

These data show how it is important to improve air quality.

«Green roofs» have a significant ability to absorb carbon dioxide. Two studies on location to determine an amount of carbon absorbed by a «green roof» surface were carried out by Getter et al. (2009). In the first study, aboveground biomass of 12 «green roofs» were analyzed. All these roofs were planted with sedums, have substrate of 2.5–12.7 cm depth and have the age of 1 to 6 years. In the second research, which lasted over two growing seasons from June 2007 to October 2008, plant roots and substrate were also studied in addition to aboveground biomass. The results of the first research were diverse. The amount of carbon absorbed by each of 12 green roofs ranged from 73 to 276 g C/m^2. Probably this spread was the result of different ages, substrate depth and irrigation regimes of studied roofs. The second research showed that aboveground biomass accumulated 168 g C/m^2, substrate accumulated 100 g C/m^2 and roots accumulated 107 g C/m^2. The author concluded that «green roofs» can become an effective means of absorbing excessive amounts of carbon dioxide of the urban atmosphere. He calculated that if all rooftops in Detroit metropolitan area were greened, they could sequester 55,252 tons of carbon [16].

4 Cities Resilience to Climate Change

In a hot season, a «green roof» surface warms up much less than a traditional roof surface, and this difference can reach 30–40 °C [17]. Owing to this, the widespread use of «green roofs» in urban planning can reduce the heat island effect and thereby compensate for the average annual increase in temperature as a result of global warming [18–20]. Using the example of Greater Manchester County (Great Britain), it was estimated that the greening of 10% of roofs in densely built-up areas will help to neutralize the effect of global warming in these areas until 2080 [21]. This suggests that «green roof» technology can be an effective tool for adapting cities to climate change.

The quantitative research made by Eleftheria Alexandri and Phil Jones permit to assess the cooling efficiency of «green roofs» in different climates. A two-dimensional dynamic model was created in C++ programming environment to describe heat and mass transfer in a typical street canyon. The problem was solved for nine different cities with different climates (from Riyadh in Saudi Arabia with desert climate to Moscow in Russia with continental climate with cool summer). The results show that «green roofs» have the greatest temperature mitigation effect for a hot and dry climate. Thus, in the case building walls and roofs were covered with vegetation the average air temperature inside the street canyon during a typical hot day would decrease by 9.1 °C for Riyadh and 3.0 °C for Moscow. So, for mild summer of Moscow, the greening of the building envelope doesn't bring to a significant improvements of the outdoors thermal comfort [22]. However, the cooling «green roof» capacity can have an important role for south regions of Russia.

5 Summary

Based on the literature review and the experimental study results, it was shown that the use of «green roof» technology will allow:

1. Improving the energy efficiency of buildings. It has been experimentally established that the greening of the public building roof can increase its thermal insulation qualities by 27–81% depending on the type of drainage element. It was also established that a type of drainage element has the key influence on a roof thermal performance in the case of thin substrate layer.
2. Reducing the concentration of airborne solid dust particles adversely affects public health and also to sequester an excessive amounts of carbon dioxide of the urban atmosphere. WHO Air Quality Guidelines shows how it is important to achieve good air quality for people health.
3. Adaptation of cities to global warming by summer temperature mitigation. This « green roofs» feature best manifests itself in a hot dry climate and can be used more effectively for south regions of Russia.

Green roofs can become a solution to a number of problems faced by urbanism today. It stimulates a lot of experimental studies devoted to them. The results of these studies will help to create a methodology for assessing the buildings environmental design for the formation of safe and developing human urban surroundings.

References

1. Pryadko, I.P., Boltayevsky, A.A.: Biosphere compatibility: human, region, technologies **1**, 10 (2014)
2. United Nations, Habitat III, New urban agenda (2017)
3. United Nations, Habitat III, Regional report of the Commonwealth of independent States region "Cities of the CIS: the way to a sustainable future"
4. Jaffal, I., Ouldboukhitine, S.-E., Belarbi, R.: A comprehensive study of the impact of green roofs on building energy performance. Renewable Energy **43**, 157–164 (2012)
5. Ahmadi, H., Arabi, R., Fatahi, L.: Curr. Environ. **10**, 908–917 (2015)
6. Zhao, M., Tabares-Velasco, P.C., Srebric, J., Komarneni, S., Berghage, R.: Effects of plant and substrate selection on thermal performance of green roofs during the summer. Build. Environ. **78**, 199–211 (2014)
7. Eumorfopoulou, E., Aravantinos, D.: Energy Build. **57**, 29–36 (2005)
8. Set of rules 131.13330.2012. Building climatology. Ministry of regional development of Russia, Moscow (2012)
9. Set of rules 50.13330.2012. Thermal performance of buildings. Ministry of regional development of Russia, Moscow (2012)
10. WHO, information bulletin. Air quality and health (2016)
11. Johnson, J., Newton, J.: Building Green, a Guide for Using Plants on Roofs and Pavement. The London Ecology Unit, London (2006)
12. Torpy, F., Zavattaro, M.: Bench- study of green wall plants for indoor air pollution reduction. J. Living Archit. **5**, 1–15 (2018)
13. Rowe, D.B.: Green roofs as a means of pollution abatement. Environ. Pollut. **159**, 2100–2110 (2011)
14. Llewellyn, D., Dixon, M.: Agric. Relat. Biotech. **4**, 338–431 (2011)
15. WHO, Air Quality Guidelines Global Update (2005)
16. Getter, K.L., Rowe, D.B., Robertson, G.P., Cregg, B.M., Andresen, J.A.: Carbon sequestration potential of extensive green roofs. Environ. Sci. Technol. **43**, 7564–7570 (2009)
17. Dalinchuk, V.S., Vlasenko, D.A., Startsev, S.A.: Innov. Dev. **4**, 12–18 (2017)
18. Li, Y., Babcock, Jr. R.W.: Green roofs against pollution and climate change. a review. Agron. Sustain. Dev. **34**, 695–705 (2014)
19. Takakura, T., Kitade, S., Goto, E.: Cooling effect of greenery cover over a building. Energy Build. **31**, 1–6 (2000)
20. Theodosiou, T.G.: Summer period analysis of the performance of a planted roof as cooling passive technique. Energy Build. **35**, 909–917 (2005)
21. Gill, S.E.: Adapting cities for climate change: the role of the green infrastructure. Built Environ. **33**, 115–133 (2007)
22. Eleftheria, A., Phil, J.: Temperature decreases in an urban canyon due to green walls and green roofs in diverse climates. Build. Environ. **43**, 480–493 (2008)

Modeling of Processes of Convective Transfer of Air Masses in the Atrium Spaces of Buildings

Nellya Kolosova$^{(\boxtimes)}$ (iD), Sergej Kolodyazhnyj (iD), Vladimir Kozlov (iD),
and Inna Pereslavceva (iD)

Voronezh State Technical University, Moscow Avenue, 14, Voronezh 394026,
Russia
belyaeva-sv@mail.ru

Abstract. The overwhelming majority of formulas for determining the mass
flow rate of a smoke ventilation system were obtained empirically when conducting experimental studies in specific conditions that did not meet the necessary and sufficient conditions of the similarity theory of heat and mass
exchanges, especially in high-ceilinged rooms. Therefore, it is necessary to
clarify the formulas for calculating the flow rate of the exhaust ventilation
system when considering the specific conditions of fire development, including
the atrium spaces of buildings. In this paper, the convective transfer of air
masses in high-ceilinged rooms is considered using a zone model of fire
development, which takes into account the change in the height of the convective column of its half-angle of opening. For high-ceilinged rooms, the
adoption of the shape of a convective column in the form of a conical
unbounded jet is correct only for the part of the column located in the lower half
of the room's height. In the process of fire development at the initial stage, when
the smoke layer beneath a ceiling fills only the upper part of the room, this
model of convective transfer of air masses determines the excessive mass flow
rate of gases and temperature in the smoke layer beneath a ceiling.

Keywords: Convective transfer · Atrium spaces of buildings · Management ·
Air masses

1 Introduction

The flow rate of the mixture of combustion products and air in a convective column,
which enters the zone of the layer beneath a ceiling, determines the speed of lowering
the smoke screen to the floor of the room. In the zone mathematical model for calculating the thermal and gas dynamic processes of a fire, one of the main assumptions
is that the shape of the convective column above the combustion source is taken as an
unbounded free-convective jet. Studies conducted in [1–3] show that the shape of a
convective column is significantly affected by the presence of enclosing building
elements - walls and floors. In this regard, the actual scientific and practical task is to
further investigate the dependence of the shape of the convective column on the shape
of the enclosing structures of the room, their geometry and location in relation to the

© Springer Nature Switzerland AG 2020
V. Murgul and M. Pasetti (Eds.): EMMFT 2018, AISC 982, pp. 146–155, 2020.
https://doi.org/10.1007/978-3-030-19756-8_15

source of ignition in order to improve the mathematical model of fire and refine the results obtained with it. In the noted papers [1–3], when studying the shape and opening angle of a convective column, a three-dimensional differential model was used to calculate the thermal and gas dynamic processes of a fire with a series of numerical experiments. In this case, the half opening angle of the cone of a convective column in a differential equation, from the solution of which the mass flow rate of the gas mixture is determined, takes an averaged constant value over the entire height of the column. In the proposed work, we study the law of changing the half-angle of a convective column in a high-ceilinged room using a fire zone mathematical model, the refinement of which is to take into account the local distribution of the half-angle of a column opening along its height.

2   Determination of the Mass Flow Rate of the Gas Mixture Along the Height of the Convective Column

As a rule, a massive evacuation of people from a room in fire takes place at the initial stage of fire development, at a time when this evacuation is most effective. It is at this stage that the parameters of the gas mixture in it are distributed very unevenly throughout the entire volume of the room [4]. The volumetric space of the room is divided into several zones, in which the values of the main parameters of the gas environment are significantly different from each other. The physical picture of the development of a fire at its initial stage determines the correctness of the use of a zone mathematical model for this stage in order to determine the magnitudes of dangerous fire factors and to consider the dynamics of their development. At ignition of combustible substances, the rising convective stream from a hot combustion products is formed over the ignition center. Having reached the ceiling of the room, the heated smoke layer spreads over the overlap and forms a layer beneath a ceiling that increases in height, which descends and completely fills the room with time. In this case, the number of zones varies, the boundaries between them are erased, the characteristics of the gas mixture in the entire volume of the room take approximately the same values. Ultimately, in the case of a voluminous fire, the smoke-filled gas mixture in the room forms one homogeneous zone.

In general, the volume of the room can be divided into any number of zones. For example, in the integral mathematical model of a fire, in which all the characteristics of a gas mixture take averaged values over the entire volume of the room, the gas mixture is considered as one zone. Splitting the volume of the room with the fire source into separate zones allows eliminating the errors of the approximate integral model, since in this case, for each of the zones, the more accurate thermophysical properties of the gas mixture are set, the corresponding ratios and equations determining the flow of characteristic processes in each of the zones are taken. The choice of the number and size of zones is determined by the purpose of the research. The criterion for selecting zones and their boundaries is the minimum difference in the values of different characteristics of the gas mixture for each zone. Zone models have a wider range of applications, in particular, to take into account the laws of hydrodynamic and thermal effects of jet flows when meeting with solid surfaces, to explore areas of self-similar and accelerated flow, critical point, etc. [5].

Like all models used, the zone ones have some inaccuracies and disadvantages. One of these inaccuracies is that with zone modeling, the shape of a convective column above the ignition center is taken in the form of a regular cone, which retains it until the room overlaps, regardless of the height of this room. But the convective column will be such a free convective jet only in an unlimited space, without taking into account the impact of enclosing building structures on its form. Therefore, to determine the true distribution of column parameters in the longitudinal and transverse directions, it is necessary to conduct scientific research confirmed by experimental studies. When using both integral and zone models of fire, to determine the main parameters of the gas mixture in each of the selected zones, additional ratios and equations are used. Considering the layer beneath a ceiling, the dynamic ratios of the thermal boundary layer on the surface of the overlap are used. To clarify some results, a differential model is used, which requires, as a rule, numerical implementation.

In the zone approach, the three-zone model is used most often, in which the volume under study is divided into three zones with different thermal characteristics. This is directly the convective column itself, the upper smoke-filled layer under the overlap, and the lower layer of cold air [4]. Such a model quite correctly describes the thermal and gas dynamics processes occurring in the initial stage of fire development, since at that time, the flame did not cover the whole room, and the source of ignition is much smaller than the size of the room. In the first initial stage of fire development, the hot, smoke-filled mixture rises to overlap, increasing in volume in the layer beneath a ceiling. At some point, the lower limit of the smoke-filled gas mixture, falling, reaches an open doorway or window opening, and the hot mixture of air with combustion products begins to leave the room. Until this time, with open doorway or window openings, unheated air flowed out of the room from the lower non-smoked area.

Determining the position of the lower boundary of the smoke-filled gas mixture under the ceiling of the room, we assume that the activation of the supply and exhaust smoke control ventilation has not yet occurred, and in the volume occupied by the heated smoke screen, there are no openings. The differential equation determining the vertical coordinate z_k measured from the floor of the room, the lower limit of the smoke-filled area beneath a ceiling, depending on the current time τ, has the form [4]:

$$\frac{dz_k}{d\tau} = -\frac{1}{\rho_0 \cdot F_c} \cdot \left(G_{cc} + \frac{Q_f \cdot (1 - \phi)}{c_p \cdot T_0} \right) \tag{1}$$

This equation expresses the law of energy conservation in the upper smoke-filled layer. Here:

F_c (m^2) - the surface area of the overlap;
ρ_0 (kg/m^3) - the density of the outside air before the fire;
T_0 (K) - the air temperature in the room before the fire;
c_p (J/(kg·K)) - the specific value of the isobaric heat capacity of the gas;
Q_f (W) - the power of heat release from the burning center;

G_{cc} (kg/s) - mass flow rate of the gas mixture entering the layer beneath a ceiling from the convection column;

φ - the coefficient of heat loss, determined by the ratio

$$\phi = \left(Q_{w1} + Q_{w2} + Q_c + Q_f + Q_o\right)/Q_f$$

Q_{w1}, Q_{w2} (W) – the power of total convective and radiant heat fluxes extending into the walls below and above the lower boundary of the upper smoke-filled layer, respectively;

Q_c, Q_f (W) – the power of total convective and radiant heat fluxes going to the overlap (ceiling) and to the floor, respectively;

Q_o (W) – the power of the heat flow leaving through open openings.

Since at the initial moment of time the upper smoke-filled layer is absent, its lower boundary coincides with h (height of the room), therefore, the initial condition at $\tau = 0$:

$$z_k = h$$

In the considered initial stage of development, the specific isobaric heat capacities and gas constants for cold air and air mixtures with combustion products differ slightly. Therefore, the same values are assumed for them. Equation (1) is solved numerically using the Runge-Kutta method, which has the fourth order of accuracy [6].

The law of conservation of mass in the layer beneath a ceiling is determined by the differential equation [4]:

$$\frac{d(\rho_2 V_2)}{d\tau} = G_{cc} \tag{2}$$

where

ρ_2 (kg/m^3) – average density of the gas mixture in the volume of the layer beneath a ceiling;

V_2 (m^3) – the entire volume of the layer beneath a ceiling.

Defining from Eq. (2) the average density in the volume of the layer beneath a ceiling, it is possible to determine the value of the average temperature in the volume of the layer beneath a ceiling from the equation of state of a mixture of ideal gases [4]:

$$p_2 = \rho_2 R T_2, \; p_2 \approx p_0 \tag{3}$$

where

p_2 (Pa) – pressure value in the upper smoke-filled layer;

p_0 (Pa) – atmospheric pressure at the floor level;

T_2 (K) - average volume temperature of the gas mixture of the layer under consideration; R(J/(kg•K)) – the gas constant.

Let's perform the calculation of the mass flow of gases from the combustion products along the height of the convective column. Assuming a uniform distribution of heat release over the height of the flame zone, a one-dimensional approximation can be applied at any time for differential equations that determine the laws of conservation of momentum and energy. Figure 1 shows the received heat exchange scheme in the elementary volume of a convective column. The letter symbols indicate the following parameters:

γ (rad) - half of the opening angle of the convective column;
z_f (m) - the height of the heat release zone (burning center);
T (K) - the average temperature of the gas mixture in the cross section of the column;
G (kg/s) - the mass flow rate of the emitted gases, which passes through the cross-sectional area of the convective column at the z coordinate.

According to the presented scheme for a mixture of gases in a dedicated elementary volume with height dz, the equation expressing the energy conservation law will be written as:

$$c_p d(TG) = c_p d(T_0 G) + \frac{Q_f \cdot (1 - \chi)}{z_f} dz \qquad (4)$$

where χ determines the fraction of the power of thermal radiation from the source of combustion, flowing into the building envelope from a convective column.

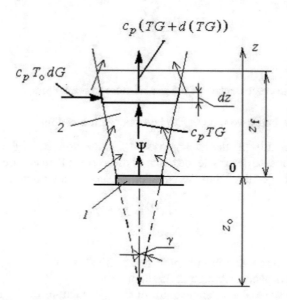

Fig. 1. Scheme of heat and mass transfer for the elementary volume of the convective column 1 – combustible agent; 2 – heat release zone.

Equation (4), in contrast to [1, 7], takes into account the fact that the mass flow rate varies along the height of the convective column. Considering the same elementary volume of a mixture of gases filled with smoke, we write the equation defining the law of conservation of momentum in the selected volume of a convective column with height dz

$$\frac{d\left(\rho v_z^2\right)}{dz} = -\frac{dp}{dz} - \rho g,$$

(5)

where

z(m) – vertical coordinate measured from the burning surface;
g (m/s^2) – the acceleration of gravity;
v_z (m/s) – the average velocity of the mixture of gases in the vertical cross section of the column (along the z axis);
ρ(kg/m^3) – the average density in the cross section of the column;
p (Pa) – the pressure in the convective section of the column;

Let's integrate Eqs. (4), (5) from 0 to the current z coordinate. After simple transformations, we come to a differential equation, from the solution of which it is possible to determine the mass flow distribution along the height of the convective column:

$$\frac{dG}{dz} = \frac{c_p \cdot z_f}{2c_p z_f T_0 G + z Q_f(1-\chi)} \cdot \left\{ T_0 G \cdot \left[2Gtg\gamma\sqrt{\frac{\pi}{F(z)}} - \frac{Q_f(1-\chi)}{c_p z_f T_0} \times \left(1 + 2ztg\gamma\sqrt{\frac{\pi}{F(z)}}\right)\right] \right.$$

$$\left. + gp_0 F^2(z) \cdot \left[\rho_0 - \frac{c_p z_f p_0 G}{R\left(c_p z_f T_0 G + 2Q_f(1-\chi)\right)}\right]\right\}$$

(6)

where

p_0 (Pa) – initial pressure at z = 0 before the fire;
$F(z)$ (m^2) – convective column cross-sectional area at height z.

The law of distribution of the average temperature in the considered section of the convective column is represented by the Eq. [4]:

$$T = T_0 + \frac{Q_f \cdot (1-\chi)}{c_p G}$$

(7)

The obtained Eqs. (6), (7), with the help of the angle γ that is into these equations, allow taking into account the change in the shape of the convective column along its height when calculating the mass flow rate of gases emitted during combustion.

3 Results

The assumption of a convective column as a free convective jet is true only up to a certain height z. This is confirmed by numerical calculations given in [8] using a three-dimensional field model for placing a cross section of 30 m × 24 m and a height of 26.3 m, with a thermal power of $Q_f = 1.3$ MW (Fig. 2). Curves 1 and 2 are obtained in accordance with the analytical dependencies presented in [4] and [9], respectively, curve 3 - according to Eq. (6). Numerical results for the field model [8]: 4 - in the cross sections of the convective column, 5 - in the cross sections of the room as a whole.

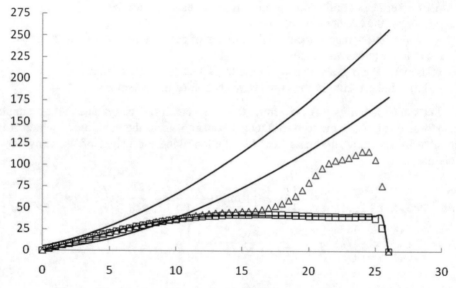

Fig. 2. Graphic dependences of the mass flow rate of the gas mixture.

G, kg/s
Graphical dependences 1 and 2 confirm the correctness of applying the approximation by a free convective jet to a relative height $\bar{z} = 0,4$ ($\bar{z} = z/h$, h is the height of the room). The graphical dependence (3) of the mass flow rate of the gas mixture through the cross section of the convective column, obtained using differential Eq. (6), coincides with the numerical calculations for a three-dimensional field model over the entire height of the room. The matching graphical dependences 3 and 4 are physically more reliable than those presented in [3, 4, 10–13], obtained under the assumption that the column propagates as an unrestricted convective jet.

We use the formula for determining the mass flow rate of a gas mixture in a convective column from [11]:

$$G = \left[7,1 \cdot \sqrt[3]{Q_f \cdot (1 - \chi) \cdot z^5} + 0,0018 \cdot Q_f \cdot (1 - \chi) \right]/10^3. \qquad (8)$$

The value of the relative consumption is determined by the equality

$$\overline{G} = G_h/G_{z*}$$

where

G_h (kg/s) – mass flow determined by the equality (8), at z = h at the level of the ceiling of the room;

G_{z*} (kg/s) – mass flow at height z*;

z* (m) – the value of the vertical coordinate, at which the constant to this height value of half the opening angle of the convective column in the formula (8) decreases to zero.

Taking the values of the original parameters z* = 1,0 m; z = 1,5 m; Q_f = 1 kW, for the value of the relative consumption, we get the result $\overline{G}$ = 1,97.

From the formula (7)

$$T - T_0 = \frac{Q_f \cdot (1 - \chi)}{G \cdot c_p}$$

Therefore, the temperature increment is inversely proportional to the mass flow rate. From this it follows that the value of the increment of the average over the column section temperature under the room overlap calculated by the formula (7) will be approximately twice the value obtained using the formulas for calculating the mass flow rate G of the gas mixture from [4, 9, 11].

When analyzing the presented results, we can conclude that the column above the ignition center in a burning room can be considered as a free unlimited jet only in the lower half of the room height. Above, such an approximation would be incorrect. And in the case when the lower boundary of the upper smoke-filled zone is in the upper part of the room, which takes place in the initial stage of fire development, the assumption of a free unrestricted jet leads to a significantly overestimated mass flow rate of the gas mixture and temperature in the smoke-filled zone under the ceiling. In addition, the exhaust fan flow rate in the mechanical system for removing smoke from the burning room is determined by the mass flow rate of the gas mixture in the convective column. If, when calculating the smoke removal system, the mass flow rate of the gas mixture is significantly higher than the actual value, then unheated air will flow into the exhaust opening along with the combustion products, preventing more volumetric removal of harmful substances and smoke, thereby reducing the efficiency of the exhaust ventilation system. In foreign literature, this phenomenon is called "plug-holing". Consequently, the formulas determining the mass flow rate of the gas mixture along the vertical axis of the convective column during combustion, which are used by many authors and are given in well-known sources [4, 10, 11, 14, etc.], require clarification and confirmation of these clarifications by appropriate experimental studies.

4 Conclusion

A mathematical zone model for the development of thermodynamic processes at the initial stage of fire development has been proposed, the refinement of which is that the height of the convective column above the fire source in the proposed ratios takes into account the change in the shape of this column under the influence of the enclosing building structures of the room, using variable half-angle of opening of the column. If the half-angle of opening is considered constant along the entire height of the convective column, then, as the considered example shows, it turns out to be an underestimate of the average volume of temperature in the upper smoke-filled layer under the ceiling of the room. If we take into account the change in the half-angle of opening along the height of the convective column, then the value of the average volume of the temperature in the upper smoke-filled zone will be twice the value obtained using well-known formulas.

References

1. Puzach, S., Abakumov, E.: Modificirovannaya zonnaya model' rascheta termogazodinamiki pozhara v atrium. Inzhenerno-fizicheskij zhurnal **80**(2), 84–89 (2007)
2. Puzach, S., Abakumov, E.: Nekotorye osobennosti termogazodinamicheskoj kartiny pozhara v vysokih pomeshcheniyah. Pozharovzryvobezopasnost' **19**(2), 28–33 (2010)
3. Park, S., Miller, K.: Random number generators: good ones are hard to find. Commun. ACM **10**(32), 1192–1201 (1988)
4. Koshmarov, Yu.: Prognozirovanie opasnyh faktorov pozhara v pomeshchenii. Uchebnoe posobie. Akademiya GPS MVD Rossii, Moscow (2000)
5. Astapenko, V., Koshmarov, Yu., Molchadskij, I., Shevlyakov, A.: Termogazodinamika pozharov v pomeshcheniyah. Strojizdat, Moscow (1988)
6. Korn, G., Korn, T.: Spravochnik po matematike dlya nauchnyh rabotnikov i inzhenerov. Nauka, Moscow (1968)
7. Than' Haj, N.: Metodika rascheta neobhodimogo vremeni ehvakuacii lyudej pri pozhare v mashinnyh zalah GEHS V'etnama v usloviyah raboty sistemy dymoudaleniya, Dis. kand. tekhn. Nauk. AGPS MCHS Rossii, Moscow (2010)
8. Puzach, S.: Metody rascheta teplomassoobmena pri pozhare v pomeshchenii i ih primenenie pri reshenii prakticheskih zadach pozharovzryvobezopasnosti. Akademiya GPSMCHS Rossii, Moscow (2005)
9. NFPA 92B. 1990 NFPA Technical Committee Reports. Technical Guide for Smoke Management Systems in Malls, Atria and Large Areas. National Fire Protection Association, Quincy, MA (1990)
10. Kumar, S., Cox, G.: Mathematical modeling of fires in road tunnels. In: 5th International Symposium on the Aerodynamics and Ventilation of Venicle Tunnels, vol. 1, pp. 61–68. Lille, France (1985)
11. NFPA 72. NationalFire Alarm and Signaling Code. National Fire Protection Association (2002)
12. Ruegg, H., Arvidsson, T.: Fire safety engineering concerning evacuation from buildings. CFPA-E Guidelines **19**, 45 (2009)

13. Welch, S., Rubini P.: Simulation of Fires in Enclosures. User Guide. Cranfield University, United Kingdom (1996)
14. Spalding, D.: Mixing and chemical reaction in steady-state confined turbulent flames. In: 13th Symposium (International) Combustion, vol. 1, pp. 649–657. The Combustion Institute, Pittsburg (1996)

Feasibility of Personal Ventilation System Use in Office Spaces

Sergei Yaremenko(ID), Igor Zvenigorodsky(ID), Dmitry Lobanov(ID),
and Roman Sheps(✉)(ID)

Voronezh State Technical University, st. 20 let Oktyabrya, 84,
Voronezh 394006, Russia
romansheps@yandex.ru

Abstract. The article presents effectiveness analysis of office space microclimate systems, which are the main thermal energy consumers. The authors conducted a significant amount of experimental studies in order to determine dynamics of basic microclimate parameters changes and concentration of carbon dioxide in the breathing zone at permanent workplaces for different periods of the year. Measurements were carried out every working day for two calendar years. The research was carried out under various operating conditions of the microclimate supply systems. As a result, it was established that none of the considered modes provides in a room temperature, humidity and gas comfort at the same time. Thus, despite the significant capital and operating costs of the microclimate support systems, the required air conditions are not provided. Currently, the cost of energy and rates are growing rapidly. Alternative energy systems development and implementation in Russia is difficult for several reasons. It is known that ventilation systems are particularly energy-intensive. It is worth noting that there is a dependence of incoming air flow rate and air quality of air exchange arrangement. The article considers the feasibility of personal ventilation system use in offices with permanent jobs.

Keywords: Ventilation system · Office spaces · Microclimate parameters

1 Introduction

Employees' mental workability and desire to work depend on air quality in offices. Heating, ventilation and air conditioning systems create and maintain comfortable microclimate. Currently, the following engineering "traditions" are carried out for most office buildings design: general exchange supply and exhaust ventilation system with mechanical motivation (top-up air exchange arrangement scheme) in conjunction with air conditioning system installation (chiller-fan coil, split-systems, VRF systems, etc.). As shown by numerous surveys, the vast majority of employees are not satisfied with air-conditioning systems work: some of them are hot, others are cold, etc. The reason for dissatisfaction is that "traditional" systems are not able to provide thermal, gas comfort at workplaces and create a high-quality air environment satisfying the needs of all employees. The fact is that the concept of "comfort" is individual.

© Springer Nature Switzerland AG 2020
V. Murgul and M. Pasetti (Eds.): EMMFT 2018, AISC 982, pp. 156–168, 2020.
https://doi.org/10.1007/978-3-030-19756-8_16

Table 1. Optimal and permissible temperature, relative humidity and air velocity norms in public buildings serviced area.

Period of year	Room name or category	Air temperature, °C		Resulting temperature, °C		Relative humidity, %		Air velocity, m/s	
		Optimal	Allowable	Optimal	Allowable	Optimal	Allowable, no more than	Optimal, no more than	Allowable, no more than
Cold	2 category	19–21	18–23	18–20	17–22	45–30	60	0,2	0,3
Warm	Rooms with the constant presence of people	23–25	18–28	22–24	19–27	60–30	65	0,3	0,5

Table 2. Classification of indoor air quality.

Class	Indoor air quality		Allowable CO_2[a], ppm
	Optimal	Allowable	
1	High	—	400 and less
2	Average	—	400–600
3	—	Allowable	600–1000
4	—	Low	1000 and more

[a]Allowable CO_2 in rooms is in excess of CO_2 in outside air.

Table 3. The effect of carbon dioxide concentration on human health and workability.

CO_2 level (ppm)	Air quality and its effect on humans
Atmospheric air 300–400 ppm	Ideal for human health
400–600 ppm	Normal air quality
600 ppm	Level recommended for bedrooms, kindergartens and schools
600–800 ppm	Single air quality complaints appear
800–1000 ppm	More frequent complaints about air quality
1000 ppm	Maximum level of standards ASHRAE and OSHA
Higher than 1000 ppm	General discomfort, weakness, headache, problems with concentration. A growing number of errors in work. Negative changes in DNA begin
Higher than 2000 ppm	May cause serious health problems. The number of errors in the work increases greatly. 70% of employees cannot focus on work
5000 ppm	Maximum permissible concentration during the 8 h working day

For offices parameters of internal microclimate are taken according to ones given in Tables 1, 2, that comply with the requirements of current regulatory documents [1–3].

For justifying the choice of the required air quality, it's necessary to take into account its impact on people health and workability given in Table 3.

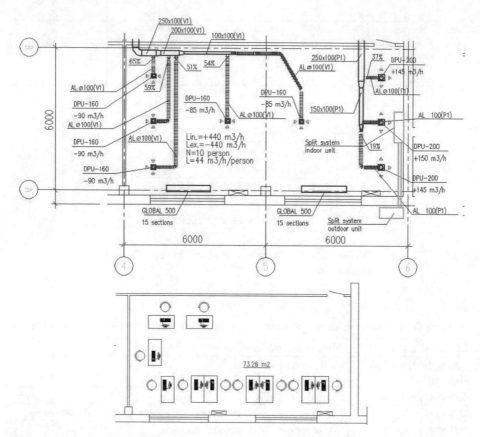

Fig. 1. Scheme of the office under study with microclimate support systems placement.

2   Materials and Methods

The authors conducted experimental studies on ventilation systems efficiency in office buildings in Voronezh. Characteristics of a room (see Fig. 1): windows orientation - South, area 73.26 m², height 2.7 m), the office is equipped with 10 permanent jobs (8-with fixed PCs, 2-with laptops). The room is equipped with systems of direct-flow supply and exhaust mechanical general ventilation (top-up air exchange arrangement scheme) and air conditioning (split system). The systems comply with the requirements and recommendations of current regulatory documents.

The measurements were carried out every working day (from 9.00 to 18.00) during two calendar years. Following modes of microclimate systems were studied:

- the first mode - mechanical ventilation system is turned off, natural one is working (windows are wide open), air conditioning/cooling system (split system) is turned off;
- the second mode - mechanical and natural ventilation systems are turned off; air conditioning/cooling system is on (split system);
- the third mode - natural ventilation does not work, mechanical ventilation and split-system works;
- the fourth mode - natural ventilation and split-system does not work, mechanical ventilation works;
- the fifth mode - ventilation and cooling systems (split system) are disabled.

According to measurement results, dynamics of changes graphs in internal air temperature, relative humidity and carbon dioxide concentration during the working day for different periods of the year were made (see Figs. 2, 3, 4, 5, 6 and 7).

Graphs show the most "typical" parameters changes of microclimate under various external and internal conditions. Also, for convenient comparison of experimental values obtained with the regulatory requirements, corresponding areas of permissible and optimal values are mentioned.

Analysis of the results showed that operating mode microclimate support systems or their combinations don't provide temperature and gas comfort in the room [4].

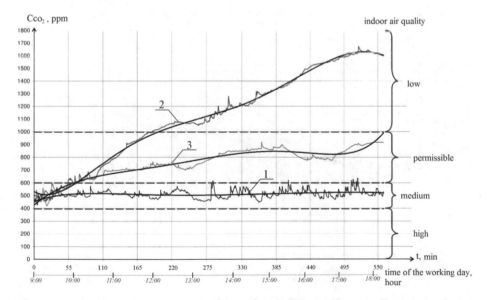

Fig. 2. Concentration of carbon dioxide (CO_2) distribution in respiratory zone during a working day in warm season.

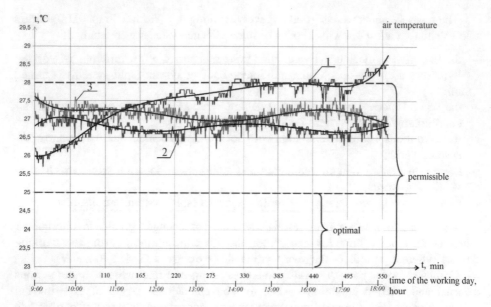

Fig. 3. Air temperature distribution (t_B) in the respiratory zone during a working day in warm season.

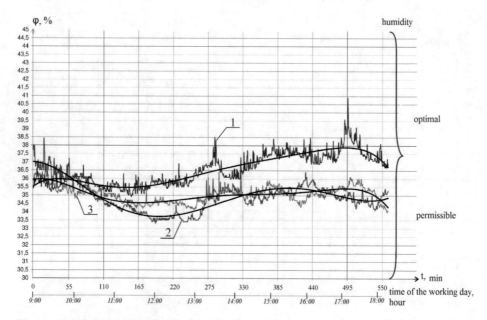

Fig. 4. Relative humidity distribution (φ_B) in breathing zone over time during warm season.

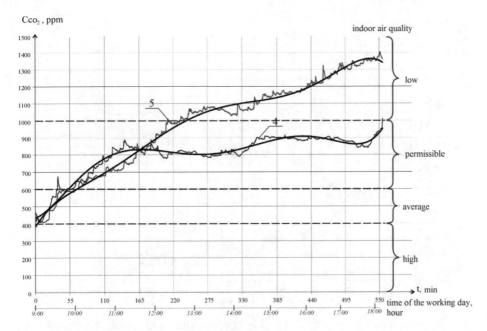

Fig. 5. Carbon dioxide concentration (CO_2) in breathing zone over time during cold season.

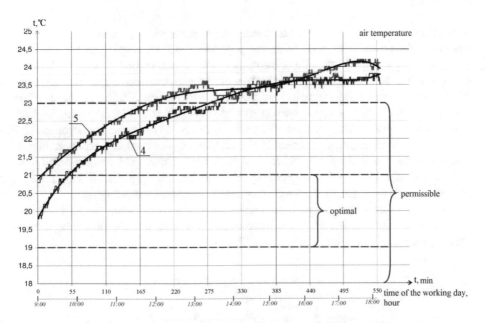

Fig. 6. Air temperature distribution (t_B) in breathing zone over time during cold season.

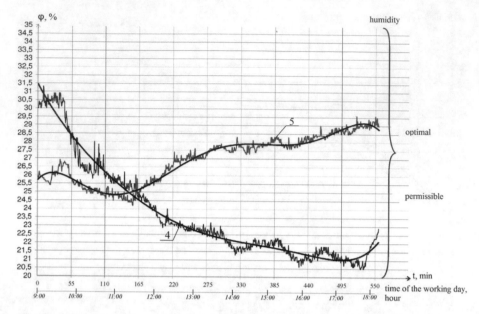

Fig. 7. Relative humidity distribution ($\varphi_в$) in breathing zone over time during cold season.

3 Results

According to modern domestic standards, air consumption should be determined separately for warm, cold periods of the year and transient conditions for the main insalubrities in a room (in the office - heat, moisture excess, carbon dioxide), taking the greatest value as a designed one [5]. The given approach does not guarantee the dilution (removal) of the estimated air flow (for a given air distribution scheme) of other insalubrities, for example, carbon dioxide (CO_2), that is considered as an indicator of air quality [1, 6]. It is known that CO_2 is about 1.5 times heavier than air, so it is not possible to organize a comfortable indoor air environment using a mixing ventilation system without any additional measures. Conducted studies justify the relevance of effective ventilation systems development and implementation, allowing to create comfortable microclimate parameters for a concrete workplace.

To create comfortable microclimate parameters in a room based on personal ventilation, firstly, it is necessary to estimate need for ventilation air on example of an office space with permanent work places (Fig. 8).

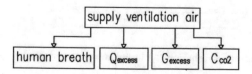

Fig. 8. Ventilation air need in office space.

Leading scientists' modern researches [7–10] suggests the need to revise air exchange standards, moreover, the required air flow abroad is significantly lower for similar premises than in the Russian Federation. In this regard, the solution of two tasks is actual: provision of standardized sanitary and hygienic conditions and a comfortable microclimate and compliance with energy saving requirements.

Let's perform an analysis of estimated expenses of fresh air for one person needs. The required ventilation air flow rate will be:

- human breath, $L_{in}^{brth} = 5$ m^3/h (with a reserve 100%);
- on heat excess from a person assimilation, $L_{in}^{cold} = 56$ m^3/h;
- carbon dioxide dilution (CO_2), emitted by a human, $L_{in}^{CO2} = 36.7$ m^3/h.

Determine connection between air flow rates:

$$\frac{L_{in}^{cold}}{L_{in}^{brth}} = 10,2 \cdots \cdot \frac{L_{in}^{CO2}}{L_{in}^{brth}} = 7,3$$

Consequently, for a human breathing needs, fresh air is required in much smaller quantities than the amount of air required for assimilating heat excess (10.2 times) or diluting CO_2 (7.3 times).

Thus, it is advisable to supply high-quality air in small quantities directly to each consumer instead of driving huge amounts of mediocre air through a room [11]. Personalized air (PA) should be supplied so that a person inhales clean, cool and dry air from the core of an air stream, where the supplied air is not mixed with the contaminated room air. An example of a personal supply air supply is shown in Fig. 9.

Fig. 9. An example of PA supply from an air stream near personal computer at a workplace in a room for mental work using PC.

In our opinion, in relation to rooms for mental work with permanent workplaces, it is required to supply fresh air directly into the breathing zone of each person in quantity necessary for breathing. Excess heat and insalubrities should be compensated for with cold (fan coil units, split systems, etc.) and absorb in installations of various designs, rather than assimilate and dilute to the level of the TLV of a working area with significant amounts of fresh air. The authors proposed a schematic diagram of required parameters ensuring of microclimate in rooms for mental work with permanent workplaces [12].

This scheme includes the device GMCP (generator of microclimate comfortable parameters), that is the development of Voronezh State Technical University staff.

Based on the foregoing, the traditional ventilation air scheme in rooms for mental work (Fig. 10) is:

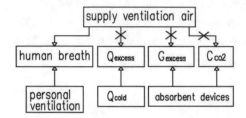

Fig. 10. Ventilation air need, taking into account the concept of required parameters of microclimate ensuring in premises for mental work with permanent workplaces.

4 Discussions

For further research, the authors implemented a personal ventilation system (Fig. 12) at the department "Housing and communal services" of FSBEI VSTU in one of the laboratories (Fig. 11).

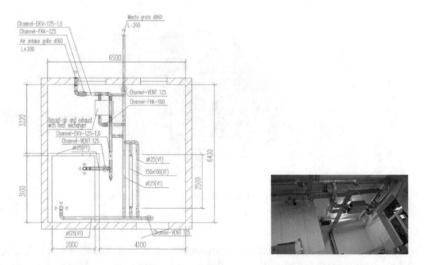

Fig. 11. Laboratory of ventilation and air conditioning systems: (a) plan diagram, (b) general view.

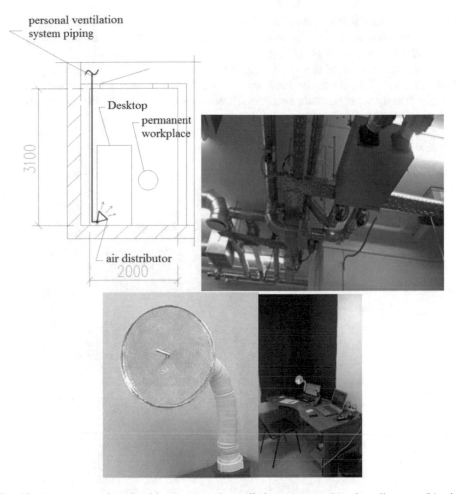

Fig. 12. A room equipped with a personal ventilation system: (a) plan diagram, (b) air distributor, (c) general view.

The studies were conducted for comparing the values of CO_2 concentration in respiratory zone with the standard and recommended values for various air exchange organization modes. A laboratory for study is a room with a permanent workplace for mental work with a PC (see Fig. 12), equipped with two laptops (one designed for fixing and plotting the dynamics of changes in air environment, the second is for performing work related to mental activity). During a working day, one person performs design work in the program AutoCad at the workplace. To ensure the required parameters of the air environment, the room has a combined ventilation system: supply and exhaust general exchange (top-up) and supply personal (see Figs. 11, 12). Supply air parameters: t = 19 °C, $\varphi_{in} = \varphi_{ex} = 22\%$, $C_{in.} = 450$ ppm. During the research, two options for ventilation organization were compared: (1) general exchange supply and exhaust ventilation. The inlet and exhaust air distribution devices of the ceiling type, type USD-L, are placed according to Fig. 11; (2) supply personal and exhaust total (from the upper zone of the room). The air intake device is located above a working table (see Fig. 12), it can be adjusted in horizontal and vertical directions with air supply directly to a person's breathing zone. In Fig. 13, the results of research are presented.

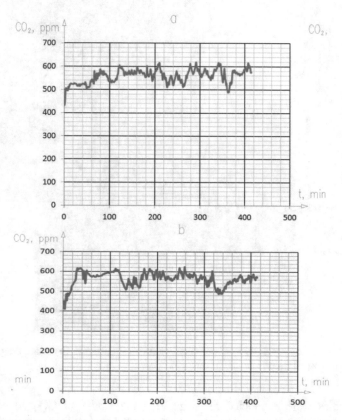

Fig. 13. Graph of CO_2 concentration changes in breathing zone for ventilation systems various operating modes: (a) general exchange supply and exhaust; (b) supply personal and exhaust total.

The research results analysis showed that to ensure the same air environment quality in breathing zone at a permanent workplace of a room for mental work at 600 ppm, 73 m^3/h are required for the first scheme, and 18 m^3/h for the second. Thus, the expediency and effectiveness of personal ventilation system in given conditions was established. Some articles about it was published in the ABOK magazine [13–15].

5 Conclusion

The use of the most common microclimate support systems often does not solve the task of providing indoor gas and temperature-humidity comfort at the same time. Solution of the problem in functioning buildings consists in the refinement (reconstruction) of the microclimate support systems with the possibility of local regulation and local devices installation to improve air environment quality. For buildings under construction it is necessary to solve the problem by organizing a ventilation system with air distribution high efficiency. Such systems can be personal, displacing other systems with installation of local means of highly efficient air purification. The authors proposed a scheme to ensure microclimate required parameters that allows:

- significantly improving inhaled air quality as it is not mixed into a room's volume and employees' productivity; decreasing the risk of allergic diseases;
- reducing both the excess heat and moisture, and the concentration of various insalubrities in a room (carbon dioxide, dust, building materials emission, etc.);
- regulating temperature and air flow rate at permanent workplaces depending on an employee individual needs;
- significantly reducing capital and operating costs (heat, cold, electricity) on microclimate maintenance systems.

The proposed solution to create and ensure comfortable microclimate parameters is promising. It should be noted that numerous studies of personal ventilation system results allow making a conclusion about the relevance of its use and the need for further field experiments on equipment and air diffusers improvement.

References

1. Russian Federation Standard. GOST 30494-2011. Residential and public buildings. The parameters of the microclimate in the premiscs
2. Russian Federation Standard. SanPiN 2.2.2/2.4.1340-03. Hygienic requirements for personal computers and organization of work
3. Russian Federation Standard. GOST R EN 13779-2007. Ventilation in non-residential buildings. Requirements for ventilation and air conditioning systems
4. Lobanov, D., Sheps, R., Portnova, N.: Experimental studies of energy efficiency of air conditioning systems in the office. Bull. Voronezh State Tech. Univ. **14**(3), 71–79 (2018)
5. Russian Federation Standard. SP 60.13330.2016. Heating, ventilation and air conditioning
6. Naumov, A., Kapko, D.: CO2: criterion of efficiency of ventilation systems. AVOK **1**, 12–21 (2015)
7. Livchak, V.: About norms of air exchange of public buildings and consequences of their overstatement. AVOK **6**, 4–10 (2007)
8. Shilkrot, E., Gubernskiy, Yu.: How much air does a person need for comfort? AVOK **4**, 4–18 (2008)
9. Kvashnin, I., Gurin, I.: To the question about regulation of the ventilation according to the CO2 content in the outer and inner air. AVOK **5**, 34–42 (2008)
10. ASHRAE 62.1-2004, 62.1-2007. Ventilation for acceptable indoor air quality
11. Fanger, O.: Quality of indoor air in the XXI century. AVOK **2**, 14–21 (2000)

12. Lobanov, D., Polosin, I.: The scheme of creation of comfortable climatic parameters in offices. Plumb. Heat. Air Cond. **2**(158), 58–61 (2015)
13. Dieckmann, J., Sooreh, A., Brodrick, J.: Personal ventilation: comfort and energy saving. AVOK **4**, 46–49 (2011)
14. Individual comfort in the workplace in the office. AVOK **4**, 54–55 (2011)
15. Naumov, A., Kapko, D.: Local air conditioning systems in office buildings. AVOK **2**, 14–19 (2012)

The Efficiency of Heat Energy Generators for Individual Thermal Points

Vladimir Papin, Roman Bezuglov$^{(\boxtimes)}$, Evgeniy Dyakonov,
Irina Denisova, Alexander Yanuchok, and Denis Dobrydnev

Platov South-Russian State Polytechnic University (NPI),
Prosveschenia Street 132, 346428 Novocherkassk, Russia
romanbezuglov@inbox.ru

Abstract. The calculation method of needs for power supply of the individual consumer is given in this article. The choice algorithms for the most effective for technical and economic reasons equipment for heating and cogeneration are given. Methods of their main characteristics are also discussed. The individual systems of heat supply reserve the centralized systems of heat supply, increasing by their reliability. Using the renewable energy resources in individual thermal points promotes decrease in consumption of fuel centralized system and decrease in emissions of products to combustion in the atmosphere. The centralized heating systems in a combination with reserve thermal points represent one of the most perspective technologies of the centralized heat supply.

Keywords: Thermal energy · Renewable power · Heat generator ·
Perspective technology

1 Introduction

A standard object for the analysis of heat supply sources is an individual house. For it, heat losses were calculated. On the basis of losses, the amount of the energy consumed in a year was expected needs of heating and hot water supply. After that, different thermal points (heat generators) were calculated: electric boiler, gas boiler, gas condensation boiler, liquid-fuel boiler, solid propellant boiler, thermal pump.

The given method of calculation of thermal protection for the building was a little simplified. If necessary, for obtaining more reliable information about heat losses of real objects, authors of article recommend to use [1]. When making a choice of the source of heat supply, emphasis is placed on the cost of equipment and its installation, as well as on the cost of used fuel or electricity. Repair costs and a technical maintenance of the equipment, increase in electricity rates and the used types of fuels were not considered in calculation.

© Springer Nature Switzerland AG 2020
V. Murgul and M. Pasetti (Eds.): EMMFT 2018, AISC 982, pp. 169–185, 2020.
https://doi.org/10.1007/978-3-030-19756-8_17

2 The Review of the Existing Situation in the Considered Subject

In the article [1], contrastive analysis of the existing systems of heat supply is made. The solutions concerning reconstruction of already existing system of heat supply are analyzed: on the basis of centralization and decentralization of sources of heat. As a result of the made analysis, it is revealed that for the accepted basic data on the value of a simple payback period, the project of decentralization of the existing system of heat supply of the city is more preferable. In article [2] actions which can provide big efficiency from use of local sources of heat supply are considered. In particular, the problem of accumulation of warmth when using decentralization is considered. It is for this purpose offered to use heating systems with the increased capacity. The economic problems concerning implementation of this solution are not considered, shortcomings of the offered scheme are not specified.

Economic aspects of decentralization are considered in article [3]. Implementation of the decentralized heat supply in the existing apartment house is considered. The research is performed by method of mathematical modeling. As a result of work an impression about cost efficiency of the roof boiler house is gained.

Nondeliveries and advantages of hostless systems of heat supply are stated in article [4] and also short characteristic of decentralization as alternative system of heat supply is given. Are specified, including, the main difficulties arising in design process of hostless systems.

In the article [5], advantages and shortcomings of hostless systems are considered and also the forecast of development of the above-stated systems is submitted in the near future.

3 Problem Definition and Basic Data

It is necessary to perform a calculation of any object: select the settlement and design the building. It is necessary to present its general view on the drawing and to define of what materials all its enclosing structures will be made. It is possible to select a real object for calculation and to take all necessary data from its technical data sheet. You should determine the key climatic parameters [6] of the heating period of the region in which there is a calculated building. Also, you should to determine parameters of an internal microclimate of the building [7].

4 Estimated Technique – A Calculation Execution Order

For this section, it is necessary to select the settlement and to design the building, that is to present on the drawing its general view, as shown in Figs. 1 and 2. You should define of what materials all its enclosing structures [8] (Fig. 2) will be made. It is possible to select a real object for calculation, and to take all necessary data from its technical data sheet. In Figs. 1 and 2 drawings of the private house (Rostov-on-Don,

Russia) are submitted. Calculation of heatlosses of this building will be given as an example to each paragraph of this section.

The scheme of building is shown in the Fig. 1.

Fig. 1. Southern facade of the building.

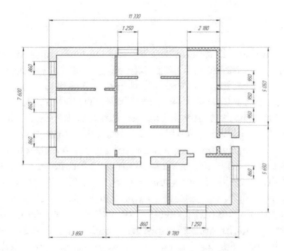

Fig. 2. Scheme of the building.

The first that needs to be made, it to determine the key climatic parameters of the heating period of the region. You should get it by the table "Climatic Parameters of the Cold Period of Year for Some Cities of Russia" according to Russian standard Construction climatology Construction Norms and Regulations 23-01-99, Moscow, 2003. For a start it is necessary to determine duration and average daily temperature of the heating period according to Russian standard Construction climatology Construction Norms and Regulations 23-01-99, Moscow, 2003. For example, for Rostov-on-Don (Russia) duration of the heating period is $z_{o.p.} = 171$ days, and average daily air temperature $t_{o.p.} = -0.6$ °C.

The most severe conditions of the cold period are described by calculated values of climatic parameters. The minimum, peak values of climatic parameters happen extremely seldom and to calculate heat supply and heat insulation of enclosing structures of buildings proceeding from them, it does not make sense. Therefore, take values with a certain security as which understand the total probability that this value of parameter (estimated) during the cold period of year will not be exceeded for estimated values. The most important climatic parameter is the temperature of fresh air. As enclosing structures of the building have thermal inertia, take the average temperature of the coldest five-day weeks, but not the coldest day for calculated value. Calculated value of outside temperature is defined also according to Russian standard Construction climatology Construction Norms and Regulations 23-01-99, Moscow, 2003, usually take value with probability 0.92. For Rostov-on-Don (Russia), for example, the calculated temperature of fresh air with coefficient 0.92 according to Russian standard Construction climatology Construction Norms and Regulations 23-01-99, Moscow, 2003 is $t_n = -22$ °C.

Further, it is necessary to define a zone of humidity of fresh air from 3 zones:

1. wet;
2. normal;
3. dry.

The zone of humidity is determined by the card of zones of humidity Russian standard Construction Norms and Regulations 23-02-2003 Thermal protection of buildings. Rostov-on-Don treats a zone of humidity 3 – dry.

For calculation of heat losses, it is necessary to determine parameters of an internal microclimate of the building. According to Russian standard GOST 30494-2011. Buildings inhabited and public. Microclimate parameters in rooms. it is possible to take the maximum value of optimum temperature for the calculated temperature of internal air $t_{vn} = 22$ °C. Relative humidity of internal air on 3 for rooms of residential buildings Russian standard Construction Norms and Regulations 23-02-2003 Thermal protection of buildings is $\varphi_{vH} = 55\%$. According to Russian standard Construction Norms and Regulations 23-02-2003 Thermal protection of buildings the selected parameters of a microclimate correspond to normal moisture conditions. Parameters of an internal microclimate of the building presented in an example are specified in Table 1. For further calculations, similar parameters of an internal microclimate of the room of the building can be accepted.

Table 1. Parameters of an internal microclimate of the building.

Temperature of internal air, t_{vn}, °C	Relative humidity of internal air, φ_{vn}, %	Moisture conditions
22	55	Normal

Depending on a zone of humidity in which there is a building and moisture conditions of its internal microclimate, operating conditions of enclosing structures are selected. Further coefficients of heat conductivity are determined by normative documents at the selected operating conditions.

4.1 Calculation of Resistance to a Heat Transfer of Enclosing Structures

Within this point it is necessary to define: heat emission coefficients on internal and outside (for winter) surfaces of enclosing structures, coefficients of heat technical uniformity and coefficients of heat conductivity of materials of components enclosing structures of the building; resistance to a heat transfer of enclosing structures.

Resistance to a heat transfer of the multilayer wall which is consisting of n layers and having a one-dimensional temperature field is defined by expression:

$$R_0^{usl} = 1/\alpha_B + \sum_{i=1}^{n} (\delta_i/\lambda_i) + 1/\alpha_n \tag{1}$$

where α_v – heat emission coefficient on an internal surface an enclosing structure;
α_n – heat emission coefficient on an outside surface of an enclosing structure;
δ_i – thickness of i layer of an enclosing structure;
λ_i – heat conductivity of i layer of an enclosing structure, is accepted according to Russian standard Joint venture 23-101-2004 Design of thermal protection of buildings.

Real barriers of modern buildings have difficult construction and, the temperature field arising in them, is also two - and three-dimensional. For such barriers the thermal resistance counted on a formula (1) is designated as conditional R_0^{usl}. The specified resistance to a heat transfer is accepted to calculated value R_0. It represents the general resistance to a heat transfer of such barrier with a one-dimensional temperature field which on average in the area has barrier with not one-dimensional temperature field Russian standard Joint venture 23-101-2004 Design of thermal protection of buildings. The given value of resistance to a heat transfer is determined by a formula:

$$R_0 = R_0^{usl} \cdot r, \tag{2}$$

where r – the coefficient of heat technical uniformity which is also a share of the specified resistance in the conditional resistance to a heat transfer of an enclosing structure. According to [5] for walls of residential buildings from a brick the coefficient of heat technical uniformity should be, as a rule, not less than 0.74 at a thickness of wall of 510 mm, 0.69 – at a thickness of wall of 640 mm and 0.64 – at a thickness of wall of 780 mm.

Having substituted expression (1) in a formula (2):

$$R_0 = r \cdot (1/\alpha_B + \sum_{i=1}^{n} (\delta_i/\lambda_i) + 1/\alpha_n) \tag{3}$$

Thus, for walls of the building designed early on a formula (3) it is necessary to count resistance to a heat transfer, as shown in an example below. Resistance to a heat transfer of windows and balcony doors are defined according to Russian standard GOST 30494-2011. Buildings inhabited and public. Microclimate parameters in rooms.

4.2   Calculation of Resistance to a Heat Transfer of Enclosing Structures

Within this point it is necessary to determine resistance to a floor heat transfer by soil [9].

Process of a heat transfer through a floor and walls on soil is quite difficult therefore on soil apply the simplified technique Russian standard Joint venture 23-101-2004 Design of thermal protection of buildings according to which the surface of a floor and walls on soil divide into zones – bands 2 m wide parallel to a joint between a floor and a wall to calculation of thermal resistance of a floor and walls, as shown in Fig. 3. Each zone has the thermal resistance which the is more, than the zone from a wall is located further. At the same time a floor is considered as continuation of a wall on soil. One zone can begin on a wall, and come to an end on a floor. The enclosing structure of the building which is not containing layers from materials with heat conductivity coefficient $\lambda = 1.2$ W/(m · °C) below is called not warmed. For each zone of not warmed floor normative values of thermal resistance are provided:

$$R_I = 2.1(m^2 \cdot °C)/W; \; R_{III} = 8.6(m^2 \cdot °C)/W;$$

$$R_{II} = 4.3(m^2 \cdot °C)/W; \; R_{IV} = 14.2(m^2 \cdot °C)/W; \tag{4}$$

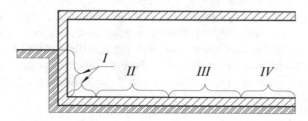

Fig. 3. Distribution of estimated zones on a surface of a floor and walls on soil.

Resistance to a floor heat transfer is determined by soil from a formula:

$$R_{soil} = (R_I \cdot A_I + R_{II} \cdot A_{II} + R_{III} \cdot A_{III} + R_{IV} \cdot A_{IV})/(A_I + A_{II} + A_{III} + A_{IV}) \tag{5}$$

where A_I, A_{II}, A_{III}, A_{IV} – the spaces occupied according to I, II, III and IV zones.

4.3   Calculation of Transmission Losses

Within this point it is necessary to calculate transmission heat losses through enclosing structures of the building and to present results in the table form.

Heat losses of the building at the expense of a heat transfer are calculated by a formula:

$$Q_{tr} = K \cdot F \cdot (t_v - t_n) \cdot n, \ W, \tag{6}$$

where F – the external surface area of barrier, m^2.
t_v – temperature of internal air of the room, according to [4] $t_v = 22\ ^\circ C$;
t_n – temperature of fresh air, according to [2] for Rostov-on-Don $t_n = -22\ ^\circ C$;
K – the heat transfer coefficient determined by a formula, $W/(m^2 \cdot {}^\circ C)$:
n – coefficient of provision of barrier of rather fresh air.

$$K = 1/R, \tag{7}$$

For the enclosing structures of the building separating the heated rooms with a temperature of internal t_v from not heated with a temperature t_s (such as $< t_s < t_v$), the coefficient of provision of barrier of rather fresh air is entered. It is necessary to increase the difference of temperatures $(t_v - t_n)$, when determining heat losses of such barrier. The coefficient of n is determined by a formula:

$$n = (t_v - t_s)/(t_v - t_n), \tag{8}$$

4.4   Calculation of Losses of Warmth on Heating of Infiltration Air

It is necessary to determine amount of heat necessary for heating of ventilating norm of air exchange, total heat losses of the building and required power of the boiler unit.
Losses of warmth are defined on heating of infiltration air [10] by expression:

$$Q_{inf} = Q_{ogr} + Q_{went}, \ W, \tag{9}$$

where Q_{org} – losses of warmth on heating of the air coming to the room through thinnesses of outside barriers. As in the building considered in an example metal-plastic windows and hermetic entrance doors are set, losses through thinnesses of enclosing structures will not be considered. In case of need accounting of this component of infiltration heat losses, calculation should be conducted according to recommendations Russian standard Joint venture 23-101-2004 Design of thermal protection of buildings;
Q_{went} – losses of warmth on heating of ventilating air, W.
Amount of heat necessary for heating of ventilating norm of air exchange is defined by expression:

$$Q_{went} = G \cdot c \cdot (t_v - t_n), \ W, \tag{10}$$

where G – design flowrate of ventilating air, kg/s;
c – heat capacity of air, $J/(kg \cdot {}^\circ C)$, $c = 1005\ J/(kg \cdot {}^\circ C)$.

The design flowrate of ventilating air is defined by expression:

$$G = (V \cdot n \cdot \rho_{av})/3600, \text{kg/s}, \tag{11}$$

where V – internal volume of the room, m^3;
ρ_{av} – air density at an average temperature of air, kg/m^3;
n – ventilation rate, h^{-1}.

According to the standards adopted in the world Russian standard Construction climatology Construction Norms and Regulations 23-01-99, Moscow, 2003 for premises with providing social norm of the square at the person more than 20 m^2, multiplicity of air exchange is accepted by equal n = 0.35 h^{-1}.

Total heat losses of the building are defined from expression:

$$Q_{ot} = Q_{tr} + Q_{inf}, \tag{12}$$

Power reserve which needs to be considered when choosing the boiler equipment for ensuring needs of hot water supply is defined from expression:

$$Q_{gvs} = n \cdot q_{gvs}, \tag{13}$$

where q_{gvs} – necessary power of hot water supply counting on one person, kW/persons, q_{gvs} = 0.25 kW/persons.
n – number of people, living in the building, persons.

Required power of the boiler unit is defined by a formula:

$$Q_k = Q_{ot} + Q_{gvs}, \tag{14}$$

5 Choice of a Heat Supply Source

The centralized system of heat supply, redundant individual thermal point, is perspective technology of the centralized heat supply. For the choice of a reserve source of heat supply in the above described technique heat losses and need of the building for warmth for the heating period were calculated. Capital and operating costs will act as the main selection term of the heating equipment. Capital expenditure will consist of the equipment cost and of its installation cost. Operating costs will include costs of the fuel used for heating or the electric power. It is necessary to select according to directories or by means of the official sites of vendors of heating equipment Russian standard GOST 30494-2011. Buildings inhabited and public. Microclimate parameters in rooms boiler units of the following types:

- electric boiler;
- gas boiler;
- gas condensation boiler;
- liquid-fuel boiler;
- solid propellant boiler;
- heat pump.

Further it is required to calculate capital and operating costs when using each type of the heating equipment and to make their comparison.

Capital costs is determined by a formula:

$$S_{cap} = C_k + C_{day} + C_{dr}, \tag{15}$$

where C_k – cost of the boiler unit;
C_{day} – cost of installation of the boiler unit;
C_{dr} – cost accompanying the equipment and its installations (a buffer tank, the soil probe of the thermal pump and others).

Operating costs are determined by a formula:

$$S_{ex}^{k.gas} = C_{top} \cdot C_{top}, \tag{16}$$

where V_{top} – fuel quantity or electric power necessary for ensuring needs of heating of the building and hot water supply;
C_{top} – cost of fuel unit or electric power.

6 The Energy Consumed for the Heating Period

Within this point it is necessary to define coefficient of irregularity of distribution of heating loading and amount of heat necessary on heating of the building during the heating period; It is also necessary to determine amount of heat necessary on hot water supply, to define total demand of the building for warmth on hot water supply and heating during the heating period.

Necessary amount of heat on hot water supply and heating of the building during the heating period will be determined by a formula:

$$E = E_{ot} + E_{gvs}, kJ, \tag{17}$$

where E_{ot} – amount of heat necessary on heating of the building during the heating period, kJ;

E_{gvs} – amount of heat necessary on hot water supply, kJ. As hot water supply is necessary during the whole year, the need for warmth on hot water supply should be calculated for 365 days.

The amount of heat necessary on heating of the building for the heating period will be defined by expression:

$$E_{ot} = 24 \cdot 3600; \ Q_{ot} \cdot z_{o.p.} \cdot K, kJ, \tag{18}$$

where Q_{ot} – design load of a heating system, kW;

K – the coefficient of irregularity of distribution of heating loading during the heating period determined by a formula:

$$K = (t_v - t_{o.p.})/(t_v - t_n),\qquad(19)$$

Amount of heat necessary is defined on hot water supply by expression:

$$E_{gvs} = V_v \cdot \rho_v \cdot c_v \cdot n \cdot z_{year} \cdot (t_{gvs} - t_{c.w.}),\, kJ,\qquad(20)$$

where V_v – the volume of the hot water necessary on one person a day, $m^3/(per\cdot day)$, according to Russian standard Construction Norms and Regulations 23-02-2003 Thermal protection of buildings $Vv = 0.105\ m^3/(per\cdot day)$;

c_v – heat capacity of water, $c_v = 4.19\ kJ/(kg \cdot {}^\circ C)$;

n – number of people, living in the building, for the building considered in an example n = 2 per.;

z_{year} – the number of days in a year, $z_{year} = 365$ days;

t_{gvs} – temperature of hot water in the line of the hot water supply, $^\circ C$, $t_{gvs} = 60\ ^\circ C$;

$t_{c.w.}$ – temperature of a cold water, $^\circ C$ $t_{c.w.} = 10\ ^\circ C$;

ρ_B – water density, kg/m^3, $t_{gvs} = 60\ ^\circ C$ and $p_v = 0,6$ MPa for $\rho_v = 983.3\ kg/m^3$.

7   Results of Calculation for Different Sources and Their Comparison

The analysis of different heat generators for supply of the consumer with heat is presented in this section.

Main Selection Terms of the Heating Equipment. For creation of a heat power complex it is necessary to decide on high power performance, the existing systems of power supply using both traditional and renewable energy resources are how effective. The main selection terms of the heating equipment are:

1. Capital cost;
2. Operating costs.

Capital expenditure in turn includes such concepts as:

1. Equipment cost;
2. Cost of its installation.

Operating costs include:

1. Expenditure for use for fuel heating;
2. Expenditure for use of the electric power.

According to directories or by means of the websites of hardware manufacturers, it is necessary to select boiler units of the following types:

1. Electric boiler;
2. Gas boiler;
3. Gas condensation boiler;
4. Liquid-fuel boiler;
5. Solid propellant boiler;
6. Heat pump.

In view of the fact that the thermal pump enters the developed heat power complex, in this point comparison of heat pumping technology with traditional technologies of heat generation on organic fuel is made.

Further it is necessary to calculate capital and operating costs when using each type of the equipment (Table 2).

Table 2. Parameters of an internal microclimate of the building

Region	Rostov region, Russia
Total area, m^2	54
Gas price, rub./m^3	5.9
Price of solid fuel, rub./m^3	1500
Price of liquid fuel, rub./l	44.99
Electric power rate, rub./kW·h	3.8
Total loss, W	5531
Need of the house for heat energy in a year, kW·h	1 825.497

In Fig. 4, the diagram of total costs the house heating when using the equipment of a different type is submitted.

Having analyzed dependences, it is possible to draw a conclusion, that the most favorable is installation of the gas boiler or the gas condensation boiler. The low cost of a boiler, high efficiency, maintainability and also low operating costs do it reasonable for application as the heating equipment for a covering of needs of the house (Table 3).

Table 3. Parameters of an internal microclimate of the building.

Region	Rostov region, Russia
Total area, m^2	117.5
Gas price, rub./m^3	6.02
Price of solid fuel, rub./m^3	1800
Price of liquid fuel, rub./l	41.9
Electric power rate, rub./kW·h	5.1
Total loss, W	33057.5
Need of the house for heat energy in a year, kW·h	78 389.61

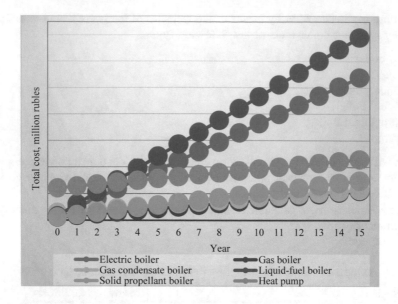

Fig. 4. Total costs of heating of the house when using the equipment of a different type.

In Fig. 5, the diagram of total costs the house heating when using the equipment of a different type is submitted.

Having analyzed dependences, it is possible to draw a conclusion that the most favorable, within the reviewed example, is installation of the gas condensation boiler. The low cost of a boiler, high efficiency, maintainability and also low operating costs do it reasonable as the heating equipment for a covering of the house needs (Tables 4 and 5).

Table 4. Parameters of an internal microclimate of the building.

Region	Krasnodar region, Russia
Total area, m^2	200.16
Gas price, rub./m^3	6.02
Price of solid fuel, rub./m^3	1800
Price of liquid fuel, rub./l	41.9
Electric power rate, rub./kW·h	5.1
Total loss, W	33057.5
Need of the house for heat energy in a year, kW·h	49102.86

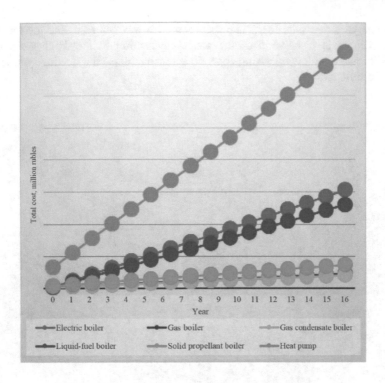

Fig. 5. Total costs of heating of the house when using the different equipment.

Table 5. Parameters of an internal microclimate of the building.

Region	Rostov region, Russia
Total area, m^2	130
Gas price, rub./m^3	6.13
Price of solid fuel, rub./m^3	8000
Price of liquid fuel, rub./l	42
Electric power rate, rub./kW·h	4.5
Total loss, W	10522.08
Need of the house for heat energy in a year, kW·h	24373

In Fig. 6 the diagram of total costs of heating of the house when using the different equipment is submitted.

Having analyzed dependences, it is possible to draw a conclusion, and the fact that the most favorable is installation of the gas condensation boiler. High efficiency, maintainability and also low operating costs do it reasonable as the heating equipment for needs of the house.

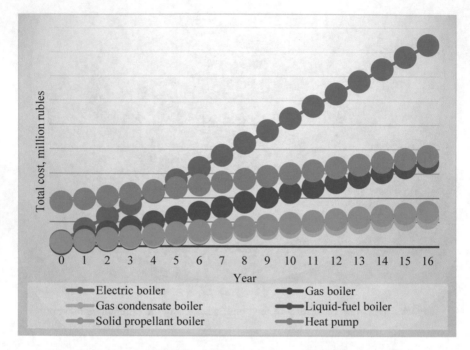

Fig. 6. Total costs of the house heating when using the different equipment.

Conclusion. Having constructed the diagram, it is possible to draw a conclusion on efficiency of each type of thermal installation. So, for example, installation of the heat pump will be favorable to this house only in case of lack of a possibility of carrying out gas or impossibility of supply of coal. However, it anyway will be very expensive action. Nevertheless it looks quite profitable as, according to the diagram, installation of the diesel generator or electric boiler are not justified at all. It should be noted that if in the nearest future there is an opportunity to carry out gas or to arrange coal delivery, then it is worth reviewing these options of heating the house as initial capital expenditure for installation of the thermal pump too big (Fig. 7 and Table 6).

In Fig. 8 the diagram of total costs the house heating when using the different equipment is submitted.

Having analyzed dependences it is possible to draw a conclusion, that the most favorable is installation of the gas condensation boiler. The low cost of a boiler, high efficiency, maintainability and also low operating costs do it reasonable as the heating equipment for a covering of needs of the house.

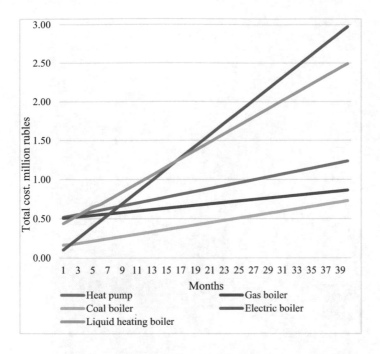

Fig. 7. Total costs of the house heating when using the different equipment.

Table 6. Parameters of an internal microclimate of the building.

Region	Rostov region, Russia
Total area, m^2	64
Gas price, rub./m^3	6.02
Price of solid fuel, rub./m^3	1800
Price of liquid fuel, rub./l	41.9
Electric power rate, rub./kW·h	5.1
Total loss, W	13600
Need of the house for heat energy in a year, kW·h	39 005.898

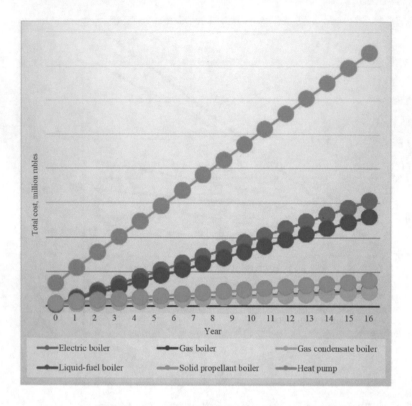

Fig. 8. Total costs the house heating when using the different equipment.

8 Conclusion

The centralized heating systems in a combination with reserve individual sources of a cogeneration is one of the most perspective technologies of the heat supply. Therefore as a reserve of the centralized system the heat and power supply autonomous sources of a cogeneration, as a rule, are considered. Thus, for creation the effective thermal point it is necessary to define using both traditional and renewable energy resources.

Acknowledgments. The article was prepared with the financial support of the Russian Federation President grant for state support of young Russian scientists - candidates of sciences MK-3537.2018.8.

References

1. Efimov, N., Papin, V., Lavrentiev, A., Dobrydnev, D., Yanuchok, A.: Heat pumps based on theoretical and experimental studies. Int. J. Mech. Eng. Technol. **9**(10), 695–705 (2018). http://www.iaeme.com/MasterAdmin/UploadFolder/IJMET_09_10_072/IJMET_09_10_072.pdf
2. Veselovskaya, E., Papin, V., Bezuglov, R.: High energy efficiency and high energy efficiency compatibility of energy sources. IOP Conf. Ser. Mater. Sci. Eng. **463** (2018). https://iopscience.iop.org/article/10.1088/1757-899X/463/2/022011/pdf. International Multi-Conference on Industrial Engineering and Modern Technologies, Vladivostok, Russian Federation
3. Efimov, N., Papin, V., Bezuglov, R.: Micro energy complex based on wet-steam turbine. Procedia Eng. **150**, 324–329 (2016). https://doi.org/10.1016/j.proeng.2016.07.022
4. Penkovskii, A., Stennikov, V., Postnikov, I.: Unified heat supply organization: mathematical modeling and calculation. Procedia Eng. **158**, 3439–3444 (2019). https://doi.org/10.1016/j.egypro.2019.01.930
5. Postnikov, I., Stennikov, V., Mednikova, E., Penkovskii, A.: Methodology for optimization of component reliability of heat supply systems. Appl. Energy **227**, 365–374 (2018). https://doi.org/10.1016/j.apenergy.2017.11.073
6. Zabolotnik, S.: The harshness of climatic conditions on the territory of Russia. Geogr. Nat. Resour. **31**(3), 251–256 (2010). https://doi.org/10.1016/j.gnr.2010.09.010
7. Castaldo, V., Pisello, A., Piselli, C., Fabiani, C., Cotana, F., Santamouris, M.: How outdoor microclimate mitigation affects building thermal-energy performance: a new design-stage method for energy saving in residential near-zero energy settlements in Italy. Renew. Energy **127**, 920–935 (2018). https://doi.org/10.1016/j.renene.2018.04.090
8. Gagarin, V., Neklyudov, A., Zhou, Z.: Determination of possible increase of R-value of enclosing structures based on the payback conditions. Procedia Eng. **146**, 100–102 (2016). https://doi.org/10.1016/j.proeng.2016.06.358
9. Wang, Y., Jiang, C., Liu, Y., Wang, D., Liu, J.: The effect of heat and moisture coupling migration of ground structure without damp-proof course on the indoor floor surface temperature and humidity: experimental study. Energy Build. **158**, 580–594 (2018). https://doi.org/10.1016/j.enbuild.2017.10.064
10. Pereira, P., Melo, C.: A methodology to measure the rates of air infiltration into refrigerated compartmentsUne méthodologie de mesure des taux d'infiltration d'air dans les compartiments réfrigérés. Int. J. Refrig **97**, 88–96 (2019). https://doi.org/10.1016/j.ijrefrig.2018.08.024

Theoretical Study of "Green Roof" Energy Efficiency

Elena Sysoeva[(✉)] and Margarita Gelmanova

NIU MGSU, Yaroslavskoye Shosse, 26, 129337 Moscow, Russia
SysoevaEV@mgsu.ru

Abstract. The article is devoted to the theoretical study of energy efficiency and a feasibility of using the technology of a "green roof" in the coatings of public buildings for large and great cities. The need to optimize energy consumption in large cities becomes a prerequisite for the improvement and implementation of innovative, energy-saving and technological systems as a method of solving the problem of energy efficiency of buildings. During creation of a "green roof" in the coating of the building, there is an increase in its energy efficiency, which is the result of reducing heat losses through the outer coating for heating in winter. A qualitative comparative analysis of the results of thermal calculations of existing non-operated insulated roofs and their coatings reorganized into the operated "green roof" is made for two public buildings. According to the results of calculations, the assessment of energy efficiency of public buildings coating when constructing a "green roof" is made. The present study shows that the construction of a "green roof" with a substrate thickness of 300 mm in the coating of two public buildings leads to an increase in heat transfer resistance of more than 5%, which leads to a decrease in energy consumption.

Keywords: Coatings of public buildings · Heat transfer resistance ·
Thermal insulation · Energy saving

1 Introduction

In the past three decades, there has been a rapid increase in the pace of technological development and, as a result, an increase in energy consumption of resources. In this regard, there is a need for rational design, construction and operation, aimed at the principles of environmental sustainability, safety and energy efficiency, which are the basis of sustainable development [1, 2]. Finding alternative ways of obtaining natural resources and saving of existing ones are global problems of the modern world.

In Russia, sustainability is determined by the stock of natural resources, which are limited and renew much slower than the development of technologies, their rational use and ensuring a decent quality of life [3, 4].

As a result, since 2009, an issue of rational energy consumption has become a priority for Russia. Since this moment, Federal laws [5, 6] which are aimed at creating the necessary conditions for reducing energy costs by 40% by 2020 has been appearing.

© Springer Nature Switzerland AG 2020
V. Murgul and M. Pasetti (Eds.): EMMFT 2018, AISC 982, pp. 186–198, 2020.
https://doi.org/10.1007/978-3-030-19756-8_18

In addition to the above, in Central parts of large cities, the percentage of public buildings can reach 90% or more.

For example, the concentration of public functional areas of Moscow, taking into account transport links, is 19.24% of the total area of the city. At the same time, in Moscow within the Garden Ring - 95.40% of the total central area of Moscow, which significantly reduces the possibility of the location of park areas (Table 1).

Table 1. Ratio of the area occupied by public functional areas to the total area of Moscow and its Central part before the expansion project of 2012.

Evaluation criterion	Total area, km^2	Area including transport links (3%)
Moscow area, S	1070.00	1037.90
The area of the public spaces of the city of Moscow, S	212.19	205.82
The area of the city of Moscow within the Garden ring, S	18.50	17.95
The area of public areas of the city of Moscow within the Garden ring, S	18.20	17.65

Thus, for the existing development of the Central parts of large cities mainly by public buildings, the best option for solving the problems of lack of green areas and rational energy consumption can be the reorganization of the coatings of existing public buildings in a multi-layer system of "green roofs".

Studies have shown that the device of intensive "green roofs" can reduce the average temperature of the air in the room by 0.7 K in summer, which will reduce costs [7]. Moreover, the results of modeling and evaluation of conventional and green coatings for a single-family house showed that the indoor air temperature in the buildings with "green roofs" decreased by 2 °C, which resulted in a 6% decrease in annual energy demand [8]. In addition, in article [9], as a result of the experimental study that compares the energy efficiency of three roofs, it was concluded about the reduction of energy consumption of "green roofs" in summer. The results of the article [10] showed that with an increase in the thickness of the substrate by 10 cm, the thermal resistance of dry clay soil increases by 0.4 m^2 k/W, but the presence of moisture in the soil has a passive effect on thermal properties of a roof, providing heat removal from the building [11]. In addition, in buildings with "green roofs", the choice of vegetation and its density can lead to a significant change in thermal insulation [12–14]. For example, scientists MacIvor, Ranalli, Lundholm found that in "green roofs", the heat transfer resistance of arid plants is higher than that one of wet-lands [15]. Also, studies have been conducted in the field of energy efficiency of buildings using "green roofs" [8, 16–22].

Currently, there are many publications on the technology of "green roof" and its advantages, but their significant part does not distinguish the impact of this technology on the energy efficiency of buildings. Conducting research in the direction of increasing the energy efficiency of buildings due to the construction of "green roofs" is relevant and promising direction for further study.

The purpose of this study is to prove the increase in energy efficiency of the coating structure of the operated "green roof" in comparison with non-operated insulated roofs.

The object of the study is the construction of the coating of some buildings of the multifunctional center in Moscow and the training center of Tula.

The subject of the study is the resistance to heat transfer.

The main tasks of theoretical research are formed:

- Implementation of thermal engineering calculation of the existing non-operated insulated roofs for two buildings;
- Implementation of heat engineering calculation of a "green roof" coating;
- Comparative analysis and assessment of the impact of "green roofs" on the energy efficiency of public buildings.

2   Materials and Methods

Research methods are based on a theoretical research, on the collection and analysis of literary and graphic materials on the subject of research, and on the study of normative documents.

Theoretical studies were carried out with the thermal calculation of the "green roofs" of two public buildings. The technique of thermal calculation is regulated by the requirements of SP 50.13330.2012 [23] and using SP 131.13330.2012 [24].

For comparative calculation, the existing insulated roof of K-2 type and intensive type of a "green roof" are taken (Table 2).

Table 2. The main comparative characteristics of non-operated and operated roofs

Characteristics of a roof	Intensive roof (operated)	Extensive roof (non-operated)
1. Thickness of the soil layer	0.2...0.6 m	0.07...0.15 m
2. Vegetation	Plants, trees, shrubs, which root system must be within the specified project height of the substrate	Only plants with horizontal root system (due to the small height of the substrate)
3. Usage	It is assumed frequent stay of people on the roof	It is assumed a very rare stay of people on the roof
4. Protective structures	The parapet height more than 1.2 m	Protective structures are not required
5. Exploitation	High level of service and care	Easy maintenance and care
6. Structure	Developed at the design stage of the building (requires calculation of loads)	Can be developed both at the design stage of the building and during operation

We will carry out thermal calculations of coatings of the building of the multi-purpose center: Moscow, the crossing of Novoyasenevsky street and Profsouznay street.

A multilayer coating system №1 using in the building is presented in layers: a corrugated sheet, TechnoNicol vapor barrier, mineral wool insulation TECHNOROOF H30 (δ = 50 mm), clonearray layer TECHNOROOF H30 KLIN (δ = 40 mm), plate PIR insulation with a covering foil (δ = 50 mm), the polymer membrane LOGICROOFV-RP (Fig. 1).

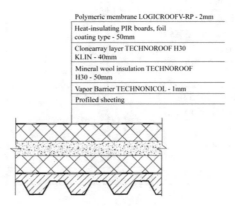

Fig. 1. Coating №1

After changing the coating №1 in the multilayer coating system №2, the composition of the layers will be as follows: profiled flooring, vapor barrier TechnoNikol, mineral wool insulation TECHNORUF N30 (δ = 50 mm), the slope layer technoruf N30 WEDGE (δ = 40 mm), insulation PIR plates with the type of coating foil (δ = 50 mm), technoelast FIX, technoelast GREEN, polyethylene film, PLANTER-life, soil substrate (δ = 300 mm), greening (Fig. 2).

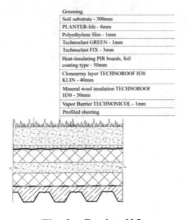

Fig. 2. Coating №2

The layers characteristics of the multilayer coating system №1 and №2 are presented in Tables 3 and 4, respectively.

Table 3. Multilayer coating system №1

Number of a layer	Type of material	Layer thickness δ, mm	Conductivity of a layer λ, W/m*C°	Thermal resistance of a layer R, $m^2*C°/W$
1	Polymeric membrane LOGICROOFV-RP	2	-	-
2	Heat-insulating PIR boards, foil coating type	50	0.022	2.27
3	Clonearray layer TECHNOROOF H30 KLIN	40	0.042	0.95
4	Mineral wool insulation TECHNOROOF H30	50	0.042	1.19
5	Vapor Barrier TECHNONICOL	1	-	-
6	Profiled sheeting	-	-	-
Thermal resistance of the enclosing structure				4.41

Table 4. Multilayer coating system №2

Number of a layer	Type of material	Layer thickness δ, mm	Conductivity of a layer λ, W/m*C°	Thermal resistance of a layer R, $m^2*C°/W$
1	Greening	-	-	-
2	Soil substrate (soil)	300	1.16	0.26
3	PLANTER-life	8	-	-
4	Polyethylene film	1	-	-
5	Technoelast GREEN	1	-	-
6	Technoelast FIX	3	-	-
7	Heat-insulating PIR boards, foil coating type	50	0.022	2.27
8	Clonearray layer TECHNOROOF H30 KLIN	40	0.042	0.95
9	Mineral wool insulation TECHNOROOF H30	50	0.042	1.19
10	Vapor Barrier TECHNONICOL	1	-	-
11	Profiled sheeting	-	-	-
Thermal resistance of the enclosing structure				4.67

Thermal resistance of the overlap of stacked homogeneous layers is defined as the sum of the thermal resistances of the individual layers according to the formula:

$$R_{w1}^{r} = \frac{1}{\alpha_B} + \frac{\delta_1}{\lambda_1} + \frac{\delta_2}{\lambda_2} + \ldots + \frac{\delta_n}{\lambda_n} + \frac{1}{\alpha_H} = \frac{1}{8.7} + 4.41 + \frac{1}{23} = 4.57 \frac{m^2 \cdot C^{\circ}}{W},$$

where α_B – a heat transfer coefficient on the inner surface of the enclosing structure, $W/(m^2 \cdot {}^{\circ}C)$; α_H – a heat transfer coefficient on the outer surface of the enclosing structure, $W/(m^2 \cdot {}^{\circ}C)$; δ_n – a thickness of the n-th layer of a fence.

Given the heterogeneity of the coating structure №1, thermal conductivity coefficient will be equal to:

$$R_{w1}^{r} = 4.57 \cdot 0.9 = 4.12 \frac{m^2 \cdot C^{\circ}}{W}$$

The normalized value of the heat transfer of the coating is:

$$R_{0,r}^{req} = 3.34 \frac{m^2 \cdot C^{\circ}}{W}$$

That is why:

$$R_{w1}^{r} = 4.12 \frac{m^2 \cdot C^{\circ}}{W} > R_{0,r}^{req} = 3.34 \frac{m^2 \cdot C^{\circ}}{W}$$

As a conclusion, it can be said that the value of the reduced resistance to heat transfer R_{w1}^{r} is much more than required $R_{0,r}^{req}$, therefore, the presented coating design fully meets the requirements for heat transfer.

The addition of new layers and adjustment of existing ones in the coating №1 was carried out according to the guide, it allowed creating the coating №2.

Thermal resistance of the overlap of stacked homogeneous layers is defined as the sum of the thermal resistances of the individual layers according to the formula:

$$R_{w2}^{r} = \frac{1}{\alpha_B} + \frac{\delta_1}{\lambda_1} + \frac{\delta_2}{\lambda_2} + \ldots + \frac{\delta_n}{\lambda_n} + \frac{1}{\alpha_H} = \frac{1}{8.7} + \frac{0.3}{1.16} + \frac{0.05}{0.022} + \frac{0.04}{0.042} + \frac{1}{23}$$
$$= 4.83 \frac{m^2 \cdot C^{\circ}}{W}.$$

Given the heterogeneity of the coating structure №1, thermal conductivity coefficient will be equal to:

$$R_{w2}^{r} = 4.83 \cdot 0.9 = 4.35 \frac{m^2 \cdot C^{\circ}}{W}$$

According to the results of comparative heat engineering calculation of non-operated coating №1 and the same coating reorganized into operated "green roof" with the addition of new necessary layers and the change (dismantling) of the former (coating №2), the value of the reduced resistance to heat transfer coating corresponds the regulatory requirements of thermal protection of enclosing structures (more than $R_{0,r}^{req} = 3.34 \frac{m^2 \cdot C°}{W}$), and, taking into account the heterogeneity of the layers, the resistance of heat perception and heat transfer, is: for covering №1 $R_{w1}^{r} = 4.12 \frac{m^2 \cdot C°}{W}$, for covering №2 $R_{w2}^{r} = 4.35 \frac{m^2 \cdot C°}{W}$, that is greater than the value of the reduced heat transfer resistance of the coating №1. Consequently, the use of a layer of soil substrate with a thermal conductivity coefficient $\lambda_B = 1.16 \frac{W}{m \cdot C°}$, according to the technology of a "green roof" in a multilayer coating system, has led not only to an increase in the reduced resistance to $0.23 \frac{m^2 \cdot C°}{W}$ but also to an increase in the energy efficiency of the building by 5.6% (Fig. 3).

We will conduct thermal calculations of the coatings of the building for the training center of specialists at the address: Tula, Shcheglovskaya Zaseka, 59.

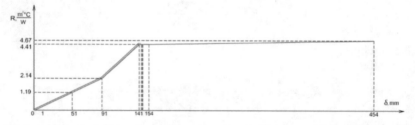

Fig. 3. Graph of heat transfer resistance versus thermal properties of coating layers 1 and 2 (blue line - coating 1 red line - coating 2)

The roof of the building is designed flat, overlaid with beam and truss metal structures.

Roof coating №3 consists of layers: beam and truss metal structures with a slope, monolithic reinforced concrete plate on profiled sheeting H75-750-0.8, reinforced with reinforcement bars A500C and A240 (δ = 150 mm), layer of Vapor Barrier, mineral wool Rockwool γ = 160 kg/m³ (δ = 200 mm), layer of Vapor Barrier, reinforced cement-sand screed on the slope (δ = 50 mm), layer of waterproofing "Unifleks" (Fig. 4).

After reorganization of coating №3, coating №4 will consist of the following layers: beam and truss metal structures with a slope, monolithic reinforced concrete plate on profiled sheeting H75-750-0.8, reinforced with reinforcement bars A500C and A240 (δ = 150 mm), layer of Vapor Barrier, mineral wool Rockwool γ = 160 kg/m³ (δ = 200 mm), technoelast FIX, technoelast GREEN, polyethylene film, PLANTER-life, soil substrate (δ = 300 mm), greening (Fig. 5).

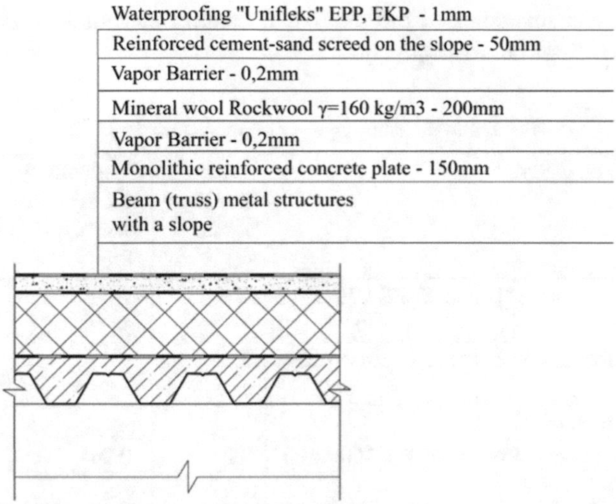

Waterproofing "Unifleks" EPP, EKP - 1mm

Reinforced cement-sand screed on the slope - 50mm
Vapor Barrier - 0,2mm
Mineral wool Rockwool γ=160 kg/m3 - 200mm
Vapor Barrier - 0,2mm
Monolithic reinforced concrete plate - 150mm
Beam (truss) metal structures with a slope

Fig. 4. Coating №3

Greening

Soil substrate - 300mm
PLANTER-life - 8mm
Polyethylene film - 1mm
Technoelast GREEN - 1mm
Technoelast FIX - 3mm
Mineral wool Rockwool γ=160 kg/m3 - 200mm
Vapor Barrier - 0,2mm
Monolithic reinforced concrete plate - 150mm
Beam (truss) metal structures with a slope

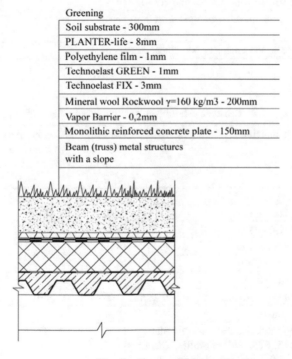

Fig. 5. Coating №4

The layers characteristics of the multilayer coating system №3 and №4 are presented in Tables 5 and 6, respectively.

Table 5. Multilayer coating system №3

Number of a layer	Type of material	Layer thickness δ, mm	Conductivity of a layer λ, W/m*C°	Thermal resistance of a layer R, m^2*C°/W
1	Waterproofing "Unifleks" EPP, EKP – 1 layer	1	-	-
2	Reinforced cement-sand screed on the slope	50	9.6	0.005
3	Vapor Barrier	0.2	-	-
4	Mineral wool Rockwool γ = 160 kg/m³	200	0.046	4.348
5	Vapor Barrier	0.2	-	-
6	Monolithic reinforced concrete plate, type B25, F75, W4 on profiled sheeting H75-750-0,8, reinforced with reinforcement bars A500C and A240	150	17.98	0.008
7	Beam (truss) metal structures with a slope	-	-	-
Thermal resistance of the enclosing structure				4.361

Table 6. Multilayer coating system №4

Number of a layer	Type of material	Layer thickness δ, mm	Conductivity of a layer λ, W/m*C°	Thermal resistance of a layer R, m^2*C°/W
1	Greening	-	-	-
2	Soil substrate (soil)	300	1.16 [6]	0.26
3	PLANTER-life	8	-	-
4	Polyethylene film	1	-	-
5	Technoelast GREEN	1	-	-
6	Technoelast FIX	3	-	-
7	Mineral wool Rockwool γ = 160 kg/m³	200	0.046	4.348
8	Vapor Barrier	0.2	-	-
9	Monolithic reinforced concrete plate, type B25, F75, W4 on profiled sheeting H75-750-0,8, reinforced with reinforcement bars A500C and A240	150	17.98	0.008
10	Beam (truss) metal structures with a slope	-	-	-
Thermal resistance of the enclosing structure				4.616

Coverage calculations are made in a similar way, taking into account the location and coating structure of the building.

The thermal conductivity coefficient λ for Tula is allowed on the basis of the operating conditions of the enclosing structures of type B (normal humidity zone and normal humidity conditions of the premises).

Thermal resistance of coating №3 with successive homogeneous layers is defined as the sum of the thermal resistances of individual layers according to the formula:

$$R_{w4}^r = \frac{1}{\alpha_B} + \frac{\delta_1}{\lambda_1} + \frac{\delta_2}{\lambda_2} + \ldots + \frac{\delta_n}{\lambda_n} + \frac{1}{\alpha_H} = \frac{1}{8,7} + 4,316 + \frac{1}{23} = 4.77 \frac{m^2 \cdot C^\circ}{W}.$$

Given the heterogeneity of the coating design №3, the thermal conductivity coefficient will be equal to:

$$R_{w3}^r = 4.52 \cdot 0.9 = 4.07 \frac{m^2 \cdot C^\circ}{W}$$

The normalized heat transfer rate is $R_{0,r2}^{req} = 3.50 \left(\frac{m^2 \cdot C^\circ}{W} \right)$

$$R_{w3}^r = 4.07 \frac{m^2 \cdot C^\circ}{W} > R_{0,r2}^{req} = 3.5 \frac{m^2 \cdot C^\circ}{W}$$

The value of the reduced heat transfer resistance R_{w3}^r is much greater than the required $R_{0,r2}^{req}$, therefore the presented design of the coating fully meets the requirements for heat transfer.

Thermal resistance of coating №4 with successive homogeneous layers is defined as the sum of the thermal resistances of individual layers according to the formula:

$$R_{w4}^r = \frac{1}{\alpha_B} + \frac{\delta_1}{\lambda_1} + \frac{\delta_2}{\lambda_2} + \ldots + \frac{\delta_n}{\lambda_n} + \frac{1}{\alpha_H} = \frac{1}{8.7} + 4.616 + \frac{1}{23} = 4.77 \frac{m^2 \cdot C^\circ}{W}.$$

Given the heterogeneity of the coating design №4, the thermal conductivity coefficient will be equal to:

$$R_{w4}^r = 4.77 \cdot 0.9 = 4.29 \frac{m^2 \cdot C^\circ}{W}$$

The value of the reduced heat transfer resistance R_{w4}^r is much greater than the required $R_{0,r2}^{req}$, therefore the presented design of the coating fully meets the requirements for heat transfer.

According to the results of comparative heat engineering calculation of non-operated coating №3 and the same coating reorganized into operated "green roof" with the addition of new necessary layers and the change (dismantling) of the former (coating №4), the value of the reduced resistance to heat transfer coating corresponds the regulatory requirements of thermal protection of enclosing structures (more than

$R_{0,r}^{req} = 3.50\,\frac{m^2 \cdot C^\circ}{W}$), and, taking into account the heterogeneity of the, the resistance of heat perception and heat transfer, is: for the coating №3: $R_{w3}^r = 4.07\,\frac{m^2 \cdot C^\circ}{W}$, for the coating №4 $R_{w4}^r = 4.29\,\frac{m^2 \cdot C^\circ}{W}$, that is greater than the value of the reduced heat transfer resistance of the coating №3. Consequently, the use of a layer of soil substrate with a thermal conductivity coefficient $\lambda_B = 1.16\,\frac{W}{m \cdot C^\circ}$, according to the technology of a "green roof" in a multilayer coating system, has led not only to an increase in the reduced resistance to $0.22\,\frac{m^2 \cdot C^\circ}{W}$ but also to an increase in the energy efficiency of the building by 5.4% (Fig. 6). This indicates the effectiveness of using the "green roof" technology to cover public buildings.

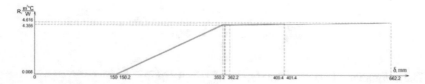

Fig. 6. Graph of heat transfer resistance versus thermal properties of coating layers 3 and 4 (blue line - coating 3, red line - coating 4)

3   Results

The results of heat engineering calculation using the "green roof" technology:

To cover the building of the multifunctional center in Moscow, the heat transfer resistance was:

$$R_{w1}^r = 4.12\,\frac{m^2 \cdot C^\circ}{W}$$

With the addition of layers to form a "green roof" on the existing coating, the heat transfer resistance was:

$$R_{w2}^r = 4.35\,\frac{m^2 \cdot C^\circ}{W}$$

To cover the building for training center of specialists in Tula, the heat transfer resistance was:

$$R_{w3}^r = 4.07\,\frac{m^2 \cdot C^\circ}{W}$$

With the addition of layers to form a "green roof" on the existing coating, the heat transfer resistance was:

$$R^r_{w4} = 4.29 \frac{m^2 \cdot C^\circ}{W}$$

4 Discussion

Thus, it is possible to identify a certain dependence in the coatings of buildings when constructing a "green roof": with an increase in the thickness of the soil layer, an increase in the numerical value of the reduced resistance to heat transfer is observed, i.e. thermal protection of the building increases. Therefore, when considering a green roof in operation, the reduced heat transfer resistance will be much higher than in the unexploited green roof due to the required greater thickness of the substrate (0.2–0.6 m > 0.07–0.15 m). When using a "green roof", it is possible to reduce economic costs in the case of a decrease in the thickness of the insulation.

Comparing the calculation results for the coatings in the two examples described above, we can conclude that the reorganization of the multilayer structure of the coating into a "green roof" can increase the heat transfer resistance by more than 5%, depending on the composition and thermal properties of the original coating layers of public buildings, which leads to an increase in the energy efficiency of the building and a reduction in economic costs due to a decrease in the thickness of the insulation.

5 Conclusions

As a result of a theoretical study, an increase in the energy efficiency of the operated "green roof" construction compared to unexploited warmed roofs by more than 5% was proved.

References

1. Berardi, U.: Sustainability assessment in the construction sector: rating systems and rated buildings. Sustain. Dev. 20(6), 411–424 (2012)
2. Ghaffarian, H.A., Dahlan, N., Berardi, U., Makaremi, N.: Sustainable energy performances of green buildings: a review of current theories, implementations and challenges. Renew. Sustain Energy Rev. 25, 1–17 (2013)
3. Ilyichev, V.A., Emelyanov, S.G., Kolchunov, V.I., Gordon, V.A., Bakaeva, N.V.: Principles of transformation of the city into a biosphere-compatible and developing person, Moscow, DIA, p. 184 (2015)
4. Ponomareva, M.A.: Energy efficient development of the region's infrastructure as a condition for its transition to sustainable development. Tomsk State Univ. Bull. 349, 149–153 (2011)

5. Federal Law "On Energy Saving and on Increasing Energy Efficiency and On Amending Certain Legislative Acts of the Russian Federation" of 23.11.2009 N 261
6. Federal Law "Technical Regulations on the Safety of Buildings and Structures" dated December 30, 2009 No. 384
7. Karachaliou, P., Santamouris, M., Pangalou, H.: Experimental and numerical analysis of the energy performance of a large scale intensive green roof system installed on an office building in Athens. Energy Build. **114**, 256–264 (2016)
8. Jaffal, I., Ouldboukhitine, S., Belarbi, R.: A comprehensive study of the impact of green roofs on building energy performance. Renew. Energy **43**, 157–164 (2012)
9. Coma, J., Pérez, G., Solé, C., Castell, A., Cabeza, L.F.: Thermal assessment of extensive green roofs as passive tool for energy savings in buildings. Renew. Energy **85**, 1106–1115 (2016)
10. Wong, N.H., Cheong, D.K.W., Yan, H., Soh, J., Ong, C.L., Sia, A.: The effects of rooftop garden on energy consumption of a commercial building in Singapore. Energy Build. **35**, 353–364 (2003)
11. Wolf, D., Lundholm, J.T.: Water uptake in green roof microcosms: effects of plant species and water availability. Ecol. Eng. **33**, 179–186 (2008)
12. Cox, B.K.: The influence of ambient temperature on green Roof R-values. Master thesis. Portland State University (2010)
13. Berardi, U., Ghaffarian Hoseini, A.H., Hoseini, G.A.: State-of-the-art analysis of the environmental benefits of green roofs. Appl. Energy **115**, 411–428 (2014)
14. Sailor, D.J.: A green roof model for building energy simulation programs. Energy Build. **40**, 1466–1478 (2008)
15. MacIvor, J.S., Ranalli, M.A., Lundholm, J.T.: Performance of dryland and wetland plant species on extensive green roofs. Ann. Bot. **107**(4), 671–679 (2011)
16. Hodo-Abalo, S., Banna, M., Zeghmati, B.: Performance analysis of a planted roof as a passive cooling technique in hot-humid tropics. Renew. Energy **39**(1), 140–148 (2012)
17. Costanzo, V., Evola, G., Marletta, L.: Energy savings in buildings or UHI mitigation? Comparison between green roofs and cool roofs. Energy Build. **114**, 247–255 (2016)
18. Heidarinejad, G., Esmaili, A.: Numerical simulation of the dual effect of green roof thermal performance. Energy Convers. Manage. **106**, 1418–1425 (2015)
19. Spala, A., Bagiorgas, H.S., Assimakopoulos, M.N., Kalavrouziotis, J., Matthopoulos, D., Mihalakakou, G.: On the green roof system. Selection, state of the art and energy potential investigation of a system installed in an office building in Athens, Greece. Renew. Energy **33**(1), 173–177 (2008)
20. Santamouris, M., Pavlou, C., Doukas, P., Mihalakakou, G., Synnefa, A., Hatzibiros, A., Patargias, P.: Investigating and analysing the energy and environmental performance of an experimental green roof system installed in a nursery school building in Athens, Greece. Energy **32**(9), 1781–1788 (2007)
21. Castleton, H.F., Stovin, V., Beck, S.B.M., Davison, J.B.: Green roofs. Building energy savings and the potential for retrofit. Energy Build. **42**(10), 1582–1591 (2010)
22. Saadatian, O., Sopian, K., Salleh, E., Lim, C.H., Riffat, S., Saadatian, E., Toudeshki, A., Sulaiman, M.Y.: A review of energy aspects of green roofs. Renew. Sustain. Energy Rev. **23**, 155–168 (2013)
23. SP 50.13330.2012 Thermal protection of buildings. Updated edition of SNiP 23-02-2003, Moscow, p. 100 (2013)
24. SP 131.13330.2012. Construction climatology. Updated version. SNiP 23-01-99 *, Moscow, p. 184 (2012)

Renewable Energy

Effect of Demand Tariff Schemes in Presence of Distributed Photovoltaic Generation and Electrical Energy Storage

Beatrice Marchi[1]([✉]) [iD], Marco Pasetti[2] [iD], and Simone Zanoni[1] [iD]

[1] Department of Mechanical and Industrial Engineering,
Universita degli Studi di Brescia, via Branze 38, 25123 Brescia, Italy
{b.marchi,simone.zanoni}@unibs.it
[2] Department of Information Engineering, Università degli Studi di Brescia,
via Branze 38, 25123 Brescia, Italy
marco.pasetti@unibs.it

Abstract. Electrical energy storage systems gathered a large interest among the energy market, since they have a key role in increasing the hosting capacity of renewables overcoming their main drawbacks (i.e. intermittency and uncertainty of the power generation). They also lead to increased share of self-consumption and improved reliability of the electric power distribution. Nevertheless, the economic feasibility of energy storage devices for distributed generation is still questionable: revenues and savings coming from the increase of self-consumption are usually not sufficient to cover the investment costs. The economic benefits are strictly dependent on the electricity tariff, which usually can be defined through a trinomial structure made up of a fixed part and two variable components depending on power and energy components. Electrical energy storage devices can be used for different purpose (e.g., load shifting and peak shaving), and the selection of the most performant is affected by the ratio between the two variable components of the tariff. The main purpose of this contribution is to investigate how the electricity tariffs could affect the diffusion of electrical energy storage systems in support of distributed generation based on renewable energy sources for different loads of end-users.

Keywords: Distributed generation · Photovoltaic · Electrical energy storage · Self-consumption · Electricity tariff schemes

1 Introduction

In recent years, Distributed Generation (DG) from Renewable Energy Sources (RESs) equipped with Electrical Energy Storage System (EESS) has received an increased attention due to the growing pressure towards a more sustainable and decarbonized energy system. EESS solutions are recognized as a key technology for overcoming, or at least mitigating, the main drawbacks of renewable energy caused by its intermittency and uncertainty since they allow to store energy and release it when needed. These devices can also increase the hosting capacity of RESs, the reliability of distribution systems, and share of self-consumption of energy prosumers. In particular, electrical

© Springer Nature Switzerland AG 2020
V. Murgul and M. Pasetti (Eds.): EMMFT 2018, AISC 982, pp. 201–215, 2020.
https://doi.org/10.1007/978-3-030-19756-8_19

storage systems are expected to play a fundamental role to support the large penetration of Electric Vehicles (EVs) in urban areas, by allowing advanced energy flow management strategies at EV charging stations [1, 2].

However, EESSs presents some issues that hinder their wide adoption: i.e., they suffer from technical limitations, entail high investment costs and pose relevant challenges on the operation of the electrical power grids. Hence, the management of EESS and the interactions with the surrounding systems is an important precondition to guarantee their economic and technical feasibility.

Battery energy storage systems (BESS), which are devices that store energy in the form of electrochemical energy, represents one of the most widely used systems for the support of distributed generation [3] thanks to their wide flexibility. To evaluate the economic feasibility of a storage system, the total costs for the installation, operation and maintenance over its entire lifetime should be considered. Operation costs consist of all the costs and revenues that occurs during the entire time of use in its lifetime [4]: i.e., costs for the energy purchased from the grid, revenues for selling back to the grid the energy produced by the RES at the corresponding feed-in-tariff, and, eventually, revenues due to feed-in tariff premiums for the increased self-consumption share.

In literature, several works exist which develop economic models for the evaluation of the savings introduced through a distributed generation equipped with a storage device. For instance, [5] proposed a methodology for the evaluation of the Levelized Cost of Electricity for utility-scale storage systems; [6] investigated the life cycle cost of three different BESS technologies, by considering investment, operation, maintenance, and disposal costs; [7] developed non-linear programming optimization models for the sizing of different storage devices in a renewable farm equipped with solar and wind power plants; [8] addressed the optimal energy storage management and sizing problem taking into account dynamic pricing of the electricity from the grid. However, all these works consider only the cost component related to the energy consumption. Currently, electricity prices can be defined through a trinomial tariff composed by a fixed component per Point of Delivery (POD), a variable component function of the maximum power taken from the grid, and a variable component function of the energy purchased from the grid. Different cost coefficients can thus affect the diffusion of BESS in support of distributed RESs, since they can increase/reduce the share of self-consumption and the peak power. For instance, [9] investigated how the Italian reform of system costs in the electricity tariffs for non-household customers (Legislative Decree No. 244/16) affects the on DG and ESS. The results of the numerical analysis revealed that, even if the presence of PV and ESS always allows the reduction of systems costs, the variations that would be introduced by the reform are variable. Generally, the bigger the PV plant with BESS, the lower is the percentage reduction of the system costs. In some cases, the variation in the system costs is even positive (i.e. leading to higher costs). Other research studies relevant for this work are focused on the technical aspects of PV and BESS. [10] proposes a PV module equipped with on-board Energy Storage Systems to properly manage the charge/discharge cycle of the ESS while implementing the Maximum Power Point Tracking of the PV source.

Furthermore, they considered that the system is capable to communicate with the external grid to allow the implementation of advanced network services, such as power support, to the distribution network. [11] defined a communication and computing architecture, based on the Virtual Power Plant (VPP) approach, able to virtualize several renewable power plants installed in the same area. The proposed architecture was deployed in the North Campus of the University of Brescia, Italy, for the monitoring of two PV plants and two different BESS. Results showed that it can simplify the monitoring from the DSO. Later, [12] presented a conceptual study on a VPP architecture, based on a service-oriented approach, for the optimal management of distributed energy resources owned by prosumers participating in Demand-Side Management (DSM) programs.

Since the benefits introduced and the return of the investment are strictly dependent on the specific electricity price considered, the objective of this work is to evaluate the impact of the electricity tariffs on DG and EESS in terms of energy balance and economic performance in the case of industrial customers. The assessment will be conduct for different values of the ratio between the cost coefficient related to the energy consumption and the one related to the peak power, and for different load profiles of the industrial users. The remainder of the article is organized as follows: Sect. 2 defines the assumptions and the objectives of the analysis and presents the models and methods. In Sect. 3 a numerical case study to illustrate the possible effect of the different electricity tariff in case of RES coupled with a BESS is presented. Finally, Sect. 4 summarizes the main findings of the present work and provides suggestions for future research.

2 Models and Methods

In the following, the models and methods used in the present analysis are described and discussed.

2.1 Assumption and Objectives

The objective of this study is to assess the impact of the electricity tariff structure on industrial users equipped with a photovoltaic (PV) plant and a battery energy storage system. We considered three industrial users with different load profiles but characterized by the same yearly energy consumption (1 GWh/year): a user continuously working on three shifts (User 1), a user working on two shifts – the first shift from 6 am to 2 pm while the second from 2 pm to 10 pm (User 2), and a user working on one shift – from 8:30 am to 16:30 pm (User 3). Detailed information about the user's load profiles considered in the present analysis are given in Sect. 2.3. For each user, we considered the presence of a PV power plant with two different configurations, with an expected yearly energy production of, respectively, the 50% and the 100% of the yearly user's load consumption. Finally, for each combination of user load profile and PV configuration, we assumed the installation of a Lithium-Ion Battery Energy Storage

System (Li-Ion BESS), with a gross capacity capable to store the average daily excess of energy produced by the PV system (computed on yearly basis). Information about the modelling of the PV and the BESS is reported in [4, 9] and some additional details are defined in Sect. 2.4. For each scenario, the energy balances within the users and the distribution grid have been computed with a simulation time step of 15 min over a time horizon of 1 year through a custom C++ code. The computed energy flows have been then used to compute the electricity costs based on the current tariff structure. To better observe the variation of the costs introduced by varying the load profiles and the cost coefficients, the acquisition and installation costs of PV and BESS systems haven't been considered, thus focusing only on currently operating systems.

2.2 Electricity Tariff Structure

In many countries, the electricity tariff is generally defined by a trinomial structure consisting of three cost components: (i) a fixed component per POD, (ii) a variable component function of the peak power taken from the grid (measured on monthly basis), and (iii) a variable component function of the energy purchased from the grid. The electricity tariff is computed as function of the fixed component α and of the variable components β and γ, as defined in Eq. 1:

$$c_e = \alpha + \beta P_{FU,max} + \gamma E_{FU} \tag{1}$$

An interesting parameter affecting the economic performance of the DG and BESS system is given by the ratio between the cost coefficient related to the energy consumption and the one related to the peak power, as defined in Eq. 2:

$$k_1 = \frac{\gamma}{\beta} \tag{2}$$

In the analysis, five scenarios characterized be different costs components, and consequently by a different value for the coefficient k_1, will be compared: (i) considers the tariff structure currently used in Italy for an industrial user ($\alpha = 1400$ €, $\beta = 65$ €/kWp, and $\gamma = 0.15$ €/kWh) to which correspond a k_1 equals to 0.023 kWp/kWh; (ii) and (iii) models the cost components offered by two different companies in the USA ($\alpha = 1679$ €, $\beta = 15$ €/kWp, and $\gamma = 0.15$ €/kWh, and $\alpha = 0$ €, $\beta = 158.4$ €/kWp, and $\gamma = 0.05$ €/kWh) to which correspond a k_1 equals to 0.01 kWp/kWh and 0.0003 kWp/kWh respectively; (iv) and (v) consider two hypothetical tariff structures characterized by a reduction of the component related to the energy consumption and by an increase of the one related to the peak power respectively ($\alpha = 1000$ €, $\beta = 100$ €/kWp, and $\gamma = 0.1$ €/kWh, and $\alpha = 1000$ €, $\beta = 1000$ €/kWp, and $\gamma = 0.1$ €/kWh) to which correspond a k_1 equals to 0.001 kWp/kWh and 0.0001 kWp/kWh.

2.3 Power Demand Profiles of the Industrial Users

The load profile was computed by considering the typical daily power demand profile for each user over a time horizon of 1 year with a time resolution of 15 min, taking into account working days, non-working days and holidays, according to the specific working schedules. During non-working days and holidays, the expected power demand was set to the baseload power, P_0, which can be defined as a function of the day peak power P_{max} by Eq. 3.

$$P_0 = k_2 P_{max} \tag{3}$$

For the case study, k_2 was set to 1 for user 1 while 4 for users 2 and 3. Furthermore, the peak power was set to 80% of the rated power. The expected load's energy consumption during one year for each i-th user, $\overline{E}_{L,Y,i}$, was then computed. The typical daily power demand profiles of users, along with the expected PV power generation profiles in the case of 100% of PV share, are depicted in Figs. 1, 2 and 3, respectively for user 1, user 2 and user 3.

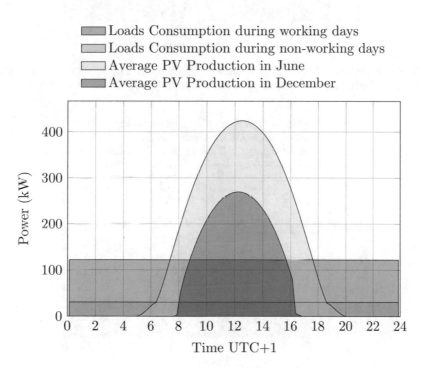

Fig. 1. Power demand profiles and expected PV production for user 1, in the case of 100% of PV share, corresponding to a nominal power of about 690 kWp.

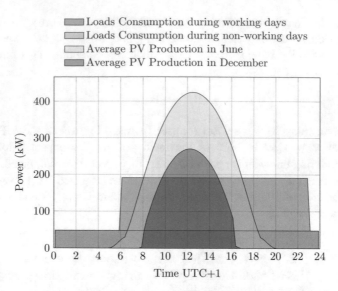

Fig. 2. Power demand profiles and expected PV production for user 2, in the case of 100% of PV share, corresponding to a nominal power of about 690 kWp.

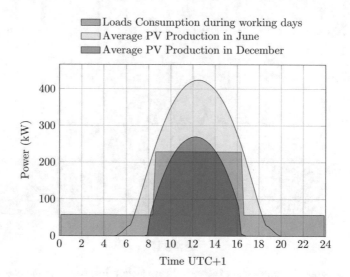

Fig. 3. Power demand profiles and expected PV production for user 3, in the case of 100% of PV share, corresponding to a nominal power of about 690 kWp.

2.4   PV and BESS Model

In the present analysis, we considered a PV power plant located in Rome (chosen as reference site because of its central location in the Italian territory) based on crystalline silicon technology, which is the most used technology for the PV plants currently installed in Italy. For each use case (i.e. for each user, with the two different PV share hypotheses), the energy balance with the distribution grid was computed by evaluating the load power demand and the PV power generation profiles. Figures 4, 5 and 6 depict the typical net power profiles during working and non-working days in June and December for user 1, user 2, and user 3 respectively with a PV share of 100%. It can be noticed how the net power is subject to relevant variation during the day, especially for user 3.

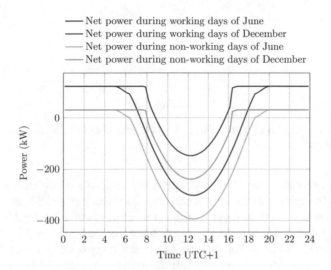

Fig. 4. Net power profiles of user 1 with a PV share of 100%. Positive values represent power taken from the grid.

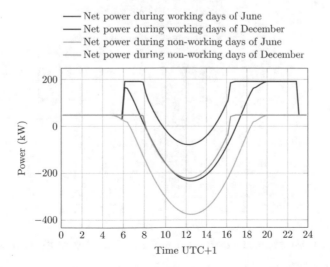

Fig. 5. Net power profiles of user 2 with a PV share of 100%. Positive values represent power taken from the grid.

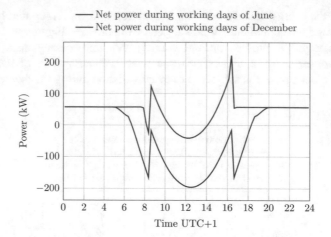

Fig. 6. Net power profiles of user 3 with a PV share of 100%. Positive values represent power taken from the grid.

The PV plant was assumed to be equipped with a BESS which nominal power was computed as the minimum value between the maximum power allowed by the whole battery pack and the maximum value of the power fed to the utility grid by the system. The efficiency of the BESS power conversion system was computed by means of the conversion efficiency curve. For each use case, the energy balance with the distribution grid (see Figs. 7, 8 and 9 for user 1, user 2, and user 3, respectively) was computed taking into account the load power demand profiles, and the PV power generation.

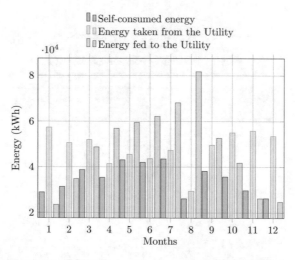

Fig. 7. Energy balance at the POD of user 1 in the case of 100% of PV share without BESS.

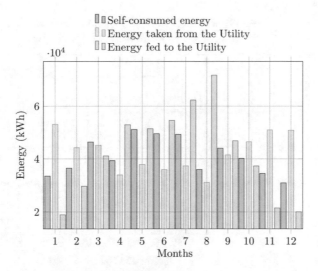

Fig. 8. Energy balance at the POD of user 2 in the case of 100% of PV share without BESS.

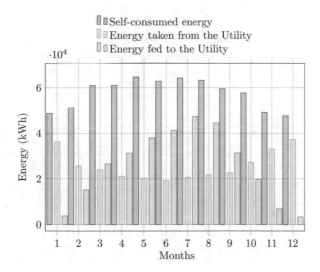

Fig. 9. Energy balance at the POD of user 3 in the case of 100% of PV share without BESS.

The typical net power profiles for the considered users with the Li-Ion system are depicted in Figs. 10, 11 and 12, respectively for user 1, user 2, and user 3.

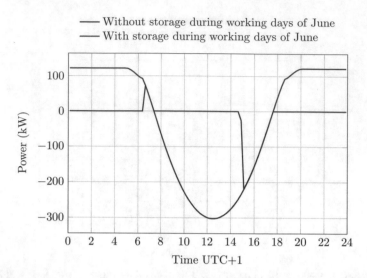

Fig. 10. Net power profiles of user 1 in the case of 100% of PV share with a 1768 kWh and 422 kWp Li-Ion BESS.

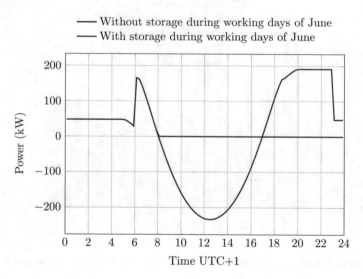

Fig. 11. Net power profiles of user 2 in the case of 60% of PV share with a 1548 kWh and 405 kWp Li-Ion BESS.

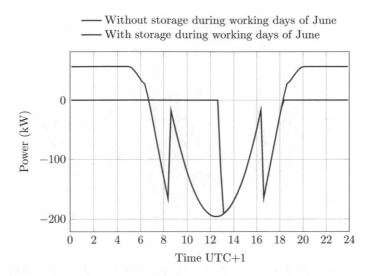

Fig. 12. Net power profiles of user 3 in the case of 60% of PV share with a 941 kWh and 225 kWp Li-Ion BESS.

Figures 13, 14 and 15 compare the self-consumption index without and with the opportunity to charge and discharge the electrical energy storage device, respectively for user 1, user 2, and user 3.

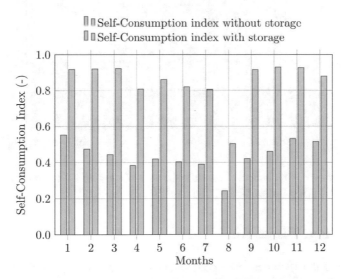

Fig. 13. Self-consumption indices of user 1 in the case of 100% of PV share without and with a 1768 kWh and 422 kWp Li-Ion BESS.

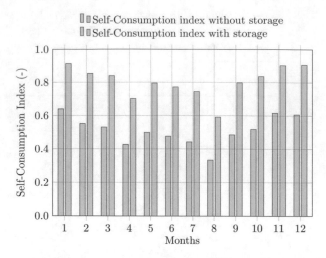

Fig. 14. Self-consumption indices of user 2 in the case of 100% of PV share without and with a 1548 kWh and 405 kWp Li-Ion BESS.

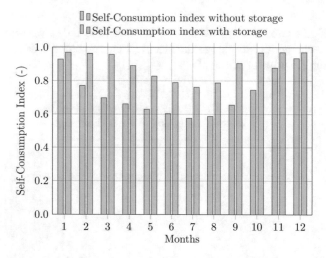

Fig. 15. Self-consumption indices of user 1 in the case of 100% of PV share without and with a 941 kWh and 225 kWp Li-Ion BESS.

3   Results and Discussion

Figures 16, 17, and 18 present the economic results obtained through the installation of the PV system and the Li-Ion BESS for user 1, user 2, and user 3 respectively. The economic results are evaluated as the variations in the total electricity costs with respect to the base case with neither PV neither BESS under the different hypotheses of the electricity tariffs. As can be expected, the introduction of both the PV and the storage device allows to highly reduce the overall electricity costs. The best performing scenario is H2 which is characterized by the lower variable component function of the peak power taken from the grid, β, and to which correspond k_1 equal to 0.01 kWp/kWh.

Shifting from the AS-IS scenario to the new hypotheses, the bigger the PV plant the higher is the reduction of the system costs. The current structure of the electricity tariff promotes users with larger PV share, especially if equipped with a BESS. This is mainly due to the reduced grid consumption for users with a PV plant and storage systems, especially for those users who can benefit from high self-consumption indices. However, if the variable cost component related to the peak power increases, the current dispatching rule aiming at maximizing the self-consumption highly reduces its convenience. For instance, shifting from H1 (current electricity tariff in Italy) to H5 (electricity tariff with the higher β), the reduction of the electricity costs is almost halved, more than halved, and reduced by 8–20% for user 1, user 2, and user 3, respectively. In these cases, a different dispatching rule should be investigated: e.g., aiming at peak shaving.

For low PV shares, the use of storages system does not significantly affect the electricity costs, especially for user 2 and 3. This result can be explained by the high self-consumptions indices of the PV configurations (see Figs. 14 and 15), that result from the low power output of the PV system, if compared to the relative high power demand of the loads. Conversely, for high PV shares, the effect of the presence of storage systems is evident. In this case, the power output of the PV plant is often higher than the power demand for all the users. Consequently, the opportunity to store the excess of energy generated by the PV and reuse it later lead to lower yearly grid consumption, and thus to lower electricity costs. However, for user 2 and 3, the installation of a BESS slightly improves the economic performance even for a high PV share, since the loads and the PV generation are almost overlapped.

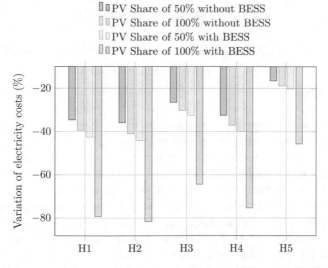

Fig. 16. Electricity cost variation with respect to the base case without PV and BESS for user 1.

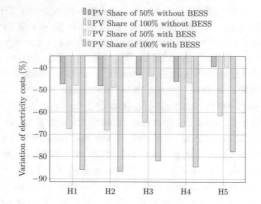

Fig. 18. Electricity cost variation with respect to the base case without PV and BESS for user 3.

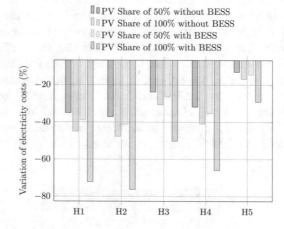

Fig. 17. Electricity cost variation with respect to the base case without PV and BESS for user 2.

4 Conclusions

In this paper, we investigated the effect of energy-based and power-based tariff schemes in presence of distributed photovoltaic generation and electrical energy storage. The variation of electricity costs was analyzed for three industrial users characterized by different working time and shifts and different load profiles but with the same yearly energy consumption (1 GWh/year). They were assumed to be equipped with a PV system and a lithium-ion BESS. The results of the numerical analysis revealed that, the installation of the PV and the storage device allows to highly reduce the electricity costs and the best performing electricity tariff is the one characterized by the lower coefficient for the cost component function of the peak power taken from the grid. Furthermore, the current structure of the electricity tariff promotes users with larger PV share, especially if equipped with a BESS. This is mainly due to the reduced grid consumption for users with a PV plant and storage systems, especially for those users who can benefit from high self-consumption indices. However, if the variable cost component related to the peak power increases, the current dispatching rule aiming at maximizing the self-consumption highly reduces its convenience. Further study may look at different price-based demand side schemes, such as Time-Of-Use (TOU), and Time-Of-Export (TOE) electricity tariff (e.g., characterized by different values of k_1), at different DG profiles (e.g., characterized by different values of k_2), and at different dispatching rules for the EES (e.g. looking at mainly peak shaving purposes).

Acknowledgments. This research activity has been partially funded by the University of Brescia as part of the research activities of the eLUX laboratory - "energy Laboratory as University eXpo".

References

1. Rinaldi, S., Pasetti, M., et al.: On the integration of E-Vehicle data for advanced management of private electrical charging systems. In: International Instrumental Measurement of Technology Conference (2017)
2. Rinaldi, S., Pasetti, M., et al.: On the mobile communication requirements for the demand-side management of electric vehicles. Energies 11(5), 1220 (2018)
3. Office of Electricity Delivery & Energy Reliability, DOE Global Energy Storage Database (2016). http://www.energystorageexchange.org/
4. Marchi, B., Zanoni, S., Pasetti, M.: A techno-economic analysis of Li-ion battery energy storage systems in support of PV distributed generation. In: Proceedings of the Summer School Francesco Turco, 13–15 September 2016
5. Obi, M., Jensen, S.M., et al.: Calculation of levelized costs of electricity for various electrical energy storage systems. Renew. Sustain. Energy Rev. 67, 908–920 (2017)
6. Marchi, B., Pasetti, M., Zanoni, S.: Life cycle cost analysis for BESS optimal sizing. Energy Procedia 113(C), 127–134 (2017)
7. Berrada, A., Loudiyi, K.: Operation, sizing, and economic evaluation of storage for solar and wind power plants. Renew. Sustain. Energy Rev. 59, 1117–1129 (2016)
8. Bortolini, M., Gamberi, M., Graziani, A.: Technical and economic design of photovoltaic and battery energy storage system. Energy Convers. Manag. 86, 81–92 (2014)
9. Marchi, B., Pasetti, M., et al.: The Italian reform of electricity tariffs for non household customers: the impact on distributed generation and energy storage. In: Proceedings of the Summer School Francesco Turco, pp. 103–109, September 2017
10. Dede, A., Della Giustina, D., Massa, G., et al.: A smart PV module with integrated electrical storage for smart grid applications. In: 2016 International Symposium on Power Electronics, Electrical Drives, Automation and Motion, SPEEDAM 2016, pp. 895–900 (2016)
11. Rinaldi, S., Pasetti, M., Ferrari, P., Massa, G., Della Giustina, D.: Experimental characterization of communication infrastructure for virtual power plant monitoring. In: Proceedings of the 2016 IEEE International Workshop on Applied Measurements for Power Systems, AMPS 2016, pp. 1–6 (2016)
12. Pasetti, M., Rinaldi, S., Manerba, D.: A virtual power plant architecture for the demand-side management of smart prosumers. Appl. Sci. 8(3), 432 (2018)

A New Photovoltaic Current Collector Optimizer to Enhance the Performance of Centralized Inverter Topologies

Ahmed Refaat🆔 and Nikolay Korovkin$^{(\boxtimes)}$🆔

Peter the Great Saint-Petersburg Polytechnic University, Polytechnicheskaya 29,
195251 Saint-Petersburg, Russia
ahmed_refaat_1984@eng.psu.edu.eg,
Korovkin_NV@spbstu.ru

Abstract. Centralized inverter topologies are the current preferred technology for medium and large-scale grid-connected photovoltaic (PV) installation because of their low cost and simplicity. However, the output power of these traditional topologies is mainly suffered from partial shading effects and mismatch between PV modules. Power losses due to shadow may reach up to 30% of total power expected, depending on PV array configuration and atmospheric conditions. This paper proposes a novel grid-connected centralized inverter topology based on a new photovoltaic current collector optimizer (CCO) to enhance the power extracted from PV array during partial shading or mismatch conditions. Computer simulation is carried out using MATLAB/Simulink in order to confirm the performance of the proposed topology. Simulation results show that the proposed topology offers an excellent steady-state response, fast dynamic response, perfect and robust tracking of the maximum power point during partial shading condition.

Keywords: Photovoltaic system · Current collector optimizer (CCO) ·
Grid-connected centralized inverter topologies · Current control VSI ·
Efficiency · PV modules mismatch · Partial shading

1 Introduction

Over recent years, the problems of energy shortage and environmental contamination have become critical research topics worldwide. Thus, the trends of employing distributed generation systems (DGSs) based on renewable energy are drawing more and more attention for reducing energy crisis and carbon emission. Among all various DGSs, solar photovoltaic (PV) systems are rapidly growing in electricity markets and are expected to continue this trend throughout the near future. Nevertheless, the unit cost of energy obtained from PV systems is still high; accordingly, current research focuses on decreasing the costs by increasing the energy production of the PV systems [1, 2]. Central inverter technology is the most common topologies of PV installation, which interfaced a large number of panels that configured in series-parallel (SP) combination to the grid. The main technical challenge for central inverter technology is the absence of a maximum power point operation for each module due to partial shading as

© Springer Nature Switzerland AG 2020
V. Murgul and M. Pasetti (Eds.): EMMFT 2018, AISC 982, pp. 216–224, 2020.
https://doi.org/10.1007/978-3-030-19756-8_20

well as bypass diodes that is used to prevent modules from hotspot effect deforms the PV array characteristics and exhibits multiple peaks, including a global and local maximum power points (MPPs). That force researchers to look for different technologies for the interconnections of the PV modules [3, 4].

The most typical solutions to partial shading effects are AC-module inverters, which utilize micro-converters for each individual PV module to adopt distributed MPP tracking (MPPT). All PV modules have the ability to be operated at each MPP, even under partial shading condition, because of the individual control for each module which improves the power efficiency of PV system. The prime downsides of AC-module inverters are high expenditure and increased system complexity since the quantity of micro-converters is proportional to that of PV modules [4, 5].

Other solutions to eliminate local MPPs and increase maximum power are using differential power processing (DPP) converters and voltage or current equalizers. A portion of the produced power of unshaded modules is transmitted to shaded ones by using DPP converters or equalizers; consequently, all the modules operate at the same voltage or close to each individual MPP. Several types of DPP converters and equalizers have been created and presented in literatures [5–14]. In fact, the number of switches in DPP converters and equalizers can demonstrate a decent insight about system intricacy due to each switch needs a controller circuit including its accessories.

Besides the system efficiency, high-quality injected power is another essential feature of grid-connected PV (GCPV) systems. To ensure compatible operation of PV systems with the electric grid, IEEE standard 929 for utility interface of PV systems recommended that PV inverters should operate at approximately unity power factor (UPF) with total harmonic current distortion less than 5%.

This paper aims at evaluating the performance of 100 kW GCPV system by using a new centralized inverter topology based on a novel photovoltaic current collector optimizer (CCO). The main objectives are to improve the power extracted from PV array during partial shading condition, avoid the misleading power loss due to local MPPs, eliminate the power loss associated with circulating currents between parallel PV generators, and inject a high quality AC current into the grid to meet the standard IEEE 929.

2 System Description

The schematic diagram of the proposed grid-tied centralized inverter based on current collector optimizer (CCO) is depicted in Fig. 1. The power circuit consists of a 100 kW PV array with CCOs, DC-link capacitor, three phase voltage source inverter (VSI), LC filter, low-frequency step-up transformer, and grid. As can be seen in the diagram, every eight modules or substrings are connected to CCO as a single stack and then these stacks may be connected in series-parallel combination to the grid through an inverter.

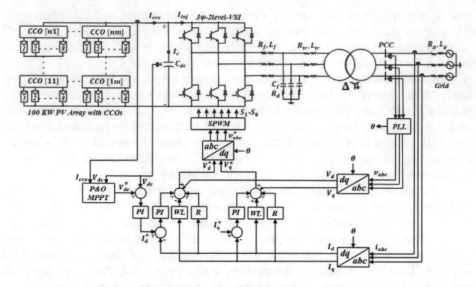

Fig. 1. The schematic diagram of the proposed topology.

2.1 Control of Current Collector Optimizer (CCO)

The main idea of the CCO is to find a way to harvest the currents from PV modules or substrings at approximately MPP voltages during shading condition. Hence, there is no need for the shaded module or substring to be short-circuited through bypass diodes. The circuit diagram of the CCO is as shown in Fig. 2. This circuit is a modified circuit that is used to collect the current from magnetohydrodynamic generator electrodes.

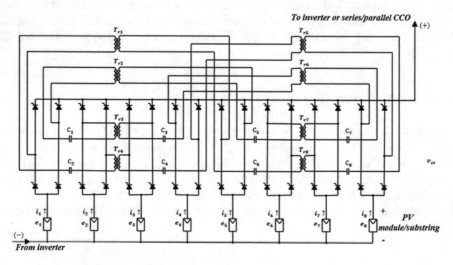

Fig. 2. Circuit diagram of the CCO.

As shown in Fig. 2, all PV generators negative terminals are connected to a common negative line while each positive terminal is connected to thyristor-bridge and then all bridges are gathered in common positive line. The H-bridges are interconnected with each other through eight capacitors ($C_1 - C_8$) and transformers ($T_{r1} - T_{r8}$) which act as self-commutation circuit and compensate the voltage difference between parallel PV generators. The bridge thyristors are controlled so that at first half cycle two diagonally opposite thyristor $T_1 \& T_4$ are forward biased while the other two thyristors $T_2 \& T_3$ are reversed biased and vice versa at the second half cycle. Therefore, the upper and lower capacitors between adjacent PV generators alternatively change their polarities every half cycle. Forced commutation of thyristors is carried out during discharge of coupling condensers.

The coupling transformers are symmetrically linked to each other concerning the power production section center. Their function is to compensate the voltage difference between PV modules; under shading condition, to a current consolidation point v_{st}. The transforms are represented by inductances and mutual coupling with approximately unity coupling factor (k) as given below.

$$\begin{cases} R_m = k \cdot \sqrt{R_{s1} \cdot R_{s2}} \\ L_m = k \cdot \sqrt{L_{s1} . L_{s2}} \quad \& \, 0 < k \leq 1 \end{cases} \tag{1}$$

Where $R_{s1}, R_{s2} \& L_{s1}, L_{s2}$ are the self-resistance and inductance of transformer windings, while R_m, L_m are mutual resistance and inductance of the transformer.

In the first half cycle, the current and voltage equations of two adjacent PV generators in a single stack CCO are given by (2), while the voltage equations at second half cycle are given by (3).

$$\begin{cases} e_1 - 2 \cdot R_s \cdot i_1 - 2 \cdot L_s \cdot \frac{di_1}{dt} - R_m \cdot i_3 - L_m \cdot \frac{di_3}{dt} - R_m \cdot i_5 - L_m \cdot \frac{di_5}{dt} - v_{C1} - v_{st,i} = 0 \\ e_2 - 2 \cdot R_s \cdot i_2 - 2 \cdot L_s \cdot \frac{di_2}{dt} - R_m \cdot i_4 - L_m \cdot \frac{di_4}{dt} - R_m \cdot i_6 - L_m \cdot \frac{di_6}{dt} - v_{C2} - v_{st,i} = 0 \\ i_{st,r} = \sum_{j=1}^{8} i_j \end{cases}$$

$$\tag{2}$$

$$\begin{cases} e_1 - 2 \cdot R_s \cdot i_1 - 2 \cdot L_s \cdot \frac{di_1}{dt} - R_m \cdot i_3 - L_m \cdot \frac{di_3}{dt} - R_m \cdot i_5 - L_m \cdot \frac{di_5}{dt} - v_{C2} - v_{st,i} = 0 \\ e_2 - 2 \cdot R_s \cdot i_2 - 2 \cdot L_s \cdot \frac{di_2}{dt} - R_m \cdot i_4 - L_m \cdot \frac{di_4}{dt} - R_m \cdot i_6 - L_m \cdot \frac{di_6}{dt} - v_{C1} - v_{st,i} = 0 \end{cases}$$

$$\tag{3}$$

Where $i_{st,r}$ is stack current, $v_{st,i}$ is stack voltage, i_j are the currents generated by PV modules, e_j are PV modules voltages and v_{Cj} the voltage across coupling capacitors.

The overall circuit current, voltage and power are given by the following equations:

$$\begin{cases} V_{dc} = \sum_{i=1}^{n} v_{st,i} \\ I_{CCO} = \sum_{r=1}^{m} i_{st,r} \\ P_{dc} = V_{dc} \cdot I_{CCO} \end{cases} \tag{4}$$

Where n and m are the number of stacks connected in series and parallel respectively.

220   A. Refaat and N. Korovkin

2.2 Control of Voltage Source Inverter (VSI)

A current controlled sinusoidal PWM (SPWM) technique is used to control the VSI in d-q synchronous reference frame with phase-locked loop (PLL) to extract the grid voltage vector angle (θ). In order to determine the MPP voltage of the CCOs, the DC link reference voltage (V_{dc}^*) is adjusted by using ordinary perturb and observe (P&O) MPPT technique. The PI controller output of the DC link voltage dynamics is used as the reference of direct current component (I_d^*). The reference command direct current can be written as:

$$I_d^* = K_P\left(V_{dc}^* - V_{dc}\right) + K_I \int \left(V_{dc}^* - V_{dc}\right)dt \tag{5}$$

To design the inner loops controller, it is acceptable to ignore the filter capacitor using the L approximation filter. This is possible because the low frequency behavior of the LCL filter is similar to an L filter as shown in Fig. 3.

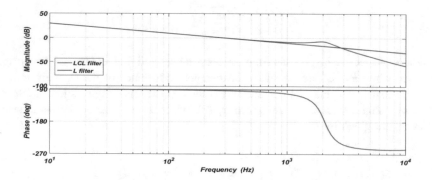

Fig. 3. Bode diagram of line filters

By adopting decoupled control, the command voltage equations by the inverter in d-q reference frame can be described as:

$$\begin{cases} V_d^* = RI_d + V_d - wLI_q + K_P\left(I_d^* - I_d\right) + K_I \int \left(I_d^* - I_d\right)dt \\ V_q^* = RI_q + V_q + wLI_d + K_P\left(I_q^* - I_q\right) + K_I \int \left(I_q^* - I_q\right)dt \end{cases} \tag{6}$$

Where V_d^* and V_q^* are the d-q reference voltages, V_d and V_q are the d-q voltages at point of common coupling (PCC), I_d and I_q are the d-q currents, ($R = R_f + R_{tr}$ & $L = L_f + L_{tr}$) are the equivalent resistance and inductance of the control loop.

In order to control the active and reactive power injected into the grid, the PLL is adjusted so ($V_q = 0$). The power equations can be written as:

$$\begin{cases} P = V_d I_d \\ Q = -V_d I_q \end{cases} \tag{7}$$

Thus, the active and reactive power can be controlled independently by controlling the dq currents. For UPF, it is required to force the reactive power to zero, thus the reference command quadrature current (I_q^*) is set to zero. Finally, command voltages are used to generate the inverter switches trigger pulses based on SPWM technique.

3 Simulation Results

In order to validate the feasibility of the proposed centralize inverter topology, a numerical model of the entire system is simulated for 100 KW PV array farms of 320 modules by using Matlab/Simulink. For the purpose of simulation, a commercial PV module, i.e. SunPower SPR-315E-WHT-D, has been selected. First of all, to demonstrate the effect of partial shading on the PV characteristics let us consider the following two scenarios. The first scenario is a conventional 10 * 32 series-parallel (SP) array with bypass diodes every 5 series modules, while in the second scenario the PV array is stacked through 2 * 4 series-parallel CCOs, which is equivalent to the first scenario. For both scenarios assume that 25% of substrings are illuminated by 0.6 kW/m^2, while the rest PV modules are fully illuminated at a constant temperature of 25 °C. Figure 4 shows the corresponding P-V characteristics of the two scenarios during partial shading and in case of clear sky (note: both scenarios have the same characteristics under a clear sky).

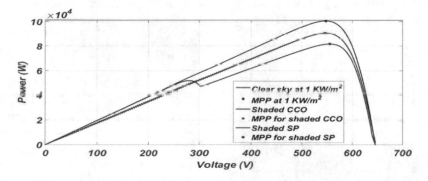

Fig. 4. The P-V characteristics of the two scenarios during partial shading and clear sky.

As can be seen from Fig. 4, in case of conventional SP array with bypass diodes, the PV characteristics exhibit multiple MPPs, which increase the probability of false tracking of MPP. On the other hand, in the case of CCO, the P-V characteristics have a unique MPP which easy to follow by a simple MPPT algorithm. The maximum powers generated in case of clear sky, shaded CCO, and shaded SP array are 100 KW, 90 KW, and 81.5 KW at MPP voltages 547 V, 545 V, and 554 V respectively. This means that 9.4% of expected power is lost in bypass diodes of conventional topologies in this shading situation. In order to investigate the effectiveness of the controllers, a dynamic simulation is done with partial shadow discussed earlier for 25% of substrings as depicted in Fig. 5(a). Figure 5(b, c) show the output power of shaded and unshaded substrings during the simulation run. As can be seen, the irradiation level dropped at 1 s

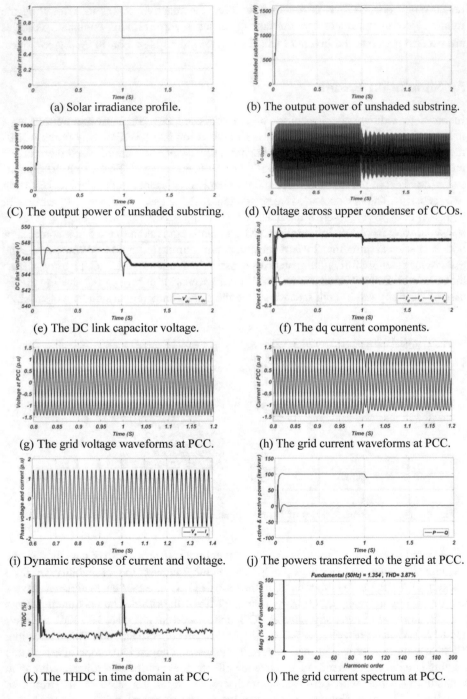

(a) Solar irradiance profile.

(b) The output power of unshaded substring.

(C) The output power of unshaded substring.

(d) Voltage across upper condenser of CCOs.

(e) The DC link capacitor voltage.

(f) The dq current components.

(g) The grid voltage waveforms at PCC.

(h) The grid current waveforms at PCC.

(i) Dynamic response of current and voltage.

(j) The powers transferred to the grid at PCC.

(k) The THDC in time domain at PCC.

(l) The grid current spectrum at PCC.

Fig. 5. Simulation results under transient condition.

and has an immediate effect on the substring output power. The voltage magnitude across coupling condensers of CCOs is dropped to compensate this situation of partial shading and collect the total power from shaded and unshaded substrings at approximately MMP voltage. The voltage across the upper condensers is as shown in Fig. 5(d), while the voltage across the lower condensers is out-of-phase with upper condensers.

The DC link voltage is monitored to verify the operation of the DC link controller. As depicted in Fig. 5(e), the DC link voltage kept tracking the MPP provided by P&O algorithm and decrease from 547 V to 545 V. The dq components of the injected current are shown in Fig. 5(f). The d-component is controlled to follow the new reference value dictated by the DC link controller at 0.9 p.u, which is dropped from 1 p. u during clear sky. The reference for the q-component, however, stayed at zero to maintain UPF operation.

The dynamic behaviors of three-phase grid voltage and grid current at PCC are shown in Fig. 5(g, h), while Fig. 5(i) illustrates the phase angle between phase current and phase voltage at PCC, where the grid current is in phase with the grid voltage (i.e. UPF). The active and reactive powers transferred to the grid at PCC are shown in Fig. 5 (j). The active power is initially at 100 kW before dropping to 90 KW after occurs of partial shading, while the reactive power injected into the grid is kept at zero Kvar. The total harmonic distortion of the current (THDC) at PCC in time domain as well as in frequency domain are shown in Fig. 5(k, l). As can be seen, the THDC less than 5% limit set by IEEE standard 929.

4   Conclusion

This paper proposed a newly centralized inverter topology based on a novel CCO. Computer simulations have been carried out to confirm the topology operation under partial shading conditions. Simulation results show that the proposed topology offers an excellent steady-state response, fast dynamic response, low total THDC, UPF operation, perfect and robust tracking of MPP. Thus, the system performance verified the standard IEEE 929. According to results, the maximum power generated by using CCOs was significantly increased compared to conventional topology with SP array under partial shading pattern. In addition, the CCO did not suffer from misleading power loss due to a unique MPP as well as the circulating currents between parallel PV generators were eliminated through the optimizer.

References

1. Bao, C., Ruan, X., Wang, X., Li, W., Pan, D., Weng, K.: Step-by-step controller design for LCL-type grid-connected inverter with capacitor-current-feedback active-damping. IEEE Trans. Power Electron. **29**(3), 1239–1253 (2014). https://doi.org/10.1109/TPEL.2013. 2262378
2. Refaat, A., Kalas, A., Daoud, A., Bendary, F.: A Control methodology of three phase grid connected PV system. Clemson University PSC 2013 (2013)

3. Hasan, R., Mekhilef, S., Seyedmahmoudian, M., Horan, B.: Grid-connected isolated PV microinverters: a review. Renew. Sustain. Energy Rev. **67**, 1065–1080 (2017). https://doi.org/10.1016/j.rser.2016.09.082

4. Kandemir, E., Cetin, N.S., Borekci, S.: A comprehensive overview of maximum power extraction methods for PV systems. Renew. Sustain. Energy Rev. **78**, 93–112 (2017). https://doi.org/10.1016/j.rser.2017.04.090

5. Uno, M., Kukita, A.: Single-switch voltage equalizer using multistacked buck-boost converters for partially shaded photovoltaic modules. IEEE Trans. Power Electron. **30**(6), 3091–3105 (2015). https://doi.org/10.1109/TPEL.2014.2331456

6. Nimni, Y., Shmilovitz, D.: A returned energy architecture for improved photovoltaic systems efficiency. In: ISCAS 2010, pp. 2191–2194 (2010). https://doi.org/10.1109/iscas.2010.5537199

7. Bergveld, H.J., Büthker, D., Castello, C., Doorn, T., De Jong, A., Van Otten, R., De Waal, K.: Module-level DC/DC conversion for photovoltaic systems: the delta-conversion concept. IEEE Trans. Power Electron. **28**, 4 (2013). https://doi.org/10.1109/tpel.2012.2195331

8. Giral, R., Carrejo, C.E., Vermeersh, M., Saavedra-Montes, A.J., Ramos-Paja, C.A.: PV field distributed maximum power point tracking by means of an active bypass converter. In: ICCEP 2011, pp. 94–98 (2011). https://doi.org/10.1109/iccep.2011.6036360

9. Qin, S., Cady, S.T., Dominguez-Garcia, A.D., Pilawa-Podgurski, R.C.N.: A distributed approach to MPPT for PV sub-module differential power processing. In: ECCE 2013, pp. 2778–2785 (2013)

10. Shenoy, P.S., Kim, K.A., Johnson, B.B., Krein, P.T.: Differential power processing for increased energy production and reliability of photovoltaic systems. IEEE Trans. Power Electron. **28**(6), 2968–2979 (2013). https://doi.org/10.1109/TPEL.2012.2211082

11. Kadri, R., Gaubert, J.P., Champenois, G.: New converter topology to improve performance of photovoltaic power generation system under shading conditions. In: International Conference on Power Engineering, Energy and Electrical Drives, pp. 1–7 (2011). https://doi.org/10.1109/powereng.2011.6036483

12. Du, J., Xu, R., Chen, X., Li, Y., Wu, J.: A novel solar panel optimizer with self-compensation for partial shadow condition. In: APEC 2013, pp. 92–96 (2013). https://doi.org/10.1109/apec.2013.6520190

13. Fernando, L., Villa, L., Ho, T., Crebier, J., Raison, B.: A power electronics equalizer application for partially shaded photovoltaic modules. IEEE Trans. Ind. Electron. **60**(3), 1179–1190 (2013). https://doi.org/10.1109/TIE.2012.2201431

14. Refaat, A., Kalas, A., Daoud, A., Bendary, F.: A control methodology of grid-connected PV system to verify the standard IEEE 929-2000. In: MEPCON 2012 (2012). https://doi.org/10.13140/rg.2.2.35699.84002

Integration of Renewable Energy Sources into Microgrid Considering Operational and Planning Uncertainties

Amir Abdel Menaem[1,2]([✉]) [iD] and Vladislav Oboskalov[1,3] [iD]

[1] Ural Federal University, Yekaterinburg, Russia
[2] Electrical Engineering Department, Mansoura University, Mansoura, Egypt
ashassan@mans.edu.eg
[3] Science & Engineering Center of Ural Branch of RAS, Yekaterinburg, Russia

Abstract. In this paper, a new integration approach for renewable energy sources (RESs), especially intermittent resources (i.e., PV and wind), is presented considering RESs planning uncertainties issues, which are size and location of units, and RESs operational uncertainties issues including operation mode of units and the stochastic nature of prime energy sources (solar irradiance and wind speed). The approach composes of three stages: uncertainties sources modelling, uncertainties expansion, and optimizing uncertainties sources. In the first stage, the nodal active and reactive power uncertainties based on size and operational mode of a renewable power unit are modelled using cumulant technique considering stochastically the solar irradiance, wind speed, and load demands. Secondly, relations of the nodal active and reactive power injections with buses voltage are established. Finally, the particle swarm optimization (PSO) algorithm is utilized to identify the optimal size and operational mode, and the best location and type of RESs for minimizing the buses voltage violations considering technical constraints. The proposed algorithm and analysis on standard 33-bus MG system is implemented using MATLAB. The proposed probabilistic approach by means of cumulant technique reduces the complexity and computational burden of RESs planning problem considering RESs planning and operation uncertainties, which is most important for any researcher in the field of RESs planning and making it very advantageous and practical.

Keywords: Microgrid · PV · Wind · Uncertainty analysis ·
Renewable energy sources · Cumulant technique ·
Probabilistic density function (PDF)

1 Introduction

The increasing deployment of weather-dependent renewable distributed generation (DG), as they are cheap and environmentally friendly energy production is forcing modulations to the planning and operation of distribution networks and developing from passive networks with unidirectional power flow supplied by the transmission grid to active distribution networks with the integration of RESs. The microgrid (MG) has been introduced as an important part of these modifications and considered as the building

© Springer Nature Switzerland AG 2020
V. Murgul and M. Pasetti (Eds.): EMMFT 2018, AISC 982, pp. 225–241, 2020.
https://doi.org/10.1007/978-3-030-19756-8_21

block of active distribution networks to facilitate the RESs integration into the power systems. The integration of RESs has increased the level of power systems uncertainties. Uncertainties can broadly be categorized into input and system uncertainties. Input uncertainties originate due to randomness related with power generation and system load demands, while the system uncertainties arise either from line parameter variations or outage of any of its components which are beyond the scope of this work. These uncertainties impose a large number of regulatory and technical issues that need to be carefully evaluated like frequency [1, 2], power quality [3], voltage [4], power systems dynamic and stability [3, 5], environment [3], and power losses [6]. voltage violation represents one of the major issues regarding RESs integration, especially in the case of MG in which Volt/Var control devices are either not be obtainable to fulfil the required functionality since the MG located at remote or rural areas or will face severe operational challenges as their interaction with the operation of DG units [7].

Therefore, this paper introduces an approach to quantify the influence of uncertain input parameters (Power injections) on the output parameters (buses voltages) in a probabilistic manner and studies how to tackle these uncertainties in the planning stage through finding an optimal integration solution. Since this impact is dependent on size, location, and the operation mode of units (lag-lead-unity power factor) as well as stochastic nature of solar irradiance, wind speed and load demands, the optimum integration aims is to identify the optimal size and operational mode, as well as the best location and type of RESs for minimizing the weighted sum of buses voltage change variances considering various technical aspects. In this context, many published reviews were carried out to do literature review about optimal sizing and placement of DGs. For example, the authors in [8–12] have concentrated on reviewing the optimization approaches utilized in DGs planning regarding decision variables, objectives, and DG type applied constraints. However, in [13–15] the authors have reviewed uncertainty modeling methods for DGs planning to illustrate both the shortcoming and validity of these methods. According to the author's knowledge, there is no study that covers all these random variables of RESs integration into MG system concurrently. This paper is structured as follows. Section 2 presents uncertainty analysis and the probabilistic problem formulation. Section 3 presents the uncertainty modeling of load demand, PV and WT generations. Section 4 introduces the PSO algorithm and solution method. Section 5 presents the case study, the proposed technique verification and discusses the results of the proposed method applied to the test system. Section 6 concludes the paper.

2   Uncertainty Analysis

An uncertainty analysis comprises three steps: uncertainty modeling, uncertainty expansion and uncertainties sources optimization. First, the uncertainty of PV and wind power generation and the stochastic behavior of loads have been considered using the probabilistic model in this work. Cumulant method (CM) combined with the Gram-Charlier Expansion theory, which was proposed in [16–18] is used. CM is executed to model the active and reactive nodal power based on linear relationships of several random input variables, while the output variables have their PDFs reconstructed by using the Gram-Charlier Expansion theory. Second, the uncertainty expansion addresses the issue of expansion the inputs uncertainties to the outputs. The CM is used to relate probabilistic data of load and RESs generation with voltage by providing a linear relationship between the cumulants of input variables (power injections) and those of output variables (buses voltages) through the power flow equations. Application of this method to load flow study is known as probabilistic load flow (PLF) study. The objective of a PLF is to determine the vulnerability of the electric system to the modeled uncertainties, which is called sensitivity analysis (SA), so that risk analysis and mitigation can be performed through optimizing uncertainties sources. The SA here uses the inverse of the power flow Jacobian matrix which can be directly acquired from the Newton-Raphson algorithm [19]. This method has been widely utilized, especially for the power-voltage sensitivity, since the propagation of the Newton-Raphson algorithm and its fast calculation of sensitivities. In probabilistic analysis, the variance or standard deviation is used to describe the result uncertainty by knowing how far the values lie from the mean value. Thus, the variance-based SA aims to understand how the variation of nodal active and reactive injections across different buses of MG affects the variance of the nodal voltage.

2.1   Probabilistic Problem Formulation

In general, the form of the power flow equations for real and reactive bus power injections, and real and reactive line flows are shown in (1) and (2), respectively.

$$
\begin{aligned}
P_i &= V_i \sum\nolimits_{j=1}^{n} V_j (G_{ij} \cos \theta_{ij} + B_{ij} \sin \theta_{ij}), \\
Q_i &= V_i \sum\nolimits_{j=1}^{n} V_j (G_{ij} \sin \theta_{ij} - B_{ij} \cos \theta_{ij})
\end{aligned}
\tag{1}
$$

$$
\begin{aligned}
P_{ij} &= -G_{ij} V_i^2 + V_i V_j (G_{ij} \cos \theta_{ij} + B_{ij} \sin \theta_{ij}, \\
Q_{ij} &= B_{ij} V_i^2 - \frac{B_{ij}}{2} V_i^2 + V_i V_j (G_{ij} \sin \theta_{ij} - B_{ij} \cos \theta_{ij})
\end{aligned}
\tag{2}
$$

Where P_{ij} and Q_{ij} are the active and reactive line flow in line ij. P_i and Q_i are injected active and reactive powers at bus i. θ_{ij} is the difference in voltage angles between bus i and j. V_i and V_j are the bus voltage magnitude at bus i and bus j, respectively. G_{ij} and B_{ij} are the real and imaginary parts of the admittance matrix of line ij, respectively. Let x be the variable vector for bus voltage magnitudes and angles, and y be the variable vector for active and reactive power injections. Equation (1) can be expressed as follows:

$$
y = g(x)
\tag{3}
$$

Where h is the bus power injection function. By using Taylor's series, expanding (3) around the operating point and omitting the second and higher order, the equations are shown as follows,

$$
\Delta x = G^{-1} \Delta y = K \Delta y
\tag{4}
$$

where Δx and Δy are the uncertainty variable vectors of x and y; K is the inverse matrix of G and is referred to as a sensitivity matrix of voltage; G is the jacobian matrix at the operating point and can be expressed as $G = \frac{\partial g(X)}{\partial X}\big|_{x=x_0}$.

Let $Y_1, Y_2, \ldots$, and Y_m be m random variables with known mean value μ_{Y_i} and variance $\sigma_{Y_i}^2$, and X has a linear relationship with the these variables $X = k_1 Y_1 + k_2 Y_2 + \cdots + k_m Y_m, k_1, k_2, \ldots$, and k_m are coefficients for X. The n-th order cumulant of X can be calculated as follow

$$\gamma_X^{(n)} = k_1^n \gamma_{Y_1}^{(n)} + k_2^n \gamma_{Y_2}^{(n)} + \cdots + k_m^n \gamma_{Y_m}^{(n)} \tag{5}$$

Where $\gamma_{Y_1}^{(n)}, \gamma_{Y_2}^{(n)}, \ldots$, and $\gamma_{Y_m}^{(n)}$ are the n-th order cumulant for random variables $Y_1, Y_2, \ldots$, and Y_m. For example, the first cumulant of X which is expected value μ_X and the second cumulant (variance σ_X^2) of X are calculated as,

$$\mu_X = \sum_{i=1}^m k_i \mu_{Y_i},$$
$$\sigma_X^2 = \sum_{i=1}^m k_i^2 \sigma_{Y_i}^2 \tag{6}$$

Based on the cumulants of the input variables $(Y_1, Y_2, \ldots, Y_m)$, the cumulants of output variable (X) are known, and then the next step is to obtain the cumulative distribution function (CDF) and probability density function (PDF) using the Gram-Charlier series expansion theory. The variable X in the expansion should be put to standard formula (mean value of zero and standard deviation of unity). The variable has to be normalized as $X^* = (X - \mu_X)/\sigma_X$, and the cumulants have to be normalized as $(\gamma_X^{(n)})^* = \gamma_X^{(n)}/\sigma_X^n$. In accordance to Gram-Charlier expansion theory, the CDF and the PDF of X can be shown as

$$F(x) = \sum_{i=0}^n \frac{c_i}{i!} \phi^{(i)}(x),$$
$$f(x) = \sum_{i=0}^n \frac{c_i}{i!} \varphi^{(i)}(x) \tag{7}$$

Where $\phi(x)$ and $\varphi(x)$ represent the CDF and PDF of the standard normal distribution, and

$$\varphi(x) = \frac{1}{\sqrt{2\pi}} e^{-x^2/2}$$

The coefficient of the Gram-Charlier Expansion c_i is obtained as a function of central moments. The central moments are the moments about the mean value and are calculated from the cumulants. The details can be found in [16–18].

3 Uncertainty Modelling of Power Injection

3.1 Probabilistic Model of Active Power from PV Generation

The probabilistic model can be founded according to the actual data to expect the distribution of PV active power generation. The authors in [20] illustrated that the active power (P_{PV}) has a linear relationship with solar irradiance as follow,

$$P_{PV} = rA\eta \tag{8}$$

Where r is the solar irradiance; A is the total area of the PV module; η is the PV generation efficiency. Thus, the PDF of PV active power generation is computed as in (9).

$$f_P(P_{PV}) = A\eta \, f_r(r) \tag{9}$$

According to [21], It has been shown that the solar irradiance PDF can be described well by a beta distribution,

$$f_r(r) = \frac{\Gamma(\alpha + \beta)}{\Gamma(\alpha) \cdot \Gamma(\beta)} \cdot (r)^{a-1} \cdot (1 - r)^{b-1} \tag{10}$$

The shape parameters a and b are calculated using the mean and standard deviation of solar irradiance as follows:

$$b = (1 - \mu_s)(\tfrac{\mu_s(1-\mu_s)}{\sigma_s^2} - 1),$$
$$a = \mu_s b/(1 - \mu_s)$$

Where μ_s, σ_s are the mean value and the standard deviation of the Beta distribution, respectively. To get the n-order moments $\alpha_{\Delta r}^{(n)}$ of the beta distribution [22], we evaluate

$$\alpha_{\Delta r}^{(n)} = \frac{\Gamma(a+n)\Gamma(a+b)}{\Gamma(a+b+n)\Gamma(a)} \tag{11}$$

Once the moments are known, the cumulant of the solar radiation are evaluated using Eq. (12).

$$\gamma_{\Delta r}^{(1)} = \alpha_{\Delta r}^{(1)},$$
$$\gamma_{\Delta r}^{(2)} = \alpha_{\Delta r}^{(2)} - (\alpha_{\Delta r}^{(1)})^2, \tag{12}$$
$$\gamma_{\Delta r}^{(3)} = \alpha_{\Delta r}^{(3)} - 3\alpha_{\Delta r}^{(1)}\alpha_{\Delta r}^{(2)} + 2(\alpha_{\Delta r}^{(1)})^3,$$

Based on Eqs. (5, 9), the cumulants of active power generation from PV are known.

3.2   Probabilistic Model of Active Power from WT Generation

It has been familiar that wind speed (v_w) in most parts of the world can be shaped by a Weibull PDF [17], thus the wind speed PDF can be written as:

$$f_v(v_w) = \frac{k}{c} \cdot \left(\frac{v_w}{c}\right)^{k-1} . exp\left[-\left(\frac{v_w}{c}\right)^k\right],$$
$$k = (\sigma_w/\mu_w)^{-1.086},$$
$$c = \frac{\mu_w}{\Gamma(1+1/k)}$$

(13)

where $\Gamma(.)$ is the Gamma function; k is the shape parameter ($k > 0$); c is the scale parameter ($c > 0$) and μ_w, and σ_w be the mean value and the standard deviation of the Weibull distribution, respectively. A piecewise linear function $P_W(v_w)$ is used to estimate the wind generator output under certain values of wind speed v_w, as shown in (14).

$$P_W(v_w) = \begin{cases} P_{w,rated}, & v_{w,rated} \leq v_w < v_{co} \\ \left(\frac{P_{w,rated}}{(v_{w,rated}-v_{ci})}\right)(v_w - v_{ci}), & v_{ci} \leq v_w < v_{w,rated} \\ 0, & otherwise \end{cases}$$

(14)

Where v_{ci}, v_{co}, and $v_{w,rated}$ are the cut-in value, the cut-out value and the rated value, respectively. Then, the PDF of P_w can be expressed in (15) as illustrated in [23].

$$f_P(P_w) = \begin{cases} [1 - (F_v(v_{co}) - F_v(v_{ci}))]\delta(P_w), & for & P_w = 0 \\ \frac{k}{d}\left(\frac{P_w-h}{d}\right)^{k-1}exp\left[-\left(\frac{P_w-h}{d}\right)^k\right], & for & 0 < P_w < P_{w,rated} \\ \left[(F_v(v_{co}) - F_v(v_{w,rated}))\right]\delta(P_w - P_{w,rated}), & for & P_w = P_{w,rated} \\ 0, & for & P_w > P_{w,rated} \end{cases}$$

(15)

Where

$$d = \frac{P_{w,rated}\mu_w}{(v_{w,rated} - v_{ci})\Gamma\left(1+\frac{1}{k}\right)},$$
$$h = -\frac{P_{w,rated}v_{ci}}{v_{w,rated} - v_{ci}}$$

The wind power variation is defined to be $\Delta P_w = P_w - P_{w0}$, where P_{w0} is the deterministic wind power generation. Estimation of n th wind active power moments $\alpha_{\Delta P_w}^{(n)}$ of the wind power variation, ΔP_w is obtained in (16) as presented in [23]. Based on (12), the n-th order cumulants of active power generation from one WT unit ($\gamma_{\Delta P}^{(n)}$) are computed.

$$\alpha_{\Delta P_w}^{(n)} = [1 - (F_v(v_{co}) - F_v(v_{ci}))](-P_{w0})^n + [(F_v(v_{co}) - F_v(v_{w,rated}))](P_{w,rated} - P_{w0})^n$$

$$+ \sum_{x=0}^{n} C_n^x (d)^x (h - P_{w0})^{n-x} \int_{(-\frac{h}{d})^k}^{(\frac{P_{w,rated} - h}{d})^k} \tau^{\frac{x}{k}} e^{-\tau} d\tau \qquad (16)$$

Where $C_n^x = (n!)/(x!(n - x)!)$ and $\int_{(-\frac{h}{d})^k}^{(\frac{P_{w,rated} - h}{d})^k} \tau^{\frac{x}{k}} e^{-\tau} d\tau$ is an incomplete

Γ function.

3.3   Probabilistic Model of Reactive Power from PV and WT Generations

PV and WT based DGs can be operated in various active and reactive power modes. The full converter synchronous machines used in wind farms and the inverter-based PV unit are allowed to absorb or inject reactive power to stabilize bus voltages. The relationship between the active and reactive power of a PV or WT unit at bus i can be expressed as follows:

$$Q_i = u_i P_i \qquad (17)$$

Where $u_i = \pm \tan(\cos^{-1} PF_i)$ or zero, and PF_i is the operating power factor of the PV or WT unit at the bus i. u_i is positive for the unit supplying reactive power (lag PF_i), negative for the unit consuming reactive power (lead PF_i) and zero for the unit doesn't supply or consume reactive power (unity PF_i). Since the cumulants of PV and WT active power generation are known, the reactive power output cumulants can be calculated using (5).

3.4   Probabilistic Model of Loads

The probabilistic models of loads may change for different conditions and customers. Therefore, the load probabilistic model requires to be set up according to the load historical data. There are in literature several prepared models for different load classes. The normal distribution is the most frequent probabilistic model for loads which is known as Gaussian distribution with zero mean and unit variance. In a Gaussian distribution, the PDF peak is at the mean value, therefore, first and second cumulants are only considered and all cumulants of order three and higher are zero [18].

4   PSO Algorithm

As an evolutionary optimization method [21], PSO considers the behavior of or bird flocking or fish schooling. This optimization algorithm provides a population-based exploration procedure in which the possible solutions are represented by many individuals moving in the search space. Each particle p is correlated with two vectors: the position vector $x_p \in \{x_p^1, x_p^2, \ldots, x_p^d\}$, and the velocity vector, $v_p \in \{v_p^1, v_p^2, \ldots, v_p^d\}$ where d is the solution space dimension. The two vectors of each particle are initialized randomly within the ranges. During optimization search process the particles change their position and velocity and try to converge in more promising region of solution. They change stochastically velocity based on the historical best position for the particle itself and the neighborhood best position at each iteration. Through iterations the particles move to an optimal or near optimal solution. The velocity and position of each particle p are updated as follows:

$$v_p^{t+1} = w^t v_p^t + C_1 rand_1 \left(Lbest_p^{1:t} - x_p^t \right) + C_2 rand_2 \left(Gbest^{1:t} - x_p^t \right)$$
$$x_p^{t+1} = x_p^t + v_p^{t+1}$$

$$(18)$$

Where w^t is an inertia weight, which is decreased linearly from 0.5 down to 0.3, in the t-th iteration. In addition, $rand_1$ and $rand_2$ are two uniformly distributed random numbers independently generated between 0 and 1. Learning factors C_1 and C_2 are constants between 0 and 2, which is, $C_1 + C_2 \leq 4$. $Lbest_p^{1:t}$ is the position with the best fitness found so far for the p th particle up to the recent iteration, and $Gbest^{1:t}$ is the best position for population recorded so far up to the recent iteration.

4.1   Objective Function

The objective of the proposed strategy is to minimize the weighted sum of buses voltage magnitude variances caused by renewable energy and demand uncertainties as shown in Eq. (19).

$$Min \sum_i^{N_{buses}} w_i \sigma_{V_i}^2$$

$$(19)$$

Where w_i the weighting factor of bus (i) is defined as follows.

$$w_i = \begin{cases} \left(\mu_{V_i} \right)^2, 1 < \mu_{V_i}(P.u) < V_{max} \\ \left(\frac{1}{\mu_{V_i}} \right)^2, V_{min} < \mu_{V_i}(P.u) < 1 \end{cases}$$

Where μ_{V_i} is the expected voltage value of bus i, and V_{min} and V_{max} are the minimum and maximum voltage limits, respectively. In this manner, the buses away limit (1 p.u) should be set a larger weight in the objective function to prevent over-voltage and undervoltage.

4.2   Constraints

There are two types of constraints, equality constraints and inequality constraints. Equality constraints consist of active and reactive power balance at each bus of the system. Renewable sources (PV and WT generators) are not dispatchable and considered as a negative load. Thus, the net total load has to be equal the power come from main grid in grid connected MG and is defined as

$$P_{main\,grid} = \sum_{i=1}^{buses\,number} (P_i^{PV} + P_i^{WT} - P_i^L) + \sum P_{losses}$$
$$Q_{main\,grid} = \sum_{i=1}^{buses\,number} (Q_i^{PV} + Q_i^{WT} - Q_i^L) + \sum Q_{losses}$$
(20)

While, inequality constraints consist of voltage profile limits, line thermal limit, substation transformer capacity limit, and number of DG limit. Finally, renewable penetration level (RPL) is defined to determine the amount of power coming from main grid as compared to the renewable generators.

$$RPL(\%) = \left(\frac{\sum_{i=1}^{buses\,number} (P_i^{PV} + P_i^{WT})}{\sum_{i=1}^{buses\,number} (P_i^L)} \right) * 100$$
$$RPL \geq RPL_{proposed}$$
(21)

4.3   Solution Strategy

For investigating the proposed methodology, the following assumptions have been made in this paper:

- It is assumed that solar irradiance and wind speed profile are the same to all buses in the system under study.
- More than one type of RES cannot be connected to the same. Due to limited accessible land bus, it is assumed that maximum number of WTs or PV arrays of each size can be allocated at each candidate bus [21] is three.
- For the simplicity, it is assumed that all units connected at any bus in the system operate at the same operation mode.
- Network parameters and configuration are considered constant.

The aim of the proposed solution strategy is to determine the locations, types, numbers of units, and operational modes of renewable DGs which lead to minimizing the weighted sum variance buses voltages. To attain this goal, the PSO is used to solve the optimization problem with mean values and standard deviation of load, solar irradiance and wind speed forecasts to define the optimal type (PV or WT), the number of units required (one, two, three), and operation mode (0.95 lag PF, 0.95 lead PF, unity PF) at each bus. With known the number of units and operating PF, the cumulants of bus active power generation from all PV or WT units is calculated based (5) and then bus reactive power generation cumulants are computed. After that, the cumulants estimation of total bus active and reactive power injections is executed. In the optimization problem, the first and second cumulants only are used. Once detecting the optimal solution, the seven cumulants of the output variables (bus voltages) are calculated. CM here is applied to 7th order cumulant which allows a better approximation to the entire PDF curve. Upon obtained, they are used to establish the output variables PDF and CDF curves. The details of proposed approach are shown in Fig. 1.

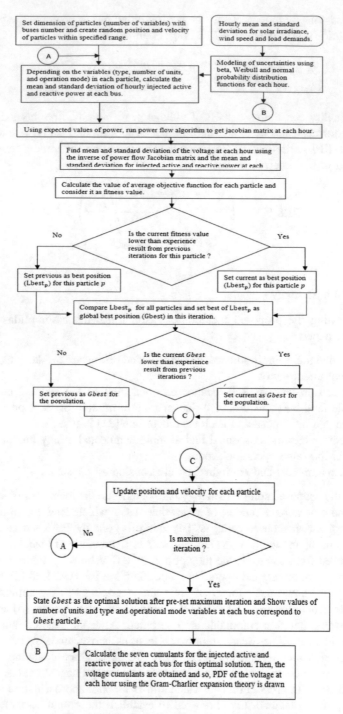

Fig. 1. Flowchart for the proposed strategy.

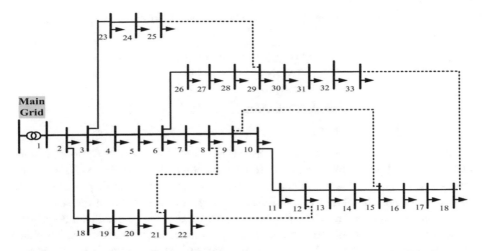

Fig. 2. MG system.

5   Case Study

The proposed technique is tested on an IEEE 33-bus MG system as shown in Fig. 2. which is radial in nature with 32 branches having total load of 3.715 + j2.3 MVA and its data can be found in [24]. The base voltage and base MVA are 12.66 kV and 100 MVA, respectively. The historical data (typical day in each season) for solar irradiance and wind speed of Mansoura city, Egypt, which is located to the north from Cairo, with latitude of 31.037933 and longitude of 31.381523 is incorporated in the study [25]. PV arrays are designed with 1000 PV modules which have 250 kW of installed capacity and WTs employed in the study have maximum capacity of 250 kW. Specification of PV module is illustrated in Table 1 and that of WT used here is shown in Table 2. The 0.95–1.05 p.u voltage magnitude is set to follow as the constraints in the optimization problem. RPL is taken 90%. The proposed methodology is carried out by Intel core i7, personal computer using MATLAB (2017a) software package.

Table 1. Specifications of PV module (polycrystalline photovoltaic panel S60PC, Solartec brand).

	Value
Rated output power	250 W
The panel area	1.62 m^2
The total number of cells connected in the panel	60
The efficiency at STC	15.43%
Number of panels	1000
Standard cell temperature	25 °C

Table 2. Specifications of wind turbine.

	Value
Rated output power	250 kW
Cut-in speed	3 m/s
Rated speed	12 m/s
Cut-out speed	25 m/s

In order to assess the accuracy of the results from CM, Monte Carlo simulation (MCS) with 10,000 samples is utilized as a comparative reference. The MCS technique comprises a recurrent chosen of the input variables value from their probability distribution and then for chosen value of these input getting the values of the state vector exactly in the same way as deterministic analysis. Finally, the probabilistic description of the state output is obtained from the results of the recurrent simulations. To obtain reliable results, thousands of MCS are usually wanted. The active power productions of each PV and WT generators are plotted in Fig. 3 for yearly averaged solar irradiance and wind speed using both MCS and CM methods. There is small difference between PDFs calculated by two methods.

Figure 4 shows the PDF curves of the voltage magnitude at bus-18 in base case (without RESs) using MCS method and CM, and in case of RESs integration in MG network. The bus-18 is chosen since having large voltage deviation. One PV unit is located randomly at buses 16, and 30 and one WT unit at buses 22, and 24 for comparison purposes. All units are operated at unity PF. According to the results, the results of the CM are in close agreement with those of the MCS technique. Moreover, CM is notably faster than MCS and so save computational time since it only needs one run of power flow program to provide the probabilistic information of the all system variables. Thus, CM can be utilized in the optimization problem.

Table 3. Optimal location, type and size of RESs in MG network.

	Type of generation	Number of installed unit	PF setting	Location in network (bus number)
Mixed solar-wind system	PV	3	Lag	2,3,4,5,19,23,24
		2	Lag	25,26
	WT	1	Lag	12,15,18,20,21,22,27,28,29,32,33

In this section of results, optimization results are shown. Table 3 illustrates the optimal solution for PV and WT power integration and Table 4 presents the annual average network performance parameters and average objective function value in comparison with base case in which only load uncertainty is found. As known the RESs integration raises the buses voltage variance but with the optimal integration of solar and wind units this increase is low. In the same time nearly more than 90% of loads are supplied from RESs which denotes drastic reduction of grid dependency with

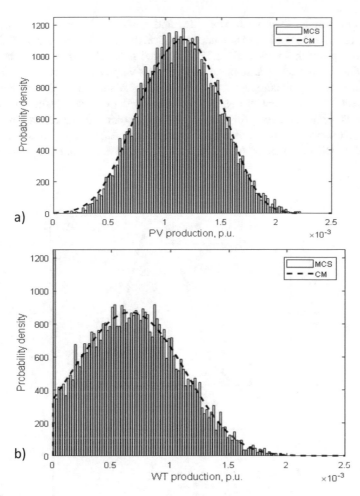

Fig. 3. (a) Hourly PDF curve of one PV generation unit; (b) Hourly PDF curve of one WT generation unit.

Table 4. Comparative study of annual average network performances.

	Objective function (p.u)	Main grid active power expected value (p.u)	Main grid reactive power expected value (p.u)	Total active power losses expected value (p.u)
Base case	6.7465e–05	0.03879	0.02609	0.001647
Mixed solar-wind system	0.00015378	0.003552	0.008599	4.2×10^{-4}

regard of constraints. Optimal mixed solar and wind has better performance indices active and reactive power losses and network security (grid dependency). With the help of proposed approach 74.5% and 71.72% reduced network active and reactive power loss and 90.84% and 54.96% reduced supplied active power and reactive power from the grid is obtained in compared with base case. Also, remarkable voltage profile improvement has been seen in Fig. 5 which shows the expected value of the voltage magnitude at different buses for yearly averaged solar irradiance and wind speed. Moreover, as shown in Fig. 6, the RESs integration reduces the probability of under voltage (V < 0.95) for bus-18 from 100% to nearly 45% in solar-wind system.

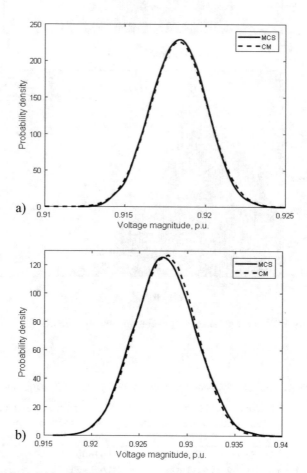

Fig. 4. (a) PDF curves of bus-18 in base case (without RES integration); (b) PDF curves of bus-18 (with RES integration).

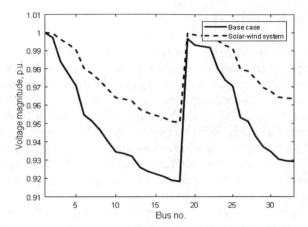

Fig. 5. Expected value of buses voltage in two cases.

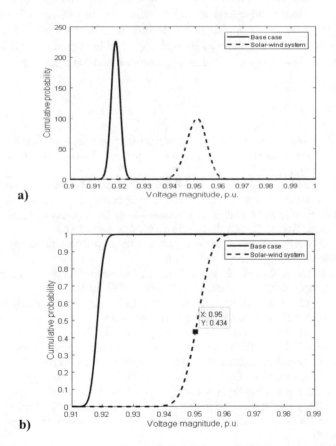

Fig. 6. (a) PDF curves of bus-18 voltage magnitude, and (b) CDF curves of bus-18 voltage magnitude in Base case and Solar-wind system.

6 Conclusions

Due to the probabilistic nature of the formulated MG planning problem results from the planning and operational uncertainties of PVs, and WTs renewable DG and uncertain load consumptions, this paper has proposed a probabilistic approach based on cumulant technique for performing uncertainty analysis. The cumulant method has been used to characterize several sources of uncertainties and relate probabilistic data of load and RESs generation in an optimal RESs-integrated MG with decision variable (voltage). In order to do an optimal RESs integration considering the stochastic bus voltages caused by the relevant uncertainties, the optimization approach (PSO) has been employed to minimize bus voltage uncertainty by finding the optimal type, size, and operation mode of RESs at each bus in respect of technical constraints such as capacity of lines and RESs penetration level. The method was implemented and tested on a 33-bus system. To show the accuracy and efficiency of the cumulant method, the results were compared with those of MCS technique with 10000 samples. The errors were found to be accepted. The simulation results show that optimal hybridization of PVs and WTs regarding size, location and operation mode assures the low increase in buses voltage uncertainties and ensures no system constraints will be violated. Moreover, large enhancement of bus voltage magnitudes and appreciable decrease of grid power consumption are resultant despite variation of resources and load patterns.

References

1. Pandey, S., Mohanty, S., Kishor, N.: A literature survey on load–frequency control for conventional and distribution generation power systems. Renew. Sustain. Energy Rev. **25**, 318–334 (2013)
2. Parmar, K., Majhi, S., Kothari, D.: Load frequency control of a realistic power system with multi-source power generation. Int. J. Electr. Power Energy Syst. **42**, 426–433 (2012)
3. Georgilakis, P.: Technical challenges associated with the integration of wind power into power systems. Renew. Sustain. Energy Rev. **12**, 852–863 (2008)
4. Chen, Z., Blaabjerg, F.: Wind farm—a power source in future power systems. Renew. Sustain. Energy Rev. **13**, 1288–1300 (2009)
5. Ulbig, A., Borsche, T., Andersson, G.: Impact of low rotational inertia on power system stability and operation. IFAC Proc. Vol. **47**, 7290–7297 (2014)
6. Soroudi, A., Ehsan, M.: A possibilistic–probabilistic tool for evaluating the impact of stochastic renewable and controllable power generation on energy losses in distribution networks—a case study. Renew. Sustain. Energy Rev. **15**, 794–800 (2011)
7. Farag, H., Abdelaziz, M., El-Saadany, E.: Voltage and reactive power impacts on successful operation of islanded microgrids. IEEE Trans. Power Syst. **28**, 1716–1727 (2013)
8. Gamarra, G., Guerrero, J.: Computational optimization techniques applied to microgrids planning: a review. Renew. Sustain. Energy Rev. **48**, 413–424 (2015)
9. Ehsan, A., Yang, Q.: Optimal integration and planning of renewable distributed generation in the power distribution networks: a review of analytical techniques. Appl. Energy **210**, 44–59 (2018)
10. Jung, J., Villaran, M.: Optimal planning and design of hybrid renewable energy systems for microgrids. Renew. Sustain. Energy Rev. **75**, 180–191 (2017)

11. Singh, A., Parida, S.: A review on distributed generation allocation and planning in deregulated electricity market. Renew. Sustain. Energy Rev. **82**, 4132–4141 (2018)
12. Fathima, A., Palanisamy, K.: Optimization in microgrids with hybrid energy systems – a review. Renew. Sustain. Energy Rev. **45**, 431–446 (2015)
13. Zubo, R., Mokryani, G., Rajamani, H., Aghaei, J., Niknam, T., Pillai, P.: Operation and planning of distribution networks with integration of renewable distributed generators considering uncertainties: a review. Renew. Sustain. Energy Rev. **72**, 1177–1198 (2017)
14. Talari, S., Shafie-khah, M., Osorio, G., Aghaei, J., Catalao, J.: Stochastic modelling of renewable energy sources from operators' point of-view: a survey. Renew. Sustain. Energy Rev. **81**, 1953–1965 (2018)
15. Aien, M., Hajebrahimi, A., Fotuhi-Firuzabad, M.: A comprehensive review on uncertainty modeling techniques in power system studies. Renew. Sustain. Energy Rev. **57**, 1077–1089 (2017)
16. Fan, M., Vittal, V., Heydt, T., Ayyanar, R.: Probabilistic power flow studies for transmission systems with photovoltaic generation using cumulants. IEEE Trans. Power Syst. **27**, 2251–2261 (2012)
17. Dadkhah, M., Venkatesh, B.: Cumulant based stochastic reactive power planning method for distribution systems with wind generators. IEEE Trans. Power Syst. **27**, 2351–2359 (2012)
18. Schellenberg, A., Rosehart, W., Aguado, J.: Cumulant-based probabilistic optimal power flow (P-OPF) with Gaussian and gamma distributions. IEEE Trans. Power Syst. **20**, 773–781 (2005)
19. Aghatehrani, R., Kavasseri, R.: Reactive power management of a DFIG wind system in microgrids based on voltage sensitivity analysis. IEEE Trans. Sustain. Energy **2**, 451–458 (2011)
20. Liu, S., Liu, P., Wang, X.: Stability analysis of grid-interfacing inverter control in distribution systems with multiple photovoltaic-based distributed generators. IEEE Trans. Ind. Electron. **63**, 7339–7348 (2016)
21. Kayal, P., Chanda, C.: Optimal mix of solar and wind distributed generations considering performance improvement of electrical distribution network. Renew. Energy **75**, 173–186 (2015)
22. Otieno, J.: From the classical beta distribution to generalized beta distributions. Master's thesis, University of Nairobi (2013)
23. Bu, S., Du, S., Wang, H., Chen, Z., Xiao, Z., Li, H.: Probabilistic analysis of small-signal stability of large-scale power systems as affected by penetration of wind generation. IEEE Trans. Power Syst. **27**, 762–770 (2012)
24. Zhang, C., Xu, Y., Dong, Z., Wong, K.: Robust coordination of distributed generation and price-based demand response in microgrids. IEEE Trans. Smart Grid **9**, 4236–4247 (2018)
25. Hassan, A., Kandil, M., Saadawi, M., Saeed, M.: Modified particle swarm optimisation technique for optimal design of small renewable energy system supplying a specific load at Mansoura University. IET Renew. Power Gener. **9**, 474–483 (2015)

Determination of Energy Costs of Wind Farms at All Life Cycle Stages

Pavel Mikheev[ID], Roman Okorokov[✉][ID], Gennady Sidorenko[ID], and Anna Timofeeva[ID]

Peter the Great St. Petersburg Polytechnic University, Polytechnicheskaya 29, 195251 St. Petersburg, Russia
roman_okorokov@spbstu.ru

Abstract. The article presents the results of calculations of energy costs at the life cycle stages of energy facilities (construction, operation and decommissioning) and the total energy costs during their life cycle. Total energy costs are widely used, along with financial costs, to assess the energy efficiency of power facilities for distributed energy resources management, including wind farms. The authors of the article have calculated the total energy costs at the life cycle stages of onshore wind power plants including the energy costs that can be compensated by the recycling and reusing of applied materials. According to the results, the energy costs of the production of wind turbines as well as foundations and cables have the greatest impact on the total energy costs during life cycle of onshore wind power plants, and therefore on their energy efficiency; the lowest energy costs are of the construction works, transportation of wind farm elements and service maintenance. It is also showed how the energy costs are changed with the increase in the installed capacity of wind turbines and the change in the technical characteristics of wind farm elements, in particular, the mass of foundations and cables.

Keywords: Wind power plants · Energy costs · Life cycle analysis · Energy efficiency · Distributed energy resources management

1 Introduction

In the world economy, the principles of resource saving and green production as well as the provision of energy services to the economy and population are generally recognized. The competitiveness of the Russian economy is mainly determined by the innovative solutions that implement the rational use of raw materials and energy resources. The supply of heat, electricity and fuel to the consumers is realized through the energy technologies (including new and innovative), which in turn also consume raw materials and energy resources [1].

At present, in foreign practice, along with financial costs, the energy costs are determined at the stages of life cycle of power facilities (construction, operation and decommissioning) to assess the energy efficiency of life cycles of electric power objects

© Springer Nature Switzerland AG 2020
V. Murgul and M. Pasetti (Eds.): EMMFT 2018, AISC 982, pp. 242–256, 2020.
https://doi.org/10.1007/978-3-030-19756-8_22

including wind power plants (wind farms) [2–4]. In this article, the energy efficiency of the life cycles of energy facilities is estimated on the basis of the energy costs calculation.

The following are taken as criteria for the energy efficiency of power facilities:

- Energy payback period (EPBP);
- Energy return on investment (EROI).

The EPBP shows the period during which the power facility compensates for the energy costs at the stages of life cycle by the energy produced, and the EROI shows how many times the energy produced by the power facility at the operation stage is greater than the energy spent during its life cycle.

The EPBP is defined as the ratio of the energy spent during the life cycle of the power facility to the energy produced at the operation stage per year [2]:

$$EPBP = E_{spent\,(lifecycle)} / E_{gen\,(year)}, \tag{1}$$

where $E_{spent}(\text{lifecycle})$ – energy spent during the all life cycle of the power facility, MWh; $E_{gen(year)}$, – energy produced at the operation stage per year, MWh/yr.

The energy spent during the life cycle of the power facility is equal [1]:

$$E_{spent\,(lifecycle)} = E_{mat} + E_{manuf} + E_{trans} + E_{inst} + E_{om} + E_{eol}, \tag{2}$$

where E_{mat} – energy spent on production of materials, MWh; E_{manuf} – energy spent on production of equipment and building constructions, MWh; E_{trans} – energy spent on transportation of equipment and materials, MWh; E_{inst} – energy spent on construction of the power facility, MWh; E_{om} – energy spent on operation and maintenance of the power facility, MWh; E_{eol} – energy spent on end of life management, MWh.

It should be noted that the recycling and secondary use of materials are considered currently instead of dismantling and utilization at the end of life stage of the power facility. In this case the energy costs that can be compensated by the recycling and secondary use of materials are determined. It is important to emphasize that in case of the energy efficiency assessment of the life cycle of the power facility, the energy costs that can be compensated by the recycling and reusing of applied materials are deducted from the total energy costs during the life cycle of the power facility.

The energy produced at the operation stage per year is equal:

$$E_{gen\,(year)} = E_{agen} - E_{aoper}, \tag{3}$$

where E_{agen} – energy produced per year at the operation stage, MWh/yr; E_{aoper} – energy spent on the own needs of the power facility, MWh/yr.

The EROI is the ratio of the energy produced at the operation stage to the energy spent during the life cycle of the power facility [2]:

$$EROI = E_{gen} / E_{spent(lifecycle)}, \tag{4}$$

where E_{gen} – energy produced at the operation stage, MWh; $E_{spent(lifecycle)}$ – energy expended during the life cycle of the power facility determined by Eq. (2), MWh.

It should be noted that at present, the energy costs during the life cycle of the power facility per kWh of produced energy is also an equally important indicator of the energy efficiency of power objects.

Energy costs during the life cycle per kWh of energy produced are determined as follows [2]:

$$E = E_{spent\,(lifecycle)} / \left(E_{gen\,(year)} \cdot L_{lifecycle} \right), \tag{5}$$

where $E_{spent(lifecycle)}$ – the same as in Eq. (4), MWh; $E_{gen(year)}$ – energy produced during one year determined by Eq. (3), kWh/yr; $L_{lifecycle}$ – life cycle of the power facility, years.

As it follows from the equations above, the energy costs at the stages of life cycle of power facilities can significantly affect the values of the total energy costs during the life cycle and, consequently, the energy efficiency of the life cycles of power facilities. In this regard, the question of how the energy costs are distributed at the stages of life cycle (construction, operation and decommissioning) of wind power plants (wind farms) is of some interest.

2 Materials and Methods

The data provided in the Vestas Wind System A/S (Denmark) research reports, the world's leading manufacturer of wind turbines are used for the energy costs calculation at the stages of life cycle of the considered wind power plants (wind farms) [5–16]. These research reports are devoted to the assessment of energy and environmental efficiency of the life cycles of land-based wind farms.

The reports deal with 12 onshore wind farms (in domestic practice, the term wind power plant (WPP) is more often used) of various rated capacity: 5 wind farms have 50 MW rated capacity, 2 wind farms – 90 MW and 5 wind farms – 100 MW. Wind turbines with a rated capacity of 3.3 MW are more commonly used, 4 out of 12 considered WPP are equipped with them. Wind turbines with a rated capacity of 3.0 MW are used at 2 WPP; wind turbines with a rated capacity of 2.6 MW – at one WPP; wind turbines with a rated capacity of 2.0 MW – at 4 WPPs and wind turbines with a rated capacity of 1.8 MW – at one WPP. The scheme of the onshore WPP is shown in Fig. 1.

The main specifications of the considered WPP and wind turbines are given below in Tables 1 and 2. The distances of transportation of raw materials and components used for the production of wind turbines and WPP elements, transportation of wind turbine and WPP elements to the places of construction and recycling adopted for calculations of energy costs as well as the distances of transportation of maintenance personnel at the operation stage for the WPP maintenance are given in Tables 3 and 4.

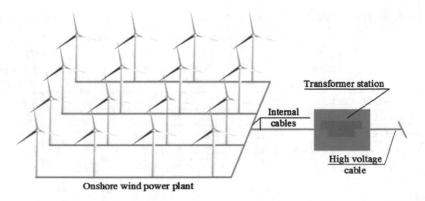

Fig. 1. Scheme of the onshore wind power plant.

Table 1. Specifications of some onshore wind power plants and wind turbines [5–23].

Description		Model, rated power and the number of wind turbines in WPP					
		V100, 1.8 MW, 28 pcs	V80, 2.0 MW, 25 pcs	V90, 2.0 MW, 25 pcs	V100, 2.0 MW, 25 pcs	V110, 2.0 MW, 25 pcs	V100, 2.6 MW 35 pcs
Wind class (IEC)		IIIA	IIIA	IIIA	IIB	IIIA	IIA
Rotor diameter, m		100	80	90	100	110	100
Swept area, m^2		7850	5027	6362	7854	9503	7854
Rotor speed, rpm		9.3–16.6	19.0[1]	9.6–17.0	13.4[1]	13.4[1]	13.4[1]
Power density, m^2/kW		4.36	2.51	3.18	3.93	4.75	3.02
Blade length, m		49	39	44	49	54	49
Tower height, m		80	80	80	80	80	80
Internal cables length (33 kV), km		138	138	138	30	30	35
High voltage cables length (110 kV), km		20	20	20	20	20	20
Mass, tons	Wind turbine	251	272	248	230	248	259
	Foundation	796	1166	789	849	963	1041
	Cables	1682	1682	1682	557	511	476

It should be noted that the distances of transportation of raw materials and components for the production of wind turbines and WPP elements and the distances of transportation of WPP elements to the places of recycling in the energy costs determination are taken the same [5–16].

Table 2. Specifications of some onshore wind power plants and wind turbines [5–23].

Description		Model, rated power and the number of wind turbines in WPP					
		V90, 3.0 MW, 30 pcs	V112, 3.0 MW, 33 pcs	V105, 3.3 MW, 30 pcs	V112, 3.3 MW, 30 pcs	V117, 3.3 MW, 30 pcs	V126, 3.3 MW, 30 pcs
Wind class (IEC)		IIA	IIIA	IA	IIA	IIA	IIIA
Rotor diameter, m		90	112	105	112	117	126
Swept area, m^2		6362	9852	8659	9852	10751	12469
Rotor speed, rpm		9.9–18.4	6.2–17.7	6.2–17.7	6.2–17.7	6.2–17.7	5.3–16.5
Power density, m^2/kW		2.12	3.28	2.60	2.99	3.26	3.80
Blade length, m		44	55	55	55	57	62
Tower height, m		80	84	72.5	84	91.5	117
Internal cables length (33 kV), km		30	32.45	30	30	30	30
High voltage cables length (110 kV), km		20	20	20	50	20	20
Mass, tons	Wind turbine	256	332	358	342	409	496
	Foundation	1041	947	1042	861	1206	1424
	Cables	590	630	590	590	590	590

Notes on Tables 1, 2:

1. Manufacturer of wind turbines is Vestas Wind System A/S (Denmark).
2. In all considered wind turbines, except for the V112 (3.0 MW) the asynchronous type generators are used, in the V112 (3.0 MW) the synchronous type generator is used.
3. All considered wind turbine towers have conical shape; the steel with the subsequent protecting coating is used for their production.
4. The fiberglass, carbon fiber and epoxy resin are used for blades production.
5. The aluminum (as a conductor), copper and polymeric materials are used for cable production.
6. WPP internal cables including wind turbines V100 (1.8 MW), V80 and V90 (2.0 MW) have section 120 mm^2, 300 mm^2 and 500 mm^2; for the rest of WPP cables cross-sections are 95 mm^2, 240 mm^2 and 630 mm^2.
7. High voltage cables have cross-section 630 mm^2 for all considered WPP.
8. The maximum speed of rotation is specified.
9. The wind turbine class is given in accordance with the IEC international classification.
10. More detailed specifications of the considered WPP and wind turbines including mass of individual elements (rotor, blades, nacelle, tower, etc.) are presented in [5–23].

According to the statistical data [4–15], the life cycle analysis (LCA) technique is used to determine the energy costs taking into account the requirements of the international standards ISO 14044 [24] and ISO 14040 [25]. The GaBi modeling and reporting software [26] is used in the energy costs determination at the stages of life cycle of the considered WPP [5–23]. During the LCA the life cycle of the power facility is divided into the stages: construction, operation, decommissioning. The scheme of LCA carrying out on the example of the wind turbine is shown in Fig. 2.

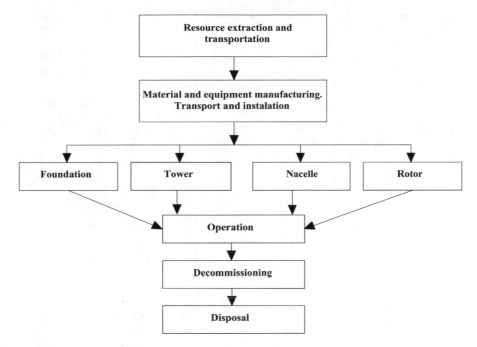

Fig. 2. LCA model of the wind turbine.

At the same time, the research reports used as a statistical base for the study do not consider the utilization, but the recycling of WPP and wind turbine elements, and determine the amount of energy that can be compensated by the recycling and secondary use of the materials. In this case, the LCA scheme is as follows below (Fig. 3).

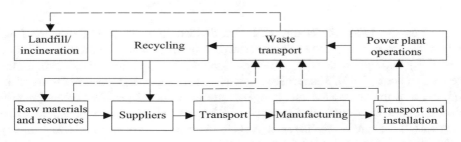

Fig. 3. LCA model with the recycling of used materials.

Table 3. Transportation distances of raw materials, WPP and wind turbine elements at the construction stage [5–16].

Model	Rated capacity MW	The number of wind turbines in WPP	Transportation of raw materials and components for production of WPP and wind turbine elements				Transportation of WPP and wind turbine elements to the site of construction		
			Transportation of raw materials		Transportation of components		WPP element	Type of transport	Distance km
			Type of transport	Distance km	Type of transport	Distance km			
V100	1.8	28	Truck	600[1]/50[2]	Truck	600	Nacelle	Truck	1025
V80	2.0	25					Rotor	Truck	1025
V90	2.0	25					Blades	Truck	600
V100	2.6	35					Tower	Truck	1100
V90	3.0	30							
V105	3.3	30						Ship	8050
V112	3.3	30							
V117	3.3	30					Other parts	Truck	600
V126	3.3	30							
V100	2.0	25	Truck	600[1]/50[2]	Truck	600	Nacelle	Truck	785
								Ship	8575
							Rotor	Truck	785
								Ship	8575
							Blades	Truck	2200
V110	2.0	25						Ship	1570
							Tower	Truck	2065
								Ship	2125
							Other parts	Truck	600
V112	3.0	33	n/a	n/a	n/a	n/a	Nacelle	Truck	1000
							Rotor	Truck	1000
							Blades	Truck	1000
							Tower	Truck	700
							Other parts	n/a	n/a

Notes:
1. Transportation distances of raw materials for the production of WPP and wind turbine elements except the raw materials for the construction of foundation.
2. Transportation distances of raw materials for the construction of foundation.

During the LCA, the energy costs are divided into the energy received from renewable resources or renewable raw materials (Primary energy from renewable raw materials) and non-renewable resources (Primary energy from resources) in the research reports [5–16]. The results of calculations of the energy costs at the stages of life cycle of the considered energy objects are given in GJ per kWh of energy produced. The energy costs at the stages of life cycle and for the production of the elements of the considered energy objects are given as a percentage of the total energy costs, the numerical values of the energy costs are not presented.

Table 4. Distances of service staff transportation at the operation stage and transportation of WPP and wind turbine elements to the place of recycling at the decommissioning stage [5–16].

Model	Rated capacity, MW	The number of wind turbines in WPP	Distance of WPP service staff transportation at the operation stage, km		Distance of transportation of WPP and wind turbine elements to the place of recycling	
			Per year	During the operation period	WPP element	Distance, km
V100	1.8	28	2160	43200	Nacelle	200
V80	2.0	25				
V90	2.0	25			Rotor	
V100	2.6	35				
V90	3.0	30			Blades	
V105	3.3	30				
V112	3.3	30			Tower	
V117	3.3	30				
V126	3.3	30			Foundation	50
V100	2.0	25	1500	30000	Nacelle	200
					Rotor	
V110	2.0	25			Blades	
					Tower	
					Foundation	50
V112	3.0	33	900	18000	Nacelle	n/a
					Rotor	
					Blades	
					Tower	
					Foundation	

The authors of the article have calculated the total energy costs (energy obtained from renewable and non-renewable resources) at the stages of the WPP life cycle based on the statistical data [5–22] of the energy costs per kWh of produced energy, the energy produced during the operation stage, the distribution of energy costs during the life cycle and the production of WPP elements (given as a percentage of the energy costs during the life cycle). The total energy costs are obtained in numerical form and include the energy costs for the production of wind turbines, foundations and cables. Also, the numerical values of the energy costs have been calculated which can be compensated by the recycling and secondary use of materials.

3 Results and Discussion

The results of calculations of the energy costs at the stages of life cycle conducted by the authors of the article are shown in Tables 5, 6, 7 and 8 as well as the total values of energy costs during the life cycle of the onshore WPP.

As follows from Tables 5, 6, 7 and 8, the energy costs at the construction stage are more than 90% of the total energy costs during the life cycle of the onshore WPP, while the energy costs for the production of wind turbines are from 49.9 to 81.5%, 4.3 ÷ 27.6% for cables, 7.9 ÷ 12.65% for foundations. The energy costs of the production of wind turbines, foundations and cables are more than 80% of the total energy costs during the life cycle of the WPP. The construction works and transportation account for 4.65 ÷ 9.7% of the total energy costs as well as 5.6 ÷ 8.8% for the service maintenance during the life cycle of the WPP. The energy costs that can be compensated by the recycling and secondary use of materials are from 17 to 26% of the total energy costs during the life cycle of the WPP. It is important to emphasize that the energy costs are negligible when the option of dismantling and recycling of the onshore WPP and wind turbine elements is considered at the stage of decommissioning. Thus, these energy costs make up 2 ÷ 3% of the total energy costs during the life cycle of power facilities according to the data presented in [5–16], i.e., they do not have practically any impact on the energy efficiency of power objects.

According to the results, it can be noted that the energy costs of the production of wind turbines as well as the production of foundations and cables are of great influence on the values of the total energy costs during the life cycle of the onshore WPP, and therefore on their energy efficiency. The energy costs of the construction works, transportation of the elements of power facilities to the construction site (even with a significant increase in the transportation distances), and service maintenance have the least influence. In this regard, the question is of certain interest how the energy costs are changed with the increase in the installed capacity of wind turbines and with the change in the technical characteristics of WPP elements, in particular, the mass of foundations and cables.

Table 5. Energy costs at the stages of life cycle of the onshore WPP.

Components of energy costs	Model, rated capacity and the number of wind turbines in WPP					
	V100, 1.8 MW, 28 pcs		V80, 2.0 MW, 25 pcs		V90, 2.0 MW, 25 pcs	
	MWh	%	MWh	%	MWh	%
Wind turbines	123817.4	55.25	104757.5	49.9	97826.25	49.15
Foundations	19817.0	8.85	22045.0	10.5	17381.25	8.75
Cables	57607.2	25.7	58050.0	27.6	55228.75	27.75
Construction works and transportation	10430.0	4.65	11481.25	5.45	18090.0	9.1
Total at stage of construction	211671.6	9 4.45	196333.75	93.45	188526.25	94.75

(*continued*)

Table 5. (*continued*)

Components of energy costs	Model, rated capacity and the number of wind turbines in WPP					
	V100, 1.8 MW, 28 pcs		V80, 2.0 MW, 25 pcs		V90, 2.0 MW, 25 pcs	
	MWh	%	MWh	%	MWh	%
Service maintenance	12516.0	5.55	13777.5	6.55	10425.0	5.25
Total at stages of construction and operation	224187.6	100.0	210111.25	100.0	198951.25	100.0
Decommissioning with recycling of materials[1]	59451.0	26.0	52815.0	25.0	50388.75	25.0
Total with recycling and secondary use of materials	164736.6	-	157296.25	-	148562.5	-

Table 6. Energy costs at the stages of life cycle of the onshore WPP.

Components of energy costs	Model, rated capacity and the number of wind turbines in WPP					
	V100, 2.0 MW, 25 pcs		V110, 2.0 MW, 25 pcs		V100, 2.6 MW, 35 pcs	
	MWh	%	MWh	%	MWh	%
Wind turbines	99151.25	65.3	107905.0	68.45	161255.5	64.5
Foundations	11897.5	7.9	12397.5	7.9	31654.0	12.65
Cables	17263.75	11.4	16286.25	10.35	16184.0	6.5
Construction works and transportation	11665.0	7.7	10507.5	6.65	22099.0	8.85
Total at stage of construction	139977.5	92.3	147096.25	93.35	231192.5	92.5
Service maintenance	11665.0	7.7	10507.5	6.65	18698.75	7.5
Total at stages of construction and operation	151642.5	100.0	157603.75	100.0	249891.25	100.0
Decommissioning with recycling of materials[1]	34995.0	23.0	31520.0	20.0	35698.25	15.0
Total with recycling and secondary use of materials	116647.5	-	126083.75	-	214193.0	-

Table 7. Energy costs at the stages of life cycle of the onshore WPP.

Components of energy costs	Model, rated capacity and the number of wind turbines in WPP					
	V90, 3.0 MW, 30 pcs		V112, 3.0 MW, 33 pcs		V105, 3.3 MW, 30 pcs	
	MWh	%	MWh	%	MWh	%
Wind turbines	138652.5	61.3	166859.55	65.15	201789.0	74.4
Foundations	28365.0	12.5	28119.3	11.0	21943.5	8.1
Cables	17500.5	7.7	18781.95	7.35	25110.0	9.2
Construction works and transportation	21888.0	9.7	19971.6	7.8	22621.5	8.3
Total at stage of construction	206406.0	91.2	233732.4	91.3	271464.0	100.0
Service maintenance	19899.0	8.8	22218.9	8.7	-	-
Total at stages of construction and operation	226305.0	100.0	255951.3	100.0	271464.0	100.0
Decommissioning with recycling of materials[1]	37989.0	17.0	52272.0	20.0	45243.0	17.0
Total with recycling and secondary use of materials	188316.0	-	203679.3	-	226221.0	20.0

Table 8. Energy costs at the stages of life cycle of the onshore WPP.

Components of energy costs	Model, rated capacity and the number of wind turbines in WPP					
	V112, 3.3 MW, 30 pcs		V117, 3.3 MW, 30 pcs		V126, 3.3 MW, 30 pcs	
	MWh	%	MWh	%	MWh	%
Wind turbines	176604.0	74.65	209742.0	79.0	233416.5	81.5
Foundations	20893.5	8.85	21238.5	8.0	22732.5	7.95
Cables	19315.5	8.15	14091.0	5.3	12351.0	4.3
Construction works and transportation	19711.5	8.35	20422.5	7.7	17901.0	6.25
Total at stage of construction	236524.5	100.0	265494.0	100.0	286401.0	100.0
Service maintenance	-	-	-	-	-	-
Total at stages of construction and operation	236524.5	100.0	265494.0	100.0	286401.0	100.0
Decommissioning with recycling of materials[1]	39420.0	17.0	61260.0	23.0	71601.0	25.0
Total with recycling and secondary use of materials	197104.5	-	204234.0	-	214800.0	-

Notes on Tables 5, 6, 7 and 8:

3. Energy costs that can be compensated by the recycling and secondary use of materials.
4. Energy costs for the WPP maintenance are negligible for the V105, V112 (3.3 MW), V117 and V126 wind turbine models in accordance with the data [5–16].
5. Tables 5, 6, 7 and 8 show the total energy costs for the element production of the energy facilities and the stages of their life cycle without splitting into energy derived from renewable resources (Primary energy from renewable raw materials) and non-renewable resources (Primary energy from resources).

The calculations of the energy costs of the production of wind turbines that are part of the considered WPP and their foundations have been made by the authors of the article. The dependencies of the energy costs on the installed capacity of wind turbines as well as the mass of foundations and cables have been built (Figs. 4, 5 and 6) on the basis of these calculations and data of the technical characteristics of WPP elements (the results given in Tables 5, 6, 7 and 8 are used).

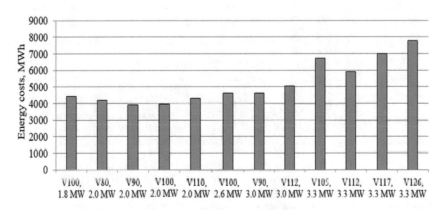

Fig. 4. Energy costs of the wind turbine production.

As can be seen from Fig. 4, the energy costs of the production of wind turbines that equal in the installed capacity and similar in specifications (Tables 1 and 2) are close in values or they have a small spread of values (V90, 2.0 MW and V100, 2.0 MW wind turbines) and, vice versa, the energy costs have a large spread of values when the wind turbines differ considerably in terms of the installed capacity (V90 and V126 wind turbines) and the technical characteristics of WPP elements.

As it follows from Fig. 5, the energy costs for the construction of foundations built in similar engineering and geological conditions have a small spread of values and, vice versa, they have a large spread of values in the case of consideration of foundations significantly different in mass. This is due to the fact that these foundations are built in different engineering conditions. In addition, they have different loads due to differences in the technical characteristics of wind turbines.

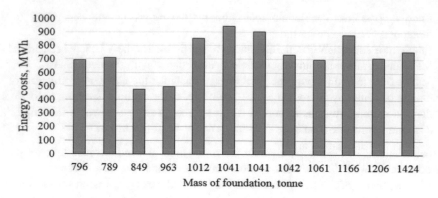

Fig. 5. Energy costs of the construction of wind turbine foundations.

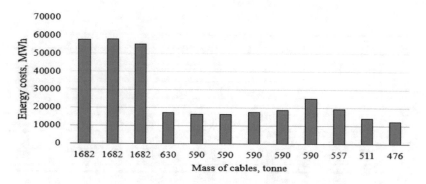

Fig. 6. Energy costs of the cable production.

The energy costs for the production of cables that equal in mass are close in values (as can be seen from Fig. 6), and vice versa. The jump in energy costs is caused by the fact that the cables for the considered WPP significantly differ (almost three times) in their mass (a large range of energy costs for the production of cables is caused by the same).

4   Conclusions

According to the results of work, the following conclusions can be drawn on the calculations of energy costs at the stages of life cycle of the onshore WPP:

6. The stage of construction accounts for more than 90% of the energy costs during the life cycle of the WPP while the energy costs for the production of wind turbines, cables and the construction of foundations account for more than 80% of the total energy costs during their life cycle. The lowest costs of energy (less than 10%) during the life cycle fall on the stage of WPP operation.

7. The energy costs of the wind turbine production account for 49.9 ÷ 81.5% of the total costs of energy over the WPP life cycle, cable production – from 4.3 to 27.6%, foundation construction – from 7.9 to 12.65% respectively. The construction works and transportation account for 4.65 ÷ 9.7%, the service maintenance – from 5.6 to 8.8% of the total energy costs during the WPP life cycle. The energy costs that can be compensated by the recycling and secondary use of materials are from 17 to 26% of the total energy costs during the WPP life cycle.

8. The energy costs of the production of wind turbines as well as the production of foundations and cables are of great influence on the values of the total energy costs during the life cycle of WPP, therefore, on their efficiency, distributed energy resources management and renewable energy commercialization. The energy costs of the construction works, transportation of the WPP elements to the construction site, and service maintenance have the least impact.

9. The energy costs of the production of wind turbines, foundations and cables that are similar in their technical characteristics and construction conditions (in considering the foundations) are close in values or have a small spread of values; conversely, they have a large spread of values for the wind turbines, foundations and cables significantly different in their technical characteristics.

References

1. Okorokov, R., Timofeeva, A.: Prospects of electric heating applying in Russian cities to ensure their sustainable development. In: MATEC Web of Conferences, vol. 170, p. 03004 (2018)
2. Sidorenko, G.I., Mikheev, P.Yu.: Otsenka energeticheskoy effektivnosti zhiznennykh tsiklov energeticheskikh ob'yektov na osnove VIE. Al'ternativnaya energetika i ekologiya **1–3**, 101–110 (2017)
3. Sidorenko, G.I., Mikheev, P.Yu.: Otsenka energeticheskoy i ekologicheskoy effektivnosti zhiznennykh tsiklov energeticheskikh tekhnologiy vozobnovlyayemoy energetiki. In: XXI International Scientific Conference Proceedings, pp. 251–259. Ekoonis Publishing, Moscow (2016)
4. Bezrukikh, P.P.: Effektivnost' vozobnovlyayemoy energetiki, mify i fakty. Vestnik agrarnoy nauki Dona **1**, 5–17 (2015)
5. Garrett, P., Ronde, K.: Life cycle assessment of electricity production from an onshore V100–1.8 MW wind plant. Vestas Wind Systems A/S, Denmark (2011)
6. Garret, P., Ronde, K.: Life cycle assessment of electricity production from a V80 – 2.0 MW Gridstreamer wind plant. Vestas Wind Systems A/S, Denmark (2011)
7. Garrett, P., Ronde, K.: Life cycle assessment of electricity production from a V90 – 2.0 MW Gridstreamer wind plant. Vestas Wind Systems A/S, Denmark (2011)
8. Razdan, P., Garrett, P.: Life cycle assessment of electricity production from an onshore V100–2.0 MW wind plant. Vestas Wind Systems A/S, Denmark (2015)
9. Garrett, P., Ronde, K.: Life cycle assessment of electricity production from an onshore V110–2.0 MW wind plant. Vestas Wind Systems A/S, Denmark (2015)
10. Garrett, P., Ronde, K.: Life cycle assessment of electricity production from an onshore V100–2.6 MW wind plant. Vestas Wind Systems A/S, Denmark (2013)

11. Garrett, P., Ronde, K.: Life cycle assessment of electricity production from an onshore V90–3.0 MW wind plant. Vestas Wind Systems A/S, Denmark (2013)
12. Souza, N., Shonfield, P.: Life cycle assessment of electricity production from a V112 turbine wind plant. Vestas Wind Systems A/S, Denmark (2011)
13. Garrett, P., Ronde, K.: Life cycle assessment of electricity production from an onshore V105–3.3 MW wind plant. Vestas Wind Systems A/S, Denmark (2014)
14. Garrett, P., Ronde, K.: Life cycle assessment of electricity production from an onshore V112–3.3 MW wind plant. Vestas Wind Systems A/S, Denmark (2015)
15. Garrett, P., Ronde, K.: Life cycle assessment of electricity production from an onshore V117–3.3 MW wind plant. Vestas Wind Systems A/S, Denmark (2014)
16. Garrett, P., Ronde, K.: Life cycle assessment of electricity production from an onshore V126–3.3 MW wind plant. Vestas Wind Systems A/S, Denmark (2014)
17. General Specification V100–1.8/2.0 MW. Document no.: 0004–3053 V04. Vestas Wind Systems A/S, Denmark (2010)
18. General Specification V80–1.8/2.0 MW. Document no.: 944411 R3. Vestas Wind Systems A/S, Denmark (2002)
19. General Specification V90–1.8/2.0 MW. Document no.: 0004–6207 V05. Vestas Wind Systems A/S, Denmark (2010)
20. General Specification V90–3.0 MW. Document no.: 950010.R1. Vestas Wind Systems A/S, Denmark (2004)
21. General Specification V112–3.0 MW. Document no.: 0011–9181 V03. Vestas Wind Systems A/S, Denmark (2010)
22. General Specification V117–3.3 MW. Document no.: 0035-1209 V02. Vestas Wind Systems A/S, Denmark (2013)
23. Catalog of Wind Turbine 3MW Platform. Document no.: 10/2015-EN. Vestas Wind Systems A/S, Denmark (2015)
24. ISO 14044. Environmental Management – Life Cycle Assessment – Requirements and Guidelines. International Organization for Standardization, Switzerland (2006)
25. ISO 14040. Environmental Management – Life Cycle Assessment – Principles and Framework. International Organization for Standardization, Switzerland (2006)
26. GaBi Product Sustainability Solution. http://www.gabi-software.com. Accessed 11 Mar 2019

Combining Phytoremediation Technologies of Soil Cleanup and Biofuel Production

Elena Elizareva[1,2]([✉]) [iD], Yulay Yanbaev[3] [iD], Nina Redkina[1] [iD],
Natalya Kudashkina[4] [iD], and Alexey Elizaryev[2] [iD]

[1] Bashkir State University, 32 Validy Street, Ufa 450076, Russia
Elizareva_en@mail.ru
[2] Ufa State Aviation University, 12 K. Marx Street, Ufa 450000, Russia
[3] Bashkir State Agrarian University, 34 50-ya Oktyabrya Street,
Ufa 450001, Russia
[4] Bashkir State Medical University, 3 Lenina Street, Ufa 450000, Russia

Abstract. The urgency of obtaining alternative fuels from renewable raw materials of plant origin has been substantiated. The possibility of combining the production of biofuels from rape with its use for phytoremediation of soils is considered. The contemporary state, problems and prospects for the development of phytoremediation of soils contaminated with heavy metals are analyzed. The main phytoremediation technologies have been characterized, such as rhizofiltration and phytofiltration, phytoextraction, phytostabilization, and phytoevaporation. It is shown that the tolerance index, translocation and bioconcentration factors are used in order to assess the effectiveness of using various plants for phytoremediation. Using these indicators, a logical function of selecting plants for phytoremediation was compiled depending on the utilization method of the generated biomass. The ability of rape to absorb metals was studied. It has been established that rape accumulates heavy metals mainly in the roots under conditions of high soil contamination level, which makes it possible to use it for plant stabilization. The feasibility of combining soil phytostabilization by using rape plant and biofuel production with careful control of the heavy metals content in it is shown.

Keywords: Bioethanol · Biodiesel · Phytoremediation · Rhizofiltration · Phytofiltration · Phytoextraction · Phytostabilization · Phytoevaporation

1 Introduction

In recent years, alternative fuels from renewable raw materials of plant origin found a wide spreading due to the price increase for traditional motor fuels, the tightening of exhaust emission standards, as well as due to the depletion of world oil reserves and the limitation of carbon dioxide emissions. At the present time, the following biofuels are most widely used: bioethanol, which is a product of processing almost any biomass, and biodiesel – a product of oil distillation obtained from oilseeds or animal fats (for rapeseed oil, these are methyl esters of fatty acids). Fuel production from renewable raw materials (continuous mass of plants and other organic matter from oils to wood) is a

© Springer Nature Switzerland AG 2020
V. Murgul and M. Pasetti (Eds.): EMMFT 2018, AISC 982, pp. 257–266, 2020.
https://doi.org/10.1007/978-3-030-19756-8_23

biocatalytic cracking. During the process, the splitting of long-chain hydrocarbons to the desired fraction of oil (diesel) fuel occurs, which is also promising in modern state [1]. Rape plant is considered the best raw material for biofuels due to its high yield and the possibility of almost waste-free use. Moreover, this culture can be cultivated in different soil and climatic zones. This article discusses the possibility of combining the production of biofuels from rape and using it for phytoremediation of soils contaminated with heavy metals. Pollution of the environment by heavy metal ions is a great danger to the biosphere. In addition to the direct toxic effects on living and plant organisms, heavy metals tend to accumulate in food chains, which increases their danger to humans. Once in the reservoirs, they are in the most dangerous ionic form for a long time, and even in a bound state (colloidal form, bottom sediments or other poorly soluble compounds) they continue to pose a potential threat for a long period of time. The increased content of heavy metals in the body leads to diseases of the cardiovascular system and causes severe allergies. In addition, heavy metals have embryotropic properties and are carcinogenic [1].

Analysis of widely used methods for removing heavy metals from such natural objects as soil cover and water bodies shows that they are associated with the formation of a large amount of toxic sludge, are expensive and difficult to perform. Therefore, searching and developing methods to extract ecotoxicants without additional burden on the environment is extremely relevant. The undoubted priority for environmental and economic efficiency is recognized by the phytoremediation method [2], which is the technology of soil and industrial wastewater purification using natural and genetically modified plants. The term is formed by a combination of two Latin words "phyto" - plant and "remedium" - to purify, restore [3]. Phytoremediation includes 4 main approaches and, is accordingly subdivided into 4 technologies: rhizofiltration and phytofiltration, phytoextraction, phytostabilization and phytoevaporation [4].

Rhizofiltration means wastewater passing through rhizofiltration facilities with hydroponically grown higher land plants. Long, fibrous and dense, covered with hairs, the root system of such plants absorbs, concentrates or precipitates heavy metals [5]. According to the same principles, wastewater is treated from heavy metal ions using higher aquatic plants (macrophytes); this method is called phytofiltration [6].

Phytofiltration of wastewater can be carried out in two ways:

(1) In the phytofiltration system by passing a stream of sewage with adjustable pH, temperature and speed through aquariums with growing macrophytes [7].
(2) The use of so-called botanical sites, which refers to a wide range of watercourses, overgrown with macrophytes in a natural way or artificially planted with them. As a rule, these are swampy areas with slower flow rates on the way to larger water bodies [8, 9].

As macrophytes become saturated with heavy metals, contaminated biomass (all or above water level) is removed or mowed.

Phytoextraction is the cultivation of specially selected plant species in contaminated areas for a certain period of time to extract heavy metals from the soil by the root system and maximize their concentration in the aboveground biomass [10, 11].

Phytostabilization (or phytorecovery) is physical and chemical immobilization of pollutants due to their sorption on roots and chemical fixation with the help of various soil additives for stabilizing toxic substances and preventing their spread by wind and water erosion. It also allows reducing the vertical migration of pollutants into groundwater. It can be used as a temporary strategy to reduce environmental risks until the selection of the most appropriate treatment technology.

Phytoevaporation is the process of adsorbing metals such as mercury and selenium from the soil by the plants, their biological transformation into a gaseous form inside the plant and their release into the atmosphere. The purification effect is due to the fact that the gaseous form of these metals is much less toxic, for example, for selenium, the toxicity decreases by 500–600 times [12, 13]. Despite the additional benefits (minimal changes in the surface being cleaned, minimal need for maintenance after planting, preventing erosion processes, no need to dispose the plant biomass), using phytoevaporation, unlike other phytoremediation technologies, makes it impossible to control the migration of pollutants entering the environment during the process. Therefore, phytoevaporation is the most controversial phytoremediation technology. Additional characteristics of phytoremediation technologies are shown in Table 1.

From Table 1 it follows that each specific phytoremediation technology involves the use of plants with certain properties. Thus, according to Baker's theory, 3 following groups of plants are distinguished by the mechanisms of metal extraction: accumulators, indicators and excludors [14]. In accumulators, the extraction of metals by roots and their transport to aerial parts are balanced, while in excludors, which do not have the ability to regulate the extraction of metals, the transport to shoots is limited. Thus, accumulators extract a large amount of metals and transport them to the aboveground part in a logarithmic relation between the metal concentration in the soil and the metals concentration in the shoots.

In addition, Baker introduced the term of hyper-accumulator, the assignment criterion for which is the following metal content in the aerial part: more than 100 mg of Cd/kg, 1000 mg of Ni or Cu/kg, more than 10,000 mg of Zn or Mn/kg of dry weight [7]. In excludors, the concentration of metals in shoots is small and constant over a wide range of soil metal concentrations until a certain value is reached, above which unlimited transport appears. Indicators reflect the concentration of metal in the soil. Thus, indicators can be effectively used for monitoring pollution with heavy metals, accumulators for phytoextraction processes, and excludors for phytostabilization. Plants used for rhizofiltration or phytofiltration processes should have properties opposite to those of accumulators, they should accumulate metals in the roots.

In order to assess the effectiveness of using various plants for phytoremediation, the following indicators are used: tolerance index, translocation and bioconcentration factors, the characteristics of which are presented in Table 2.

Table 1. Characteristics of phytoremediation technologies.

Heading level	Plant selection	Cleaning object
Rhizofiltration and phytofiltration	(1) For phytofiltration - a combination of various macrophyte types (floating, partially or completely submerged) for cleaning all layers of the water flow [14] (2) For rhizofiltration - land plants that create an extremely large contact surface with the medium being cleaned due to an extensive root system. Such plants also must be tolerant to metals [15] Selecting the pants according to translocation properties is determined by the method of disposal of the resulting contaminated biomass	Rhizofiltration is especially effective for the remediation of large wastewater volumes containing relatively low concentrations of various heavy metals Thus, for example, according to [16], regulatory wastewater treatment can be achieved by rhizofiltration if the initial concentration of heavy metals docs not exceed 20 MCL for copper, 5... 6 MCL for zinc and cadmium, 2 MCL for manganese and cobalt
Phytoextraction	Tolerance to high concentrations of metals. The ability to absorb and accumulate high concentrations of several metals or their particular forms simultaneously. Efficient translocation of metals from the root system to aboveground biomass [17]. High growth rate, large biomass, deeply growing root system High resistance to plant diseases and pests. The ability to grow with the use of conventional farming	Applicable for the remediation of large land areas, the contamination of which does not extend to great depths. In addition, high concentrations of metals can be lethal to plants, so the contamination degree should be low or medium [18] The soil surface should be free from obstacles such as fallen trees or stones, and be characterized by topography that allows the use of agricultural machinery
Phytostabilization	The plants or substances they secrete should have the ability to stabilize pollutants in the soil by binding them with lignin on the cell wall ("lignification"), absorption by soil humus using plant or microbial enzymes ("humification"), binding by organic substances or using other mechanisms [19]. Absence (or low level) of translocation of pollutants from root biomass to aboveground. High growth rate, dense aboveground and root biomass, tolerance to metals	Most effective for fine soils with high organic content. Most often used for large areas with low or medium pollution
Phytoevaporation	Some macrophytes have a good ability to convert easily volatile metals into gaseous form. In addition, it seems quite effective to use woody vegetation with a developed root system, long life expectancy and intensive production of bedding from fallen leaves, which contributes to the availability of metals in the soil. Genetically modified plants are used for phytoevaporation of mercury, developing such plants for phytoevaporation of arsenic is in progress [20]	It is recommended to use this technology far from populated areas and in places with meteorological conditions that facilitate rapid decomposition of volatile substances [21, 22]

Table 2. Indicators for assessing the performance of various plants for phytoremediation.

Performance indicators	Formula	Gradation
Tolerance to metals	Tolerance index: $TI = \frac{M_{Me}}{M_c} \times 100\%$, where M_{Me} is the dry weight of a plant biomass grown with the addition of metals, g; M_c is the dry weight of a control plant biomass grown in Hoagland solution, g The tolerance index is calculated for shoots (STI), roots (RTI) and the whole biomass in total (BTI)	TI > 100% - stimulating effect; TI = 100% - no effect; TI < 100% - the inhibitory effect of the analyzed concentrations of heavy metals on plant growth; TI = 50% is the minimum desirable volume of biomass when grown on a polluted environment [23]
Metal translocation inside the plant	Translocation factor: $TF = \frac{C_s}{C_r}$, where C_s is the concentration of metal in shoots, mg/g; C_r - metal concentration in the roots, mg/g where is the concentration of metal in shoots, mg/g;	The value of TF < 1 indicates the accumulation of metals mainly in the roots, TF > 1 – in the shoots
The ability to accumulate metals (individually and mixed)	Bioconcentration factor: $BCF = \frac{C_{pl}}{C_{sol}}$, where C_{pl} is the concentration of the metal in the plant, mg/g; C_{sol}- concentration of metal in solution, mg/l Bioconcentration factor is calculated separately for shoots (BCF$_s$) and roots (BCF$_r$)	A BCF > 1000 value is a criterion for classifying a plant as a proper accumulator [24]

Using the indicators given in Table 2, the phytoremediation potential of rapeseed on real soils adjacent to the territory of metallurgical enterprises was investigated.

2 Materials and Methods

Soils adjacent to the territory of three following enterprises were selected for the research: Karabashmed (at a 1.5 km distance from the plant), Satkinskiy chuguno-plavil'nyy zavod (at a 2 km distance from the plant) and Uchalinskiy gorno-obogatitel'nyy kombinat (1.2 km from the plant). Soil samples were taken using the "envelope" method from the upper humus-containing horizon. A sample taken on the territory of the winter garden of the Bashkir State Agrarian University (sample Ufa) was chosen as a background soil test.

Rape was grown on selected soils in a greenhouse for 90 days. Mineral fertilizers in the form of nitroammophoska were applied to all variants of the experiment (including the control ones) at the rate of N120P120K120.

The grown biomass was weighed in each experiment. In the experiment with soils from Satka, seed germination did not occur. It should be noted that in the experiment with soils from Karabash, the mass of the aboveground part of the grown plants was significantly less than in the experiments with the Uchaly and Ufa soils. This is probably due to the high contamination level of these soils with heavy metals.

The concentrations of heavy metals (manganese, iron, copper, zinc) in the soil and plant samples were determined by atomic absorption spectrometry (AAS). The metal ion content was calculated in milligrams per kilogram of dry weight (mg/kg).

3   Results

The heavy metals (HM) content in soils before and after growing plants is shown in Table 3.

Table 3. Gross concentrations of HM in the studied soil samples before and after growing plants and approximate permissible concentrations (APC), mg/kg.

Location of soil collection	Fe		Zn		Cu		Mn	
	Before	After	Before	After	Before	After	Before	After
Ufa	22484	12016	30	31	24	15	590	385
Uchaly	33438	33625	393	330	637	181	742	770
Karabash	48750	25437	3388	2551	10720	2829	423	505
Satka	29938	–	59	–	74	–	854	–
APC	25000[a]		55		33		1500	

Notes: [a]Percentage abundance in the crust; the excess of the APC is highlighted in the table.

It can be seen from Table 3 that the content of heavy metals in the background sample (Ufa) in all respects does not exceed the approximate permissible concentration of gross forms. The manganese content does not exceed the standards in all soils.

Comparing the gross concentrations of iron, zinc and copper before and after the cultivation of rape shows that the HM concentration in soil samples taken in the city of Uchaly and Karabash did not reach APC values. The copper content in soils decreased the most, 1.5 times in Ufa, 3.5 times in Uchaly, 3.8 times in Karabash. The concentration of zinc also decreased: 1.2 times in Uchaly, 1.3 times in Karabash, but practically did not change in Ufa. The iron content fell 1.8 times in the soils of Ufa and Karabash, but remained on the same level in Uchaly. The manganese content in the soil of Ufa decreased by 1.5 times, but slightly increased in the soils from Uchaly and Karabash. Seeds had been sown in the soil of Satka didn't grow. The dried rapeseed biomass grown in the soil of Karbash was not enough for analysis. Probably, the level of soil contamination with heavy metals turned out to be too high.

The content of HM in rapeseed is presented in Table 4. The calculation results of the translocation and bioconcentration factors are shown in Table 5.

Table 4. The accumulation of heavy metals by rape seed.

Location of soil collection	Plant part	Content of dry matter, mg/kg			
		Fe	Zn	Cu	Mn
Ufa	Leaves	430	25.38	3.78	23.7
	Roots	603	28.6	2.61	11.2
Uchaly	Leaves	247	154	9.34	29
	Roots	768	298	16.9	32

Table 5. Performance indicators of using the various plants for phytoremediation.

Indicators	Ufa				Uchaly			
	Fe	Zn	Cu	Mn	Fe	Zn	Cu	Mn
TF	0.71	0.89	**1.45**	**2.12**	0.32	0.52	0.55	0.91
BCF_s	0.02	0.86	0.16	0.04	0.01	0.39	0.01	0.04
BCF_r	0.03	0.97	0.11	0.02	0.02	0.76	0.03	0.04

According to the data obtained, rape is prone to translocation of copper and manganese to shoots with a low HM content in soil. Zinc, iron and cadmium mainly accumulate in the roots. With a higher level of soil contamination, all metals accumulate in the underground part of the plant.

4 Discussion

Based on the data obtained, the options for the rape use for phytoremediation and options for the possible use of the resulting biomass are considered. Using the indicators given in Table 2, let us set the efficiency criterion for using one or another plant for the purposes of phytoremediation in the form of a logical function $E = f(BTI, TF, BCF_s, BCF_r)$ shown in Fig. 1.

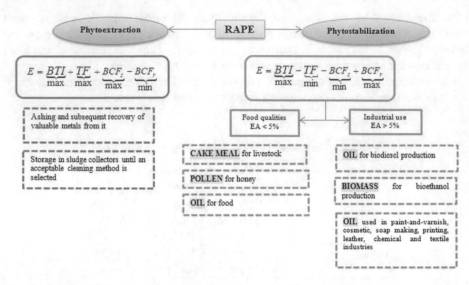

Fig. 1. Efficiency function of phytoremediation technology, depending on the method of biomass utilization.

The indicator included in the formula with the sign "+" should tend to the maximum, with the sign "−" to the minimum. It should be noted that E function is not a mathematical relationship and is only qualitative. Nevertheless, this approach may be useful in substantiating the choice of a specific plant for a particular phytoremediation method with the most suitable method of utilizing the resulting biomass.

One of the main effectiveness components of any environmental protection technology is zero waste. Phytoremediation technologies lead to the formation of biomass contaminated with heavy metals, so the possibility of its utilization is an important task. The definition of further biomass use after phytostabilization for food or technical purposes depends on the content of erucic acid (EA) in rape seeds, which belongs to omega-9-unsaturated fatty acids. Rape with an erucic acid content of less than 5% is used for food purposes such as the production of oils, margarines, mayonnaise, confectionery fats, etc. When an erucic acid content is more than 5%, it is used for technical purposes like soap making, production of fuel and lubricants, plastics, in the paint-and-varnish, metallurgical, printing, cosmetic industry, in the production of biodiesel.

According to the "Energy Strategy of Russia for the period until 2030" the expansion of the production and use of new fuel types derived from various biomass types is one of the priorities of scientific and technological progress in the energy sector [25]. In this regard, the option of obtaining biofuels from rapeseed is preferred.

The following advantages of biofuel obtained from rape should be noted:

– almost complete biodegradability;
– eco-friendly production;
– reduction of emissions of hydrocarbons, soot, nitrogen oxides due to lowering the combustion temperature;

- extremely low content of sulfur compounds;
- the absence of polycyclic aromatic hydrocarbon-carcinogens in the exhaust gases of the engine;
- adaptability to transportation and storage at the gas stations;
- renewability.

The results of domestic studies and the experience of foreign firms indicate that mixed biofuels based on rapeseed oil or rapeseed biodiesel help to save oil fuels, improve the environmental performance of diesel engines and solve a number of social problems when using rape processing by-products [1, 17]. However, it should be noted that the distribution of heavy metals in the organs of agricultural crops is selective and decreases in the following order: leaves > stalks > roots > fruit coat > seeds. Despite this, biofuel obtained from rapeseed used for phytostabilization of soils is necessary to be carefully checked for the HM content, in order to prevent secondary pollution.

5 Conclusions

The results of experiments on cultivating the rapeseed on real soils contaminated with HM and selected in the influence zone of metallurgical production testify to the effectiveness of using rapeseed for phytostabilization.

An algorithm for selecting the most efficient method of using rapeseed biomass is proposed. The biomass is formed during the process of soil phytostabilization, as a logical function that takes into account the values of the tolerance index, translocation factor, bioconcentration factor and erucic acid content. It is shown that one of the most promising areas of biomass utilization is the production of biofuels subject to strict control of heavy metals content in it.

Acknowledgements. We gratefully acknowledge financial support from the RFFI № 17-44-020574 grant.

References

1. Kashevarov, N.I., Osipova, G.M., Danilov, V.P.: Rape - a source of environmentally-friendly fuel. Siberian J. Agric. Sci. **3**, 89–97 (2008)
2. Elizareva, E.N.: Toxic effects of heavy metals. In: Current Issues of University Science Proceedings, pp. 110–120. BSU, Ufa (2016)
3. Ol'shanskaya, L.N., Sobgajda, N.A., Russkix, M.L., Valiev, R.Sh., Aref'eva, O.A.: Phytoremediation energy-saving technologies in solving the problems of hydrosphere pollution. Innov. Theor. Pract. **2**(9), 166–172 (2012)
4. Pulford, I.D., Watson, C.: Phytoremediation of heavy metal-contaminated land by trees - a review. Environ. Int. **29**, 529–540 (2003)
5. Tarushkina, Y.A., Ol'shanskaya, L.N., Mecheva, O.E., Lazutkina, A.S.: Higher aquatic plants for wastewater treatment. Ecol. Ind. Russ. **5**, 36–39 (2006)
6. Vlyssides, A., Barampouti, E.M., Mai, S.: Heavy metal removal from water resources using the aquatic plant Apium nodiflorum. Commun. Soil Sci. Plant Anal. **36**, 1–7 (2005)

7. Kamal, M., Ghaly, A.E., Mahmoud, N., Côté, R.: Phytoaccumulation of heavy metals by aquatic plants. Environ. Int. **29**, 1029–1039 (2004)
8. Constructed Wetlands for Industrial Wastewater Treatment and Removal of Nutrients. https://www.researchgate.net/publication/313059365_Constructed_Wetlands_for_ Industrial_Wastewater_Treatment_and_Removal_of_Nutrients. Accessed 21 Jan 2019
9. Paz-Alberto, A.M., Sigua, G.C.: Phytoremediation: a green technology to remove environmental pollutants. Am. J. Clim. Change **2**, 71–86 (2013)
10. Elizaryev, A.N.: Assessment of anthropogenic impact on hydro-ecological regime of water bodies. In: Abstract of Ph.D. thesis, Sankt-Peterburg (2007)
11. Marchiol, L., Assolari, S., Sacco, P., Zerbi, G.: Phytoextraction of heavy metals by canola (Brassica napus) and radish (Raphanus sativus) grown on multicontaminated soil. Environ. Pollut. **132**, 21–27 (2004)
12. Schnoor, J.L.: Phytoremediation of soil and groundwater. Technology evaluation report, Ground-Water Remediation Technologies Analysis Center, Pennsylvania (2002)
13. Lasat, M.M.: Phytoextraction of metals from contaminated soil: a review of plant/soil/metal interaction and assessment of pertinent agronomic issues. J. Hazard. Subst. Res. **2**, 5–25 (2000)
14. Cao, A., Cappai, G., Carucci, A., Muntoni, A.: Selection of plants for zinc and lead phytoremediation. J. Environ. Sci. Health **4**(39), 1011–1024 (2004)
15. Arthur, E.L., et al.: Phytoremediation - an overview. Crit. Rev. Plant Sci. **24**, 109–122 (2005)
16. Qu, R.L., Li, D., Du, R., Qu, R.: Lead uptake by roots of four turfgrass species in hydroponic cultures. HortScience **38**(4), 623–626 (2003)
17. Gomes, H.I.: Phytoremediation for bioenergy: challenges and opportunities. Environ. Technol. Rev. **1**(1), 59–66 (2012)
18. Kim, I.S., Kang, K.H., Johnson-Green, P., Lee, E.J.: Investigation of heavy metal accumulation in Polygonum thunbergii for phytoextraction. Environ. Pollut. **126**, 235–243 (2003)
19. Robinson, B., Fernández, J.E., Madejón, P., Marañón, J.E., Murillo, J.M., Green, S., Clothier, B.: Phytoextraction: an assessment of biogeochemical and economic viability. Plant Soil **249**, 117–125 (2003)
20. Lasat, M.M.: Phytoextraction of toxic metals: a review of biological mechanisms. J. Environ. Qual. **31**, 109–120 (2002)
21. Suresh, B., Ravishankar, G.A.: Phytoremediation – a novel and promising approach for environmental clean-up. Crit. Rev. Biotechnol. **24**(2–3), 97–124 (2004)
22. Afanasev, I., Volkova, T., Elizaryev, A., Longobardi, A.: Analysis of interpolation methods to map the long-term annual precipitation spatial variability for the Republic of Bashkortostan, Russian Federation. WSEAS Trans. Environ. Dev. **10**(1), 405–416 (2014)
23. Ali, N.A., Berna, M.P., Ater, M.: Tolerance and bioaccumulation of cadmium by Phragmites Australis grown in the presence of elevated concentrations of cadmium, copper, and zinc. Aquat. Bot. **80**, 163–176 (2004)
24. Kvesitadze, G.I. et al.: Metabolism of anthropogenic toxicants in higher plants. Science, Moscow (2005)
25. Energy Strategy of Russia for the period up to 2030. http://www.energystrategy.ru/projects/ es-2030.htm. Accessed 20 Dec 2018

Model of Application of Alternate Energy Sources When Reloading Fossil Fuels

Igor Zub⑩, Victor Shchemelev⑩, and Yuri Ezhov$^{(\boxtimes)}$⑩

Admiral Makarov State University of Maritime and Inland Shipping,
Dvinskaya Street 5/7, 198035 Saint-Petersburg, Russia
ezhovye@gumrf.ru

Abstract. The demand for energy is constantly growing: analysts forecast that it will increase by 50% by 2040. New deposits being discovered and the elaboration of new methods of hydrocarbons extraction will allow using these resources for a long time. Nevertheless, most countries have begun to think about using solar power. This interest is caused not only by environmental problems, but also by the fact that annually 178.000 TW of solar radiation reaches our planet. To use solar energy, two methods – thermal and photovoltaic – are used. The first one involves heating up used liquid, the second one implies the conversion of sunlight into electrical energy. Containers for storing fossil fuels are installed at oil terminals. This hydrocarbon can be crude or refined oil in a form of heavy (mazut) or light (diesel) fuel. Solar energy can be used at oil terminals both for heating hydrocarbons (in order to reduce viscosity to operating parameters), and for storing electrical power. The use of solar energy will improve energy efficiency and the environmental component of a terminal. The paper presents the functioning of an oil product heating system as a sequence of events based on cause-effect relations. For a formal description of the logical structure of processes of solar energy use, the authors proposed a solution using the Petri net machine. Prerequisites for the use of Petri nets are: the presence of random and deterministic components in the process of generation of solar energy; the possibility of representing the conversion of solar energy in the form of sets of technologically homogeneous parallel processes; situational control related to the randomness of the possibility of converting solar energy. With the help of the model and by using the Petri net, conflict situations were found that could be solved by situational control. As practice shows, such kind of local situations is solved within the framework of automated control.

Keywords: Oil loading terminals · Solar panels · Environmental ecology · Energy efficiency · Modeling · Petri nets

1 Introduction

Analysts forecast that by 2040 the demand for energy will grow by almost 50% [1]. At the same time, the consumption of solar power will increase. The total power of solar radiation reaching our planet within one year is 178,000 TW (terawatts) (1,78 * 1014), which is 15,000 times more than all power consumed by Earth population per year. The Sun transmits to the Earth such an amount of energy during a couple of hours that the

© Springer Nature Switzerland AG 2020
V. Murgul and M. Pasetti (Eds.): EMMFT 2018, AISC 982, pp. 267–277, 2020.
https://doi.org/10.1007/978-3-030-19756-8_24

population uses for a whole year [2]. The energy illumination of solar radiation is usually measured in W/m2. In 2017, 140 MW of renewable energy sources (RES) were introduced in Russia, of which more than 100 MW are solar power plants [3].

In the fuel-and-power sector, there is a decrease in the production of power generated with the use of hydrocarbon fuels. This is (among other issues) due to the reduction of solar power cost owing to the growth of the efficiency of solar modules, which entails a steady increase in solar energy consumption [4]. However, today the growth of solar energy production is possible only with the support of the government [5].

The use of solar panels (SP) as an alternative source of energy that is contained in power plants providing the work of port equipment allows not only saving hydrocarbon resources, improving energy efficiency, reducing harmful emissions from electrical and thermal plants, but also improving environmental safety. The use of solar power contributes to reducing the potential danger of emergency situations, decreasing operating costs due to the lack of mechanical devices at a SP. Another positive factor is the reduction of losses during the transportation of electrical power, since SPs are installed directly at or nearby the consumer.

There are two ways of using solar power: thermal (SP_t) and photovoltaic (SP_f). In the first case, the working liquid is heated by means of mirror solar installations; in the second case, sunlight is converted into electrical power by means of semiconductor materials.

The choice of the method of obtaining solar power can be determined using a model that has an economic effect as the criterion.

The use of solar energy is carried out using SP based on semiconductor photo-voltaic elements. When using SP_t, power is obtained directly from a solar panel.

200–300 watts of power can be obtained from 1 m^2. The energy generation in the SP E_{sb} is determined by the formula [6]:

$$E_{sb} = Ins \times P_{sp} \times \eta / P_{ins} \tag{1}$$

where Ins is insolation per square meter, kW $\times$ h, Psp is the nominal power of a SP, W, η is the total efficiency of electric current transmission through wires, Pins is the maximum insolation power per square meter of the earth's surface, kWh/m2.

The intensity of direct solar radiation Iβ reaching a SP, located at any angle β to the horizon during a clear day will be equal to [7]:

$$I_\beta = I_M \times \cos \xi, \tag{2}$$

where ξ is the angle between the direction to the Sun and the normal to the bevel face, IM is the intensity of the direct radiation reaching the Earth.

Investment costs for installation of solar panels are determined by the formula [8]:

$$C_{inv} = \frac{NIns_{stc.} Cm^2}{Ins_f Cm^2} + C_{tr} + C_{iw} \tag{3}$$

where N is the required power of solar panels, W; Ins_f – insolation for the studied region, W/m^2; $Ins_{st\ c}$ - insolation for standard conditions, W/m^2; Nm^2 - power of solar panels from 1 m^2, W/m^2; Cm^2 - price of 1 m^2 of solar panels, rub.; C_{tr} - the cost of transportation, rub.; C_{iw} - the cost of installation work, rub.

The annual economic effect is determined by the formula:

$$E_{SP} = C_{red} - C_{inv} \frac{r\%}{100\%} - \frac{C_{inv}}{P_d} - C_{main} - T_{pr} - C_{re} \qquad (4)$$

where C_{red} – reduced production costs per unit of a product using the basic version of energy supply, rubles; C_{main} – maintenance costs, rubles; T_{pr} – property tax, rubles; C_{re} – costs for retraining employees, rubles; P_d – depreciation period, years; r – the period of operation.

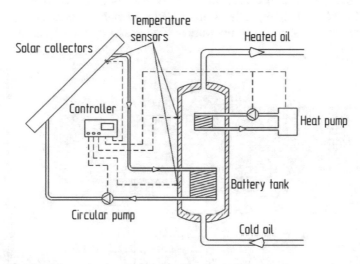

Fig. 1. Schematic circuit of the system for heating petroleum products.

The choice of *SP* is a complex both economic and technical problem, the solution of which can be found with the use of the model. To pay off investments, it is necessary to maximize the efficiency of the energy complex of an oil terminal (E_{fOT}): $E_{fOT} \rightarrow max$.

The following symbols should be introduced:

W – an option of the use of solar power:

$$W = \{SP_t, SP_f\} \qquad (5)$$

W depends on the type of fossil fuel cargo being reloaded. For the natural gas loading terminal, *SPs* will be used for energy accumulation and solar collectors (*SC*) for generating thermal energy. At oil terminals, both thermal and photovoltaic *SPs* can be used. A circuit diagram of the production and use of thermal energy is shown in Fig. 1. To generate electricity at oil terminals, *SPs* based on thin-film silicon can be used. Such *SPs* have a wide spectrum of absorbed solar radiation [2]. Mechanical flexibility can be attributed to the peculiarities of such SPs, so they can be installed on a roof or upper zones of an oil storage, which allows maximizing the area of a SP, or install them on free areas of a terminal (Fig. 2). Effective power (P) of a SP depends on the generated voltage: $P = U_{max}I_{max}$. Under certain illumination, the maximum power take-off ($P_{max} = U_{max} \cdot I_{max}$) from a SP occurs only if the resistance of the external load satisfies the relation $R_{\mu} = U_{max}/I_{max}$. Under constant load, a change in illumination of a working surface of a panel leads to a mismatch between a SP and an external load, and the power take-off will be lower than the maximum possible values.

SP_t is used to heat liquid petroleum products to a temperature when the pumped hydrocarbon load reaches viscosity (v_{opt}), for which the optimal performance of the pump P_{pump} is calculated and which depends on the type of oil product being loaded. Light and dark petroleum products are reloaded at terminals in accordance with GOST 32511-2013 (EN 590: 2009) "Diesel fuel EURO. Technical conditions" (entered into force as the national standard of the Russian Federation on January 1, 2015).

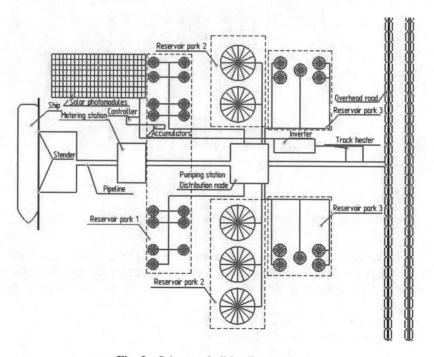

Fig. 2. Scheme of oil loading terminal.

The elaboration of a simulation algorithm, as the formalization of the process of functioning of the production system, must be preceded by the formalization of the structure of states of the system itself. The process of using solar energy to generate thermal energy used for heating the oil product to the required viscosity is considered as a structurally organized set of subsystems (solar collector, solar panels and external power supply) interconnected by control loops. To optimize the functioning of the oil product heating system, it is advisable to distinguish the following subsystems: the heating loop subsystem (HL) and the oil product subsystem itself (OP), formulating optimization models and their implementation algorithms for each of them. In order to formalize the technical aspects, we introduce the following symbols:

Parameters of the heating circuit:

- P_{SC} - power of a solar collector, kW;
- S_{acc} - accumulator area, m^2;
- V_{wl} - working liquid volume, m^3;
- t_{wl} - working liquid temperature, °C;
- c_{wl} - specific heat capacity of the working liquid, kcal/(kg × °C);
- λ - heat conductivity coefficient, (m × h × °C);
- h - heat transfer coefficient, kcal/(m^2 × h × °C);
- $grad\ c$ - heat transfer gradient;
- R - region of SP installation;
- D_s - number of sunny days per year in the region of SP installation.

The following model is considered:

$$HL = \{P_{SC}, S_{akk}, V_{wl}, t_{wl}, c_{wl}, \lambda, h, gradc, R, D_s\}. \qquad (6)$$

The following symbols are introduced for the petroleum product subsystem:

- c_{oil} – specific heat capacity of hydrocarbon, kcal/(kg × °C);
- V_{oil} – volume of fossil fuel cargo in storage, m^3;
- t_{oil} – working fluid temperature, °C. For t_{wl}, a limit is set: $t_{min} < t_{oil} < t_{cav}$, where t_{min} is the minimum temperature specified in the characteristics of the pump, t_{cav} is the temperature at which cavitation appears;
- t_{env} – temperature of environment or liquid, °C;
- ρ – the density of hydrocarbons, kg/m^3;
- a – thermal diffusivity, m^2/s, $a = \lambda/c\rho$.

$$OP = \{c_{oil}, V_{oil}, t_{oil}, t_{env}, \rho, a\}. \qquad (7)$$

To implement the project on solar power production (PSPP), it is necessary to consider the model represented by the system:

$$\Pi\Gamma C\Theta = \begin{cases} P_{CK}, S_{akk}, V_{wl}, t_{wl}, c_{wl}, \lambda, h, graadc, R, D_s \\ c_{oil}, V_{oil}, t_{oil}, t_{env}, \rho, a \end{cases} \qquad (8)$$

The amount of SP_t should be sufficient for heating the volume (V) of the fossil fuel cargo stored in a tank. When there is a lack of power received from a SP, it is collected from external sources. The decision (d) that will be made should be optimal: $d \rightarrow opt$.

The functioning of oil product heating system is represented as a sequence of events based on cause-effect relations (Fig. 3). For a formal description of the logical structure of solar power use processes, the authors proposed a solution using the Petri net machine [9, 10]. Prerequisites for the use of Petri nets are: the presence of random and deterministic components in solar energy production; possibility of representing the solar power conversion as the sets of technologically homogeneous parallel processes; situational control related to the randomness of the possibility of converting solar energy.

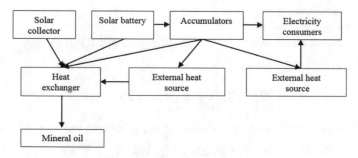

Fig. 3. Scheme of solar power usage at the terminal.

Petri nets are used to simulate the diagnostics of overloading equipment [11], assess the safe operation of reloading equipment at the container terminal [12], workflow management processes [13], as well as in many other areas of engineering and information technology.

The purpose of modeling with the Petri net (Fig. 4) is to identify conflict situations. Conflict situations imply a position in the Petri net having access to two or more transitions. The main task of the modeling with Petri nets is the verification of the process it simulates. Each execution must be completed, since the execution of a new operation is not allowed until the previous operation is finished. Depending on the choice, only one transition can be fulfilled, because starting a transition removes a chip from its position, so other transitions cannot be started. When each position of the network contains only one chip in any network marking reachable from the initial marking, then the Petri net is considered safe, i.e. the safety of the Petri net reflects the impossibility of starting a new operation until the previous one is completed. Logical connections of events (temporary, causal) are assigned based on a model representation of the work of oil product heating system with the use of Petri nets.

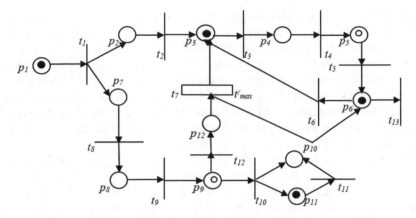

Fig. 4. Model of solar energy use at the terminal.

where p_1 – solar power; p_2 – solar collector; p_3 – heat exchanger; p_4 – oil product layer bordering by heat exchanger; p_5 – oil product temperature; p_6 – source of external energy; p_7 – solar panels; p_8 – electric power; p_9 – rechargeable batteries; p_{10} – power consumers; p_{11} – external electrical network; p_{12} – thermal energy; t_1 – absorption of solar power; t_2 – conversion of solar power into heat; t_3 – heat transfer agent pumping by the loop; t_4 – heat transfer to petroleum product; t_5 – inclusion of external source of power; t_6 – transfer of power from external source to heat exchanger; t_7 – transfer of heat energy to heat exchanger; t_8 conversion of solar power into electrical power; t_9 – battery charging; t_{10} – transmission of electricity to consumers and to external network; t_{11} – transmission of electricity from external source; t_{12} – conversion of electrical power into heat, t_{13} – transfer of heat energy to other consumers.

Solar energy (chip in position p_1), if there is any, is absorbed by the solar collector (transition t_1). If solar energy is absent, the transition t_1 does not work, so the oil is heated by the energy of an external source (position p_6) and the electrical power of accumulators (position p_9). Under favorable weather conditions, solar energy is supplied to the solar collector (position p_2) and solar panels (position p_7), where it is converted into heat (transition t_2) and electrical power (transition t_8). From the converted solar energy, heat transfer occurs in the heat exchanger (position p_3) to the transfer medium. Position p_3 has a chip that shows the presence of transfer medium. The t_3 transition (pumping the transfer medium by the contour) will work only if there are two chips: heat energy and transfer medium. Heat from the heat exchanger is transferred to the petroleum product layer bordering by heat exchanger (position p_4) and then heat is transferred to other oil product layers, which occurs due to heat propagation and gravitational heat exchange due to the difference in density of heated and cold oil products (transition t_4). The temperature of the oil product (position p_5) t_{oil} should satisfy the inequality: $t_{min} < t_{oil} < t_{cav.}$ Position p_5 has a built-in network (Fig. 5). When the temperature reaches its maximum value (position r_1) (Fig. 3), a signal is given to turn off electrical power supply from solar batteries (transition π_2, which is marked as t_{max}). Transition t_7 is marked as t'_{max} , which means that these markers complement each other and all transitions that are marked this way can be triggered simultaneously [14]. Considering the simulation model as a hierarchy of nested networks, the assumption can be made that at the top level of the hierarchy all or some of chips are used to designate the downstream cascade of private networks describing the implementation of individual functions. An important feature of the proposed approach to modeling is the description of each piece by means of its own "nested" (according to terminology [14]) network.

When the oil product temperature t_{min} (position r_2) (Fig. 5) is signaled to turn on the power from the external source (transition π_3), the chip enters the position r_3. The external energy source (position p_6) is switched on (transition t_5), then energy is supplied (transition t_6) to the heat exchanger (position p_3). In position p_6 there is a chip – the presence of external thermal energy, if it is absent the transition t_6 will not work. In the absence of solar energy (chip in position p_1), the oil product will be heated by batteries (position p_9), the electrical energy of which is converted into thermal energy (transition t_{12}), and thermal energy from an external source (position p_6).

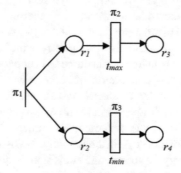

Fig. 5. Nested net of position p_5.

r_1 – temperature of oil product has reached the maximum value; r_2 – oil product temperature below the permissible value; r_3 – the signal to turn off the solar panels filed; r_4 – signal to turn on the external power source is on; π_1 – heating oil; π_2 – the signal is given to turn off the power supply from solar panels; π_3 – the signal is given to turn on the power from the external source.

While solar energy is converted into heat, solar power in solar panels (position p_7) is converted into electrical power (transition t_8). Electric power (position p_8) is supplied for charging (transition t_9) of rechargeable batteries (position p_9). Electricity is transmitted (transition t_{10}) from rechargeable batteries (position p_9) to electric loads of the terminal (position p_{10}), when they are disconnected - to the external electrical network (position p_{11}). Position p_{11} has a chip that indicates the presence of electrical power in the external network. If there is no chip in the p_{11} position, the t_{11} transition will not start.

Positions p_6 and p_9 are conflicting. Conflicts in positions p_6 and p_9 are resolved through situational control. As practice shows, such a local situation should be solved within the framework of automated control [15]. When transition k_2 having mark t''_{max} is triggered (Fig. 6) the transition t_7 is simultaneously triggered with mark t'_{max}. (Figure 4). In this case, electricity will flow to transition t_{10} and further to position p_{10}, providing consumers with electricity from rechargeable batteries, and to an external network when consumers are disconnected. When battery is out of charge, consumers will receive electricity from the external electrical network (position p_{11}).

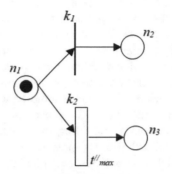

Fig. 6. Nested net of position p_9.

where n_1 - charged batteries; n_2 - electric power to the converter; n_3 - electricity to consumers; k_1 - transfer of electricity to the converter; k_2 - the transfer of electricity to consumers.

The analysis of the Petri net shows that the net is safe. The security of the Petri net reflects the impossibility of a new operation starting until the previous one is completed. There are certain conflicts in the net (positions p_6 and p_9). Conflict is resolved through automated control.

The net is alive. Transition in a Petri net is called alive if it can work when there is attainable markup. Attainability is considered as a sequence of network transitions.

The second stage of simulation process is the selection of a SP model:

$K = \{1, 2, ..., k\}$ - the set of producers of SP, $k > 1$.

$M = \{1, 2, ..., n\}$ - array of SP models of manufacturer K, $n > 1$. The choice of SP is made in terms of power of accumulated energy and geometric dimensions.

DL is the designated lifetime of the SP. As the service life increases, annual costs (C_a) for maintaining SP's working condition increase, and depreciation charges (DC) reduce or remain constant. In this case, economically justified DL will be determined by the year when total costs for maintaining the SP's working condition are minimal (saddle point of the chart):

$$DL = DC + C_a \rightarrow \min. \tag{9}$$

When using *SPs* for heating the pumped cargo, it is necessary to provide the installation of a circulation pump that will circulate the hydrocarbons through a *SP*.

After the parameters of *SP* are determined, the issue of financing the purchase is resolved: $F = \{F, LF, I\}$, where F is own funds, LF is loan funds, and I is investment.

Considering the development strategy of the terminal, funds are going to be invested for a certain planned time (T). Moments of time (t_i) of transition to new technologies are associated with losses (P). The trajectory $x(t_i) = x_i(t)$.

The use of solar power for heat-generating and power-generating sets enables increasing the energy efficiency of an oil terminal, reducing the negative impact of emissions into the atmosphere during the production of heat and electrical power with using exhaustible fuel sources. Solar panels are used as solar energy receivers (to convert solar radiation into electrical power) and solar collectors (to convert solar radiation into thermal energy). Solar radiation receivers are installed on terminal areas, which are not occupied by production equipment, and on roofs of oil storage tanks. The paper has not covered the optimization aspects, neither economical nor technical, since the authors had not posed such a task, which should be a subject of a special study.

The simulation model constructed using Petri nets shows that solar panels can work with solar collectors simultaneously, providing the production equipment of an oil-loading terminal with thermal and electrical power.

In the model, two conflict situations were identified that are solved depending on a situation by means of automated control. Unused electrical power is accumulated in batteries or transferred to an external network and can be used by consumers that do not refer to production equipment. When load on heat exchangers is absent, thermal energy is also given to external network for use by other consumers. The choice solar receiver model depends on the place of its installation at the terminal, the region of installation, the source of funding and the strategic plans for the development of the terminal.

References

1. Makarov, A.A., Mitrova, T.A., Kulagin, V.A.: Dolgosrochnyy prognoz razvitiya energetiki mira i Rossii. Ekonomicheskiy zhurnal Vysshey shkoly ekonomiki. **2**, 172–204 (2012). (in Russian)
2. Kashkarov, A.: Solnechnyye batarei i moduli kak istochniki pitaniya. Sovremennaya elektronika **5**, 8–15 (2015). (in Russian)
3. Informatsionnoye agentstvo RNS (in Russian). https://rns.online/energy/Rossiiskie-kompanii-vveli-v-ekspluatatsiyu-v-2017godu-140-MVt-vozobnovlyaemoi-generatsii-2018-01-15. Accessed 12 Jan 2019
4. Usachov, A.M.: Analiz dinamiki mirovoy industrii solnechnoy energetiki. Internet-zhurnal « NAUKOVEDENIYe » vol. 7/4 (2015). (in Russian). https://doi.org/10.15862/10evn415
5. Sedash, T.N.: Vozobnovlyayemyye istochniki energii: stimulirovaniye investitsiy v Rossii i za rubezhom. Rossiyskiy vneshneekonomicheskiy vestnik **4**, 94–97 (2016). (in Russian)
6. Il'in, R.A., Davydenko, A.I.: Fotoelektricheskiye preobrazovateli kak nezavisimyy istochnik elektroenergii na sobstvennyye nuzhdy proizvodstvennykh predpriyatiy. Simvol nauki **11**, 27–31 (2015). (in Russian)

7. Uskov, D.A., Otmakhov, G.S., Semonov, Y.A.A., Daurov, A.V.: K voprosu otsenki solnechnoy energii. Nauchnyy zhurnal KubGAU **124**, 1388–1402 (2016). (in Russian)
8. Sysoyeva, M.S., Pakhomov, M.A.: Metodika otsenki ekonomicheskoy effektivnosti innovatsionno-investitsionnykh proyektov v oblasti vnedreniya al'ternativnykh istochnikov energii. Sotsial'no-ekonomicheskiye yavleniya i protsessy **9**(31), 151–155 (2011). (in Russian)
9. Bratchenko, N.Y., Yakovlev, S.V.: Primeneniye setey Petri dlya analiza protsessov upravleniya urovnem obsluzhivaniya sistem upravleniya uslugami svyazi. Uspekhi sovremennogo yestestvoznaniya **5**, 96 (2007). (in Russian)
10. Lomazova, I.A.: Vlozhennyye seti Petri: modelirovaniye i analiz raspredelonnykh sistem s ob"yektivnoy strukturoy. M.: Nauchnyy mir, 208 p. (2004). (in Russian)
11. Zub, I.V., Yezhov, Y.Y., Sokolov, S.S.: Imitatsionnaya model' na osnove setey Petri kak sredstvo diagnostiki peregruzochnoy tekhniki. Rechnoy transport (XXI vek) **4**, 52–58 (2016). (in Russian)
12. Yezhov, Y.Y., Zub, I.V.: Model' otsenki bezopasnoy ekspluatatsii peregruzochnoy tekhniki na konteynernom terminale. Vestnik Gosudarstvennogo universiteta morskogo i rechnogo flota imeni admirala S. O. Makarova **5**(39), 50–61 (2016). (in Russian)
13. van der Aalast, V., van Khey, K.: Upravleniye potokami rabot: modeli metody i sistemy. Per. s angl. Bashkina, V.A., Lomazovoy, I.A., Pod red. Lomazovoy, I.A. FIZMATLIT, Moscow, 316 p. (2007). (in Russian)
14. Lomazova, I.A.: Vlozhennyye seti Petri: modelirovaniye i analiz raspredelonnykh sistem s ob"yektivnoy strukturoy. Nauchnyy mir, Moscow, 208 p. (2004). (in Russian)
15. Zub, I.V., Yezhov, Y.Y.: Modelirovaniye funktsionirovaniya transportnogo terminala vlozhennymi setyami Petri. Vestnik Gosudarstvennogo universiteta morskogo i rechnogo flota imeni admirala S. O. Makarova **2**(36), 41–48 (2016). (in Russian)

Transportation Engineering

Interoperability of Railway Infrastructure in the Republic of Serbia

Luka Lazarević⑩, Zdenka Popović$^{(\boxtimes)}$⑩, and Nikola Mirković⑩

Faculty of Civil Engineering, University of Belgrade,
Bulevar Kralja Aleksandra 73, 11 120 Belgrade, Serbia
zdenka@grf.bg.ac.rs

Abstract. The competitiveness of the railway transport is still limited by the differences in terms of infrastructure, rolling stock, transport technology, maintenance strategy, signaling systems, electrification systems, safety regulations, and speed limits. These circumstances force international trains to stop at state borders. Interoperability of railway infrastructure in the Republic of Serbia should allow safe and uninterrupted movement of trains. This paper presents the technical measures applied on the Serbian railway infrastructure in order to interoperate with the European railway network. Basic parameters corresponding to the essential requirements for infrastructure subsystem were presented. Furthermore, paper considers the interoperability constituents and experience regarding reconstruction and modernization on Corridor X through Serbia.

Keywords: Railway · Interoperability · Infrastructure · Technical regulation · Sustainable development

1 Introduction

Railway transport cannot provide door-to-door services, and this is the biggest limitation for railway competitiveness in the modern transport market. Time and money are lost because the railway systems and networks are not harmonized. Today, the obvious potential of the railway transport (reduction of CO_2 emissions and other ecological advantages, energy savings, relief of road traffic, transport of large quantities of goods over long distances, rail share in modal split, etc.) remains unused throughout Europe and the world.

In accordance with European transport policy, the establishment of an open transport market free of technical barriers should guarantee higher service quality and allow the railways to go ahead with new investments.

In order to attract more passenger and freight customers and consistently satisfy their requirements, more innovative and cost-effective ways need to be identified and implemented to increase punctuality, safety and capacity, improve performance at a system level and remove barriers to seamless intermodal transport and railway interoperability.

Interoperability is defined as the capability to operate on any stretch of the rail network without any difference. The formal definition of interoperability is given in the Interoperability Directive 2008/57/EC, stating that it represents "the ability of the rail system to allow the safe and uninterrupted movement of trains which accomplish the required levels of performance" [1].

© Springer Nature Switzerland AG 2020
V. Murgul and M. Pasetti (Eds.): EMMFT 2018, AISC 982, pp. 281–289, 2020.
https://doi.org/10.1007/978-3-030-19756-8_25

In the EU, Technical Specifications for Interoperability (TSIs) are law and define the technical and operational standards which have to be met in order to satisfy *the essential requirements* and to ensure the *interoperability* of the European railway system. The essential requirements could be summarized as safety, reliability and availability, health, environmental protection, technical compatibility and accessibility.

The Republic of Serbia is not an EU member (Fig. 1, left), but it passed the Law on Interoperability [2] in order to enable the harmonization of national regulations with the European regulations in the field of railway transport.

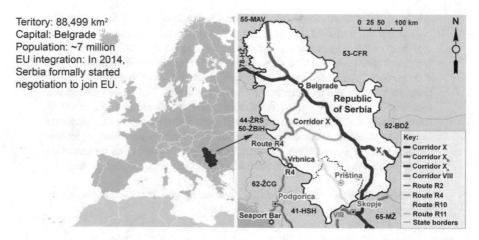

Fig. 1. General information about Serbia (left) and European corridors and routes through the Republic of Serbia (right).

Two European traffic corridors pass through the Republic of Serbia: The Danube waterway Corridor VII and the road-railway Corridor X, as well as three European routes: Route 4, Route 10 and Route 11 (Fig. 1, right).

Efficient and sustainable railway transport is essential for the development of the Republic of Serbia [3, 4]. Today, railway transport holds the second place in terms of volume of freight transport (after waterway transport) and passenger transportation (after road transport). According to the official data of the Statistical Office, the Republic of Serbia has 3809 km of tracks and 546 railway stations, including 865 km of tracks on European Corridor X (Fig. 1, right).

2   Implementation of Interoperability Requirements

TSIs are applicable to the whole EU railway network and railway networks in other countries that have passed laws on the interoperability of the railway system. TSIs cover one or more railway sub-systems (Table 1). The sub-systems considered in this paper are highlighted in gray in Table 1.

Table 1. The covering of railway sub-systems by TSIs.

Covering only one railway sub-system/Date of entry into force/		Covering more than one railway sub-system/Date of entry into force/
Locomotive & passenger rolling stock subsystem (LOCPAS TSI) /01/01/2015/	Rolling stock subsystem (freight wagons) (WAG TSI) /01/01/2014/	Safety in railway tunnels (SRT TSI) /01/01/2015/
Operation and traffic management subsystem (OPE TSI) /01/07/2015/	Control command and signaling subsystem (CCS TSI) /05/07/2016/	
Infrastructure subsystem (INF TSI) /01/01/2015/	Energy subsystem (ENE TSI) /01/01/2015/	Persons with reduced mobility (PRM TSI) /01/01/2015/
Telematics applications for freight subsystem (TAF TSI) /01/01/2015/	Telematics applications for passenger services subsystem (TAP TSI) /08/12/2013/	Rolling stock – Noise (NOI TSI) /01/01/2015/

The relevant technical specifications for *Infrastructure subsystem* refer directly to the railway infrastructure (INF TSI) [5] and indirectly to the safety in railway tunnels (SRT TSI) [6] and to the infrastructure which is adapted to the needs of persons with reduced mobility (PRM TSI) [7]. General hierarchical scheme for harmonization of legislation in the area of railway interoperability is shown in Fig. 2.

Fig. 2. The hierarchical scheme for harmonization.

Based on European legislation, Serbian laws and technical regulations constitute a legal framework for the implementation of interoperability of railway system, including railway infrastructure. The interoperability requirements apply to the railway infrastructure on the Corridor X through the Republic of Serbia. When favorable circumstances are created, interoperability requirements will be applied across the entire Serbian railway network.

3 Interoperability Requirements for the Railway Infrastructure

Basic parameters of the *Infrastructure subsystem* are based on the essential requirements defined in INF TSI [5]. Table 2 presents the basic parameters of the *Infrastructure subsystem* and Serbian practical experience on the railway Corridor X.

Table 2. Serbian experience relating to the implementation of the interoperability requirements.

Basic parameters [5] on Corridor X	Requirements according to INF TSI and their fulfillments
Structure gauge	GC structure gauge which enables intermodal transport without restrictions is applied on the international railway lines
Distance between track centers	4.0 m–4.5 m in accordance with the design speed on railway lines
Max. gradient	Max. gradient is 12% on Corridor X
Min. radius of horizontal curve	Min. radius according to the design speed, but not less than R = 300 m (max. design speed is up to 200 km/h)
Min. radius of vertical curve	Min. radius according to the design speed, but not less than R_v = 2000 m
Nominal track gauge	1435 mm
Cant	Max. cant D = 150 mm for ballasted track (recommended value is 120 mm to reduce maintenance costs)
Equivalent conicity	This request is satisfied with the use and maintenance of the nominal track gauge 1435 mm, installation of standard types of rails [8] with inclinations 1:40, as well as maintenance of wheels and track geometry
Rail profile for plain line	60E1 and 49E1 rail profiles are applied [8]
Rail inclination	1:40
Design of switches and crossings	The Infrastructure Manager defines the limit values for the geometry of switches and crossings in the Maintenance Plan
Use of swing nose crossings	Crossings with swing nose are not applied on Corridor X (design speeds are up to 200 km/h)
Max. unguided length of fixed obtuse crossings	The request is satisfied with the installation of standard types of crossings during reconstruction and construction of new tracks
Track resistance to vertical loads	Vertical loads are in accordance with [9]

(continued)

Table 2. (*continued*)

Longitudinal track resistance	Calculation of the railway superstructure takes into account the braking, acceleration and temperature force [10]. Braking force includes the effect of magnetic brake during emergency braking
Lateral track Resistance	Lateral loads are in accordance with [9]
Resistance of new bridges to traffic loads	European standards [11, 12] and national standards [10, 13] define the vertical loads including dynamic effects, centrifugal force, virtual lateral force, longitudinal forces due to braking and acceleration, and the impact of track deformations under load from the traffic
The immediate action limit for longitudinal level	European standards [14–16] define basic parameters of the track geometry, the methods for measurement and tolerances. Based on these parameters, the Maintenance Plan is defined by the Infrastructure Manager before putting the railway line in service
The immediate action limit for track twist	
The immediate action limit of track gauge as isolated defect	
The immediate action limit for cant	
The immediate action limit for switches and crossings	
Usable length of platforms	Usable length of platform is 400 m or min. 120 m for local railway stops
Platform height	Platform height is 550 mm above the upper edge of the rail head
Platform offset	According to the structure gauge
Track layout along platforms	Tracks along platforms should be designed straight. If this is not possible, larger curve radius shall be applied according to the speed of transit trains
Max. pressure variations in tunnels	Maximum pressure change is 10 kPa during train passing through the tunnel. It is applied for new tunnels and speeds over 200 km/h
Effect of cross winds	Measures taken to achieve traffic safety in critical wind conditions: local reduction of train speeds during winds, and/or installation of wind protection equipment on critical sections of the railway line
Ballast pick-up	This requirement applies to railway lines on Corridor X with max. design speed V = 200 km/h and it is the open point on Corridor X in Serbia
Equivalent conicity in service	The both parameters are essential for maintenance plan, which is developed by the Infrastructure Manager before putting the line in service according to the law. The tolerances for the track gauge, the rail inclination and the condition of the rail head are defined according to the required equivalent conicity
Maintenance rules	

3.1   Requirements for Interoperability Constituents

According to [5], interoperability constituents are the rail, the rail fastening systems, and track sleepers.

Infrastructure Manager defines the procedure for selecting the type and quality of rail steel in accordance with INF TSI [5], relevant standard [8], and adopted maintenance strategy.

Fastening system is assembly of components which secures a rail to the supporting structure and retains it in the required position, whilst permitting any necessary vertical, lateral and longitudinal movement. Standard series EN 13146 [17–25] defines test methods for fastening systems. Fastening system should be defined with type, pad stiffness, clamping force and longitudinal restraint. Moreover, fastening system has to be compliant with type of track sleeper [26–33].

According to [17], the longitudinal rail restraint is defined as the minimum axial force, applied to a rail secured to a sleeper by a fastening assembly, causing non-elastic slip of the rail through the fastening system. For general application in plain line, this value shall be at least 7 kN (for speed equal or lower than 250 km/h). A method for determining if the fastening system meets these requirements at the type approval testing stage is given in [17].

Figure 3 shows free space required for installation of rail fastening assembly.

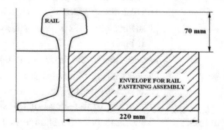

Fig. 3.   Envelope for rail fastening installation.

3.2   Requirements According to PRM TSI

Person with disabilities and person with reduced mobility represent any person who has a permanent or temporary physical, mental, intellectual or sensory impairment which, in interaction with various barriers, may hinder their full and effective use of transport on an equal basis with other passengers or whose mobility when using transport is reduced due to the age. PRM TSI [6] shall apply to the infrastructure, operation and traffic management, telematics applications and rolling stock subsystems according to [1]. PRM TSI shall apply to new railway infrastructure or rolling stock subsystems of the rail system and not apply to existing infrastructure or rolling stock of the rail system, which is already placed in service.

Essential requirements for the infrastructure sub-system should be applied to parking facilities for persons with disabilities and persons with reduced mobility, obstacle-free route, doors and entrances, floor surfaces, highlighting of transparent obstacles, toilets and baby-nappy changing facilities, furniture and free-standing devices, ticketing, information desks and customer assistance points, lighting, visual information (signposting, pictograms, printed or dynamic information), spoken information, platform width and edge of platform (Fig. 4), end of platform, boarding aids on platforms, and level track crossing at stations.

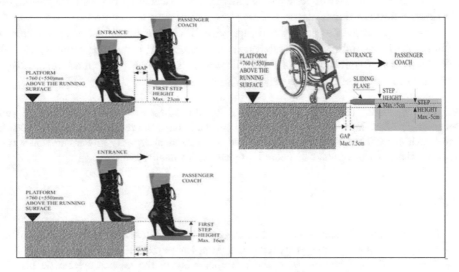

Fig. 4. Accessibility for all passenger categories.

3.3 Requirements According to SRT TSI

SRT TSI [7] concerns the following subsystems: control-command and signaling (CCS), infrastructure (INF), energy (ENE), operation (OPE), and rolling stock (locomotives and passenger units - LOC&PAS).

The purpose of SRT TSI is to define a coherent set of tunnel specific measures for the infrastructure, energy, rolling stock, control-command and signaling and operation subsystems, thus delivering an optimal level of safety in tunnels in the most cost-efficient way. It shall permit free movement of vehicles, which are in compliance with this TSI to run under harmonized safety conditions in railway tunnels.

Especially for tunnels, an alternative to the conventional ballasted track is the slab track (also called ballastless track). In slab track, the ballast is replaced by supporting concrete slab or asphalt slab. There are two principle solutions for slab tack structure with: (a) concrete slab on hydraulically bonded layer, and (b) asphalt slab on hydraulically bonded layer.

4 Conclusion

This paper presents and explains the general hierarchical scheme for the harmonization of legal and technical regulations for railway system. In addition, the basic parameters corresponding to the essential requirements for infrastructure subsystem are presented.

The paper considers the interoperability constituents (rail, rail fastening systems and track sleepers) and experience regarding reconstruction and modernization on Corridor X through Serbia.

It was pointed out the necessity of applying Technical Specifications for Interoperability (TSI) for infrastructure subsystem [5], people with reduced mobility [6], as well as safety in railway tunnels [7]. These TSIs should be applied for railway infrastructure on all of the European railway corridors.

Acknowledgement. This work was supported by the Ministry of Education, Science and Technological Development of the Republic of Serbia through the research project No. 36012: "Re-search of technical, technological, staff and organizational capacity of Serbian Rail-ways, from the viewpoint of current and future European Union requirements".

References

1. European Parliament and Council of EU: Directive 2008/57/EC on the interoperability of the rail system within the community. Official J. Eur. Union (2008)
2. The Republic of Serbia: Law on the interoperability of the railway system (2018). (in Serbian: Zakon o interoperabilnosti železničkog sistema)
3. Popović, Z., Lazarević, L., Vukićević, M., Vilotijević, M., Mirković, N.: The modal shift to sustainable railway transport in Serbia. In: MATEC Web of Conferences 106, paper 05001 (2017)
4. Vilotijević, M., Vukićević, M., Lazarević, L., Popović, Z.: Sustainable railway infrastructure and specific environmental issues in the Republic of Serbia. Tehnički Vjesnik. **25**(2), 516–523 (2018)
5. European Commission: The technical specification for interoperability relating to the 'infrastructure' subsystem of the rail system in the European Union. Official J. Eur. Union (2014)
6. European Commission: The technical specification for interoperability relating to accessibility of the Union's rail system for persons with disabilities and persons with reduced mobility. Official J. Eur. Union (2014)
7. European Commission: The technical specification for interoperability relating to 'safety in railway tunnels' of the rail system of the European Union. Official J. Eur. Union (2014)
8. CEN/TC 256: EN 13674-1 - Railway applications - Track - Rail - Part 1: Vignole railway rails 46 kg/m and above (2017)
9. CEN/TC 256: EN 14363 - Railway applications - Testing and Simulation for the acceptance of running characteristics of railway vehicles - Running Behaviour and stationary tests (2018)
10. Mirković, N., Popović, Z., Pustovgar, A.P., Lazarević, L., Zhuravlev, A.V.: Management of stresses in the rails on railway bridges. FME Trans. **46**(4), 636–643 (2018)
11. CEN/TC 250: EN 1991-2 - Eurocode 1: Actions on structures - Part 2: Traffic loads on bridges (2010)

12. CEN/TC 250: EN 1990 - Eurocode - Basis of structural design (2005)
13. DIN: Fachbericht 101 - Einwirkungen auf Bruecken. Deutschland (2003)
14. CEN/TC 256: EN 13848-1 - Railway applications - Track - Track geometry quality - Part 1: Characterisation of track geometry (2008)
15. CEN/TC 256: EN 13848-2 - Railway applications - Track - Track geometry quality - Part 2: Measuring systems - Track recording vehicles (2006)
16. CEN/TC 256: EN 13848-5 - Railway applications - Track - Track geometry quality - Part 5: Geometric quality levels - Plain line, switches and crossings (2017)
17. CEN/TC 256: EN 13146-1 - Railway applications - Track - Test methods for fastening systems - Part 1: Determination of longitudinal rail restraint (2014)
18. CEN/TC 256: EN 13146-2 - Railway applications - Track - Test methods for fastening systems - Part 2: Determination of torsional resistance (2012)
19. CEN/TC 256: EN 13146-3 - Railway applications - Track - Test methods for fastening systems - Part 3: Determination of attenuation of impact loads (2012)
20. CEN/TC 256: EN 13146-4 - Railway applications - Track - Test methods for fastening systems - Part 4: Effect of repeated loading (2014)
21. CEN/TC 256: EN 13146-5 - Railway applications - Track - Test methods for fastening systems - Part 5: Determination of electrical resistance (2012)
22. CEN/TC 256: EN 13146-6 - Railway applications - Track - Test methods for fastening systems - Part 6: Effect of severe environmental conditions (2012)
23. CEN/TC 256: EN 13146-7 - Railway applications - Track - Test methods for fastening systems - Part 7: Determination of clamping force (2012)
24. CEN/TC 256: EN 13146-8:2012 - Railway applications - Track - Test methods for fastening systems - Part 8: In service testing (2012)
25. CEN/TC 256: EN 13146-9 - Railway applications - Track - Test methods for fastening systems - Part 9: Determination of stiffness (2011)
26. CEN/TC 256: EN 13481-1 - Railway applications - Track - Performance requirements for fastening systems - Part 1: Definitions (2012)
27. CEN/TC 256: EN 13481-2 - Railway applications - Track - Performance requirements for fastening systems - Part 2: Fastening systems for concrete sleepers (2012)
28. CEN/TC 256: EN 13481-3:2012 - Railway applications - Track - Performance requirements for fastening systems - Part 3: Fastening systems for wood sleepers (2012)
29. CEN/TC 256: EN 13481-4 - Railway applications - Track - Performance requirements for fastening systems - Part 4: Fastening systems for steel sleepers (2012)
30. CEN/TC 256: EN 13481-5 - Railway applications - Track - Performance requirements for fastening systems - Part 5: Fastening systems for slab track with rail on the surface or rail embedded in a channel (2012)
31. CEN/TC 256: ENV 13481-6 - Railway applications - Track – Performance requirements for fastening systems - Part 6: Special fastening systems for attenuation of vibration (2006)
32. CEN/TC 256: EN 13481-7 - Railway applications - Track - Performance requirements for fastening systems - Part 7: Special fastening systems for switches and crossings and check rails (2012)
33. CEN/TC 256: EN 13481-8 - Railway applications - Track - Performance requirements for fastening systems - Part 8: Fastening systems for track with heavy axle loads (2006)

Assessment of Sleeper Stability in Ballast Bed Using Micro-tremor Sampling Method

Luka Lazarević[1]($\boxtimes$) (iD) and Dejan Vučković[2]

[1] Faculty of Civil Engineering, University of Belgrade,
Bulevar kralja Aleksandra 73, 11 120 Belgrade, Serbia
llazarevic@grf.bg.ac.rs
[2] Faculty of Mining and Geology, University of Belgrade,
Đušina 7, 11 000 Belgrade, Serbia

Abstract. The quality of sleeper support plays an important role in static and dynamic behavior of ballasted track. Sleepers in track should be well supported in vertical, lateral and longitudinal direction. Irregular vertical support (along z-axis) leads to significant increase in dynamic forces in the wheel-rail contact (increase up to 80% comparing to the sections with regular sleeper support). Irregular lateral support (along y-axis) reduces track lateral stability and increases the susceptibility to track buckling. Longitudinal support (along x-axis) is important for the behavior of the continuously welded rails. In the previous research conducted by the authors, application of micro-tremor analysis for determination of vertical sleeper support was considered. This paper investigates its further application for assessment of sleeper support in lateral and longitudinal direction. The obtained results are presented and discussed.

Keywords: Sleeper · Track · Sleeper support · Ballast bed · Micro-tremor

1 Introduction

Sleeper is the track element which transfers the traffic load in vertical and longitudinal direction (along z-axis and y-axis respectively in Fig. 1), as well as the load from traffic and temperature changes [1] in longitudinal direction (along x-axis in Fig. 1). Therefore, compaction of ballast around and below the sleeper plays an important role in static and dynamic behaviour of the track.

Differential settlement of the ballast bed is an inevitable process that often causes the occurrence of unevenly supported and unsupported sleepers. This leads to ballast bed degradation and track geometry deterioration. In addition, differential settlement of ballast bed at the bearing zone of sleepers leads to weakening of lateral support.

In some cases, more than 50% of the sleepers could be considered as unevenly supported [3]. Research conducted by Li and Sun showed that unevenly supported sleepers usually occur consecutively over 1–4 m of track [4], which was confirmed in the previous researches by the authors [5, 6]. Ballast-sleeper interaction contributes significantly to the dynamic behavior of the railway track [7]. Therefore, one or several consecutive poorly supported sleepers would lead to significant increase in dynamic forces. For example, dynamic forces in the wheel-rail contact could be up to 80%

© Springer Nature Switzerland AG 2020
V. Murgul and M. Pasetti (Eds.): EMMFT 2018, AISC 982, pp. 290–299, 2020.
https://doi.org/10.1007/978-3-030-19756-8_26

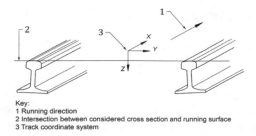

Key:
1 Running direction
2 Intersection between considered cross section and running surface
3 Track coordinate system

Fig. 1. Track axes according to the European standard EN 13848-1 [2].

higher comparing to the sections with well supported sleepers [7]. On the other hand, if several sleepers in a row have reduced lateral and longitudinal support, this could lead to track buckling [8]. Regarding previous considerations, it is obvious that investigation of sleeper support in field presents an important task for each Infrastructure Manager (IM).

There are different methods which could be used to identify the state of sleeper support in vertical direction. In general, all of these methods could be considered as direct or indirect. Direct methods provide the results related directly to the support conditions. On the other hand, indirect methods imply measuring parameters that are correlated to the sleeper support conditions, therefore providing the basis for indirect conclusions.

On the other hand, regarding the in-field application, investigation of sleeper support could be performed either by continuous or manual (non-continuous) methods. Continuous methods imply application on network level using adequately equipped inspection vehicle. Non-continuous methods are applied on section level using manually propelled or portable devices. However, both methods should be used:

– continuous method at first, in order to locate the sections with expected poor sleeper support and
– manual method afterwards, in order to thoroughly investigate these sections and schedule necessary maintenance activities.

Direct methods for investigation of sleeper support imply ballast bed scanning using acoustic waves [5], ultrasound [9], or ground penetrating radar (GPR) [10, 11]. Indirect methods imply measuring either sleeper vibrations under traffic or micro-tremor (ambient microvibrations), vertical acceleration, velocity, deflection, or other parameters related to track or vehicle that could be correlated to the sleeper support conditions [5, 6, 12–17].

From the aspect of engineering application, each method has its advantages and disadvantages. Therefore, responsibility of IM is to choose the method that would provide reliable results, thus providing the basis for creating the adequate maintenance plan.

This paper considers application of micro-tremor sampling method for identification of track sections with poor support along either vertical, lateral or longitudinal direction.

2   Application of Micro-tremor Sampling Method

Micro-tremor presents continuous ambient microvibrations which transfer from soil to the surrounding structures [18]. Measuring and interpreting micro-tremor presents useful geophysical tool which can be applied in many research areas [19, 20], including railway infrastructure [5, 6].

In the previous research [5, 6] by the authors, it was proven that poorly supported sleepers have significantly higher amplitudes along z-axis comparing to well supported sleepers. Further analysis of the obtained results showed that amplitudes in lateral direction (along y-axis) could be used for assessment of track lateral stability. If several sleepers in a row have poor lateral support, track buckling could occur on this track section [8]. This paper presents the results of micro-tremor sampling and identification of ballasted track sections with poor vertical and lateral sleeper support.

2.1   Measurement Section

Micro-tremor sampling on sleepers was previously conducted on the rail line Belgrade - Vrbnica, which is the part of international railway route that connects Belgrade railway junction on Pan-European Corridor X with seaport Bar in Montenegro (Route R4 in Fig. 2). This railway line, with length of around 287 km, realizes about 8% of total number of tonne-km and about 2% of total number of passenger-km on the entire Serbian railway network [6].

Fig. 2. Location of measurement section on the railway line Belgrade - Vrbnica (Route R4)

Measurement section was chosen according to methodology previously described in [6]. Measurement section stretches from kilometer point 57 + 300 to 57 + 350. Track on this section consists of concrete sleepers, locally known as IM2 sleepers, at axial distance 0.65 m (sleeper length 2.4 m and mass 190 kg), rigid rail fastening system (K type) and 49E1 rail profile.

2.2 Measurement System

Measurement system TM-3C400 (Fig. 3), developed by the company Centre for Nondestructive Testing and Geophysics (CNTG), was used for micro-tremor sampling. This system presents 24-bit acquisition system, which consists of three-channel GPS synchronized digitizer, with frequency range from 0.1 Hz to 600 Hz and triaxial Geo-Space sensor. Obtained data is stored to the PC's hard drive in real time. In addition, the system has autonomous high capacity power supply, which allows 48 h of continuous work.

Key:
1 - TM-3C400
2 - Triaxial sensor
3 - PC connector
   (for data gathering and storing)

Fig. 3. Measurement system TM-3C400 for micro tremor sampling

Table 1 shows the acquisition parameters for micro-tremor sampling on sleepers. The applied methodology is the result of generally accepted principles and experience in this research field. Software package AcDat was used during acquisition. All obtained data was processed afterwards using Fast Fourier Transformation.

Table 1. Acquisition parameters for micro-tremor sampling

No.	Acquisition parameters	Values
1.	Sampling length	60 s
2.	Total number of samples	24000
3.	Sampling frequency	400 Hz
4.	Sampling time	0.0025 s
5.	Number of sampling axes	3
6.	Amplitude resolution	24-bit

Sampling time of 60 s was chosen in order to obtain necessary data for reliable reconstruction recorded signal specter. On the other hand, sampling time was small enough, that measurement could be conducted in a short time. For this reason, ambient vibrations can be considered as a coherent excitation for all sampling points.

2.3 Applied Disposition for Micro-tremor Sampling

The first task before conducting the measurements was to choose placement for triaxial sensor on the sleeper. General distribution of stress under sleepers shows that maximum stress occurs in the area of rail seat [21]. Therefore, this area would be subjected to ballast settlement and loss of support, whilst the middle part of the sleeper is not relevant for measurements. Since we could not place the sensor on the rail seat, we positioned it right next to the fastening system on the inside stretch of the track (Fig. 4). This sensor placement had practical purpose, because placing the sensor on the outside stretch of track would demand removal of ballast from the end of each sleeper, which would not provide relevant data for y-axis component of micro-tremor.

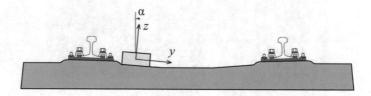

Fig. 4. Concrete sleeper shape and sensor position (with inclination angle)

As it can be observed from Fig. 4, sensor is under inclination angle α, which equals approximately 18°. Hence, measured displacement amplitudes along z-axis and y-axis would be about 5% less than real values according to Eq. (1), which could be neglected in the result analysis.

$$\Delta = (1 - cos\alpha) \cdot 100 = (1 - cos18°) \cdot 100 \approx 5\% \tag{1}$$

Figure 5 presents disposition for micro-tremor sampling on measurement section (general scheme). Sampling was performed on 40 sleepers in a row, which means 40 min of overall sampling time (according to sampling length from Table 1).

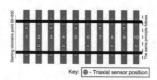

Fig. 5. Disposition for micro-tremor sampling on the measurement section (general scheme)

3   Results of Micro-tremor Sampling

Figure 6 shows the examples of measured relative displacement amplitudes (in all of three directions) on sleepers 14 and 25. In addition, Fig. 7 shows the view of sleepers 14 and 25 in field. Measurements showed that highest displacement amplitudes occurred for frequency range between 10 Hz and 20 Hz [5, 6].

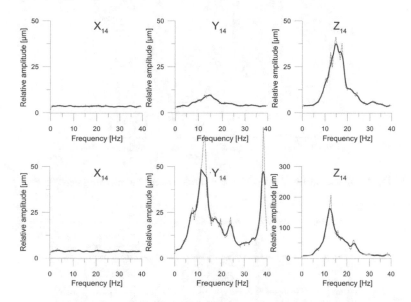

Fig. 6. Examples of obtained specters of relative amplitudes on sleepers 14 and 25

Fig. 7. The view of sleepers 14 and 25 on the measurement section

Results of micro-tremor sampling and data processing provided the insight into displacement amplitudes. In the previous research [5, 6], three ranges of relative displacement amplitudes along z-axis were adopted, which correspond to three support conditions:

- good support (relative amplitudes up to 50 µm);
- medium support (relative amplitudes from 50 µm to 150 µm);
- poor support (relative amplitudes from 150 µm to 300 µm).

However, these limits cannot be applied in general, since they were intrinsic for the chosen measurement section. Furthermore, they are not applicable for displacement amplitudes neither along y-axis nor along x-axis. After analyzing the specter on each sampling position, we separated highest relative displacement amplitudes and aligned these values along the measurement section. Afterwards, obtained diagrams for z-, y- and x-axis were filtered using Running Average function with 3 points (Figs. 8, 9 and 10).

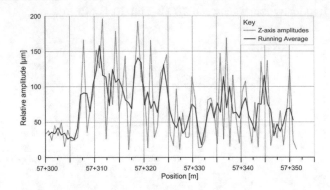

Fig. 8. Change of highest z-axis amplitudes along the measurement section

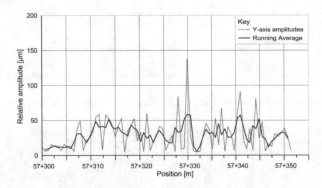

Fig. 9. Change of highest y-axis amplitudes along the measurement section

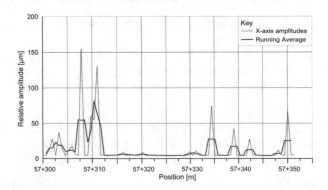

Fig. 10. Change of highest x-axis amplitudes along the measurement section

According to the obtained displacement amplitudes along y- and x-axis, following three ranges of relative displacement amplitudes were adopted (similar as the above mentioned ranges):

- good support: relative amplitudes up to 30 μm;
- medium support: relative amplitudes from 30 μm to 90 μm;
- poor support: relative amplitudes from 90 μm to 180 μm.

These ranges could be used for displacements along both x-axis and y-axis, since we obtained almost the same range of amplitudes for two axes.

Regarding the adopted ranges for each axis, it was determined that the measurement section has:

- 37 sleepers with good, 36 sleepers with medium and 7 sleepers with poor vertical support (according to displacements along z-axis);
- 47 sleepers with good, 31 sleepers with medium and 2 sleepers with poor lateral support (according to displacements along y-axis);
- 72 sleepers with good, 6 sleepers with medium and 2 sleepers with poor longitudinal support (according to displacements along x-axis).

It is important to state that the loss of sleeper support along z-axis and y-axis could not be determined visually in the early phase of development, which is shown by the number of sleepers with medium support. Sleepers with poor support are recognized by damaged ballast particles and the loss of ballast on the lateral side of sleepers.

The loss of sleeper support along x-axis occurred on several sleepers on the measurement section. This phenomenon is related to the insufficient compaction of ballast between the sleepers.

4   Conclusion

The modal shift to sustainable railway transport increases the requirements that must be met by Infrastructure Manager (IM) in order to reduce the negative impacts on the environment [22], to extend the life cycle of the railway infrastructure [23–25] and to optimize maintenance costs. Accumulated traffic and temperature load leads to inevitable ballast bed degradation and track geometry deterioration. IM has to define maintenance policy and strategy in order to ensure continuous condition monitoring, which provides the basis for planning and conducting of short term and long term maintenance.

Poorly supported sleepers influence the reduce of service life of ballast bed and con-tribute to the track geometry degradation. Therefore, these sleepers need to be detect-ed and appropriate maintenance activities performed as early as possible. This approach ensures the possibility of reducing the overall maintenance costs.

There are many methods which could be used for investigation of sleeper support, which could be either direct or indirect and used continuously or manually. Each method has its advantages and disadvantages related to the field application and reliability of the results. However, IM has to choose the optimal method by considering:

- investigation level (entire network or single section),
- reliability of the obtained results,
- time to set up and carry out the investigation,
- measurements costs and
- portability of the equipment and the necessity for the track closure.

This paper presents micro-tremor sampling on sleepers, which is non-destructive indirect method. Micro-tremor presents continuous ambient microvibrations that transfer from surrounding soil to the railway structure. In the previous research [5, 6] by the authors, it was proven that poorly supported sleepers in the ballast bed have higher vertical displacement amplitudes. This paper presents further results of micro-tremor analysis in lateral and longitudinal direction, as well as discussion of the obtained results.

Acknowledgement. This work was supported by the Ministry of Education, Science and Technological Development of the Republic of Serbia through the research project No. 36012: "Re-search of technical, technological, staff and organizational capacity of Serbian Rail-ways, from the viewpoint of current and future European Union requirements".

The authors are grateful to company Centre for Non-destructive Testing and Geo-physics (CNTG) for organizing and conducting field investigations.

References

1. Mirković, N., Popović, Z., Pustovgar, A.P., Lazarević, L., Zhuravlev, A.V.: Management of stresses in the rails on railway bridges. FME Trans. **46**(4), 636–643 (2018)
2. CEN/TC 256: EN 13848-1 - Railway applications - Track - Track geometry quality - Part 1: Characterisation of track geometry (2010)
3. Augustin, S., Gudehus, G., Huber, G., Schunemann, A.: Numerical model and laboratory tests on settlement of ballast track. In: System Dynamics and Long-term behaviour of Railway Vehicles, Track and Subgrade, pp. 317–336. Springer, Berlin (2003)
4. Li, Z.F., Sun, J.G.: Maintenance and cause of unsupported sleeper. China Railway Build **2**, 15–17 (1992)
5. Lazarević, L.: Assessment of track geometry quality using fractal analysis of measured data. Ph.D. Thesis, University of Belgrade, Faculty of Civil Engineering, Serbia (2016)
6. Lazarević, L., Vučković, D., Popović, Z.: Assessment of sleeper support conditions using micro-tremor analysis. Proc. Inst. Mech. E Part F: J. Rail Rapid Transit. **230**(8), 1828–1841 (2016)
7. Kaewunruen, S., Remennikov, A.M.: Investigation of free vibrations of voided concrete sleepers in railway track system. Proc. Inst. Mech. E Part F: J. Rail Rapid Transit. **221**(4), 495–507 (2007)
8. Lim, N.H., Park, N.H., Kang, Y.J.: Stability of continuous welded rail track. Comput. Struct. **81**(22), 2219–2236 (2003)
9. De Bold, R.: Non-destructive evaluation of railway trackbed ballast. Ph.D. Thesis, University of Edinburgh, Scotland (2011)
10. Silvast, M., Nurmikolu, A., Wiljanen, B., Levomaki, M.: An inspection of railway ballast quality using ground penetrating radar in Finland. Proc. Inst. Mech. E Part F: J. Rail Rapid Transit. **224**(5), 345–351 (2010)

11. Kathage, A., Niessen, J., White, G., Bell, N.: Fast inspection of railway ballast by means of impulse GPR equipped with horn antennas. In: Proceedings of the Eight International Conference "Railway Engineering 2005", London, UK (2005)
12. Kim, D.S., Kim, S.D., Lee, J.: Easy detection and dynamic behavior of unsupported sleepers in high speed ballasted track. In: Seventh World Congress on Railway Research, Montreal, Canada (2006)
13. Brajović, Lj., Malović, M., Popović, Z., Lazarević, L.: Wireless system for sleeper vibrations measurements. Commun. Sci. Lett. Univ. Žilina 16(4), 21–26 (2014)
14. Pinto, N., Ribeiro, C.A., Gabriel, J., Calcada, R.: Dynamic monitoring of railway track displacement using an optical system. Proc. Inst. Mech. E Part F: J. Rail Rapid Transit. 229 (3), 280–290 (2015)
15. Puzavac, L., Popović, Z., Lazarević, L.: Influence of track stiffness on track behaviour under vertical load. Promet – Traffic Transp. 24(5), 405–412 (2012)
16. Berggren, E.: Dynamic track stiffness measurement - a new tool for condition monitoring of track substructure. Licentiate Thesis, KTH Royal Institute of Technology, Sweden (2005)
17. Sadeghi, J.: Field investigation on vibration behavior of railway track systems. Int. J. Civ. Eng. 8(3), 232–241 (2010)
18. Sanchez-Sesma, F.J., Rodriguez, M., Iturraran-Viveros, U., Luzon, F., Campillo, M., Margerin, L., Garcia-Jerez, A., Suarez, M., Santoyo, M.A., Rodriguez-Castellanos, A.: A theory for microtremor H/V spectral ratio: application for a layered medium. Geophys. J. Int. 186(1), 221–225 (2011)
19. Delgado, J., Lopez Casado, J., Giner, J., Estevez, A., Cuenca, A., Molina, S.: Microtremors as a geophysical exploration tool: applications and limitations. Pure. appl. Geophys. 157(9), 1445–1462 (2000)
20. Papanova, Z., Papan, D., Kortis, J.: Microtremor vibrations in the soil experimental investigation and fem simulation. Commun. Sci. Lett. Univ. Žilina 16(4), 41–47 (2014)
21. Sadeghi, J.: Field Investigation on dynamics of railway track pre-stressed concrete sleepers. Adv. Struct. Eng. 13(1), 139–151 (2010)
22. Popović, Z., Lazarević, L., Vukićević, M., Vilotijević, M., Mirković, N.: The modal shift to sustainable railway transport in Serbia. In: MATEC Web of Conferences 106, paper 05001 (2017)
23. Popović, Z., Lazarević, L., Brajović, L., Vilotijević, M.: The importance of rail inspections in the urban area - aspect of head checking rail defects. Procedia Eng. 117, 596–608 (2015)
24. Popović, Z., Brajović, Lj., Lazarević, L., Milosavljević, L.: Rail defects head checking on the Serbian railways. Tehnički Vjesnik 21(1), 147–153 (2014)
25. Popović, Z., Radović, V.: Analysis of cracking on running surface of rails. Gradevinar 65(3), 251–259 (2013)

Modeling of Multi-agent Voltage Control in Distribution Electric Networks of Railways

Evgeny Tretyakov$^{(\boxtimes)}$ (iD), Vasily Cheremsin (iD),
and Grigory Golovnev (iD)

Omsk State Transport University, Marksa 35, 644046 Omsk, Russia
eugentr@mail.ru, cheremisinvt@gmail.com,
grishantiy@gmail.com

Abstract. The introduction of adjustable reactive power compensation devices in distribution electric networks of railways opens up new opportunities for increasing their efficiency through the use of group-based voltage control methods based on the agent-based approach. Multi-agent voltage control allows obtaining new results related to the possibility of self-organizing agents - active elements of the electric network, which leads to an increase in the reliability of power supply and power quality. Modeling of considered multi-agent control systems on classical models of system dynamics is difficult because of the complex interaction of agents due to their individual utility goals, the presence of logical operations, and the event nature of the processes. State diagrams of agents were developed for modeling multi-agent voltage control using reactive power sources in distribution electric networks of railways in the Anylogic environment. The modeling of the voltage control in the test electric network when changing the mode parameters is done. The obtained modeling results show the validity of approaches to voltage stabilization using multi-agent control methods and the possibility of their practical implementation on the basis of modern equipment.

Keywords: Modeling · Sources of reactive power · Voltage stabilization ·
Agent-based approach · State diagram

1 Introduction

The introduction of adjustable devices for reactive power compensation (CD) in the distribution electric networks of railways opens up new opportunities for increasing their efficiency through group-based voltage control methods based on the agent-based approach. A significant part of publications on multi-agent control of electric network modes, including those with elements of distributed generation, energy accumulators, is devoted to the development of concepts and subsystems of such control, in which the results of modeling are presented for individual components, [1–4].

Modeling the considered multi-agent control systems on classical models of system dynamics is difficult because of the complex interaction of agents due to their individual utility goals, the presence of logical operations, and the event nature of the processes. When calculating the parameters of the modes of distribution networks, it is

© Springer Nature Switzerland AG 2020
V. Murgul and M. Pasetti (Eds.): EMMFT 2018, AISC 982, pp. 300–306, 2020.
https://doi.org/10.1007/978-3-030-19756-8_27

often assumed [5–8] that the electric network is static, all data values are known constants, and the actual load change over time is considered taking into account several different discrete cases. The implementation of the approach based on a combination of traditional methods of system dynamics and agent-based modeling will solve these problems.

2 Theoretical Part

To create agent-based models, specialized software products have been developed, for example, NetLogo, StarLogo, Repast Simphony, Eclipse AMP, JADE, Jason and others [7–10], many of which are based on the FIPA specification [11]. However, these agent-based platforms require special programming skills, so their widespread use by researchers in wide areas of knowledge is limited. One of the convenient tools for research on modeling agent-based systems is the AnyLogic software product, which does not yet have packaged libraries for the power engineering.

The model of multi-agent voltage control in the distribution electric network of the railways in Anylogic can be represented as a well-known description of steady-state modes [12] and characteristics of local agents and coordinator agents in the form of state diagrams, ontology, algorithms of interaction and coordination [13, 14].

In the task in question, the active element controllers—devices for reactive power compensation, and the coordinator agents—the voltage controllers of the electric network section, act as local agents.

Local controllers have their own rules of behavior, and their joint work creates a complexity of the model, the emergent properties of which determine the behavior of the voltage control system in the electric network as a whole. For the best behavior of the specified multi-agent control system, local agents should be involved, which have maximum efficiency for performing a specific task, possessing the rule of self-organization under external and internal influences [14]:

$$J = \sum_{v=1}^{n} \sum^{min,} q_v \rightarrow max \qquad (1)$$

where q_v – assessment of the efficiency of execution of the action by the agent; n - the number of agents, among the set of actions of which there are all actions ensuring the achievement of the target task.

Coordination of local agents is carried out on the basis of the principle of "auction", which consists in choosing the best offers for control purposes among local agents. The auction is held iteratively until all the tasks are distributed among local agents in the best way (Fig. 1). The main actions of agents: the formation by agents of the price array (k_Q), the selection of the most effective agents, the notification of agents about the completion of the task, the exclusion of the task from the price arrays of agents.

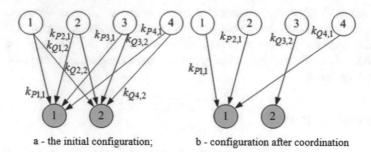

a - the initial configuration;   b - configuration after coordination

Fig. 1. Coordination of agents.

Agents are selected on the basis of the ranking of their efficiency estimates (in this case, the ratio of the "cost" of the task implementation to the voltage sensitivity of the electric network buses to the injection of reactive power of CD - k_Q/b_{jk}).

The voltage sensitivity of the electric network busses to the injection of the reactive power of CD of the local agent bjk is determined on the basis of the corresponding elements of the Jacobian matrix [12].

The amount of injections of CDs of local agents is determined as a result of solving the problem, taking into account the known limitations of the mode parameters:

$$\Delta U_D = \sum\nolimits_{j=1}^{N} \left(k_P \Delta P_j + k_Q \Delta Q_j \right) \to min, \qquad (2)$$

where k_P, k_Q – coefficients for active (if available) and reactive power; $\sum_{j=1}^{N} \left(a_{jk}\Delta P_j + b_{jk}\Delta Q_j \right) = \Delta U_D$.

Modeling of multi-agent voltage control in distribution electric networks consists in the integration of calculations of the mode parameters and simulation of the work of local agents, coordinator agents.

To describe the ontology, state diagrams and specified limitations of the mode parameters are used, on the basis of which all knowledge that the agent needs for both individual work and interaction with other agents is described [14].

A provisional list of communication protocol commands in a multi-agent voltage control system:

- informing agents about the readiness and the end of negotiations;
- request the value of the "cost" of voltage control;
- nswer with the value of the "cost" of task implementation by the agent;
- prior consent to work;
- refusal to work;
- notification of the consent of all agents to work;
- notification of the refusal of some agents;
- confirmation of consent and transition to work;
- rejection of prior consent and return to the original work.

The voltage control algorithm due to the coordinated generation of active and reactive power in the electric network is shown in Fig. 2.

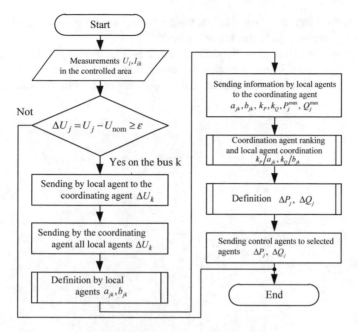

Fig. 2. The voltage control algorithm.

Let us consider the implementation of multi-agent voltage control modeling in the distribution electric networks of railways in the AnyLogic software product using the example of a fragment of a 10 kV electric network shown in Fig. 3.

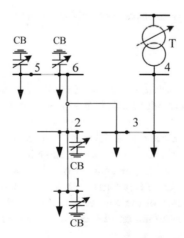

Fig. 3. The electric network fragment.

The calculation of the steady-state regimes for a given topology and the parameters of the electric network equivalent circuit was performed at each time step. In the modeling, twenty-four steps were taken, which can be increased to the level of detail of the time schedule in a few minutes or seconds.

To describe the behavior of the considered agents in AnyLogic, the state diagrams of the local agent of CD and the coordinator agent are developed on the basis of the presented algorithm and the principles of their coordination. They are shown in Fig. 4.

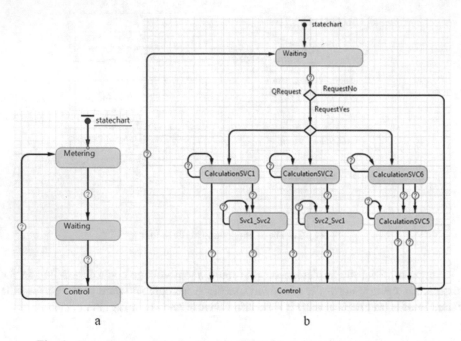

Fig. 4. State diagram of the local agent of CD (a) and the coordinator agent (b)

3   Practical Part

Figures 5 and 6 show the modeling results of multi-agent voltage control on buses 1 and 2 in the test electric network based on the presented approaches in the AnyLogic program.

The simulation results indicate the efficiency of the modeling of multi-agent voltage control in the electric network. According to the modeling conditions, the tolerance on buses is set to ±6%, the reactive power limit of CD is 400 kvar. The compensation device on the bus 2 of the test electric network is activated only if CD on the bus 1 cannot provide for stabilization of the voltage level within the specified limits, which is based on the principles of coordination of local agents presented above and solving the optimization problem (2).

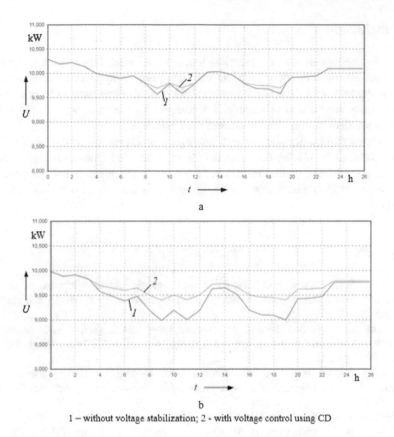

a

b

1 – without voltage stabilization; 2 - with voltage control using CD

Fig. 5. The results of the voltage modeling on the bus 2 (a), on the bus 1 (b) of the electric network.

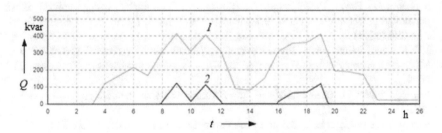

Fig. 6. Reactive power generated by CD for voltage stabilization: 1 – on the bus 1; 2 – on the bus 2.

4 Conclusion

The research results showed the practical feasibility of modeling voltage control in the distribution electric networks on the basis of the presented approach. The integration in one software product of the components of the system dynamics and agent behavior makes it possible to holistically model the behavior of the multi-agent voltage control system in the distribution electric networks of railways.

Of considerable scientific interest are also the issues of modeling multi-agent voltage control in the electric networks in case of failure and limitations in the operation of one or several CDs, given limitations of the mode parameters, and the presence of distributed generation with active consumers, which is the subject of further research by the authors.

References

1. Ismoilov, S.T., Fishov, A.G.: Modeling and analysis of the effectiveness of voltage regulation in an electrical network with distributed generation. Prob. Transp. Siberia Far East **2**, 302–305 (2014)
2. Niknam, T., Zare, M., Aghaei, J.: Scenario-based multiobjective volt/var control in distribution networks including renewable energy sources. IEEE Trans. Power Del. **27**(4), 2004–2019 (2012)
3. Karbalaei, F., Shahbazi, H.: Quick method to solve the optimal coordinated voltage control problem based on reduction of system dimensions. Electron. Power Syst. Res. **1420**, 310–319 (2017)
4. Morattab, A., Akhrif, O., Saad, M.: Decentralised coordinated secondary voltage control of multi-area power grids using model predictive control. IET Gener. Transm. Distrib. **11**, 4546–4555 (2017)
5. Alobeidli, K., Moursi, S.: Novel coordinated secondary voltage control strategy for efficient utilisation of distributed generations. IET Renew. Power Gener. **8**(5), 569–579 (2013)
6. Yassami, H., Bayat, F., Jalilvand, A., Rabiee, A.: Coordinated voltage control of wind-penetrated power systems via state feedback control. Int. J. Electr. Power Energy Syst. **93**, 384–394 (2017)
7. JAVA Agent DEvelopment Framework. http://jade.tilab.com. Accessed 11 Feb 2019
8. MASwarm Agent Platform Homepage. http://navizv.github.io/MASwarm. Accessed 11 Feb 2019
9. NetLogo Agent Platform Homepage. http://ccl.northwestern.edu/netlogo. Accessed 11 Feb 2019
10. Repast Suite Homepage. http://repast.sourceforge.net, last accessed 2019/02/11
11. Foundation for Intelligent Physical Agents (FIPA) Homepage. http://www.fipa.org. Accessed 11 Feb 2019
12. Pochaevets, V.S.: Automated Control Systems for Power Supply Devices for Railways. UMC ZDT Publications, Moscow (2003)
13. Tretyakov, E.A.: The method of controlling the power supply system of railways. Patent RF, no. 2587128, Russia (2016)
14. Rassel, S., Norvig, P.: Artificial Intelligence: a Modern Approach. Williams Publisher, Moscow (2006)

Oscillation Process of Multi-support Machines When Driving Over Irregularities

Sergey Ovsyannikov[1]([⊠]) [iD], Evgeniy Kalinin[2] [iD],
and Ivan Koliesnik[2] [iD]

[1] Belgorod State Technological University Named After V.G. Shukhov,
Kostyukova str. 42, 308012 Belgorod, Russia
ovsannikov.si@bstu.ru
[2] Kharkiv Petro Vasylenko National Technical University of Agriculture,
Alchevski str. 44, Kharkiv 62002, Ukraine

Abstract. The article considers a mathematic model which describes multiaxial machine movement while moving over irregularities. Generalized coordinates are: vertical and angular movement of center of gravity in lateral plane. It was discovered during the modeling process that a sufficient oscillation decay is achieved much earlier than under extreme aperiodic motion. Solid oscillation dampening for unsprang mass can be expected at the rate of 0.5 of relative dampening coefficient. Obviously, the coefficient of aperiodic vertical oscillations must be 0.16 for passenger and 0.29 for freight automobiles. Thus, vertical oscillations of multi-support machines must increase while angular oscillations dampen. If the length of irregularity is big, multi-support machine behaves like a mass on a spring. If the length of irregularity is set, its impact on vertical and particularly angular oscillations declines with increased number of supports. Under sufficient rate of damping, body oscillations become steady, practically on the third and sometimes on the second irregularity. It is shown that motion smoothness of a multi-support machine is higher than of a biaxial machine. The difference is greater under transition from two to three axes. If the number of supports increases, effect from suspension softening declines. Some decline in angular oscillation damping with increased number of supports caused insignificant increase in the range of oscillations over the next irregularities.

Keywords: Multi-support machines of the oscillation process ·
Driving over irregularities

1 Introduction

Proper analysis of oscillations and choice of parameters play an extremely important role for automobile or track-type machine as well as for state and productivity of the driver [1]. Nowadays, the theory of suspension is quite fully developed for biaxial cars [2–4]. The base for theoretical research and calculations is that distribution of spring mass with enough accuracy allows to consider body oscillations on the front and back suspensions independent from each other. This gives a formal advantage – when choosing the generalized coordinates of vertical transitions of the body above the front

© Springer Nature Switzerland AG 2020
V. Murgul and M. Pasetti (Eds.): EMMFT 2018, AISC 982, pp. 307–317, 2020.
https://doi.org/10.1007/978-3-030-19756-8_28

and back wheels to divide the system of four differential equations, which describes oscillations of the automobile in lateral plane, into two systems with two equations each [5].

However, for multi-axial machines such simplifications do not give the desired effect. It is known that angular oscillations of the frame are particularly frequent in multi-axial machines, so it makes sense to view angular transition of the very frame as the generalized coordinate. Theory of oscillations and cushioning of machines was studied by P.M. Volkov, N.S. Piskunov, D.A. Popov and other researchers, whose works are described in the works [6, 7]. Though the cushioning theory development is not anyhow connected to the theory of suspended systems, it is reasonable to consider some conditions from unified positions which are interesting in relation to suspension development of both wheeled and track-type machines.

This work [8] describes a mathematic model of oscillations of wheeled transportation-technological systems, in work [9] a model of machine-tractor aggregate with considered function conditions is described. [10] describes modeling in the system of road-tractor-driver with the account of smoothing features of track-movers. But the mentioned above works do not consider the effect of springing degree and the impact of aperiodic perturbing factors on body oscillations of multi-axial means of transport.

2 Methods and Materials

A systematic approach based on modern methods of theoretical research was used to the solution of this works problematic. The authors utilized modern mathematical methods of differential equation solutions, strain-energy and probable methods, frequency method for ranking of dynamic systems, methods of analysis of complex movement of mechanism and systems. Software based analytical modelling was utilized for the study of multiaxial machine stability.

3 Results

Figure 1, a - schematic of a multiaxial machine as an oscillating system; Fig. 1, b – equivalent oscillating system with 3 or 4 supports.

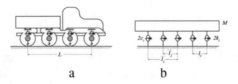

a   b

Fig. 1. Oscillating system a – wheeled multiaxial automobile; b – multi-support machine

Let's imagine a multi-support machine as an oscillating system that is slightly simplified compared to the initial conditions – unsprang mass is not accounted for. A more detailed analysis reveals that such a simplification is suitable in the areas of low frequency resonance, an area which determines the smoothness of movement in the first place. If the unsprang mass is not accounted for, then the rigid element of suspension and tyres can be replaced with a cushioned element of equivalent rigidity.

We will examine the oscillating system (Fig. 1b) using vertical movements of center of mass as unified coordinates Z_o and angular movement in lateral plane as α. In line with the set admissions, we consider oscillations to be low, characteristics of all elements of suspension to be linear and the length between supports to be equal $l_i = const$ Oscillations of a multi-support machine are described the same equations that describe biaxial automobile with unified coordinates Z_o and α:

$$\begin{cases} \ddot{Z}_o + 2h_z\dot{Z}_o + \omega_z^2 Z_o + \eta_{zk}\dot{\alpha} + \eta_{zc}\alpha = Q_z \\ \ddot{\alpha} + 2h_\alpha\dot{\alpha} + \omega_\alpha^2\alpha + \eta_{\alpha k}\dot{Z}_o + \eta_{\alpha c}Z_o = Q_\alpha \end{cases} \tag{1}$$

Relation coefficients η_z η_α condition the connection between vertical and angular movement and equal to:

$$\eta_{zk} = \frac{2}{M}\sum_1^j k_i l_i; \eta_{\alpha k} = \eta_{zk}\frac{1}{\rho^2};$$

$$\eta_{zk} = \frac{2}{M}\sum_1^j k_i l_i; \eta_{\alpha c} = \eta_{zc}\frac{1}{\rho^2};$$

In these equations, the sum includes the total number of supports j, and distances from center of mass to the support l_i have opposite operators depending on the position of support in relation to the center of mass. That is why we can use the following values, in particular with a symmetrical scheme of the machine

$$\sum k_i l_i = 0 \text{ and } \sum c_i l_i = 0.$$

In that case the system (1) breaks down into two independent equations:

$$\begin{aligned} \ddot{Z}_o + 2h_z\dot{Z}_o + \omega_z^2 Z_o = Q_z \\ \ddot{\alpha} + 2h_\alpha\dot{\alpha} + \omega_\alpha^2\alpha = Q_\alpha \end{aligned} \tag{2}$$

Let's notice that vertical and angular movements of the frame are described equations of the same type. This allows for introduction of uniformity into the method of quantification and calculation of both types of oscillations.

Coefficients in the left part of the Eq. (2) determine the flow of proper oscillations. Damping of vertical oscillations h_z and their frequency ω_z equal to:

$$h_z = \frac{\sum k_i}{M} \text{ and } \omega_z = \sqrt{\frac{2\sum c_i}{M}}. \tag{3}$$

In the case of angular oscillations for damping h_α and own frequency ω_α, we have following equations:

$$h_\alpha = \frac{\sum k_i l_i^2}{M \rho^2} \text{ and } \omega_\alpha = \sqrt{\frac{2 \sum c_i l_i^2}{M \rho^2}}. \tag{4}$$

We know that it is better to estimate an oscillating system using relative coefficient (aperiodic coefficients) which, in our case, equal to:

$$\psi_z = \frac{h_z}{\omega_z} \text{ and } \psi_\alpha = \frac{h_\alpha}{\omega_\alpha}. \tag{5}$$

Result:

$$\psi_z = \frac{h_z}{\omega_z} = \frac{2 \sum k_i}{M} \sqrt{\frac{M}{2 \sum c_i}} = \frac{2 k_i i}{\sqrt{2 c_i M}}, \tag{6}$$

because $\sum k_i = j k_i$ and $\sum c_i = j c_i$.

In the case of relative damping for two wheels, we have:

$$\psi_k = \frac{h_k}{\omega_k}.$$

When stationary, sprung mass's own frequency ω_k, a damping h_k of unsprang mass equal to:

$$h_k = \frac{2 k_i}{m_i} \text{ and } \omega_k = \sqrt{\frac{2 c_{pi}}{m_i}},$$

m_i – mass of the left and right wheels.
Therefore

$$\psi_k = \frac{2 k_i}{\sqrt{2 c_{pi} m_i}}. \tag{7}$$

Let's calculate the relation of relative coefficients of damping of sprung and unsprang mass. Using Eqs. (6) and (7) we can write down:

$$\psi_z = \psi_k \sqrt{j \frac{m_i}{M}}. \tag{8}$$

Accordingly, to Eq. (8) the maximal value of relative damping should be different for passenger and freight automobiles with different value of working load. It is obvious that $\psi_k > \psi_z$. That is why ψ_z should be chosen according to the influence of dampers on the movement of unsprang mass ψ_k. When $\psi_k = 1$, movement of unsprang mass becomes aperiodic: wheels are going to get worse at following road irregularities and will lose traction more often in result.

Considering jm/M for a passenger car to be 10 and for the freight car to be 3, we will find that the oscillation of unsprang mass in comparison with sprung mass will stop ($\psi_k = 1$) when ψ_z, is corresponding to 0,32 and 0,58. Although sufficient damping is achieved much sooner than during aperiodic movement. Reliable damping can be expected starting with $\psi_k = 0,5$ for unsprang mass. It is established that ψ_z should be 0,16 passenger automobiles and 0,29 freight automobiles. These values are very close to the values derived from experimental data: $\psi_z = 0,15 \div 0,20$ and $\psi_z = 0,25 \div 0,30$ respectively for both types.

Perturbations in the right side of initial Eqs. (1) are composed from the sum of interactions that are created by the movement of the wheels over each irregularity.

For vertical motions we have:

$$Q_z = \sum_1^j Q_{zi} = \frac{2}{M}\left(\sum_1^j k_i \dot{q}_i + \sum_1^j c_i q_i\right). \tag{9}$$

In case of a single irregularity each summand is corresponding to a particular time period in which force Q_{zi} acts. Assuming that irregularity profile is changing according to $q = q_0(1 - \cos vt)$, for the first wheel pair we will have:

$$Q_{z1} = 0 \text{ at } t = 0 \text{ and } t \geq t_v, \text{ where } t_v = \frac{S}{v};$$
$$Q_{z1} = q_0[\omega_{z1}^2(1 - \cos vt)h_{z1}v \sin vt] \text{ at } 0 < t < t_v;$$
$$Q_{z2} = 0 \text{ at } t \leq t_i \text{ and } t \geq (t_v + t_i) \text{ where } t_i = \frac{l_i}{v};$$
$$Q_{z2} - q\{\omega_{z2}^2[1 - \cos v(t - t_i)] - h_{z2}v \sin v(t - t_i)\};$$
$$Q_{zj} = 0 \text{ at } t \leq \frac{L}{v} \text{ and } t \geq (t_v + \frac{L}{v});$$
$$Q_{zj} = q_0\left\{\omega_{zj}^2[1 - \cos v(t - \frac{L}{v}) - h_{zj}v \sin v(t - \frac{L}{v})]\right\}$$

Respectively, for the planar angular oscillations, we have:

$$Q_\alpha = \sum_1^j Q_{\alpha i} = \frac{2}{M\rho^2}\left(\sum_1^j k_i l_i \dot{q}_i + \sum_1^j c_j \bar{l}_i q_i\right). \tag{10}$$

This equation opens similar to Q_z. It is worth noting that rule of signs for these sums is different. For Q_z all the summands are of the same sign. For Q_α summands corresponding to supports and located on opposite sides of the center of mass have opposite signs. An example for a quad axial automobile:

$$Q_z = Q_{z1} + Q_{z2} + Q_{z3} + Q_{z4}$$
$$Q_\alpha = Q_{\alpha 1} + Q_{\alpha 2} - Q_{\alpha 3} - Q_{\alpha 4}$$

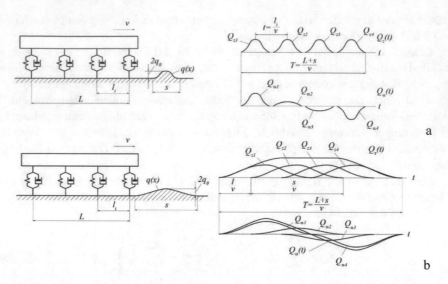

Fig. 2. Plan of movement over short (a) and long (б) single irregularity and input disturbance Q_z and Q_α

These values are illustrated on Fig. 2.

Formula for a function of Q_z and Q_α significantly depends on the relation of s and l_i. For short irregularities, when $s < l_i$ movement of irregularity excites a series of impacts j to the automobile body. This is the effect of an explicit multi-axial machine peculiarity: movement over a single irregularity leads to a series of j periodic irregularities.

Movement over long irregularities (when $s > l_i$) leads to impacts from different supports combining into a single impact, with its duration described by $T = \frac{s+L}{v}$. So another peculiarity of a multi-support machine is that a long irregularity with measures s and $2q_o$ is substituted by a given irregularity with length $s + L$ and height q_n.

Let's assume that are no dampers in order to illustrate the behavior of functions Q_z and Q_α with different irregularity length. In that case we can write down:

$$Q_z \approx \frac{2cj}{M} \sum_1^j q_i = \frac{2cj}{M} \bar{q}_n^z,$$

$$Q_\alpha \approx \frac{2cj}{M\rho^2} \sum_1^j \bar{l}_j q_j = \frac{2cj}{M\rho^2} \bar{q}_n^\alpha.$$

Figure 3 depicts change of value $\bar{q}_z^n$, and therefore Q_z for a symmetrical quad axial automobile with curves on Fig. 3, b given for comparison, are derived from biaxial automobile.

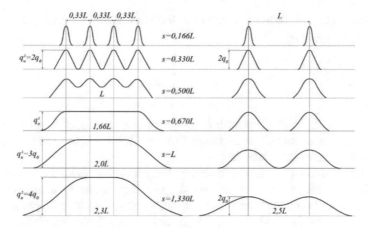

Fig. 3. Perturbation functions $\bar{q}_z^n$, caused by movement of quad-axial automobile over a single irregularity (a) and biaxial automobile (b).

For the purpose of approximate computer modelled measurement of perturbation functions, values of q_z^n were calculated for different number of supports $j = 3 - 8$ (Fig. 4a) in function l or s/L ($s = 1000$ in all cases).

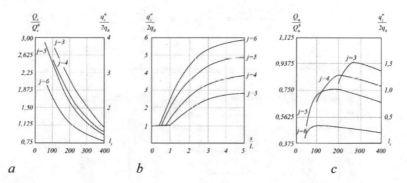

Fig. 4. Height of irregularities relegated to actual height (*a*), it's correlation with s/L (b) and normalized irregularity height (*в*) at different number of supports j

Composite diagram (Fig. 4b) illustrates that starting with the length of irregularities $s = (2,5 - 3,0)L$ the height of virtual irregularity approaches asymptotic value $2q_a j$ with 3 or 4 supports.

However, despite such increase of q_{in}^z, value of Q_z decreases because the rigidity of cushioned element of each support drops with the increase of number of supports. That is how Q_z/Q_z^o ratio accounts for both overlap of impact of separate supports and for decrease in suspension rigidity that is expected with an increase of number of supports.

Q_α function should change according to another law with the increase in length. The longer the irregularity while $s > l$ is strongly expressed, the lower should q_n^α be. At

$\frac{s}{l_i} \to \infty$ we get $q_n^\alpha \to 0$. To test that statement with an electro modulating device, we got the following functions:

$$\bar{q}_n^\alpha = \sum_1^j \bar{l}_i q_i.$$

Figure 4 illustrates q_n^α value with different number of supports. All of the curves have their maximums roughly at the interval of $s = (1,3 - 2,0)L$. After further increase of the length, q_n^α and Q_α values decrease.

That data allows us to establish a connection between unfavorable combination of the length of the irregularity and movement speed with operating conditions.

Area of possible combination of s and v_a is shaded in Fig. 5. This area is limited by the interval of operating speeds $v_{\min} - v_{\max}$ and the maximum length of irregularity $s_{\max}$, that is not yet deemed exceptional. The most unfavorable irregularities would be those that are so short that the following is true:

$$\frac{3,6s}{v_a} = \frac{2\pi}{\omega} \text{ or } s = \frac{2\pi v_a}{3,6 \omega}, \tag{11}$$

Or so long, that the following is true:

$$\frac{3,6(s+L)}{v_a} = \frac{2\pi}{\omega} \text{ or } s = \frac{2\pi v_a}{3,6 \omega} - L. \tag{12}$$

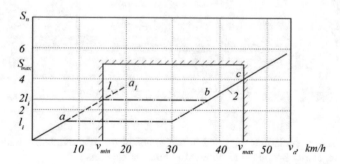

Fig. 5. Dependence of combination of operational s and v_a with resonance modes

Data on Fig. 6 illustrates a quad axial automobile ($L = 4,5$ m; $l_i = 1,5$ m; $\omega = 10$ sec-1). Line 1 is drawn for the resonance with short irregularities and makes sense only when $0 < s < l_i$, although these values do not match with operating speed and therefore have no practical meaning. Long irregularities that cause the effect corresponding to them conform to $s > 2l_i$ (Line 2). Only s and v_a, lying in segment bc, may have operational values. It is important to notice that longer irregularities will have more impact on the machine according to the statements above. That is why the increase in movement speed over irregularities will produce bigger oscillations even if the height of the irregularities stays the same.

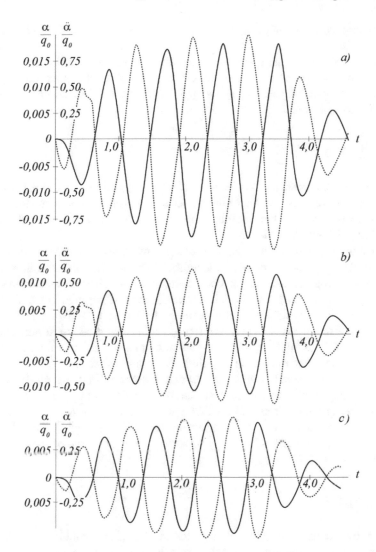

Fig. 6. Angular movement and accelerations, induced by movement over 4 irregularities of automobiles with various number or supports: a – biaxial; b – triaxial; c – quad axial

Both analyzed harmonicas will be induced when lengths of irregularities lie in $l_i < s < 2l_i$ so it is much harder to predict what oscillations will be more unfavorable then.

Q_z and Q_α functions were taken as perturbating actions, moreover they were repeated for four consecutive times. Even washboard road surface is an exception. It is much more common to meet two-three irregularities of similar size. That is why we used an excessive amount of four. Under the impact of irregularities, the body of the automobile experienced unsteady oscillations, that will approach steady oscillations with sufficient damping. Free converging oscillations are registered after the end of the impact of perturbation.

Examples of skewed angular movements and angular acceleration relegated to the height of irregularities for machines with different numbers of support are depicted on Fig. 6. For comparison reasons ω_z and ψ_z values were set equal and the mode of oscillations to resonance.

4   Discussion

The study of characteristics of oscillations of multi-support machines lead to a number of practical conclusions. As an example, let's take some recommendations on test methodology of multi-support machines's movement smoothness that are derived from the specifics of their oscillations.

To ensure the harshest testing environment for when tests are carried out in artificial conditions, it is important to decide on a number of parameters: the length, number and frequency of irregularities, movement speed, the sufficient measurements for an objective decision on suspension quality. The interval of irregularities that cause significant oscillations correlates to the segment bc of Line 2 (Fig. 5.) If all of the irregularities of that interval were of the same height, then the longest one would have the bigger effect. This implies that it is sufficient to choose s and v_a values in point B on Fig. 5 for the means of running only a limited number of tests.

In order to understand how exactly shock-absorbers limit the oscillation of the automobile body while moving over subsequent irregularities, it is better to have a number of irregularities. Irregularities that are spaced to close to each other will not provide sufficient oscillation, so irregularities should be spaced further. Q_z and Q_α functions of single impacts will add up to form a periodic curve of bigger amplitude when the length is L.

5   Conclusion

Consequently, the increase in irregularities length $(s > 2L)$ should increase the vertical oscillations of multi-support machines and decrease angular ones. A multi-support machine acts like a mass on a spring when the length of irregularity is big. When irregularity's length is fixed (or when s/L) its impact on vertical and angular oscillations decreases with the number of supports.

We separate two distinct cases based on the results of computer modeling:

1. With a sufficient damping of $\psi \geq 0,25$, body oscillations become set after third, sometimes even after second irregularity. This conclusion has an important theoretical value as it justifies the calculation for constrained oscillations which is simpler and better developed than calculation for unconstrained oscillation mode.

2. Aperiodic coefficients ψ_z and ψ_α provide good enough first estimate on efficiency of damping for oscillations of different types (free, induced, unsteady). Aperiodic coefficients of Eq. (5) are very useful for engineering evaluation of appropriate suspension damping. They are quite simple and allow to account for the influence of shock-absorber characteristics, their number and location in the automobile as well as separately account for vertical and angular oscillations of the body.

Acknowledgement. The article is prepared in the framework of development of the Base University on the basis of BSTU named after V.G. Shukhov.

References

1. Ovsyannikov, S.I.: Method for calculating walking tractor performance considering operator's energy consumption IOP Conf. Series Mater. Sci. Eng. **327**, 052025 (2018). https://doi.org/10.1088/1757-899x/327/5/052025
2. Grzegozek, W.: Modelowanie dynamiki samochodu przy stabilizujacym sterowaniu silami hamowania. Politech. Krakowska im Tadeusza Kosciuszki, Krakow, (2000)
3. Zukurov, A.M.: Fundamentals of Movement of Pulling and Transporting Vehicles. URGEUS Publsihment, Shahti (2005)
4. Alexandrov, E.E., Bech, M.V.: Automated design of dynamic systems with Lypaunov function. Kharkov, Osnova (1993)
5. Tverskoy, V.M.: Automobile Dynamics. KMI, Kurgan (1995)
6. Rotenberg, R.V.: Automobile Suspension and Movement Smoothness. Machinostroenie, Moscow (1972)
7. Blinov, E.I.: Dynamic and Energetics of Wheeled Machines. Machinostroenie, Moscow (2005)
8. Voloshin, Y.L.: Mathematic models of oscillations of wheeled transport and pulling and transporting vehicles. Tractors Agric. Mach. **6**, 37–43 (2007)
9. Vazhenin, A.N., Aryutov, B.U., Passin, A.V.: Dynamic model of MTA with factored operating conditions. Tractors Agric. Mach. **9**, 21–23 (2007)
10. Nosov, S.V., Kindyuhin, Y.Y.: Modeling of road-tractor-driver system with factoring of tire substructure smoothing. Tractors Agric. Mach. **10**, 12–15 (2009)

Development of the Logistical Model for Energy Projects' Investment Sources in the Transport Sector

Olga Kalinina[1(✉)] [iD], Snezhana Firova[1] [iD], Sergey Barykin[1] [iD], and Irina Kapustina[2] [iD]

[1] Peter the Great St. Petersburg Polytechnic University, Polytechnicheskaya 29, 195251 St. Petersburg, Russian Federation
olgakalinina@bk.ru
[2] Selectel Ltd., Str. Tsvetochnaya 21, lit. A, 196084 Saint-Petersburg, Russian Federation

Abstract. This article is dedicated to the research on the principles of modelling the investment sources structure in innovation regarding to the complicated scientific area spreading over energy strategies in the transport sector. In this paper, arguments to the statement concerning the systemic nature of energy projects open innovations are presented together with the principles of evaluating the effectiveness of innovation and investment projects. The study develops an allocation model for investment sources in innovation and investment projects, taking into account the systemic nature of open innovations and its complexity. The observed model involves designing a transport model to optimize the structure of investment sources in innovation projects in terms of the minimum payback period of an innovation project.

Keywords: Energy strategies · Transport sector · Open innovations · Investment sources · Resources logistical (Allocation) model · Innovations and investment projects

1 Introduction

In accordance with the Energy sector strategy developed by the European bank for reconstruction and development (*EBRD*) since 2006, it has invested EUR 8.6 billion in 172 projects. Computational analysis of the investment sources structure implies the development of methods for effective allocation of the investment sources and testing the results basing on the analysis of empirical and factual information concerning the assessment of the contribution of innovation and investment projects (*IIP*) in the company's economic development. One could define the logistically developed model for energy projects' investment sources in the transport sector as allocation model for calculating the amount of funds being invested into specific energy project. According to the valid opinion of the Russian scientists Silkina G. and Shevchenko S., S. V. Shevchenko further, conceptual areas in the world practice that are largely exposed to the idea of increasing the company's openness in terms of innovation [1].

© Springer Nature Switzerland AG 2020
V. Murgul and M. Pasetti (Eds.): EMMFT 2018, AISC 982, pp. 318–324, 2020.
https://doi.org/10.1007/978-3-030-19756-8_29

The systemic exchange of knowledge in the process of developing open innovations might be claimed. The argument concerning systemic exchange of knowledge in innovative companies was put forward in the research of G. Chesbro (Head of the Haas Business School at the University of California, Berkeley) who was the first to classify innovations as autonomous and systematic within which it is possible to create partnerships [3]. The remark itself made by Henry Chesbro and Professor David Tees refers to the systemic nature of technological innovation which, according to the scientists, involves the innovative development of the related systems and products. Basing on this, Chesbro and Tees come to a conclusion that the centralized control over the implementation of innovations is valid, due to their diversity and large scale, which is economically effective within large corporations. However, in their discussions Chesbro and Tees conclude the systematic exchange of knowledge in the process of partnerships. We believe it is the high complexity of the innovation process that determines the need for systematic interaction between the *IIP* participants (stakeholders) in the development of open innovations at the current time. In the knowledge-based economy the complexity of organizing the management structure lies in the fact that hidden knowledge accumulated in the form of skills and personal developments of specialists, as well as corporate traditions of technical culture cannot be considered separately from the individual holders of this knowledge (employees of a particular company), in contrast to the codified knowledge (specifications recorded in industry standards and development norms) that can be transferred from one organization to another within the group of companies without any significant loss of information quality.

In 1990, the article by Gary Hamel - visiting Professor of strategic and international management at the London School of Business, and K. Prahalad - Professor of corporate strategy and international business at the University of Michigan, called "Basic Competencies of a Corporation" introduced the term "basic competence" (meaning a combination of individual technologies and production skills that form the foundation for the entire set of the company's product lines), and the term "ability" (meaning capacity to manage dealers through training and support of the dealer network, sales, space planning and maintenance, as well as the ability to design a product that is manifested in a continuous and simultaneous planning and testing processes that go separately from the implementation) [7]. With the advent of such concepts as core competencies and competition on the basis of abilities in the 1990s the emphasis has shifted from the external environment of a company to the internal components covering the importance of skills, collective learning, and ability of management to control these factors. According to George Stock, Philip Evans and Lawrence Shulman (Boston Consulting Group), the differences between abilities and competencies are rather profound: basic competence focuses on technological and production experience in terms of specific stages of the value chain, while abilities have a broader base that covers the entire value chain of a company [2]. According to the Stoke-Evans-Shulman conceptual approach, competencies and abilities characterize the "behavioral" aspects of the strategy unlike the traditional structural model.

In general, the systematic nature of open innovations stems from the very essence of the knowledge-based economy. The term "knowledge-based economy" (economy based on knowledge) was introduced in 1996 as a concept of economy based on knowledge, referring to the developed economies that directly rely on production, distribution and use of knowledge and information [6]. At the same time, the interactive innovation model based on interaction of producers and consumers in the process of the formalized (codified knowledge) and non-formalized knowledge (tacit knowledge) exchange, replaces the traditional linear innovation model. In our opinion, the systematic nature of open innovation should be first taken into account in the models of investment sources allocation in the *IIP*.

We agree with the statement made by G. U. Silkina and S. V. Shevchenko, according to which the open innovation model was formed in the process of overcoming negative market trends that were causing a massive reduction in the time horizons of R&D spending, thus, minimizing the payback period of the *IIP*. Developing the approach of G. U. Silkina and S. V. Shevchenko further, it becomes possible to suggest new principles of investment sources allocation for innovative projects:

1. in the *IIP* evaluation due attention should be paid to the models of interaction between investors and innovation companies, taking into account the asymmetry and non-verifiability of information in the field of innovation and investment projects;
2. higher value of improvements in business models in comparison to innovations in the field of product development;
3. shift the major focus to investing in applied developments based on external and internal ideas (from the company's viewpoint);
4. investment in intellectual developments that are classified as the most appropriate for the company's business model.

2 Materials and Methods

The following statement was formulated as hypothesis for the study: In our opinion, the innovation and investment process is characterized by both its systematic nature and complexity of behavioral aspects that are shown due to the influence of the interaction between the *IIP* stakeholders (that is primarily determined by the investor–innovator interaction).

It is possible to develop an economic and mathematical model of the investment sources allocation in order to ensure that the tasks concerning innovation and investment projects are being implemented on the criterion of the minimum values of the payback period for investment in the *IIP*. The mathematical apparatus for solving such tasks is developed [4] and does not concern the linear programming models. Let us denote by "P" the matrix of size "$m \times n$", where at the position "i", "j" is the value "ρij".

$$P = \left\{ \begin{array}{cccc} \rho_{11} & \rho_{12} & \cdots & \rho_{1n} \\ \rho_{21} & \rho_{22} & \cdots & \rho_{2n} \\ \vdots & \vdots & \rho_{ij} & \vdots \\ \rho_{m1} & \rho_{m2} & \cdots & \rho_{mn} \end{array} \right\} \tag{1}$$

The element of the matrix "ρij" indicates the payback period assuming the return of investments directed from the i funding source to the j innovation and investment project of a company. The set of tasks provided by *R&D* can be performed by any innovation and investment project (*IIP*) with a payback period of "ρij". Using the transport scheme, it is possible to create a matrix of the payback period indicators. The target function reflects the company's efforts to allocate investment sources in the innovative projects that most effectively perform the tasks. It is necessary to make a plan for allocation of the flow of investment sources in which the payback period within a set of projects in the task would be minimal.

Let the "ϕ_{ij}" be the amount of investment resources of the i source (funding directed for the implementation of the j task from the budget of the i source). The total volume of resources that can be directed from the i source amounts to "μ_i", $i = 1, 2, ...,$ m; the size of investment needs within some j task is "v_j", $j = 1, 2, ..., n$.

Investment resources are allocated to projects that have completed the tasks in the shortest period of time possible. The system presupposes competition among innovation and investment projects that perform a set of tasks in the shortest period of time. When the longest of all *IIPs* completes, the figures indicating its payback time should be as minimal as possible to provide allocation of investment resources. In this case, the target function "f" according to the criterion of the minimum payback period should be equal to the maximum value of all the payback period values, meaning:

$$f(\varphi_{ij}) = \max_{\varphi_{ij} > 0} \rho_{ij} \to min, \tag{2}$$

where the maximum is taken only from those "ρ_{ij}", that correspond to the positive payback period values. The task under consideration should be written as follows:

$$\begin{cases} \sum_{j=1}^{n} \phi_{ij} = \mu_i, & i = 1,...,m \\ \sum_{l=1}^{m} \phi_{ij} = v_i, & j = 1,...,n \\ \sum_{i=1}^{m} \mu_i = \sum_{j=1}^{n} v_j, & i = 1,...,m; \ j = 1,...,n \\ \varphi_{ij} \geq 0, & i = 1,...,m; \ j - 1,...,n \end{cases} \tag{3}$$

Table 1. Transport table for allocation of investment sources within *IIP*.

Sources of investment	Investment resources "ϕ_{ij}" from the i source ($i = 1, 2, ..., m$) to the j IIP ($j = 1, 2, ..., n$)						Total volume of investment in R&D, Rub.
	1	2	...	j	...	n	
1	ϕ_{11}	ϕ_{12}	...	ϕ_{ij}	...	ϕ_{1n}	μ_1
2	ϕ_{21}	ϕ_{22}	...	ϕ_{ij}	...	ϕ_{2n}	μ_2
...	...	...	...	...	...	...	...
i	ϕ_{i1}	ϕ_{i2}	...	ϕ_{ij}	...	ϕ_{in}	μ_i
...	...	...	...	...	...	...	...
m	ϕ_{m1}	ϕ_{m2}	...	ϕ_{mj}	...	ϕ_{mn}	μ_m
Needs, Rub.	v_1	v_2	...	v_j	...	v_n	x

The meaning of the constructed algorithm for solving the task observed is to stop at the smallest possible value of $\rho ij \ \omega$ at which some solution still can be found within the process of solving an economic and mathematical model of the investment sources allocation. The task on allocation of investment sources within the *IIP* is solved via iterations (Table 1).

The transport table for the allocation of investment resources by the *IIP* can be presented as follows (Fig. 1).

The economic and mathematical model developed allows allocating investment resources depending on the payback period of the *IIP*.

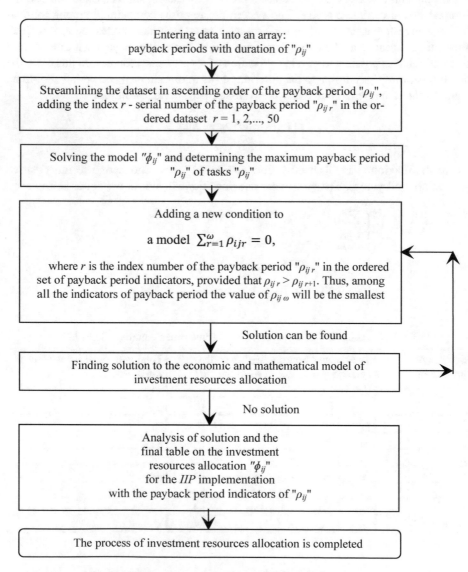

Fig. 1. Block diagram of allocation of investment resources.

3 Results

The problem of allocation of investment sources within *IIP* can be written as follows:

$$f(\varphi_{ij}) = \max_{\substack{\varphi_{ij} > 0}} \rho_{ij} \rightarrow \min) \tag{4}$$

subject to the following:

$$
\begin{cases}
\varphi 11 + \varphi 12 + \varphi 13 + \varphi 14 + \varphi 15 + \varphi 16 + \varphi 17 + \varphi 18 + \varphi 19 + \varphi 1,10 = 0.19; \\
\varphi 21 + \varphi 22 + \varphi 23 + \varphi 24 + \varphi 25 + \varphi 26 + \varphi 27 + \varphi 28 + \varphi 29 + \varphi 2,10 = 0.18; \\
\varphi 31 + \varphi 32 + \varphi 33 + \varphi 34 + \varphi 35 + \varphi 36 + \varphi 37 + \varphi 38 + \varphi 39 + \varphi 3,10 = 0.205; \\
\varphi 41 + \varphi 42 + \varphi 43 + \varphi 44 + \varphi 45 + \varphi 46 + \varphi 47 + \varphi 48 + \varphi 49 + \varphi 4,10 = 0.21; \\
\varphi 51 + \varphi 52 + \varphi 53 + \varphi 54 + \varphi 55 + \varphi 56 + \varphi 57 + \varphi 58 + \varphi 59 + \varphi 5,10 = 0.215; \\
\varphi 11 + \varphi 21 + \varphi 31 + \varphi 41 + \varphi 51 = 0.113; \\
\varphi 12 + \varphi 22 + \varphi 32 + \varphi 42 + \varphi 52 = 0.088; \\
\varphi 13 + \varphi 23 + \varphi 33 + \varphi 43 + \varphi 53 = 0.11; \\
\varphi 14 + \varphi 24 + \varphi 34 + \varphi 44 + \varphi 54 = 0.091; \\
\varphi 15 + \varphi 25 + \varphi 35 + \varphi 45 + \varphi 55 = 0.1; \\
\varphi 16 + \varphi 26 + \varphi 36 + \varphi 46 + \varphi 56 = 0.105; \\
\varphi 17 + \varphi 27 + \varphi 37 + \varphi 47 + \varphi 57 = 0.085; \\
\varphi 18 + \varphi 28 + \varphi 38 + \varphi 48 + \varphi 58 = 0.109; \\
\varphi 19 + \varphi 29 + \varphi 39 + \varphi 49 + \varphi 59 = 0.095; \\
\varphi 1,10 + \varphi 2,10 + \varphi 3,10 + \varphi 4,10 + \varphi 5,10 = 0.104; \\
\sum_{r=1}^{\omega < 50} \rho_{ijr} = 0, r = 1,2,\ldots,50; \\
\varphi ij \geq 0, i = \overline{1,5}; j = \overline{1,10}.
\end{cases}
\tag{5}
$$

The total investment needs of five *IIP* are fully met by the amount of 1.0 million rubles, if the equality is implemented:

$$\sum_{i=1}^{m} \varphi_{ij} = \sum_{j=1}^{n} \varphi_{ij} \tag{6}$$

The maximum payback period of the *IIP* receiving investment resources from the *i* source for implementation of j *IIP* will be 3.6 years.

The results of solving the economic and mathematical model can be interpreted as the investment sources allocation within the *IIP*, which have different characteristics both in the context of investment sources, as well as conditions for implementation.

4 Discussion

The allocation model for energy projects' investment sources in the transport sector is considered for the purpose to cover the features of logistically developed model for calculating the funds invested to the wide range of energy projects. Further discussion of the issue dedicated to the research on the principles of modelling the investment sources structure in innovation may be associated with such feature as the systemic nature of open innovations are presented, together with the principles of evaluating the effectiveness of innovation and investment projects.

The practical application of the conditions for a certain structure of investment sources in an innovative project to be developed can be considered using the example of the task on allocation of investment sources within several *IIPs*.

The developed model of resource allocation can be applied not only within the open innovation theory, but also in practical and applied problems in the concept of the sharing economy, the main approaches to which were considered by I. Alexandrov and M. Fedorova [5].

5 Conclusion

In the course of the study, the hypotheses put forward earlier were tested. The innovation and investment process is characterized by both, its systematic nature and complexity of behavioral aspects that is primarily determined by the investor – innovator interaction regarding to energy strategies in the transport sector.

Thus, in our opinion, first of all, the contribution of the innovation and investment project in terms of increasing the company's economic value can be estimated basing on the payback period of capital investments in the implementation of innovations, that allows to assess the effectiveness of investing resources in the development of a particular area. Consequently, it is necessary to allocate funding for innovation and investment projects that participated in specific tasks.

References

1. Silkina, G.U.: Innovative processes in the digital economy. Information and communication drivers. In: Silkin, G.U., Shevchenko, S.U. (ed.) Publishing house of the Polytechnic University, St. Petersburg (2017)
2. Stoke, G., Evans, P., Shulman, L.: Competition based on abilities. New rules of corporate strategy. Collection of articles: Corporate strategy: Translation from English, 2nd edition. United Press LLC, Moscow (2009)
3. Henry, C., Tees, D.: When virtual is justified. Organization for innovations. Collection of articles: Strategic alliances: Translation from English. Alpina Business Books, Moscow (2008)
4. Chernyak, A., Novikov, V., Melnikov, O., Kuznetsov, A.: Math for Economists Based on Mathcad. BHV-Petersburg, St. Petersburg (2003)
5. Aleksandrov, I., Fedorova, M.: International Science Conference SPbWOSCE-2017 Business Technologies for Sustainable Urban Development: Strategic planning of the tourism development in small cities and rural territories as a tool for the development of the regional economy. MATEC Web Conference, vol. 170, p. 01011 (2017)
6. OCDE, The Knowledge-Based Economy, Paris (1996). http://www.oecd.org/dataoecd/51/8/1913021.pdf
7. Prahalad, C.K., Hamel, G.: Core competence of the corporation. Harvard business review, pp. 1–15 (1990)

Modeling of Railway Track Sections on Approaches to Constructive Works and Selection of Track Parameters for Its Normal Functioning

Alexey Loktev[1] , Vadim Korolev[1] , Irina Shishkina[1(✉)] ,
Lidia Chernova[1] , Pavel Geluh[1] , Alexander Savin[2] ,
and Daniil Loktev[3]

[1] Russian Open Academy of Transport of the Russian University of Transport,
Chasovaya str. 22/2, 125190 Moscow, Russia
`shishkinaira@inbox.ru`
[2] Far 2JSC Railway Research Institute (JSC VNIIZhT),
3rd Mytischinskaya str., 10, 129626 Moscow, Russia
[3] Moscow State University of Civil Engineering (MGSU),
Yaroslavskoye Shosse, 26, 129337 Moscow, Russia

Abstract. The article deals with the districts of the transition zone from the ballast sub-basement to the constructive works, observations of these areas and their research. The analysis of observations of the process of accumulation of residual deformations on the approaches to constructive works are considered. A mathematical model to determine the optimal parameters of the elements of the track bed structure and the roadbed is considered.

Keywords: The permanent deformation · District of variable rigidity ·
Settlement of track · Deformation · Elastic sediment · Grading of track ·
Recursive equations

1 Introduction

In the process of development of railway infrastructure and construction of lines with high-speed and high-speed train traffic, more and more attention is paid to the transitional areas to constructive works, where the ballast-free base is operated, on which special designs of transition areas are not applied. In these poorly districts, there is the greatest accumulation of path geometry disorders and residual deformations and more intensive wear of the elements of the track bed structure. All these processes occur as a result of a sharp change in the stiffness of the track during the transition from the approaches of the elevated approach to the bridge or tunnel and vice versa. Unlike conventional sections of the road, transition zones require more attention and increase the amount of work on the running-maintenance.

These processes can be determined by the following reasons, such as the accumulation of sediment compacted (underconsolidation) of the ballast under the assembled rails and sleepers and a draught of the upper top embankment. An important role

© Springer Nature Switzerland AG 2020
V. Murgul and M. Pasetti (Eds.): EMMFT 2018, AISC 982, pp. 325–336, 2020.
https://doi.org/10.1007/978-3-030-19756-8_30

in the stability of the path in front of constructional works plays a violation of drainage behind the walls of the abutments as a result of blockage of drains, which are not cleaned in a timely manner and as a result of this there is waterlogging of the soil at the border of the main site of the roadbed, which also affects the appearance of subsidence directly behind the abutments. In the process of accumulation of residual deformations in the sub-basement, there is an increase in the elastic subsidence of the path under the influence of dynamic loads from the rolling stock and in the process of operating of sites of variable stiffness leads to a progressive accumulation of residual deformations and disruption of the stable track condition [1–4].

The consequence of the accumulation of residual deformations are:

– the appearance of deformations in the boundary of the main site of the roadbed in the form of ballast recesses (lodges);
– damage to bridge superstructure of a bridges during the occurrence of increased dynamic loads at the time of failure of the locomotive when entering the bridge;
– gaps between the sleeper and ballast, which cause hidden tremors when the train and the higher the speed of the train, the stronger the tremors.

An important factor in the development of deformations in the transition section is the deflection of the rail, which has a significant impact on the magnitude of the dynamic effect of the passage of the rolling stock on the track bed structure. Rail deformation occurs due to uneven precipitation on the transitional section of the track and requires special attention. Construction of speed lines and high-speed passenger-carrying lines, as well as increasing the axial load of cars make the most serious approach to the study and study of areas of variable stiffness. Various designs of such sections are being developed with the possibility of using them on lines with train speeds of 160, 200, 250 km/h and more. The problems of creating an optimal design in the area of the track interface with constructional works are solved [5].

The purpose of this design is to prevent the appearance of sediment path within the boundaries of areas of variable stiffness from the dynamic effects of the rolling stock, as well as increasing the resistance to horizontal shifts to maintain the design geometry of the track thereby reducing the cost of the current content [6].

Transition section with variable stiffness, as a rule, are arranged on the approaches to bridges, tunnels and viaducts with a ballast-free track design, where the indications of the track measuring car are marked by periodic or permanent disorders of the rail track. The device plots a variable stiffness needed to ensure a smooth and stable transition of stiffness of track with roadbed on an artificial construction to obtain reduce the dynamic effects of rolling stock on the track and maintaining the integrity and stability of roadbed.

The most common construction sites of variable stiffness, used to date, are:

– design phase of a variable stiffness reinforced with geogrids;
– construction of the section of the transition path of reinforced concrete boxes;
– construction with sub-ballast reinforced concrete slabs;
– construction of the transition area with the use of gabions.

In addition to the considered variants of the device sections of variable stiffness, which are designed for the device in front of artificial structures (bridges, tunnels, overpasses and viaducts), there are sections of the path on which the device is necessary transition zone. These portions include transitions with conventional ballast track on the way ballastless design. The design of the transition zone of the substructure of the track on the concrete base to the track on the ballast is a track with a reinforced design of the sub-basement. The minimum thickness of the reinforced structure of the sleeper base is 1.20 m. The problem of transitions from ballast to ballast track can be solved by strengthening the track bed structure on the approach to the ballast-free segment by homogenizing the ballast prism with binders and laying additional rails inside the track. Strengthening the track bed structure is due to the bonding of ballast prism particles and the use of flange rails. As the ballast-free structure approaches, the depth and area of the ballast material is increased. Due to this, the rigidity of the track bed structure and its modulus of elasticity gradually increases. The use of flange rails is designed to reduce the deformability of the track of the traditional design and to ensure maximum integrity of the path geometry. The device of transitional sites with variable rigidity on the existing railway lines is more expedient to make during carrying out capital repairs of a way or at the time of carrying out strengthening and reconstruction of constructive works. Due to the complexity of the construction of sections of variable stiffness, which temporarily require the closure of the track, which is necessary for the preparation of project documentation, a technological "occupation" has been laid down to carry out these works. The duration of such occupations depends on the selected technological process of work in the project. If possible, the construction of sections of the alternating occupation is advisable to perform immediately on both sides of the constructive works to reduce the closing time of the track [7–10].

In projects on the device of sites of variable rigidity, it is necessary to provide mainly the mechanized way of performance of works for the maximum exception of manual work from process of works, to apply advanced technologies of work to reduce the construction period and the duration of technological "occupations".

2 Research Technique

In order to select the optimal design of the variable stiffness section, it is necessary to perform field studies of the dynamic behavior of the transition zone during regular trains on the section of the operating railway track. Nineteen engineering structures on double-track and single-track sections of the track, including reinforced concrete bridges (9 pieces), metal bridges (10 pieces) were chosen as objects of observation.

In the process of visual inspection during the observation period of defects of the subgrade (floats, flaps, longitudinal and transverse cracks on the slope and edge, the presence of ballast plumes, depths) was not found. The ball was shot exaggerated longitudinal profiles made from the boundaries of the cabinet walls of abutments in each side of the stretch for 200 m on the left and right thread of the track. The first 50 m of the road from the bridge mark, were taken through 2.5 m, and the remaining 150 m every 5 m. Measurement accuracy up to 1 mm. Throwing off of profiles elevation was carried out quarterly during the year.

Precipitation measurement of elastic track is made using a theodolite 4T30P under passing freight trains. On the rail web were glued to mark with a millimeter scale, and in the process the deflection of the rail under load was carried out the measurement.

Also, the amount of elastic precipitation was determined by the gap formed between the pin hammered under the rail and the sole of the rail after the passage of the freight train. Analyzing the results obtained in the course of field measurements of the roadbed sections on the approaches to constructive works, it is easy to notice that the greatest precipitation of the path occurred in the period March–July. It is during this period that the strength characteristics of the roadbed are reduced by more than 50% due to spring thawing. Maximum track precipitation was detected mainly in the zone from 10 m to 30 m (from the cabinet wall) before the bridge or overpass abutment which reached more than 20 mm per year. In the area of the abutment of the bridge, the track precipitation is reduced. The results of measurements of elastic precipitation also indicate that the largest deformation of the sub-basement occurs in the border zone before the abutment. The data of track measuring tapes collected for the period January-December (working passes) were analyzed. The predominant defects of the transition zones are distortions and subsidence of the second degree, other deviations from the norms are less common. The process of development of subsidence begins in the period of debauchery, when there is thawing of the roadbed and its strength properties are reduced. Further, the processes of instability development either fade or continue throughout the warm period.

In recent years, the increasing use in the diagnosis of the roadbed has become a high-speed georadar sensing in combination with laser scanning. High-speed georadar sensing makes it possible to cover a large polygon in order to detect defects of the roadbed. The result of these observations is to identify areas of the roadbed with increased moisture saturation and deformability with the determination of the degree of development. Laser scanning shows the degree of narrowing of the main site of the roadbed, the size of the edge, the overstated slope. On the basis of these measurements, it is possible to identify the primary signs of defects of the roadbed.

Having the results of the passes of track measuring cars in the form of degrees of faults of the geometry of the track and subsidence, we cannot see the full picture of what is happening in the sub-basement, and at the same time it is difficult to track the beginning of the development of defects of the roadbed. Giordano survey complements the system diagnose the track and roadbed. On one of the studied objects, according to the results of surveys, the development of the defect in the form of ballast deepening was recorded in the period from March to November. This defect of the roadbed was not visible in the analysis of track measuring tapes. This can be explained by the fact that after the working passage of the track recording car, the elimination of the identified disorders occurs, the way it is straightened in the plan and profile by filling the rubble and its subsequent compaction thereby hiding the development of subsidence under the ballast bed. Analyzing the GPR data, we also see the places of moisture accumulation in the boundary of the subgrade. In our case, on all approaches to the constructive works observed by us, there is an accumulation of moisture from the foundations of the bridge. This indicates a violation of drainage from the surface of the subgrade, thereby contributing to the liquefaction of the upper layers of the soil of the roadbed, which in turn leads to deformations and subsidence.

During the analysis of the results of the survey on the experimental sites, it is safe to say that the sediment track on the approaches to constructive works is present on each side. To a greater extent, the sediment path develops in front of the bridges in the course of trains and to a lesser extent behind them. In order to choose the correct variant of the construction of the variable stiffness section, it is necessary to consider the process of interaction of the wheel and rail on the basis of a mathematical model in order to determine the optimal parameters of the railway track.

There are many methods and computational algorithms to solve the problem of calculating structures, their elements and entire structures for different types of applied load. The most relevant from the point of view of fundamental research on the mechanics of a deformable solid body, and from the point of view of practical applications in engineering problems of calculation of the track, methods allowing to take into account not only the dynamics in the application of the load, but also the dynamics in changes in the deformation and force characteristics of the track form.

The most commonly available methods are divided into two large groups: research, in which the desired values are described as functions of known quantities, coordinates and time, and numerical, in which the desired values are described by linear dependencies on known characteristics. Each class of tools acquires its positive and negative properties. In the research solution of modern multiparameter problems, almost always there is a need to use special conditions governing the use of the data. Linearization of functions, which is best suited in the case of numerical methods, significantly expanding the number of required values and equations that unite them, and, despite the simplicity of the derived systems, their mathematical solution requires significant computational resources. Among the numerical methods, the methods of finite and boundary elements are distinguished, thanks to a relatively simple implementation using modern programming languages. Parametric dependences of the final characteristics of the dynamic effects obtained by analytical methods, enable employees of research and design institutes to study the impact of structural elements on the final result of the calculation, and the practical application of numerical methods in software systems make it possible to effectively perform the most complex tasks, leading to the result by means of graphical or tabular dependencies.

In our case, we would like to consider an approach to the construction of a mathematical model based on obtaining recurrent equations, the solution of which will allow us to obtain dependencies for the required quantities beyond the upper limits of the maximum gap. Here, the beam method for cylindrical waves-strips, the appearance of which occurs under the dynamic action of a solid-state drummer and a plastic target, is applied and refined. The problems based on the dynamic action of a solid body on a flat linear structure are solved with the help of the wave theory based on the distribution of the wave planes of the maximum and minimum gap in the contact target. As a solution tool used ray method, which was applied Achenbach'om and Reddy and improved for dynamic contact Loktevs problems.

In continuation of the interaction of the wheel, considered as a solid and the track bed structure, a contact region of radius r0 appears on the rail and quasi-longitudinal and quasi-transverse waves arise from its edges, the regions of which are the surfaces of a strong rupture. Note that in a flat element of the surface of a strong rupture are cylindrical surfaces-strips, forming which are parallel to the normal to the middle surface, i.e., z axis, and guides, located in the middle plane, are circles, expanding at normal speeds. In this mathematical model, the rail is considered as an elastic orthotropic plate, and as a material having the ability to have a cylindrical anisotropy.

The dynamic operation of a circular elastic orthotropic Ufland-Mindlin plate with cylindrical anisotropy in the polar coordinate system is described by equations extracted from the equations given in [1], by taking into account the inertia of the cross sections rotation and the deformation of the transverse shear, and in the case of the axisymmetric problem, the wave characteristics do not depend on the angle θ:

$$D_r\left(\frac{\partial^2 y}{\partial r^2} + \frac{1}{r}\frac{\partial \varphi}{\partial r}\right) - D_\theta \frac{\varphi}{r^2} + hKG_{rz}\left(\frac{\partial w}{\partial r} - \varphi\right) = -\rho \frac{h^3}{12}\frac{\partial^2 \varphi}{\partial t^2} \tag{1}$$

$$KG_{rz}\left(\frac{\partial^2 w}{\partial r^2} - \frac{\partial \varphi}{\partial r}\right) + KG_{rz}\frac{1}{r}\left(\frac{\partial w}{\partial r} - \varphi\right) = \rho \frac{\partial^2 w}{\partial t^2} \tag{2}$$

$$C_r\left(\frac{\partial^2 u}{\partial r^2} + \frac{1}{r}\frac{\partial u}{\partial r}\right) - C_\theta \frac{u}{r^2} = \rho h \frac{\partial^2 u}{\partial t^2} \tag{3}$$

$$C_k\left(\frac{\partial^2 v}{\partial r^2} + \frac{1}{r}\frac{\partial v}{\partial r} - \frac{v}{r^2}\right) = \rho h \frac{\partial^2 v}{\partial t^2} \tag{4}$$

$$D_k\left(\frac{\partial^2 \psi}{\partial r^2} + \frac{1}{r}\frac{\partial \psi}{\partial r} - \frac{\psi}{r^2}\right) - KhG_{\theta z}\psi = -\rho \frac{h^3}{12}\frac{\partial^2 \psi}{\partial t^2} \tag{5}$$

To find the dynamic deflection, as well as the contacting force, we find the transverse displacement w(t), which is a component in the system of Eqs. (1) and (2). The remaining equations represent independent subsystems that can be solved separately. The functions found from the relations (3–5), do not affect the desired and investigated dynamic parameters, based on this, we consider the following system of Eqs. (1 and 2). To find the designated (desired) functions included in the Eqs. (1–5), it is necessary to perform the expansions in power series of spatial coordinates and time. But this technique is based not only on the use of decomposition of the required quantities in a series, but also on the use of geometric and kinematic compatibility conditions proposed by Thomas [2] and recorded for physical components in the following form:

$$G\left[\frac{\partial Z_{,(k)}}{\partial s}\right] = -[Z_{,(k+1)}] + \frac{\delta[Z_{,(k)}]}{\delta t} \tag{6}$$

The following expression is used to move from the jump of the derivative of the function Z in coordinate to the jump of the derivative of the desired function in time of the highest order:

On the boundary of the contact area of the wheel and rail at $r = r0$, the expression can be written as a decomposition only in time:

$$Z(x_\alpha, t) = \sum_{\alpha=1}^{2} \sum_{k=0}^{\infty} \frac{1}{k!} [Z_{,(k)}]_{t=0} t^k \tag{7}$$

Next, we find the unknown functions W and Qr (in the form of ray series intervals) from the boundary conditions:

$$W = \sum_{\alpha=1}^{2} \sum_{k=0}^{\infty} \frac{1}{k!} X_{w(k)}^{(\alpha)} (y_\alpha)^k H(y_\alpha) \tag{8}$$

$$Q_r = KG_{rz}h \sum_{\alpha=1}^{2} \sum_{k=0}^{4} \frac{1}{k!} \left(-X_{w(k)}^{(\alpha)} G^{(\alpha)-1} + \frac{\delta X_{w(k-1)}^{(\alpha)}}{\delta t} G^{(\alpha)-1} - \omega_{\varphi(k-1)}^{(\alpha)} \right) (y_\alpha)^k H(y_\alpha) \tag{9}$$

With the aim of finding invariant values of integration, to solve the problem of dynamic interaction of wheelset and track bed structure, which was simulated using a flat type element Uflyand-Mindlin to model the behavior of interaction uses a buffer that can be presented in the form of elastic, viscoelastic and elastoplastic elements. For these three elements (respectively), the dependences of the emerging contact force at the interaction point on the track movement and on the mechanical characteristics of the materials used are considered:

$$P(t) = E_1(a(t) - w(t)) \tag{10}$$

$$P(t) = E_1(a - w) - \frac{E_1}{\tau_1} \int_0^t (\dot{a} - \dot{w}) e^{-\frac{t-t'}{\tau_1}} dt' \tag{11}$$

$$\alpha = \begin{cases} bP^{2/3}, & \frac{dP}{dt} > 0, P_{max} < P_1 \\ (1+\beta)c_1 + (1-\beta)Pd, & \frac{dP}{dt} > 0, P_{max} > P_1, \\ b_f P^{2/3} + \alpha_p(P_{max}), & \frac{dP}{dt} < 0, P_{max} > P_1, \end{cases} \tag{12}$$

As a result, we obtain expressions to determine the dynamic deflection:

$$
\begin{aligned}
\widetilde{w}(\tilde{t}) = \frac{\widetilde{E}\,\widetilde{V}}{\pi\widetilde{h}}\Bigg[& \frac{\tilde{t}^3}{6} - \left(1 + \frac{G^{(2)}}{G^{(1)}}\right)\frac{\tilde{t}^4}{12} + \left[3\left(1 + \frac{G^{(2)}}{G^{(1)}}\right)^2 - \frac{\widetilde{E}}{\pi\widetilde{h}}(\tilde{m}+2)\right]\frac{\tilde{t}^5}{120} \\
& + \left[\frac{2\widetilde{E}}{\pi\widetilde{h}}\left(1 + \frac{G^{(2)}}{G^{(1)}}\right)(\tilde{m}+4) - 4\left(1 + \frac{G^{(2)}}{G^{(1)}}\right)^3 + \frac{1}{4}\left(1 + \frac{G^{(2)3}}{G^{(1)3}}\right)\right. \\
& \left. + \frac{12G^{(2)2}}{\widetilde{h}^2 G^{(1)2}}\left(1 - \frac{G^{(2)3}}{G^{(1)3}}\right)\left(1 - \frac{G^{(2)2}}{G^{(1)2}}\right)^{-1} - \left(\frac{E_\theta}{E_r} - 1\right)\frac{G^{(2)}}{G^{(1)}}\left(1 - \frac{G^{(2)}}{G^{(1)}}\right)^{-1}\right]\frac{\tilde{t}^6}{720}\Bigg]
\end{aligned}
\tag{13}
$$

In this study, one of the main is a method based on the representation of unknown quantities in the form of series expansion by spatial coordinate and time, as well as on the use of compatibility conditions for the transition from the derivative of an unknown quantity by spatial coordinate to the derivative of the same quantity by time. Also, the required functions are given, which are the solution of the dynamic problem, in the form of ray series segments, the coefficients of which are jumps of time derivatives of the required functions of different orders, and the variable is the time elapsed since the arrival of the wave at a given point in the track construction.

Systems of recurrent differential equations for the orthotropic elastic element are derived from equations describing the dynamic behavior of plane elements using compatibility conditions. In the course of solving these systems, the coefficients of the ray series are determined up to arbitrary constants up to the fifth term of the series.

When merging solutions outside the contact area and inside it, the values of the integration constants included in the coefficients of the required quantities expansions are obtained. Analytical expressions for the dynamic deflection (subsidence) and contact force arising at the interaction of the mounted wheels and the track are obtained. These expressions include parameters of dynamic interaction (composition speed and load on the mounted wheels) and mechanical characteristics of interacting bodies. These expressions are ready for use in real engineering problems, it remains only to substitute the parameters determined by field measurements or numerical modeling.

Further, using the results of engineering-geological studies, we determine the mechanical characteristics of the roadbed and the ground under it, and then determine the parameters of the path from the action of the dynamic load from the mounted wheels passing along the rail. The obtained values of dynamic deflections (displacements) and stresses are compared with the given data of the track recording car.

3 Results and Discussion

Using the relations (11) for dimensionless dynamic deflection, we construct graphs of the dynamic characteristics (dynamic deflection and interaction force) versus time.

Figure 1 shows the dependence of the dynamic deflection of the track bed structure on time for different values of the ratio E_θ/E_r, which are indicated by the numbers in the curves, other parameters in the calculations take the following values: $\tilde{m} = 25$, $\tilde{h} = 1$, $\tilde{E} = 1.3 \times 10^{-6}$, $\tilde{V} = 8.8 \times 10^{-3}$. The influence of anisotropic properties of the embankment on the characteristics of dynamic action is considered: as the ratio decreases, E_θ/E_r the deflection-drawdown increases (see Fig. 1) to a certain value at $E_\theta/E_r < 1$; as the ratio increases, E_θ/E_r the deflection decreases, since the last member of the series segment (11) decreases with the ratio $E_\theta/E_r > 1$.

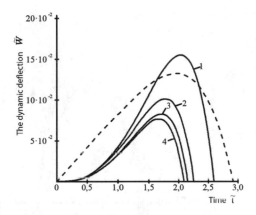

Fig. 1. Dynamic deflection versus time for different ratio values E_θ/E_r.

Figure 2 shows the precipitation of the embankment, based on the assumption that it has transversally isotropic properties for different shear modules in the direction perpendicular to the plane of the embankment, i.e. according to the proposed model of constructing a wave pattern of changes in parameters from the dynamic action. For different ratios of the shear modulus in the direction perpendicular to the rail thread and the deformation modulus in the direction of the thread, G_{rz}/E_r the ratio values are given in figures at curves, at $G_{rz}/E_r = 0.54$ the ballast prism has isotropic properties. Figure 2 shows that an increase in the value of the ratio of modules G_{rz}/E_r leads to a decrease in the deflection-precipitation.

The influence of these elastic characteristics on the dynamic precipitation of the upper structure of the path has the same character as the influence of the rate G_{rz}/E_r in the previous figure, but at the same time, some distinctive features are visible. For example, it can be seen that the change $G_{\theta z}$ slightly increases the deformation time of the track, and the change in the ratio $G_{\theta z}/E_r$ leads to a greater spread of the maximum precipitation values compared to the graphs of the previous figure. That is, the shear modulus in the plane θ_z affects the final dynamic draft more strongly than the shear modulus in the RZ plane. The shear modulus $G_{r\theta}/E_r$ has less effect on the value of dynamic precipitation than shear modulus in other planes, but at the same time, this effect is quite noticeable (although initially this effect is not obvious) and in the design of the railway track with small precipitation can be recommended, including vertical reinforcement, to increase the shear modulus in the plane $R\theta$.

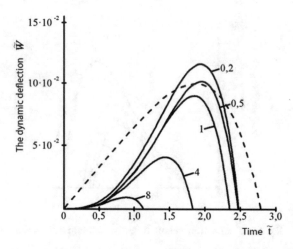

Fig. 2. Dynamic deflection versus time for different ratio values $G_{r\theta}/E_r$.

Comparing the graphics according to the dynamic precipitation can pick up the mechanical characteristics of the embankment c anisotropic properties (by using different designs for the transitional sections of variable stiffness). Thus, to achieve a decrease in precipitation, to the level necessary for the movement of the composition at a given speed, as well as the maximum force at which there will be no defects in the roadbed.

The work also identified the movement of the embankment base in the area of its contact with the residual soil.

Figure 3, a show that with a decrease in the elastic modulus E_r and E_θ the maximum value of the dynamic deflection increases, in addition, E_r the greatest impact on the deflection. With a decrease in the values of the shear modulus, the deflection increases, while, as shown in Fig. 3b, more than other parameters affect the dynamic deflection of the target.

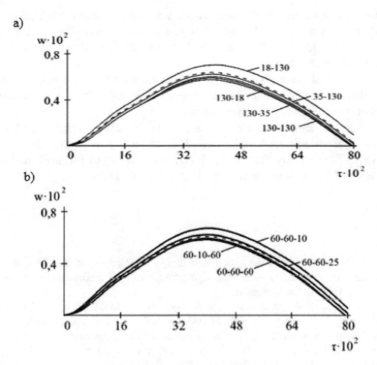

Fig. 3. The dependence of the dynamic deflection at the point of impact time for various values: (a) modulus of deformation, (b) the shear moduli.

By reducing G_{rz}, $G_{\theta z}$ maximum deflection and the time during which the deflection is equal to zero, increases, and the decrease in time, related to the zero deflection increases more intensively, i.e. the restoration of the roadbed will be slower. Thus, analyzing the obtained theoretical dependences for different values of the mechanical parameters of the roadbed with the data obtained as a result of the calculation, it can be seen that changing the mechanical parameters of the soil it is possible to both increase and decrease the deformation of the railway track. The best approximation and the truth of the obtained data of mathematical modeling and experimental studies make it possible to apply the obtained methods in solving the main problems of ensuring the stability of the behavior of Railways in operation and in the design of high-speed railway lines.

4   Conclusions

By analyzing the results of analytical calculation and graphical comparison for dynamic deflection (precipitation) and the interaction force for the railway track embankment, which has anisotropic elastic properties, we can draw the following conclusions:

1. to minimize dynamic effects from the railway equipment on the track6 it is necessary to consider the individual design sections of the variable stiffness after carrying out a detailed survey and analysis of statistical observations on the passages of diagnostic cars;
2. using numerical methods in solving dynamic problems, makes it possible to determine the optimal values of the parameters of the railway track in the area of variable stiffness during the dynamic load;
3. to increase the speed of the rolling stock on the site of variable stiffness up to 140 km/h at an axle load of 22 tons, it is necessary to provide the modulus of elasticity of the railway track in the direction of the rail to 190 MPa, and along the sleepers to 140 MPa, in the vertical direction to 90 MPa.

References

1. Instructions for the use of structural and technical solutions of transition areas on the approaches to constructive works for high-speed combined traffic areas. Approved. order No. 2754p from 12 December 2013
2. New construction of transitional sections from the embankment to the bridge. P760/4. Approved by the meeting of the OSJD Commission for infrastructure and railway equipment from 7 to 10 November 2005, the Committee of the OSJD, Warsaw
3. Loktev, A.A., Sycheva, A.V.: Sci. Technol. Transp. **4**, 4 (2013)
4. Loktev, A.A., Sycheva, A.V.: Path Track Facil. **7**, 3 (2014)
5. Loktev, A.A., Sycheva, A.V., Zaletdinov, A.V.: In the collection: mathematics, informatics, natural science in economics and society proceedings of the all-Russian scientific conference, vol. 27 (2014)
6. Loktev, A.A., Talashkin, G.N., Stepanov, K.D.: Transp. Russ. Fed. **2–3**(63–64), 5 (2016)
7. Loktev, A.A.: Mech. Compos. Mater. Struct. **11**(4), 14 (2005)
8. Loktev, A.A.: Her. Voronezh State Univ. Ser.: Physics. Math. **1**, 6 (2007)
9. Loktev, A.A.: Lett. J. Tech. Phys. **33**(16), 5 (2007)
10. Loktev, A.A.: Appl. Math. Mech. **72**(4), 6 (2008)

Calculation of Heat Distribution of Electric Heating Systems for Turnouts

Boris Glusberg[1] , Alexey Loktev[2] , Vadim Korolev[2] ,
Irina Shishkina[2](✉) , Diana Alexandrova[2] ,
and Dmitri Koloskov[2]

[1] Railway Research Institute, Third Mytishchenskaya 10,
107996 Moscow, Russia
[2] Russian Open Academy of Transport of the Russian University of Transport,
Chasovaya str. 22/2, 125190 Moscow, Russia
shishkinaira@inbox.ru

Abstract. The article is devoted to ensuring reliable operation of railroad switches in winter conditions for Russian Railways. On the basis of the conducted research, the generalized data on observations of work of various systems of ensuring reliable work of railroad switches in winter conditions are presented.

Keywords: The railroad switch · Methods and systems of snow removal · Ensuring reliable operation of railroad switches · Snow removal efficiency · Innovative systems and methods

1 Introduction

In conditions of low temperatures, causing icing and snow drifts of Railways and railroad switches, the problem of keeping the track in a state that ensures uninterrupted and safe passage of trains in the winter, remains relevant in many countries. The solution of this problem requires improving the reliability of technical means of railway transport, primarily technical means of infrastructure. In recent years, especially due to the instability of climatic factors, the countries participating in the international Union of Railways (UIC) have faced the problem of ensuring the smooth operation of technical means of railway infrastructure in winter conditions. In the last 5 years, in winter, on the roads of Russia, Central and Northern Europe, due to heavy snowfalls, there were breaks in the movement. The main reason is the disruption of the railroad switches. This issue is also being given serious attention in our country. Figure 1 shows the data on the volume and costs of improving traffic safety on the railroad switches for 2016–2020.

© Springer Nature Switzerland AG 2020
V. Murgul and M. Pasetti (Eds.): EMMFT 2018, AISC 982, pp. 337–345, 2020.
https://doi.org/10.1007/978-3-030-19756-8_31

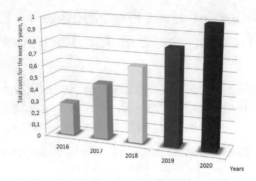

Fig. 1. Schedule of planned investments to ensure reliable operation of the railroad switches in winter conditions for 2016–2020.

The switch economy of the main Railways of Russia has more than 160 thousand railroad switches. Figure 2 shows the shares of distribution systems for reliable operation of the railroad switches in winter conditions on Russian Railways.

The climate of Russia is such that for a large part of the year, on the roadbed and stations is snow cover. There are roads where the time during which the earth is covered with snow exceeds 8 months, so the problem of improving the reliability of the railroad switches in the winter for Russia is very relevant.

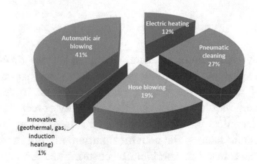

Fig. 2. Distribution schedule of systems to ensure reliable operation of the railroad switches in winter conditions on Russian Railways.

2 Research Technique

In the framework of solving the problem of ensuring reliable operation of switches in winter conditions, OAO "VNIIZHT" in 2013 was performed pioneering work on the evaluation of existing systems for the protection of railroad switches in winter from the adverse effects of snow and ice for the Russian Railways.

As a result of the work done, the life cycle cost of the six main systems was estimated: electric, gas and geothermal heating, pneumatic cleaning, air blowing automatic and hose. When forming an assessment of the effectiveness of each of the systems, following were taken into account: the cost of their purchase, installation, energy, fuel, repair costs, replacement of system components, maintenance and maintenance. In addition, failures were taken into account, both in the operation of systems and in the operation of the railroad switches due to failures of elements of snow removal systems. Taking into account the indicators for each of the systems made it possible to assess the effectiveness of their work and to collect the necessary material to determine the cost of their life cycle related to one railroad switch.

The work carried out allowed to formulate the following statement of the problem, which includes three stages:

1. Formation of an approach to the organization and testing, ensuring comparability of results and assessments of systems to ensure trouble-free operation of the railroad switches in winter conditions.
2. Obtaining representative estimates of the work of existing and innovative systems.
3. Development of technical requirements for systems and subsystems to ensure trouble-free operation of the railroad switches in winter conditions, based on the experience of their application.

With the aim of continuation of the work by employees of the Department "Transport building" Russian university of transport MIIT in the initiative order was the calculation of heat thermal transfer system for electric heating of the railroad switches. The purpose of the calculation was: getting the required heating power depending on the ambient temperature, the environment for the two heating options: one heating Element or two heating Elements, laid on the side of the wisp, obtaining the temperature field of the rail and heating Elements, obtaining the temperature dependence of the Heating element on the ambient temperature for different heating options.

Description of the heated object: rail R-65, installed in the switch gear, which was placed outdoors. The input for the thermal calculation was taken:

4. Temperature parameters: the calculation was carried out at different ambient temperatures $Toc = 20\ °C;\ -40\ °C;\ -50\ °C$.
5. Geometric dimensions of R-65 rail: rail height $H_p = 180$ mm, neck height $H_p = 105$ mm, head width $B_p = 75$ mm, sole width $B_p = 150$ mm, neck thickness $e_p = 18$ mm.
6. The Thermophysical parameters of the material of the rail: body material-carbon steel, thermal conductivity $\lambda_{corp} = 45\ W/(m \cdot K)$, heat capacity $Sm_{corp} = 460\ j/(kg \cdot K)$, the density of $_{rcorp} = 7800$ kg/m^3.
7. SEO Parameters: body material of heating element – steel, the thermal conductivity of the case of PETN $\lambda_\kappa = 45$ W/(m $\cdot$ K) material lived Tena – tungsten, the conductivity of PETN lived $\lambda_{Dg} = 173\ W/(m \cdot K)$ material of core insulation of heating element – magnesium oxide, the thermal conductivity of magnesium oxide $\lambda_{MgO} = 18\ W/(m \cdot K)$, the length of heating element L = 4.1 m, cross-section dimensions of Ten (a x b) = (12 × 6) mm, thickness of wall of the Heater $T_{st} = 1$ mm.

The calculation took into account the fastening of the heating Element to the rail. The design of the Heaters and their attachment to the rail as shown in Fig. 3.

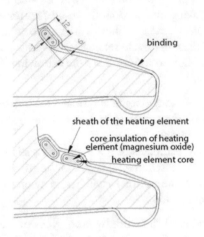

Fig. 3. Diagram of heating Elements attachment to the rail.

The calculation was carried out for two cases: the rail and wit are partially covered with snow, in the gap between the rail and wit is also snow, and the rail and wit are not covered with snow. The following assumptions were made in the calculation: the climatic parameters of the location of the object, the diameter of the tungsten core of the heating Element is taken to be 1 mm.

The calculation was carried out for the case of heating the rail with one heating Element laid on the rail from the rail tongue. The calculation of the temperature field of the rail is carried out using the software package ElCut.

At the same time, calculation models are built: model No. 1 (without snow, with one heater), model No. 2 (with snow, with one heater), presented in Fig. 4, model No. 3 (without snow, with two heaters), model No. 4 (with snow, with two heaters).

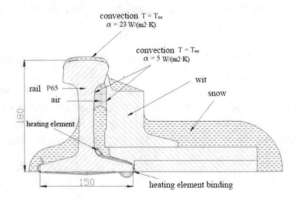

Fig. 4. Calculation model №2 for the case of heating the rail with snow heating element.

In the calculation model the following boundary conditions were set:

1. To take into account the convective heat transfer between the outer surface of objects and the environment in the calculation model, the ambient temperature is set Tos, °C and the convective heat transfer coefficient $\alpha_{nar} = 23\,W/(m^2 \cdot K)$.
2. On the surfaces in the gap between the tip and the rail is also given the condition of convective heat transfer, the coefficient of convective heat transfer is taken in $\alpha\,inside = 5\,W/(m^2 \cdot K)$.
3. In the calculation model, the heating power is set as the volume heat output power, W/m^3, in the veins of the heating Element.

3   Results and Discussion

Figures 5, 6, 7 and 8 show the temperature fields of the rail obtained by the design models №№1-4 at an ambient temperature of minus 50 °C.

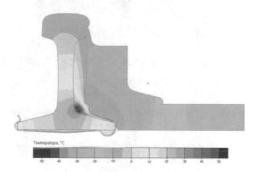

Fig. 5. Temperature field of the rail, calculated according to the calculation model №1: one heater, no snow.

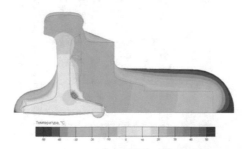

Fig. 6. Temperature field of the rail, calculated according to the calculation model №2: one heater, with snow.

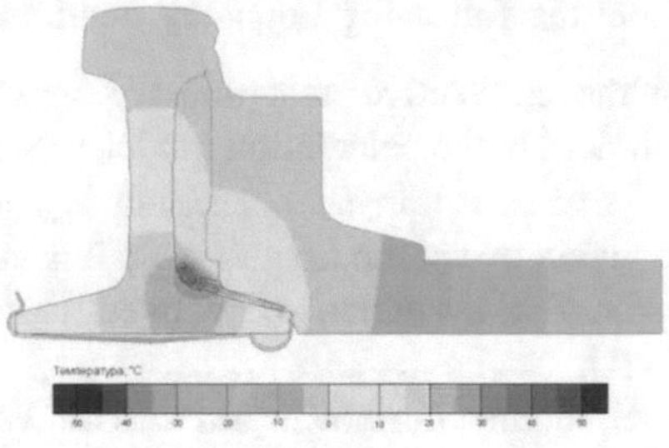

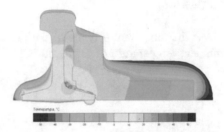

Fig. 7. Temperature field of the rail, calculated according to the calculation model №3: two heaters, no snow.

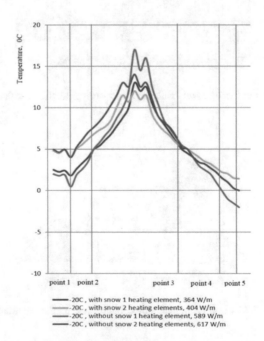

Fig. 8. Temperature field of the rail, calculated according to the calculation model №4: two heaters, with snow.

Fig. 9. Temperature distribution on the side surface of the rail at ambient temperature minus 20 °C.

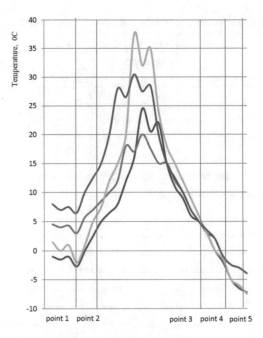

Fig. 10. Temperature distribution on the side surface of the rail at ambient temperature −50 °C.

Graphs of temperature distribution on the side surface of the rail are shown in Figs. 9, 10. The graph shows the characteristic points of the lateral surface of the rail (shown in Fig. 11). The heating power was selected so that the temperature at point 3 (the middle of the rail) was equal to + 5 °C.

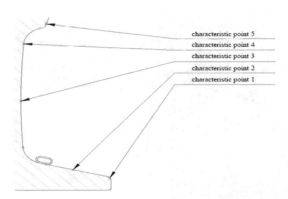

Fig. 11. Characteristic point calculation results.

The obtained dependences of the required heating power on the ambient temperature are shown in Table 1 and Fig. 12.

Table 1. Obtained dependences of the required heating power on the ambient temperature.

| | Required heating power, W/m | | | |
| | One heater | | Two heaters | |
Temperature of the environment, °C	With snow	Without snow	With snow	Without snow
−50	801	1272	889	1435
−20	364	539	404	617
0	73	108	81	123

The graph of the heater core temperature dependence on the ambient environment temperature for the selected heating options is shown in Figs. 13, 14.

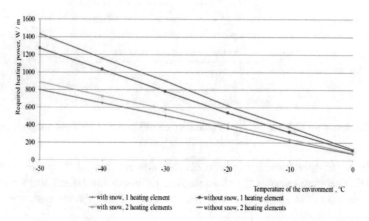

Fig. 12. Dependence of the required heating power on the ambient environment temperature.

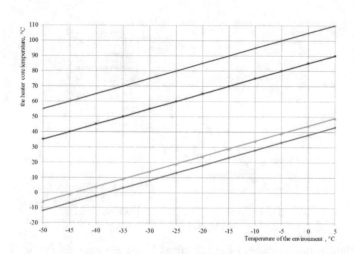

Fig. 13. Diagram of heater core temperature dependence on ambient environment temperature.

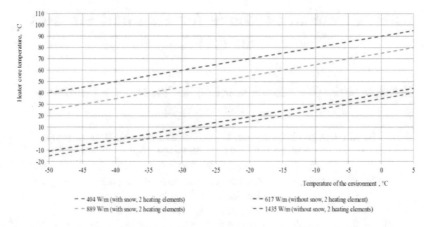

Fig. 14. Graph of temperature of a core of the heater from ambient environment temperature.

4   Conclusions

Dependences of the required power of sea on the basis of tubular heaters on the ambient environment temperature during heating of the rail switch gear (Table 1 and Fig. 12) are obtained in the calculation, as well as the temperature fields of the rail during heating (Figs. 5, 6, 7, 8, 9, and 10) and the maximum temperature of the heater core at different heating options (Figs. 13, 14) are estimated. Due to the design of the wit, considered in the calculation, when using two heaters, the contact of the heater and the wit fasteners is observed, as a result of which there is an outflow of heat. Department of "Transport construction" Russian university of transport MIIT continues to work on improving the various systems of snow removal on the railroad switch. The Department of "Transport construction" Russian university of transport MIIT is also ready to conduct research in conjunction with other experts within the UIC, to monitor the various systems to ensure reliable operation of the railroad switches in the winter, and also for a more accurate and complete assessment, to determine the effectiveness of their use in different operating conditions, as well as to improve existing systems to ensure reliable operation of switches in winter conditions.

References

1. Korolev, V.V.: Modern and advanced design of railway path for different operating conditions the Works of JSC "VNIIZHT" INTEKS, Moscow, pp. 138–148 (2013)
2. Korolev, V.V.: Conference on Innovative Methods of Design, Construction, Diagnostics and Maintenance of Infrastructure, Subiit, Chita, pp. 82–87 (2013)
3. Glusberg, B.E., Korolev, V.V.: The application of different systems snow removal on railroad switches. In: 6th Scientific-Practical Conference Introduction of Modern Structures and Technologies in Track Facilities. ITC "Aint", MIIT, Moscow, pp. 76–86 (2013)
4. Glusberg, B.E., Korolev, V.V.: Way and track facilities, vol. 3, pp. 26–30 (2014)

Measurement Methods for Residual Stresses in CWR

Nikola Mirković[1](✉) ⓘ, Ljiljana Brajović[1] ⓘ, Miodrag Malović[2] ⓘ,
and Petr Vnenk[3] ⓘ

[1] Faculty of Civil Engineering, University of Belgrade, Bulevar kralja
Aleksandra 73, 11120 Belgrade, Serbia
{nmirkovic, brajovic}@grf.bg.ac.rs
[2] Faculty of Technology and Metallurgy, University of Belgrade, Karnegijeva 4,
11120 Belgrade, Serbia
office@malovic.in.rs
[3] University of Pardubice, Studentská 95, 532 10 Pardubice, Czech Republic
petr.vnenk@upce.cz

Abstract. Continuously welded rails are a part of the track structure since the 1950s. Total stresses in continuously welded rails include residual stresses generated during the manufacturing process as well as the service life of rails. The control of rail stresses is very important for traffic safety. Measurement methods for residual stress in rails are presented in this paper. Since destructive measurement methods are unsuitable for use in the railway track, the modern non-destructive methods were particularly presented and discussed. The advantages and disadvantages of X-ray diffraction, ultrasonic, magnetic and electromagnetic methods are particularly explained and considered.

Keywords: Railway · Residual stress · Strain gauge · X-ray diffraction · Ultrasonic measurement · Magnetic measurement

1 Introduction

Residual stresses exist in rails regardless of the external influence (temperature changes, railway traffic load, track/bridge interaction etc.). The residual stresses in new rails are generated during the manufacturing process of hot-rolled steel, including the uneven cooling, roller straightening and leveling rail profiles. The manufacturing technology directly influences the residual stresses in the new rails [1]. The mass of Vignole rail is concentrated in the rail head and the middle of the rail foot (Fig. 1), affecting the cooling speed of rail profile and uneven distribution of residual stresses in rail profile. The amount of the residual stresses in the rails can be reduced by the modifications in the manufacturing process. The residual stress changes during the rail life cycle [2]. The complex distributions of residual stresses in used rail according to [3] and in new rail according to [4] are shown in Fig. 2. During track laying and rail maintenance, the rail welding process (the both flash butt and aluminothermic welding) induces additional residual stresses (due to differential expansion and contraction in

© Springer Nature Switzerland AG 2020
V. Murgul and M. Pasetti (Eds.): EMMFT 2018, AISC 982, pp. 346–355, 2020.
https://doi.org/10.1007/978-3-030-19756-8_32

metal) which contribute to the total level of tensile and pressure stresses in continuously welded rails (CWR).

Fig. 1. Uneven cooling of rail profiles (left) and bending of the rails in cooling bed (right)

Furthermore, the actual value of the residual stress depends on the traffic loads and affects the stability of the track with CWR (in particular on the track sections on the bridge and in narrow curves). As shown in Fig. 3, the residual stresses are a part of basic tensile stress in used rail during service life of railway track with CWR [5].

In the engineering calculations, the residual stress value of 80 MPa is commonly used for CWR during railway service life [5, 6]. Given that residual stresses in CWR might significantly influence the traffic safety (risk of the rail break [5, 7], track stability [8], railway superstructure design, and life span of rails in track [9, 10]), determining the current value of the stresses in CWR is of great importance for engineering practice. Contrary to previous engineering practice, the residual stress in rail foot up to 250 MPa for all steel grades is defined by European standard [4]. Moreover, the same standard prescribes the destructive method using electrical strain gauge for measurement of rail residual stress [4]. This method is presented and discussed in this paper. Since destructive measurement methods are unsuitable for use in railway track with CWR, the non-destructive measuring methods are presented and discussed in this paper.

Figure 4 presents dangerous corrosion-fatigue crack in the middle of rail foot which was developed under the influence of cyclical traffic loads and tensile residual stresses in rail foot.

2 Destructive Saw-Cutting Measurement Method

The European standard [4] specifies 23 rail profiles of Vignole railway rails of 46 kg/m and greater linear mass, for conventional and high speed railway track usage. Moreover, nine pearlitic steel grades are specified covering a hardness range of 200 HBW to 440 HBW and include non-heat treated non alloy steels, non-heat treated alloy steels, and heat treated non alloy steels and heat treated alloy steels. This European standard prescribes a destructive measuring method for residual stresses in foot of Vignole rail.

The description of the measurement method is presented in accordance with [4]. The residual stresses are determined on a sample of a rail with length of 1 m. As presented in Fig. 5, the electrical strain gauge is attached to the underside of the rail foot: in the middle of the Vignole rail profile, and in the longitudinal direction.

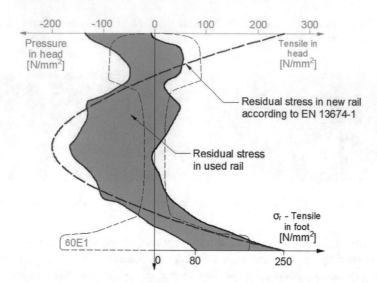

Fig. 2. Complex distribution of residual stresses in Vignole rail profile

The gauge is 3 mm long, encapsulated type, and its gauge factor accuracy must be better than 1%. It measures average strain along its length. The first strain measurements are taken in that state. After that, two saw cuts are performed in order to remove a 20 mm thick slice from the middle of the rail (Fig. 5). A second set of the strain gauge measurements is taken on that rail slice as relaxed strains. The relieved strain is estimated from the differences between the first and the second sets of measurements. That value, with changed sign, is multiplied by 2.07×105 MPa in order to calculate residual stress [4].

However, this measurement method itself may cause the change in the three dimensional residual stress in the rail due to the slice extraction.

3   Non-destructive Measurement Methods

There are methods for nondestructive assessment of stress in steel. These measurement methods could be suitable for laboratory use only or could be used for in situ measurements.

3.1   XRD Method

The most used non-destructive method for evaluating residual stresses is X-ray diffraction (XRD) method. It is based on the interactions of the wave front of the X-ray beam, and the crystal lattice of investigated material. In real ferromagnetic materials, there are domains with different orientation of crystallographic planes. Figure 6 shows principle for measurement of 2θ angle between the directions of incident and diffracted rays.

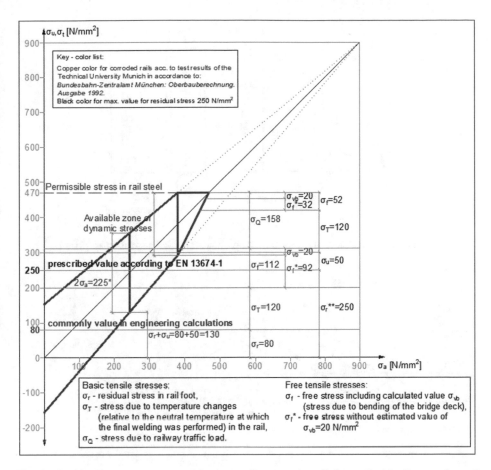

Fig. 3. Residual stress as a part of basic tensile stress in rail foot - Smith diagram for rail 60E1/900 on the track section on the bridge

Fig. 4. Corrosion-fatigue crack in rail foot [11]

When X-ray beam of wavelength λ incidents on the material atoms at angle θ to the atomic planes (Fig. 6) the diffracted rays from the different planes at distance $d = d_{\psi\phi}$,

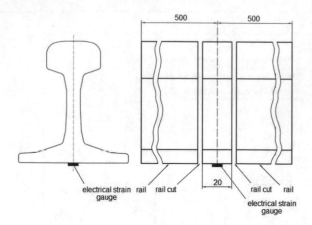

Fig. 5. The one-meter long rail sample and rail slice according to [4]

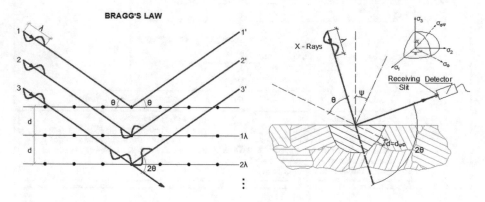

Fig. 6. Principle of measurement by XRD method

will interfere at the detector and thus have maximum values in detected signals for angle θ obeying Bragg's law (1), where z is an integer.

$$2 \cdot d \cdot \sin\theta = z \cdot \lambda \tag{1}$$

The angle equals 2θ between the directions of incident and diffracted rays d can be estimated by practical measuring. Since orientation of crystallographic planes is not parallel to the material surface, angle ψ (between incident ray and a normal to the material surface) and an azimuth angle ϕ have to be measured in order to define the beams directions connected to main coordinates of the material primary strains and stresses σ_1, σ_2 and σ_3 (Fig. 4). At XRD graph of signal intensity versus 2θ, maximum values correspond to distances $d_{\psi\phi}$ of various directions [12].

In the case of residual stress, there is a difference between the distance d$\psi\phi$ and corresponding distance d0 of non-stressed material. By changing the incident angle ψ, wavelength of X-ray or using two detectors, the strain and the residual stress $\sigma\phi$ could be evaluated even without known d0. This is cheap and quick method for obtaining surface biaxial residual stresses in small measuring volumes. For the high intensity of residual stresses, the nominal accuracy is 20 MPa for steel. The main disadvantage is small penetration depth (up to 30 μm) and unknown elastic constants of steel crystal lattice for all directions. In order to calibrate system and apply appropriate modeling for residual stress estimations, the same slice samples are measured with more accurate and deeper laboratory destructive and non-destructive methods.

Furthermore, Synchrotron X-ray and Neutron diffraction methods are also based on Bragg's law.

Synchrotron X-ray Method

Synchrotron X-ray method provide very intense x-rays beams with million times higher energy than laboratory based x-ray systems and, for that reason, have a depth of penetration about 20 mm in steel components. It enables obtaining 3D maps of the strain distribution to millimeter depth in inspected components. Measurement is fast, and high magnitude residual stresses could be measured with nominal accuracy of 30 MPa. However, this technique is strictly laboratory, with limited size of investigated components. Also, it does not need stress-free specimen for calibration.

Neutron Diffraction Method

In 1993, neutron diffraction method was recommended by the European Rail Research Institute (*ERRI*) as measurement method for residual stress measurements on naturally hard and two head-hardened rails (D 173/RP42 Report, [13]).

Neutron diffraction method is based on larger wavelength λ, so the penetration depth is larger than in XRD (about 60 mm in steel). This method enables the detection of changes in atomic lattice spacing due to the stress. In order to calculate absolute stress values, stress free sample ("d_0" sample) should be measured for calibration. With this method, tri-axial residual stress components and stress gradients could be measured. The accuracy for steel is 30 MPa. Unfortunately, this method could only be used in laboratory. Moreover, this method is very expensive and it is not available in all countries.

3.2   Ultrasonic Method

The speed of ultrasonic waves that travel through a material is affected by the direction and magnitude of the present stresses. Based on this acoustoelastic effect, ultrasonic waves in a frequency range 2–10 MHz are used for measurement of applied and residual stresses. The average velocity of ultrasonic waves is measured along the chosen path. This method is most sensitive when the chosen path and motion direction of material particles is parallel to the direction of stress [14].

Different types of waves can be employed, but the most commonly used are the longitudinal critically refracted (L_{cr}) waves [14], which travel just beneath and parallel to the specimen surface. The measurement equipment for the time of flight (TOF) measurement consists of one transmission and one or more receiving ultrasonic probes fitted in the plexiglass wedge at fixed distance is presented in Fig. 7. The L_{cr} waves are passing fixed length L for TOF $t = L/v$ in stressed material, and for $t_0 = L/v_0$ in stress-free material, where v and v_0 are corresponding wave velocities. In the range of elasticity, the average residual stress along the path compared to stress–free material σ is given by Eq. (2).

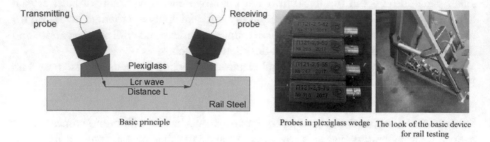

Basic principle   Probes in plexiglass wedge   The look of the basic device for rail testing

Fig. 7. Basic principle of ultrasonic method, probes and device for rail testing

$$\sigma = \frac{1}{K \cdot t_0} \cdot (t - t_0) \tag{2}$$

The acoustoelastic constant K depends on the type of waves, elastic properties of the material and direction of wave propagation. It should be determined using appropriate calibration tests on the same or similar samples. For increase in stress of 10 MPa, TOF difference is about 10 ns for the rail steel.

The speed of ultrasonic waves also depends upon temperature and microstructure changes in the steel. Changes in temperature can be corrected by simultaneous temperature measurements. The acoustic methods enable accurate tri-axial high residual stresses measurements at penetration depth up to 150 mm. Required stress-free reference, sensitivity to microstructural changes, average stress measuring over relatively large gauge volume and difficulty to specify spatial resolution are disadvantages of the method. It is suitable for in situ stress detection in CWR [15, 16]. The total (thermal and residual) stresses could be measured in unloaded CWR at a temperature that is different than the rail neutral temperature. On the other hand, the residual stress in CWR could be measured at its neutral temperature.

3.3   Magnetic and Electro-Magnetic Measurements

The most of the magnetic characteristics of ferromagnetic materials are influenced by mechanical stresses. In the non-stress ferromagnetic material each domain is magnetized in one direction and during magnetizing-demagnetizing process the volume and magnetic orientation of domains are changing. The magnetic flux density or induction vector B in a material is changing under applied magnetic field in cycling process, presented with magnetic hysteresis loop in Fig. 8 [17].

The characteristic values on graph B_r (remanence) and H_c (coercivity), permeability $\mu = B/H$ and others could be evaluated as non-stressed material parameters. In case of applied or residual stresses in material, the boundaries between domains called domain

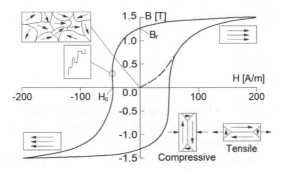

Fig. 8. Magnetic domain alignment during magnetization and under stress

walls are moving [17], and the magnetization orientations of the domains are changing. Under tensile stresses, the domains with the same magnetization orientation as stress direction are enlarging and thus increase magnetization in that direction. The compressive stresses enlarge domains with transverse magnetization orientation and increase magnetization in the direction perpendicular to stress directions. So, the residual stresses lead to change in the magnetic characteristics, and the hysteresis loop. This is the foundation of one of mobile type instruments, MAPS (Magnetic Anisotropy and Permeability System), for residual stress measurements [17] designed for manual probe manipulation that measures simultaneously a large number of magnetic parameters from hysteresis loop. The penetration depth of measurements is between 0.1 mm and 5 mm, controllable by the frequency of applied magnetic field. Spatial resolution is between 5.2 mm and 15.5 mm, depending on a probe type. Obtaining complete biaxial stress measurement lasts less than 2 min. This system has to be calibrated against known stress levels for the investigated material [17].

The MBN (Magnetic Barkhausen Noise) method is based on Barkhausen effect. Figure 8 shows that changing of B is happening in discrete steps (Barkhausen jumps). Those sudden changes in B could be measured with pick up coils above the surface of a material, thus the typical voltage pulses in the signal are obtained. The variation of MBN signals shape, the number and values of peaks, the RMS (root mean square value) of the overall MBN signal over a number of cycles indicate the changes of magnetic material structure due to the stress. If the frequency of applied alternating field frequency is higher than 10 Hz the depth of this measurements is 20 μm, while for low frequency (0.1–1 Hz) the penetrating depth is of order of 1 mm and can be used for evaluation of subsurface stresses. The ACSM (alternating current stress measurements) technique of stress measurement is based on ACFM (Alternating Current Field Measurement) technique of defect detection. The coil with alternating current induces currents in the metal surface that are unidirectional and uniform in the non-stressed material without defects. These induced currents produce the magnetic field above the surface, which is measured in two directions. Small changes measured in the strength and the direction of those magnetic fields by array of magnetic sensors or sensing coils could be related to changes in stress state of a material and ACSM output signal is almost linear under applied stress. This is non-contact, rapid technique, sensitive to both tensile and compressive stresses and does not need special surface preparation. It is better for stress changes measurements than for absolute stress measurements [18, 19].

4 Conclusion

Because of higher speeds and increased axle and traffic loads, measurement of residual stresses in new rails, as well as in continuously welded rails (CWR), has great importance on modern railways in recent years. Residual stresses in rails could significantly influence the risk of the rail break, track stability, railway superstructure design, as well as the life span of track with CWR. For now, there is no standard method for testing the residual stresses in CWR. In accordance with European standard [4], residual stresses in rail shall be estimated by destructive cutting method in laboratory. This method is not applicable to control the rail residual stress in the track with CWR. On the other hand, the prescribed value of residual stress in rail foot up to 250 MPa could adversely affect the traffic safety (especially in CWR on the bridges or transition zones [5]). If the current residual stress values in CWR could be measured, the level of total stresses in rails and its effects on the traffic safety could be predicted.

This paper presented standard destructive method prescribed in [4]. Unfortunately, this method is not applicable for CWR. Moreover, several non-destructive methods for measurement of rail residual stresses were presented.

The X-ray diffraction method is based on interaction of the X-ray beam and the crystal lattice of the material of the investigated rail. The method is fast, but it lacks higher accuracy. Another limitation of this method is the penetration range, as it is able to identify residual stress only in the surface layer with the thickness of 30 μm. The ultrasonic method is based on measurement of the velocity of ultrasonic waves going through the investigated rail material. Key issues to cope with while using this method are the precise TOF measurement, as the measured times are usually very low on short distances, and extract the influence of the thermally induced stress, as it is a very common factor in field measurements. Finally, the methods using magneto elastic effect are based on magnetization changes of magnetic domains in the investigated material, which are oriented in various directions. The changes in magnetization orientation of magnetic domains in the ferromagnetic materials due to presence of residual and applied stresses are the base of various magnetic and electromagnetic methods. Practical portable instruments, based on measuring the changes in hysteresis loop parameters of rail steel or the Barkhausen noise detection and analysis, enable residual stresses mapping. ACFM method is also adapted from defect identification to residual stress measurements in rails. Those methods are very promising for rapid practical measurements and models for interpretation of measuring results and residual stress estimation are constantly improving. All presented methods have a great potential. However, interpretation and correct evaluation of the results is a subject of ongoing research by the authors.

Acknowledgements. This work was supported by the Ministry of Education, Science and Technological Development of the Republic of Serbia through the research project No. 36012:

"Research of technical-technological, staff and organization capacity of Serbian Railway, from the viewpoint of current and future European Union requirements".

References

1. Nejad, R.M., Shariati, M., Farhangdoost, K.: Three-dimensional finite element simulation of residual stresses in UIC60 rails during the quenching process. Therm. Sci. **21**(3), 1301–1307 (2017)
2. Popović, Z., Lazarević, L., Brajović, L., Gladović, P.: Managing rail service life. Metalurgija **53**(4), 721–724 (2014)
3. European Rail Research Institute: Verbesserung der Kenntnis der Kräfte im lückenlose Gleis (einschließlich Weichen): Schlußbericht (ERRI D 202/RP 12), Utrecht (1999)
4. CEN: EN 13674-1 - Railway applications – Track – Rail - Vignole railway rails 46 kg/m and above, Brussel (2017)
5. Mirković, N., Popović, Z., Pustovgar, A., Lazarević, L., Zhuravlev, A.: Management of stresses in the rails on railway bridges. FME Trans. **46**(4), 636–643 (2018)
6. DIN: Fachtbericht 101 - Einwirkungen auf Bruecken, Deutschland (2003)
7. Popović, Z., Radović, V.: Analysis of cracking on running surface of rails. Gradjevinar **65** (3), 251–259 (2013)
8. Popović, Z., Lazarevic, L., Vatin, N.: Railway gauge expansion in small radius curvature. Proc. Eng. **117**(1), 841–848 (2015)
9. Popović, Z., Lazarević, L., et al.: The importance of rail inspections in the urban area - aspect of head checking rail defects. Proc. Eng. **117**(1), 596–608 (2015)
10. Popović, Z., Lazarević, L., et al.: The modal shift to sustainable railway transport in Serbia. In: MATEC Web of Conferences, vol. 106, paper 05001 (2017)
11. UIC: UIC - Merkblatt Nr. 712 – Schienenfehler, Paris (2002)
12. Fitzpatrick, M., Fry, A., et al.: NPL Measurement Good Practice Guide No. 52. Determination of Residual Stresses by X-ray Diffraction. NPL Teddington, UK (2005)
13. ERRI Committee D 173/RP42: Rail rolling contact fatigue - Residual stress measurements on naturally hard and two head-hardened rails by neutron diffraction, Utrecht (1993)
14. Li, Z., He, J., Teng, J., Wang, Y.: Internal stress monitoring of in-service structural steel members with ultrasonic method. Materials **9**, 223–240 (2016)
15. Szelazek, J.: Monitoring of thermal stresses in continuously welded rails with ultrasonic technique. Electromagn. Nondestruct. Test. **3**(6) (1998)
16. Vangi, D., Virga, A.: A practical application of ultrasonic thermal stress monitoring in continuous welded rail. Exp. Mech. **47**, 617–623 (2007)
17. Buttle, J., Moorthy, V., Shaw, B.: NPL Measurement Good Practice Guide No. 88 - Determination of Residual Stresses by Magnetic Methods. NPL, Teddington (2006)
18. Lugg, M., Topp, D.: In: Proceedings of ECNDT 2006, Berlin, Germany, pp. 1–14 (2006)
19. Utrata, D., Strom, A., et al.: In: Thompson, D.O., Chimenti, D.E. (eds.) Review of Progress in Quantitative Nondestructive Evaluation, pp. 1683–1691. Springer, Boston (1995)

Evolutionary-Functional Approach to Transport Hubs Classification

Oksana Pokrovskaya[1] and Roman Fedorenko[2(✉)]

[1] Siberian Transport University, Dusi Kovalchuk Street, 191, 630049
Novosibirsk, Russia
insight1986@inbox.ru
[2] Samara State University of Economics, Sovetskoi Armii Street, 141, 443090
Samara, Russia
fedorenko083@yandex.ru

Abstract. The purpose of the article is to characterize an evolutionary-functional approach to the classification of transport hubs. This approach is a part of the theory of terminalistics, which is proposed to be included in the practice of transport science. We have developed a pyramid of terminalistics and a hierarchy of its objects, taking into account the complexity and integration of the service. We propose to use the four-stage adapted Rodrigue-Notteboom model when analyzing the operation of railway transport hubs. In modern conditions, transport hubs have long performed the role of full-fledged multi-functional logistics entities that implement a wide range of not only transport, storage, but also consulting, customs, distribution, expeditionary and other services. However, so far transport science has not offered a full-fledged transport hubs' classification reflecting the entire rich functional potential of these complex logistic systems. The evolutionary-functional approach to the development of transport hubs as multimodal logistics entities is the basic theoretical position of terminalistics. This approach reflects a whole range of parameters that are important for transport hubs analyses: infrastructure base, geography, service orientation, features of regional development and development of logistics solutions. The authors presented transport hubs' classification according to the stage of their evolutionary-functional development. The basis of the classification is a sign of the perfection of logistic solutions when interacting with the parties of logistics.

Keywords: Evolutionary-functional approach · Transport hubs · Terminalistics · Logistics entity · Logistics service · Multiplicative effect

1 Introduction

The issues of origin, formation and subsequent transformation (evolution) are fundamental in the design of technical equipment, work technology, organization mechanism and inter-element interaction of such complex systems as transport hubs (TH). The role of TH in the delivery of goods is difficult to overestimate. The efficiency of the entire transport and logistics system of the country depends on their uninterrupted work and harmonious interaction of the modes of transport.

© Springer Nature Switzerland AG 2020
V. Murgul and M. Pasetti (Eds.): EMMFT 2018, AISC 982, pp. 356–365, 2020.
https://doi.org/10.1007/978-3-030-19756-8_33

Organization of effective transport hubs' work is a complex, multidimensional task. Unfortunately, it is usually solved unilaterally. Traditional approaches research economic and geographical position, track development, integration with the locality. All other issues of warehouse infrastructure, technical equipment, interaction of modes of transport and others are solved "pointwise" locally. There are still no comprehensive approaches to the study of the design and development of transport hubs, integrating knowledge of logistics, economic geography, system theory and many other disciplines related to the activities of TH. In modern conditions, transport hubs have long been fulfilling the role of full-fledged multi-functional logistics entities that implement a wide range of services including not only transport, storage, but also consulting, customs, distribution, expeditionary and other [9]. However, until now, transport science has not proposed a full-fledged classification of transport hubs, reflecting all their rich functional potential. This situation is associated with the prevalence of the traditional understanding of the transport hub as a junction of various types of transport, communications hub, infrastructure unit, but not as a logistics entity. The paradox of the situation is that transport hubs of various types of transport, while performing logistic functions, do not have a current logistical approach to the classification, reflecting the diversity of modern transport and logistics activities [14, 15]. Obviously, an approach aimed at transport hubs as an object of study is necessary for a full-fledged, comprehensive study of transport logistics. The above determines the choice of the topic of this work. The aim of the work is to develop the theoretical foundations of transport hubs' development from the point of view of terminalistics as a new scientific direction.

2 Materials and Methods

The methodological basis for the development of alternative classification and the hierarchy of transport hubs was the result of Russian and foreign classifications' analysis. We used methods of logistics, synergistic, transport geography, cluster and system analysis. In order to achieve our goal in this research, we have analyzed a number of original terms that reflect the authors' TH typology. While analyzing the literature, we have found that most authors refer to transport hubs as territorial entities, located at the nodal points of transport system. As a result, researchers unite different types of transport hubs in one term. New terminology is necessary for theoretical support of a unified approach to the definition of the hierarchical positions of transport hubs, as well as their classification and functions. We have used evolutionary-functional approach to establish the conceptual aspect of the new classification system. The new approach integrates all of the existing approaches while focusing on the functionality of each type of transport hubs. While developing this approach, we performed a thorough analysis of scientific literature in the field of transport, logistics and economics. It showed that today's scientific trends show mutual distribution, or integration of scientific research, that is to say that the research has an interdisciplinary nature.

3 Results

3.1 Characteristics of Existing Approaches to Transport Hubs Typology

We considered the difference between the evolutionary-functional approach of terminalistics and the traditional one, using the definitions and typologies of technical specifications, as well as the place and composition of the approaches available in the transport literature. Traditionally, researchers considered transport hubs from the point of view of structural implementation in specific geographical conditions.

Currently, there are three following approaches to the typology of transport hubs in accordance with their definitions:

1. Railway approach, in which the transport hub is considered an object with complex road development and a road junction, as well as single geographical object itself;
2. Commercial, in which the transport hub is a system of elements, a local logistic object;
3. Economic and logistic, in which the transport hub is a logistics service provider.

The key features of these classifications are track development, technical equipment, design, logistic role in the delivery system, economic and geographical position. Unfortunately, all the signs are scattered, there is no unified approach to the development of transport hubs. For example, the railway aspect of understanding and typology of transport hubs covers only such areas as technical equipment, track development and economic and geographical position.

Thus, the increasing complexity of the range of logistics services, a variety of approaches to transport hubs organization, the increase in the quality requirements for them objectively require that all knowledge be transformed into an independent industry - terminalistics.

3.2 Proposed Terminology

This paper presents some theoretical concepts of terminalistics - the logistics of transport hubs and terminal networks [13], in particular the evolutionary-functional approach to the development of transport hubs as multimodal logistics facilities. Terminalistics uses the following approaches to the study of transport hubs and terminal networks: process, synergistic, systemic, logistic, cluster, evolutionary. We consider in detail the evolutionary approach to the development of the transport hub functional. In our analysis, we proceed from the assumption that the hub is based on a railway junction. The term "terminalistics" comes from the merger of the words terminal + logistics = terminalistics. This is the logistics of terminal networks and transportation hubs. This is the science of organization, design, management, structure and configuration of networks of cargo terminals. It studies issues of the number and location of hubs, the functional and technological composition, as well as transport, infrastructure, integration, economic and environmental components of the regional terminal networks.

We offer a pyramid of terminalistics with the hierarchy of objects reflecting the complexity and integration of the service (see Fig. 1).

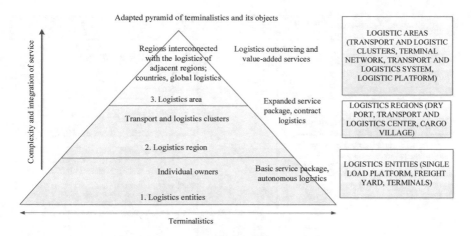

Fig. 1. The pyramid of terminalistics and its objects.

The logistics area is a set of interconnected logistic areas, providing integration into transport corridors and building a stable in composition and complexity of transport and logistic service of the terminal network.

3.3 Evolutionary-Functional Approach to the Development of Transport Hubs

In the general case, the development of any transport hub as logistics entities goes through the following stages of evolution (See Fig. 2):

Fig. 2. Enlarged evolution of the essence of transport hubs.

1. Road junction where, during the development of intra-hub interaction, conditions are formed for the provision of additional service and transition to a new stage;
2. A docking station for means of transport where interaction becomes inter-hub and the service allows servicing complex cargo delivery systems;
3. Multimodal transport and logistics entity, which provides comprehensive end-to-end service for customers, rolling stock, cargo («seamless technology»).

In particular, for railway transportation hubs, we propose to adapt the Rodrigue-Notteboom model [20, 21] taking into account the theoretical foundations of terminalistics (Fig. 3).

The development of the hub (hubs network) takes into account the transformation of internal processes:

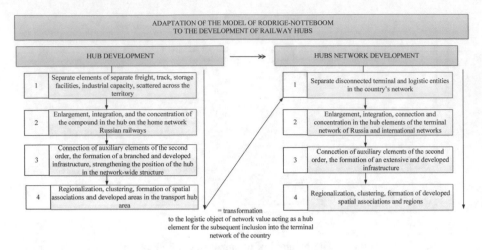

Fig. 3. Adaptation of the Rodrigue-Notteboom model to railway transportation hubs.

1. Disconnected existence of separate elements (entities);
2. Concentration - integration of entities in the hub;
3. Building up infrastructure (connecting support elements, building up extended infrastructure support);
4. Regionalization with the subsequent exit to a new level ("connection" to local and global transport and logistics systems).

We offer an evolutionary-functional approach to the analysis of the development of transport hubs according to implementation of transport and logistics services. In this scheme, the gradation of the logistics service and orientation to the service object is integrated. The stages of grading the service level are standard service = >> extended service = >> unique service.

In addition to improving the level of service, there is a gradual shift in the key orientation of the service: the internal environment = >> the external environment = >> client (See Fig. 4). We have developed a classification of transport hubs on the stage of their evolutionary-functional development (see Fig. 5). A distinctive feature of the proposed classification of transport hubs is the synthesis of cluster [23] and synergistic [17] approaches. The development of hubs is viewed from two sides: as a logistics entity and a self-organizing complex system. The basis is a sign of perfection of logistic solutions in the interaction of the parties of logistics. Thus, the stage of generation corresponds to the stage of chaos (chaotic development) of transport hubs. As noted in the central unit, the existing basis serves as the infrastructure basis, the corresponding development of which can take place in the form of reconstruction or new construction. Conditions are created for the formation of center to attract labor, financial and other resources. However, the element-by-element development is disordered, and the links of the technological interaction are poorly developed.

Focus on service

↓

STAGE 1. Increasing the track development, storage and carrying capacity. Priority travel development. Service orientation to large industrial enterprises. The main work is the shipment and acceptance of goods in the regional and interregional connections on the orders of individual large enterprises. Standard service.
Reference point - the internal environment.

↓

STAGE 2. Modernization of technical equipment, comprehensive mechanization and automation. Technical equipment for multimodal transport. Own warehouse infrastructure. Priority development of storage capacity and technical equipment. Focusing on comprehensive customer service. Expanding the range of services provided by the pack. Construction of delivery chains. The expansion of the customer base. Attracting new members to the structure. Increasing the investment attractiveness of objects and the region as a whole. Extended Service. **Reference point - the external environment.**

↓

STAGE 3. Customer focus. Quality service in integrated delivery chains. Offering a unique service in one place. The complexity of the structure and layout of the site. orientation to the virtual information support of the service, providing online control. **Reference point– client.**

Fig. 4. Staged development of transport hubs.

The synergetic approach involves the study of external disturbances (fluctuations), which are a condition for the subsequent qualitative change and the transition to a higher level of order and development.

We propose to correlate stages of maturity in the cluster approach with the stage of bifurcation in the synergistic approach. The two central blocks reflect the stages of growth and sustainable growth to the level of 2- and 3-PL, respectively. There is a gradual expansion of participants number and the range of services. As a result, the first manifestations of the multiplicative effect are observed. The difference between the growth stage and the stage of sustainable growth is the focus of integrated development - the territory in the first case and the service in the second case, respectively. For a synergistic approach, this corresponds to the stage of bifurcation. Bifurcation in the theory of synergetic is considered a change in the steady state system operation.

The main property of the bifurcation point is the unpredictability of the subsequent development. There may be an irreversible change in both forward and reverse. In Fig. 5, this is illustrated by the dotted arrow, which returns to the stage of chaos. If the stage of sustainable growth is successfully passed, then the final stage of maturity becomes qualitative development (as opposed to the initial quantitative). It corresponds to the third stage in a synergistic approach - the order stage.

The stagnation stages of the cluster approach correspond to the order stage of the synergistic approach. There is a cycle of development associated with a new appeal to quantitative growth, different from the previous stage in its scale. At this stage, integration or metamorphosis may occur in a larger object.

Integration implies mutually equal penetration, merger or absorption by a larger object. The result of the development at this stage is the formation of a large facility and infrastructure conditions for accessing the city chain => region => transport

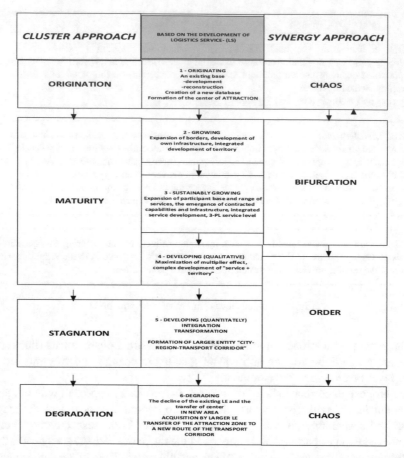

Fig. 5. Classification of transport hubs by stages of evolutionary-functional development.

corridor. The stages of degradation in the cluster approach correspond to chaos in the synergistic approach. The main load of transport hubs here is transferred to a new area of local or third-party origin, or to a new promising point of interaction between modes of transport.

4   Discussion

The development of modern transport and logistics services market has long put before science and practice the issue of creating a unified approach to the typology of transport hubs, the format of which and the range of services is changing and expanding. At the same time, the theory is lagging behind such a rapid development of the terminal warehouse business. The situation is complicated by the fact that a large number of participants work in modern supply chains - customers, freight forwarders, logistics providers, warehouses and transport operators, carriers [22]. The organization of effective interaction of individual participants in the process of product distribution [10] is becoming increasingly important in building an effective logistics chain. At the same time, each group of participants has its own slang, its own approach to the essence of transport hubs, which makes it difficult to find a "common language" in the delivery process. A single information field is required to facilitate the interaction of participants in the transportation process [5]. First, the unified transport hubs classification, alternative to the existing one, can become such a theoretical support and condition for the unity of the information field.

The first attempts to systematize the types and build the transport hubs hierarchy (the term "logistics center" was used in the original sources) were made in the 1990s:

1. Commission on Transport and Logistics of the European Union - dividing transport hubs by status, taking into account their location;
2. Scientific and Practical Center NeLoc - according to the degree of interaction of transport hubs in the Baltic [12];
3. The European Association of Freight Villages "Europlatforms" on the basis of the analysis of terminal networks investigated the essence and characteristics of European transport hubs;
4. The United Nations Economic Commission for Europe (UNECE) investigated the role of transport hubs in intermodal transport [8].

The Institute of Logistics and Supply Chain Management (ILSCM, Australia) singled out a gradation: cargo village, inland port, cargo hub, logistic city [16]. According to the most common approach in the European Union, there are:

1. transport terminals (concentration points);
2. logistics centers;
3. freight villages with a wide range of services;
4. logistics hubs (gateways, sea and air ports) [2].

The approach to the 5-level hierarchy of large transport hubs in accordance with their geographical coverage and capacity is based on a spatial-scale attribute and shows the great importance of transport hubs in the organization of cargo delivery [13]. Notteboom and Rodrigue divided transport hubs by their location and functions [21]. Rimienė and Grundey proposed the 3-level hierarchy, taking into account the volume role of transport hubs in the transport chain [19]. However, there is no clear boundary between the selected levels. Rodrigue [20] based its 3-level hierarchy of transport hubs on the zoning of intermodal freight traffic. Higgins and Ferguson developed the 3-level hierarchy of transport hubs, in which its type is determined by the distance from the sea [11].

Russian transport science classifies transport hubs according to the nature of operational work [1], cargo processing technology [3], geographical location [6], nature of logistics services [7], geometric outline [4] and control system [18]. For the domestic and foreign approach to the classification, it is characteristic that they clearly distinguish two key approaches to the essence of transport hubs – the physical representation (cargo terminal), as the lower stage of their development (dominates in Russia), and the virtual (information center), as their highest stage development (dominates abroad). All classifications have floating criteria with minor differences. As a result, the same object can simultaneously end up in several classes according to different classifications. Our terminology is the theoretical basis for the classification and serves as a tool for identifying the transport hubs type. Evolutionary-functional approach is characterized by the assessment of the functional stage of development of the logistics service of transport hubs according to the criteria of stability, structuredness, development of infrastructure and range of services. The application of this approach to the classification of transport hubs may be required in various studies.

5 Conclusions

The article studies an evolutionary-functional approach that can be used as a methodology for studying the characteristics of the origin and development of transport hubs. We adapted the theory of Rodrigue J.-P. and developed a pyramid of terminalistics and the hierarchy of its objects, taking into account the complexity and integration of the service. The result of the research is the definition of the logic of the functional evolution of transport hubs and their role in the general theory of terminalistics.

For the classification of railway transport hubs, we propose to use the four-stage adapted Rodrigue-Notteboom model. Based on the analysis of the existing classification and evolutionary approaches to the development of transport hubs, we have developed our own classification by service orientation, which integrates the gradation of the logistics service and orientation to the service object.

During the study, we used the integration of cluster and synergistic approaches. The result of the study was the development of a classification of transport hubs in the six stages of their evolutionary-functional development. Evolutionary-functional approach can be a tool for determining the stage of logistic development of transport hubs. It reflects the whole range of parameters important for transport hubs logistics: infrastructure base, geography, service orientation, features of regional development and development of logistics solutions (PL).

Acknowledgments. The reported study was funded by RFBR and FRLC according to the research project 19-510-23001.

References

1. Apatsev, V., Efimenko, Y.: Railway Stations and Hubs. Marshrut, Moscow (2014)
2. Baumann, L., Behrendt, F., Schmidtke, N.: Applying Monte-Carlo simulation in an indicator-based approach to evaluate freight transportation scenarios. In: Proceedings of the International Conference on Harbor Maritime and Multimodal Logistics M&S, pp. 45–52 (2017)
3. Beltiukov, V., et al.: Evaluation of effectiveness of separating layers in railroad track structure using life cycle cost analysis. Proc. Eng. **189**, 695–701 (2017)
4. Bosov, A., Khalipova, N.: Formation of separate optimization models for the analysis of transportation-logistics systems. East. Eur. J. Enterp. Technol. **3**(3–87), 11–20 (2017)

5. Bubnova, G., Efimova, O., et al.: Information technologies for risk management of transportation – logistics branch of the "Russian Railways". In: MATEC Web Conference, vol. 235 (2018). https://doi.org/10.1051/matecconf/201823500034

6. Chislov, O., Trapenov, V.: Rational practicing methods of location logistics and warehouse transportation centers and distribution traffic flows in large urban agglomerations. Vestnik Rostovskogo Gosudarstvennogo Universiteta Putey Soobshcheniya 1(61), 87–97 (2016)

7. Dybskaya, V.: Promising directions for the logistics service providers development on the russian market in times of recession. Transp. Telecommun. 19(2), 151–163 (2018)

8. Erkayman, B., Gundogar, E., Yilmaz, A.: An integrated fuzzy approach for strategic alliance partner selection in third-party logistics. Sci. World J. (2012). http://dx.doi.org/10.1100/2012/486306

9. Fedorenko, R., Persteneva, N., Konovalova, M., Tokarev, Y.: Research of the regional service market in terms of international economic activity's customs registration. Int. J. Econ. Financ. Issues 6(5), 36–144 (2016)

10. Fedorenko, R., Zaychikova, N., et al.: Nash equilibrium design in the interaction model of entities in the customs service system. Math. Educ. 11(7), 2732–2744 (2016)

11. Higgins, C., Ferguson, M.: An exploration of the freight village concept and its applicability to Ontario. McMaster University, Hamilton (2011)

12. Kondratowicz, L., NeLo, C.: Planning of logistics centres. Department of Scientific Publications of the Maritime Institute, Gdansk (2003)

13. Kurova, A.: The role of logistics centres in supply chain management. European Applied Sciences: modern approaches in scientific researches, Papers of the 1st International Scientific Conference, vol. 2, pp. 230–231 (2012)

14. Pokrovskaya, O.: Chi terminelistica reale come una nuova direzione scientific. Ital. Sci. Rev. 1(34), 112–116 (2016)

15. Pokrovskaya, O.: Terminalistics as the methodology of integrated assessment of transportation and warehousing systems. In: MATEC Web of Conferences, vol. 216 (2018). https://doi.org/10.1051/matecconf/201821602014

16. Pretorius, M.: Logistical cities in peripheral areas. University of the Free State, Pert (2013)

17. Prigogine, I., Stengers, I.: Order Out of Chaos: Man's New Dialogue With Nature. Heinemann, London (1984)

18. Prokofieva, T.: Development of Logistic Infrastructure of Murman Transport Hub and Organization of Interaction of "Rzd" with Sea Ports with Use of Progressive Logistics Technologies. RISK: Resources, Information, Supply, Competition, vol. 1, pp. 14–20 (2016)

19. Rimienė, K., Grundey, K.: Logistics centre concept through evolution and definition. Eng. Econ. 4(54), 87–95 (2007)

20. Rodrigue, J., Comtois, C., Slack, B.: The Geography of Transport Systems. Taylor & Francis e-Library, New York (2006)

21. Rodrigue, J., Notteboom T.: Inland terminals within North American and European supply chains. Transp. Commun. Bull. Asia Pac. 78, 21–26 (2009)

22. Šakalys, R., Batarliene, N.: Research on intermodal terminal interaction in international transport corridors. Proc. Eng. 187, 281–288 (2017)

23. Yap, M., Luo, D., Cats, O., Van Oort, N., Hoogendoorn, S.: Where shall we sync? Clustering passenger flows to identify urban public transport hubs and their key synchronization priorities. Transp. Res. Part C-Emerg. Technol. 98, 433–448 (2019). https://doi.org/10.1016/j.trc.2018.12.013

Information Technologies in the Area
of Intersectoral Transportation

Valeriy Zubkov[1]([⊠]) [iD], Nina Sirina[2] [iD], and Oleg Amelchenko[3] [iD]

[1] Production Infrastructure Department of «FGK» plc, Yekaterinburg, Russia
zubkovvvl973@gmail.com
[2] Ural State University of Railway Transport, Kolmogorova Street 66,
620034 Yekaterinburg, Russia
[3] Computation Centre in Chelyabinsk, Structural Unit of the Main Computation
Center of "Russian Railways" plc Branch in Chelyabinsk, Chelyabinsk, Russia

Abstract. This article is devoted to the implementation issue of transport functionality in the segment of freight transportation in Russia. The focus is made on the search of efficient ways of development and implementation of the railway transport system potential in the area of intersectoral freight transportation, the idea of which lies in the consolidation of information and technology systems of parties of freight flow in the current as well as future transport infrastructures. The authors of this article have developed and now propose to implement the automated system named "Electronic service of complex transport services". This system is meant to create the integral information space, in which each participant of the united transportation process in the field of intersectoral cooperation is provided with a multifaceted variety of options for effective conduct of business and optimal expenditures and investments for execution or purchasing of complex transport services. The introduction of the automated system increases data exchange between carrier companies and service receivers, with regard to shipping requests, selection of an optimal and economically viable means of transportation for specific conditions as well as providing result in the form of a commercial offer for the request in question.

Keywords: Automated system · Complex transport service ·
Information space · Module · Transport and economic potential

1 Introduction

Forecasting and planning of freight transportation in the interaction space of various means of transport using information systems as tools of transport and economic potential of Russia, is the priority scientific and applied task, aimed at development of transport and business processes, taking into account the interdependence of goods production, their consumption and the very need for transport services.

The transport and economic potential in the Russian Federation is represented by anticipated and actual production volumes and the need for freight flow correspondence both within a specific region and between regions by means of road, railway, sea and inland water transport. The forecast of cargo base and freight volume relies on strategic

© Springer Nature Switzerland AG 2020
V. Murgul and M. Pasetti (Eds.): EMMFT 2018, AISC 982, pp. 366–375, 2020.
https://doi.org/10.1007/978-3-030-19756-8_34

analysis and planning, which include a variety of options of predictive estimates and conditions of internal and external regulatory impact, as well as country's economic development indices with due regard for information and technology production processes and freight flow correspondence. Potential efficiency is assured through analysis and calculation of anticipated production volumes [1, 2]. This potential allows to reveal possible impact on transport infrastructure in case of freight flow changes, its modernization and renovation, to efficiently plan anticipated freight in the united transport system with regard for required delivery time, traffic capacity characteristics, possible motion speed and narrow (barrier) areas. Transport and economic potential contributes to sustainable functionality of the transport sector, but, as research has shown, does not result in efficient development of the united transport system in the field of intercommunication of various means of transport and transportation, does not lead to implementation of universal analysis, accounting and planning methods of intersectoral transportation. As a result, there lacks a uniform complex technology process among parties of the transportation process and a uniform informational space of industry-specific means of transport and receivers of services. This eventually leads to poor planning and inefficient implementation of transport services, noncompetitive performance of certain means of transport, irrational use of transport infrastructure, baseless overpricing of transport services and overcharge of added value of finished products. The authors of this article propose an alternative efficient way of development and implementation of the railway transport system potential [3] in the field of intersectoral freight transportation. The authors have developed and offered to introduce the automated system "Electronic service of complex transport services (ES CTS)".

The main target of creation and introduction of the automated system is the creation of the united information space [4]:

– systems of collection, accounting, analysis and processing of statistic information, anticipated and actual volumes of freight transportation of various means of transport;
– systems of collection, accounting, analysis and processing of anticipated and actual capacities of carriers to cover needs for freight transportation;
– systems of forecasting, planning and monitoring of freight flow correspondence within the united transport system of the country;
– systems of cooperation of administrative authorities of the united transport system with participants of the complex transport services market [4, 5].

The main target is achieved through execution of the following tasks:

1. Transparency of anticipated and actual volumes of production and freight transportation.
2. Minimal time and financial expenditure of receivers of transport services, their affordability.
3. Increase in quality of complex transport services [5, 6].
4. Introduction of the information technology, its integration in the logistics processes to enable online orders of the main and additional freight transportation services.
5. Procedure simplification of internal and external documentation of a desired transportation by a user in the united informational space.

Implementation of these priority tasks [7] by building of the automated system ES CTS means well-balanced comprehensive development of the statistical data accounting system, use of forecasting methods to analyze the needs of country's regions and population for transport services [8], production volume dynamics, development of transport infrastructure components [8, 9] and also analysis and modelling of development options of the united transport system.

2   Structure and Methods of System Construction

The automated ES CTS functions in the united informational space on the basis of universal database and systems of regulatory and reference information. Data exchange is carried out based on the approved regulations of data exchange [10].

ES CTS allows users to get freight quotes by specified criteria online as well as submit an application for a complex freight transport service.

The automated system executes the following:

1. Informing an ES CTS operator about a customer's need for transportation (sending information letters).
2. Estimation options of preliminary shipment cost (transportation by various or several means of transport).
3. Distance and delivery time calculation (transportation by various or several means of transport).
4. Finding an optimal means of transport for shipment.
5. Selection of transport and logistics companies which provide services for the shipment in question.
6. Providing, storage and editing of regulatory documents, required for quality cooperation with regard to providing services under the contract.
7. Opportunity for the system's operator to process shipping requests from cargo owners (customers).

ES CTS envisages an opportunity to integrate parameters from other information systems operative [10, 11] in the country's information space (subject to approval). Cooperation is carried out upon user's request with structured messages exchange protocol in the distributed computer environment SOAP, which keeps data consistency and does not hinder normal operation.

The software is fully located on the server and does not require installation on the user's workstation of any kind of programs. The software and data store are located on the server [12] and envisage transfer of report forms into documents Office Microsoft Excel, PDF. It is designed for standard commercial computers, connected by standard data transmission facilities through application software, developed within task creation.

The main task for the receiver of transport services "Preparation of a shipping request" is based on the principle of displaying on the workstation of the output forms required by the user. The minimal required input operational parameters for the task "Preparation of a shipping request" are as follows:

- station, departure point;
- station, destination point;
- date of shipment;
- cargo name;
- weight, volume and number of cargo units;
- type of shipment;
- required type of rolling stock, transport vehicle.

According to duties and tasks performed by the user, each workstation in the task "Preparation of a shipping request" envisages the following access permissions:

1. Workstation of the process executive: reports viewing, getting information and operational tasks of a business unit.
2. Workstation of operational personnel – viewing of incoming requests, well-timed reacting on requests and complaints of users.
3. Workstation of a system's user: preliminary calculation of shipping price, preparation of a shipping request, complaints about met obligations, getting informed about current organizational documentation, carrying out tests for the rules of load fastening and positioning.
4. Workstation of a system administrator: user administration, altering, updating of regulatory and reference information for building a database.

The system builds on the principle of workstation organization of a user of transport services to enable an opportunity, while staying focused on the working process, to make a preliminary price calculation of the required transportation, including additional terminal or warehouse services, receipt of cargo, delivery on the principle of "door to door" [8], application for purchasing of a transport service, stage tracking of freight transportation.

The Fig. 1 shows organizational and functional scheme of the automated system ES CTS.

To achieve desired functionality, the following program components and informational and functional modules have been outlined:

- "User registration in the system" module;
- "Personal account of a user" module;
- "Intersystem cooperation" module, which includes the following components;
- "Request for regulatory and reference information" component (information from tables: classifier of interstations, classifier of railways, classifier of roads, classifier of cargos, classifier of rolling stock types, classifier of transport vehicles, classifier of shipment types, container capacity, transport vehicles capacity, classifier of very large cargo containers;
- "Request for testing connection with other information systems" component;
- "Inquiry calculation of transport rate" component;
- "Delivery time calculation for two stations" component;
- "User Authorization" module;
- "Restore of lost data for user authorization" module;
- "Providing the user with operational information about preliminary price of freight transportation according to specific parameters" module;

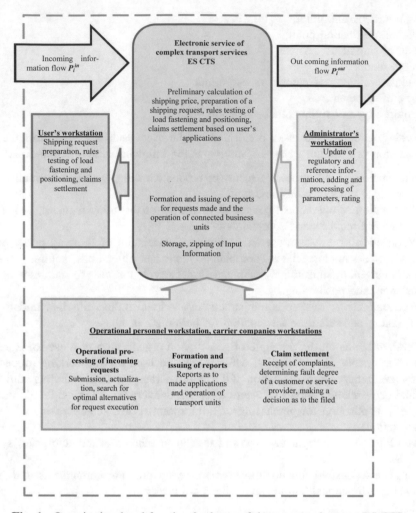

Fig. 1. Organizational and functional scheme of the automated system ES CTS.

- "Preparation of a shipping request" module;
- "Uploading of information from other information systems to the ES CTS data base" module;
- "Automated workstation of a ES CTS operator" module.

The Fig. 2 shows the structure of the automated system ES CTS.

The Module "User registration in the system" allows users of the information system to perform registration procedure in ES CTS. It displays a registration form with data-entry field, checks correct input of information, informs user of occurred errors while filling in registration data. It conducts preliminary checking of a registering user by input of information from a graphic object, sending registration confirmation to the given e-mail address, inserts input information in the data base table upon successful

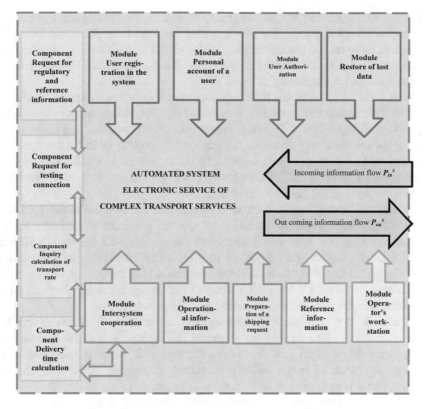

Fig. 2. Structure of the automated system ES CTS.

completion of data-entry field. It informs the user about successful registration procedure, encrypts accounting data in the ES CTS, protects data-entry fields from sql-
injections by screening of input information.

The module "Personal account of a user" enables users of the information system to
view status of shipping requests, to prepare such requests, to edit personal data, to
apply for guidance. It displays specified form with source data, checks data-entry field
for correct input, including mandatory fields, informs the user of occurred errors by
filling in registration data, allows the user to track status of sent requests, to view the
list of commercial offers for those requests, inserts input data in the database table upon
successful completion of the data-entry field by the user.

The module "Intersystem cooperation" is designed for solving the complex of tasks
of data exchange and control when making requests for shipment. It is basically an
interface of intersystem cooperation of the ES CTS with other information systems in
terms of request execution for getting regulatory and reference information, preliminary
price of shipment, distance and delivery time. This module is a back-office subsystem
within the ES CTS, which performs intercommunication tasks. The intersystem
cooperation module is a web-service, which uses SOAP protocol.

The module "User's authorization" is responsible for the authorization procedure of a user in the ES CTS system. It displays accounting data-entry field, conducts search and matches input data to the information in the ES CTS database, informs the user of mismatch of input data and data stored in the ES CTS database, protects data entry-field from sql-injections by screening of input data.

The module "Restore of lost data for user authorization" provides the user with an opportunity to execute restore of forgotten or lost data for authorization in the ES CTS. It displays a dialogue box with e-mail entry field which the user indicated by registration in the information system for the purposes of restore of lost data. It conducts search and matching of input data to the information stored in the ES CTS database, informs the user of mismatch of data given in the restore lost data entry field to the information stored in the ES CTS database.

The module "Providing the user with operational information about preliminary price of freight transportation according to specific parameters" is responsible for providing the system's user with information about estimated distance, preliminary time and price of cargo delivery according to specified criteria of shipment. It displays a dialogue box with data-entry field for shipment criteria [11] in the form of a drop-down list of prompts in the data-entry fields, estimated distance, preliminary time and price of cargo delivery according to specified criteria of shipment.

The module "Preparation of a shipping request" enables the users of the information system to make a request for shipment. It displays a dialogue box with data entry field for shipment criteria. It provides the user with estimated distance, preliminary time and price of cargo delivery according to specified criteria of shipment, informs the user whether the request have been registered.

The shipping request can be made either in the personal account or in the main form of the information system. However, the link "Preparation of a shipping request" is available to all users. The module "Uploading of reference data" is responsible for uploading of regulatory and reference information from database tables of other information systems and uploading of ES CTS database in a table. The module contains a range of procedures and collaborative functions with the intersystem cooperation module with regard to getting data as to required parameters. In case of emergencies in the data import and export procedures, the module returns an error code and its description.

The module "Automated workstation of ES CTS operator" is an ES CTS subsystem, which enables an operator to perform operations over made requests for shipment, to set the information system, prepare and send replies for the requested advice. It provides the ES CTS with a fully functional workstation and all the necessary mechanisms for performing required operations. It allows the user to track shipping requests, to process made requests, to track requests for customer support, to set the system, to upload necessary regulatory documents in the system. It sends e-mails to the given addresses containing information about execution of requests and conducts control over the current operation.

The automated information system developed by the authors carries the following features, which differentiate it from other systems: statistic data accounting and analysis of production volumes and freight flow correspondence, incorporation of data from different information systems in one, adaptation of ES CTS in the united information space of inter-sectoral transportation, online preparation of documents required for shipment, payments for the complex transport service by any convenient method, online performance assessment of a loading master. The above-mentioned features characterize the system as whole new compared to other automated systems.

3   Results

The results of ES CTS use on the railway service testing ground are presented on Fig. 3.

As it can be observed, the use of the automated system of the electronic service of complex transport services in the united information space of intersectoral freight transportation contributes to an increase in efficiency of planning by 12.7%, increase of rationalization level of transportation by 5.7%, reduction in cargo execution time by 15.8%. There is an increase by 4.3% in the freight flow volume of the railway transport in the category of interregional transportation and 4.1% of the road transport in the category of regional transportation. The overall speed of delivery is increased by 27%.

The conducted estimates of transport service quality show the financial positive effect as the overall decrease in costs of intersectoral transportation by 3.2% has been achieved and the costs of additional transport services have been reduced by 1.7%.

As a result, the adaption of the united transport system to the existing conditions by means of consolidation of information systems of the parties of the united transportation process with the use of automated system ES CTS assures reduction in added

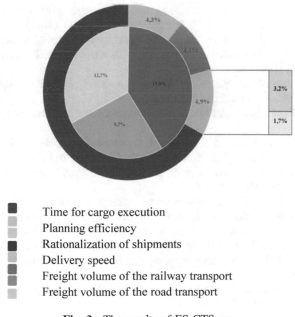

Time for cargo execution
Planning efficiency
Rationalization of shipments
Delivery speed
Freight volume of the railway transport
Freight volume of the road transport

Fig. 3. The results of ES CTS use.

value growth rate of the finished product. The decrease of financial losses of transport service receivers, in its turn, leads to creation of competitive environment in the market of railway transportation.

The current computer program has been registered by the Federal service for intellectual property of the Russian Federation.

4 Conclusions

The use of the ES CTS program implements the project of united information space and allows building of the united transport system of the Russian Federation with universal, adaptive and consolidating intercommunication and reduction in unproductive losses in the functioning field of several means of transport in the area of intersectoral transportation.

The received results are tools for analysis, accounting and information processing of anticipated and actual production volumes and freight flow correspondence.

The use of this system enables to achieve positive effects in the main direction of competitiveness and focus on customer, namely transparency of pricing of transport service and one-stop service principle.

The possible use of the current system in information technologies in the sector of international freight transportation is a subject for further research.

References

1. Galkin, G., Zubkov, V., Sirina, N.: Model of complex transport service as a prospect of freight transportation development. Transp. Ural. 1(56), 7–11 (2018)
2. Sirina, N., Zubkov, V.: Modernization of transport business processes. In: 9th International Scientific and Practical Conference-Transport Infrastructure of the Siberian Region, Irkutsk, pp. 134–137 (2018)
3. Misharin, A., Evseev, O.: Updating of the transport strategy of the Russian Federation for the period 2030. Transp. Russ. Fed., 4–13 (2013)
4. Zubkov, V.: Developing a transportation management model. Transp. Ural. 1(40), 12–17 (2014)
5. Lapidus, B., Macheret, D., Miroshnichenko, O.: On increasing the productivity of resource use and efficiency of the railways. Railw. Econ. 6, 12–22 (2011)
6. Zubkov, V.: Concept of interaction of infrastructure regional boards on railway borders. Transp. Ural. 2(33), 18–21 (2012)
7. Zubkov, V.: Developing a transportation management model. Management of the transportation process based on the theory of forward-thinking systems. Intelligent control systems in railway transport: collection of scientific articles, pp. 138–140. NIIAS (2012)
8. Mitrofanov, V., Grishina, T., Feofanov, A.: Management of automated technological systems and modeling of efficiency when making decisions. Tekhnologiya Mashinostroeniya 8, 43–45 (2015)
9. Armand, V., Zheleznov, V.: Bar Codes in Information Processing Systems. http://www.retail.ru/biblio. Accessed 20 Sept 2018

10. Henry, E.: Precision apiculture: development of a wireless sensor network for honeybee hives. Masters of Science, McGill University, Montreal, Quebec, Canada (2016)
11. Zhou, C., Yao, K., Jiang, Z., Bai, W.: Research on the Application of NoSQL Data base in intelligent Manufacturing. In: Wearable Sensors and Robots. LNEE, pp. 423–434 (2017)
12. Agarwal, R., Dhar, V.: Editorial big data, data science, and analytics: the opportunity and challenge for is research. Inf. Syst. Res. 3(25), 443–448 (2014)

Concurrent Intelligent Transport Systems Based on Neuroprocessor Devices

Vitaliy Romanchuk(✉)

Ryazan State University named for S. Yesenin,
Svobody ave. 46, 390000 Ryazan, Russia
v.romanchuk@365.rsu.edu.ru

Abstract. The principles of developing the intelligent transport systems based on neuroprocessors are exposed in the paper. Neuroprocessor devices are characterized by high processing speed, low price and low power consumption. They can be used for image processing and real-time car recognition. The efficient parallelization algorithms have been developed in order to ensure high processing speed. The algorithms are based on the rational separation of many homogeneous neural network operations.

Keywords: Intelligent transport systems · Neuroprocessor system · Efficiency analysis · Neuroprocessor

1 Introduction

At the present time, intelligent transport systems are being actively developed. Systems that use innovative developments in modeling transport systems and regulating traffic flows are called intelligent. They provide end users with greater information and security, as well as qualitatively increase the interaction level between traffic participants, comparing to the conventional transport systems. Generally, such systems are transport modeling systems and means of controlling traffic flows. In order to regulate traffic flows, software controls and parallel processing of data from video cameras on the server or directly from the traffic lights are used. At the same time, the option with information processed on the server is more efficient due to the decision making considering traffic at multiple points, not at one point.

Then the problem of creating intelligent traffic control systems comes down to the development of high-speed intelligent pattern recognition algorithms and decision making based on the number of cars at intersections and roads. In order to ensure high real time calculation speed, it is rational to use specialized computer equipment based on neuroprocessor devices. Various companies actively produce neurochips, for example NTC Module, Qualcomm, IBM, Toshiba, Human Brain Project, KnuEdge Inc., Analog Devices, Texas Instruments and others [1]. In 2014, IBM Research unveiled the TrueNorth chip consisting of a million digital neurons and 256 million synapses that are part of 4,096 synaptic nuclei. During the capability demonstration, the chip recognized cars, cyclists and pedestrians on a video at a speed of 100 times faster and consumed 1000 times less energy than existing computing devices based on the

© Springer Nature Switzerland AG 2020
V. Murgul and M. Pasetti (Eds.): EMMFT 2018, AISC 982, pp. 376–382, 2020.
https://doi.org/10.1007/978-3-030-19756-8_35

von Neumann architecture. The developed neurochips have limited performance, so usually cluster, multiprocessor systems based on neurochips are used. Thus, in order to solve the problem of developing intelligent transport systems, it is proposed to use neurocomputer devices that implement matrix calculations. The main task is to create efficient signal and image processing algorithms that can be executed on neural calculators. It is also necessary to take into account the possibility of parallelism, considering the architectural features of calculators, which allows reaching high performance.

2 Hardware Base of Intelligent Transport Systems

Let us consider one of the options for using a set of hardware and software tools for developing and implementing deep neural nmDL networks on the neuroprocessor 1879VM6Ya.

The 1879VM6Ya processor was developed by ZAO STC Modul (Moscow) and is a high-performance vector-matrix microprocessor with the original NeuroMatrix® architecture and ultra-large command words, high-performance digital signal processor. The architecture is based on using the new generation VLIW/SIMD processor core NMC4. The processor contains two processor cores NMPU0 and NMPU1, each one includes a RISC processor and a vector coprocessor. The first 64-bit coprocessor is designed to perform vector-matrix operations on integer data of variable length from 1 to 64 bits. The second 64-bit coprocessor is designed for floating point vector operations. The RISC processor is 32 bits wide; bit length of commands is 32 and 64 bits; capacity is 1000 MIPS (3000 MOPS). The vector-matrix coprocessor of integer arithmetic has a programmable data length from 2 to 64 bits (64 bits is a packed word length); one basic operation – integer matrix multiplication; simultaneous execution of 2 saturation functions. Its performance (MAC – Multiplication and Accumulation per step) is 2 MAC for 32-bit data; 4 MAC for 16-bit data; 24 MAC for 8-bit data; 80 MAC for 4-bit data; 224 MAC for 2-bit data.

In order to implement deep learning on a neuroprocessor device, software that implements a deep convolutional network for the MC121.01 module has been developed. DLE (Deep Learning Engine) is software for the MC121.01 module, implementing a deep convolutional network in real time, the topology and parameters of which are set by the configuration file. At the moment, there various architecture types are supported. The software includes a library of optimized functions for programming deep neural networks on an 1879VM6Ya – DLCL processor. In order to ensure the effective implementation of a convolutional neural network, DLE uses the DLCL (Deep Learning Compute Library), a library of optimized functions that accelerate the work of deep neural networks on the 1879VM6Ya platform. The library implements the primitives most frequently used in deep neural networks (convolutional layers with ReLu, pooling, normalization, etc.). DLE during initialization will create the necessary network implementation using these primitives and the deep neural network configuration file, which is obtained from the DLDT loader. At the same time, various optimization algorithms for working with memory and convolution kernel are used for maximum performance. This frees the user from the need to optimize the placement of various architectures on the hardware platform. The user can also create his own deep neural network topology by using neural network primitives from the DLCL library. In order to do this, the file with a neural network topology, which is compatible with the DLDT package loader, needs to be provided. The DLE program will generate the required network for real-time operation on the basis of this file. In addition to the features described, it is permissible to use neural network primitives from the DLCL library in user applications for solving arbitrary application problems.

3 Parallelism Organizing Method in Intelligent Transport Systems Based on Neuroprocessors

The problem of developing intelligent transport systems based on neuroprocessors can be represented as an artificial neural network (ANN), which needs to be emulated using specialized hardware, the neurocomputer system. ANN can be set in the form of a tuple of parameters and characteristics:

1. The set of ANN inputs $Net_X = \{net_{x1}, \ldots .net_{xi}, \ldots, net_{xn}\}$, where each input net_{xi} is characterized by a length capacity $NetX_{R_i}$ and type $NetX_{N_i}$.
2. The set of ANN outputs $Net_Y = \{net_{y1}, \ldots .net_{yi}, \ldots, net_{yn}\}$, where each output net_{yi} is characterized by a length capacity $NetY_{R_i}$ and type $NetX_{N_i}$.
3. The set of neurons $Net_N = \{net_{n1}, \ldots .net_{ni}, \ldots, net_{nc}\}$, each of which must be emulated to solve some problem $Z^{(j)}$.
4. The set of weight ratios $W = \{w_1, \ldots .w_i, \ldots, w_{nc}\}$
5. ANN topology. There is currently no formal procedure for determining the ANN topology (the number of layers, the number of neurons in each layer, the type of connections). Due to this fact the heuristic methods are used based on previous experience with similar neural network tasks [1].

All topologies can be divided into two groups: networks with feedback and networks without feedback. The following types of topologies can be distinguished: multilayer perceptron, recurrent and associative networks, self-organizing networks, radial-basis networks.

The ANN topology can be described as a set of connections between neurons, which define the dependencies between neurons of the ANN:

$$N_E = \{N_{11}, N_{12}, \ldots, N_{1nc}, \ldots, N_{ncnc}\},$$

which can be represented as a matrix of connections between neurons, inputs and outputs of the ANN.

$$N_E \rightarrow MNet = [MNet_{ij}] \text{ of length } (nc + xn + yn) \times (nc + xn + yn).$$

This matrix shows the ANN topology, the number of ANN layers, the presence or absence of bypass connections, the transfer functions of neurons.

Let us introduce the concept of a precedence relation $\prec$, which is defined as follows: $\forall n_i, n_j \in N, n_i \prec n_j$, the neuron n_j uses the n_i neuron's output data as input, so the n_j

neuron emulation cannot be performed without prior emulation of the n_i neuron. The precedence relations between neurons are determined depending on the availability of ANN connections, which is a set of connections $N_E = \{N_{11}, N_{12}, \ldots, N_{1nc}, \ldots, N_{ncnc}\}$.

Each element of the matrix can take values $MNet_{ij} = \{'-', 'n', 'y'\}$:

- $'n'$ ('no') - there is no precedence relation $\neg(n_i \prec n_j)$ between subroutines n_i and n_j $(j > i)$;
- $'y'$ ('yes') - there is a precedence relationship between subroutines n_i and n_j;
- $'-'$ - there can be no precedence relationship between neurons.

In the *MNet* matrix, all elements below the main diagonal equal $'n'$, since there are no feedbacks in the perceptron.

Let us assume that $Z^{(j)}$ is some j-th neural network problem, which is implemented on a neuroprocessor platform.

Let us introduce a set of operations $O = \{O_1, O_2, \ldots, O_i, \ldots, O_{nc}\}$. Taking into account the introduced set of operations, the neural network problem is a tuple consisting of operations $O_1, O_2, \ldots, O_m, \ldots, O_{nc}$ of length $nc = |Z^{(j)}|; j = \overline{1, N}$, i.e.:

$$Z^{(j)} = <O_1, O_2, \ldots, O_m, \ldots, O_{nc}> ; j = \overline{1, N}$$

Thus, the set of operations $O_1, O_2, \ldots, O_m, \ldots, O_{nc}$ corresponds to the number of emulated ANN neurons, since a certain operation O_l is a mathematical model of a formal neuron:

$$O_l = f\left(\sum_{m=1}^{n} a_m x_m + a_0\right) \tag{1}$$

For some operations O_i and O_j the precedence relation $\prec$ is also applicable and defined as follows: $\forall O_i, O_j \in O, O_i \prec O_j$ - the operation uses the output data O_i of the operation as input data O_j, so the operation O_j cannot be executed without a preliminary executed operation O_i.

The purpose of the parallelization algorithm for computations is to redistribute the tuple of operations $Z^{(j)} = <O_1, O_2, \ldots, O_m, \ldots, O_{nc}> ; j = \overline{1, N}$ and organize them in p tuples of operations:

$<O_1, O_2, \ldots, O_{nc}> \rightarrow <O_{11}, O_{12}, \ldots, O_{1nc}>, \ldots, <O_{p1}, O_{p2}, \ldots, O_{pnc}>$ for their subsequent most rational displaying in the microcode: $<O_{11}, O_{12}, \ldots, O_{1nc}>$, $\ldots, <O_{p1}, O_{p2}, \ldots, O_{pnc}> \rightarrow <MK_1, MK_2, \ldots, MK_K>$

Let us introduce the segment of operations term *SO*, which implies a tuple of operations O, there is no precedence relationship between those operations:

$$SO_i = <O_{i1}, O_{i2}, \ldots, O_{inc}> ; O_{i1}\neg \prec O_{i2}\neg \prec \ldots \neg \prec O_{inc}. \tag{2}$$

The precedence relation $\prec$ for operations is defined as follows: $\forall O_i, O_j \in O, O_i \prec O_j$- the operation uses O_j the output of the neuro microcode O_i as input data. That is, the command O_j can be executed if and only if the operation O_i has already been executed.

Then the entire operation tuple implementing the problem $Z^{(j)}$ can be divided into a tuple of segments defined as (2): $<O_1, O_2, \ldots, O_m, \ldots, O_{nc}> \rightarrow <SO_1, SO_2, \ldots, SP_p>$, where $SO_1 \prec SO_2 \prec \ldots \prec SP_p$.

Let us consider the ratio of two arbitrarily taken operations O_k and O_l. The concept of equality of operations $O_k = O_l$ is introduced. The same functional purpose, which is an emulation of a neuron, is meant.

Then the statement is true that the ratio E of two arbitrarily taken microcodes O_k and O_l, satisfying (1), is an equivalence relation.

The equivalence relation is a binary relation, for which the following conditions are fulfilled: reflexivity, symmetry, transitivity. Let us consider each of these conditions:

1. Any operation O_k that satisfies (1) and is a reflection of the neuron model can run parallel to itself. This is implied by the architecture of the neural network itself and the neurocomputer, i.e. $O_k E O_l$. Thus, the reflexivity condition of operations is valid.
2. If the operation O_k is equal to the operation O_l, then the operation O_k is equal to the operation O_l and can be performed simultaneously with it, which is assumed according to the paradigm of neurocomputing, i.e.: $\forall O_k, O_l : O_k E O_l \rightarrow O_l E O_k$

Therefore, the condition of symmetry of two operations is satisfied.

If the operation O_k is equal to the operation O_l, and the operation O_l equals O_m, then the command O_k functionally equals O_m and parallel to the command O_m, that is the following: $\forall O_k, O_l : O_k E O_l, O_k E O_l \rightarrow O_k E O_l$.

Thus, the transitivity condition of the operations is true.

The division into segments consists of the following steps:

1. Filling the matrix of precedence relations depending on the neural processor type. Only the functional block area of the neural processor is considered in the matrix for solving this problem, limited by the precedence relations of the input and output data.
2. For each column value $MNet_i = \{n_1, n_2, \ldots, n_{nc}\}$ of the selected matrix area, the following conditions are considered:
   a. If for n_i all the column values equal 'n', then the operation O_i corresponding to the emulation of the neuron n_i can be placed in any (current) segment SO_j:
   $MNet_{ij} = $ 'n'; $i = \overline{1, N} \rightarrow SO_j = SO_j + O_i$, where N is the dimension of the $MNet$ matrix.
   b. all $n_i, i = \overline{1, nc}$ having the same location of values 'y' in the columns of the selected matrix area are placed in a separate segment SO_j: $MNet_{ij} = $ 'y';$\rightarrow SO_j = SO_j + \{O_i\}; i = \overline{1, nc}$.
   c. If the solution of the problem requires feedback, which is expressed in preceding relation between the last and the first neuron of the ANN, the value 'y' in the lower left corner of the matrix $MNet_i = \{n_1, n_2, \ldots, n_{nc}\}$ is not taken into account. $\max k(SO_i = $ 'y') $> \max k(SO_j = $ 'y'); $k = \overline{1, N} \rightarrow SO_j \prec SO_i$.
3. Having a set of segments $\{SO_1, SO_2, \ldots, SP_p\}$, the preceding relationships can be defined as follows:

– if the position of the lowest value 'y' in the matrix is greater for SO_i segment operations than the position of the lowest value 'y' for SO_j segment operations, then the preceding relationship $SO_j \prec SO_i$ is true: having a tuple of segments connected by precedence relationships, it is necessary to split the segments for programming and mapping an operation into multiple processor microcodes, in order to isolate a tuple of operations performed per processor cycle as follows:

– if the number of operations $|SO_i|$ in the SO_i segment is greater than the maximum number of emulated neurons in the neuroprocessor (MAC), then it is necessary to divide this segment into two, while the precedence relationships between them will not be indicated: $|SO_i| > MAC : SO_i = SO_{i1} \bigcup SO_{i2}$.

As a result, a tuple of segments obtained, each of which can be represented as a microcode (or a tuple of microcodes) with a maximum parallelism level in the execution of multiple operations per cycle. That is, using the programming language according to the chosen neuroprocessor, the tuple of segments is obtained:

$$<O_{11}, O_{12}, \ldots, O_{1nc}>, \ldots, <O_{p1}, \ldots, O_{pnc}> \rightarrow <SO_1, SO_2, \ldots, SO_L>.$$

Thus, the results of the study are as follows:

1. Mathematical apparatus for regulating traffic flows depending on real-time traffic.
2. Parallel algorithms for the regulation of traffic flows.
3. Software for regulating traffic flows.

4 Conclusion

The use of neuroprocessor devices as a hardware base for real-time image processing is proposed. Efficient paralleling algorithms have been developed in order to develop intelligent transport systems. Practical studies using the developed software have shown that the speed of the neuroprocessor program increases by 5–14% due to the application of the parallelization method. The increase value depends on the class of the problem and the homogeneity degree of operations. Speed increases due to the rational separation of many homogeneous neural network operations.

Acknowledgments. The research was carried out within the frame-work of the assignment for the performance of public works in the sphere of scientific activity within the framework of the initiative scientific project of the state task of the Ministry of Education and Science of the Russian Federation No. 2.9519.2017/BC on the topic "Technologies for parallel processing of data in neurocomputer devices and systems".

References

1. Singh, H., Lee, M.-H., Lu, G., Kurdahi, F.J., Bagherzadeh, N., Filho, E.M.: MorphoSys: an integrated reconfigurable system for data-parallel and computation-intensive applications. IEEE Trans. Comput. **49**(5), 465–481 (2000). https://doi.org/10.1109/12.859540
2. Romanchuk, V.A.: The method of optimization of neuro-based concurrent operations in neurocomputers In: IOP Conference Series: Materials Science and Engineering (2017). https://doi.org/10.1088/1757-899x/177/1/012033
3. Goswami, S., Chakraborty, S., Saha, H.N.: An univariate feature elimination strategy for clustering based on metafeatures. Int. J. Intell. Syst. Appl. (IJISA) **9**, 20–30 (2017)
4. Barabash, O., Kravchenko, Y., Mukhin, V., Kornaga, Y., Leshchenko, O.: Optimization of Parameters at SDN Technologie Networks. Int. J. Intell. Syst. Appl. (IJISA) **9**, 1–9 (2017)
5. Bottou, L.: Stochastic gradient descent tricks. In: Neural Networks: Tricks of the Trade, pp 421–436 (2012)
6. Niu, C.R.F., Retcht, B., Wright, S.J.: Hogwild! a lock free approach to parallelizing stochastic gradient descent In: Neural Information Processing Systems (2011)
7. Izeboudjen, N., Larbes, C., Farah, A.: A new classification approach for neural networks hardware: from standards chips to embedded systems on chip. Artif. Intell. Rev. **41**, 491–534 (2014)
8. Kolinummo, P., Pulkkinen, P., Hämäläinen, T., Saarinen, J.: Parallel implementation of self-organizing map on the partial tree shape neurocomputer. Neural Process. Lett. **12**, 171–182 (2000)
9. Romanchuk, V.: Mathematical support and software for data processing in robotic neurocomputer systems. In: MATEC Web of Conferences, vol. 161 (2018). https://doi.org/10.1051/matecconf/201816103004

The Method of Optimum Design of Energy Saving Electromagnets for Levitation Magnetic Transport

Anna Balaban⬤, Yury Bakhvalov⬤, Valeriy Grechikhin$^{(\boxtimes)}$⬤,
and Julia Yufanova⬤

Platov South-Russia State Polytechnic University (NPI),
Prosvesheniya Str. 132, 346428 Novocherkassk, Russia
vgrech@mail.ru

Abstract. The article proposes a method of optimal design of electrical devices, based on solving conditionally-correct inverse problems. The method is used to design electromagnets of levitation magnetic transport. Permanent magnets are built into the design of electromagnets to increase energy saving. The effectiveness of the method is due to the use of the hierarchy of mathematical models. In the first, second, and third stages, inverse problems are solved analytically, and in the fourth, numerically. Such an approach makes it possible to determine the initial approximations of the desired parameters with sufficiently high accuracy and to reduce the total time to solve the problem. At the fourth stage, the finite element method is used to solve direct problems of calculating the magnetic field and forces, and the gradient descent method is used to solve optimization problems. Reduction of time for solving the problem is achieved by transformation of constraints into objective functions and sequential minimization of these functions over a limited number of variables. An algorithm for solving the inverse problem is given. The results of the computational experiment showed a high efficiency of the method in the design of levitation electromagnets with permanent magnets. The method can be used in the design and other electrical devices.

Keywords: Magnetic transport · Energy efficiency · Permanent magnet · Inverse problems · Optimization

1 Introduction

At present, permanent magnets (PM) are embedded in electromagnets of various applications in order to save electricity. Therefore, the urgent task is to develop an effective method for the optimal design of such devices. The method is based on solving inverse problems. This approach is one of the promising areas in the design, identification and diagnosis of technical and biological objects. Next are used only conditionally correct inverse problems, according to Tikhonov [1].

Transport in the life of modern society plays an extremely important role. One of the directions in the development of transport systems is the replacement of a wheel-rail node in rail transport with a non-contact electromagnetic suspension. A new type of

© Springer Nature Switzerland AG 2020
V. Murgul and M. Pasetti (Eds.): EMMFT 2018, AISC 982, pp. 383–391, 2020.
https://doi.org/10.1007/978-3-030-19756-8_36

ground transportation with magnetic levitation of crews involves the use of a linear traction electric drive and it is capable of speeds up to 500 km/h [2–5]. In the article, the proposed method is used to design electromagnets for levitation of magnetic transport. The method is a modification of the method proposed earlier by the authors of this article [6–10].

2   Materials and Methods

The design of the W-shaped electromagnet (EM) (Fig. 1); required levitation force Fl; permissible error of calculating the levitation force δF; air gaps δg; the maximum average magnetic induction in the rods EM Bδ, which ensures the effective operation of the gap regulator; permissible error of calculation of magnetic induction δB; length of PM along 0z axis L; permissible error of calculating the size and magnetomotive force (mmf) δ; type of PM and its magnetization are known.

It is required to determine the geometrical dimensions of EM: a, h, l, b_{fr}, h_{pm} and mmf iw, at which the values F_l and B_δ are determined with known allowable errors δ_F and δ_B, and the mass EM (without ferrorails) is minimal.

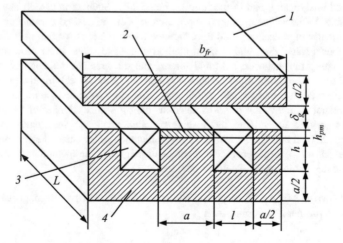

Fig. 1. Energy-saving EM cross-section: 1 – ferrorail; 2 – PM; 3 – current coil; 4 – core of EM.

The mathematical model of the problem consists of an optimization model and a model for electromagnetic calculations. The first model includes the mass (goal function), determined by the formula

$$M_{em} = \left[2hS_p + \left(\frac{S_w}{h} + \frac{S_p}{L}\right)S_p\right]\rho_{st} + 2S_wL\rho_{cu} \qquad (1)$$

where ρ_{st}, ρ_{cu} – steel and copper density; S_p – central pole area, $S_p = aL$; S_w – coil window area, $S_w = hl = iw/jk_f$, j – coil current density, k_f – filling factor of the window by the copper.

The model contains constraints arising from the formulation of the problem

$$F_l \leq F_l^{(k)} \leq F_l(1 + \delta_F); \ \ B_\delta \leq B_\delta^{(k)} \leq B_\delta(1 + \delta_B) \tag{2}$$

$$\left| a^{(k+1)} - a^{(k)} \right| \leq \delta; \ \ \left| h^{(k+1)} - h^{(k)} \right| \leq \delta; \ \ \left| l^{(k+1)} - l^{(k)} \right| \leq \delta; \ \ \left| iw^{(k+1)} - iw^{(k)} \right| \leq \delta \tag{3}$$

In Eqs. (2) and (3) the superscript «k» means that the value was obtained at the k-th step of the problem solving algorithm.

A model for electromagnetic calculations is presented in the form of a hierarchy of models: at the first stage, only air gaps are known. Therefore, the equivalent magnetic circuit of EM for model 1 is of the form shown in Fig. 2.

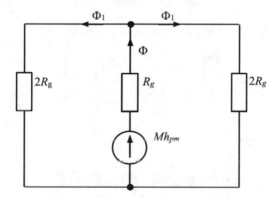

Fig. 2. Magnetic circuit of model 1.

Kirchhoff's laws for the magnetic circuit (Fig. 2), taking into account its symmetry, are written in the form

$$\Phi - 2\Phi_1 = 0; \ \ \Phi R_g + \Phi_1 2R_g = Mh_{pm} \tag{4}$$

where Φ – magnetic flux in the central core of the EM core, $\Phi = B_\delta S_p$; Φ_1 – magnetic flux in the side rods, $\Phi_1 = 0.5\Phi$; R_g – gap magnetic resistance, $R_g = \delta_g/(\mu_0 S_p)$; μ_0 – magnetic constant, $\mu_0 = 4\pi \cdot 10^{-7}$ H/m; Mh_{pm} – mmf of PM, M – magnetization of PM.

On the other hand, is performed $\oint \vec{H} d\vec{l} = Mh_{pm}$, where does the formula for the first approximation h_{pm}

$$h_{pm}^{(1)} = \frac{B^* 2\delta_g}{\mu_0 M - B^*} \tag{5}$$

where B^* – EM magnetic induction value, which provides mmf PM, $B^* = 0,5B_\delta$.

The information obtained at the first stage is used to build model 2.

Taking $iw^{(1)} = Mh_{pm}^{(1)}$ is received $S_w^{(1)} = h^{(1)}l^{(1)} = iw^{(1)}/jk_f$.

The force of attraction EM for two gaps is determined using an approximate formula $F_l = B_\delta^2 k_s S_p/\mu_0$, from where

$$S_p^{(1)} = \frac{\mu_0 F_l}{B_\delta^2 k_s}, \quad a^{(1)} = \frac{S_p^{(1)}}{L} \tag{6}$$

where k_s – coefficient of buckling of the magnetic field in the gap.

From Eq. (1) follows, if known S_w and S_p then mass of the EM depends only on h. The value of the parameter h, is determined by minimizing $M_{em}(h)$.

Necessary and sufficient condition for the extremum of the function M_{em} are

$$\frac{dM_{em}}{dh} = 0; \frac{d^2 M_{em}}{dh^2} > 0.$$

When performing the differentiation $M_{em}(h)$, it is obtained

$$\left.\begin{aligned} \frac{dM_{em}}{dh} &= \left(2S_p - \frac{S_p S_w}{h^2}\right)\rho_{st}; \\ \frac{d^2 M_{em}}{dh^2} &= \frac{S_p S_w}{h^3}\rho_{st} > 0. \end{aligned}\right\} \tag{7}$$

From Eq. (7) follows

$$h^{(1)} = \sqrt{0,5S_w^{(1)}}, l^{(1)} = S_w^{(1)}\big/h^{(1)}. \tag{8}$$

The width of ferrorail is determined by the formula

$$b_{fr}^{(1)} = 2.4\left(l^{(1)} + a^{(1)}\right) \tag{9}$$

The equivalent magnetic circuit for building model 2 is shown in Fig. 3.

Kirchhoff's laws for the magnetic circuit (Fig. 3), taking into account its symmetry, are written in the form

$$\left.\begin{aligned} \Phi - 2\Phi_1 &= 0; \\ \Phi\left(R_g + R_c\right) + \Phi_1\left(R_{fr} + 2R_g + R_s + R_h\right) &= Mh_{pm} \end{aligned}\right\} \tag{10}$$

where R_c – magnetic resistance of central rod of the EM core, $R_c = \dfrac{h + 0.5a}{\mu_0 S_p}$; R_{fr} –

magnetic resistance of a section of the ferrorail, $R_{fr} = \dfrac{l + 0.75a}{\mu_{fr} 0.5 S l_p}$; R_s – lateral rod magnetic resistance, $R_s = \dfrac{h+0.5a}{\mu_{fr}0.5S_p}$; R_h – magnetic resistance of the horizontal section

of the EM core, $R_h = \dfrac{l+0.75a}{\mu_{fr}0.5S_p}$.

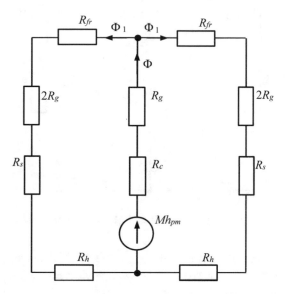

Fig. 3. Magnetic circuit of model 2

Considering that $\Phi = B_\delta S_p$, $\Phi_1 = 0{,}5\Phi$ from (10) second approximation is obtained h_{pm}

$$h_{pm}^{(2)} = \frac{2B_\delta}{\mu_0}\left(\delta_g + \frac{h^{(1)} + l^{(1)} + 1.25a^{(1)}}{\mu_{fr}}\right) \tag{11}$$

where $\mu_{\phi r}$ – relative magnetic permeability, determined by the main magnetization curve of a selected steel grade for a given value B_δ.

The value $h_{pm}^{(2)}$ and consequently, $Mh_{pm}^{(2)}$ are used to refine the values of all the parameters of the problem using the above formulas.

The equivalent magnetic circuit of EM is shown in Fig. 4 for building model 3, which allows for given values of all parameters and Mhpm to clarify the value of iw, which provides the required value of Bδ. Then the parameters of the problem are specified.

The three models discussed above provide a fairly good approximation for all the parameters of the problem, reducing the total time to solve it by analytically solving inverse problems.

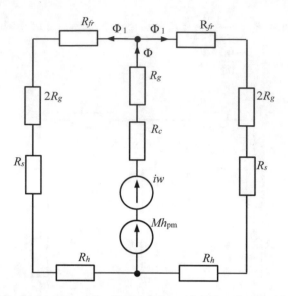

Fig. 4. Equivalent magnetic circuit of EM with PM and coil with current

Model 4, consisting of equations that describe the EM magnetic field and optimization model, allows numerical (using the gradient descent method) determination of all parameters with the required accuracy.

The calculation of the magnetic field and force interactions is performed by the finite element method.

The optimization problem formulated above can be solved using the Lagrange indefinite multipliers method [11]. However, this method is ineffective, since it leads to the problem of solving a nonlinear system of equations of high dimensionality.

Next, is used the method proposed by us, based on the transformation of constraints Eq. (12) into objective functions and the successive minimization of these functions over a limited number of variables.

The following system of objective functions is introduced:

$$\left. \begin{array}{c} J_1 = \left[F_l^{(k)} - (1+\delta_F)F_l\right]^2; \\ J_2 = \left[B_\delta^{(k)} - (1+\delta_\delta)B_\delta\right]^2; \\ J_3 = M_{em}. \end{array} \right\} \tag{12}$$

Assumptions are accepted: J1 is a function of Sp only, which follows from Eqs. (2) and (3), J2 is a function of iw only (or Sw). When known Sp, and Sw, then J3 is a function of h only. Therefore, we will use relations Eqs. (11), (12), which minimizes Mam. Therefore, relations Eqs. (11), (12) are used that minimize Mem.

The algorithm of numerical iterative refinement of the desired variables, that is, the solution of the inverse problem, using model 4 consists of the following steps.

1. The solution of the direct problem of calculating a stationary magnetic field and forces by the finite element method, taking $a = a^{(k)}$, $h = h^{(k)}$, $l = l^{(k)}$, $iw = iw^{(k)}$, $b_{fr} = b_{fr}^{(k)}$, $h_{pm} = h_{pm}^{(k)}$. Determination of the force $F_l^{(k)}$, the average magnetic induction in the gap $B_\delta^{(k)}$.

2. Calculation the values $J_1^{(k)}$ and $J_2^{(k)}$.

3. Check for compliance with conditions $J_1^{(k)} \le \delta_F^2$; $J_2^{(k)} \le \delta_B^2$, as well as conditions (Eq. (3)).

4. If the conditions of step 3 are satisfied, then the solution is obtained. If not, go to the next step.

5. Determination of the following approximations of the desired parameters using the gradient descent formulas

$$S_p^{(k+1)} = S_p^{(k)} - \lambda_{S_p}^{(k)} \frac{\partial J_1}{\partial S_p^{(k)}}; iw^{(k+1)} = iw^{(k)} - \lambda_{iw}^{(k)} \frac{\partial J_2}{\partial iw^{(k)}},$$

where $\lambda_{S_p}^{(k)}$; $\lambda_{iw}^{(k)}$ is method steps. Partial derivatives $\partial J_1 / \partial S_p^{(k)}$ and $\partial J_2 / \partial iw^{(k)}$ are determined numerically.

6. Calculation $k+1$ approximations of the unknown quantities by formulas

$$a^{(k+1)} = S_p^{(k+1)} \Big/ L; \quad S_{ok}^{(k+1)} = iw^{(k+1)} / jk_f; \quad h^{(k+1)} = \sqrt{0,5S_w^{(k+1)}}; \quad l^{(k+1)} = S_w^{(k+1)} / h^{(k+1)}; \, b_{fr}^{(k+1)} = 2.4(a^{(k+1)} + l^{(k+1)}).$$

7. Move to step 1.

3 Results and Discussion

Initial data for designing of the EM: levitation force (the force of attraction EM to ferrorails) $F_l = 30 \cdot 10^3$ N; allowable error of force calculation $\delta_F = 0.01$; air gap $\delta_g = 0.015$ m; length of EM is $L = 1$ m; current density $j = 3 \cdot 10^6$ A/m^2; coefficients $k_b = 0.8$; $k_f = 0.7$; relative magnetic permeability of steel 1010: $\mu_{fr} = 1660$ when $B_\delta = 0.75$ T; steel density $\rho_{st} = 7800$ kg/m^3; copper density $\rho_{cu} = 8900$ kg/m^3; allowable size calculation error and mmf $\delta = 0.01$. $B_\delta = 0.75$ T, $B_\delta^* = 0,5 B_\delta$ T; $\delta_B = 0.01$, as well as the value of magnetization $M = 950$ kA/m for the selected type of PM.

The results of the calculations are given in Table 1.

Thus, the values of the required parameters with a given error are obtained, at which EM, having a minimum mass, provides the required values of lift and magnetic induction with permissible errors.

Analysis of the results shows that the use of PM in the EM design allows reducing the power consumption by 2.7 times with the mass of the electromagnet Mem not exceeding 275 kg.

Table 1. The results of the calculations.

Parameters	Model 1	Model 2	Model 3	Model 4 (after 16 iteration)
h_{pm}, m	0.0140	0.0142	0.0142	0.0142
iw, A	13000	13350	12300	9171
S_p, m^2	0.084	0.084	0.084	0.099
a, m	0.084	0.084	0.084	0.099
S_w, m^2	$6.33 \cdot 10^{-3}$	$6.43 \cdot 10^{-3}$	$5.86 \cdot 10^{-3}$	$4.37 \cdot 10^{-3}$
h, m	0.056	0.057	0.054	0.047
l, m	0.112	0.113	0.103	0.093
b_{fr}, m	0.377	0.379	0.355	0.461
M_{em}, kg	291	289	280	275

Thus, the values of the required parameters with a given error are obtained, at which EM, having a minimum mass, provides the required values of lift and magnetic induction with permissible errors.

Analysis of the results shows that the use of PM in the EM design allows reducing the power consumption by 2.7 times with the mass of the electromagnet Mem not exceeding 275 kg.

4   Conclusions

The article proposes a method of optimal design of electrical devices, the effectiveness of which is due to the following – using a hierarchy of mathematical models. In the first, second and third stages, inverse problems are solved analytically, in the fourth - numerically. Such an approach makes it possible to determine the initial approximations of the desired parameters with sufficiently high accuracy and to reduce the total time to solve the problem.

The finite element method is used to solve direct problems of calculating the magnetic field and forces at the fourth stage, and the gradient descent method is used to solve optimization problems. Reducing the time to solve a problem here is achieved by converting constraints to objective functions and sequential minimization of these functions over a limited number of variables. One of the objective functions (mass) is minimized analytically, the rest are numerically.

A comparison of the proposed method with a method in which a multicriteria task is reduced to a single-criterion introduction of a complex criterion showed its acceptable accuracy (the error does not exceed 1%), as well as reducing the problem solving time by about two times. The use of PM in EM has reduced the power consumption by 2.7 times. The application of the method is demonstrated on a specific task, however, it can be used in the design and other electrical devices. The advantage of the method is the introduction of the device mass minimization procedure into the algorithm. The method can be used in both design and identification problems.

Acknowledgments. The reported study was funded by RFBR according to the research project № 18-01-00204.

References

1. Tikhonov, A.N., Arsenin, V.Y.: Solution of Ill-Posed Problems. Winston & Sons, Washington (1977)
2. Murai, M., Tanaka, M.: Magnetic levitation (Maglev) technologies. Jpn. Railway Transp. Rev. **25**, 61–67 (2000)
3. Schach, R., Jehle, P., Naumann, R.: Transrapid und Rad-Schiene- Hochgeschwindigkeits-bahn: Ein gesamtheitlicher Systemvergleich. Springer, Heidelberg (2006)
4. Lee, H.W., Kim, K.C., Lee, J.: Review of Maglev train technologies. IEEE Trans. Magn. **42**(7), 1917–1925 (2006)
5. Liu, Z., Long, Z., Li, X.: Maglev Trains: Key Underlying Technologies. Springer, Heidelberg (2015)
6. Bakhvalov, Y., Gorbatenko, N.I., Grechikhin, V.V., Balaban, A.L.: Optimal design of energy saving electromagnetic systems using solutions of inverse problems. J. Fundam. Appl. Sci. **8**(3S), 2505–2513 (2016)
7. Bakhvalov, Y.A., Gorbatenko, N.I., Grechikhin, V.V., Yufanova, A.L.: Design of optimal electromagnets of magnetic-levitation and lateral-stabilization systems for ground trans-portation based on solving inverse problems. Russ. Electr. Eng. **88**(1), 15–18 (2017)
8. Balaban, A.L.: Method of the optimal design of electric engineering systems electromagnetic actuators on the basis of the inverse problems solving. Russ. Electromech. **4**, 34–39 (2017)
9. Balaban, A.L., Bachvalov, Y., Pashkovskiy, A.V., Yufanova, Ju.V: Method for solving inverse problems of optimal design of electrical devices. Russ. Electromech. **5**, 23–31 (2018)
10. Balaban, A.L., Bachvalov, Yu.A., Grechikhin, V.V.: Optimization of actuating elements of transport control systems with magnetic levitation based on the solution of inverse problems. In: MATEC Web of Conferences, vol. 226, 04006 (2018)
11. Nocedal, J., Wright, S.J.: Numerical Optimization. Springer, New York (1999)

Improvement of Fuel Injection Process in Dual-Fuel Marine Engine

Vladimir Gavrilov[1]([⊠]) , Valeriy Medvedev[2] ,
and Dmitry Bogachev[1]

[1] Admiral Makarov State University of Maritime and Inland Shipping,
Dvinskaya street 5/7, 198035 Saint-Petersburg, Russia
gavrilov@vg5647.spb.edu, dougle305@gmail.com
[2] State Marine Technical University of Saint-Petersburg, Lotsmanskaya,
3, St. Petersburg 190121, Russian Federation
office@smtu.ru

Abstract. The paper notes that the main problem of fuel delivery in a dual-fuel marine engine lies in the difficulty of efficient pilot fuel injection. Irrationally organized, it may result in undesirable knocks of the combustion process and cylinder cut-outs (misfires). Excessive pilot fuel consumption is likelihood. Therefore, the goal of this work is to determine the direction and means of improving the pilot fuel delivery. Fuel delivery is considered as perfect, at which reliable auto-ignition of the air-fuel mixture is ensured with a minimum cyclic dose of pilot fuel. Rational organization of the pilot fuel injection process cannot be ensured through the use of standard engine fuel equipment, which is designed to work in diesel mode. The paper proposes principles for solving this problem, outlines the main provisions of the methodology for preliminary calculation of the pilot fuel injection process, and makes several suggestions on choosing the main parameters of special fuel equipment. Using the results of the current research will ensure a highly efficient operation of a dual-fuel engine in both diesel and gas-diesel mode.

Keywords: Diesel and gas-diesel modes · Pilot fuel · Fuel atomization ·
Knock · Fuel equipment

1 Introduction

Currently, natural gas, due to its relatively low cost, high calorific value, low sulfur content and favorable composition of exhaust gases, is becoming one of the main fuels in transport [1]. Along with its advantages, gas fuel has several significant drawbacks: complexity of storage and transportation, increased fire and explosion hazard, toxicity and others.

In order to capitalize on the advantages of natural gas and diminish its drawbacks, dual-fuel engines have become popular in recent years. These engines burn either liquid fuel (diesel mode) or gas with auto-ignition of small amount of pilot fuel (gas-diesel mode). It should be noted, that using of dual-fuel engine is flexible since either of the fuel types can be chosen, depending on the availability and price. The

© Springer Nature Switzerland AG 2020
V. Murgul and M. Pasetti (Eds.): EMMFT 2018, AISC 982, pp. 392–399, 2020.
https://doi.org/10.1007/978-3-030-19756-8_37

engine can be easily switched over to diesel mode in case of its malfunction in gas-diesel one and back.

Principles of high-efficient operation of dual-fuel engine vary in different modes. The list of these principles can be long. In particular, we propose to change the actual compression ratio by changing the position of the closing point of the intake valves during the compression stroke [1]. Different injection of liquid diesel fuel should be organized in the two modes under consideration - in diesel mode, when a full cycle portion of fuel is injected in the cylinder, and in gas-diesel mode, in which a small amount of ignition fuel is fed, which is usually a few percent of the total cyclic portions. It should be admitted that neither the principles of the organization of work of the relevant fuel equipment, nor its design are currently sufficiently developed. Therefore, the goal of this stage of our research is to determine the direction of improving the process of injection of liquid pilot fuel, to develop the main provisions of the process calculation method and select the main parameters of special fuel equipment to ensure highly efficient operation of a dual-fuel engine in both diesel and gas-diesel modes.

2 Specific Requirements to the Fuel Injection Process

First of all, basic requirements to the process of liquid fuel delivery to the engine should be stated, which are vital for the case under consideration. These requirements to the operation of the engine in diesel mode are well known [2]. To ensure proper quality of mixing and combustion in this case, it is necessary to have high quality atomization of the fuel (so-called "fineness" and "homogeneity" of atomization), compliance with the range of fuel jets and the nature of the mass distribution of fuel to the distribution of mass and velocity of air in combustion chamber, as well as its configuration. The quality of mixing is largely determined by the parameters of the fuel injection process, the design and parameters of the fuel equipment, in particular, the fuel atomization pressure, the diameter and number of nozzle holes, the diameter and speed of the high pressure fuel pump plunger.

High-quality mixture formation when injecting pilot fuel in the gas-diesel mode means, first of all, obtaining a sufficiently "powerful" source of auto-ignition when the engine is operating in a wide range of loads, and with the lowest possible cyclic dose of fuel. The most important specific requirement for the organization of the injection process of the pilot fuel should be considered to ensure the sustainability of ignition and the absence of knock of the mixture of gas and liquid fuels.

Let us clarify the described requirement. During knock, a complex consisting of a shock wave and following exothermic chemical reactions propagates through the cylinder charge at supersonic speed, causing excessive mechanical (shock) loads on the main parts of the cylinder-piston and crank-connecting rod groups of the engine, up to their destruction.

The presence or absence of the knock depends on many factors. To illustrate the complexity of the organization of detonation-free combustion, on the one hand, and sustainable ignition, on the other hand, consider the influence of the most important factor - the excess air coefficient for combustion, α. The figure shows the dependency of some characteristics and location of characteristic zones of the combustion process on α in a gas engine at various levels of the average effective pressure p_{me}, MPa in it (Fig. 1).

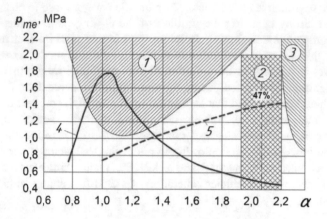

Fig. 1. Dependency of some characteristics and location of characteristic zones of the combustion process on excess air coefficient for combustion in the gas engine (according to Wärtsilä): *1* – knock zone; *2* – best operational parameters zone (Performance factor of the engine is 47%); *3* – zone of «impossible combustion» ; *4* – zone borders, in which exhaust of NO_x does not exceed 1 g/(KW·h); *5* –nature of engine performance factor change.

It can be seen from the figure that the knock-free, "stable" and sufficiently effective operation of the engine with the current level of average effective pressure is observed only in a relatively narrow range of $1.9 \leq \alpha \leq 2.2$. If α is less than 1.9, the engine works with knocks, and if it is more than 2.2, there is a risk of misfiring. Note that the effect of only factor (α) is highlighted here and the influence of many other important variables, such as the degree of compression of the cylinder charge, is not taken into account. Of course, the presented results in quantitative form should not be used in the ignition process in gas-diesel engine. However, in this engine, the picture quality can be similar.

3   Current Problems

The provision of high-quality mixture formation when applying to the dual-fuel engine ignition fuel is hampered by the circumstances described below. There is a low quality atomization in the case of injection of the fuel when using standard equipment. Therefore, the well-known problem of ensuring the required quality is faced, due to the fact that the cyclic dose of fuel by mass is a very small part of its nominal cycle supply, which results in the lack of the injection pressure before the nozzle holes. The increase in the proportion of pilot fuel in a mixture with gas (as a mean to improve the quality of atomization) significantly reduces the benefits from the use of gas fuel. Another extremely important circumstance, which does not allow recommending the use of standard fuel equipment for ignition fuel injection, is related to its well-known property

- a very significant degree of uneven distribution of the supplied fuel over the working cylinders, which can be increased by an order of magnitude and more than nominal cyclic feed. Due to the increase in the degree of irregularity in the transition to small cycle feeds, so-called "misfires" occur in part of the cylinders of a multi-cylinder engine. The latter circumstance in the case of gas diesel is unacceptable.

To solve the problems of the improvement of fuel injection in a dual-fuel marine engine, we have planned the following tasks:

– development of principles for the organization of the delivery of pilot fuel in a dual-fuel engine, ensuring stable ignition and knock-free combustion of the mixture;
– development of a methodology for calculating the parameters of fuel equipment in order to ensure a highly efficient injection of pilot fuel in a dual-fuel engine.

4 Fundamental Solutions

Let us outline the possible ways of meeting fundamentally different requirements. On the one hand, it is necessary to ensure maximum power of the auto-ignition source of the pilot fuel. On the other hand, reliable auto-ignition should be ensured with a minimum dose of pilot fuel. In order to solve these problems, we briefly describe the process of auto-ignition. As it is known, auto-ignition happens in the fuel jet zone, in which there is a concentration of fuel vapor close to the stoichiometric fuel-air ratio (in the zone corresponding to the excess air ratio for fuel combustion close to unity: $\alpha \approx 1.0$). This zone is formed during the development of the fuel jet in its so-called "shell". The centers of self-ignition are formed at some distance from the nozzle holes, and the flame spreads over the envelope zone to the jet front and the nozzle with high speed. This is the general picture of the auto-ignition process.

To solve the problems posed, means, that are similar to those used in high-quality mixture formation in a diesel engine, can be used. It is known that the mixture formation in a diesel engine is a very complex, multifactorial process. In practice, various criteria for the quality of individual mixing are used [2]. For example, there are known requirements of providing the maximum possible area of the "surface" of the fuel jet. Apparently, this area is in some way connected with the volume of the mentioned above zone with $\alpha \approx 1.0$, therefore, the criterion under consideration can be used in our case. Note that in the case of the organization of pilot fuel injection, the question of the ratio of the fuel jet range and the distance from the nozzle to the wall of the combustion chamber at this stage of the study can now be left open.

The area of the "surface" of the fuel jet is determined by its so-called range and the angle of the cone. Therefore, one can use known formulas for calculating the indicated quantities to estimate the required parameters of the pilot fuel jet.

In relation to the conditions of the ship's medium-speed diesel engine, an empirical formula for the range l_c, mm has been obtained [2].

$$l_c = 230 p_{j\,\text{max}}^{0,15} d_{\text{c.o}}^{0,546} \rho_a^{(0,092\,\lg\,\tau - 0,409)} \tau^{0,375}, \tag{1}$$

where p_{jmax} – is the maximum fuel injection pressure, MPa; $d_{c.o}$ is the diameter of the nozzle hole of the burner, mm; ρ_a – the average density of the air charge of the cylinder in the fuel injection phase, kg/m³; τ – time, ms.

The choice of the diameter $d_{c.o}$ in a complex way is connected with the following important values: (a) with the total cross-sectional area of all nozzle holes intended for delivering pilot fuel (this area, in turn, depends on the desired injection pressure level); (b) with the number of nozzle holes, which, together with their diameter, determines the mentioned total cross-sectional area and, accordingly, the injection pressure, as well as the total surface area of the fuel jets; (c) with the desired duration of the pilot fuel injection phase, determined by the specified total cross-sectional area of the nozzle holes and injection pressure. Some recommendations for choosing the listed values are set out below.

Formula (1) is valid within the following limits of independent variables: $p_{jmax} = 76.5 \ldots 117$ MPa; $d_{c.o} = 0.45 \ldots 0.55$ mm; $\rho_a = 23 \ldots 40$ kg/m³; $\tau = 1.0 \ldots 8.0$ ms.

The angle of the jet cone ψ_c can be calculated by the well-known method, developed on the basis of the criterion equations A.S. Lyshevsky. This method, as well as the method for calculating the range, is applicable to the conditions of the so-called "cold jet". To take into account the actual conditions of the engine on the temperature of the cylinder charge, the dependencies obtained by V.P. Lazurko and A.N. Finogenov can be used:

$$\psi_c = \left(\frac{T}{300}\right)^{0,245} \psi_{c\,300}, \psi_c = \left(\frac{T}{300}\right)^{0,245} \psi_{c\,300} \tag{2}$$

where T is the charge temperature, K; $l_{c\,300}$ is the range of the jet at a temperature of 300 K; $\psi_{c\,300}$ is the angle of the jet cone at a temperature of 300 K.

Along with the above-mentioned maximum possible surface area of the fuel jet, it is necessary to have high quality atomization of pilot fuel in order to ensure the required high quality of the mixture, when provoking auto-ignition in a gas-diesel engine. The atomization quality is characterized, in particular, by the average diameter of the atomized fuel. There are at least six types of average diameters that can be used to represent atomization quality. Due to the fact that the surface area and volume of droplets are of great importance for evaporation and combustion, in such cases the so-called Sauter average diameter d_{32} is usually used. It is the diameter of the droplet, which has a ratio of surface area to volume corresponding to the actual atomization.

There are various methods for estimating the parameters of fuel atomization by diesel injectors [3], starting with the simplest methods based on the Tanasawa formula and including the relatively frequently used methods of V.A. Kutovoy and A.S. Lyshevsky. The analysis shows that the applied mathematical models give slightly different results. This is explained by the differences in the conditions and means of experimental research used by the authors in obtaining empirical dependencies. However, a sufficiently adequate response of the model to the qualitative change of significant factors in the description of atomization is sufficient to solve these problems. The mentioned models generally reflect the influence of the main factors. It is important to note that the most significant influence on the atomization parameters (at a constant fuel viscosity) is exerted by the injection pressure and the diameter of the nozzle holes. The ratio of the length of the nozzle channel to its diameter, surface tension and fuel density affect less. The density factor of the air charge of the engine cylinder showed itself indefinitely. Therefore, the most important tasks of the design of fuel equipment for the implementation of the pilot fuel injection are a reasonable choice of the diameter of the nozzle holes and the provision of the necessary injection pressure.

5   Recommendations

In addition to the general considerations outlined above, a few more specific recommendations regarding the organization of pilot fuel injection should be considered. In particular, it is necessary to answer two questions. Firstly, until what time point of the development of the fuel jet is it necessary to evaluate its range and the angle of the cone? Secondly, how should the correlation between the injection pressure $p_{j\,cp}$ and the total flow area of the nozzle (diameter $d_{c.o}$ and number of nozzle holes $i_{c.o}$), as well as the volume flow of fuel by the high-pressure pump, which is determined by the speed v_{Π} and diameter d_{Π} of the plunger pump, be considered?

When solving the first question, one should bear in mind that in the presented case it is necessary to ensure high quality of mixing not during the usual long burning period of a full-cycle fuel dose when the engine is running in diesel mode, but only at least during the auto-ignition delay and, partly in the rapid combustion phase (when the fuel that has evaporated during the delay period burns). To estimate this point in time, one of the known formulas for calculating the duration of the auto-ignition delay period τi can be used.

Based on the O.M. Todes formula, which is an approximate solution of the differential equation for the auto-heating of the mixture, according to the results of our experiments, which were carried out on a ship's medium-speed diesel engine of the type DN 23/30 (2-stroke engine), the following dependence of the delay period τ_i, s was obtained:

$$\tau_i = 3,4 \times 10^{-6}(T/p)^{0,5}\exp(E/R_\mu T), \qquad (3)$$

where T and p are temperature, K and pressure, MPa in the engine cylinder at the time of the start of fuel injection; $E = 23500$ is apparent activation energy, J/mol; $R_\mu = 8.314$ J/(mol·K) is the universal gas constant.

When solving the second question, one should first select the cyclic dose of pilot fuel q_u, sm^3 and set the average injection pressure, Pa ($p_{j\,cp} \approx 0,7p_{j\,max}$), as well as the desired injection duration $\varphi_{B\Pi}$, expressed in degrees of rotation of the engine crankshaft, ° r.e.c. and, at a minimum, the appropriate duration τ_i.

The discussed values are interconnected by a well-known formula for estimating the total flow area of the nozzle, sm^2:

$$f_c = \frac{0,06 n q_{\text{ц}}}{\varphi_{\text{вп}} \mu_c \sqrt{\left(\dfrac{2}{\rho_f}\right)\left(p_{f\ \text{ср}} - p_{a\ \text{ср}}\right)}},$$

(4)

where n is the engine speed, min^{-1}; μ_c is nozzle discharge coefficient (it is recommended to take $\mu_c = 0.6 \ldots 0.7$); ρ_f is fuel density, kg/m^3; $p_{a\ \text{ср}}$ is the average charge pressure of the working cylinder of the engine during the injection period, Pa.

Taking into account the calculated value f_c, the number and diameter of the nozzle holes of the injector are determined $i_{\text{c.o}} \times d_{\text{c.o}}$, focusing, on the one hand, on achieving the maximum possible total surface area of the fuel jets, on the other hand, on ensuring high quality fuel atomization, is largely determined by the diameter of the nozzle holes.

Furthermore, it is important to solve the problem of determining the design parameters of the high-pressure fuel pump, which provide the required cyclic dose of pilot fuel $q_{\text{ц}}$ and its injection pressure $p_{j\ \text{ср}}$ at the engine design modes. These values depend mainly on the working area of the plunger $f_{\text{П}}$ (the diameter of the plunger) and its average speed $v_{\text{П.ср}}$ during the fuel injection. In the experiment of fuel equipment of a diesel engine of type ChN 30/38 (4-sroke engine), with a certain actual range of variables, an almost proportional dependence was found:

$$p_{f\ \text{ср}} \sim \left(\frac{f_{\text{П}} v_{\text{П.ср}}}{f_c}\right).$$

(5)

From consideration of this dependence, it follows that at a sufficiently high injection pressure of pilot fuel $p_{f\ \text{ср}}$ (ensuring the required atomization quality) and with a small total flow area of the nozzle holes f_c (determined by the need to obtain high atomization quality and smallness of the cyclic dose of fuel), the high-pressure fuel pump should have a small plunger diameter and probably reduced speed $v_{\text{П.ср}}$. We believe that there are various design solutions that meet these requirements, including the equipment of a dual-fuel engine with a separate additional fuel system for injecting pilot fuel or a high-pressure fuel pump with a double plunger designed for operation with significantly different cyclic doses of the injected fuel [4]. To obtain a reduced speed of the plunger $v_{\text{П.ср}}$, the following device can be used to change the actual compression ratio in the engine when changing the type of fuel [1]. This device is a gas distribution mechanism, the shaft of which contains two sets of cams and can change its position in the axial direction. This allows to put into work the right set of cam jaws when changing the type of fuel. The change in gas distribution phases occurring at the same time (angle of inlet valves closing, °r.e.c.) entails a corresponding change in the actual compression ratio. In order to provide the possibility of reducing the speed $v_{\text{П.ср}}$ during the switch over to the gas-diesel mode on the gas-distributing shaft, it is possible to similarly place two sets of cams of the high-pressure fuel pump drive, one of which allows to obtain a reduced speed of the pump plunger movement.

Of course, the article deals with only a small part of technical problems and their solutions in the direction of improving the fuel injection processes in a dual-fuel marine engine. For example, it is necessary to solve the problem of coking (loss of the flow section) of inoperative nozzle holes designed to deliver a full cyclic dose of fuel after the engine has been switched to gas-diesel mode.

6 Conclusion

When solving relevant problems of improving the process of fuel injection in a dual-fuel engine, the organization of highly efficient mixture formation when injecting pilot fuel into the working cylinder is most important.

At the same time, it is necessary to obtain, on the one hand, reliable auto-ignition of pilot fuel, provided by sufficient power of the auto-ignition source, on the other hand, the minimum possible cyclic dose of this fuel.

The specific of the pilot fuel injection is that it is vital to ensure the maximum possible surface area of the fuel jet (determined by its range and angle of the cone) during the auto-ignition delay of the combustible mixture. It is also necessary to obtain a high quality atomization of fuel.

The main difficulty lies in the fact that the use of standard fuel equipment, designed for engine operation in diesel mode, does not provide a sufficiently high quality of mixture formation when the pilot fuel is injected in gas-diesel mode. Therefore, when creating a dual-fuel engine, for each of these modes, the choice of a set of fuel equipment parameters should be substantiated: the diameter and number of nozzle holes of the injector, the diameter, the active stroke and the speed of the high-pressure fuel pump plunger. These parameters determine the key parameters of the fuel jets, the duration of the fuel injection and the quality of its atomization.

References

1. Zhou, S., Gao, R., Feng, Y., Zhu, Y.: Evaluation of Miller cycle and fuel injection direction strategies for low NOx emission in marine two-stroke engine. Int. J. Hydrogen Energy **42**(31), 20351–20360 (2007)
2. Stoumpos, S., Theotokatos, G., Boulougouris, E.: Marine dual fuel engine modelling and parametric investigation of engine settings effect on performance-emissions trade-offs. Ocean Eng. **157**, 376–386 (2018)
3. Sun, X., Liang, X.: Influence of different fuels physical properties for marine diesel engine. Energy Procedia **142**, 1159–1165 (2017)
4. Noor, C.W.M., Noor, M.M., Mamat, R.: Biodiesel as alternative fuel for marine diesel engine applications: a review. Renew. Sustain. Energy Rev. **94**, 127–142 (2018)

Treatment Process of Container Cargo in the Form of Open-End Queue System

Igor Rusinov$^{(\boxtimes)}$ (iD)

Admiral Makarov State University of Maritime and Inland Shipping,
Dvinskaya Street 5/7, 198035 Saint-Petersburg, Russia
rusinovia@gumrf.ru

Abstract. The process of handling container cargo at the terminals in the form of an open open-end queue system (QS) is investigated. Based on the classical QS theory, we can assume that the flow of ships to the port is a regular stream of events, in which events follow one after another, strictly according to schedule, at equal intervals of time. In a real-life setting, the approach moments of the courts are irregular streams of events.

Keywords: Container terminals · Deterministic models · Probabilistic models · Ships · Berths · Queue system (QS)

1 Introduction

Container shipping is the most convenient and economical way to transport goods. The design of a large-tonnage container provides safe transportation of goods in any combination of sea, river and land modes of transport. At the same time, cargo handling processes as a whole are intensified, costs in ports are minimized, monitoring is simplified. The main gateway hub during the transportation of container cargo is the terminal. Cargo arrives at the terminal by rail and vehicular transportation as export and transit traffic. All the imports on sea and river vessels also arrive at the terminal. The total transit time of cargo delivery and the cost of the whole transportation depend on the terminal operation.

2 Discussion

It can be assumed that any terminal contains cargo processing channels, which in particular cases are berths, sea or river ones. During the design, construction and operation processes of terminal complexes, there is a need to determine the quality indicators of cargo handling processes. These indicators characterize the average ship waiting time and their average time of stay at the terminal. Also, the indicators, which are proportional to these, like average values of the number of vessels in the queue or terminal, should be determined. These indicators characterize the quality of the services provided, as waiting in a queue leads to substantial expenditures for shipping

© Springer Nature Switzerland AG 2020
V. Murgul and M. Pasetti (Eds.): EMMFT 2018, AISC 982, pp. 400–409, 2020.
https://doi.org/10.1007/978-3-030-19756-8_38

companies. As a result, despite an increase in cargo turnover, an excessive increase of the channel load factor becomes economically unprofitable. At the same time, the expenses increase with an increase in the number of idle ships waiting for the clearance of channels.

As is known, deterministic models are the most convenient in practical application for the calculations of cargo handling. The flow of certain ships to the sea trade port or to a separate terminal is assumed to be a regular flow of events, in which the approaches follow one after the other, strictly according to schedule, at equal intervals of time. However, we know that deterministic models have their drawbacks. They do not indicate the specific nature of the port or terminal. Especially since in reality the moments of approach of ships to the ports are irregular streams of events.

The processes occurring during the loading/unloading of ships transform from one state to another at random times. At the same time, the number of ships in the queue and the number of occupied channels are constantly changing. The transition of processes from one state to another occurs at the moments when a new ship approaches the receiving buoy, or when one of the berths becomes available. The system contains $n + 1$ of countable set of states: $E_0, E_1, E_2, \ldots, E_n$, where n is the number of ships in the system. The ships in the queue and ships moored under cargo operations are considered.

It is known that the total flow of moments of the ships approaching the port can be considered as the flow sum of ships belonging to different shipping companies. It can be MAERSK, MSC, CMA-CGM, OOCL, almost anyone, any shipowner.

According to the QS theory, it is known that with the superposition of a large number of ordinary stationary flows, a flow is obtained with virtually any after-effect. That flow is arbitrarily close to the stationary Poisson (simplest) one. It is as if the treatment time of ships is subject to the exponential distributive law. The man-made assumptions about the Poisson flow of ship arrival and the exponential distribution of cargo handling time in containers allow us to use the framework of a queuing system (QS) to describe the production processes occurring in ports and terminals. The use of QS framework allows us to describe the ship treatment process in a port or at a terminal using linear differential equations and present formulas for probabilistic quality rates of processes in an analytical form. However, the use of existing open-end queue models in order to determine the probabilistic rates of treated ships is not advisable, since these models do not always adequately describe the handling operations under real-life conditions of the port. Thus, the classical queuing theory assumes the study of a multi-channel system, and the number of devices is equal to the number of channels. Each channel can be serviced by one device, independently of other channels (a queuing system without mutual assistance). In addition, channels can serve all free devices or part of free devices (QS with full or partial mutual assistance). The probabilities of system transitions from the E_n state to E_{n-1} states, i.e., the probability of serving a single query depends on the number of service channels that are running. The resulting service intensity in the n-th state is determined according to the principle of linear superposition, i.e. is equal to the total intensity of all service devices and is a multiple of the predicted intensity of a single device μ_0. Thus, the resulting service intensity in this case cannot exceed $S\mu_0$, therefore $\mu_p \leq S\mu_0$, and the intensity of service with one device μ_0 does not change depending on the state of the QS. In addition, the service process is considered unpredictable and unmanageable, since the number of applications that will be sent to the system in the near future is unknown to the dispatcher of the QS. The dispatcher cannot, depending on the status of the QS, change the intensity of the service devices.

3 Materials and Methods

In real-life conditions of the port or terminal, the ship treatment is not adequate to these assumptions. That is why a centralized container treatment system, managed by the port dispatcher, is considered. The dispatcher determines the queue order and the fleet treatment sequence. He also performs the distribution of human and technical resources among all berths. If necessary, when the queue is getting longer, the dispatcher draws additional resources, thereby significantly increasing the fleet treatment intensity at individual berths. However, sometimes the resulting intensity of the system becomes less than the total intensity of individual technical means due to the work front limitations of the handling operations. Thus, under real-life conditions, the resulting intensity of fleet treatment usually is not a multiple of the average treatment intensity μ_0. In some cases, it may exceed the value of $S_\mu 0$.

The peculiarities of the container cargo treatment system at the port require the applied analytical models to take into account the possibility of changing the treatment intensity of individual berths, depending on the general state of the QS. However, the classical queuing system operates with constant intensity values of individual recovery devices. Let us start with a container terminal, which includes S berths, the simplest flow of ships (applications) with intensity λ approaches the receiving buoy. The design treatment flow by each berth is μ_0. However, the resulting intensity of the process may vary depending on its state. The resulting intensity of the cargo treatment in the state $E_n = r_n\mu_0$, where r_n is the treatment intensity factor can be either an integer or a fractional number. As a rule, when all the berths are occupied, i.e. $n \geq S$, it is assumed that $r_n = const$ (usually $r_n = r_{max}$).

Supposing that the possibility of the system "to jump" through a state (for example, from E_n to E_{n+2} or from E_{n-1} to E_{n+1}, bypassing the state E_{n+1} and E_n, respectively), for a small period of time dt can be neglected as the value of higher order of smallness.

The value r_n characterizes the resulting intensity of cargo treatment in the state E_n, r_n can be both an integer and a fractional number, and in some cases exceed the value of S.

A visual representation of the Markoff cargo handling process is achieved by a square intensity matrix of the simplest flows of approaching ships and cargo handling. In the case when the state of the system is $n + 1$, then the intensity matrix has the following form:

$$\lambda = \begin{array}{c} \\ \end{array} \begin{array}{cccccccc} E_0 & E_1 & E_2 & E_3 & \cdots & E_{n-1} & E_n & E_{n+1} \\ \end{array}$$

$$\lambda = \left[\begin{array}{cccccccc} 0 & \lambda & 0 & 0 & \cdots & 0 & 0 & 0 \\ r_1\mu_0 & 0 & \lambda & 0 & \cdots & 0 & 0 & 0 \\ 0 & r_2\mu_0 & 0 & \lambda & \cdots & 0 & 0 & 0 \\ 0 & 0 & r_3\mu_0 & 0 & \cdots & 0 & 0 & 0 \\ \vdots & \vdots & \vdots & \vdots & \cdots & \vdots & \vdots & \vdots \\ 0 & 0 & 0 & 0 & \cdots & 0 & \lambda & 0 \\ 0 & 0 & 0 & 0 & \cdots & r_n\mu_0 & 0 & \lambda \\ 0 & 0 & 0 & 0 & \cdots & 0 & r_{n+1}\mu_0 & 0 \\ \end{array} \right] \begin{array}{l} E_0 \\ E_1 \\ E_2 \\ E_3 \\ \vdots \\ E_{n-1} \\ E_n \\ E_{n+1} \end{array} \quad (1)$$

The intensity matrix corresponds to an oriented state graph, in which the size of the edge connecting two successive states is equal to the corresponding intensity of the transition from one state to another. The indicated state graph is presented in the figure.

Fig. 1. The indicated state graph.

The required properties of the intensity matrix and the state graph, which allow the Poisson processes to procced in the QS, are analyzed. It is known that this requires all event flows, which transfer the system from one state to another, to be Poisson or simplest, so the elements of the intensity matrix (1) do not change over time. Thus, the factors r_n ($n = 1, 2, 3, \ldots$) can take any integer or fractional positive values, but should remain constant. Such a matrix is usually called the simplest matrix, and the corresponding random process is the simplest Markoff process. Similarly, the state graph (see Fig. 1), where each edge corresponds to constant intensity values, will be called the simplest state graph.

The stochastic transition matrix can be compiled based on the state graph. It should be considered that, as is shown above, due to the "jump" absence event v can be only one of the three events $E_n, E_{n-1}, E_n, E_{n+1}$ ($v = n - 1, n, n + 1$). Accordingly, each row and each column of a stochastic matrix must not contain more than three nonzero entries. According to the aforesaid, the probabilities of the system transition from one state to another are presented as follows:

$$\begin{aligned} E_0 &\to E_0 & P_{00} &= 1 - \lambda t \\ E_0 &\to E_1 & P_{01} &= \lambda t \\ E_1 &\to E_0 & P_{10} &= r_1\mu_0 t \end{aligned}$$

$$1 \leq n \leq S \begin{cases} E_n \to E_n & P_{nn} = 1 - (\lambda + r_n\mu_0)dt \\ E_n \to E_{n+1} & P_{n,n+1} = \lambda dt \\ E_n \to E_{n-1} & P_{n,n-1} = r_n\mu_0 dt \end{cases}$$

$$n \geq S \begin{cases} E_n \rightarrow E_n & P_{nn} = 1 - (\lambda + r_s\mu_0)dt \\ E_n \rightarrow E_{n+1} & P_{n,n+1} = \lambda dt \\ E_n \rightarrow E_{n-1} & P_{n,n-1} = r_s\mu_0 dt \end{cases}$$

Then the stochastic transition matrix can be represented as a matrix:

$$J(t) = \begin{bmatrix} & E_1 & E_2 & E_3 & E_4 & \cdots & E_{n-1} & E_n \\ 1-\lambda dt & \lambda dt & 0 & 0 & \cdots & 0 & 0 \\ r_1\mu_0 dt & 1-(\lambda+r_1\mu_0 dt) & \lambda dt & 0 & \cdots & 0 & 0 \\ 0 & r_2\mu_0 dt & 1-(\lambda+r_2\mu_0 dt) & \lambda dt & \cdots & 0 & 0 \\ \vdots & \vdots & \vdots & \vdots & \cdots & \vdots & \vdots \\ 0 & 0 & 0 & 0 & \cdots & 1-(\lambda+r_{n-1}\mu_0 dt) & \lambda dt \\ 0 & 0 & 0 & 0 & \cdots & r_n\mu_0 dt & 1-(\lambda+r_n\mu_0 dt) \\ 0 & 0 & 0 & 0 & \cdots & 0 & r_{n+1}\mu_0 dt \end{bmatrix}$$

$$(2)$$

The matrix rows correspond to the state $E_n(t)$, and the columns correspond to $E_{n+1}(t)$. Having conducted the transposition on the left and right sides of the expression $P^T(t+dt) = P^T(t)J(t)$, the expression $\vec{P}(t+dt) = J^T(t)\vec{P}(t)$ is obtained, where $J^T(t)$ is the matrix transposed to the transition matrix $J(t)$, $\vec{P}(t)$ is a column vector of dimension probabilities $n * 1$. Put the expression (2) in the form $\vec{P}(t+dt) = [J^T(t) - E_n]\vec{P}(t) + P(t)$, transfer $P(t)$ to the left side, then: $\vec{P}(t+dt) - \vec{P}(t) = [J^T(t) - E_n]\vec{P}(t)$, dividing the left and right parts (2) by dt, the following is obtained: $\vec{P}'(t) = R\vec{P}(t)$, where $R = \frac{1}{dt}[J^T(t) - E_n]$ is the matrix, which has the following form:

$$R = \begin{bmatrix} P_0 & P_1 & P_2 & \cdots & P_{n-1} & P_n & P_{n+1} & \\ -\lambda & r_1\mu_0 & 0 & \cdots & 0 & 0 & 0 & P_0 \\ \lambda & -(\lambda+r_1\mu_0) & r_2\mu_0 & \cdots & 0 & 0 & 0 & P_1 \\ 0 & \lambda & -(\lambda+r_2\mu_0) & \cdots & 0 & 0 & 0 & P_2 \\ 0 & 0 & \lambda & \cdots & 0 & 0 & 0 & P_3 \\ \cdots & \cdots & \cdots & \cdots & \cdots & \cdots & \cdots & \cdots \\ 0 & 0 & 0 & \cdots & -(\lambda+r_{n-1}\mu_0) & r_n\mu_0 & 0 & P_{n-1} \\ 0 & 0 & 0 & \cdots & \lambda & -(\lambda+r_n\mu_0) & r_{n+1}\mu_0 & P_n \end{bmatrix}$$

The first $n+1$ differential equations of a cargo handling system can also be represented as follows:

$$P_0'(t) = -\lambda P_0(t) + r_1\mu_0 P_1(t)$$
$$P_1'(t) = \lambda P_0(t) - (\lambda+r_1\mu_0)P_1(t) + r_2\mu_0 P_2(t)$$
$$\cdots$$
$$P_n'(t) = \lambda P_{n-1}(t) - (\lambda+r_n\mu_0)P_n(t) + r_{n+1}\mu_0 P_{n+1}(t), n = 1,2,3\ldots$$

$$(3)$$

It's as if the cargo handling process is Markoff, random and ergodic. According to this, after a sufficiently long period of time (theoretically when $t \rightarrow \infty$), the probabilities of the cargo handling system states practically do not depend on the system's state at the initial moment of time at $t = 0$ and does not depend on the time interval itself. Such an assumption is possible, since all the event flows transferring the system from one state to another are the simplest, so all elements of the intensity matrix (1) are constant values.

Based on Eq. (3), the dynamic modes of cargo handling system are analyzed.

The container terminal operation mode, in which the probability P_n of the system being in the state n does not depend on time, is called the stationary mode. Thus, the container terminal has the limit stationary modes. The characteristics of these modes do not depend on the state of the terminal at the initial moment of time, but they depend on the adopted sequence of fleet treatment, or on the distribution of the container terminal's resources.

I order to determine the probability values of system's individual states in stationary modes, it is necessary to equate the values of the derived states to 0, i.e. left parts of the equation system (3).

Spelling out the details of the equation system, after transferring one of the items to the left in each equation, the following is obtained:

$$r_1\mu_0 P_1 = \lambda P_0$$
$$r_2\mu_0 P_2 = (\lambda + r_1\mu_0)P_1 - \lambda P_0$$
$$r_3\mu_0 P_3 = (\lambda + r_2\mu_0)P_2 - \lambda P_1$$
$$\cdots$$
$$r_n\mu_0 P_n = (\lambda + r_{n-1}\mu_0)P_{n-1} - \lambda P_{n-2}$$

Let us introduce the following notation $\psi = \frac{\lambda}{\mu_0}$ and call it the reduced density of the arriving flow of ships. The correlation expression between the stationary probability values of the individual states is as follows:

$$P_1 = \frac{1}{r_1}\psi P_0$$
$$P_2 = \frac{1}{r_2}\psi P_1 = \frac{1}{r_1 r_2}\psi^2 P_0$$
$$P_3 = \frac{1}{r_3}\psi P_2 = \frac{1}{r_2 r_3}\psi^3 P_0 \qquad (4)$$
$$\cdots$$
$$P_n = \frac{1}{r_n}\psi P_{n-1} = \frac{1}{\prod_{i=0}^{n} r_i}\psi^n P_0$$

Using expression (4), the normalizing condition is obtained:

$$P_0 \left[\sum_{n=0}^{\infty} \frac{1}{\prod_{i=0}^{n} r_i} \psi^n \right] = 1,$$

where r_0 is taken equal to 1.

Accordingly, the probability expression for the zero state of the system, i.e., the probability that all the berths will be free at the arriving time of the ship to the port, is determined using the following expression:

$$P_0 = \frac{1}{\sum_{n=0}^{\infty} \frac{1}{\prod_{i=0}^{n} r_i} \psi^n} \tag{5}$$

Let us assume that for $n \geq S'$, the resulting intensity of the system is $r_{max}\mu_0$. Usually $S' = S$ and is equal to the number of berths. However, in some cases, S' can be either more or less than the number of berths S. Hereafter, $S' = S$ if unbespoken. Then:

$$P_n = \frac{P_0}{\prod_{i=0}^{n} r_i} \psi^n,$$

$$P_{S+d} = P_0 \frac{\psi^S}{\prod_{n=1}^{S} r_n} \left(\frac{\psi}{r_{max}}\right)^d,$$

$$\lim_{t \to \infty} p_{S+d} = 0$$

The normalizing condition can be exposed as follows:

$$P_0 \left[\sum_{n=0}^{S-1} \frac{\psi^S}{\prod_{n=1}^{S} r_n} + \frac{\psi^S}{\prod_{n=1}^{S-1} r_n} \sum_{d=0}^{\infty} \left(\frac{\psi}{r_{max}}\right)^d \right] = 1 \tag{6}$$

The second sum (6) represents the sum of the geometric progression members. Then it follows:

$$P_0 = \frac{1}{\sum_{n=0}^{S-1} \frac{\psi^S}{\prod_{i=0}^{n} r_i} + \frac{\psi^S}{\prod_{n=1}^{S} r_n \left(1 - \frac{\psi}{r_{max}}\right)}} \tag{7}$$

Expression (7) is the most common expression for the probability of case, when during the arrival time of the ship at the port all berths are free. Depending on the order of ship treatment (without mutual assistance, with full or partial mutual assistance, as well as in cases when the ship treatment intensity by individual berths varies according to the system state), the factors ri, and therefore the elements of the intensity matrix and transitions will change. Accordingly, the expressions for the probabilities of the system's states will change. But all of them can be considered as special cases of expressions (4) and (7). The stationary mode of the system exists only when the following condition is met:

$$\frac{\psi}{r_{max}} < 1$$

Considering the case of $r_{max} = S$ (as it often occurs in real-life conditions), the existence condition of a stationary mode can be exposed in the following form:

$$\varphi = \frac{\psi}{S} \leq 1$$

where φ is the load factor of the fleet treatment system.

According to these conditions, the average number of treated ships per unit of time by all S berths must be greater than the average number of ships entering the port or terminal per unit of time. If this condition is not met, the number of places in the queue is considered to be infinite, and there will be no stationary mode in the terminal. When $\psi \geq r_{max} (\varphi \geq 1)$, the process moves indefinitely towards states with a larger number d, and the queue will increase indefinitely. Then, when $\psi \geq r_{max} (\varphi \geq 1)$, for any finite d $\lim\limits_{t \to \infty} P_{S+d} = 0$, so the system must necessarily pass the state $S+d$ and not return to it. Therefore, when $r_{max} = \psi$, as can be seen from (5) and (6), the probabilities P_0 and P_n are equal to 0. When $r_{max} < \psi$, these expressions lose their physical meaning.

Let us determine the average number of ships in the queue.

$$\bar{d} = \sum_{n=S+1}^{\infty} (n - S)P_n = \sum_{d=1}^{\infty} dP_{S+d} = P_S \sum_{d=1}^{\infty} d\left(\frac{\psi}{r_{max}}\right)^d$$

where $d = n - S$ is the number of ships in queue:

$$P_S = P_0 \frac{\psi^S}{\prod\limits_{i=1}^{S} r_i}$$

Then the expression for the average number of ships in the queue is obtained as follows:

$$\bar{d} = P_0 \frac{\psi^S}{\prod\limits_{i=1}^{S} r_i} \frac{\frac{\psi}{r_{max}}}{\left(1 - \frac{\psi}{r_{max}}\right)^2} = P_0 \frac{\psi^{S+1}}{\prod\limits_{i=1}^{S-1} r_i (r_{max} - \psi)^2}$$

The average number of ships at the terminal is determined as follows:

$$\bar{d}_\Sigma = \sum_{n=0}^{\infty} nP_n = \sum_{n=0}^{S-1} nP_n + \sum_{n=S+1}^{\infty} nP_n$$

Then the average total number of ships at the terminal:

$$\bar{d}_\Sigma = \bar{d} + S - \sum_{n=0}^{S} (S-n) \frac{\psi^n}{\prod_{n=1}^{S} r_n} P_0$$

It should be noted that, regardless of the ψ and S values, the average number at the terminal is equal to the sum of the average number of treated ships and the average number of ships in the queue, i.e. $\bar{d}_\Sigma = \bar{d}_o + \bar{d}$.

Accordingly, the average number of ships being treated:

$$\bar{d}_o = S - \sum_{n=0}^{S} (S-n) \frac{\psi^n}{\prod_{n=1}^{S} r_n} P_0$$

The average waiting time for a ship in the queue and the average total staying time at the terminal are determined using Little's formulas.

The average time of ship delay in anticipation of treatment due to the busy berths:

$$\bar{T}_o = \frac{\bar{d}}{\lambda} = \frac{P_0}{\lambda} \frac{\psi^{S+1}}{\prod_{i=1}^{S-1} r_i (r_{max} - \psi)^2} = \frac{P_0}{\mu_0} \frac{\psi^S}{S \prod_{i=1}^{S-1} r_i (1-\varphi)^2}$$

The average total staying time of the ship in the terminal:

$$\bar{T}_\Sigma = \frac{\bar{d}_\Sigma}{\lambda} = \frac{\bar{d}}{\lambda} + \frac{S}{\lambda} + \frac{1}{\lambda} \left[S - \sum_{n=0}^{S} (S-n) P_n \right] = \bar{T}_o + \frac{1}{\lambda} \left[S - \sum_{n=0}^{S} (S-n) P_n \right]$$

The obtained probabilistic models of cargo handling processes allow us to make a probabilistic analysis of the characteristics of these processes in the general case, taking into account the possible dependence of the intensity factor r_n on the cargo handling discipline.

4   Results

Calculations made on the basis of the above mentioned expressions allow the container terminal dispatcher to practically solve the problems about the planned flow rate of ships in a certain period of time, taking into account the allowable length of the queue and a fairly high through-put. If the intensity of approaching ships increases during the specified period in comparison with the planned intensity, the dispatcher transfers part of the ships to another terminal or draws on additional resources.

5   Conclusion

The main scientific result of the paper is the development of the classical queue system theory, taking into account the real-life conditions of the container terminal, which means the possibility of changing the intensity value of the service using individual devices (berths), depending on the state of the QS.

References

1. Gnedenko, B.V., Kovalenko, I.N.: Vvedenie v teoriiu massovogo obsluzhivaniia. Nauka, Moscow (1987). (in Russian)
2. Kuzin, L.T.: Osnovy kibernetiki. Energiia, Moscow (1979). (in Russian)
3. Rusinov, I.A.: Obrabotka i khranenie refrizheratornykh gruzov na spetsializirovannom terminale. SPbII RAN, Saint-Petersburg (2005). (in Russian)
4. Rusinov, I.A.: Formalizatsiia i optimizatsiia protsessov pererabotki refrizheratornykh gruzov na spetsializirovannykh terminalakh. Politekhnika, SPb (2008). (in Russian)
5. Rusinov, I.A., Zubarev, Iu.Ia.: Pererabotka konteinernykh gruzov. Politekhnika, SPb (2009). (in Russian)
6. Rusinov, I.A.: Formalizatsiia, identifikatsiia i optimizatsiia protsessov pererabotki konteinernykh gruzov na spetsializirovannykh terminalakh. Diss. na soisk. uch. st. d.t.n. Sankt-Peterburgskii gosudarstvennyi universitet vodnykh kommunikatsii, Saint-Petersburg (2010). (in Russian)
7. Rusinov, I.A.: Konteinery i terminaly. Politekhnika, Saint-Petersburg (2011). (in Russian)
8. Rusinov, I.A.: Logiko-veroiatnostnoe modelirovanie protsessov pererabotki konteinernykh gruzov s ogranichennym ozhidaniem obsluzhivaniia. Ekspluatatsiia morskogo transporta. GMA 2(66), 279–285 (2011). (in Russian)
9. Rusinov, I.A.: Osobennosti formalizatsii protsessov pererabotki konteinernykh gruzov v usloviiakh nepolnoi i nechetkoi apriornoi i operativnoi informatsii sostoianiia terminala. Ekspluatatsiia morskogo transporta. GMA 1(65), 218–224 (2012). (in Russian)
10. Rusinov, I.A.: Nechetkaia formalizatsiia granichnoi zadachi identifikatsii protsessov pererabotki konteinernykh gruzov v klasse polinomialnykh modelei. Ekspluatatsiia morskogo i rechnogo transporta. MGA 2, 79–86 (2012). (in Russian)
11. Rusinov, I.A., Kuznetsov, A.L., Kitikov, A.N.: Ogranicheniia pri raschete morskogo fronta metodami teorii massovogo obsluzhivaniia. Ekspluatatsiia morskogo transporta ezhekvartalnyi sb. nauchn. 1(71), 3–6 (2013). (in Russian)
12. Rusinov, I.A., Gavrilova, I.A., Bersenev, A.I.: Protsessy pererabotki gruzov pri konteinernykh perevozkakh: matematicheskoe modelirovanie, kriterii optimalnosti. MATEC Web Conf. 239, 03011 (2018)

Maintenance of Operating Devices of Dredgers and Ship Unloaders Throughout the Operational Lifetime

Yuri Yezhov$^{(\boxtimes)}$ ⓘ, Aleksey Bardin ⓘ, and Vladimir Sidorenko ⓘ

Admiral Makarov State University of Maritime and Inland Shipping,
Dvinskaya Street 5/7, 198035 Saint-Petersburg, Russia
ezhovye@gumrf.ru

Abstract. Technical operation of dredgers and ship unloaders is aimed at organizing and managing the process of maintaining equipment in working condition. Improving the efficiency of maintenance is performed by changing the work organization principle in favor of preventive maintenance according to the actual condition of equipment. This change is based on upgrading the system of diagnostic of components and mechanisms. To diagnose the condition of equipment, they implement a strategy of permanent monitoring of the actual state of components and mechanisms throughout the entire operational lifetime of equipment. The article describes the method and the order of activities that provide early warning of potential failures or emergencies of work equipment. The proposed approach significantly extenuates disadvantages of traditional maintenance methods, increasing the efficiency, reducing the costs of crashes and unplanned downtime, providing in real time non-destructive control over the state of the equipment without stopping the machines. The implementation of the method increases the productivity of service personnel and provides optimization of production asset management.

Keywords: Technical maintenance · Efficiency · Ship unloader · Reliability · Working capacity · Monitoring · Diagnostics

1 Introduction

To achieve the stable working condition of equipment, the strategy of scheduled preventive maintenance (SPM) has been traditionally used, which implies strict regulation of frequency of servicing equipment mechanisms. This way of ensuring efficiency was based on the correlation of the equipment operation time and the probability of a failure. The SPM strategy ensures the consistent running of repair services while ensuring an adequate level of reliability. Nevertheless, the so-called "Industry 4.0" is gaining more and more power following the globalization process, while the service of technical maintenance and repair based on SPM strategy turns out to be inefficient. Following SPM strategy is expensive, since the cost of its implementation exceeds the production cost.

© Springer Nature Switzerland AG 2020
V. Murgul and M. Pasetti (Eds.): EMMFT 2018, AISC 982, pp. 410–418, 2020.
https://doi.org/10.1007/978-3-030-19756-8_39

The use of SPM strategy often entails the following claims:

- Not demanded works (unnecessary works on maintenance of obviously serviceable easily accessible components, moreover some components and mechanisms are replaced before the exhaustion of their supposed service life);
- Risks of deterioration of the equipment condition (any unwarranted intrusion into the equipment implies possible risk of personnel errors, which can lead to additional post-repair failures and equipment repairs);
- Excessive staffing, high labor intensity with low labor productivity;
- Nor using working capital for parts and materials in stock.

"Industry 4.0" enables the use of new strategies of maintenance and repair based on interrelation of diagnostic features, which being registered allow talking confidently about emerging defects in work equipment [3, 4]. Measuring, monitoring and analysis of such symptoms will allow informing of defects, their development and making recommendations aimed at ensuring reliable and operational condition of the equipment. The idea of upgrading maintenance systems implies minimizing costs by applying methods for assessing the state of components and mechanisms by non-destructive testing approaches in addition to the methods used to maintain equipment in working condition [1, 5, 6]. In order to increase the efficiency of maintenance of working equipment of Dredgers and Ship Unloaders (DSU), the following requirements for the new strategy were outlined:

1. Simplicity and clarity in use;
2. Increase in predictability of failures and breakdowns of working equipment of dredgers and ship unloaders;
3. Reduction of the number of failures in components and mechanisms;
4. Reduction of unreasonable expenses;
5. Improvement of the quality and control of maintenance and repair work.

The present article is focused on the description of the method of maintenance throughout the operational lifetime of equipment. Studies show that the cost of maintenance and repair for different service strategies may vary significantly [1].

The authors propose the method based on the use of stationary systems of Actual State Monitoring (ASM) of equipment. ASM is widely used in many industries, [2] and has proven to be effective. However, in Russian sand extraction enterprises, such tools are very rarely introduced. The advantage of the stationary type ASM is the simplicity, reliability, the possibility of permanent monitoring and transmission of data on measurement results throughout the entire life cycle of equipment [3, 4].

The use of ASM systems does not contradict the classical approach of SPM. ASM systems conversely complement it, make maintenance and repair more efficient due to both more accurate prediction of technical condition of certain components and mechanisms and analysis of the consequences of possible failures [5].

ASM system allows identifying the majority of recognizable defects by recording diagnostic signs and parameters, signaling that there are defects that develop and can lead to equipment breakdown [6].

The use of ASM as part of DSU will ensure the long-term operation of equipment and timely awareness of the state of equipment, thereby increasing the efficiency of the services responsible for the maintenance and repair due to prompt decision-making (Figs. 1 and 2).

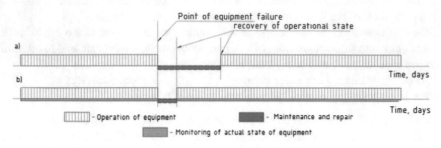

Fig. 1. Methods of maintenance and repair: (a) Maintenance and repair after failure; (b) Maintenance and repair after failure applying actual state monitoring

According to data from various sources (SIEMENS, METSO, etc.), when switching to the service based on the actual condition, the savings are on average more than $ 5 per year per unit of power (1 kW) [3].

With reference to the considered installations of SG, equipped with unique and expensive equipment of the rotor type of high power (>1500 kW), the authors recommend using stationary systems of the remote ASM. The state of equipment is recommended to be recorded by measuring the parameters of the following processes: vibration and temperature. To record changes in parameters over time, only stationary sensors and modules should be used. Measurements take place automatically at a fixed interval, for example, $0.5 \div 1$ s.

2 Method

In the basic version of technical diagnostics of equipment, the authors recommend using vibration and thermal monitoring sensors, using among others the methods of organoleptic diagnostics. The use of this recommendation and measurement results of working equipment of DSU will allow to detect and predict the development of incipient defects with a high degree of confidence (up to 90–95%) (Fig. 2).

Methods of vibration diagnostics are the most simple and common. Vibration diagnostic methods can detect a wide range of equipment defects, such as: imbalance, weakening of supports, shaft misalignment, violation of structural rigidity, breakage of anchor bolts, etc. [5, 6].

Figure 2 shows specific steps of the process of component wear. Area 1: For the running-in process, no increase in wear rate is observed; Area 2: For the process of regular wear, there is no correlation between the wear rate and vibration rate; Areas 3–4: Increased wear and vibration rates are peculiar to the process of enhanced wear.

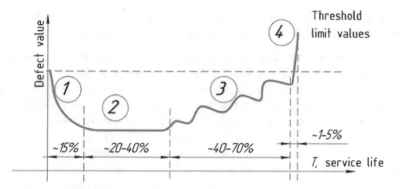

Fig. 2. The dependence of wear on the level of vibration. Area 1 – running-in process; Area 2 – process of regular wear; Areas 3-4 – process of enhanced wear.

ASM system provides information on the actual state of equipment in the required amount to ensure the permanent monitoring throughout the operational lifetime. When building ASM system, the minimum required number of sensors is used that record the state of working equipment and are able to define and record the technical condition in real time. Such sensors are installed in places most susceptible to wear (Fig. 3).

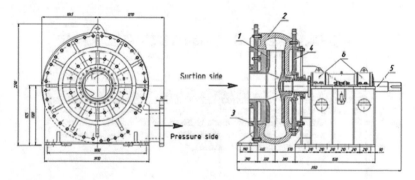

Fig. 3. Dredging pump of ship unloader. 1 – Impeller; 2 – Body frame; 3 – Front wall; 4 – Back wall; 5 – Drive shaft; 6 – Drive shaft bearings.

"Sensors" wirelessly transmit measurement results to the "Control Unit". The "Control Unit" records and preprocesses the received information recorded by sensors (Level I). After processing the received data via wireless communication channels, the results of diagnostics and primary analysis are transmitted to the "Organization Server" (equipment owner). On "Servers of the Organization" the data is stored/archived and then analyzed, according to the developed algorithms for the functioning of the maintenance and repair, the data obtained from the "Control Units". After final processing of information on "Servers of the Organization", the "Information Output" of monitoring results and recommendations for maintenance and repair in a convenient form are made (for example, on a monitor or a printer of an employee who is responsible for the technical condition of equipment) (Control Level II).

- "Sensors" are the measuring tool designed to transmit a signal of measurement information in a form convenient for further use and processing.
- "Control Unit" is the computing device used as part of the system for monitoring the actual condition of the equipment controls the process of collecting, preprocessing and analyzing incoming information about the actual condition of the equipment and transmitting it to the "Organization Server".
- "Server of Organization" is a software and hardware complex based on a specialized computer of enhanced reliability, ensuring the collection, storage, and transmission of information on the actual state of equipment to the user station.
- "Information Output" is a complex based on general-purpose computers designed to display information about the actual condition of equipment, in the required quantity and quality to ensure visualization of its technical condition.

A qualitative assessment of the type of technical condition is formed depending on the compliance of the obtained results with the technical condition of the equipment to one of the ranges shown in Fig. 4. Visualization of the analysis results is recommended to be divided into four information areas by three conventional lines: alerts (green), warnings (yellow) and alarms (red) (see Fig. 4).

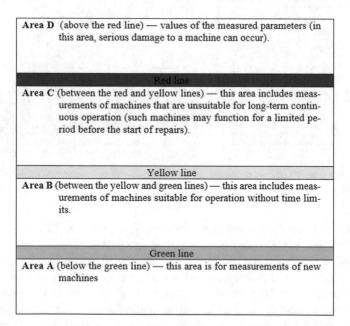

Fig. 4. Areas of visualization of the results of analyzing ASM systems.

Information zones characterize the technical state. The use of visualization tools to the results of the analysis of monitoring the state of equipment allows to improve the efficiency and quality of maintenance and repair, employee productivity and optimize labor costs to maintenance of equipment in working condition [3].

Among the necessary features of ASM systems, the following ones are distinguished:

1. The ability to process Big Data without distortion or loss of information;
2. Multi-channel communication, with the possibility of synchronous recording of signals from various measured physical processes;

The ability to display measurement results and analysis in a convenient form.

Functioning of ASM in the maintenance system can be divided into several serial stages showed in Fig. 5.

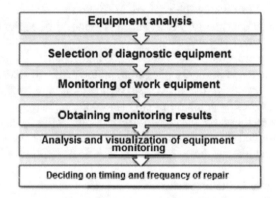

Fig. 5. Stages of ASM functioning.

Description of operation stages.

3. Equipment analysis. (Analysis of the failures of the equipment serviced by the SPM and Run-To-Failure strategies. Classification of equipment into groups (based on similarity of functioning and failures) for their transfer to ASM.
4. Selection of diagnostic equipment (for each group, diagnostic parameters and stationary measuring equipment are determined according to the accepted classification).
5. Monitoring of work equipment (with the help of stationary equipment, selected parameters characterizing the state of components and mechanisms are measured).
6. Obtaining monitoring results (primary recording of results in "control units" with subsequent transfer to the "organization server").
7. Analysis and visualization of equipment monitoring results. (The final analysis and visualization in industrial control system of the Organization. From the analysis of the history, one can learn the patterns of occurrence of faults) (Fig. 6).

8. Deciding on the timing and frequency of equipment repair (Drawing up acts on the structure and frequency of maintenance and repair, to ensure the effective operation of dredgers and ship unloaders).

3   Results

The introduction of maintenance system with the use of remote monitoring entailed the following positive results:

- Information about the technical condition of the equipment produced by non-destructive testing;
- The number of unpredictable equipment failures decreased (more than 80%);
- The total number of failures decreased (more than 15%);
- Reducing the cost of maintaining equipment performance (up to $20 \div 25\%$);
- Reducing the cost of owning the equipment (more than 25%);
- Allowed to reduce the time for repair and maintenance of equipment (up to 60%);
- Increases staff productivity ($\approx 30\%$);
- Allows predicting the volume of maintenance and repair;
- Repairs should be made primarily of equipment located in the area of risk of failure;
- Optimization of assets of the enterprise (spare parts and works are planned in advance, more favorable purchase prices) (more than 25%);
- Reduction of operating costs throughout the equipment life cycle (more than 15%);
- Among the identified difficulties in implementing ASM are the following:
- The need for additional investments to technical refitting of equipment and personnel;
- The cost of training/retraining staff;
- There is a possibility of errors due to the difference in measurement conditions (when using mobile ASM).

Within assessing the condition of the equipment, the integral part/result of the use of ASM system is the process of visualization throughout the life cycle. Visualization (the result of a range of activities) is shown in Fig. 5.

Example of visualization of the ASM system functioning for the rotary type of equipment.

As one can see, the vibration levels on the bearings of the electric drive mechanisms are located in areas A and B (Figs. 4 and 6). Vibration levels in areas A and B are indicative for new machines and machines suitable for operation without a time limit.

The diagnosis of rotary mechanisms is based on the fact that each defect has its own diagnostic features. Defects of the mechanisms are observed in the distribution of the signal power, in the form of a time realization, in the spectral picture of the auto-spectra, in the distribution of various additional vibration parameters. Most of the defects of the rotor mechanisms manifest in the spectra by harmonics peaks or "bursts". Some types of defects manifest in the form of changes in the general background in the high-frequency ranges (defects of lubricants in nodes, cavitation, resonances, etc.).

For materials with a high degree of absorption, the change in the overall thermal pattern at places where fatigue cracks begin to form is indicative when the deformation is almost completely transformed into heat. Thus, the use of ASM systems can provides control over the state of mechanisms without stopping machines in real time and allows for the planned maintenance according to the technical condition.

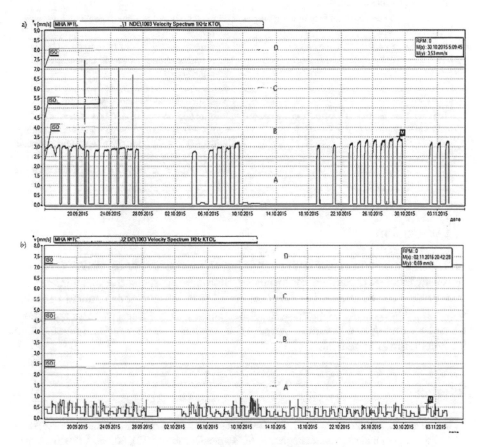

Fig. 6. Example of the output of information from sensors measuring the vibration velocity of the electric drive bearing of a pump of ship unloader [7] (a) yellow level (Area B) – signs of origin of the defect; (b) green level (Area A) – no defect.

4 Discussion

The goals of improving maintenance were achieved.

Monitoring the actual state allows maintenance and repair to be more clear, transparent and simple, allowing one to improve its efficiency and predictability. The monitoring process is easy to digitize, which allows increasing the level of maintenance automation. To obtain reliable results of the state of equipment, the use of stationary ASM is recommended (Table 1).

Table 1. Summary table of the research results.

No	Tasks and goals set	Results
1.	Simplicity and clarity in use	Standardization of maintenance staff actions. Ability to automate management
2.	To increase the predictability of failures and breakdowns of the working equipment of dredgers and ship unloaders	The number of unpredictable equipment failures has decreased (to 0%)
3.	Reducing the number of failures in components and mechanisms	The total number of failures decreased by more than 25%
4.	Reduction of unreasonable expenses	Optimization of assets of the enterprise (spare parts and work are planned in advance) (more than 30%) Reduction of operating costs throughout the equipment life cycle (more than 15%)
5.	Improve the quality and control of maintenance and repair work	Information on the technical condition of the equipment is obtained by non-destructive testing Repairs are made primarily of equipment located in the area of failure risk Allows predicting the volume of maintenance and repair

5   Conclusion

The use of strategy of actual state monitoring showed not only its viability, but also significant efficiency. Further, by means of thorough approach the maintenance and repair process can be arranged as the coordinated work of several departments and then be integrated with EAM (Enterprise Asset Management) technologies. The application of EAM technologies will further increase the efficiency of not only maintenance, but also an enterprise in general.

References

1. Cherkaoui, H., Huynh, K.T., Grall, A.: On the assessment of performance and robustness of condition-based maintenance strategies. IFAC PapersOnLine **49**(12), 809–814 (2016). https://doi.org/10.1016/j.ifacol.2016.07.874
2. Zub, I.V.: Vestnik Gosudarstvennogo universiteta morskogo i rechnogo flota imeni admirala Makarova S. O. **5**(39), 50–61 (2016). https://doi.org/10.21821/2309-5180-2016-8-5-50-61. (in Russian)
3. Uhlemann, T.H.J.: The digital twin: realizing the cyber-physical production system for industry 4.0. Procedia Cirp. **61**, 335–340 (2017). https://doi.org/10.1016/j.procir.2016.11.152
4. Jardine, A.K.S., Lin, D., Banjevic, D.: Mech. Syst. Signal Process. **20**(7), 1483–1510 (2006). https://doi.org/10.1016/j.ymssp.2005.09.012
5. Lee, J.: Prognostics and health management design for rotary machinery systems—reviews, methodology and applications. Mech. Syst. Signal Process. **42**, 314–334 (2014). https://doi.org/10.1016/j.ymssp.2013.06.004
6. SadaShiva, B.: Vibration monitoring mathematical modelling and analysis of rotating machinery. Int. J. Adv. Res. Electr. Electron. Instrum. Eng. **2**, 1325–1329 (2013)

Monitoring Systems of Ship Power Plants During Operation

Vladimir Zhukov⑩, Artem Butsanets$^{(\boxtimes)}$⑩, Sergey Sherban⑩,
and Vladimir Igonin⑩

Admiral Makarov State University of Maritime and Inland Shipping,
Dvinskaya Street 5/7, 198035 Saint-Petersburg, Russia
butsanetsaa@gumrf.ru

Abstract. The paper is devoted to the problem of creating a universal integrated monitoring system for marine diesel engines. A review of the existing systems for diagnosing the monitoring of ship engines is made, and their disadvantages are shown. The structure of diagnostic parameters and the selection method of measuring instruments are proposed. It is shown that the main objects to be monitored during the monitoring are the diesel engine workflow, the condition of the cylinder-piston group parts, the condition of the engine oil, vibration indicators of the engine, and systems serving the main engine. When creating monitoring systems, it is proposed to maximize the use of elements of the engine electronic control system and information obtained from the sensors of this system. The paper outlines the steps for the practical implementation of the developed system, it is shown that its implementation will reduce the operating costs of shipping companies and increase the safety of navigation.

Keywords: Ship power plants · Diesel engines · Monitoring systems · Diagnostic methods · Diagnostic parameters · Choice of measuring instruments

1 Introduction

The most important condition for the success of any shipping company is to reduce operating costs without compromising the quality of provided transportation services (timeliness, reliability and security of the delivery of passengers or goods). The main share of operating costs is directly related to ship power plants: the cost of fuel and engine oil, the cost of maintenance and repair of the ship power plant elements during operation. Thus, ensuring an economical and reliable operation of the ship power plant gives shipping companies an additional competitive advantage. The main elements of ship power plants, determining their environmental, economic and resource indicators are diesel engines used on ships as the main and auxiliary engines.

Improving methods for assessing the technical condition of ship diesel engines and creating modern systems for monitoring the state of main and auxiliary engines under operating conditions are effective means of combating environmental pollution with power plants of operated vessels, as well as ensuring their economical and reliable operation [1, 2].

© Springer Nature Switzerland AG 2020
V. Murgul and M. Pasetti (Eds.): EMMFT 2018, AISC 982, pp. 419–428, 2020.
https://doi.org/10.1007/978-3-030-19756-8_40

Monitoring the workflow of ship diesel engines makes it possible to assess the processes quality of mixing and combustion in engine cylinders, which determine the economic and environmental indicators of ship power plants. The methodology for monitoring the diesel engine workflow under operating conditions is described in sufficient detail in [3, 4], however, in these works, insufficient attention is paid to technical means of monitoring. The works [5, 6] are devoted to methods and algorithms for assessing the state of the cylinder-piston group of diesel engines, but they do not pay attention to the methods of transmitting and storing the received information.

Known methods for the control of diesel engines as engine oil [7, 8]. In [9], a diagnostic method was proposed using sensors that record the presence in the engine oil of wear products, water, soot, corrosion, sulfur content, and also determine the viscosity of the oil and the degree of aeration. For the detection of wear particles in engine oil it is proposed to use induction sensors [10]. The development of a lubricant flow measurement methodology is described in [11].

The use of vibroacoustic methods as a means of indivisible diagnostics of power plant elements is described in [12, 13]. The work [14] is devoted to the diagnostics of bearing assemblies during operation.

To assess the technical condition of diesel engines, it is proposed to use the analysis of exhaust gases [15] and the state of diesel engine individual components [16, 17].

Perspectives are studies of the management capabilities of ship diesel engines using methods for predicting their state during operation [18].

A number of studies solve the problems of using modern methods and means of complex diagnostics of diesel engines using automated systems, both in real time and periodically (upon request) [19, 20]. Work is underway on the use of wireless engine diagnostics systems [21]; systems of ship mechanisms and means of controlling emissions into the atmosphere are being designed to meet future emission standards [22]. In [23], the efficiency of using neural networks to determine the threshold values corresponding to the limiting state of diesel engines is shown.

In the process of monitoring ship diesel generators in accordance with the requirements of the RMRS Rules, electrical quantities and power quality should be subject to control. Studies aimed at developing new instruments for monitoring the parameters of electrical energy in the process of monitoring auxiliary marine diesel generators are presented in [24]. The same author conducts work on the analysis of the impact of the competence of ship crews on the likelihood of emergency situations associated with the state of equipment and power quality [25].

An overview of existing diesel engine monitoring systems is given in [26–28], in which it is noted that most of the systems for monitoring marine diesel engines are a single software and hardware system that records parameters and partially calculates the workflow in real time.

Analysis of the existing systems for diagnosing ship diesel engines allows us to draw the following conclusions:

- the majority of systems solve particular problems of diagnostics, to ensure comprehensive complex monitoring of ship power plants in accordance with the requirements of classification societies, it is necessary to develop a single measuring and information complex that provides for obtaining and storing information about the current and future state of main and auxiliary diesel engines, as well as providing information to the crew, shipowner, classification society;

- leading manufacturers of marine diesel engines are focused on diagnostics of their own engines, there are no universal systems that can be used to equip a wide range of vessels;
- favorable prospects for the creation of monitoring systems are due to the increased use of electronic control and regulation systems in the construction of modern diesel engines; it is advisable to use the information obtained by sensors of such systems to assess the state of individual systems and the engine as a whole.

The aim of this work was to develop general principles for creating a universal integrated system for monitoring ship diesel engines and determining the diagnostic parameters of the monitoring system for diesel power plants of ships.

2   Method and Results

On the basis of operating experience of ship power plants, testing and thermal control of main and auxiliary engines and in accordance with the requirements of the RMRS for the monitoring systems of the mechanism technical condition, the structure of diagnostic parameters and means for their determination were proposed.

2.1   Effective Engine Performance

The measured values are calculated engine power, the specific effective fuel consumption. The required accuracy of measurements and the corresponding accuracy class of instruments shall be established by the classification society or the shipowner. To correctly determine the effective performance of the engine, it is necessary to take into account the actual operating conditions (weather conditions, navigation area, ship loading, etc.). To obtain this information, the monitoring system must incorporate additional sensors (Table 1).

Table 1. Effective engine performance.

Diagnostic parameter	Determination method	Measurement tool
Torque, N·m	Method of elastic deformation	Tensor-resistive, inductive, agnetoelastic sensors
Crankshaft rotational frequency, min^{-1}	Contact, contactless	Electromagnetic, electronic, stroboscopic tachometers
Fuel consumption, kg/h	Mass, volume	Mechanical, optical, electrical flow meters

2.2 Quality of Workflow

Diagnosing a diesel workflow is the most difficult procedure in monitoring due to the high rate of change of parameters characterizing the process, but sensors with the required speed are now developed and mass-produced (Table 2).

Table 2. Quality of workflow

Diagnostic parameter	Determination method	Measurement tool
Maximum pressure of compression, MPa	Workflow indexing	Strain gauge, piezoelectric, optical, capacitive sensors
Maximum pressure of combustion, MPa	Workflow indexing	Strain gauge, piezoelectric, optical, capacitive sensors
Maximum temperature of the cycle, °C	Workflow indexing	Strain gauge, piezoelectric, optical, capacitive sensors
Pressure rise rate $(dp/d\varphi)$, MPa/°p.kv	Workflow indexing	Strain gauge, piezoelectric, optical, capacitive sensors
The angle corresponding to the maximum pressure of combustion, °p.kv	Workflow indexing	Strain gauge, piezoelectric, optical, capacitive sensors
The pressure on the compression line at a point of 12° to TDC, MPa	Workflow indexing	Strain gauge, piezoelectric, optical, capacitive sensors
The pressure on the expansion line at a point of 36° after TDC, MPa	Workflow indexing	Strain gauge, piezoelectric, optical, capacitive sensors
The angle of advance of the start of combustion, °p.kv	Workflow indexing	Strain gauge, piezoelectric, optical, capacitive sensors

Based on the results of the workflow indexing, the average indicator pressure, specific indicator fuel consumption, cylinder indicator power are calculated. When conducting the indexing, it is also necessary to determine the fuel injection parameters: the start of fuel injection, the angle of fuel injection duration, maximum fuel injection pressure.

2.3 The State of the Cylinder-Piston Group

The wear of the cylinder-piston group parts limits the resource indicators of the engine and determines the timing of the maintenance work (Table 3).

Table 3. The state of the cylinder-piston group.

Diagnostic parameter	Determination method	Measurement tool
The breakthrough of gases from the combustion chamber kg/s	Mass	Mass, electronic flow meters, gas meters
Compression end pressure p_s, MPa	Workflow indexing	Strain gauge, piezoelectric, sensors
Engine oil consumption G_M, kg/h	Mass	Mass, electronic flow meters
Vibration in the area of cylinder covers (vibration velocity v_c, mm/s)	Vibroacoustic	Vibration meters-analyzers

2.4 Condition of Lubricating Oil and Coolant

The quality of engine oil and coolant, characterized by their physicochemical properties, has a significant effect on engine reliability. Reasonable prolongation of the use of motor oils and material coolants ensures a reduction in operating costs (Table 4).

Table 4. Condition of lubricating oil and coolant

Diagnostic parameter	Determination method	Measurement tool
Kinematic viscosity of oil, m^2/s	Capillary, rotary	Mechanical, electrical sensors
The content of impurities in the oil, mg/kg	Spectrographic	electronic sensors
The physical and chemical properties of the coolant	Physical, chemical	Analyzers
The content of pollution in the coolant, mg/kg	Optical	Electronic sensors

2.5 The State of the Systems Serving the Main Diesel

The good condition and trouble-free operation of the fuel injection systems, lubrication, cooling and pressurization are essential conditions for reliable and economical operation of the engine. In this regard, the monitoring of major systems should be given special attention (Table 5).

When determining the list of diagnostic and monitoring parameters, it is necessary to take into account that the lack of parameters can lead to errors during monitoring, and their redundancy can lead to excessive complication and cost of the monitoring system. The choice of the optimal number of diagnostic parameters is a separate scientific task.

To select a specific type of sensors and other measuring instruments from among those produced commercially, the principles described in [29, 30] should be used. This will minimize costs when creating monitoring systems with specified functionality.

Table 5. The state of the systems serving the main diesel.

Diagnostic parameter	Determination method	Measurement tool
Fuel injection pressure, MPa	Workflow indexing	Strain gauge, piezoelectric, sensors
The advance angle of fuel injection, °p.kv	Workflow indexing	Strain gauge, piezoelectric, sensors
Pressure behind the fuel priming pump, MPa	Contact	Mechanical, electric sensors
Oil pressure, MPa	Contact	Mechanical, electric sensors
Oil temperature at the engine outlet and at the engine inlet, °C	Contact	Mechanical, electric sensors
Coolant temperature at engine outlet and at the engine inlet, °C	Contact	Mechanical, electric sensors
Coolant flow rate, kg/s	Contact	Mechanical, electric sensors
Charge pressure, MPa	Contact	Mechanical, electric sensors
Charge air temperature, °C	Contact	Mechanical, electric sensors
Pressure before the turbine, MPa	Contact	Mechanical, electric sensors
Rotor speed, min^{-1}	Contactless	Electric, electronic, optical sensors

3 Discussion

An experimental system for monitoring the state of a diesel engine is being developed based on the proposed concept. When developing the system, the possibility of its scalability is taken into account: the ability to connect from several sensors for small vessels, to a complete monitoring system of ship indicators. This allows you to make the system universal and expands the possibilities of its application.

The developed monitoring systems for ship diesel include: micro-PC (or router with data transfer function), graphic monitor (touch pad), web interface, analog-digital converters, sensors of pressure, temperature, torque, speed, flow meters, gas analyzers, smoke meters and etc., a device for express analysis of the condition (engine oil, coolant).

The data obtained as a result of diagnosis can be stored on the MicroPK for a specified time, then transferred to the central server for permanent storage and/or study. In addition, they are displayed on a graphic display for the mechanics of the engine room. Basic functions for mechanics are supposed to be, such can be: topping up oil, coolant, partition of internal combustion engines, etc.

At the first stage, it is supposed to use open source software. The view of the system interface is shown in Fig. 1.

Fig. 1. An example of the interface of the developed system.

The presence of the laboratory with the engine SKL 3VD14.5/12-2 allows us to carry out bench tests of the system. This work was supported by a real transport company (shipping company «Navigator»), which is confirmed by the concluded cooperation agreement.

In case of a positive result of the pilot tests, the shipowner is ready to provide a river-sea class tanker for carrying out field tests.

At present, work continues on detailing the list of devices, selecting the necessary components for the effective operation of an experimental monitoring system and choosing the optimal operating modes of the equipment (Fig. 2).

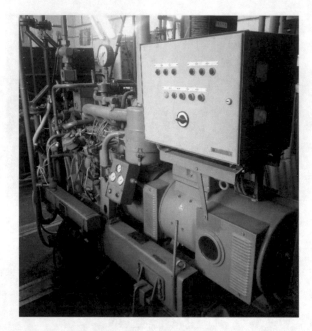

Fig. 2. Engine SKL 3VD14,5/12-2 for bench testing of the monitoring system.

4 Conclusions

1. The problem of creating an integrated universal monitoring system of ship power plants is relevant.
2. The development of monitoring systems should be carried out taking into account the requirements of the Rules of the classification societies (Russian Maritime Register of Shipping).
3. When developing monitoring systems, it is necessary to maximize the use of sensors and converters installed in modern electronic engine control systems.
4. To install additional sensors, a rational list of diagnostic parameters must be scientifically substantiated and the brands and models of measuring instruments should be determined taking into account their technical level.
5. For the transmission of information obtained in the process of monitoring ship engines to remote users, modern means of communication can be used.

References

1. Russian Maritime Register of Shipping. Rules for the classification and construction of ships. Part VII Mechanical installation. Saint-Petersburg, p. 72 (2016)
2. Klyuev, V.V., Parkhomenko, P.P., Abramchuk, V.E., et al.: Technical Means of Diagnosis: A Handbook, p. 672. Mashinostroenie, Moscow (1989)

3. Ivanovsky, V.G., Varbanets, R.A.: Monitoring of the working process of ship diesel engines in operation. Intern. Combust. Engines **2**, 138–141 (2004)

4. Polovinka, E.M., Yakovenko, A.Yu.: Improving the monitoring of diesel engines. Am. Sci. J. **20–1**, 22–33 (2018)

5. Ageev, E.V., Kudryavtsev, A.L., Sevostyanov, A.L.: Algorithm for diagnosing cylinder-piston group using a technical endoscope. World Transp. Technol. Mach. **1**(36), 116–122 (2012)

6. Snarsky, S.V., Gafiyatullin, A.A., Kulakov, A.T.: Methods for determining the residual life of a car diesel engine in on-board diagnostics. Sci. Thought **3**, 210–221 (2017)

7. Wakiru, J.M., et al.: A review on lubricant condition monitoring information analysis for maintenance decision support. Mech. Syst. Signal Process. **118**, 108–132 (2019)

8. Zhukov, V.A., Butsanets, A.A.: Monitoring of engine diesel engine oil status during operation. Collection of theses of the national scientific and practical conference. Saint-Petersburg, GUMRF them. adm. C.O. Makarova, pp. 30–32 (2018)

9. Zhu, X., Zhong, C., Zhe, J.: Lubricating oil conditioning sensors for online machine health monitoring–a review. Tribol. Int. **109**, 473–484 (2017)

10. Du, L., Jiang, Z.: An integrated ultrasonic-inductive pulse sensor for wear debris detection. Smart Mater. Struct. **22**(2), 025003 (2012). https://doi.org/10.1088/0964-1726/22/2/025003

11. Delvigne, T., et al.: A new methodology for on line lubricant consumption measurement. SAE Technical Paper, №. 2005-01-2172 (2005)

12. Reshetov, A.A.: Information technology to improve the efficiency of vibration monitoring of energy-mechanical equipment. Instrum. Syst. Manage. Control Diagn. **12**, 50–59 (2013)

13. Geng, Z., Chen, J., Hull, J.B.: Analysis of engine vibration and design of an applicable diagnosing approach. Int. J. Mech. Sci. **45**(8), 1391–1410 (2003)

14. Vakharia, V., Gupta, V.K., Kankar, P.K.: A comparison of feature ranking techniques for fault diagnosis of ball bearing. Soft. Comput. **20**(4), 1601–1619 (2016). https://doi.org/10.1007/s00500-015-1608-6

15. Henningsson, M., et al.: Dynamic mapping of diesel engine through system identification. In: Identification for Automotive Systems, pp. 223–239. Springer, London (2012). https://doi.org/10.1007/978-1-4471-2221-0_13

16. Kurbakov, I.I., Pankov, A.I., Karpov, V.N., Kurbakova, M.S., Ladikov, S.A.: Analysis of indivisible methods for diagnosing turbochargers. In Proceedings of the Energy-Efficient and Resource-Saving Technologies and Systems, pp. 224–228. Institute of Mechanics and Energy, Saransk (2016)

17. Sergeev, K.O., Pankratov, A.A.: Diagnostics of high speed marine diesel injectors. Bull. Astrakhan State Tech. Univ. Ser. Mar. Eng. Technol. **1**, 50–58 (2017)

18. Karlsson, M., et al.: Multiple-input multiple-output model predictive control of a diesel engine. IFAC Proc. Vol. **43**(7), 131–136 (2010). https://doi.org/10.3182/20100712-3-de-2013.00003

19. Lazakis, I., et al.: Advanced ship systems condition monitoring for enhanced inspection, maintenance and decision making in ship operations. Transp. Res. Procedia **14**, 1679–1688 (2016). https://doi.org/10.1016/j.trpro.2016.05.133

20. Lazakis, I., Dikis, K., Michala, A.L.: Condition monitoring for enhanced inspection, maintenance and decision making in ship operations. In: 2016 PRADS 13th Triennial Proceedings of the 13th International Symposium on Practical Design of Ships and Other Floating Structures. Technical University of Denmark (DTU), Denmark (2016). ISBN 978-87-7475-473-2

21. Michala, A.L., et al.: Wireless condition monitoring for ship applications. Smart The Royal Institution of Naval Architects, Ship Technology London (2016)

22. Balland, O., Erikstad, S.O., Kjetil, F.: Concurrent design of vessel machinery system and air emission controls to meet future air emissions regulations. Ocean Eng. **84**, 283–292 (2014). https://doi.org/10.1016/j.oceaneng.2014.04.013

23. Basurko, O.C., Uriondo, Z.: Condition-based maintenance for medium speed diesel engines used in vessels in operation. Appl. Thermal Eng. **80**, 404–412 (2015). https://doi.org/10.1016/j.applthermaleng.2015.01.075

24. Mindykowski, J., Tarasiuk, T.: Development of DSP-based instrumentation for power quality monitoring on ships. Measurement **43**(8), 1012–1020 (2010). https://doi.org/10.1016/j.measurement.2010.02.003

25. Mindykowski, J.: Impact of staff competences on power quality-related ship accidents. In: 2018 International Conference and Exposition on Electrical and Power Engineering (EPE). IEEE (2018). https://doi.org/10.1109/icepe.2018.8559646

26. Okunev, V.N., Burkov, D.E.: Modern systems of diagnostics and monitoring of the technical condition of marine diesel engines. Marine Radio Electron. **3–4**(33–34), 40–43 (2010)

27. Danilyan, A.G., Chimshir, V.I., Razinkin, R.A., Naidenov, A.I.: Improvement of technical diagnostics systems for low-speed ship diesel engines. Young Sci. **2**(82), 138–142 (2015). https://moluch.ru/archive/82/14613/

28. Soloviev, A.V.: Monitoring systems for ship diesel engines in operation. Bull. Astrakhan State Tech. Univ. Ser. Marine Equip. Technol. **1**, 87–92 (2018)

Data Processing Model in Hierarchical Multi-agent System Based on Decentralized Attribute-Based Encryption

Andrey Nyrkov⬤, Yulia Romanova⬤, Konstantin Ianiushkin$^{(\boxtimes)}$⬤,
and Izolda Li⬤

Admiral Makarov State University of Maritime and Inland Shipping,
Dvinskaya Street 5/7, 198035 Saint-Petersburg, Russia
kayahidden@gmail.com

Abstract. Multi-Agent and distributed systems are widely used in military defense systems, logistics, navigation and motion control systems of unmanned aerial vehicles and other systems. Such systems are complex and include many different types of agents, the number of which can vary in real time. Elements of such a system require proper management of access roles and security keys during the process of sharing sensitive information. In this paper, we tested the applicability of decentralized attribute-based access control for communication in multi-agent systems. By implementing one of the algorithms, the operation speed and practical benefits of attribute-based encryption were evaluated, and also the encryption rate on the synchronization frequency of the multi-agent system was estimated.

Keywords: Distributed systems · Attribute-based access control ·
Attribute-based encryption · Multi-agent system · Multi-authority

1 Introduction

1.1 Multi-agent Systems and Access Policies

Consideration of multi-agent systems begins with the definition of the agent as the main system element. The agent is a specific entity that is able to collect available data, process it, send it via the information transfer channels, and make decisions on its own. The set of such agents that can be organized both into a peer-to-peer flat and into a multi-peer (but not strictly defined) system, and having some common goal, is called a Multi-Agent System (MAS).

Multi-agent systems are used to model traffic for the needs of transport, military defense systems, logistics systems, navigation systems, motion control systems of unmanned aerial vehicles, and other systems.

In the role of such intelligent agents can be either physical individual devices (machines and sensors), or a software agent, which is a piece of code that performs the appropriate actions to interact with the environment. Software agents can run on different physical nodes or on a single computing device in a virtual environment [1–6].

© Springer Nature Switzerland AG 2020
V. Murgul and M. Pasetti (Eds.): EMMFT 2018, AISC 982, pp. 429–438, 2020.
https://doi.org/10.1007/978-3-030-19756-8_41

Decentralized systems or, more precisely, security problems of information transmission in systems are considered. In the systems agents can join and leave the system, communication channels can also appear and disappear, and encrypted messages can be transmitted through known open communication channels to everyone who can decipher them. The main advantage of this architecture is fault tolerance. In the case of the main node failure in a traditional information system, the entire system will fail, while the MAS will be able to continue working even if a few nodes fail.

1.2   Decentralized Attribute-Based Encryption and Access Control

It is possible to build an access control scheme in distributed networks using attribute-based encryption (ABE). Attribute-based encryption is a subtype of functional encryption. The first ABE scheme was proposed by Sahai and Waters in 2005 [5].

In 2006, in the study of Goyal [6] an ABE scheme with a key-policy attribute-based encryption access rule was proposed. The encryption algorithm is different from the original version of ABE at the stages of generating private keys and at the stage of decryption. The agent's private key is generated according to the requested access structure. In each agent's key, an access tree is encrypted indicating the attribute set values. The consistency between key and data attribute values is checked. If the packet attributes satisfy the agent's key, then it can decrypt the message. This approach is called Key Policy (KP-ABE). The keys are given to agents by a trusted center, it also verifies the authenticity of attribute values, which means that agents actually possess them.

In 2007, an ABE encryption scheme with a ciphertext-based access rule was proposed by Bethencourt et al. in their research [7]. Access control is performed in a manner similar to KP-ABE, however, the data access rule is not contained in the agent's private key, but in the encrypted data itself (ciphertext), and the agent's private key at the same time corresponds to the set of attributes. If the attributes contained in the agent's private key correspond to the ciphertext access structure, the agent can decrypt the data. The ciphertext method (Ciphertext Policy ABE, CP-ABE), which is different because the access tree is encrypted into a data packet, and the agent's key includes verification attributes, has the following four major functions:

- Setup - generates the public key and the secret master key associated with it;
- Encrypt - encrypts the file using the public key and the specified policy;
- Keygen - creates a private key with attributes that is associated with the secret master key and its public key. This requires a secret master key;
- Decrypt - decrypts the encrypted file only if the private key attributes satisfy the policy and the private key is associated with the public key that was originally used for encryption. It requires an encrypted file and a private key;

The problem, which arises when constructing data transfer schemes for multi-agent systems by the means of above-mentioned schemes, is that a single and common trusted center is needed in these schemes. In addition, the rules for constructing policies limited the organization of the system; only so-called monotone structures were allowed, in which it is impossible to describe attributes with denied access.

An option for expanding the algorithm in the case of non-monotonic access structures was proposed in 2007 in the research of Ostrovsky et al. [8]. Along with each element, its negation is added to the attribute space, being essentially a new attribute. With its help, the total number of possible attributes is doubled compared with the classical scheme. In other aspects the principle of the algorithm remains the same.

This scheme type has significant drawbacks. For example, if it is necessary to explicitly indicate the negation of all attributes in the access structure, by which it is not supposed to open access to the data, then as a result, the ciphertext can contain a large number of attribute negations that have no practical value for decrypting the data. As a result, the size of the ciphertext is greatly increased. In some papers [9], it was noted that new attributes may appear in the system as a result of such encryption, which will require that encryption to be restarted.

For a self-organizing system without a specific coordination center, it is necessary to consider the concept of a decentralized encryption scheme based on attributes, which consists of a set of trusted centers (authorities), a trusted initialization system, and agents. The centers themselves, in turn, may act as agents of a higher level in the hierarchy tree.

Attribute-based decentralized encryption is a special form of multi-level ABE schemes, in which no common coordination center is required other than creating uniform initial parameters. One of the most interesting schemes of that type was presented in the work of Rao and Dutta [9], which was taken as a basis for the study.

The scheme described by them allows complete implementing the attribute-based access control policy (ABAC)—a model of access control to data based on analyzing the rules for attributes of objects, possible operations with them and the environment corresponding to the query [10]. Attribute-based access control systems provide mandatory and selective access control. The scheme allows implementing more simple role-based access control (RBAC). Herewith, the access rights of the system agents to objects are grouped taking into account the specifics of their use, i.e. forming roles [11].

For a variety of agents of the same type, role model elements, which can be implemented as part of attribute-based encryption, are suitable. The formation of roles is intended to define clear rules for access control.

2 Model and Methods

The purpose of the study is to determine the feasibility of implementing the above-mentioned system, as well as determining the requirements and restrictions imposed on multi-agent systems in connection with the use of decentralized CP-ABE encryption.

For the implementation of such system, the following model was designed: after the system starts, attribute authority (AA) trusted centers are registered on the authentication server (SA), generating a shared secret key. After the AA key is generated, the authentication and authorization procedures are performed. The AA issuing centers conduct the same operations for agents. Thus, each node of the system and all attribute authorities have common initialization parameters. Therefore, access rights are differentiated by using attribute-based encryption, since all messages between devices are encrypted with private keys based on the access structure.

The access structure will be based on policies. Standard for access control based on XACML attributes (eXtensible Access Control Markup Language) [12] will be adhered in order to describe policies and access control. The main concepts of XACML are rules, policies, rule-combining-algorithms, attributes (of subject, object, actions and environmental conditions), obligations and advices. The central element is the rule, which contains the goal, effect and conditions [12].

The ABAC model management structure consists of the following 4 basic mechanisms that are key nodes in the implementation of policies [12]:

The policy decision point (PDP) evaluates digital policies (DP) and metapolicy (MP) and decides on giving the access to an object, and also mediates between DP and MP;

The policy enforcement point (PEP) implements PDP solutions in response to a request from a subject for access to an object protected by an established policy.

The policy information point (PIP) provides the PDP with the necessary attribute data for further evaluation.

The policy administration point (PAP) helps with the creation of the policy, contains information about it. Various characteristics of agents can be considered as attributes, depending on the overall scheme hierarchy. The scheme of distributing keys from attribute authorities to agents should be described, since in a dynamic environment attributes may change, and it becomes necessary to obtain a new secret key in response to new changes. Let us describe this process using the Otway–Rees protocol [8], assuming that attribute authorities have the ability to authenticate a new set of attributes, and their transfer is initiated by the agent itself. An example of such an interaction system is presented in the diagram (see Fig. 1).

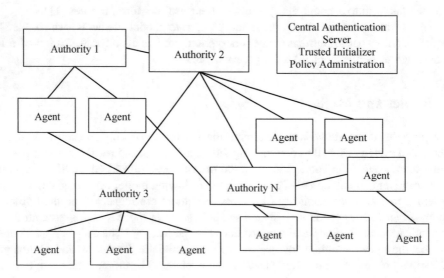

Fig. 1. An example of a modeled communication scheme between agents and attribute authorities.

In such a system, the main bottleneck will be the synchronization speed of the agents transmitting encrypted data, which requires computational power and time to encrypt.

When it comes to managing a multi-agent system, synchronization is either a consistent change, or a match, or an approximation of the parameter's values for a certain number of agents. Often such a goal is formulated as reaching a consensus or agreeing on certain characteristics. In practice, synchronization algorithms can be used in such areas as parallel computing, power supply systems, sensor systems and robotics [13].

The model of interaction between the system's agents can be represented as an oriented graph $G = (V, E)$, where $V = [1 : N]$ is the set of vertices, $E \subset V \times V$ is the set of edges, $N_k = \{i \in V : \exists (i, k) \in E, k \in V\}$ is the set of neighboring nodes of the node k.

For the simplest case, it was shown that the following simple linear system comes to a consensus [14]:

$$\dot{x}_k(t) = \sum\nolimits_{i \in N_k} (x_i(t) - x_k(t)), x_k(0) = z_k \in \mathbb{R} \tag{1}$$

It was proved that for connected graphs there exists a unique equilibrium point $x^* = (\alpha, \ldots \alpha)$ [15].

The used computing systems do not operate continuously, but with a certain frequency. Therefore, in order to use the algorithm in practice, it is necessary to perform a discretization of the formula in the computing system. Among other things, the sampling interval can be determined based on the speed of the data encryption scheme. In order to determine these characteristics of the system, the implementation of the described model was created and investigated using the available means. To simulate real agents for experimental purposes, the use of their software simulations based on the existing platform Java Agent Development Framework is planned [16].

Access control is based on a set of pre-formed rules that will determine the scope of the entire system as a whole. The rule set simply lists all the rules included in this policy. The result of the policy calculation is one particular value, as in the case with the rules. To get this value from the calculation results of all the rules, the rule combination algorithm specified by the policy according to the XACML standard is applied. The scheme used by the standard will be distributed among the components of the multi-agent system as shown in the diagram (see Fig. 1). The creation of security policies for a multi-agent system is a separate topic, which will not be considered in the framework of the study, but a policy must be created to calculate the policies. The management component is responsible for the creation of policies, which is called the policy administration point (PAP) in the standard. Another data source needed by the PDP for computing is the source of attribute values, called the policy information point (PIP).

The indicated elements will fulfill the following roles in the developed scheme:

Central Authentication Server (CA): The CA is responsible for setting global public parameters, processing and developing policy-based access rules. Moreover, it is responsible for registration requests from AA. The CA generates the unique global agent identification (ID) for each agent and a unique identification (AID) for each AA in the system. CA does not participate in any attribute operations.

Attribute Authority (AA): Each AA is an independent center that is responsible for managing the attributes of agents. According to the role or identification of the agent, the corresponding AA can grant the rights, revoke and update the attributes of the agent in their domain. Each attribute is associated with one AA, but each AA can control an arbitrary number of attributes. Each AA generates its own public key and issues secret keys associated with attributes to agents.

Agent: Each agent is assigned a global identifier (ID), and it contains a set of attributes. Agents will request their secret keys associated with attributes from the corresponding AA. All agents are free to download encrypted data and decide whether to try to decrypt the data or not. Only the agents with attributes satisfying access policy can successfully decrypt the encrypted data using their secret keys. The agent also acts as a data source. Before distributing data, the agent firstly determines the attribute access policy, and then encrypts the data in accordance with this policy. The agent does not rely on AA and CA to control data access. Agents with different attributes get different decryption rights, and therefore, access control occurs within the crypto-graphic system.

The algorithms described in the research of Rao and Dutta [9] were implemented for the cryptographic protection described in the model, as well as for the imple-mentation of differentiating the access rights according to the attribute model.

An attempt to implement such an algorithm already existed, but we were unable to launch the implementation found using the Charm environment [17]. We developed our own implementation using Charm and related applicative cryptographic packages, as well as performance measurement tools. System booting and profiling were per-formed using the AMD CodeXL package [18].

Several experiments were conducted in order to investigate the correctness and performance of implementations. A basic work scenario of determining and comparing various characteristics of the circuit is studied in the first experiment. In the following experiments, the impact of policies, number of attributes, and also the impact of the size of keys on system performance were analyzed. Each experiment that required changing the system parameters was performed 1000 times, the results were averaged. Each case considered had a different input value (different policy) and agent config-uration, measurements are performed in several runs to limit the mutual influence of the measurements.

In order to represent the boolean expression, obtained by processing attribute-based access policies in XACML, in the form of an LSSS (Linear Secret Sharing Scheme) matrix, the algorithm proposed by Lewko and Waters [19] was used. The only change is that the formulas should be presented in disjunctive normal form [9]. Let us consider a boolean expression as an access tree, where the internal "AND" and "OR" are nodes, and the leaves correspond to attributes. Then, use $(1, 0, \ldots, 0)$ as a sharing vector for the LSSS matrix. Let us start with designating the tree root node by the vector (1) (vector of length 1). Then we go down the levels of the tree, marking each node with a vector determined by the vector of the parent node. We maintain a global counter variable C, which is initialized by one.

The LSSS access matrix (2) was constructed and used for the first experiment.

$$\begin{pmatrix} 1 & 1 & 0 \\ 0 & -1 & 1 \\ 0 & 0 & -1 \\ 0 & -1 & 0 \end{pmatrix} \tag{2}$$

The experiment bench was a computer with a Core i5-8400 processor running at 2.8 GHz. The experiments were carried out on a virtual machine running Fedora Linux 29. The measured parameter is the time required to perform all calculations with a full system load. The system status was monitored using the AMD CodeXL package. For the first series of experiments, the duration of each separate step of the cryptographic scheme algorithm was measured. The measurement error was calculated using the Kornfeld method and is 0.012 s.

3 Results

The most important metric is the execution time of each cryptographic operation. The execution time of six basic cryptographic operations, which are performed by CA, AA or by the agent itself were checked: system initialization, initialization of attribute authorities (AA), generation of decryption keys, encryption and decryption of data. The measurement results are presented in the graph on Fig. 2. Updating policy is the longest operation, as it requires both encryption and decryption of data.

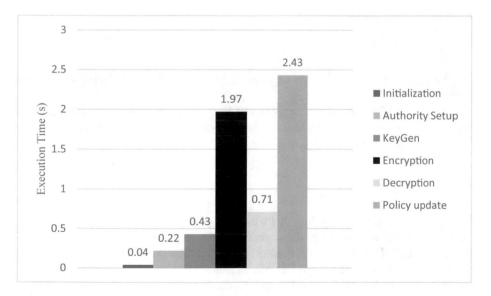

Fig. 2. The average time of completing the steps of the described algorithm.

The following graph shows the change in the execution time of encryption and decryption operations, depending on the size of the policies (see Fig. 3). As was predicted in the analytical assessment, the execution encryption time linearly depends on the number of attributes defined by the policy. However, the used scheme showed a constant time to perform the decryption operation. This can be an important parameter for choosing this encryption scheme, if the system agents have to decrypt much more information from surrounding source agents than to encrypt for sending. The average decryption time for this configuration was 0.73 s.

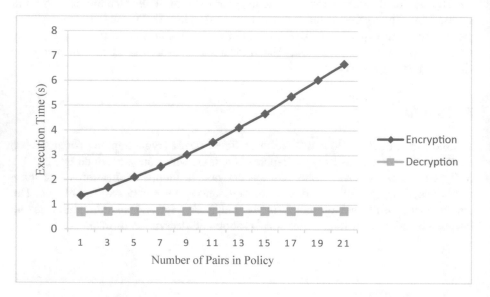

Fig. 3. The average time to perform encryption and decryption operations.

The access rules used for the experiment were compiled in a disjunctive normal form.

The length of the agent's composite key linearly depends on the number of attributes assigned to the agent. During the experiment, the size of the memory allocated for the longest key was about 2 KB.

4   Conclusion

The agent communication system that we developed on the basis of the decentralized attribute encryption scheme has shown its efficiency. In order to build an access control system, the use of attribute encryption was proposed. The secret key distribution model in the conditions of a large number of devices and trusted centers was given. The ABAC-based access control scheme supports data privacy and access policies and uses the LSSS flexible matrix access structure. Role model elements for the large number of agents, which are lower in the hierarchy, allows not saving computing resources along with reducing the number of attributes, which directly affects the speed of data encryption. The advantages of attribute encryption certainly include flexible configuration of access structures, ABE allows encrypting messages between all system agents. However, at the same time, a situation where attribute values of each of them are not secret can cause confidentiality risks for specific application areas.

References

1. Lewis, F.L., Zhang, H., Hengster-Movric, K., Das, A.: Cooperative Control of Multi-agent Systems. Springer, London (2014)
2. Nyrkov, A., Belousov, A., Sokolov, S.: Algorithmic support of optimization of multicast data transmission in networks with dynamic routing. Mod. Appl. Sci. **9**, 162 (2015)
3. Zhilenkov, A., Nyrkov, A., Chernyi, S., Sokolov, S.: Simulation of in-sensor processes in the sensor – object system type when scanning the elements of underwater communication lines with a probe beam. Int. Rev. Model. Simul. (IREMOS) **10**(15), 363–370 (2017)
4. Nyrkov, A., Sokolov, S., Zhilenkov, A., Chernyi, S.: Complex modeling of power fluctuations stabilization digital control system for parallel operation of gas-diesel generators. In: 2016 IEEE NW Russia Young Researchers in Electrical and Electronic Engineering Conference (EIConRusNW), St. Petersburg, pp. 636–640 (2016)
5. Sahai, A., Waters, B.: Fuzzy identity-based encryption. In: Cramer, R. (ed.) Advances in Cryptology – EUROCRYPT 2005. EUROCRYPT 2005. Lecture Notes in Computer Science, vol. 3494. Springer, Heidelberg (2005)
6. Goyal, V., Pandey, O., Sahai, A., Waters, B.: Attribute-based encryption for fine-grained access control of encrypted data. In: ACM CCS (2006)
7. Bethencourt, J., Sahai, A., Waters, B.: Ciphertext-policy attribute-based encryption. In: IEEE Symposium on Security and Privacy (SP 2007), Berkeley, CA, pp. 321–334 (2007)
8. Ostrovsky, R., Sahai, A., Waters, B.: Attribute based encryption with non monotonic access structures. In: Proceedings of the 14th ACM Conference on Computer and Communications Security (CCS 2007), pp. 195–203. ACM, New York (2007)
9. Rao, Y.S., Dutta, R.: Decentralized ciphertext-policy attribute-based encryption scheme with fast decryption. In: De Decker, B., Dittmann, J., Kraetzer, C., Vielhauer, C. (eds.) Communications and Multimedia Security, CMS 2013. Lecture Notes in Computer Science, vol. 8099. Springer, Heidelberg (2013)
10. Karp, A.H., Haury, H., Davis, M.H.: From ABAC to ZBAC: the evolution of access control models. Technical report HPL-2009-30, HP Labs (2009)
11. Ferraiolo, D., Kuhn, D.R.: Role-based access control. In: NIST-NSA National (USA) Computer Security Conference, pp. 554–563 (1992)
12. Rissanen, E., Brossard, D., Slabbert, A.: Distributed access control management – a XACML-based approach. In: Baresi, L., Chi, C.H., Suzuki, J. (eds.) Service-Oriented Computing. ServiceWave 2009, ICSOC 2009. Lecture Notes in Computer Science, vol. 5900. Springer, Heidelberg (2009)
13. Chen, Y., Lu, J., Yu, X., Hill, D.J.: Multi-agent systems with dynamical topologies: consensus and applications. IEEE Circuits Syst. Mag. **13**(3), 21–34 (2013)
14. Olfati-Saber, R., Murray, R.: Consensus problems in networks of agents with switching topology and time-delays. IEEE Trans. Autom. Control **49**, 1520–1533 (2004)
15. Olfati-Saber, R., Fax, J., Murray, R.: Consensus and cooperation in networked multi-agent systems. Proc. IEEE **95**, 215–233 (2007)

16. JAVA agent development framework homepage. http://jade.tilab.com. Accessed 15 Jan 2019
17. Charm: a framework for rapidly prototyping cryptosystems. https://github.com/JHUISI/charm. Accessed 15 Jan 2019
18. CodeXL. https://github.com/GPUOpen-Tools/CodeXL. Accessed 15 Jan 2019
19. Lewko, A., Waters, B.: Decentralizing attribute-based encryption. In: Paterson, K.G. (ed.) Advances in Cryptology – EUROCRYPT 2011. EUROCRYPT 2011. Lecture Notes in Computer Science, vol. 6632. Springer, Heidelberg (2011)

Development and Analysis of Diagnostic Models of Electrical Machine Windings

Evgeniy Bobrov, Aleksandr Saushev$^{(\boxtimes)}$,
Aleksandr Monahov, and Aleksandr Chertkov

Admiral Makarov State University of Maritime and Inland Shipping,
5/7, Dvinskaya Street, Saint Petersburg 198035, Russia
saushev@bk.ru

Abstract. It is shown that the reliability of electrical machines is mainly determined by the state of insulation of electrical windings. Most often, the stator windings break down. It has been established that the main causes of the loss of operability of the windings of electrical machines are their breakage, short circuit and decrease in insulation resistance. The equivalent circuits of the windings are considered, which make it possible to compose mathematical models and solve the main problems of technical diagnostics—assessment of the state of the winding and the search for a possible defect. The conducted experiments and the carried out analysis confirmed the correctness of the choice of equivalent circuits for the windings of electrical machines necessary for solving the problems of their technical diagnostics.

Keywords: Winding · Electrical machine · Equivalent circuit ·
Diagnostic model

1 Introduction

An analysis of the literature [1–3] shows that the greatest proportion of motor failures are faults in the stator windings (interturn faults, wire breaks, etc.). This conclusion is confirmed by numerous studies [1] showing the percentage ratio of common damage to induction motors: damage to the stator elements - 38%; damage to the rotor elements - 10%; damage to bearing elements - 40%; other damages - 12%.

In [4], a more detailed analysis of the main sources of development of damages to an induction motor is given: overload or overheating of the stator of the electric motor - 31%; interturn fault - 15%; bearing damage - 12%; damage to the stator windings or insulation - 11%; uneven air gap between the stator and the rotor - 9%; motor operation on two phases - 8%; breakage or loosening of the rod mounting in the rotor windings - 5%; loosening the stator windings - 4%; motor rotor imbalance - 3%; shaft misalignment - 2%.

It is obvious that all possible malfunctions of electrical machines can be divided into mechanical and electrical.

Mechanical faults include: incorrect operation of bearings; damage to the rotor shaft (anchor); loosening of brush holder fingers; the formation of deep wear outs on the surface of the reservoir and slip rings; loosening the mounting of the poles or stator core to the frame; cracks in bearing shields or in the frame, etc.

© Springer Nature Switzerland AG 2020
V. Murgul and M. Pasetti (Eds.): EMMFT 2018, AISC 982, pp. 439–447, 2020.
https://doi.org/10.1007/978-3-030-19756-8_42

Electrical faults include: interturn faults; breaks in the windings; insulation breakdown; the unsoldering of winding connections with a collector; incorrect polarity of the poles; wrong connections in coils, etc.

2 Analysis of the Winding of an Electrical Machine as an Object of Diagnosis

The main reasons for the loss of performance of the windings of electrical machines are: breakage due to high mechanical stresses during winding; breakage due to high temperature stresses during winding; breakage due to high mechanical stresses caused by thermal expansion of the frame; breakage due to corrosion in exposed areas of insulated wire; breakage due to electrolysis in the exposed areas of insulated wire; short circuit of turns or layers due to mechanical or thermal breakdown of wire insulation; insulation breakdown due to overvoltage or external wire insulation; decrease in insulation resistance due to deterioration of its insulation properties on temperature, humidity, pressure, time [5–7]. The analysis performed allows us to conclude that the reliability of electrical machines is largely determined by the reliability of its windings, which, in turn, largely depends on the state of the insulation.

Statistical studies of EPRI (Electrical Power Research Institute, USA) show that more than 37% of failures and breakdowns of synchronous generators and electric motors are associated with defects in electrical insulation of stators and rotors. The main mechanisms of insulation aging during the operation of electrical machines are temperature exceedance, cyclic loads, abrasion of the semiconducting coating of coil sections or rods as a result of vibration, separation of insulation, leakage from contamination.

Damage to the insulation of working electrical machines is subject to four main factors: thermal, electrical, mechanical, and atmospheric. Insulation of windings is subject to the influence of these factors in varying degrees. The intensity of impacts depends on the operating voltage, operating modes, and many other design and operational reasons. Under operational conditions, the intensity of impacts is often unstable. The dynamics of the effects affects the aging process of insulation.

The main effect on insulation is thermal effect. Heat is generated as a result of electrical losses in conductors when current flows, steel losses and mechanical losses. The amount of overheating depends on the class of insulation of the machine winding and the overload during its operation. With increasing temperature decreases the mechanical strength of the insulation. With temperature expansion of the insulation material, its structure is weakened, air cavities increase, internal mechanical overloads and overvoltages occur. Especially large mechanical stresses occur in rigidly coupled insulation systems with significantly different thermal expansion coefficients.

In the process of aging of the insulation, decay products can accumulate, leading to the formation of gas bubbles, which reduces the breakdown voltage of the insulation and mechanical strength. Thermal aging makes the insulation vulnerable to mechanical stress. When mechanical strength or elasticity is lost, the insulation is not able to withstand the usual conditions of vibration or shock, moisture penetration, and the difference in thermal expansion and contraction of copper, steel, and other materials. Exposure to heat over time leads to loosening of the coil mounts, wedges, slotted gaskets, and other fasteners, which makes the winding susceptible to damage due to relatively weak mechanical effects.

When electric motors are started up, the current values are usually 5–7 times higher than the nominal values, and significant electrodynamic forces are generated in the winding, which are proportional to the square of the current, which causes deformation of the windings, displacement of turns in the presence of mounts loosening and damage to the turn insulation. The resulting center of destruction of the turn insulation progresses to a short circuit between the turns and leads to a short circuit on the frame.

The mechanical forces to which the windings of rotating electrical machines are exposed are the result of centrifugal forces, inrush currents, vibrations, and mechanical shock.

During the operation of electric motors, the casing and motor windings are vibrated due to unbalance of the rotor (anchor), bearing wear and other reasons. In the initial period of operation, the vibration of electric motors does not significantly affect the reliability of the windings, since the impregnating varnish cements the winding. Subsequently, with the change in the mechanical properties of insulation materials on impregnating varnishes due to thermal aging, the effect of vibrations on the reliability of the windings and lead wires becomes noticeable.

Often, insulation damage is a result of its interaction with the environment. In the most difficult conditions is insulation exposed to atmospheric effects and, in particular, the action of moisture, which depends on the ambient temperature.

The presence of moisture in the pores significantly reduces the mechanical and electrical strength of the insulation, enhances the effect of ionization phenomena, accelerates the aging of some materials (for example, cellulose), increases dielectric loss, which reduces the service life and increases the accident rate of electrical machines. Moisture is especially dangerous when it is combined with contaminated insulation surfaces or with chemically active and conductive substances, which is typical, for example, for dry cargo and bulk cargo ships.

Thus, the aging processes of the insulation of electrical machines are interconnected and activate each other. The destruction of the insulation occurs gradually, and it begins due to the process of thermal aging.

As a conclusion, let us consider the main ways to improve the reliability of the windings of electrical machines:

improvement of the thermal state of electrical machines by switching to a higher class of heat resistance of insulation, equalizing the temperature of individual parts of the machine due to the optimal choice of loads, developing cooling systems, applying overload protection;

development and implementation of measures to reduce vibrations both inside the machine and vibrations of the external affecting environment;

improving the quality of components and materials, for example: the introduction of modern impregnating varnishes with high heat resistance, the use of special wires with durable and elastic insulation, reducing the rigidity of the winding wires, etc.;

expanding the automation of technological and production processes and ensuring high quality of operations;

systematic study of the operating conditions and operational reliability of electrical machines, determining the nature, causes and laws of the distribution of failures.

3 Mathematical Model of the Winding

The equivalent circuit of the winding of an electrical machine must ensure with the required accuracy the solution of the main problems of technical diagnostics - determination of the technical state of the winding and the search for a defect, if the operating conditions are not met. To substantiate the choice of a winding equivalent circuit, it is sufficient for a specific type of electrical machines to derive the equation for the transient response in the winding when a test signal is applied to its input, and to determine the parameters of the equation experimentally. Most often, rectangular voltage pulses with steep waves are used as test signals. During transient responses resulting from the supply of pulsed voltages to the input of the windings of electrical machines, the latter behave simultaneously as circuits with distributed and lumped parameters. As a result, depending on the task, transient responses in the windings should be considered on equivalent circuits with either distributed or lumped parameters.

In the general case, the windings of electrical machines with respect to the frame are complex non-linear circuits. However, as shown by special studies [8], the windings of electrical machines can be considered linear circuits, the parameters of which do not depend on the amplitude of the incident wave. When considering impulse processes in the windings, the latter can be represented as long lines, which can be replaced by valuable L-shaped circuits. An example of such a circuit is presented in Fig. 1, where the complex resistances Z_2, Z_4, ..., Z_n take into account the frame winding parameters, and the complex resistances Z_1, Z_3, ..., Z_{n-1} – turn parameters. In this case, the output pulse signal $\dot{U}_1$ is fed to one of the winding outputs (terminal 1) and the machine frame (terminal 2), and the signal $\dot{U}_n$ is removed from the other winding output (terminal 3) relative to the machine frame (terminal 4).

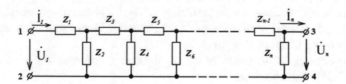

Fig. 1. Recurrent circuit of the replacement of the electric machine winding.

Using the methods of analyzing electrical circuits, for example, the method of loop currents or by recording the input resistance, it is possible to determine the type of transfer function of this circuit.

To select the equivalent circuit of the winding of an electric machine, we use the principle of sequential complication of the mathematical model [9], which involves choosing the simplest model at the first stage of the iterative search process, and, if it is inadequate, moving on to the next, more complex model. The analysis of literature shows that, in the first approximation, a winding equivalent circuit can be represented as a second-order dynamic link.

Let us consider the proposed approach on the example of tachometer generator of TMG-30 type. Studies were performed on the excitation windings of these machines. The equivalent circuit is shown in Fig. 2. It consists of two identical links connected in series. This is explained by the fact that the excitation winding of the TMG-30 machine consists of two sections located at different poles and connected in series.

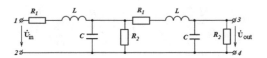

Fig. 2. The equivalent circuit of the TMG-30 tachometer generator.

A more compact scheme for replacing the excitation winding of such a machine can be represented as a quadrupole, as shown in Fig. 3.

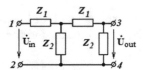

Fig. 3. Simplified equivalent circuit of the TMG-30 tachometer generator

Here, $Z_1(p)$ and $Z_2(p)$ — operator resistances of a single-link circuit:

$$Z_1(p) = R_1 + pL; \quad Z_2(p) = \frac{R_2}{R_2 cp + 1}$$

It can be shown that the voltage value at the output of the equivalent circuit when applying voltage spike to its input will look like:

$$U_{out}(p) = U_{in}(p) \frac{Z_2(p)^2}{Z_1(p)[2Z_2(p) + Z_1(p)] + Z_2(p)[Z_1(p) + Z_2(p)]} \tag{1}$$

After substituting the values of $Z_1(p)$ and $Z_2(p)$ and corresponding transformations, we obtain

$$U_{out}(p) = U_{in}(p)[c^2 L^2 (p^4 + A_1 p^3 + A_2 p^2 + A_3 p + A_4)]^{-1},$$

where auxiliary complex resistances A_1, A_2, A_3, A_4 have the following form:

$$A_1 = \frac{2R_1}{L} + \frac{2}{R_2 C},$$

$$A_2 = \frac{R_1^2}{L^2} + \frac{4R_1}{R_2 CL} + \frac{1}{R_2^2 C^2} + \frac{3}{CL},$$

$$A_3 = \frac{2R_1^2}{R_2 CL^2} + \frac{2R_1}{R_2^2 C^2 L} + \frac{3R_1}{CL^2} + \frac{3}{R_2 C^2 L^2},$$

$$A_4 = \frac{R_1^2}{R_2^2 C^2 L^2} + \frac{1}{C^2 L^2} + \frac{3R_1}{R_2 C^2 L^2}.$$

As a result of the field experiment, the parameters of the excitation winding section were determined, which turned out to be equal to:

$$R_1 = 360 \text{ Ohm}, \quad L = 2 \text{ mH}, \quad R_2 = 0,5 \cdot 10^6 \text{ Ohm}, \quad C = 70 \cdot 10^{-12} \text{ F}.$$

Substituting these values into expression (1) and dividing both its parts by the input operator voltage $U_{in}(p)$, we obtain the transfer function in the form

$$\frac{U_{out}(p)}{U_{in}(p)} = \frac{1,9600003 \cdot 10^{20}}{p^4 + 0,57503 \cdot 10^5 p^3 + 22,2655 \cdot 10^9 p^2 + 6,16398 \cdot 10^4 p + 0,511306 \cdot 10^{20}}$$

The roots of the denominator of this function are

$$\left.\begin{array}{l} p_1 = -U_1 + jV_1 = -1,4376976 \cdot 10^4 + j5,02679 \cdot 10^4 \\ p_2 = -U_1 - jV_1 = -1,4376976 \cdot 10^4 - j5,02679 \cdot 10^4 \\ p_3 = -U_2 + jV_2 = -1,4374524 \cdot 10^4 + j13,60042 \cdot 10^4 \\ p_4 = -U_2 - jV_2 = -1,4374524 \cdot 10^4 - j13,60042 \cdot 10^4 \end{array}\right\}.$$

The system response to a single step for the resulting transfer function will take the form:

$$U_{out}(p) = \frac{F_1(p)}{pF_2(p)} = [c^2 L^2 p(p - p_1)(p - p_2)(p - p_3)(p - p_4)]^{-1}$$
$$= [c^2 L^2 p(p^2 + 2U_1 p + U_1^2 + V_1^2)(p^2 + 2U_2 p + U_2^2 + V_2^2)]^{-1},$$

where $F_1(p)$ and $F_2(p)$ – operator expressions in the form of polynomials, where degree of $F_1(p)$ not higher than degree of $F_2(p)$. In addition, the polynomial $F_2(p)$ does not have a root equal to zero.

To go from the image of $U_{out}(p)$ to the original $U_{out}(t)$, we use the decomposition theorem:

$$U_{out}(t) = \frac{1}{c^2 L^2} \left[\frac{F_1(0)}{F_2(0)} + \sum \frac{F(p_i)}{p_i F_2^1(p_i)} e^{p_i t} \right]$$

$$= \frac{1}{c^2 L^2} \left[\frac{1}{(p_{1,2})^2 (p_{3,4})^2} + \frac{e^{(-U_1 + jV_1)t}}{(-U_1 + jV_1)F_2^1(p_1)} + \frac{e^{(-U_1 - jV_1)t}}{(-U_1 - jV_1)F_2^1(p_2)} \right. \tag{2}$$

$$\left. + \frac{e^{(-U_2 + jV_2)t}}{(-U_2 + jV_2)F_2^1(p_3)} + \frac{e^{(-U_2 - jV_2)t}}{(-U_2 - jV_2)F_2^1(p_4)} \right].$$

Here, in general terms

$$F_2^1(p) = (p - p_2)(p - p_3)(p - p_4) + (p - p_1)(p - p_3)(p - p_4)$$
$$+ (p - p_1)(p - p_2)(p - p_4) + (p - p_1)(p - p_2)(p - p_3)$$

For the case under consideration

$$F_2^1(p_1) = (p_1 - p_2)(p_1 - p_3)(p_1 - p_4) = 2(A + jB)jV_1;$$

$$F_2^1(p_2) = (p_2 - p_1)(p_2 - p_3)(p_2 - p_4) = -2(A - jB)jV_1;$$

$$F_2^1(p_3) = (p_3 - p_1)(p_3 - p_2)(p_3 - p_4) = 2(C + jD)jV_2;$$

$$F_2^1(p_4) = (p_4 - p_1)(p_4 - p_2)(p_4 - p_3) = -2(C + jD)jV_2.$$

Let us perform transformations of expression (2). We do this in pairs, first for the second and third members, and then for the fourth and fifth.

The expression for the second and third members will look like:

$$\frac{e^{(-U_1 + jV_1)t}}{(-U_1 + jV_1)2jV_1(A + jB)} - \frac{e^{(-U_1 - jV_1)t}}{(-U_1 - jV_1)2jV_1(A - jB)} = e^{-U_1 t} \frac{e^{jV_1 t} R e^{j\varphi_1} - e^{-jV_1 t} R e^{-j\varphi_1}}{j2}$$

$$= Re^{-U_1 t} \frac{e^{j(V_1 t + \varphi_1)} - e^{-j(V_1 t + \varphi_1)}}{j2} = R_I e^{-U_1 t} \sin(V_1 t + \varphi_1),$$

where $\frac{1}{(-U_1 + jV_1)V_1(A + jB)} = K + jM = R_I e^{j\varphi_1}$; $\frac{1}{(-U_1 - jV_1)V_1(A - jB)} = K - jM = R_I e^{-j\varphi_1}$.

Performing similar transformations for the fourth and fifth members, we get:

$$\frac{e^{(-U_2 + jV_2)t}}{(-U_2 + jV_2)F_2^1(p_3)} - \frac{e^{(-U_2 - jV_2)t}}{(-U_2 - jV_2)F_2^1(p_4)} = R_{II} e^{-U_2 t} \sin(V_2 t + \varphi_2),$$

where $\frac{1}{(-U_2+jV_2)V_2(c+jD)} = R_{II}e^{j\varphi_2}$; $\frac{1}{(-U_2-jV_2)V_2(c-jD)} = R_{II}e^{-j\varphi_2}$.

Thus, after transformations, the output voltage $U_{out}(t)$ will be:

$$U_{out}(t) = 1 + \frac{R_I}{c^2L^2}e^{-U_1t}\sin(V_1t+\varphi_1) + \frac{R_{II}}{c^2L^2}e^{-U_2t}\sin(V_2t+\varphi_2).$$

After substituting the corresponding numerical values, we get:

$$U_{out}(t) = 1 - 1{,}213156\,e^{-1{,}4377\times10^4t}\sin(5{,}027\cdot10^4t+77{,}64°)$$
$$+ 0{,}1717576\,e^{-1{,}4374\times10^4t}\sin(13{,}600\cdot10^4t+83{,}97°).$$

To compare the type of transient processes obtained by calculation on the basis of the chosen model of the equivalent circuit of an electrical machine winding and experimentally (according to the scheme of Fig. 3), we compare their damping decrements χ and the natural oscillation periods Tn. The results of the comparative analysis are given in the Table 1.

Table 1. Results of a comparative analysis of the results of calculation and experiment.

Calculated values	Experimental values
$1/\chi_c = 0.34$	$1/\chi_e = 0.38$
$Tn_c = 115$ (µs)	$Tn_e = 120$ (µs)

Let's determine the relative error of the obtained results:

$$\delta_1 = \frac{1/\chi_e - 1/\chi_c}{1/\chi_e}\cdot100 = 3{,}44\%; \; \delta_2 = \frac{Tn_e - Tn_c}{Tn_e}\cdot100 = 4{,}12\% \;.$$

The analysis shows that the relative errors do not exceed 5%. This circumstance allows us to conclude about the possibility of using the chosen equivalent circuit for solving the considered problems of technical diagnostics of the winding of an electric machine.

4   Conclusion

Electrical machines are the most important element of electric motor drives, which are widely used in industry and transport. The reliability of electrical machines is largely determined by the technical condition of their windings and, above all, the condition of the winding insulation. Diagnosing the technical condition of the windings of electrical machines is the most important way to increase their reliability. The equivalent circuit of a winding of an electrical machine, considered in this paper, allows with high confidence solving the problem of assessing the state of windings during operation of an electric motor and the problem of finding a defect in case of loss of its performance due to a sudden or gradual failure.

References

1. Petukhov, V.: Diagnostics of the state of electric motors. The method of spectral analysis of current consumption. Electr. Eng. News **1**(31), 23–28 (2005)
2. Fedotov, M.M., Tkachenko, A.A.: On the issue of building fault diagnosis systems for asynchronous electric motors. Electr. Eng. Electromechanics **2**, 59–61 (2006)
3. Manohar, V.G., Kumar, P.: Comprehensive predictive maintenance of electrical motors in Indian nuclear power plants. Int. J. Nucl. Power **17**, 1–3 (2003)
4. Shevchuk, V.A., Semenov, A.S.: Comparison of asynchronous motor diagnostic methods. Int. Student Sci. J. **3**(4), 419–423 (2015)
5. Sedunin, A.M., Afanasyev, D.O., Sidelnikov, L.G.: Diagnostic methods for asynchronous motors. OOO TestService, Perm (2012)
6. Sidelnikov, L.G., Afanasyev, D.O.: Review of methods for monitoring the technical condition of asynchronous motors in operation. Perm. J. Petrol. Mining Eng. **7**, 127–137 (2013)
7. Saushev, A.V., Gasparyan, K.K.: Prediction of the state of electrical systems based on information technology. International Research Journal: a collection of the results of the International Research Journal Conference VII. IRJ, Yekaterinburg, vol. 2(1), pp. 93–95 (2015)
8. Meshgin, H., Milimonfared, J.: Effects of air-gap eccentricity on the power factor of squirrel cage induction machines. In: International Conference on Electrical Machines (ICEM-2002), Old St. Jan Conference Center, Belgium (2002)
9. Saushev, A.V.: Planning of an Experiment in Electrical Engineering. Petersburg State University of Water Communications, Saint-Petersburg (2012)

Parametric Identification of Electric Drives Based on Performance Limits

Aleksandr Saushev$^{(\boxtimes)}$ ⓘ, Svyatoslav Antonenko ⓘ,
Alexey Lakhmenev ⓘ, and Aleksandr Monahov ⓘ

Admiral Makarov State University of Maritime and Inland Shipping,
5/7, Dvinskaya Street, Saint Petersburg 198035, Russia
saushev@bk.ru

Abstract. Identification methods for the dynamic systems are considered with respect to electric drives. It is shown that electric drives are complex electromechanical systems, the mechanical part of which is multimass and contains elements with elastic connections. The features of these systems that require considering the gaps in gear units when solving modeling and identification problems; elasticity arising in the shafts; stiffness change of elastic elements in couplings. It is noted that the identification problem of electric drives is a subproblem of managing the state of electromechanical systems. The basis of methods and algorithms for its solution is a set is information about the performance limits of the electric drive. It is shown that identification of the controlled parameters of the electric drive requires using statistical methods based on the expansion of the pulse transient functions to a series of the orthonormal Laguerre transformed system of functions. The method has been proposed in order to assess the condition of the electric drive during operation. This method involves setting the performance limits as a set of boundary points. A distinctive feature of the method is the high identification accuracy.

Keywords: Electric drive · Performance limits · Working capacity ·
Identification · State assessment

1 Introduction

The identification of a technical object is usually assumed as defining the structure and parameters of its analytical model. The parameters ensure the maximum closeness of the model output values and the object itself with the same input effects [1]. The problem of identifying dynamic systems, including the electric drives, usually comes down to defining the system model's operator, which transforms its input actions into output values. Both deterministic and statistical methods for processing observational data are used for the purposes of identification [1].

The parametric identification problem regarding electric drives is solved in the framework of the relevant and currently developing scientific field, the purpose of which is to ensure the capacity and increase the efficiency of electromechanical systems (EMS) at all life cycle stages by parametric control of their state.

© Springer Nature Switzerland AG 2020
V. Murgul and M. Pasetti (Eds.): EMMFT 2018, AISC 982, pp. 448–458, 2020.
https://doi.org/10.1007/978-3-030-19756-8_43

In order to solve the problems regarding parametric control of the electric drive's state, which is the most important EMS, the necessary condition is to obtain information about the performance limits [2]. The problem of building and analyzing the performance limits plays an important role in the process of designing and operating electric drives and their elements. Information about the boundary of this area allows to successfully solve the problem of parametric synthesis, determination and predicting of the technical state of EMS.

The most urgent problems regarding parametric identification of electric drives are considered in the paper. They are solved during the drive's operation, like the problem of calculating the parameters of the electric drive elements in real time, as well as recognizing the electric drive's state based on the calculated values of its parameters. In this case, it is assumed that the working conditions in the form of a system of inequalities are defined for the electric drive, which determines the allowable limits of variation of its parameters.

2 Methods and Materials

The electric drive is a complex electro mechanic system designed to set in motion the operating member of the working machine (OMWM) and control this movement. The elements of the electric drive are as following: an electrical energy converter (EEC), an electromechanical converter (EMC), a mechanical transducer (MT) and a control system of the electric drive (CSED), which includes an interface device, an information device (ID) and a control device (CD) (Fig. 1).

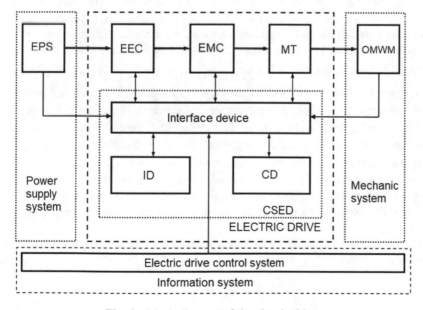

Fig. 1. Block diagram of the electric drive.

Figure 1 also represents the EPS – electric power source, OMWM – operating member of the working machine and the electric drive control system, which is external to the drive system of a higher hierarchical level.

The exploitation of general industrial electric drives is usually characterized by the following features, the presence of random factors determining the external environment and technological process; limited range of changes in design and performance parameters that affect quality indicators; the continuity of the main processes in the drive. Electric drives of transport machines and mechanisms are characterized by variable and complex operating conditions, large circuit diversity, lack of statistical information about the condition of their elements, and self-containment. The following factors should be taken into account for modeling and identifying most of electric drives: gaps in gears; elasticity in the shafts; change of the inertia moments in gear rims; stiffness change of elastic elements in elastic couplings. It should also be considered that the mechanical part of an electric drive generally is a multi-mass branched mechanical circuit with connections with parameters needed to be identified [3–5].

The state of the drive at any fixed time point is characterized by a certain set or vector of parameters, such as following:

input parameters $u = (u_1, u_2, \ldots, u_g, \ldots, u_e)$, which characterize the stimulus commands u(t) and observed at the system inputs. There are control commands, which characterize the operating modes of the drive, and test commands, which take place in the modes of adjustment and technical diagnostics;

external parameters $V = (v_1, v_2, \ldots, v_\rho, \ldots, v_f)$, which characterize the properties of the external environment affecting drive's functioning;

internal parameters $X = (X_1, X_2, \ldots, X_i, \ldots, X_n)$, which characterize the state of the drive's components and are also called primary parameters. They include both the parameters of the elements themselves (resistances, inductances, capacitances, masses, inertia moments, stiffness of elastic connections), and functions of these parameters that have a certain physical meaning (gain rates, time constants, mass ratio coefficients);

internal parameters $u^v = (u_1^v, u_2^v, \ldots, u_i^v, \ldots, u_e^v)$, $Z^v = (Z_1^v, Z_2^v, \ldots, Z_j^v, \ldots, Z_c^v)$ which characterize the input and output signals of electrical devices included as elements, respectively $v = \overline{1, h}$, h is the number of elements in the electric drive;

output parameters $Y = (Y_1, Y_2, \ldots, Y_j, \ldots, Y_m)$, characterizing the properties of the drive, which are interesting for the consumer. These are functional parameters, i.e. the functional correlations of the drive's phase variables $Z = (Z_1, Z_2, \ldots, Z_g, \ldots, Z_d)$, $d = c \cdot v$ and the parameters that are the boundary values of external variables. The efficiency of the system is preserved between them. These parameters usually indicate the quality of the drive. They include, for example, the speed control range, efficiency, transient time.

The connection of the drive's outputs with its inputs can be represented by the equation

$$Y = F(X, u, Z, V, t), \tag{1}$$

where F is a communication statement. By the form of the operator F, the elements of the electric drive can be divided into the following: continuous and discrete, linear and nonlinear, stationary and non-stationary, deterministic and stochastic, with lumped parameters and with distributed parameters. The elements of the electric drive with continuous and deterministic operator F are considered.

3   Identifying the Primary Parameters of the Electric Drive

An analysis of the existing methods for identifying dynamic systems [6] showed that fairly simple methods of identification in the time and frequency domains can be used to estimate the values of the primary parameters of the drive's elements. Increasingly widespread use is found in the statistical identification in the basis of orthogonal functions. As applied to drive control systems, the method of calculating the expansion coefficients of their impulse transition functions in a series using a system of orthonormal transformed Laguerre functions should be used [7]. In this case, the pulse transient function of the electric drive is represented by the model as a following series:

$$\omega(\tau) = \sum_{j=0}^{\infty} \beta_{j\alpha} l_{j\alpha}(\tau) \tag{2}$$

where $\beta_{j\alpha}$ is the expansion coefficient, $l_{j\alpha}(\tau)$ are Lagger basis functions.
    Based on the formula (2), the following can be obtained:

$$\beta_{k\alpha} = \int_{0}^{\infty} \omega(\tau) l_{k\alpha}(\tau) d\tau \tag{3}$$

The formulas allowing us to determine unknown parameters of the electric drive can be obtained using coefficients $\beta_{j\alpha}$:

$$\beta_{k\alpha} = \sqrt{2\alpha} \sum_{j=0}^{k} \frac{C_k^j}{j!} (2\alpha)^j W^{(j)}(\alpha); \quad W^{(j)}(\alpha) = \frac{d^j}{dp^j} W(p)_{p=\alpha} \tag{4}$$

Then, the expressions for the first four $\beta_{j\alpha}$ coefficients of the expansion (4) are:

$$\begin{cases} \beta_{0\alpha} = \sqrt{2\alpha} W(\alpha); \\ \beta_{1\alpha} = \sqrt{2\alpha}[W(\alpha) + 2\alpha W'(\alpha)]; \\ \beta_{2\alpha} = \sqrt{2\alpha}[W(\alpha) + 4\alpha W'(\alpha) + 2\alpha^2 W''(\alpha)]; \\ \beta_{3\alpha} = \sqrt{2\alpha}[W(\alpha) + 6\alpha W'(\alpha) + 6\alpha^2 W''(\alpha) + (4/3)\alpha^3 W'''(\alpha)], \end{cases} \tag{5}$$

Where W', W'', W''' – the first, second and third derivative of the transfer function, respectively.

Suppressing the index α with the coefficients $\beta_{j\alpha}$, the expressions for the derivatives can be obtained, provided that k = 0, 1, 2, 3:

$$
\begin{cases}
W(\alpha) = \frac{1}{\sqrt{2\alpha}} \beta_0; \\
W'(\alpha) = \frac{1}{2\alpha\sqrt{2\alpha}} (-\beta_0 + \beta_1); \\
W''(\alpha) = \frac{1}{2\alpha^2\sqrt{2\alpha}} (\beta_0 - 2\beta_1 + \beta_2); \\
W'''(\alpha) = \frac{3}{4\alpha^3\sqrt{2\alpha}} (-\beta_0 + 3\beta_1 - 3\beta_2 + \beta_3).
\end{cases}
\tag{6}
$$

The method of determining the functional correlation between the expansion coefficients of the pulse transient function $\beta_{j\alpha}$ and the parameters of the electric drive lies in finding the correlations between these parameters and the derivatives of the transfer function. According to this, at $p = \alpha$, and after the subsequent representation of the derivatives $W^{(k)}(\alpha)$ through the coefficients $\beta_{j\alpha}$ in accordance with formula (6), the following can be obtained:

The DC motor of independent excitation was considered as an example. It is the most important element of the electric drive. The equations describing an electric motor are as follows:

$$
\begin{cases}
U = iR + L\frac{di}{dt} + c\omega; \\
M = ci; \\
J\frac{d\omega}{dt} = M - M_r
\end{cases}
\tag{7}
$$

where c is the machine constant, Ua is the armature winding voltage, iя is the armature current, Lя is the armature winding inductance, Ra is the armature winding active resistance, M is the electromagnetic moment of the machine, ω is the angular velocity, J is the rotor inertia moment of the motor, Mc - the moment of resistance on the motor shaft, taken to be zero.

In the operator form, the entries of Eq. (7) will take the following form:

$$
\begin{cases}
U(p) = (R + Lpi(p)) + c\omega(p); \\
M(p) = ci(p); \\
Jp\omega = M - M_r.
\end{cases}
\tag{8}
$$

A structural diagram of the electric drive be compiled on the basis of above mentioned equations, and its transfer function was determined according to the control action U_a.

$$
W(p) = \frac{K}{1 + \tau_1 p + \tau_2 p^2},
$$

where $K = 1/c$, $\tau_1 = T_m = \frac{RJ}{c^2}$, $\tau_2 = T_e T_m = \left(\frac{L}{R}\right) \cdot \left(\frac{RJ}{c^2}\right) = \frac{LJ}{c^2}$; T_e, T_m – the electromagnetic and electromechanical constants of the electric motor, respectively. Let us express the parameters K, τ_1, τ_2 of the transfer function in terms of the $\beta_{j\alpha}$ coefficients. The following notations are introduced:

$$\begin{cases} 1 + \tau_1\alpha + \tau_2\alpha^2 = A_0; \\ \tau_1 + 2\tau_2\alpha = A_1; \\ 2\tau_2 = A_2, \end{cases} \tag{9}$$

After excluding the coefficients τ_1, τ_2 from the system of Eqs. (9):

$$2A_0 - 2\alpha A_1 + \alpha^2 A_2 = 2 \tag{10}$$

The equality from the expression of the transfer function:

$$W(p)(1 + \tau_1 p + \tau_2 p^2) = K.$$

By successively differentiating this expression and replacing p with α, we obtain as follows:

$$\begin{cases} W(\alpha)A_0 = K; \\ W'(\alpha)A_0 + W(\alpha)A_1 = 0; \\ W''(\alpha)A_0 + 2W'(\alpha)A_1 + W(\alpha)A_2 = 0; \\ W'''(\alpha)A_0 + 3W''(\alpha)A_1 + 3W'(\alpha)A_2 = 0. \end{cases} \tag{11}$$

The desired parameters of the electric drive can be expressed through auxiliary functions A_0, A_1, A_2 as follows:

$$K = A_0 W(\alpha); \quad \tau_1 = A_1 - \alpha A_2; \quad \tau_2 = A_2/2. \tag{12}$$

After considering a system of linear algebraic equations:

$$\begin{cases} W'(\alpha)A_0 + W(\alpha)A_1 = 0; \\ W''(\alpha)A_0 + 2W'(\alpha)A_1 + W(\alpha)A_2 = 0; \\ 2A_0 - 2\alpha A_1 + \alpha^2 A_2 = 2, \end{cases}$$

and solving this system with respect to the unknown functions A_0, A_1, A_2, the following expressions (the argument for the derivatives is omitted for short) obtained as follows:

$$A_0 = \frac{2W^2}{F[\alpha, W, W']}; \quad A_1 = \frac{-2WW'}{F[\alpha, W, W']}; \quad A_2 = \frac{2\left[2(W')^2 - WW''\right]}{F[\alpha, W, W']};$$

$$F[\alpha, W, W'] = 2W^2 + 2\alpha WW' - \alpha^2 WW'' + 2\alpha^2 W'^2.$$

Substituting the determined values of the functions A_0, A_1, A_2, into the formula (12) and taking into account the system of Eqs. (6), we obtain:

$$K = \frac{\beta_0^3}{\sqrt{2}\alpha \cdot F[\beta_j]}; \quad \tau_1 = \frac{3\beta_{01}^2 - 5\beta_0\beta_1 + 2\beta_0\beta_2 + 2(\beta_1 - \beta_0)}{\alpha F[\beta_j]};$$

$$\tau_2 = \frac{-\beta_0 + \beta_1 + \beta_0^2 - 2\beta_0\beta_1 + \beta_0\beta_2}{\alpha^2 F[\beta_j]}$$

In these equalities $F[\beta_j] = 4\beta_0^2 + (\beta_0 - \beta_1)^2 - \beta_0(3\beta_0 - 4\beta_1 + \beta_2)$.

The fourth equation of system (11) can be used to find the interconnection of the coefficients $\beta_{j\alpha}$.

4 Identifying the State of a Drive

By analogy with the EMS, the concept of stamina of the electric drive is injected. Stamina is considered as the approximation degree of the vector X_t of the actual drive's state to its maximum limit value X_{lim}. The set of maximum permissible values of the vector X_{lim} is determined by the limit of the drive's stamina area. The approximation degree of a vector X_t is determined by the distance from its end to the nearest boundary point of this area [8, 9].

A point on the boundary of the stamina area is expressed through $\mathbf{X}_\Gamma = [X_{\Gamma 1}, X_{\Gamma 2}, \ldots, X_{\Gamma h}]$. The minimum distance between the vector of primary parameters $X_t = [X_{1t}, X_{2t}, \ldots, X_{nt}]$ and the vector $\mathbf{X}_\Gamma$ over all values of the boundary points will determine the stamina of the drive. If the nominal values of the primary parameters $X_n = [X_{1n}, X_{2n}, \ldots, X_{nn}]$ are determined, then the nominal stamina ρ_n can also be determined. In this case it is convenient to express stamina in relative units $\lambda(X)$:

$$\rho = \min_{[X_b]} \sqrt{\sum_{i=1}^{n} (X_i - X_{bi})^2}, \quad \rho_o = \min_{[X_b]} \sqrt{\sum_{i=1}^{n} (X_{io} - X_{bi})^2}, \quad \lambda(X) = \rho/\rho_o.$$

Such definition of stamina allows taking into account both external and internal conditions of the drive's workability [10].

The stamina area G is prescribed in the space of internal parameters of the electric drive X and is determined by its workability conditions:

$$Y_{jmin} \leq Y_j = F_j(\mathbf{X}) \leq Y_{jmax}, j = \overline{1, m};$$
$$Z_{gmin} \leq Z_g = F_g^v(\mathbf{X}) \leq Z_{gmax}, d = \overline{1, g}; \qquad (13)$$
$$X_{imin} \leq X_i \leq X_{imax}, i = \overline{1, n}.$$

The diagnostic parameters in the paper are the primary parameters X. This is because the primary parameters determine the state of the drive. In addition, it is possible to determine stamina, which is the most important indicator of the electric drive, only in the

X parameter space. In order to assess the state of the drive in the space of primary parameters, it is necessary to establish whether the measured values of these parameters are in the stamina area or not.

The stamina area can be specified as a set of boundary points or in the analytical form. In order to set the stamina area with an array of boundary points, methods and algorithms implementing them have been developed [2], based on a discrete and continuous search for the coordinates of these points. For the technical implementation of methods for searching and storing boundary points, a sufficiently large memory amount is required. At the same time, the computational capabilities of modern computers make it possible to implement the developed algorithms for searching the boundary points with a sufficiently large number of primary parameters. To identify the state of the drive, it is preferable to represent the stamina area in the set form of boundary points coordinates.

In order to solve this problem, the sequential iteration method can be used, which allows to assess the state of the electric drive. This method involves successive analysis of all stamina area boundary points and calculation of distances in the Euclidean space Rn between each boundary point and the current point Rt, which determines the drive's state at a given time. The shortest distance, provided that a point is in the stamina area, will determine the workability of the drive. The method requires large computational costs and time expenditures. When the number of diagnostic parameters n > 2, the use of this method to assess the drive's state is inexpedient.

In the absence of information characterizing the quantitative correlation between the primary and output parameters of the electric drive, in order to recognize the belonging of the point Rt to the stamina area, a full-scale experiment and corresponding measuring equipment are objectively required (13). Analysis has shown that it is possible to synthesize alternative, more convenient and effective recognition rules. For the considered case, when the stamina area is defined as a set of boundary points, one should use the methods of mechanical and electrical analogies [9], which are designed to solve the problem of parametric optimization of electrical systems.

The most general case is considered, when the stamina area has an arbitrary shape. The optimal internal point R0, while lacking the information about the process of changing the values of the drive's primary parameters, is determined during the search by the objective function, which has the following form [9]:

$$F = \frac{1}{N} \sum_{k=1}^{N} \left(1 \Big/ \sum_{i=1}^{n} (R_i - X_{ik})^2 \right), \tag{14}$$

where N is the total number of specified stamina boundary points; n is the number of primary parameters; X_{ik} – coordinates of the stamina boundary points.

Let us consider the algorithm for solving the problem. The values of the function F are calculated at different time points t, first at the points R_0 (if this value is not known a priori) and Rt. If $F(R_0) < F(R_t)$, then the point is outside the stamina area. Otherwise, additional analysis is required. If the function F monotonously decreases from the value $F(R_0)$ to the value $F(R_t)$ on the line segment connecting the points R_0 and R_t, then the studied point belongs to the stamina area and the drive is in a operating condition.

In order to verify this condition, it is enough to calculate the value of the function F at the point R_1 with coordinates different from the ones of R_t point by the value ΔX_i in the direction to the R_0 point. If $F(R_1) < F(R_t)$, then the point is in the stamina area. Indeed, at the point R_r, the function F undergoes a discontinuity (Fig. 2) and its value tends to infinity. Between the points R_0 and R_r and the points R_t and R_r, the function F is monotonously increasing when moving from the R_0 to the R_r and from the R_t to the R_r, as follows from expression (14). Thus, if the condition $F(R_1) < F(R_t)$ is not satisfied, then there is a contradiction to the original statement.

Further development of the method lies in reducing the time spent on its technical implementation. If the stamina area is convex, and the coordinates of the point R_0, which are optimal according to the stamina criterion, are known, these costs can be reduced by excluding boundary points from consideration. In this case they are redundant. The essence of the proposed algorithm is as follows. Mutually perpendicular planes H_1, H_2, ..., H_n are carried out through the point R_0, dividing the stamina area into 2^n subareas $(G_1, G_2, ..., G_2^n) \in G$. Then, the subarea $G_i \in G$ is determined, to which the point R_t belongs. For this purpose, the sign of each of the coordinates is determined: $(\text{sign} X_i > 0) \lor (\text{sign} X_i < 0), i = \overline{1, n}$, and the belonging of the R_t point to some subarea G_i is identified on the basis of the obtained information.

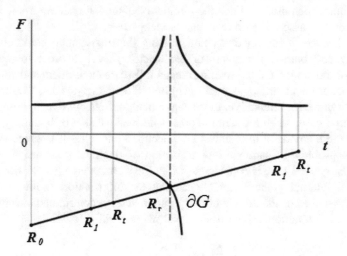

Fig. 2. Form of the F-criterion for recognizing the drive's status.

The boundary point of stamina area, which is most closely located to the R_t, and which determines the drive's state at the considered moment of time, is called the nearby point. It is possible to prove [9] that if the stamina area G in the n-dimensional space R^n of primary parameters, the R_0 point, which determines the maximum possible stamina l_0, and the point R_t corresponding to the drive's state at a given time are specified, then the nearby boundary point R_t^Γ does not belong to the subarea $G_i \in G$, the point coordinates $A_1, ..., A_N$ of which are opposite in sign to the R_t coordinates.

The considered algorithm allows to significantly reduce the number of boundary points of the stamina area that are used for analysis.

Quite often, the diagnostic parameters of the electric drive are the gain rates and time constants of its control system. Periodic identification of the values of these parameters and the evaluation of the drive's state by the considered methods allow increasing its operational reliability and efficiency.

5 Results and Discussion

The results obtained allow us to continuously calculate the values of diagnostic parameters, determine the stamina and evaluate the state of almost any electric drive, for which the stamina area is defined.

A distinctive feature of the considered algorithm for the identification of electric drives based on Lagger filters is its high accuracy due to the use of scaling factors for the decomposition of the function.

The proposed algorithm for assessing the state of the electric drive ensures high accuracy of identification. Information on the boundary of the drive's stamina area is required for its application, presented as an array of boundary points. To obtain such information, special methods, algorithms and software have been developed.

6 Conclusions

The process of managing the state of electromechanical systems involves solving a whole complex of analysis and synthesis problems. This includes problems of methodological support of this process, problems of structural and parametric synthesis, monitoring and forecasting the state of the system. The most important problem is to identify the state of electric drives during operation. Modern electric drives are complex electromechanical systems and find the widest application in industry and transport. Using information on the boundary of the stamina area for the purpose of identifying electric drives guarantees high reliability of the problem's solution and the fundamental possibility of calculating the stamina.

References

1. Alekseev, A.A., Korablev, Iu.A., Shestopalov, M.Iu.: Identifikatsiia i diagnostika sistem. Akademiia, Moscow (2009). (in Russian)
2. Saushev, A.V.: Oblasti rabotosposobnosti elektrotekhnicheskikh system. Politekhnika, SPb (2013)
3. Kuznetsov, N.K., Perelygina, A.Iu., Kononenko, R.V.: Identifikatsiia parametrov i modelirovanie dinamiki trekhmassovoi mekhatronnoi sistemy. Vestnik Irkutskogo gos. tekhnich. un-ta 3(43), 6–12 (2010). (in Russian)
4. Petrov, V.L.: Identifikatsiia modelei elektromekhanicheskikh sistem s ispolzovaniem s ispolzovaniem spektralnykh metodov analiza v bazisakh nepreryvnykh ortonormirovannykh funktsii. Mekhatronika, avtomatizatsiia, upravlenie 10, 29–36 (2003). (in Russian)

5. Rodkin, D.I., Romashikhin, Iu.V.: Energeticheskii metod identifikatsii elektromekhanich-eskikh ustroistv i system, Izvestiia VUZov i energeticheskikh obieedinenii SNG. Energetika **3**, 10–20 (2011). (in Russian)
6. Saushev, A.V., Troian, D.I.: Identifikatsiia elektroprivodov portovykh peregruzochnykh mashin. Vestnik GUMRF im. adm. Makarova S.O. **5**(33), 169–183 (2015). (in Russian)
7. Nivin, A.E., Saushev, A.V., Shoshmin, V.A.: Sintez ortogonalnykh filtrov pri statisticheskoi identifikatsii dinamicheskikh system. Priborostroenie **10**, 5–11 (2013). (in Russian)
8. Saushev, A.V., Kuznetsov, S.E., Karakayev, A.B.: System approach to ensure performance of marine and coastal electrical systems during operation. IOP Conf. Ser. Earth Environ. Sci. **194** (8), 082037. https://doi.org/10.1088/1755-1315/194/8/082037
9. Saushev, A.V.: Parametricheskii sintez elektrotekhnicheskikh ustroistv i system. GUMRF im. Adm. Makarova, S. O., Saint-Petersburg (2013). (in Russian)

Electro-Mobile Installation of the Vessel with Cascade Electric Converter

Fedor Gelder[ID], Izolda Li[ID], Aleksandr Saushev[✉][ID],
and Dmitry Semenov[ID]

Admiral Makarov State University of Maritime and Inland Shipping, 5/7,
Dvinskaya Street, Saint Petersburg 198035, Russia
{saushev, ep-gumrf}@bk.ru

Abstract. A perspective electric propulsion installation of high-capacity vessel is considered. The analysis of the existing structures of the electric propulsion systems is carried out. The advantages and disadvantages of the construction and operation of the existing systems of electric propulsion systems are considered. The idea of using a cascade frequency converter in the electric propulsion installation of the vessel was proposed, which will eliminate the cumbersome and heavy matching transformer from the electric drive power circuit. Various variants of the design of electric propulsion installation of a vessel with a cascade electric converter are presented. Recommendations for the implementation of the proposed structure are suggested. The advantages and the expected effect of the implementation of an electric propulsion vessel with a cascade electric converter are presented.

Keywords: Electric propulsion installation · Structural scheme ·
Electric drive · Generator set · Cascade frequency converter · Cell ·
Single-phase frequency converter · Switching frequency ·
Modulation frequency · Voltage quality · Reliability ·
Synthesized voltage levels

1 Introduction

At the moment, the Russian and world shipbuilding trends have seen a tendency to build special types of large capacity vessels using electric propulsion systems. The use of such systems allows obtaining the best performance and energy characteristics. The most full of the advantages of ships with electric propulsion systems are reflected in [1, 2]. The most frequently used electric propulsion system vessels are: icebreakers, tugs, ferries, fishing vessels, as well as ships and boats of the Navy. So for the development of the Northern Sea Route and the development of new Arctic territories on The Baltic plant today manufactures three nuclear icebreakers of project 22220 with a domestic electric propulsion system with an installed capacity of 60 MW (3 screws of 20 MW each) [3, 4]. In 2015 JSC CKB "Iceberg" in conjunction with the FSUE "Krylovsky State Scientific Center" in the framework of the federal tafgered program "Development of the Civil marine technology for 2009–2016 years" developed a conceptual design of the atomic icebreaker "Leader" with the total installed power

© Springer Nature Switzerland AG 2020
V. Murgul and M. Pasetti (Eds.): EMMFT 2018, AISC 982, pp. 459–465, 2020.
https://doi.org/10.1007/978-3-030-19756-8_44

120 MW for propeller screws (4 screws of 30 MW each) [5]. The decision on the timing and order of further design and manufacturing stages should be taken by the Russian government in 2019. There is a development of both partial and full electric propulsion systems for the needs of the Navy, and the power of such systems are in the ranges from units to hundreds of MW. The growth of the installed power of the electric propulsion systems of such vessels requires the development, design and implementation of both new modern element base and new circuit design solutions at building the ship's electric propulsion complex [6].

The difference in the use of electric propulsion systems on vessels, ships and submarines is that this object simultaneously produces, distributes, converts and consumes electrical energy, and then converts it into mechanical energy for propulsion of the vehicle.

2 Materials and Methods

Classical structures of vessel electric propulsion systems, as a rule, have the following construction scheme (Fig. 1).

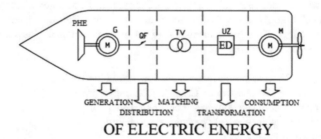

Fig. 1. Classical structure of a vessel electric propulsion system

This structure (Fig. 1) consists of a primary heat engine (PHE), electric generator, switchgear, matching transformer, frequency converter and propeller motor. Each of the elements of such structure is made in full capacity of the propulsion complex and contains a large number of complex expensive and large in mass and size of equipment. In the practical implementation of such a structure of the electric propulsion system, the question of feasibility as well as the possibility of placing electrical equipment, which sometimes is not always possible to place in areas given by the designer, is particularly acute. At this point in time for the Navy, the auxiliary systems of electric propulsion have received the greatest distribution. When using such a structure in civil shipbuilding, the criteria of mass and dimensions are usually not so significant, but important, and the main criterion is the cost of electrical equipment and the entire electric propulsion system.

Such a structure (Fig. 1) of the construction of an electric propulsion complex is typical of almost all modern shipbuilding systems with electric propulsion systems. A common feature of this structure is the use of a frequency-controlled drive with an electric alternating current machine. As an electric converter of such a structure, various schemes of frequency converters are used, constructed using power semiconductor elements. The growth of the installed power of electrical transducers and electric propulsion systems is limited by the switching capacity of the currently produced power electronic semiconductor components. In order to increase the installed power of the electric propulsion system of vessels and ships, it is crushed by the number of propulsion units, the number of independent windings of the rowing electric motors, parallel connection of electrical converters operating on a common load, etc. To increase the unit installed power of such frequency converters and improve the quality of the synthesized voltage to power the rowing electric motor, they are usually performed multi-level [7, 8], and the circuitry of the frequency converter itself and the control system of such a converter are much more complicated. The disadvantages of such a frequency converter include large dimensions, weight, and low reliability when using it, since the failure of any of the elements of the power structure leads to the failure of the entire electrical converter. The disadvantages of the used powerful frequency converters include the limited switching frequency of power electronic fully controlled semiconductor components, as a rule, not exceeding 400 Hz. This circumstance has a significant impact on the quality of synthesizing voltages used to control electric motor, which significantly affects its performance and energy characteristics.

3   Research Result

To eliminate the drawbacks of the well-known structure of building the propulsion complex of the vessel, a different approach will be offered. It is proposed to exclude from the classical scheme of the propulsion system complex distribution and matching elements and reduce the structure of the proposed complex to the form of generation - transformation - consumption. As a conversion element, it is proposed to use a cascade electric converter, which allows improving the quality of the synthesized voltage, as well as increasing the reliability, both of the electric converter itself and of the entire electric propulsion installation. Thus, the solution of the task is to use a cascade electric converter circuit with its single-phase frequency converters powered from the isolated windings of the electric generating unit. At the same time, an expensive, large-sized (size is comparable to the size of the executive rowing electric motor) and a heavy matching voltage transformer is excluded from the structure, which is the main disadvantage of a cascade electric frequency converter during its general industrial use.

Cascade frequency converter circuits have been known for a long time [9–13] and have gained wide acceptance in the industry as an electrical converter for powerful and high-voltage AC drives. In more detail, the scheme and operation of a cascade electric converter is presented in [9–13] and does not require additional explanation.

Consider the construction of a structural scheme of an electric propulsion complex and its operation. The Fig. 2 shows the simplest structure of a vessel propulsion complex with a cascade electric converter.

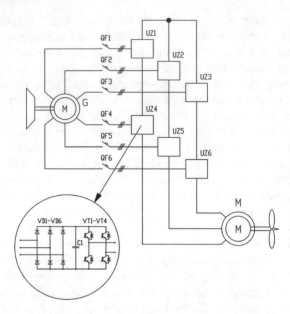

Fig. 2. Structure of a vessel propulsion complex with a cascade electrical converter and one multiple-winding generating unit

In the diagram Fig. 2 electric power generator contains such a number of multi-phase electrical windings galvanically isolated from each other, the number of which is equal to the number of single-phase frequency converters of which consists the cascade electric converter. The number of single-phase frequency converters is a multiple of the number of phases of the rowing electric motor and depends on the required nominal voltage and the number of levels synthesized by the voltage. Such an organization of the power source, whose role is played by a multiple-winding electrical generator, makes it possible to exclude the matching transformer from the electric propulsion complex circuit.

The advantage of the proposed structure is the fact that it can be implemented using high-speed gearless primary heat motors, thereby greatly improving the weight and size and energy characteristics of the propulsion system. In addition, the proposed electric propulsion installation scheme allows realizing electric transmission on the vessel of practically unlimited power, since the voltage for powering the rowing electric propulsion motor is gained from voltages of low-voltage single-phase frequency converters that have a simple circuit design and a high degree of reliability. Another advantage of the proposed structure is its modularity, which provides the flexibility to build an electric propulsion unit with different designs and different numbers of electrical energy generators, propeller motors, with different nominal voltage levels, both an electric alternator and a rowing electric motor. Such a structure of a cascade electric converter allows synthesizing almost sinusoidal voltage to power the rowing electric motor. Despite the fact that the switching frequency of power fully controlled keys in each single-phase frequency converter is limited by their functionality, it is possible to shift the voltage instantaneously synthesized by each of the single-phase frequency converters. In this case, the equivalent modulation frequency of the voltage applied to the windings of rowing electric motor is increased multiple of the number of single-phase frequency converters installed in each of the phases of the cascade electric converter. An increase in the equivalent modulation frequency leads to a reduction in losses, both in the electric converter and in the rowing electric motor. The proposed structure of a cascade electric converter allows reducing the voltage rise rate (dU/dt) on the windings of the rowing electric motor and helps to avoid resonances of electro-motive processes occurring in the electric drive. In case of failure of one of the single-phase frequency converters, its output terminals can be shunted in one of the ways, while the serviceable single-phase frequency converters of the cascade frequency converter can continue to work and synthesize the required voltage levels of the cascade electric converter. As an electric machine drive propeller can be used any electric machine with alternating current.

In order to maximize the load of the primary heat engines for different modes of vessel operation, the electric propulsion complex of the vessel can be made according to a scheme with several generator units Fig. 3.

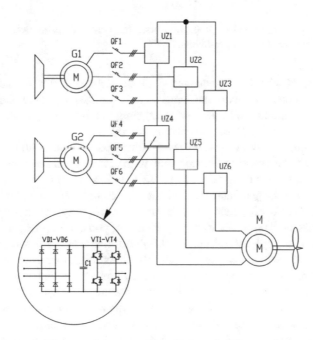

Fig. 3. The structure of the propulsive complex of a vessel with cascade converter with several generator units

Another variation of the proposed structure with a cascade frequency converter can be offered a scheme of an electric propulsion complex implemented using standard commercially available electric generators (Fig. 4).

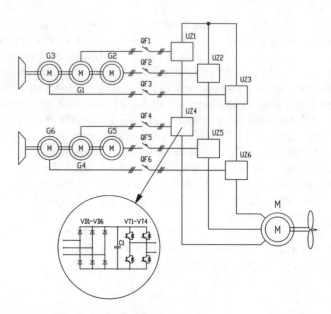

Fig. 4. The structure of the vessel propulsion complex with a cascade electric converter and commercially available electrical generators

The number of possible instantaneous phase voltage levels at the output of a cascade electric converter with the same rated voltage of single-phase frequency converters and the number of series-connected single-phase frequency converters equal N is determined according to dependence $3 + 2*(N-1)$.

In order to improve the quality of the synthesized voltage at output of a cascade electric converter, the vessel propulsion unit with a cascade electric converter, whose structural diagrams are presented in Figs. 2, 3 and 4, can be designed so that the isolated windings of the generator are made at a different level of nominal voltage [14]. Using synchronous electric generators with an electromagnetic excitation system, using the structural diagrams presented in Fig. 3. Figure 4, you can implement an operational change in the voltage level of the single-phase frequency converters voltage. In addition, this circuit will allow you to smoothly charge the storage capacitors installed in the direct current link of single-phase frequency converters, which will eliminate the need to use an additional charging circuit. Using the proposed schemes (Figs. 3 and 4), and different power levels of single-phase frequency converters, the number of potential instantaneous phase voltage levels is related to the number of single-phase frequency converters N in each phase of a cascade frequency converter according to $2(N + 1) + 1$.

4   Conclusion

In conclusion, it should be noted that the proposed structure is universal, competitive capable and the performed analysis shows that it has the best weight and size and energy characteristics in comparison with the known structures of electric propulsion systems.

References

1. Darenkov, A.B., Miryasov, G.M., Titov, V.G., Ohotnikov, M.N., Umyarov, D.V.: The rowing electrical installation. Tutorial of Nizhny Novgorod State Technical University named R. E. Alekseeva, p. 219. Nizhny Novgorod (2014). (in Russian)
2. Verevkin, V.F.: Trends in the development of electric propulsion of ships. Eurasian Union Scientists **1–2**(22), 38–39 (2016). (in Russian)
3. Dyakova, A.A., Dyakov, O.D.: The Northern Sea Route: perspectives of the sea innovations. Young Scientists **44**, 235–239 (2018)
4. Vershinin, V.I., Mahonin, S.V., Parshikov, V.A., Khomyak, V.A.: Algorithm of control of the universal atomic icebreaker rowing electric drive project 22220. Proc. Krylov State Res. Center **4**(382), 95–102 (2017). (in Russian)
5. Kashka, M.M., Smirnov, A.A., Golovinsky, S.A., Vorobev, V.M., Ryzhkov, A.V., Babich, E.M.: Prospects for the development of an atomic icebreaking fleet. Ecol. Econ. **3**(23), 98–107 (2016). (in Russian)
6. Lazarevsky, N.A., Khomyak, V.A., Samoseiko, V.F., Gelver, F.A.: Structural diagrams of rowing equipment, analysis and development prospects. Shipbuilding **3**(802), 44–47 (2012). (in Russian)
7. Filatov, V.: Two- and three-level inverters on IGBT. Promising Solutions Power Electron. **4**, 38–41 (2012). (In Russian)
8. Mikheev, K.E., Tomasov, V.S.: Analysis of energy indicators of multi-level semiconductor converters of electric drive systems. Sci. Tech. J. Inform. Technol. Mech. Optics **1**(77), 46–52 (2012). (in Russian)
9. Pronin, M.V., Vorontsov, A.G., Gogolev, G.A., Osipova, L.I.: A marine electric propulsion system with poly-phase permanent magnet synchronous motor under full fnd partial-phase operation, pp. 227–232 (2011)
10. Milosha, I.V., Korotaev, A.D.: Development of a cascade type frequency converter for a pump of an external pump. Bull. PNIPU **7**, 105–114 (2013). (in Russian)
11. Lazarev, G.B.: High-voltage transducers for variable frequency electric drives. Construction of various schemes. Electr. Eng. News **2**(32), 30–36 (2005). (in Russian)
12. Burdasov, B.K., Nesterov, B.B., Fedotov, Y.B.: Multi-level and cascade frequency converters for high voltage AC drives. Electron. Sci. J. **5**, 2–15 (2015). "Apriori Series: Natural and Technical Sciences", (in Russian)
13. Schavelkin, A.A.: Cascade multi-level frequency converters with improved energy characteristics. Applied science magazine «Technical electrodynamics» Power electronics and energy efficiency, Part 1, pp. 65–70. Kiev (2010). (in Russian)
14. Khakimyanov, M.I., Shabanov, V.A.: Multi-level frequency converter with differentiated voltage levels and bypass semiconductor switches (in Russian), RU 2510769, application 201214848 1/07 of 11.14.2012 (2012)

Reduction of the Electromagnetic Torque Pulsations in a Valve-Inductor Machine

Veniamin Samoseiko$^{(\boxtimes)}$ (iD), Aleksandr Saushev$^{(\boxtimes)}$ (iD),
Tatiana Knish (iD), and Eduard Shiryaev (iD)

Admiral Makarov State University of Maritime and Inland Shipping,
5/7, Dvinskaya Street, Saint Petersburg 198035, Russia
{ep-gumrf, saushev}@bk.ru

Abstract. The principle of operation of inductor machines is based on the inductance pulsations of the stator windings in time with the pulsations of the currents feeding them. It is known that with the sinusoidal nature of the inductance pulsations of the stator phase windings, caused by the rotor rotation, the constant component of the electromagnetic moment is produced by the harmonic, whose frequency coincides with the pulsation frequency of the inductances, as well as the zero component. The remaining current harmonics create only pulsations of the electromagnetic moment. Therefore, it is advisable to synthesize the design of the machine so as to ensure the condition of constancy of the electromagnetic moment, achieving the sinusoidal nature of the inductance pulsations of the stator phase windings, and controlling the machine to carry out sinusoidal currents with a constant component. The paper discusses design solutions that allow the pulsation form of the inductances of the stator phase windings of the valve-inductor machines to be approximated to a sinusoidal form.

Keywords: Valve-inductor machine · Shape of teeth · Pulsations ·
Sinusoidal nature · Electromagnetic moment · Inductances

1 Introduction

The machine, the magnetic core of which has a gear stator and a rotor, has been known for a long time as a stepping motor [1]. The first publications about stepper motors refer to the beginning of the last century. The stator windings of stepper motors are powered by current pulses, under the influence of which the teeth are magnetized and the rotor rotates by a certain amount, called the step [2].

New life of the machine with a gear stator and rotor received with the development of power valve electronics. In modern Russian literature, they are known as valve-inductor motors [3, 4]. An example of a magnetic system of a valve-inductor machine with a gear stator and a rotor is shown in Fig. 1. Valve-inductor motors with a gear stator and a rotor are devoted to a large number of both domestic [5–7] and foreign [8–10] papers, in which it is believed that the stator windings are powered from the valve switch in the form of current pulses. The main disadvantage of valve-inductor motors is the high pulsations of the electromagnetic moment [11, 12]. Reducing the

© Springer Nature Switzerland AG 2020
V. Murgul and M. Pasetti (Eds.): EMMFT 2018, AISC 982, pp. 466–480, 2020.
https://doi.org/10.1007/978-3-030-19756-8_45

pulsations of the electromagnetic moment leads to a decrease in power losses and vibro-noise characteristics of electric drives. Therefore, reducing the pulsations of the electromagnetic torque of the engine is an urgent task.

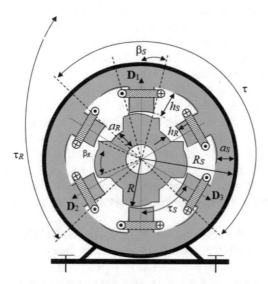

Fig. 1. Magnetic system of the machine with the number of teeth on the stator $Z_1 = 6$ and the number of teeth on the rotor $Z_2 = 4$ (the number of phases of the stator $m = 3$, the number of pole pairs p = 1)

2 Formulation of the Problem

The principle of operation of inductor machines is based on the inductance pulsations of the stator windings in time with the pulsations of the currents feeding them. It can be shown [13] that with the sinusoidal nature of the inductance pulsations of the stator phase windings, caused by the rotor rotation, the constant component of the electromagnetic moment is produced by the harmonic, whose frequency coincides with the frequency of the inductance pulsations, and the zero component. The remaining current harmonics create only pulsations of the electromagnetic moment. Therefore, it is advisable to synthesize the design of the machine in such a way as to ensure the condition of the constancy of the electromagnetic moment with the sinusoidal nature of the inductance pulsations of the stator phase windings. The control of the machine in this case should be carried out by sinusoidal currents with a constant component. The paper discusses design solutions that allow the pulsation form of the inductances of the stator phase windings of the valve-inductor machines to be approximated to a sinusoidal form.

A distinctive feature of this work as applied to inductor electric machines, the magnetic core of which has a gear stator and rotor, is the position that sinusoidal currents with a constant component flow through the windings, the frequency of which coincides with the inductance pulsation frequency of the stator phase windings. Ways to reduce the pulsations of the electromagnetic moment in a valve-inductor machine are considered.

2.1 Improving the Harmonic Composition of the Magnetic Conductivities of the Air Gap Due to the Choice of Geometrical Ratios of Teeth

The magnetic core of the machine stator has the shape of a hollow cylinder. In describing the geometry of the stator and rotor, the polar coordinate system is mainly used. The basic value in this coordinate system has the concept of a polar geometric angle G between two radial rays. The main values of the angle $G \in [0, 2 \cdot \pi]$. Many dimensions of the machine are indicated in geometric angles. As the base value of the polar radius, the value of the inner radius of the stator R is used. Linear dimensions can be given in relative units, as the ratio of the actual linear size to the baseline - the inner radius of the stator R. Dimensions in relative units are marked with an upper symbol *.

The outer radius of the cylindrical stator magnetic core is denoted by R_S. On the inner side of the stator cylinder there are Z_1 grooves (teeth). The depth of the groove (tooth height) is indicated by the letter h_S or in relative units (ru) $h_S^* = h_S/R$. The number of teeth of the stator $Z_1 = 2 \cdot p_1$ is even, where $p_1 = 1, 2, 3, \ldots$ is the number of pairs of stator teeth. The ratio of the angular size of the stator tooth β_S to the angular tooth division of the stator τ_S is hereinafter referred to as the duty ratio of the stator teeth.

The gear rotor of the machine has Z_2 grooves (teeth). The depth of the groove (tooth height) is denoted by the letter h_R or in relative units (ru) $h_R^* = h_R/R$. The number of teeth of the stator $Z_2 = 2 \cdot p_2$ is even, where $p_2 = 1, 2, 3, \ldots$ is the number of pairs of teeth of the rotor. The ratio of the angular size of the stator tooth β_R to the angular tooth division of the stator τ_R is hereinafter referred to as the duty ratio of the rotor tooth.

The greatest common divisor of the number of teeth of the rotor and stator is the number of pole pairs p. The number of poles is even: $2 \cdot$ p. The ratio of the number of teeth

$$\frac{Z_2}{Z_1} = \frac{n}{m}$$

forms an irreducible fraction, where m is an integer, called the number of stator phases; n is the number of pairs of the rotor teeth on the pole division of the stator (coefficient of electrical reduction).

In classic AC machines, the number of pairs of rotor teeth and pairs of stator poles are the same. In a jet machine, there are n pairs of rotor teeth per stator pole. Therefore, under the electrical angle we will understand the angle of repeatability of electrical processes

$$\gamma = p_2 \cdot G = n \cdot p \cdot G = \frac{Z_2}{2} \cdot G$$

Thus, the electric period corresponds to the rotation of the rotor into two tooth divisions.

All angular dimensions of the machine can be expressed in electric angles. So, for example, angular tooth divisions of the stator and rotor in electric angles

$$\tau_R^* = p_2 \cdot \tau_R = \pi; \quad \tau_S^* = p_2 \cdot \tau_S = \pi \cdot \frac{n}{m},$$

and the electric angular size of the stator tooth

$$\beta_S^* = p_2 \cdot \beta_S = \tau_S^* \cdot \gamma_S = \pi \cdot \frac{n}{m} \cdot \gamma_S.$$

The view of the magnetic conductivity function of the winding on the angle of rotation of the rotor is shown in Fig. 2a [5]. Note that the symmetry of the magnetic conductivity function of the air gap from the angle of rotor rotation is ensured when the rotor teeth have a duty ratio of $\gamma_R = 1/2$. Then the electric angular size of the rotor tooth

$$\beta_R^* = \pi/2 \tag{1}$$

To describe the magnetic conductivities pulsations of the air gap, we introduce a periodic function

$$\mathrm{lr}(\gamma, \beta_S^*) = \frac{\mathrm{ass}(2 \cdot \gamma + \beta_S^*) - \mathrm{ass}(2 \cdot \gamma - \beta_S^*)}{2 \cdot \mathrm{ass}(\beta_S^*)} \tag{2}$$

where $\mathrm{ass}(t) = 2 \cdot \arcsin(\sin(t))/\pi$. The type of function (2) is shown in Fig. 2b.

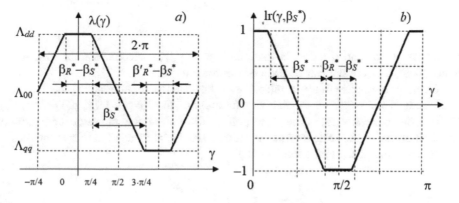

Fig. 2. Dependence of the function type on the angle of the rotor rotation: a - magnetic conductivity of widding, b - $\mathrm{lr}(\gamma, \beta_S^*)$ **Error! Reference source not found.**

Using the function $Lr(\gamma, \beta_S^*) = lr(\gamma, \beta_S^*)$, the magnetic conductivity of the winding will be written in the following form:

$$\lambda(\gamma) = \Lambda_{00} + \Lambda_m \cdot Lr(\gamma, \beta_S^*) \tag{3}$$

where Λ_{00} and Λ_m are the average value and amplitude of the first harmonic of the winding magnetic conductivity.

The function $lr(\gamma, \beta_S^*)$ can be represented by a Fourier series.

$$lr(\gamma, \beta_S^*) = \sum_{k=1}^{\infty} \frac{A_v}{v^2} \cdot \cos(2 \cdot v \cdot \gamma) \tag{4}$$

where $v = 2 \cdot k - 1$;

$$A_v = (-1)^{k-1} \cdot \frac{4}{\pi} \cdot \frac{\sin(v \cdot \beta_S^*)}{\beta_S^*} \tag{5}$$

harmonic amplitude coefficients.

The sum of the constant component and the first harmonic of the winding magnetic conductivity

$$\lambda_{i(1)}(\gamma) = \Lambda_{00} + A_1 \cdot \Lambda_m \cdot \cos(2 \cdot \gamma + 2 \cdot \rho \cdot i) \tag{6}$$

where

$$A_1 = \frac{4}{\pi} \cdot \frac{\sin(\beta_S^*)}{\beta_S^*}$$

amplitude coefficient of the main harmonic; $\beta_S^* = p_2 \cdot \beta_S$—electric angular size of the stator tooth; p_2 - the number of pairs of the rotor teeth.

Consider the choice of the angular size of the stator tooth, minimizing the distortion coefficient of the magnetic conductivity function of the air gap $Lr(\gamma, \beta_S^*) = lr(\gamma, \beta_S^*)$, defined by expression (4). The distortion coefficient of the magnetic conductivity function of the air gap is defined as the ratio of the actual value of higher harmonics to the actual value of the first harmonic [10]

$$k_d = \frac{1}{A_1} \cdot \sqrt{\sum_{k=2}^{\infty} \frac{A_v^2}{v^4}} \tag{7}$$

where $v = 2 \cdot k - 1$; A_v is the harmonic amplitude of order v, defined by expression (5). The graph of the distortion coefficient dependence of the function $Lr(\gamma, \beta_S^*)$ on the electric angular size of the stator tooth β_S^* is shown in Fig. 3.

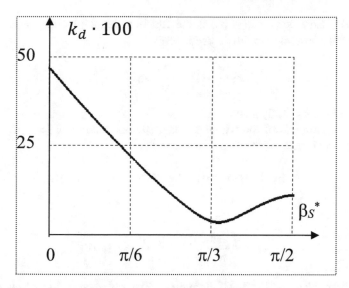

Fig. 3. Distortion coefficient dependence of the function (4) on the electric angular size of the stator tooth

Note that the distortion coefficients of the function $Lr(\gamma, \beta_S^*)$ with the values of the parameters $\beta_S^* = 1.082$ and $\beta_S^* = \pi/3$ are slightly different. Therefore, you can take $\beta_S^* = \pi/3$. When $\beta_S^* = \pi/3$ harmonics are multiples of 3.

Thus, the electric angular size of the stator tooth, minimizing higher harmonics, can be taken

$$\beta_S^* = \frac{2}{3} \cdot \beta_R^* = \frac{\pi}{3} \tag{8}$$

At the same time the geometric angular size of the stator tooth

$$\beta_S = \frac{2}{3} \cdot \beta_R = \frac{2 \cdot \pi}{3 \cdot Z_2} \tag{9}$$

and duty ratio of stator teeth

$$\gamma_S = \frac{m}{3 \cdot n} = \frac{Z_1}{3 \cdot Z_2} \tag{10}$$

The magnetic conductivity of the winding (3) with $\beta_S^* = 1.082 \approx \pi/3$ will have the best harmonic composition

$$\lambda(\gamma) = \Lambda_{00} + \Lambda_m \cdot Lr(\gamma) \tag{11}$$

where $Lr(\gamma) = Lr(\gamma, \pi/3) = lr(\gamma, \beta_S^*)$ is the function defined by the expression (4) at $\beta_S^* = \pi/3$, with harmonic amplitude coefficients:

$$A_v = (-1)^{k-1} \cdot \frac{12}{\pi^2} \sin\left(\frac{v \cdot \pi}{3}\right) \tag{12}$$

where $v = 2 \cdot k - 1$; $k = 1, 2, 3 \dots$.

The main harmonic of the windings magnetic conductivities $i = 1, 2, \dots, m$ as a function of the electric angle of rotor rotation

$$\lambda_{i(1)}(\gamma) = \Lambda_{00} + A_1 \cdot \Lambda_m \cdot \cos(2 \cdot \gamma + 2 \cdot \rho \cdot i),$$

where

$$A_1 = \frac{4 \cdot \sin(\beta_S^*)}{\pi \cdot \beta_S^*} = \frac{6 \cdot \sqrt{3}}{\pi^2} = 1,053 \tag{13}$$

amplitude coefficient of the main harmonic. The obtained value of $A_1 = 1.053$ is somewhat overestimated, since it does not take into account the smoothing of the magnetic conductivity function due to the buckling of the magnetic flux. Therefore, when calculating, you can take $A_1 = 1$.

Relationships (9) and (10) allow the design to choose the angular size of the stator tooth β_S and the duty ratio of the stator teeth γ_S depending on the number of teeth on the stator and rotor so as to improve the harmonic composition of the magnetic conductivities due to the geometry of the magnetic core in the cross-sectional plane of the machine.

2.2 Improving the Harmonic Composition of the Magnetic Conductivities of the Air Gap Due to the Displacement of the Rotor Parts

Let the cylindrical rotor consist of two parts of equal length, displaced relative to each other by an electric angle α_d. Let us consider the influence of the shift of two parts of a cylindrical rotor on the form of the pulsations of the magnetic conductivities of the air gap during the rotation of the rotor.

The function characterizing the magnetic conductivities of the air gap in this case will take the following form:

$$Lr(\gamma, \alpha_d) = \frac{lr\left(\gamma - \alpha_d/2, \beta_S^*\right) + lr\left(\gamma + \alpha_d/2, \beta_S^*\right)}{2}, \tag{14}$$

where $lr(\gamma, \beta_S^*)$ is a function defined by the expression (2) or (4); β_S^* is the electric angular size of the stator tooth.

This function can be represented by a Fourier series.

$$Lr(\gamma, \alpha_d) = \sum_{k=1}^{\infty} \frac{A_v(\alpha_d)}{v^2} \cdot \cos(2 \cdot v \cdot \gamma), \qquad (15)$$

where $v = 2 \cdot k - 1$;

$$A_v(\alpha_d) = (-1)^{k-1} \cdot \frac{4 \cdot \sin(v \cdot \beta_S^*) \cdot \cos(2 \cdot v \cdot \alpha_d)}{\pi \cdot \beta_S^*}. \qquad (16)$$

The distortion coefficient of the function $Lr(\gamma, \alpha_d)$ is determined by expression (7). When $\beta_S^* = \pi/3$ and $\alpha_d = 0.288$, it reaches the minimum value of $k_d = 01.01506$. The graph of the distortion coefficient of the function (15) versus the electric angle of displacement of the rotor parts α_d is shown in (Fig. 4a).

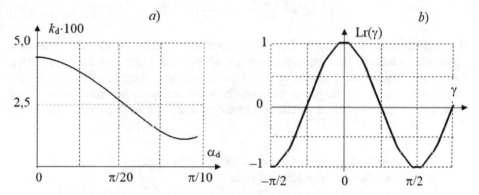

Fig. 4. Graps of the functions: a - distortion coefficient of the function $Lr(\gamma, \alpha_d)$ versus the electric angle of displacement of the rotor parts α_d; b - $Lr(\gamma)$ at displacement of two parts of rotor on the electric angle $\alpha_d = \pi/10$

If the function (15) we take $\alpha_d = \pi/10 = 0.314$, when it will take form

$$Lr(\gamma) = Lr(\gamma, \pi/10) = \sum_{k=1}^{\infty} \frac{A_v}{v^2} \cdot \cos(2 \cdot v \cdot \gamma), \qquad (17)$$

where $v = 2 \cdot k - 1$;

$$A_v = (-1)^{k-1} \cdot \frac{12}{\pi^2} \sin\left(\frac{v \cdot \pi}{3}\right) \cdot \cos\left(\frac{v \cdot \pi}{5}\right) \qquad (18)$$

The distortion factor of this function is $k_d = 01.01607$. The graph of the function (17) is shown in Fig. 4b.

Note that for $\alpha_d = \pi/10$ odd harmonics multiple of five disappear: $v = 5, 15,\ldots$. Using the function $Lr(\gamma)$, the magnetic conductivities of the windings are written as follows:

$$\lambda(\gamma) = \Lambda_{00} + \Lambda_m \cdot Lr(\gamma)/2.$$

The main harmonic of the magnetic conductivities of the phase windings $i = 1, 2, \ldots,$ m as a function of the electric angle of the rotor rotation

$$\lambda_{i(1)}(\gamma) = \Lambda_{00} + \Lambda_m \cdot A_1 \cdot k_d \cdot \cos(2 \cdot \gamma + 2 \cdot \rho \cdot i)$$

where A_1 is the amplitude coefficient of the main harmonic without displacement of the rotor parts (13); $k_d = \cos(\alpha_d)$ is the coefficient of rotor parts displacement. When $\alpha_d = \pi/10$, the amplitude coefficient of the main harmonic

$$A_1 \cdot k_d = \frac{6 \cdot \sqrt{3}}{\pi^2} \cdot \cos\left(\frac{\pi}{10}\right) = 1.001. \tag{19}$$

The displacement of two parts of the rotor equal in length by the electric angle $\alpha_d = /10$ allows to remove from the representation of the magnetic conductivities near the Fourier harmonics multiples of five and thus significantly improve the form of the inductance function of the phase winding from the angle of the rotor rotation. The distortion coefficient of the magnetic conductivity function at an angle of displacement of $\alpha_d = \pi/10$ is $k_d = 0.01607$.

2.3   Improving the Harmonic Composition of the Magnetic Conductivities Due to the Bevel of the Rotor Teeth

Let us suppose that the rotor teeth are beveled at an angle α_b, and consider the influence of the bevel angle on the coefficient of distortion of the magnetic conductivity of the air gap. When the bevel of the rotor teeth along the length of the cylinder l at an angle α_b, the function characterizing the magnetic conductivities of the air gap will take the following form:

$$Lr(\gamma, \alpha_b) = \frac{1}{\alpha_b} \cdot \int_{-\alpha_b/2}^{\alpha_b/2} lr(\gamma - x, \beta_S^*) \cdot dx, \tag{20}$$

where $lr(\gamma, \beta_S^*)$ is a function defined by expression (2) or (4).

This function can be represented by a Fourier series.

$$Lr(\gamma, \alpha_b) = \sum_{k=1}^{\infty} \frac{A_v}{v^2} \cdot \cos(2 \cdot v \cdot \gamma),$$

where $v = 2 \cdot k - 1$;

$$A_v = (-1)^{k-1} \cdot \frac{4 \cdot \sin(v \cdot \alpha_b) \cdot \sin(v \cdot \beta_S^*)}{\pi \cdot v \cdot \beta_S^* \cdot \alpha_b}. \tag{21}$$

The distortion factor of the function $Lr(\gamma, \alpha_b)$ is determined by expression (7). When $\beta_S^* = \pi/3$ and $\alpha_b = 0.622$, it reaches the first minimum value of $k_d = 0.00483$. The graph of the distortion coefficient of the parameter α_b is shown in Fig. 5a. If in expression (20) we take $\alpha_b = \pi/5 = 0.628$, then $k_d = 0.00485$ and function (20) takes the form shown in Fig. 5, b.

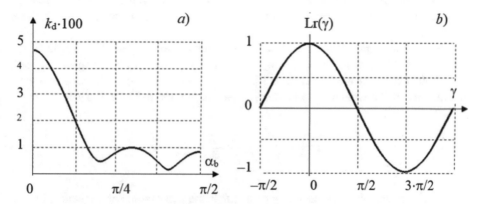

Fig. 5. Graphs: a - distortion coefficient of the function $Lr(\gamma, \alpha_b)$ at $\beta_S^* = \pi/3$ at the function of the bevel angle α_b; b - function $Lr(\gamma)$ at the bevel angle α_b

With the help of the $Lr(\gamma)$ function, the magnetic conductivities as a function of the electric angle are written in the following form:

$$\lambda(\gamma) = \Lambda_{00} + \Lambda_m \cdot Lr(\gamma)/2.$$

The main harmonic of the magnetic conductivities of the phase windings $i = 1, 2, \dots m$ as a function of the electric angle of the rotor rotation

$$\lambda_{i(1)}(\gamma) = \Lambda_{00} + \Lambda_m \cdot A_1 \cdot k_b \cdot \cos(2 \cdot \gamma + 2 \cdot \rho \cdot i),$$

where A_1 is the amplitude coefficient of the main harmonic without the rotor bevel (13); $k_b = \sin(\alpha_b)/\alpha_b$ is the coefficient of the rotor bevel. When $\alpha_b = \pi/5$ the amplitude coefficient of the main harmonic

$$A_1 \cdot k_b = \frac{30 \cdot \sqrt{3}}{\pi^3} \cdot \sin\left(\frac{\pi}{5}\right) = 0.985. \tag{22}$$

The bevel of the rotor teeth along the length of the cylinder l at the angle α_b significantly improves the form of the magnetic conductivity function with the angle of the rotor rotation. The distortion coefficient of the magnetic conductivity function at the bevel angle $\alpha_b = \pi/5$ is $k_d = 0.00485$. Note that the amplitude coefficient of the main harmonic is quite close to unity.

2.4   Improving the Harmonic Composition of the Magnetic Conductivities Due to the Cylinder Base Geometry of the Rotor Magnetic Core

In this section we consider the magnetic conductivities formation of the stator windings by changing the air gap under the stator winding tooth $\delta(\gamma)$ so that the function $Lr(\gamma)$ is sinusoidal.

The electric angle of the rotor rotation γ is related to the geometric angle of the rotor rotation G by the relation $\gamma = p_2 \cdot G$, where p_2 is the number of the rotor teeth pairs. In this case, the electric angular size of the tooth $\beta_S^* = p_2 \cdot \beta_S$. Then the magnetic conductivity of the winding can be determined by the approximate formula

$$\lambda(\gamma) = \int_{-\beta_S^*/2}^{\beta_S^*/2} \frac{\mu_0 \cdot R \cdot l}{2 \cdot p_2 \cdot \delta(x+\gamma)} \cdot dx, \tag{23}$$

where μ_0 is the magnetic permeability of air; R is the inner radius of the stator; l is the calculated length of the magnetic core.

Let us suppose that the magnetic conductivity of the winding pulses according to a sinusoidal law

$$\lambda(\gamma) = \Lambda_{00} + \Lambda_m \cdot \cos(2 \cdot \gamma) \tag{24}$$

Let us determine how the air gap should change in order for the magnetic conductivity of the winding to correspond to expression (24). For this, it is necessary to solve the integral Eq. (23). It is easy to verify that the solution to Eq. (23) is a function of the form

$$\delta(\gamma) = \frac{\delta_d \cdot \delta_q}{(\delta_d + \delta_q) \cdot \cos^2(\gamma)}, \tag{25}$$

where δ_d is the air gap along the longitudinal axis d; δ_q is the air gap along the transverse axis q. Thus, the dependence of the air gap under the stator and rotor teeth, defined by formula (25), provides a sinusoidal nature of the magnetic conductivity pulsations.

The rotor radius is a function of the electric angle γ

$$r(\gamma) = R - \frac{\delta_d \cdot \delta_q}{(\delta_d + \delta_q) \cdot \cos^2(\gamma)} \tag{26}$$

where R is the inner radius of the stator. The rotor radius can be written as a function of the geometric angle: $r(p_2 \cdot G)$, где $G = \gamma/p_2$ is the geometric angle of the polar coordinate system; p_2 - the number of the rotor teeth pairs. The graph of the function r $(p_2 \cdot G)$ from the angular coordinate G is shown in Fig. 6.

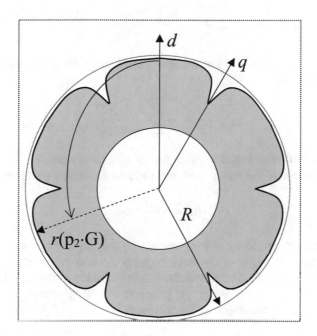

Fig. 6. Graph of the function $r(p_2 \cdot G)$ from the angular coordinate G with $p_2 = 3$

If the sinusoidal dependence of the winding magnetic conductivity on the rotation angle of rotor is provided by the size of the air gap, then the angular size of the stator tooth β_S can be not associated with the harmonic composition of the winding magnetic conductivity and choose it from other considerations. If the air gap between the stator and the rotor changes in accordance with the expression (25), then the average and amplitude magnetic conductivities of the winding:

$$\Lambda_{00} = \frac{\mu_0 \cdot R \cdot l \cdot (\delta_d + \delta_q) \cdot \beta_S^*}{4 \cdot \delta_d \cdot \delta_q \cdot p_2} \tag{27}$$

$$\Lambda_m = \frac{\mu_0 \cdot R \cdot l \cdot (\delta_d + \delta_q) \cdot \sin(\beta_S^*)}{4 \cdot \delta_d \cdot \delta_q \cdot p_2}$$

Maximum and minimum magnetic conductance of the winding:

$$\Lambda_{dd} = \Lambda_{00} + \Lambda_m = \frac{\mu_0 \cdot R \cdot l \cdot (\delta_d + \delta_q) \cdot (\beta_S^* + \sin(\beta_S^*))}{4 \cdot \delta_d \cdot \delta_q \cdot p_2} \tag{28}$$

$$\Lambda_{qq} = \Lambda_{00} - \Lambda_m = \frac{\mu_0 \cdot R \cdot l \cdot (\delta_d + \delta_q) \cdot (\beta_S^* - \sin(\beta_S^*))}{4 \cdot \delta_d \cdot \delta_q \cdot p_2}$$

You can assess the quality of the reactive machine by the coefficient of transverse magnetic conductivity.

$$k_q = \frac{\Lambda_{qq}}{\Lambda_{dd}} = \frac{\beta_S^* - \sin(\beta_S^*)}{\beta_S^* + \sin(\beta_S^*)},$$

where $\beta_S^* = p_2 \cdot \beta_S$ is the electric angular size of the stator tooth.

The angular size of the stator tooth we write in the following form:

$$\beta_S^* = \pi \cdot \frac{n}{m} \cdot \gamma_S,$$

where m is the number of the stator phases; n is the number of rotor teeth pairs per stator phase; γ_S - duty ratio of the stator teeth. To obtain the minimum value of the transverse magnetic conductivity coefficient, it is advisable in the machine to have the number of rotor teeth pairs per stator phase $n = 1$. In this case, the average and amplitude magnetic conductance of the winding:

$$\Lambda_{00} = \frac{\mu_0 \cdot R \cdot l \cdot (\delta_d + \delta_q) \cdot \pi \cdot \gamma_S}{4 \cdot \delta_d \cdot \delta_q \cdot m \cdot p} \tag{29}$$

$$\Lambda_m = \frac{\mu_0 \cdot R \cdot l \cdot (\delta_q + \delta_d) \cdot \sin(\pi \cdot \gamma_S/m)}{4 \cdot \delta_d \cdot \delta_q \cdot p}$$

where $p = p_2$ is the number of pole pairs of the machine.

An example of a machine magnetic core with the number of phase windings $m = 3$, the number of pole pairs $p = p_2 = 2$ and the air gap between the stator and the rotor, changing in accordance with the expression (26), is shown in Fig. 7.

Thus, imparting to the teeth of the rotor a special form allows to obtain a sinusoidal dependence of the inductance pulsations of the phase windings on the angle of the rotor rotation.

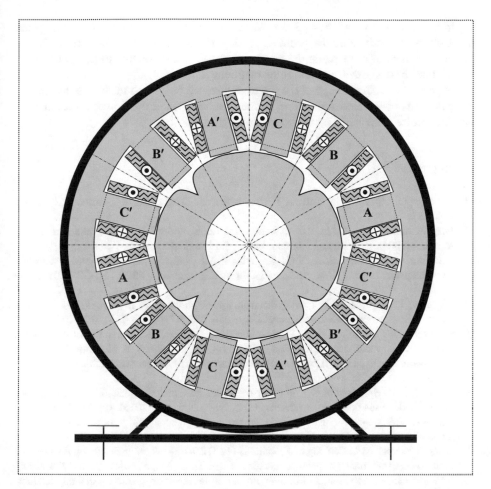

Fig. 7. Example of a machine magnetic core with the number of phase windings $m = 3$, the number of pole pairs $p = 2$ and the air gap between the stator and the rotor, changing in accordance with the expression (26)

3 Conclusion

The principle of inductor machines operation is based on the inductance pulsations of the stator windings in time with the pulsations of the currents feeding them. With the sinusoidal nature of the inductance pulsations of the stator phase windings, caused by the rotor rotation, the constant component of the electromagnetic moment is produced by a harmonic whose frequency coincides with the frequency of the inductance pulsations, and also the zero component. The remaining current harmonics create only pulsations of the electromagnetic moment. Therefore, it is advisable to synthesize the design of the machine so as to ensure the condition of the constancy of the electromagnetic moment, achieving the sinusoidal nature of the inductance pulsations of the stator phase windings, and controlling it with sinusoidal currents with a constant component.

It is possible to impart a sinusoidal character to the pulsations of the phase inductances by: optimizing the geometric ratios of the teeth of the stator and the rotor; the displacement of rotor parts; the bevel of the rotor teeth relative to the stator teeth; the cylinder base geometry of the rotor magnetic core.

Improving the curve shape of the phase inductance pulsations leads to a decrease in the pulsations of the electromagnetic moment, a decrease in the power losses and the vibration-noise characteristics of the electric drives.

References

1. Kopylov, I.P.: Electric Machines: A Textbook, 5th edn.. Higher School, Moscow (2006). (in Russian)
2. Ivanov-Smolensky, A.V.: Electric machines. In: Textbook for Universities, 2nd edn., 2 vol., Moscow (2004). Recycled and supplemented, (in Russian)
3. Petrushin, A.D., Shcherbakov, V.G., Kashuba, A.V.: Optimization of magnetic system of a valve-inductor motor. News universities. Electromechanics. **60**(1), 20–27 (2017). (in Russian)
4. Vaganov, M.A., Goryushkin, I.A.: Electromagnetic moment of a valve inductor motor. News of St. Petersburg Electrotechnical University "LETI", pp. 67–71 (2012). (in Russian)
5. Samoseiko, V.F., Saushev, A.V.: Perspectives of Reluctance Machines in the Electric Drive, Science in the modern information society XII, vol. 1: Proceedings of the Conference. North Charleston, 19–20 June 2017, vol. 1-North Charleston, SC, USA. CreateSpace, pp. 66–69 (2017)
6. Temirev, A.P., Shakurin, A.S., Temirev, A.A.: Analysis and synthesis of a valve-inductor motor for deep-sea underwater vehicles. Lick, Novocherkassk (2013). (in Russian)
7. Shabaev, V.A.: Analysis of the criteria for the technical-economic optimum of the valve-inductor engines use. Electr. Eng. **4**, 44–51 (2008). (in Russian)
8. Lamghari-Jamal, M.I., Fouladgar, J., Zaim, E.H., Trichet, D.: A magneto–thermal study of a high–speed synchronous reluctance machine. IEEE Trans. Magn. **42**(4), 1271–1274 (2006)
9. Kolehmainen, J., Ikaheimo, J.: Motors with buried magnets for medium-speed applications. IEEE Trans. Energy Convers. **23**(1), 86–91 (2008)
10. Bal, V.B., Tang, T.A.: The Electromagnetic torque of the inductor machine, no. 11, pp. 63–66 (2012). (in Russian)
11. Samoseiko, V.F., Belousov, I.V., Saushev A.V.: International Russian Automation Conference, RusAutoCon (2018). https://doi.org/10.1109/rusautocon,.8501699
12. Krasovsky, A.B.: Investigation of torque pulsations of the valve-inductor engine when adjusting the average torque in the low-speed zone. Electr. Eng. **5**, 2–8 (2017). (in Russian)
13. Samoseiko, V.F.: Theory of Electric Drive Control. Saint-Petersburg, Elmore (2007). (in Russian)

Cyber Security on Sea Transport

Maria Kardakova$^{(\boxtimes)}$ ⓘ, Ilya Shipunovⓘ, Anatoly Nyrkovⓘ,
and Tatyana Knyshⓘ

Admiral Makarov State University of Maritime and Inland Shipping,
Street Dvinskaya, 5/7, St. Petersburg 198035, Russia
m.v.kardakova@ya.ru

Abstract. Automation and digitization of the transport industry is a global priority. In place of the traditional "pilot" vehicle control comes the technology of using autonomous "crewless" control. The developments of various crewless vehicles, trains, aircraft, and ships are carried out. Crewless vehicles for various economic assessments will give a significant increase in the economy, not only the transport industry. However, in the absence of the pilot (crew), various situations arise that lead to a violation of transport safety. One of the main problems can be considered cybersecurity on vehicles of any type. In this paper, an analysis of the cybersecurity status of information systems on the maritime "crewless" vessels of the new generation is carried out. Data for the study were taken from open sources. The analysis is based on the Maritime Cyber-Risk Assessment model. The least protected systems that are most valuable for cybercriminals are identified: an automatic identification system, global navigation satellite systems, and sensor systems.

Keywords: Crewless vessels · Cyber risks · Risk assessment ·
Rolls-Royce AAWA · YARA Birkeland · Mayflower Autonomous Ship · AIS ·
GNSS

1 Introduction

No field of human activity today cannot do without the use of modern information technology. The introduction of new information technologies not only makes life easier for employees, but also ensures safety [1–3]. Transportation is not an exception, the development of information systems in this area contributes to the safety of navigation, increasing environmental protection, reducing the workload of crew members and reducing the negative impact of the human factor [4–8]. Despite this, the problem of safety of navigation is relevant, it is due to numerous factors and is exacerbated every year. Emergencies, incidents and accidents with passenger and merchant vessels happen more often due to the negligence of crews, the deterioration of technical support and unpreparedness for informational attacks [1, 8–10]. To date, the world has embraced the trend of creating and using autonomous transport. One of the most important areas in the development of autonomous transport is the automation of sea transportation [11, 12]. The creation of autonomous (robotic, unmanned, crewless) vessels are engaged in almost all the leading countries of the world, including Russia

© Springer Nature Switzerland AG 2020
V. Murgul and M. Pasetti (Eds.): EMMFT 2018, AISC 982, pp. 481–490, 2020.
https://doi.org/10.1007/978-3-030-19756-8_46

[5, 11]. According to the assurances of the head of USC A. Rakhmanov, the first samples of vessels with automated course holding and mooring processes have already been launched and are being tested [13]. An important aspect in the creation of autonomous vessels is the legality of their use in maritime transport. A number of organizations, such as SARUMS and AAWA, are engaged in creating the necessary regulatory and legal framework [14]. To date, there are various assessments of the economic efficiency of this class vessels. However, they ignore potential losses due to cyber and cyber-physical attacks. This situation arose as a result of the inability to comprehensively assess the risks and vulnerabilities related with autonomous shipping, which uses a combination of systems consisting of the latest developments in the field of autonomous and traditional marine systems [15]. It is worth noting that with the increase in the interaction level of traditional maritime transport with coastal infrastructure elements, the possibilities for managing crewless vessels are increasing. But this factor also increases the number of potential cyber attacks on vessels, including crewless ones.

To ensure cybersecurity, it is necessary to identify the most significant models of violators and threats to cyber resources, to establish a list of cyber resources of the vessel, the most susceptible to cyber attacks [9]. The purpose of this work is to identify violators, identify possible cyber threats, assess cyber risks, and identify the most vulnerable systems on crewless vessels.

2   Materials and Methods

As a method of more accurate assessment of cyber threats for crewless vessels, we use the Maritime Cyber-Risk Assessment (MaCRA) model, which, based on current data on traditional shipping, allows drawing parallels with autonomous shipping and assuming the potential of cyber vulnerabilities on autonomous vessels. The initial data for it will be the characteristics of the three autonomous vessels closest to the implementation of projects. YARA Birkeland (Y). The operation of the first autonomous zero-emission container ship in the world was planned to begin in the second half of 2018, shipping products from the Porsgrunn plant in Brevik and Larvik in Norway. While this period was postponed to 2019. Mayflower Autonomous Ship (M) is an autonomous research vessel, the first voyage of which is announced for 2020. Rolls Royce AAWA (R) - multipurpose vessel. In the reduced crew mode, it will be put into operation in 2020, and full autonomy should be achieved by 2035. Let us analyze the possible cybersecurity risks of these autonomous vessels. It is clear that many cyber-attacks on modern sea transport data are not published by carriers in order to preserve customers. Another important factor is that today, in the fleet, insufficient crew training in the field of cybersecurity takes place and some cyber attacks were mistaken for a manifestation of the so-called "human error" [1, 16, 17].

It should be noted that the MaCRA model is based on the laws governing the construction of risk assessment models for traditional vessels. In this paper, this model is expanded so that it can be applied to crewless vessels. To do this, we will evaluate according to three criteria. Each criterion will be represented by one of the axes of the MaCRA model. The s-axis is the influence of technological systems.

The e-axis is the simplicity of the system for the attacker.

The r-axis is the attack value for the attacker.

Despite the fact that this model is most often represented in two or three-dimensional measurements, it is much more complicated, since all three axes are functions of the following attributes:

$$\text{Attacker}\ (a) = (a_{\text{direction}}, a_{\text{target}}, a_{\text{profil}}, a_{\text{resources}}) \tag{1}$$

$$\text{Target}\ (t) = (t_{\text{vulnerability}}, t_{\text{consequences}}, t_{\text{type}}, t_{\text{resources}}) \tag{2}$$

These attributes are directly related to each other, since attack directions are always chosen from possible target vulnerabilities and are acceptable to an attacker if with the help of his resources on one system or another it is possible to cause the effect he needs [17]. Do not forget that the resources of both sides (targets - to protect, attacking - to conduct an attack) to improve the accuracy of risk assessment should be considered comprehensively. The s-axis is designed to store a set of marine systems used in the fleet, their technological vulnerabilities (weaknesses) and the possible negative consequences of their work:

$$\text{s-axis} = f_{\text{vulnerability}}(a_{\text{direction}}, t_{\text{vulnerability}}, t_{\text{consequences}}) \tag{3}$$

In this case, we will consider specialized systems that are planned to be used on crewless vessels. For crewless, factors such as the ship's environment (other vessels) at sea, transported cargo and current location should be taken into account. This is due to the traditional threats of piracy. Pirates today are increasingly using cyber attacks in their raids. YARA assumes movement on two routes with a deviation from the coastline of not more than 12 miles. In order to determine the level of autonomy of the vessel, we will use the determining SAE base for autonomous machines. The data are presented in Table 1. This model assumes 5 levels. According to the information currently available, the current prototype can be attributed to level 3, and the vessel to which YARA is aiming will have levels 5 or 4, depending on the conditions of autonomous movement.

Rolls-royce recently launched a new project called Advanced Autonomous Waterbourne Application (AAWA) to research autonomous vessels. On the new vessels it is planned to install a number of new sensors along with the traditional AIS. For software development, especially in the field of object recognition, they collaborate with Google.

Unlike the previous two projects, MAS aims to fully automate a vessel with limited routes. To this end, MAS includes in its project a wide range of technologies such as: new sensors, new communicators, small autonomous drones for collecting samples, new propulsion systems using renewable energy sources, etc. After analyzing the data of these projects, you can create a list of systems that will be on almost all future autonomous vessels.

Table 1. Separation of autonomy, values for the attacker and simplicity in conducting an attack in five levels.

Level	Autonomy of vessels	Stimulus for the attacker	Simplicity/resources
1	Minimum crew required for most or all ship operations	Practically not interesting for an attacker	State: advanced persistent threats, requires a national level of resources
2	Partial automation with crew for performing simple tasks, for example, advanced autopilot	Small value for an attacker	Corporate: advanced attacks that require significant resources
3	Conditional autonomy, potential crew intervention	Medium to moderate value for the attacker	Professional: moderate attack with significant investment in resources
4	Complete autonomy, crew rarely needed	Valuable for attacker and third parties	Basic attack: minimal skills or resource used
5	Complete autonomous operation of the vessel in all potential settings	Extremely valuable for the majority of players	Lack of skills, ready-made exploits are used

This list is shown in Fig. 1. In particular, the two-dimensional mapping of the MaCRA model illustrates the connection between technological systems and the possible negative consequences in the case if system data will be attacked.

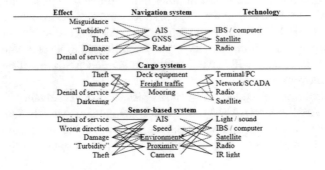

Fig. 1. Possible effects from attacks on various systems of an autonomous ship.

Let's take a closer look at AIS, GNSS, VDR, sensor systems and systems for mooring.

AIS is a multifunctional information and technical system, the equipment of which is installed on vessels and in coastal services in order to ensure the safety of shipping and automate the exchange of navigation information. AIS in the VHF range transmits messages to the maritime radio service containing information about the vessel, cargo, coordinates, course, destination, etc. The transmission frequency, for most vessels, is measured in the range of 2–10 s [18].

One of the key problems determining the effectiveness of the practical AIS use is the display of AIS information. This problem is not finally resolved, nor is it regulated by normative documentation or standards on AIS, there are only general requirements [9, 19]. On modern vessels, AIS data along with information from other sensors can be displayed on integrated navigation systems (INS) displays.

Using vulnerabilities in the AIS system, it is possible to intercept communications between vessels, send SOS signals, create nonexistent vessels on the map, send collision reports, disable AIS on the vessel, etc.

The GNSS system is a satellite system designed to determine location, time and speed around the world. It is based on GPS and GLONASS. This system includes user receivers and one or more constellations of satellites, management infrastructure and ground segments.

There are drawbacks in the GLONASS and GPS navigation systems, namely: they do not allow to ensure the accuracy necessary for safe navigation of vessels in narrow spaces with restricted freedom of vessel maneuvering and on approaches to ports; inability to provide prompt notification of system malfunctions. To eliminate these shortcomings and improve the accuracy of the readings of the GLONASS and GPS systems, a differential mode of operation is used to expand their functionality. This mode is the most promising for ensuring safe navigation in narrows and approaches to ports, and also allows solving a number of special navigation tasks [18].

The VDR (Voyage Data Recorder) system is designed for recording voyage data and is analogous to the "black box" used in aviation. The objectives of this system are: recording the vessel's voyage information, which includes technical and course data, voice recordings from the captain's bridge, saving this information to determine the circumstances in the event of an emergency. These devices are usually not protected from the possibility of changing or deleting the information received by a member of the crew. There is also the possibility of remote access through the network, which may affect the operation of the system [19]. On the crewless vessels, it can be considered that there is no one to change the data.

To ensure the mooring (docking) of the vessel, it is planned to use an automated docking system. This system operates using a radio channel. The vulnerability of the radio channel creates a serious problem, since this system is a tasty morsel for attackers s aimed at stealing cargo [20, 21]. On MAS and YARA vessels, various cranes and winches are included in the loading/unloading system, taking control of which can harm not only the cargo, but also the objects surrounding the vessel. Such attacks are unlikely on traditional vessels, as the mooring, loading and unloading process is controlled by the team. For crewless vessels such attacks are quite possible.

The failure of the sensory systems on the vessel with the crew will not create any particular problems in the management of the vessel, since the crew can operate with what they see. However, crewless vessels rely mainly on their sensors, so their decommissioning may also be the target of an attack.

The second axis of the MaCRA model simulates the ease of carrying out an attack on a vessel using vulnerabilities. Lightness depends on the attacker's resources and protection installed on the attack object.

$$\text{e-axis} = f_{\text{lightness}} \left(a_{\text{type}}, t_{\text{type}}, a_{\text{resources}}, t_{\text{resources}} \right) \tag{4}$$

Typically, crew experience and qualifications affect cyber risks, but for crewless vessels this is not significant. It should be borne in mind that the crewless vessel for cyber security has mainly technological means, and the attackers also have human resources, which simplifies attacks on the electronics of such vessels. Therefore, the autonomous protection of the vessel must exceed the combined capabilities of the attackers.

The third axis models the ultimate value of the attack. It is always necessary for an attacker to evaluate the result in order to decide whether to spend resources on a given attack.

$$\text{r-axis} = f_{\text{value}} \left(a_{\text{type}}, t_{\text{type}}, a_{\text{goal}}, t_{\text{consequences}} \right) \tag{5}$$

The existing model MaCRA classifies attackers into various types. However, for consideration of autonomous vessels, it is necessary to allocate only groups of intruders for whom this topic is relevant and considering autonomous vessels as a potential target. MaCRA models incentives for attackers using a five-level system (these types are presented in Table 1).

Terrorists (T): the actions of terrorists are aimed at undermining the sovereignty of the state, demonstrating the ineffectiveness of the ruling group actions, and also often have religious overtones.

Competitors (C): the actions of competitors are aimed at creating financial difficulties for a firm engaged in the same type of activity, in order to take this company out of business or take a leading position.

Activists (A): they can still be called "Hacktivists", directing their actions towards promoting political ideas, protecting human rights, ensuring freedom of information and freedom of speech.

Criminals (R): the actions of criminals are aimed at obtaining material gain, through theft, blackmail, smuggling.

3   Results

Processing the parameters of possible attackers and possible equipment for potential targets (YARA, MAS, AAWA) using the MaCRA model makes it possible to estimate cyber risks (Tables 2 and 3).

Table 2. Assessment for AIS, Cargo, GNSS

			AIS					Cargo				GNSS		
			I	II	III	IV	V	I	II	IV	V	I	III	V
A	S	Y	0-1	2-3	3-4	1	1	0-2	0-5	0-1	0-3	1	3-4	1
		M	0-1	2-3	3-5	1	0-1	0-2	0-5	0-1	1-2	1	3-5	1-2
		A	0-1	2-3	3-4	1	0-3	0-2	0-5	0-1	0-3	1	3-4	0-3
	E	Y	2-5	3-4	3-5	3-4	3-5	3-4	2-5	2-3	4	2-5	3-5	3-5
		M	2-4	3-4	3-5	3-4	3-5	3-4	2-5	2-3	4	2-4	3-5	3-5
		A	1-3	3-4	2-4	3-4	2-3	2-5	2-3	2-3	1-3	2-4	2-4	2-3
R	S	Y	1-2	2-5	3-5	1	1	0-3	0-5	0-1	0-3	1-2	3-4	1
		M	1-2	2-5	3-5	1	4	0-3	0-5	0-1	1	1-2	3-5	2-4
		A	1-2	3-5	3-5	2	0-3	0-3	0-5	0-1	0-3	1-2	3-4	0-3
	E	Y	2-5	3-4	3-5	3-4	3-5	3-4	2-4	2-4	2-5	2-5	3-5	3-5
		M	2-4	3-4	3-5	3-4	3-5	3-4	2-4	2-4	2-5	2-4	3-5	3-5
		A	1-3	3-4	2-4	3-4	2-4	2-4	2-4	2-4	2-3	1-3	2-4	2-4
C	S	Y	3-4	2	3-5	2	2-4	0-3	0-2	0-4	0-5	3-4	3-5	2-4
		M	3-4	2	4-5	2	0-1	0-3	0-2	0-4	1-2	3-4	2-5	2-4
		A	3-4	2	3-5	2	2-4	0-3	0-2	0-4	0-5	3-4	3-5	2-5
	E	Y	4-5	3-4	3-5	3-4	2-5	4-5	3-5	3-5	4-5	4-5	3-5	2-5
		M	3-5	3-4	3-5	3-4	2-5	3-5	3-5	3-5	4-5	4-5	3-5	2-5
		A	2-4	3-4	2-4	3-4	2-4	2-5	3-5	3-5	2-5	4-5	2-4	2-5
T	S	Y	3-5	2	2-4	2	3-5	0-4	0-2	0-3	0-5	3-5	2-4	3-5
		M	3-5	2	3-5	2	0-1	0-3	0-2	0-4	1-2	3-5	3-5	2-4
		A	3-5	2	2-4	2	3-5	0-3	0-2	0-4	0-5	3-5	2-4	3-5
	E	Y	4-5	3-4	3-5	3-4	2-5	4-5	3-5	3-5	4-5	4-5	3-5	2-5
		M	3-5	3-4	3-5	3-4	2-5	3-5	3-5	3-5	4-5	4-5	3-5	2-5
		A	2-4	3-4	2-4	3-4	2-4	2-5	3-5	3-5	2-4	4-5	2-4	2-5

Table 3. Assessment for mooring, radar and sensor systems

			Mooring			Radar			Sensor				
			I	II	V	I	II	IV	I	II	III	IV	V
A	S	Y	1-2	2-5	0-3	1	2	1-2	0-1	2-3	3-4	1	0-3
		M	1-2	2-5	0	1	2	1-2	0-1	2-3	3-5	1	1-2
		A	1-2	2-5	0-3	1	2	1-2	0-1	2-3	3-4	1	0-3
	E	Y	3-4	3-4	3-4	3-5	2	3	3-5	2-3	3-5	3-4	4
		M	3-4	3-4	3-4	2-3	2	3	3-5	2-3	3-5	3-4	4
		A	3-4	2-3	1-2	2	3	2-4	2-3	2-4	3-4	2-4	3
R	S	Y	1-3	3-5	0-3	1	2-3	1-2	1-2	1	3-4	1	0-3
		M	1-3	3-5	0-1	1	2-3	1	2	1-2	1	3-5	1
		A	1-3	3-5	0-3	1	2-3	1-2	1-2	1	3-4	1	0-3
	E	Y	3-4	3-4	1	3-5	2	3	3-5	2-4	3-5	3-4	2-5
		M	3-4	3-4	1	2-3	2	3	3-5	2-4	3-5	3-4	2-5
		A	2-4	3-4	2-3	1-2	2	3	2-4	2-4	2-4	3-4	2-4

(*continued*)

Table 3. (*continued*)

			Mooring			Radar			Sensor				
			I	II	V	I	II	IV	I	II	III	IV	V
C	S	Y	1-2	1-2	0-5	3-5	2	1-2	3-4	2	3-5	2	0-3
		M	1-2	1-2	1	3-5	2	1-2	3-4	2	4-5	2	2-4
		A	1-2	1-2	0-5	3-5	2	1-2	3-4	2	3-5	2	0-3
	E	Y	3-4	3-4	3-4	3	2-3	3	4-5	2-4	3-5	3-4	4-5
		M	3	3-4	3-4	3	2-3	3	3-4	2-4	3-5	3-4	4-5
		A	2-3	3-4	1-2	2	2-3	3	3-4	2-3	2-4	3-5	1-3
T	S	Y	3-5	1-2	0-5	1	2	2-5	3-5	2	1-2	2	0-3
		M	1-2	1-2	1	1	2	2-5	3-5	2	2-5	2	2-4
		A	1-2	1-2	0-5	1	2	2-5	3-5	2	1-2	2	0-3
	E	Y	2-5	3-4	3-4	3	2-3	2-3	4-5	2-4	3-5	3-4	4-5
		M	2-3	3-4	3-4	3	2-3	2-3	3-4	2-4	3-5	3-4	4-5
		A	2-3	3-4	1-2	2	3-2	3-2	3-4	2-4	2-4	3-4	1-3

In these tables, for each type of intruder, an assessment of the vulnerability and ease of attack on the systems under consideration for each system is given by type of attack, where:

Damage-I, DOS-II, Misguidance-III, "Turbidity"-IV, theft-V.

Figure 2 presents the results of an assessment of the benefits of hacking a particular system for an attacker, obtained using the MaCRA model. Systems located in the lower left area have minimal value for an attacker, and attacks on them are the most complex. Systems that fall into the upper right area have the greatest value, attacks on these systems are the simplest.

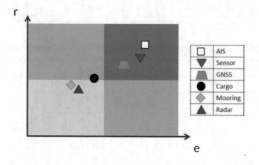

Fig. 2. The final two-dimensional mapping of the model.

4   Discussion

Analyzing the result, we note that special attention should be paid to the protection of AIS systems, GNSS and sensors, increasing their security against cyber attacks. For safe navigation, these systems must be designed so that the value of their hacking decreases for the attacker. It is safer to design multilayer systems with redundant duplication of control nodes. This will increase the cyber systems resistance of autonomous vessels to attacks and cause great difficulty for the attackers.

It is worth noting that the estimates were based on the information published so far about the autonomous vessels under consideration in this work and should be adjusted after their launch (2020), when the equipment installed on them becomes known. In future work, it is planned to expand and refine the current model in accordance with the new data.

5   Conclusions

In conclusion, we note that autonomous vessels will undoubtedly be able to make a revolution in the field of maritime transport. But it is important to remember about ensuring cyber security on them. After all, everyone knows that with the growth of human cyber capabilities, the possibilities of cybercriminals are growing.

References

1. Sokolov, S.S.: The safety assessment of critical infrastructure control system. In: Proceedings of the 2018 IEEE International Conference "Quality Management, Transport and Information Security, Information Technologies", pp. 154–157 (2018). https://doi.org/10.1109/ITMQIS.2018.8524948
2. Chernyi, S., Zhilenkov, A., Sokolov, S., Nyrkov, A.: Algorithmic approach of destabilizing factors of improving the technical systems efficiency. Vibroeng. Procedia 13, 261–265 (2017). https://doi.org/10.21595/vp.2017.19003
3. Nyrkov, A.P., Abramova, K.V., et al.: Technologies of safety in the bank sphere from cyber attacks. In: 2018 IEEE Conference of Russian Young Researchers in Electrical and Electronic Engineering (EIConRus), pp. 102–104 (2018). https://doi.org/10.1109/EIConRus.2018.8317040
4. Reshnyak, V., Sokolov, A., et al.: Inland waterway environmental safety. J. Phys: Conf. Ser. 1015(4), 042049 (2018). https://doi.org/10.1088/1742-6596/1015/4/042049
5. Nyrkov, A., Pavlova, L., Sokolov, S., et al.: Development and analysis of models for handling by marine transport. Far East J. Electron. Commun. 17(5), 1253–1264 (2017). https://doi.org/10.17654/EC017051253
6. Nyrkov, A., Maltsev, V.: Mathematical foundations and software simulation of stress-strain state of the plate container ship. J. Marit. Res. XIII(I), 47–52 (2016). https://doi.org/10.1007/978-3-319-33581-0_24
7. Bukharmetov, M.R., Kuznetsov, D.G., et al.: Robust method for protecting electronic document on waterway transport with steganographic means by embedding digital watermarks into images. In: Proceedings of the 16th Conference on Reliability and Statistics in Transportation and Communication, pp. 507–514 (2017). https://doi.org/10.1016/j.proeng.2017.01.097
8. Nyrkov, A., Goloskokov, K., Koroleva, E., et al.: Mathematical models for solving problems of reliability maritime system. In: Advances in Systems, Control and Automation. LNEE, vol. 442 (2018). https://doi.org/10.1007/978-981-10-4762-6_37

9. Nyrkov, A.P., Zhilenkov, A.A., Sokolov, S.S., et al.: Hard- and software implementation of emergency prevention system for maritime transport. Autom. Remote Control **79**, 195 (2018). https://doi.org/10.1134/S0005117918010174

10. Nyrkov, A.P., Karpina, A.S., et al: Providing the integrity and availability in the process of data transfer in the electronic documents management systems of transport-logistical clusters. In: Proceedings of the 2nd International Conference on Industrial Engineering, Applications and Manufacturing, ICIEAM (2016). https://doi.org/10.1109/ICIEAM.2016.7910915

11. Sokolov, S., Zhilenkov, A., Nyrkov, A., Chernyi, S.: The use robotics for underwater research complex objects. In: Computational Intelligence in Data Mining. AISC, vol. 556 (2017). https://doi.org/10.1007/978-981-10-3874-7_39

12. Nyrkov, A., Chernyi, S., et al.: The effective optimization methods of port activity on the basis of algorithmic model. Int. J. Electr. Comput. Eng. **7**(6), 3578–3582 (2017). https://doi.org/10.11591/ijece.v7i6

13. https://news.rambler.ru/weapon/38708781-pervye-bespilotnye-suda-poyavyatsya-v-rossii-s-2018-goda. Accessed 21 Jan 2019

14. Sokolov, S.S., Glebov, N.B., Boriev, Z.V., et al.: Formation of common legal framework for biometric data security based on contradictions in international legislation. IOP Conf. Ser. Earth Environ. Sci. **194**(2) (2018). https://doi.org/10.1088/1755-1315/194/2/022039. art. no. 022039

15. Veselkov, V., Vikhrov, N., Nyrkov, A., et al.: Development of methods to identify risks to build up the automated diagnosis systems. In: 2017 IEEE NW Russia Young Researchers in Electrical and Electronic Engineering Conference (EIConRusNW), Saint-Petersburg, pp. 598–601 (2017). https://doi.org/10.1109/EIConRus.2017.7910625

16. Nyrkov, A.P., Katorin, Y.F., Gaskarov, V.D., et al.: Aggregation process for implementation of application security management based on risk assessment. In: 2018 IEEE NW Russia Young Researchers in Electrical and Electronic Engineering Conference (EIConRusNW), Saint-Petersburg, pp. 98–101 (2018). https://doi.org/10.1109/EIConRus.2018.8317039

17. Nyrkov, A.P., Novoselov, R.O., Glebov, N.B., et al.: Databases problems for maritime transport industry on platform highload. In: Proceedings of the 2018 IEEE International Conference "Quality Management, Transport and Information Security, Information Technologies", pp. 132–135 (2018). https://doi.org/10.1109/ITMQIS.2018.8525058

18. Balduzzi, M.: AIS Exposed: Understanding Vulnerabilities & Attacks 2.0 (video), Black Hat Asia (2014)

19. Nyrkov, A.P., Belousov, A.S., Sokolov, S.S.: Algorithmic support of optimization of multicast data transmission in networks with dynamic routing. Mod. Appl. Sci. **10**(5), 162–176 (2015). https://doi.org/10.5539/mas.v9n5p162

20. Zetter, K.: Hackers Could Heist Semis by Exploiting This Satellite Flaw, Wired, 30 July 2015

21. https://news.1k.by/events/Stydentyi_Tehasskogo_Yniversiteta_ypravlyali_yahtoii_s_pomoshciy_yyazvimosti_GPS-24655.html. Accessed 13 Feb 2019

Energy-Saving Driving of Heavy Trains

Natalya Ryabchenok$^{(\boxtimes)}$ (ID), Tatyana Alekseeva (ID),
Leonid Astrakhancev (ID), Nikolaj Astashkov (ID),
and Vladimir Tikhomirov (ID)

Irkutsk State Transport University, 664074 Irkutsk, Russia
astranal@mail.ru

Abstract. The paper is aimed at studying the energy processes in the electric traction drive and in the contact network in order to reduce energy losses and increase the performance of electric traction of trains. The calculation method is based on a refined law of energy saving in an electromagnetic field using computer simulation and the FFT spectral analysis of a nonsinusoidal voltage, current. The objects of research are the contact networks of alternating and direct current, the well-known pulse power controllers, and the controllers of the input electrical resistance of the traction electric drive, proposed in the paper. Research has proven the cause of unsatisfactory energy efficiency and a reduction in the speed of transport work - a reduction in the duration of the use of the electrical potential of a contact network for electric traction of a train. The inductance of the AC contact network and the inductance of the traction transformer cause the input converter of the AC electric locomotive to work in the short circuit mode. Therefore, considerable loss of electrical energy occurs in this equipment. A further increase in the carrying and traffic capacity of the railway with energy-saving electric traction can be realized by applying increased voltage in DC contact networks and electric semiconductor variators to control collector and three-phase asynchronous motors on the traction rolling stock. Engineering tasks to improve the electric traction of trains can be successfully solved by specialists who are free from the idea of isolating the main harmonic components of voltage and current from the entire spectrum of harmonics and from the task of developing compensators for shear power and distortion power.

Keywords: Electric locomotive · Traction electric drive ·
Traction electric motor · Voltage · AC and DC current · Power controller

1 Introduction

The intensification of work aimed at increasing the carrying and traffic capacity of the railway is due to the increasing demand for the transportation of raw materials and finished products of enterprises, the growth of foreign orders for the transit transport of goods. The effect of the experienced organizational technologies of transport work is achieved due to the unjustified intensive wear of the track equipment, locomotive facilities, and the traction power supply system. The need for a comprehensive technical improvement of the traction electric rolling stock (ERS) equipment and the traction

© Springer Nature Switzerland AG 2020
V. Murgul and M. Pasetti (Eds.): EMMFT 2018, AISC 982, pp. 491–508, 2020.
https://doi.org/10.1007/978-3-030-19756-8_47

power supply system to improve the performance of railways follows from the results of pilot driving of heavy, articulated trains and operating high-speed passenger ERS [1, 2].

Solving industry problems is possible through the use of the results of modern scientific theories, the introduction of the achievements of scientific and technological progress, and the use of engineering innovation. The energy efficiency of train traction using AC and DC networks is assessed in the paper. For comparison, the energy characteristics of traction ERS with pulsed and continuous use of the electric potential of contact networks for train traction are investigated. The calculations were performed according to the method based on the updated law of energy conservation in the electromagnetic field [3].

2   State of the Issue. Setting the Research Task

The theory of energy processes [4, 5], which is used in educational institutions in Russia and abroad, in electrical circuits with semiconductor devices is based on isolating the main voltage harmonic and the first current harmonic from the spectrum of harmonic components of the nonsinusoidal voltage and current at the converter input. With the permissible error, this method allowed us to simplify the calculation of the active power at the input of semiconductor converters with various pulse methods of controlling power semiconductor devices (PSD). The calculation method, which is not correct by mathematical rules, is later accompanied by the introduction into the known power balance [6, 7] of the so-called shear power and distortion power, which are commonly considered as reactive powers. The resulting unsatisfactory assessment of the energy efficiency of pulsed power controllers is taken by experts as a guide to the development of additional equipment to compensate for shear power and distortion power [8–11]. Unfortunately, in the educational process of educational institutions, the attention of future specialists is not drawn to the reduction of the duration of irreversible conversion of electrical energy into another kind of energy by reactive elements of electrical circuits [12, 13]. With the help of an improved law of energy conservation in the electromagnetic field [3], new energy characteristics have been developed.

The proposed energy characteristics take into account the reduction of the irreversible conversion of electric energy into another type of energy due to the nonconductive state of semiconductor devices of converters and reactive elements of the electric power system.

3   Research Method

The balance of power at the input of the controller, taking into account the reduction in the duration of irreversible conversion of electric energy into a different type of energy by reactive elements and PSD of power controller (1)

$$\sqrt{S_G^2 - \Delta S^2} = \sqrt{P^2 + Q^2}, \tag{1}$$

where

- ΔS – part of the total power SG at the input of the controller, which is not used for the irreversible conversion of electric energy into another type of energy and for energy exchange in an electrical circuit;
- P – the total active power of the harmonic components of the voltage and current at the input of the power controller;
- Q – the total reactive power of the harmonic components of the voltage and current at the input of the power controller.

The calculation and measurement of the components of the proposed power balance (4) can be performed using Fourier transform, analytical calculations, mathematical modeling, and control and management instruments in practice.

To calculate the power ΔS, which is formed at the input of the controller, and which takes into account the electric potential of the energy source during the nonconductive state of PSD of the controller

$$\Delta S = \sqrt{\sum_{k=0}^{n} U_{pk}^2 \cdot I_k^2} = U_P \cdot I_{in}, \qquad (2)$$

where

- U_{pk} – the effective voltage of the k-th harmonic component at the input of the power controller during the nonconductive state (pause) of PSD;
- I_k – the effective current of the k-th harmonic component at the input of the power controller; k – the number of the harmonic component;
- n – the number of the last accounted harmonic;
- U_P – the effective voltage at the input of the power controller during the nonconductive state (pause) of PDS;
- I_{in} – the effective current at the input of the power controller.

Active power at the input of the controller characterizes a part of electrical energy that is irreversibly converted to a different type of energy in the load and to thermal energy losses in the power circuit of the traction electric drive of ERS

$$P = U_{c0} \cdot I_0 + \sum_{k=1}^{n} U_{ck} \cdot I_k \cdot cos\varphi_k, \qquad (3)$$

where

- U_{c0} – the constant component of the voltage at the input of the power controller during the conductive state of PSD;
- U_{ck} - the effective voltage of the k-th harmonic component at the input of the power controller during the conductive state of PSD;
- I_0 - the constant component of the current at the input of the power controller during the conductive state of PSD;
- φ_k – the phase angle of the k-th current harmonic relative to the same k-th harmonic of the voltage at the input of the power controller.

Reactive power at the input of the controller, which characterizes part of the electrical energy consumed for energy exchange between the energy source and the reactive elements of the electrical circuit

$$Q = \pm\sqrt{\sum_{k=1}^{n} U_{ck}^2 \cdot I_k^2 \cdot sin^2\varphi_k}. \tag{4}$$

Argument of full power and input electrical resistance $\varphi_{\sum}$ of a power controller with a load

$$\varphi_{\sum} = arctg\left[\frac{\pm\sqrt{\sum_{k=1}^{n} U_{ck}^2 \cdot I_k^2 \cdot sin^2\varphi_k}}{U_{c0} \cdot I_0 + \sum_{k=1}^{n} U_{ck} \cdot I_k \cdot cos\varphi_{k,}}\right]. \tag{5}$$

Using Eqs. (1–5) and mathematical modeling in the Simulink environment of the MATLAB program, it is possible to get an estimate of the dynamics and energy efficiency of a traction electric drive of one section of the «Yermak» electric locomotive when operating in the middle of an inter-station zone with duplicate power supply of the contact network.

4 Research Results

Set-up, start-up and acceleration of the electric drive is performed at the nominal moment of resistance on the shafts of the traction electric motors (TEM) NB-514B, constant control angle $\alpha = 74.52$ electrical degrees of the thyristors of the reversible converter (RC) with acceleration to a steady speed of 5 km/h (Fig. 1).

During set-up, start-up and acceleration of an electric locomotive up to a steady speed of 5 km/h, for traction and acceleration of the train (Fig. 2), the contact network is loaded with a current 1.6 times higher than the steady-state current and is 59 A.

During set-up, start-up and acceleration of ERS with RC, the maximum current in the secondary winding of the traction transformer reaches $I_2 = 3536$ A and decreases according to aperiodic law to $I_2 = 2547$ A The thyristor control angle is changed by a pulse-phase control system manually or by an automatic control system ranging from π to 0 electrical degrees, and in this case $\alpha = 74.52$ electrical degrees. Switching angles of the thyristors of RC $\gamma_1 = 0.4$, $\gamma_2 = 15.7$ electrical degrees can be determined using oscillograms of voltages and currents in the primary and secondary windings of the traction transformer (Fig. 3).

The total power at the input of the traction drive $S_G = U_{tp} \cdot I_{in} = 27250 \cdot 35.51 = 967.65$ kVA is calculated using the readings of the instruments (Fig. 1) to measure the effective nonsinusoidal voltage at the current collector U_{tp} and the effective nonsinusoidal current in the primary winding of the traction transformer I_{in}. The effective value of the voltage in the primary winding of the traction transformer during the conductive state of the RC thyristors (Fig. 4) was measured using the RMS device and is shown on the display $U_c = 20338$ V (Fig. 1).

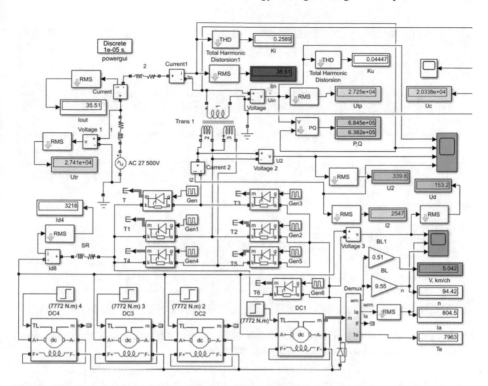

Fig. 1. Mathematical model of traction electric drive with RC of one section of a freight electric locomotive.

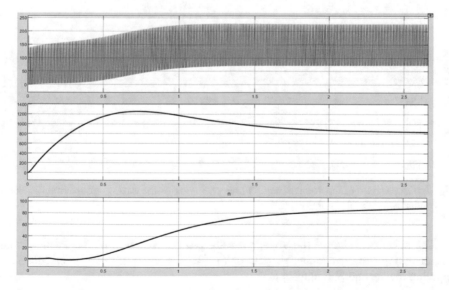

Fig. 2. Oscillograms of voltage (first oscillogram), current (second oscillogram) in windings and shaft rotation speed n (third oscillogram) of TEM during set-up, start-up and acceleration of a train.

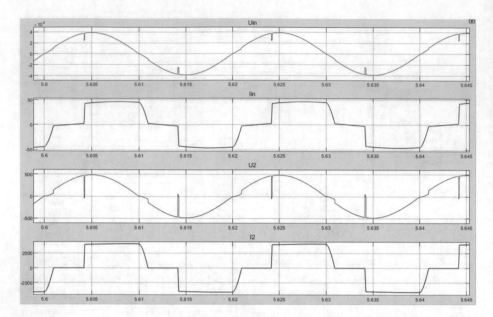

Fig. 3. Oscillogramms of voltage U_{in}, current I_2 in the primary and U_2, I_2 in the secondary winding of the traction transformer during RC operation.

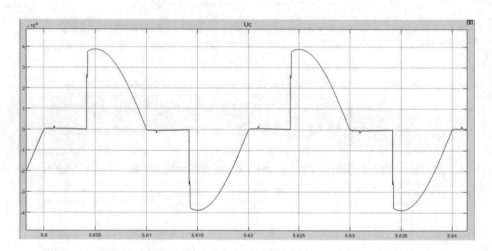

Fig. 4. Oscillogram of the voltage in the primary winding of the traction transformer U_c during the conductive state of the RC thyristors.

The effective value of the voltage on the primary winding of the traction transformer during the nonconductive state of the RC thyristors U_P can be calculated by applying the second Kirchhoff's law

$$U_P = \sqrt{U_{tp}^2 - U_C^2} = \sqrt{27250^2 - 20338^2} = 18136 \text{ V}.$$

Power ΔS, which is formed at the input of the controller due to the nonconductive state of the RC thyristors, can be calculated by the formula (2)

$$\Delta S = U_P \cdot I_{in} = 18136 \cdot 35,51 = 644,01 \text{ kVA}.$$

To calculate the active and reactive power in the power balance (1), we can use the FFT analysis using oscilloscopes and the powergui block (Fig. 1).

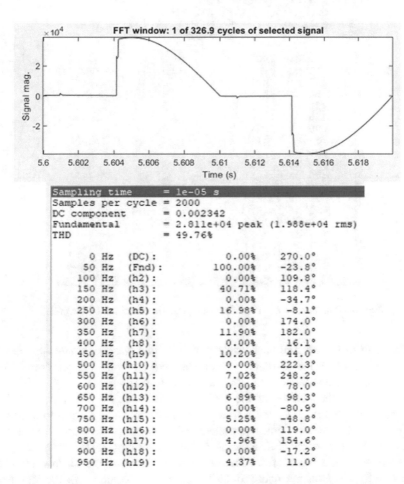

```
Sampling time        = 1e-05 s
Samples per cycle = 2000
DC component        = 0.002342
Fundamental         = 2.811e+04 peak (1.988e+04 rms)
THD                 = 49.76%

    0 Hz  (DC):          0.00%    270.0°
   50 Hz  (Fnd):       100.00%    -23.8°
  100 Hz  (h2):          0.00%    109.8°
  150 Hz  (h3):         40.71%    118.4°
  200 Hz  (h4):          0.00%    -34.7°
  250 Hz  (h5):         16.98%     -8.1°
  300 Hz  (h6):          0.00%    174.0°
  350 Hz  (h7):         11.90%    182.0°
  400 Hz  (h8):          0.00%     16.1°
  450 Hz  (h9):         10.20%     44.0°
  500 Hz  (h10):         0.00%    222.3°
  550 Hz  (h11):         7.02%    248.2°
  600 Hz  (h12):         0.00%     78.0°
  650 Hz  (h13):         6.89%     98.3°
  700 Hz  (h14):         0.00%    -80.9°
  750 Hz  (h15):         5.25%    -48.8°
  800 Hz  (h16):         0.00%    119.0°
  850 Hz  (h17):         4.96%    154.6°
  900 Hz  (h18):         0.00%    -17.2°
  950 Hz  (h19):         4.37%     11.0°
```

Fig. 5. Results of spectral analysis of periodic functions of Uc and I_{in}.

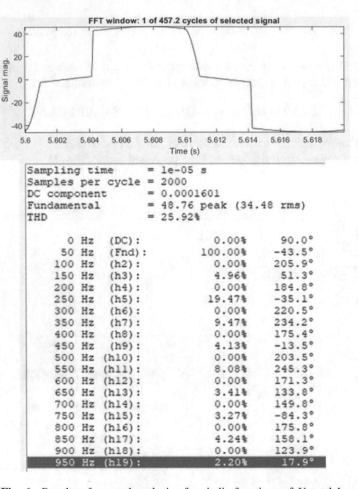

Fig. 6. Results of spectral analysis of periodic functions of Uc and I_{in}.

As a result of the spectral analysis of nonsinusoidal curves Uc and I_{in}, the values and initial phases of same harmonic components were obtained (Figs. 5 and 6). The effective value of the voltage of the first harmonic is $Uc_1 = 19880$ V, $f_1 = 50$ Hz, the effective value of the current of the first harmonic is $I_{in1} = 34.48$ A, $f_1 = 50$ Hz. According to the formula (3), the active power is calculated taking into account the phase angle of the same harmonic components of the current with respect to the voltage $P = 686.41$ kW; the calculation error is 0.3%. According to the formula (4), the reactive power at the input of the traction electric drive of the electric locomotive section $Q = 231.74$ kvar is calculated. With the help of formula (5), the argument of total power, the input electrical resistance of the traction drive $\varphi_\Sigma = 18.4°$, which indicates the active-inductive nature of the input electrical resistance, is calculated. The power factor of the traction drive $K_M = 686.41/967.65 = 0.71$.

The power balance (1) is checked, and the calculation error is determined.

$$\sqrt{S_G^2 - \Delta S^2} = \sqrt{P^2 + Q^2};\ \sqrt{967,65^2 - 644,01^2} = \sqrt{686,41^2 + 231,74^2}.$$

722,217 kVA $\approx$ 724,474 kVA, the calculation error is 0.3%.

The use of the electrical potential of an AC contact network for train traction is reduced due to the nonconductive state of the RC thyristors, the current switching by the RC thyristors, and the active-reactive nature of the input electric resistance of the electric traction drive, which are the **reasons** for the decrease in the energy efficiency of electric train traction.

To estimate the power and energy losses, one can calculate the efficiency factor (EF). The total power on the shafts of 4 TEM $Ps = 306.2$ kW was determined taking into account the moment of resistance on the shaft of each TEM 7772 N·m and the shaft rotation speed of 94 rpm (Fig. 1). The efficiency factor of the traction motors ηTEM is calculated taking into account the voltage U_d and the current Id in the inverted current loop

$$\eta_{TEM} = \frac{P_s}{U_d \cdot I_{d4}} \cdot 100\% = \frac{306,2}{153,2 \cdot 3218} \cdot 100\% = 62\%.$$

The efficiency factor of RC η_{RC}

$$\eta_{RC} = \frac{U_d \cdot I_{d4}}{P} \cdot 100\% = \frac{153,2 \cdot 3218}{686,41} \cdot 100\% = 72\%.$$

High losses of electrical energy in RC are due to the operation of the traction transformer, which is a constructive element of RC, in the short circuit mode during the switching of the current by the thyristors. The efficiency factor of the traction motor with RC $\eta = 44.6\%$. The active power losses in the AC contact network are 3.152 kW, and the active power losses in RC and traction transformer of ERS are 192.2 kW. Thus, in a traction transformer and RC of ERS, the active power loss is 98%, and the active power loss in an AC contact network with duplicate power supply is 2% of the total active power loss in these devices.

5   Recommendations for the Use of Research Results

The power balance and proposed energy characteristics (1–5) showed that scientific work should focus on the development of power controllers with control of the size and maintaining the active nature of the input electrical resistance of the ERS traction electric drive instead of developing compensating devices for shear power and distortion power.

The papers [15, 16] set forth the principles of operation of innovative power controllers for the traction electric drive of ERS with input electrical resistance control, ranging from infinitely large $Z_{in} = U_{tp}/I_{in} \approx \infty$ to Z_{in} depending on the traction and speed mode of ERS. For comparison, the calculation of the energy characteristics of the traction electric drive of the electric locomotive section with the replacement of RC (Fig. 1) with a power controller with controlled input electrical resistance by mathematical modeling (Fig. 7) is performed.

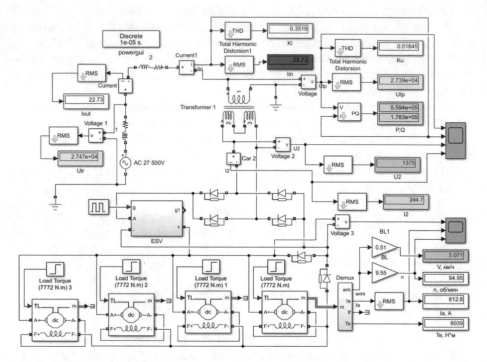

Fig. 7. Mathematical model of a traction electric drive with controlled input electrical resistance of one section of a freight electric locomotive.

On ERS of AC with a traction transformer that practically runs at idle speed before starting the drive. Therefore, the value of its input electrical resistance is large. With the supply of voltage U_{tp} to the primary winding of the transformer, from the sections of the secondary winding, the total voltage of the sections U_2 is applied to the input of the rectifier assembled on the diodes. The intermediate energy storage device in the ESV (Electric Semiconductor Variator) unit is charged to the amplitude value of the voltage U_2, the rectifier diodes are locked, and the input electrical resistance of the traction electric drive is $Z_{in} = U_{tp}/I_{in} \approx \infty$. When feeding rectangular control pulses to the input g of the ESV unit, energy is taken from the intermediate storage device, and the pulse voltage from terminals +, - of the ESV unit is applied to the windings of four TEMs (Figs. 5 and 6). The voltage on the intermediate energy storage device decreases, and the unlocking and locking of the rectifier diodes shifts from amplitude to the moments of the transition of the alternating voltage curve through zero. A current flows from the secondary winding of the traction transformer to charge the energy storage device $I_2 = 394.7$ A. The input electrical resistance of the traction electric drive decreases to $Z_{in} = U_{tp}/I_{in} = 1375/394.7 = 3.8484$ O. In the set-up, start-up and acceleration mode of ERS with ESV, the maximum current from the secondary winding of the traction transformer for charging the intermediate storage device is $I_2 = 707$ A and decreases according to the oscillatory damped law to $I_2 = 394.7$ A, which is 5–6 times less than the current in the set-up, start-up and acceleration mode of EPS with RC.

Total power at the input of the traction electric drive $S_G = U_{tp} \cdot I_{in} = 27.39 \cdot 22.73 = 622.57$ kVA. Since the rectifier is assembled on diodes, in this mode of operation, the electrical potential of the contact network is used for electric train traction, for the operation of a traction transformer in the short circuit mode during current switching in the rectifier diodes, and for a change in the magnetic field energy of the traction transformer. During the switching of the current in the rectifier diodes $\gamma_2 = 10.8°$, the electrical potential of the contact network at the input of the traction electric drive is used for operation of the traction transformer in the short circuit mode, therefore, $\Delta S = 0$ (2).

To calculate the active and reactive power in the power balance (1), we can use the FFT analysis using an oscilloscope and the powergui block (Fig. 7). As a result of the spectral analysis of the nonsinusoidal U_{tp} and I_{in} curves, the magnitudes and initial phases of the same harmonic components were obtained (Fig. 8).

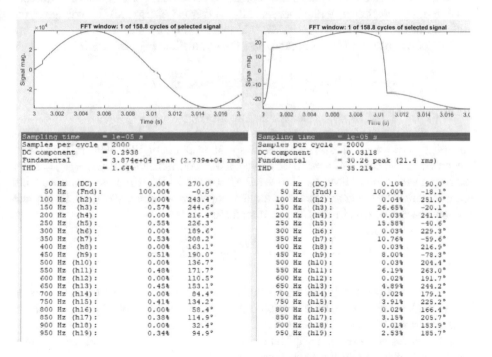

Fig. 8. Results of spectral analysis of periodic functions U_{tp} and I_{in}.

The effective value of the voltage of the first harmonic is $U_{TP1} = 27390$ V, $f_1 = 50$ Hz, the effective value of the current of the first harmonic is $Iin_1 = 21.40$ A, $f_1 = 50$ Hz (Fig. 7).

According to the formula (3), the total active power of the harmonic components taken into account (Fig. 8) is calculated taking into account the phase angle of the same harmonic components of the current with respect to the voltage $P = 558.7$ kW. According to the formula (4), the reactive power at the input of the traction electric drive of the section of the electric locomotive $Q = 177.4$ kvar is calculated. On the instrument indicator P, Q (Fig. 7), the active power is 559.4 kW, the reactive power is 178.3 kvar, the calculation error does not exceed 0.1%. With the help of formula (5), the argument of total power is calculated, the input electrical resistance of the traction electric drive $\varphi_\Sigma = 17,7\,^\circ$, which indicates the active-inductive nature of the input electrical resistance. The power factor of the traction electric drive $K_M = 558.7/622.57 = 0.897$.

The power factor of the traction electric drive with ESV is 20.8% higher than the power factor of the traction electric drive with RC. In ERS with RC, for irreversible conversion of electrical energy into another type of energy, the voltage on the current collector during the conductive state of the RC thyristors $U_C = 20338$ V is used, numerically equal to the geometric sum of effective values taken into account odd harmonic components of voltage and current (Figs. 5 and 6). For irreversible conversion of electrical energy into mechanical and thermal energy, all the voltage at the collector of ERS with ESV $U_{TP} = 27390$ V is used (Fig. 8).

The power balance (1) is checked, and the calculation error is determined.

$$\sqrt{S_G^2 - \Delta S^2} = \sqrt{P^2 + Q^2};\, 622,57 = \sqrt{558,7^2 + 177,4^2}.$$

622,57 kVA $\approx$ 586,20 kVA, the calculation error is 5,8%.

The traction and high-speed operation mode of the electric locomotive section does not differ from the previously considered operation mode of ERS with RC, therefore the power at the shafts of four TEMs $Rs = 309.1$ kW, efficiency factor $\eta_{TEM} = 62\%$. Active power of four TEMs $P_d = 309.1/0.62 = 498.53$ kW, which is consumed from the ESV unit. The efficiency factor of the developed power controller η_{ESV}, taking into account the power loss in the traction transformer

$$\eta_{ESV} = \frac{P_d}{P} \cdot 100\% = \frac{498,53}{558,7} \cdot 100\% = 89,2\% .$$

The active power loss in the AC contact network is 1.292 kW, and the active power loss in the ESV and in the ERS traction transformer is 60.264 kW. Thus, in a traction transformer and ESV of ERS, the active power loss is 95%, and the active power loss in the AC contact network with duplicate power supply is 5% of the total active power loss in these elements of the traction electric drive.

The current at the ESV output is the total current of four TEMs and is $I_d = 4 \cdot 812.8 = 3251.2$ A. ESV has the property of an electric variator. If the power loss in the ESV is neglected, then the voltage at the ESV output is less than the voltage at the ESV input $U_2 = 1375$ V, therefore the current at the input of ESV $I_2 = 394.7$ A is 8 times less than the current at the output of ESV I_d. Efficiency factor of a traction electric drive with ESV $\eta = 55\%$ and higher by 18.9% then efficiency factor of a traction electric drive with RC due to reducing current, reducing power losses in the traction transformer, in the rectifier of the developed power controller, and excluding the smoothing reactor from the electric circuit of the electric drive. The effective current in the contact network is reduced by 36% when the ERS is operating with the developed power controller compared to the current when ERS is operating with RC.

To study the energy efficiency of driving a train weighing 7100 tons, the conditions for grip of wheels of ERS are checked. If the wheel pressure on the rails is 25 ts, then taking into account the grip limit for AC electric locomotives at an hourly speed of 45–50 km/h, the traction force F_{gr} = 849.3-8 843.7 kN is realized by the three-section ERSs previously considered. The tractionof a train with a three-section electric locomotive 3ES5K in the middle of a contact network with duplicate power supply (Fig. 9) is performed at a speed of 42.95 km/h, which is limited by the voltage U_{tp} = 24.97 kV on the ERS current collector.

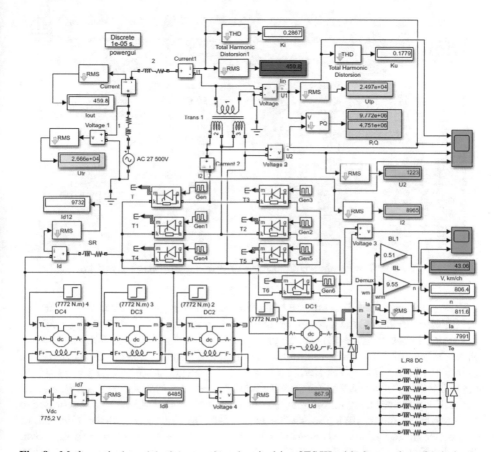

Fig. 9. Mathematical model of the traction electric drive 3ES5K with the traction of train in the middle of a contact network with duplicate power supply.

The total power at the input of the traction electric drive $S_G = U_{tp} \cdot I_{in} = 24970$ $459.9 = 11483.70$ kVA is calculated using the readings of the instruments (Fig. 9) to measure the effective nonsinusoidal voltage at the current collector U_{tp} and the effective nonsinusoidal current in the primary winding of the traction transformer I_{in}. Power $\Delta S = 0$ at the input of RC. To calculate the active and reactive power in the power balance (1), it is possible to use the FFT analysis with the help of an oscilloscope and the powergui block (Fig. 9). As a result of the spectral analysis of nonsinusoidal voltage and current curves, the magnitudes and initial phases of the same harmonic components were obtained (Fig. 10).

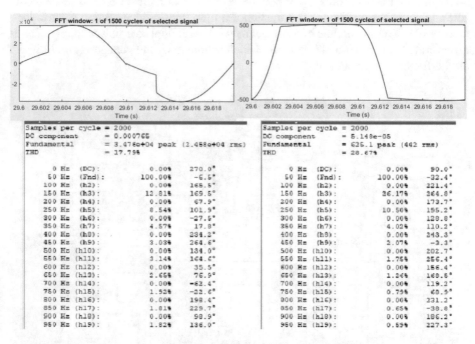

Samples per cycle = 2000			Samples per cycle = 2000		
DC component = 0.000765			DC component = 5.148e-05		
Fundamental = 3.476e+04 peak (2.458e+04 rms)			Fundamental = 625.1 peak (442 rms)		
THD = 17.79%			THD = 28.67%		
0 Hz (DC):	0.00%	270.0°	0 Hz (DC):	0.00%	90.0°
50 Hz (Fnd):	100.00%	-6.5°	50 Hz (Fnd):	100.00%	-32.4°
100 Hz (h2):	0.00%	165.5°	100 Hz (h2):	0.00%	221.4°
150 Hz (h3):	12.81%	169.5°	150 Hz (h3):	26.17%	264.8°
200 Hz (h4):	0.00%	67.9°	200 Hz (h4):	0.00%	173.7°
250 Hz (h5):	8.54%	101.9°	250 Hz (h5):	10.50%	155.2°
300 Hz (h6):	0.00%	-27.5°	300 Hz (h6):	0.00%	128.8°
350 Hz (h7):	4.57%	17.8°	350 Hz (h7):	4.02%	110.2°
400 Hz (h8):	0.00%	234.2°	400 Hz (h8):	0.00%	243.3°
450 Hz (h9):	3.03%	264.6°	450 Hz (h9):	2.07%	-3.3°
500 Hz (h10):	0.00%	134.0°	500 Hz (h10):	0.00%	202.7°
550 Hz (h11):	3.14%	164.6°	550 Hz (h11):	1.75%	256.4°
600 Hz (h12):	0.00%	35.5°	600 Hz (h12):	0.00%	156.4°
650 Hz (h13):	2.66%	76.9°	650 Hz (h13):	1.24%	169.6°
700 Hz (h14):	0.00%	-62.4°	700 Hz (h14):	0.00%	119.2°
750 Hz (h15):	1.52%	-22.6°	750 Hz (h15):	0.79%	68.5°
800 Hz (h16):	0.00%	193.4°	800 Hz (h16):	0.00%	231.2°
850 Hz (h17):	1.81%	229.7°	850 Hz (h17):	0.65%	-38.8°
900 Hz (h18):	0.00%	98.9°	900 Hz (h18):	0.00%	186.2°
950 Hz (h19):	1.82%	136.0°	950 Hz (h19):	0.59%	227.3°

Fig. 10. Results of spectral analysis of periodic functions U_{tp} and I_{in}.

The effective value of the voltage of the first harmonic is $U_1 = 24580$ V, $f_1 = 50$ Hz, the effective value of the current of the first harmonic is $I_1 = 442$ A, $f_1 = 50$ Hz. According to the formula (3), the active power is calculated taking into account the phase angle of the same harmonic components with respect to the voltage $P_1 = 9797.8$ kW, $P_3 = 98.5$ kW, $P_5 = 44.4$ kW, $P_7 = -12.3$ kW, $P_9 = -1.0$ kW, $P_{11} = 2.9$ kW, $P_{13} = -1.5$ kW, $P_{15} = 1.1$ kW, $P_{17} = -1.3$ kW, $P_{19} = 1.2$ kW, and the total active power of the harmonics taken into account is $P = 9929.90$ kW. According to the formula (4), the reactive power at the input of the traction electric drive of 3 sections of the electric locomotive $Q_1 = 4694.3$ kvar, $Q_3 = 350.74$ kvar, $Q_5 = -86.725$ kvar, $Q_7 = 15.65$ kvar, $Q_9 = -6.74$ kvar, $Q_{11} = 5.22$ kvar, $Q_{13} = -3.3$ kvar, $Q_{15} = 3.49$ kvar, $Q_{17} = -0.24$ kvar, $Q_{19} = 0.07$ kvar is calculated. The total reactive power of the harmonics taken into account is $Q = 4707.00$ kvar. Using formula (5), the argument of total power and the input electrical resistance of the traction drive $\varphi_\Sigma = 25,4^o$ is calculated, which indicates the active-inductive nature of the input electrical resistance. The power factor of the traction electric drive $K_M = 9929.90/11483.70 = 0.86$.

The power balance (1) is checked, and the calculation error is determined

$$\sqrt{S_G^2 - \Delta S^2} = \sqrt{P^2 + Q^2}; 11483, 70 = \sqrt{9929, 90^2 + 4707, 00^2}.$$

11483,70 kVA $\approx$ 10989,03 kVA, the calculation error is 4,3%.

The use of the electric potential of the AC contact network for train traction is reduced due to the switching of the current by the RC thyristors, the operation of the traction transformer in the short circuit mode during the switching of the current by the RC thyristors, and the active-reactive nature of the input electric resistance of the traction electric drive. Efficiency factor of the traction transformer and RC

$$\eta_{TT+RC} = \frac{U_d \cdot I_{d12}}{P} \cdot 100\% = \frac{867(4 \cdot 811, 6 + 6485)}{9929, 9} \cdot 100\% = 85\%.$$

The loss of active power in the traction transformer and RC is 2.8 times greater than the loss of active power in the AC contact network. The efficiency factor of the traction electric motors $\eta_{TEM} = 93\%$, and the efficiency factor of the traction electric drive of an electric locomotive is 79% and decreases due to the unsatisfactory performance of the power controller. Due to the decrease in voltage at current collectors when 3–4 sectional electric locomotives with RC are operating and driving heavy, articulated trains, the average service speed of train does not exceed 43–46 km/h, the performance of electric traction decreases even with traction power supply by the system 2×25 kV.

The energy performance indicators of the traction electric drive do not improve in this traction and speed mode when using the power controller developed at Irkutsk State Transport University (IrGUPS), since the rectifier assembled on diodes works almost like RC. Due to the inductance of the AC contact network and the inductance of the ERS traction transformer, the switching angle of the RC thyristors and the diode switching angle when using ESV is 45 electrical degrees. During this time, the electrical energy from the power supply system is spent on the operation of the ERS traction transformer in the short circuit mode.

6 The Promising Direction of Research Work on Driving Heavy and Articulated Trains

From the power balance and the proposed energy characteristics (1–5), it follows that the full and continuous use of electric potential is achieved using a high-voltage DC contact network.

The implementation of the energy-saving train traction (Fig. 11) is made of a 3-section DC electric locomotive with the ESV power controller. When the voltage in the contact network of direct current is 27.5 kV, the operation of a 3-section electric locomotive with a nominal load, the voltage at the traction substation does not decrease, and the voltage on the ERS current collector $U_{tp} = 26.52$ kV ensures simultaneous traction of the second train at a speed not lower than 50 km/h. The speed of a DC electric locomotive is 14% higher than that of an AC electric locomotive. The loss of active power in ESV when operating in a DC contact network is reduced by 62%, and the efficiency of the power controller is 94.5%. The loss of active power in a DC contact network is reduced by 27% compared with the loss of active power in an AC contact network. Since using a three-phase rectifier on a traction converter station, it is possible to convert an alternating line voltage of 27.5 kV to a rectified voltage of 37.1 kV, the efficiency of train traction is improved with the help of a DC ERS.

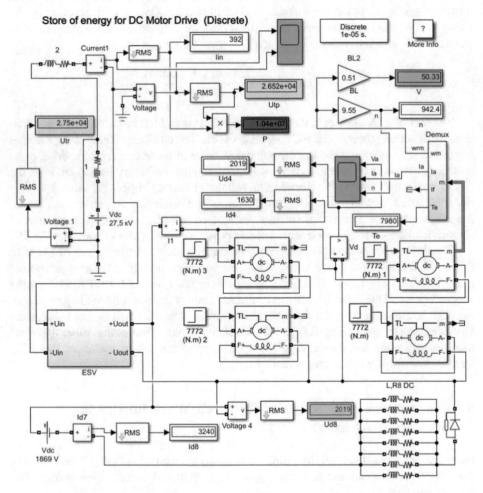

Fig. 11. Mathematical model of the electric traction drive of a DC electric locomotive with the voltage of 27.5 kV.

7 Conclusion

1. To limit the starting currents, for smooth starting and acceleration of the traction electric drive of ERS, the voltage on the electric motors is reduced by pulsed power controllers. Due to the nonconductive state of the power semiconductor devices of the pulse controllers, the energy efficiency of the electric traction of a train deteriorates.
2. Spectral analysis of the nonsinusoidal voltage and the current in the primary winding of the ERS traction transformer has shown that the active and reactive power are generated by 3, 5, 7, 9 and other harmonic components of voltage and current, and not only by the main harmonics.
3. Single-phase rectifiers, which are used on the AC ERS as an input converter, reduce the duration of use of the electric potential of the AC contact network for train traction due to the operation of a traction transformer in a short circuit mode during switching of current by power rectifier devices. The voltage at the current collector of ERS decreases, the speed of the train is limited. The efficiency factor of the electric drive is reduced due to significant loss of active power in the ERS pulse controller.
4. Continuous and full use of the electric potential of the DC contact network is realized with the help of the ESV power controller due to the smooth change of the input electrical resistance of the traction electric drive. Energy saving driving of heavy, articulated trains with high speed is achieved by reducing the load of locomotive by the contact network current.
5. The proposed method for calculating the energy characteristics of the ERS traction electric drive allows establishing the reasons for the unsatisfactory energy efficiency of the traction rolling stock and developing promising technical solutions for driving heavy, articulated trains and implementing high-speed freight traffic on the railway.

References

1. Report on the research work "Tyagovo-energeticheskiye ispytaniya elektrovoza 2ES7 v usloviyakh Vostochno-Sibirskoy zh.d." (in Russian). AO "VNIIZHT", Moscow (2016)
2. Gapanovich, V.A.: Sokhranit' nabrannyye tempy innovatsionnogo razvitiya. http://www.vostok1520.com/
3. Ryabchenok, N.L., Alekseyeva, T.L., Yakobchuk, K.P., Astrakhantsev, L.A.: Utochnennyy zakon sokhraneniya energii. http://www.rusnauka.com/42_PRNT_2015/Tecnic/5_202603.doc.htm
4. Mayevskiy, O.A.: Energeticheskiye kharakteristiki ventilnykh preobrazovateley. Energiya, Moscow (1978)
5. Zinov'yev, G.S.: Pryamyye metody rascheta energeticheskikh pokazateley ventilnykh preobrazovateley. Publishing House of the Novosibirsk State University, Novosibirsk (1990)
6. Burkov, A.T.: Elektronika i preobrazovatelnaya tekhnika. UMC ZDT, Moscow (2015)
7. Chaplygin, E.E., Panova, O.S.: Teoriya moshchnosti v silovoy elektronike (in Russian). MEI, Moscow (2013)

8. Seo, S.-S., Han, S.-W.: A study on control algorithm of thyristor dual converter system for DC electricity supply of urban railway. J. Electr. Eng. Technol. **14**(2), 979–983 (2019)
9. Prasuna, P.V., Rama Rao, J.V.G., Lakshmi, C.M.: Int. J. Eng. Res. Appl. (IJERA) **2**(4), 2368–3376 (2013)
10. Jenella, S., Radj Kumar, V.: Power electronics and renewable energy system. In: Proceedings of ICEPERES, pp. 225–236 (2014)
11. Hwang, J., Lim, S., Choi, M., Kim, M.: Reactive power control method for grid-tie inverters using current measurement of DG output. J. Electr. Eng. Technol. **14**(2), 603–612 (2019)
12. Umov, N.A.: Izbrannyye sochineniya (in Russian). M.-L.: Gostekhizdat (1950)
13. Poynting, J.H.: On the transfer of energy in the electromagnetic field. Philos. Roy. Soc. **175**, 343–361 (1884)

Technical and Economic Criteria for Selection of Transformer Power in Distribution Circuits

Akilya Galimova[✉][iD], Anna Novikova[iD], and Elena Strizhakova[iD]

Samara State Technical University (Samara Polytech),
224, Molodogvardeyskaya str., 443001 Samara, Russia
akilya@mail.ru

Abstract. When selecting transformer power at the design stage, it is necessary to quantify the reliability and service life of the electrical equipment, as well as the cost efficiency of the construction of power supply system. This issue is especially relevant for 6–10 kV distribution circuits, since a significant part of the substations of this voltage class is on the balance sheet of consumer who bear the costs for the design, construction and operation of these substations. The service life of a transformer, as well as high and low voltage devices of substations, depends on the voltage class, operating conditions, the operation of electrical power equipment in normal and emergency modes, and the load factor of a transformer in various operating modes. The paper considers the technical and economic criteria for selection of an oil-immersed transformer for outdoor unitized transformer substations with a voltage of 6–10 kV, the reasons of reducing its service life, the influence of the load factor of a transformer on the wear of insulation and the reduction of the service life of its safe operation. The methods and results of calculation of the relative service life of a transformer depending on the load factor are given. The criteria for selection of the load factor are determined depending on the capacity, structure and type of consumers of electricity.

Keywords: A transformer of 6–10 kV distribution · Service life ·
Load factor of a transformer β

1 Introduction

Calculation and selection of the equipment for a transformer substation is one of the main sections of the project of power supply systems. Thus, the transformer is the main device whose parameters depend on the power of the projected power supply system and on the operation modes of this system.

Some aspects of the selection of a power transformer for outdoor unitized transformer substations of 6–10 kV distribution circuits are considered in the paper. Usually, oil-immersed transformers with ONAN or ONAF cooling systems are used in such power supply systems. They belong to the group of distribution transformers and perform a reducing function. The power of such devices does not exceed 2500 kVA, voltage class up to 35 kV inclusive.

© Springer Nature Switzerland AG 2020
V. Murgul and M. Pasetti (Eds.): EMMFT 2018, AISC 982, pp. 509–516, 2020.
https://doi.org/10.1007/978-3-030-19756-8_48

During design, the voltage class and the rated load power to be connected to the substation are taken as basic criteria for selection of a power transformer. If consumers of 1st or 2nd reliability categories are connected to the power supply system, then the transformer operation in emergency mode is also checked. For power supply to consumers of 3rd category, the power of a transformer must not be less than the rated load power in accordance with the selection condition.

An integrated technical and economic approach is needed to solve the problem of selection of transformer power. This paper considers a parameter that, according to GOST 11677-85 [1], is not included in the list of technical parameters and characteristics included in the transformer's passport - the service life of a transformer. Generally, the design and installation of electrical networks and a transformer substation for the power supply to an enterprise are one of the significant expenditure items of the electricity consumer, especially in energy-intensive technologies and industries. Therefore, it is important to know the economic and technical efficiency of the costs for construction of power supply system. At the same time, the passport data of electrical equipment, which are of interest to energy specialists and electricians, practically do not contain information for specialists of other industries, i.e. for consumers.

2   Materials and Methods

In fact, the service life of a transformer cannot be clearly estimated. One can talk about the expected service life - some conventional value taken for a continuous constant load at normal ambient temperature and nominal operating conditions. The load that exceed the nominal one and the ambient temperature higher than the design one lead to accelerated wear of insulation and involve a certain degree of risk. In addition, the actual service life also depends on exceptional influences such as overvoltage, short circuits occurring in the network, and emergency overload. The probability of failure-free operation under such influences depends on the degree of impact (duration and amplitude) of emergency modes and overvoltage, transformer design, moisture content in insulation and oil. In the mode of permissible overload, the temperature of the various parts of a transformer increases. This also reduces the period of failure-free operation of a device. At the same time, other electrical equipment of the substation - switches, current transformers, as well as cable glands and cable sealing ends are also under a negative influence. If external influences often depend on the operation of the power system as a whole, then the normal operating modes can be controlled directly by the consumer. The main parameter that significantly affects the service life of a transformer and the electrical equipment as a whole is the load factor of a transformer β [3]. It determines the ratio of the operating load current to the rated current of a transformer or the electricity consumption to the rated power of a transformer. The load factor also affects the efficiency of a transformer. The permissible operating time of a transformer in overload conditions also depends on the degree of load of a transformer operating in steady-state mode.

It is known that the efficiency of a transformer depends on magnetic core losses and losses in the windings. Magnetic losses are constant; losses in the windings can vary. They depend on the square of the current flowing through the windings. Such losses are also called load losses. Therefore, the load factor affects the losses in a transformer and, consequently, on the efficiency [4]. The maximum value of the efficiency is observed at the optimum load factor when the losses in the windings will be equal to the no-load losses, that is:

$$\Delta P_0 = \beta^2 \cdot \Delta Pk \tag{1}$$

Then the load factor:

$$\beta_{O\Pi T} = \sqrt{\frac{\Delta P_0}{\Delta Pk}} \tag{2}$$

For distribution transformers, the maximum efficiency value occurs when the load factor $\beta = 0.4 \div 0.5$. In practice, when selecting transformer power, one often guided by the economic load factor which is equal to $\beta = 0.6 \div 0.7$.

The main normative document regulating the permissible loads and overloads of oil-immersed transformers is GOST 14209-97 (IEC 354-91) "Loading guide for oil-immersed power transformers" [2]. The document provides guidance on the definition of technically valid load modes for power transformers in terms of permissible temperatures and thermal wear. The calculation of the temperature in different parts of the windings is based on the thermal characteristics of different groups of transformers combined by power and cooling system. Each group has its limit values for permissible overloads.

The paper considers a group of distribution transformers with a power of no more than 2500 kVA. This is the most common class of transformers which is used for power supply to enterprises of nearly all industries, institutions of various fields of activity, and household load. Substations with distribution transformers are mainly on the balance sheet of an enterprise, and often controlled by specialists who are not related to energy. Therefore, it is important to ensure the high reliability of electrical devices at the design stage of the power supply system.

In distribution transformers with natural oil cooling (type of ON) in steady-state mode, the temperature of the most heated point and the metal parts in contact with the insulation should not exceed 140 °C. The temperature at which the normal service life of the transformer is ensured is 98 °C. Exceeding the temperature for every 6 °C more than 98 °C increases the wear rate of insulation in 2 times (the so-called 6-degree rule) and, consequently, significantly reduces the service life of the transformer. It is known that in 43.6% of cases, the failure of transformers with a power of up to 2500 kVA occurs due to insulation failure [6]. The main component influencing the temperature of the most heated point is the load factor β.

It should be noted that the calculation of the temperature of the most heated point is carried out in accordance with the load curve of the consumer for two modes - steady-state thermal mode and unsteady mode when the load changes during the time interval [2]. However, it is advisable in the design stage to perform a calculation for the load that determines the output power or loads in the emergency mode for enterprises that have consumers of 1st or 2nd categories, since these are the modes of maximum load of the transformer. Then the temperature in the steady-state thermal mode for transformers with ON cooling will be determined by the following equation:

$$\Theta_h = \Theta_a + \Delta\Theta_{hr}\left[\frac{1+R\beta}{1+R}\right] + H_{gr} \cdot \beta^y \qquad (3)$$

Where:

- Θ_a – ambient temperature, °C;
- $\Delta\Theta_{br}$ – increase of the oil temperature at the bottom of the winding, °C;
- R – ratio of short-circuit losses to no-load losses at rated load;
- β – load factor;
- H_{gr} – temperature gradient of the most heated point, °C;
- y – exponent of winding.

According to GOST 14209-97, when calculating for transformers with the ON cooling system, it is necessary to apply the thermal characteristics presented in Table 1.

Table 1. Thermal characteristics of transformers with the ON cooling system

Θ_a, C°	$\Delta\Theta_{hr}$, C°	R	H_{gr}, C°	y
20	78	5	23	1.6

Substituting the characteristics and coefficients listed in the table, we obtain the dependence of the maximum temperature of the most heated point on the load factor of the transformer:

$$\Theta_h = 65 \cdot \beta + 23 \cdot \beta^{1.6} + 33 \qquad (4)$$

On the basis of calculated values of the temperature of the most heated point, it is possible to determine certain parameters of thermal wear of insulation, for example:

- the relative rate of thermal wear of insulation

$$V = A \cdot 2^{\frac{\Theta_h - 98}{6}}, \qquad (5)$$

where A – constant, corresponds to the service life at t = 0°, A = 0.112;

- the service life is determined by the exponential ratio of Montsinger

$$Service\ life = e^{-\rho\Theta h} \qquad (6)$$

where ρ – constant.

The value of the constant in this equation depends on many factors: proportion of mixture of raw materials, chemical additives, environmental parameters (contents of moisture, free oxygen in the system), and others. However, it is possible to take a certain constant value of the coefficient in the range of permissible temperatures from 80 °C to 140 °C. According to [5], $\rho = 0.1155$ °C^{-1}.

So far there is no single and simple criterion for estimating the service life of the transformer, which can be used to quantify the useful life of the transformer. One can make comparisons based on the wear rate of insulation. Nevertheless, for experts in various industries who are not related to electrical engineering, which are mainly consumers of electricity, it is difficult to estimate the prospects for using electrical equipment, the reliability of its operation, and the quality of electrical power in terms of the rate of wear of insulation. Therefore, when choosing options for the power supply system and the power of electrical equipment, it is important to rely on quantitative criteria.

3 Results

A relative change in the service life of a transformer N from the load factor is proposed in the paper as a quantitative criterion, i.e. how the service life of a transformer increases or decreases depending on the load in comparison with the base value β. The base value of the load factor is the value that provides the temperature of the most heated point of 98 °C. The calculations are based on the dependencies and coefficients contained in GOST 14209-97 for a class of distribution transformers with natural oil cooling and with the power up to 2500 kVA.

The results of the calculation are shown in Table 2.

Table 2. Calculation of the relative change in the service life of a transformer from the load factor

Load factor, β	Temperature of the most heated point, °C	Relative change in the service life of a transformer, N
0.4	64.0	49.0
0.45	69.0	29.0
0.5	73.0	18.0
0.55	78.0	11.0
0.6	82.0	6.0
0.65	87.0	4.0
0.7	91.0	2.0
0.75	96.0	1.2
0.768	*98.0*	*1.0*
0.8	101.0	0.7
0.85	106.0	0.4
0.9	111.0	0.22
0.95	116.0	0.12
1.0	121.0	0.07

The calculations showed that the temperature of 98 °C is provided when a transformer is loaded by 76.8%. which corresponds to the value N = 1. The adopted economic load factor β = 0.6 ÷ 0.7 corresponds to the increase in the service life by 6 ÷ 2 times. If the consumer of electricity does not plan to increase production volumes, it is enough to focus on the economic load factor when selecting the power of transformer for a substation. If a transformer substation is designed for the power supply to new buildings, it is advisable to use the load factor β = 0.4 ÷ 0.5. which corresponds to the operating mode of a transformer with the maximum efficiency. In the course of time, new buildings increase electricity consumption by developing infrastructure. And there is also a problem of availability of a free territory for a new substation, especially in the conditions of urban area development.

Figure 1 shows the graphical dependence of the relative change in the service life of a transformer N on the load factor β. The dependence is exponential. so even a slight decrease in the load factor can significantly increase the service life of an oil-immersed transformer.

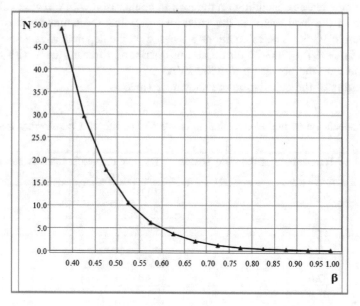

Fig. 1. Dependence of the relative change in the service life of a transformer N on the load factor β.

One of the most important parameters of the transformer's efficient operation is the efficiency which depends on the loss of electrical power. As is known, the power losses in the transformer windings, the so-called load losses, also depend on the load factor β [7, 8]. The calculations of the power losses and the efficiency for oil-immersed transformers of TM series with the power from 100 to 2500 kVA showed that the decrease in the load of a transformer to (45–50)% increases the efficiency by (1–2)%.

Moreover, the greater the power of a transformer, the greater the increase in the efficiency.

Let's consider the economic aspect of transformer power selection. The cost of a substation is formed from the cost of the electrical equipment of the high and low voltage switchgear. the facility or the modular building. and the transformer. Since electric devices of high and low voltage have high parameters for electrodynamic and thermal resistance to short-circuit currents, manufacturers produce substations for power ranges, for example up to 400 kVA and install transformers from 63 kVA to 400 kVA. The cost of a transformer itself is about 20% of the cost of the entire substation. Therefore, increasing the power of a transformer by one tap will not lead to a significant increase in the cost of the substation, but it will significantly increase the service life of a transformer.

4 Discussion

As a result of the analysis of the influence of the load factor on the service life of a transformer and according to calculations. The following conclusions can be formulated:

- the load factor of a transformer has a significant effect on the service life of an oil-immersed transformer and the electrical equipment of the substation;
- for power supply to low-power consumers, the load factor should be chosen on the basis of its economic value of 0.6–0.8;
- when selecting a transformer for power supply to power-intensive enterprises with the prospect of further development. And also, for power supply to new buildings, it is advisable to choose the load factor corresponding to the maximum efficiency of 0.4–0.5;
- when the load of a transformer is close to its rated power, the service life is significantly reduced since the dependence of the relative change in the service life of a transformer is exponential;
- the decrease in the load factor also leads to the decrease in power losses in the transformer windings and the increase in the efficiency;
- the final decision on selection of transformer power should be taken by the designer in conjunction with the consumer on the basis of the technical and economic criteria for estimating the projected power supply system.

References

1. GOST 11677-85: Power transformers. General technical specifications
2. GOST 14209-97 (IEC 354-91): Loading guide for oil-immersed power transformers
3. Galimova, A.A.: Criteria for choosing the load factor of a power transformer for the design of substations of distribution circuits. Izvestiya Vysshikh Uchebnykh Zavedenii. Problems of Energy, pp. 5–6 (2013)

4. Makarov, E.F.: Handbook of electrical networks 0.4 - 35 kV and 110 - 1150 kV, vol. 2, p. 688 (2004). Ed. Papyrus Pro
5. Kish, L.: Heating and cooling of transformers. Moscow, Energy, 208 p. (2007)
6. Tsirel. Y.A., Polyakov, V.S.: Operation of power transformers in power plants and in electrical networks, Saint-Petersburg, Energoatomizdat (2004)
7. Kopylov, I.P.: Electric Machines: Textbook for High Schools, 5th edn, p. 607. Vysshaya Shkola, Moscow (2006)
8. Tulchinskaya, Y.I.: Evaluation of the effectiveness of application of transformer with low load factor. Electron. Sci. J. Oil Gas Bus. 5 (2012)

Microelectronics Devices Optimal Design Methodology with Regard to Technological and Operation Factors

Sergey Meshkov$^{(\boxtimes)}$ ⓘ, Mstislav Makeev ⓘ, Vasily Shashurin ⓘ,
Yury Tsvetkov ⓘ, and Boris Khlopov ⓘ

Bauman Moscow State Technical University, 105005 Moscow, Russia
sb67241@mail.ru

Abstract. Microelectronic devices are the basis of sensor devices used to measure temperature, humidity, pressure, movement or acceleration, light, chemical elements concentration, etc. The aim of this work is to create a methodology of microelectronic devices integrated design, which, with the use of the existing group production technology, will achieve the optimal linkage of the main device quality indicators (target values, mass-availability) under given production conditions and reliability under given operating conditions and given restrictions on device purpose indicators. An additional stage of design and technological optimization, which receives the device design parameters from the circuit and engineering design stages, data on their technological errors from the technological production preparation stage and data on the kinetic of design parameters and purpose indicators of the device under given operating conditions, can be introduced into the traditional design scheme to solve the assigned task. As a result of the optimization, corrections for the device nominal design parameters maximizing the objective function are determined. The probability of device purpose indicators destination within the imposed limits or the probability of performing the specified functions during the operation time are used as the objective function. The proposed methodology is illustrated by the example of microelectronic SHF radio signals mixer and rectifier with a resonant-tunneling diode as a nonlinear element.

Keywords: Microelectronics device · Key quality indicator ·
Design and technological optimization · Technological errors · Degradation ·
Reliability · Radio signals mixer · Rectifier · Resonant-tunneling diode

1 Introduction

The current practice of designing of microelectronic devices includes systems engineering, circuit and engineering design, technological process preparation stages. The device structure is determined at the system engineering stage. The electrical circuit and circuit elements parameters that determine the functional parameters of the device Y (purpose indicators) are the results of circuit design stage. The design implementation of circuit solutions and design parameters X, which define the required parameters Y, are determined at the design stage. The technological route is developed, the

© Springer Nature Switzerland AG 2020
V. Murgul and M. Pasetti (Eds.): EMMFT 2018, AISC 982, pp. 517–523, 2020.
https://doi.org/10.1007/978-3-030-19756-8_49

parameters of technological operations are determined and the necessary technological equipment is designed at the technological process preparation (TPP) stage.

This methodology does not take into account the fact that microelectronic devices design implies the use of group technology production methods (often planar), if the device construction elements are located on the same substrate. The high correlation of the device's parameters formed simultaneously within the common substrate during the structural elements production operations is a feature of group technologies. On the other hand, the location of similar structural elements on a common substrate is the reason for the high correlation of aging and degradation processes occurring under the influence of destabilizing factors during device operation. This leads to the fact that the existing linear design procedure does not allow achieving devices line optimal technological efficiency and reliability indicators. One of the most important characteristics of manufacturability is the serial production availability, which is quantitatively described by the yield probability under given production conditions. The problem of ensuring the micro devices serial production availability was described in [1] and developed in [2–7] using the example of SHF hybrid integrated circuits (HIC). In these works, it was shown that the known methods of ensuring the devices serial production availability (such as methods of complete, incomplete and group interchangeability that operates with constructive tolerance as a tool of ensuring the serial production availability and accuracy) when using group treatment technologies are not operational due to the high correlation of technological errors of devices parameters within a common substrate. Application of the adjustment procedure in the transition to microsize is associated with a sharp production cost increasing and also may be impossible due to design and technological limitations. The problem of ensuring device reliability in the specified operating conditions is not considered. The aim of this work is to create a methodology of microelectronic devices integrated design, which with the use of the existing group production technology, will achieve the optimal linkage of the main device quality indicators (target values, mass-availability) under given production conditions and reliability under given operating conditions.

2   Methods, Results and Discussion

The initial concept is based on the proposition that device specified quality indicator assurance instruments (variable parameters) must, on the one hand, allow their independent variation, and on the other hand, be independent from the group technological production process, which is assumed to be fixed and unchanged. These parameters are the values of the design parameters X, and, consequently, the purpose indicators Y.

The design methodology has an integrated nature under these conditions, because it aimed on creating a device that, on the one hand, is according to the requirements of the technical specification for the purpose indicators, and on the other hand, it has the optimal technological series production cost within the existing technology and failure time under the specified operating conditions. All three groups of instrument quality indicators are interrelated. Moreover, the device y-percentile time to failure can be considered as the probability of falling devices line random purpose indicators into the given constraints (yield probability), determined after operating time t.

The tasks of analyzing and ensuring the stated reliability of radio electronic devices are solved traditionally using the sudden failures model. The device reliability is determined by the reliability indicators of its elements; therefore, device reliability R is equal to the multiplication of the reliability of the device elements R_i:

$$R = R_1 * R_2 * \ldots R_n. \tag{1}$$

This formula is valid in the case if the reliabilities R_i are independent. The condition of R_i independence is satisfied for a device consisting of discrete electronic components. In this case, R_i is the probability of failure-free operation of one element.

The device elements located within the common substrate are exposed to impact of destabilizing factors (high temperature, ionizing radiation, etc.) during operation. These factors accelerate destructive processes in devices structure. All elements located within the substrate are subjected to destabilizing effects simultaneously. This is the reason for the correlation of the degradation rates of various elements structures and hence elements functional parameters. On the other hand, degradation rates correlation of various elements structures on the substrate is caused by the similarity of the physical processes occurring under the influence of external factors and leading to failure. Therefore, the probabilities R_i are not independent in formula (1). Thus, the probability R should be considered as conditional and formula (1) for its calculation is not applicable. The gradual failures reliability model is more physically and adequate [8–13]. This model uses the aging processes regularities and takes into account their probabilistic nature, representing a sequence of phenomena and events that lead to failure.

The task of reliability ensuring of microelectronic devices line produced with application of group technology set in the following way. The objective function is the probability of performing the specified functions by devices line $P_\phi^n(t, \bar{Y}, \bar{\sigma}, \bar{\Delta})$, where $\bar{Y}$ - the device purpose indicators vector $(\bar{Y} = q(\bar{X}))$, $\bar{X}$ - the device design parameters vector; t - operating time, $\bar{\sigma}$ - the device design parameters dispersion vector (technological accuracy); $\bar{\Delta}$ - the vector of tolerance limits of device purpose indicators. P_ϕ^n is determined as:

$$P_\phi^n = \frac{\sum_i \Delta P_\Gamma^i P_\phi^i}{P_\Gamma}, \tag{2}$$

where P_Γ – devices yield probability:

$$P_\Gamma = \int_{Y_{MIN}}^{Y_{MAX}} f(Y)dY, \tag{3}$$

where $f(Y)$ – the joint probability density function of the device purpose indicators.

The illustrations are shown in Figs. 1 and 2, where Y_{NOM} – the device purpose indicator nominal value; Y_{MIN}, Y_{MAX} – tolerance limits $\bar{\Delta}$ of device purpose indicators; P_ϕ^i - the probability of performing specified functions with a parameter Y^i; P_ϕ^{MIN} - minimum tolerance probability of the device performing the specified functions; P_Γ^i - the probability of the parameter Y^i falling into the elementary interval ΔY^i.

A function $P_\Gamma\left(t, \overline{Y}, \overline{\sigma}, \overline{\Delta}\right)$ can be used as the target function if the function $P_\Phi(Y)$ is not specified.

Therefore:

- Optimization criterion - $max\, P_\phi^n\left(t, \overline{Y}, \overline{\sigma}, \overline{\Delta}\right)$ or $max\, P_\Gamma\left(t, \overline{Y}, \overline{\sigma}, \overline{\Delta}\right)$.
- Variable parameters - nominal values $\overline{X}$.
- Limitations $- \overline{\Delta} = const, \overline{\sigma} = const, \overline{X} \in O_k, \overline{X} \in O_T, P_\Gamma(t=0) \geq P_\Gamma^{min}$, where O_k, O_T – design and technology limits.

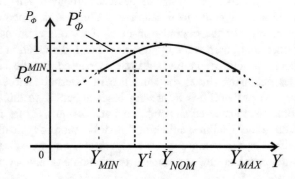

Fig. 1. The dependence of the device performing specified functions probability on the purpose indicator deviation from the nominal value.

The new nominal values $\overline{X}$ or Y_{OPT}, to which the maximum values P_ϕ^n или P_Γ are correspond, are obtained as a result of solving the optimization problem. The graphical interpretation of the solved problem using the criterion $max\, P_\Gamma\left(t, \overline{Y}, \overline{\sigma}, \overline{\Delta}\right)$ is shown in Fig. 3. The graphical representation of the problem is similar if using $max\, P_\phi^n\left(t, \overline{Y}, \overline{\sigma}, \overline{\Delta}\right)$ criterion.

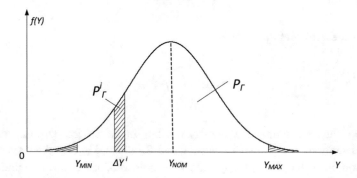

Fig. 2. The devices yield probability determination.

This problem is complex, since it allows maximizing the yield probability or the probability of performing specified functions by devices line at t = 0. Thus, the task of ensuring serial production availability (manufacturability), as well as the task of ensuring reliability (the device's performance indicators must fit the specified requirements) when introducing a time factor and taking into account the device elements degradation process under the influence of external factors are solved.

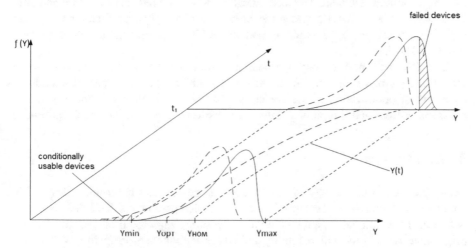

Fig. 3. Graphical representation of the optimization problem using the criterion $max P_\Gamma$ $\left(t, \overline{Y}, \overline{\sigma}, \overline{\Delta}\right)$.

The scheme of the described methodology is shown in Fig. 4. An additional stage of design and technological optimization is added to the traditional linear design process diagram. At this stage, the parameters X and Y synthesized at the circuit design and engineering design stages are received, as well as technological errors data from the stage of technical preparation of production and data on the design $X(t)$ and functional $Y(t)$ device parameters kinetics under given operating conditions. As a result of design and technological optimization, optimizing corrections to the nominal values ΔX_{opt} and ΔY_{opt} (they maximizes the yield) or the probability of performing specified functions by devices line are determined. One of these criteria goes into the category of constraints, when the other is maximized.

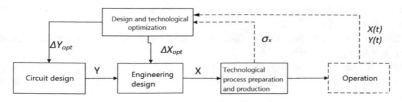

Fig. 4. Complex design methodology scheme.

The problem is solved with the use of simulation methods, including stochastic modeling of technological errors and physical degradation processes of the device elements functional parameters in the specified operating conditions and based on the results of studies of the device elements degradation processes under the influence of external factors.

The described methodology was developed on the example of microelectronic SHF radio signals mixer and rectifier, in which the resonant-tunnelling diode (RTD) was used as a nonlinear element. The methodical basis was the results of the research of the RTD degradation processes under the influence of temperature factor and ionizing radiation, carried out by the authors and the RTD diagnostic model based on these results [14–18].

The described methodology approbation shows that the yield can be increased by 10–15%, and y-percentile time to failure – by 10–30% in specified operating conditions by means of optimization of the device design parameters with the unchanged technological process and the purpose indicators conformance with the specifications.

3   Conclusions

Group production technology of microelectronic devices are characterized by a high correlation of technological errors of the device design parameters. This does not allow, within the conventional design methodology, ensuring high requirements for the serial production availability and reliability indicators without adjusting the existing production technology.

The methodology of complex design is proposed, that allows providing the optimal balance between serial production availability and reliability in the specified operation conditions under given restrictions on the purpose indicators of sensor devices used to measure temperature, humidity, pressure, movement or acceleration, light, chemical elements concentration, etc.

The research work was supported by Ministry of Education and Science of the Russian Federation under state task № 16.1663.2017/4.6.

References

1. Bushminsky, I.P., Morozov, G.V.: SHF Circuits Technological Design. Bauman University Publishing House, Moscow (2001)
2. Gudkov, A.G., Leushin, V.Yu., Meshkov, S.A., Popov, V.V.: Application of complex technological optimization for monolithic microwave circuits designing. In: 2008 18th International Crimean Conference - Microwave & Telecommunication Technology, pp. 535–536. IEEE (2008)
3. Gudkov, A.G., Popov, V.V.: GIC and MIC Reliability and Quality Increasing. Autotest LLC, Moscow (2012)
4. Koziel, S., Bandler, J.W.: Rapid yield estimation and optimization of microwave structures exploiting feature-based statistical analysis. IEEE Trans. Microw. Theory Tech. 63(1), 107–114 (2015)

5. Singhee, A., Rutenbar, R.A.: Why quasi-monte carlo is better than monte carlo or latin hypercube sampling for statistical circuit analysis. IEEE Trans. Comput.-Aided Des. Integr. Circ. Syst. **29**(11), 1763–1776 (2010)
6. Koziel, S.: Simulation-Driven Design Optimization and Modeling for Microwave Engineering. World Scientific, Singapore (2013)
7. Kabir, H., et al.: Smart modeling of microwave devices. IEEE Microw. Mag. **11**(3), 105–118 (2010)
8. Birolini, A.: Reliability Engineering: Theory and Practice. Springer, Heidelberg (2013)
9. Blischke, W.R., Murthy, D.N.P.: Reliability: Modeling, Prediction, and Optimization. Wiley, London (2011)
10. Gorjian, N., et al.: A Review on Degradation Models in Reliability Analysis in Engineering Asset Lifecycle Management. Springer, London (2010)
11. Chekanov, A.S.: Calculations and Ensuring the Reliability of Electronic Equipment. KNORUS, Moscow (2012)
12. Pronikov, A.S.: Parametric Reliability of Machines. Bauman University Publishing House, Moscow (2002)
13. Yang, Y., et al.: Design for reliability of power electronic systems. In: Power Electronics Handbook, pp. 1423–1440. Butterworth-Heinemann, Oxford (2018)
14. Makeev, M.O., Ivanov, Y.A., Meshkov, S.A., Sinyakin, V.Y., Ivanov, A.I.: Estimation of thermal destruction of AuGeNi ohmic contacts of resonant tunneling diodes on the basis of nanoscale AlAs/GaAs heterostructures. In: 2015 5th International Workshop on Computer Science and Engineering: Information Processing and Control Engineering, WCSE 2015-IPCE, pp. 260–265. The SCIence and Engineering Institute, LA (2015)
15. Makeev, M.O., Meshkov, S.A., Sinyakin, V.Yu.: State diagnostics of RTD based on nanoscale multilayered AlGaAs heterostructures. J. Phys.: Conf. Ser. **741**(1), 012176 (2016)
16. Makeev, M.O., Meshkov, S.A., Sinyakin, V.Yu.: Prediction of electronic nanodevices technical status and reliability based on analysis of their performance parameters kinetics under the influence of external factors. In: MATEC Web Conference, vol. 129, p. 03019 (2017)
17. Makeev, M.O., Sinyakin, V.Yu., Meshkov, S.A.: Reliability prediction of RFID passive tags power supply systems based on RTD under given operating conditions. In: MATEC Web Conference, vol. 224, p. 02095 (2018)
18. Makeev, M.O., Sinyakin, V.Yu., Meshkov, S.A.: Reliability prediction of radio frequency identification passive tags power supply systems based on A3B5 resonant-tunneling diodes. In: 2018 International Russian Automation Conference (RusAutoCon 2018), pp. 850–1748. IEEE (2018)

Assessment of Energy Intensity of the Drive for Traction Power Supply System

Vladislav Nezevak$^{(\boxtimes)}$ ⓘ, Vasily Cheremisin ⓘ,
and Andrey Shatokhin ⓘ

Omsk State Transport University, Marx ave. 35, 644046 Omsk, Russia
NezevakWL@mail.ru, CheremisinVT@gmail.com,
Shatohin_ap@mail.ru

Abstract. The article discusses questions about using the electric energy storage units in the DC electric traction system. As placements of the electric energy storage units, there were chosen the posts of sectionalization on the areas of the electric rolling stock's maintenance using the regenerative braking in the conditions of the way's mountain profile. In the research, there was used a method of simulation modeling of the interaction between the electric rolling stock and the electric traction system on the example of one of the Transsiberian railway's areas. There were given the results of the statistical processing of the received graphs of the change of the storage units' voltage and workload. There was shown the influence of the electric energy storage units to the main energy parameters of the electric traction system's work. There were received the frequency distributions for the electric energy volumes, the duration of the episodes of the storage unit's work, the assessment of the energy intensity in the modes of charge and discharge, there was done the assessment of the energy intensity, the capacity and the electric energy storage unit's operation time. Based on the data obtained, there were considered the basic provisions of the methodology for determining of the electric energy storage unit's capacity and energy intensity in the electric traction system subject to the cyclicity of work, the conditions for the regimes change of the charge and discharge, the energy intensity's effective usage, and the interaction variants in the information system of the energy parameters' monitoring.

Keywords: Electric energy storage unit · Electric traction system ·
Post of sectionalization · Work conditions · Daily load schedule ·
Energy intensity's determination · Work episode's duration ·
Frequency distribution

1 Introduction

The increase of the electric power systems' work efficiency is associated with the development of the electrical energy storage's systems, allowing doing the aligning of the electrical load's graph and the reservation in the systems of the electricity supply, to provide the effective work of the renewable sources of the electric energy and different transport systems. The researches of the electrical energy storage's systems allow identifying the all features of theirs using in the electro power systems. The wide

© Springer Nature Switzerland AG 2020
V. Murgul and M. Pasetti (Eds.): EMMFT 2018, AISC 982, pp. 524–538, 2020.
https://doi.org/10.1007/978-3-030-19756-8_50

distribution of the electric energy storage units in the transport systems is caused by the cyclicity of work of the transport units of any purpose and type, such as urban rail and railless transport, subway and railway transport [1–5]. The usage of the storage units at present is considering as the onboard systems of the electric rolling stock, and the stationary devices in the electric traction system [6–10]. A number of outstanding issues, which is bonded with the assessment of the electric energy storage unit's necessary power and energy consumption in the electric traction system, determine the task's urgency of the accounting of the storage unit's specificity of work and the formation of the basic provisions of the methodology and their usage in the electric traction system.

2   The Setting the Task

The expected effects from the electric energy storage unit's usage compel to consider the areas with a predominance of the freight service as the most energy-intensive. The efficiency of the electric energy storage unit's usage on the post of sectionalization (PS) is achieved by reason of the voltage level's increase on the buses of the PS, the power consumption level's reduction; the conditions' improvement of the regenerative braking's usage, the reduction of the load factor of the traction substations' power units etc. The effects by the storage unit's usage are determining by the conditions and its operating modes. The stated fact determines the necessity of the assessment of the energy parameters of the storage unit's work in the electric traction system for the freight service, which is attending with the peak loads [11–13].

The freight service determines the appearance of the traction substations' peak loads (TS – traction substations). The calculation of the TS' load TS allows assessing the energy parameters of the electric traction system's work (the values of the currents and voltages, the power consumption volumes, the specific consumption etc.). Besides the given parameters it is necessary to do the analysis of the load's daily schedule and the voltage on the TS and PS' buses. The specified task can be solved based on results of the instant schemes' calculation of the electric traction system, which allow receiving the graphs of the traction load and voltage. The calculation's results are using further for the determination of the storage unit's main energy parameters.

In the light of the above-stated, there is relevant a question of the assessment of the storage unit's capacity and energy intensity subject to the specific of the electric traction system's work.

3   The Simulation Modeling

The assessment of the main energy parameters of the electric storage unit's work we will receive on the example of the one of the Transsiberian railway's areas with the way's mountain profile (III-rd type) [14]. The simulation modeling is performing for the conditions of the implementation of the movement of trains' random graph, for which the dimensions in the freight service for the examined area are 50 pairs. The traction calculations in the freight service were done for trains with the following weights (taking into account the weights of locomotives) – 4200, 9300, 12000 t (Table 1). The traction calculation's results were received for areas with the length of 181.9 km in the software complex KORTES, which is used in the holding of RZD for the electric traction system's electric calculations.

Table 1. The main results of the traction calculations

Direction	Locomotive's series	Weight, t	Consumption, kWh	Recovery, kWh
Uneven	EP2K	1065	3726.3	–
Even	EP2K	1065	3069.2	–
Uneven	ED4m	629	2209	159.0
Even	ED4m	629	1827.9	147.6
Uneven	2ES6	4200	8956.2	1646.8
Even	2ES6	4200	6824.4	2124.2
Uneven	2ES10	12300	24031.4	1820.5
Even	2ES10	12300	17396.4	2695.6

The simulation modeling allows doing mathematical processing of results [15] of the instant calculations and solving the following tasks:

1. to assess the voltage's change on the PS during the day – minimum, maximum, average and to receive the frequency distribution over the ranges:

$$U_{\min}, U_{\max}, \bar{U} = \frac{\sum\limits_{i=1}^{n} U_i}{n}, p_1(U_1), p_2(U_2), \ldots p_n(U_n), \tag{1}$$

where U – the voltage (minimum, maximum, average); n – the number of the measurements; p_i – the i-th probability density by the i-th range.
For the traction load the assessment (1) is doing like the voltage's assessment;

2. to assess the change of the energy's return and transmission on the PS with the electric energy storage units, to receive the frequency distribution of the energy volume over the ranges:

$$W_{\text{ret}} = \sum\limits_{i=1}^{N_{\text{ret}}} W_{\text{ret}\,i}, W_{\text{tran}} = \sum\limits_{j=1}^{N_{\text{tran}}} W_{\text{tran}\,i}, p_1(W_{\text{ret(tran)}1}), p_2(W_{\text{ret(tran)}2}), \ldots, p_n(W_{\text{ret(tran)}n}), \tag{2}$$

where $W_{\text{ret(tran)}}$ – the volume of the returned to the buses of the PS and the transmissed into the contact network electricity, respectively; N_{ret}, N_{tran} – the number of the episodes of the energy's return and the transmission on the PS, respectively.

The episodes of the energy's return and the transmission are the cases of the capacity's continuous transmission in the directions from the contact network to the uses of the PS and from the PS to the contact network, c. The power flow's direction is determined by the current's direction I. The purpose of the assessment of the energy consumption's volumes on the episodes is the determination of the electric energy storage unit's energy intensity, which is necessary for work in charge and discharge mode. Taking into account the traction load's specific the accounting of the episodic work in one mode or another allows, based on the statistical processing, determining the necessary the electric energy storage unit's energy intensity, the conditions of its work in charge and discharge mode.

On the Fig. 1 there is presented the graph's fragment of the total traction load's change of the substation. On the given graph, the time intervals t1 and t3 correspond the episodes' duration of the energy transmission to the contact network, and the time interval t2 – the episodes' duration of the energy return from the contact network to the buses of the TS. The current direction's change is a sign of the episode's ending, herewith the interval of the traction load's absence corresponds the standby mode. The work's episodes are considering as continuous, if the current direction's change is not happened. By the change of the direction or by being in standby mode, when the current is zero, there is doing a transition to the next work's episode in the appropriate mode.

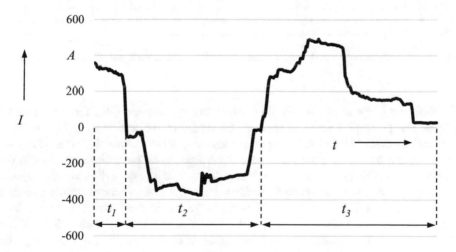

Fig. 1. The graph's fragment of the TS' traction load.

3. the assessment of the episodes' duration of the power's flow in the different modes (energy's transmission and return– respectively t_1, t_3 and t_2) and the distribution over the ranges:

$$t_{\text{ret}} = \sum_{i=1}^{N_{\text{ret}}} t_{\text{ret}}, t_{\text{tran}} = \sum_{i=1}^{N_{\text{tran}}} t_{i\,\text{tran}}, p_{i\,\text{ret}}(t_{i\,\text{ret}}), p_{i\,\text{tran}}(t_{i\,\text{tran}}). \tag{3}$$

For the simulation modeling with the purpose of the influence's assessment of the electric energy storage unit's work on the PS on the traction load's modes and the main energy parameters of the electric traction system, there are accepted the following conditions: the type of the main ways' catenary – 2MF100+M120+A185; the converting units – twelve pulse of the sequential type. On the Fig. 2 there is given the layout of the electric energy storage units (EESU) on the two PS in the electric traction system of the considering.

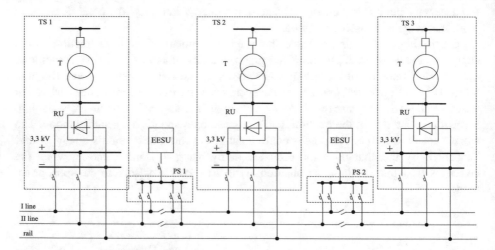

Fig. 2. The scheme's fragment of the electric traction system with the electric energy storage unit.

With the purpose of the full similarity's implementation [16], the model of the electric energy storage unit was replaced by model of the rectifier-inverter converter, the similar model of the electric energy storage unit, which allows doing modeling and assessing the necessary energy and capacity volumes for both directions by the different work's episodes. For the modeling there was accepted the «hard» external converter's feature. As the conditions for the electric energy storage unit's transition from the rest mode to the discharge and charge mode there was accepted a level achievement of the voltage on the buses of the PS, respectively, below 3200 V and above 3600 V. The values' scope of the setpoints of the transition to the charge and discharge mode has significant influence on the storage unit's energy intensity in the electric traction system [17, 18] and in the electricity supply's systems of the different transport systems [19–21], determining the energy exchange processes. The similar calculations using the similar substitution schemes show the adequacy of the work's similarity of the rectifier-inverter converter and the electric energy storage unit in the electric traction system.

The calculation with the instant schemes allows giving the daily schedule of the load and voltage at the connection points of the electric traction system with a 1 min step. On the Figs. 2 and 3 there is given the daily schedule of the change of the electric energy storage unit's voltage and traction load on the PS 1 and PS 2 km of the area, respectively (Fig. 4).

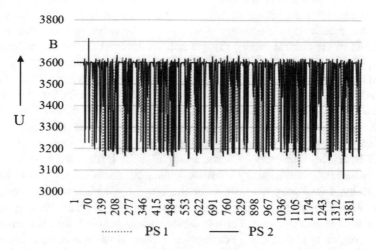

Fig. 3. The voltage's graph on the buses off the PS 1 and PS 2 with EESU.

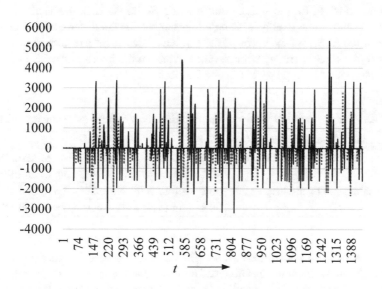

Fig. 4. Traction load's graph of the EESU on the PS on PS 1 and PS 2.

4 The Received Results

By the statistical processing results of the values of the electric energy storage unit's traction load on the PS 1 and PS 2 there was received, that the most frequency of the current's monitoring of the electric energy storage unit for the PS 1 and PS 2 is noticed in the charge mode in the values' range of 100–0 A. the most frequency of the voltage's monitoring on the buses of the PS at the level of 50% is fixed in the range of 3500–3600 V. In the discharge mode, the most monitoring's frequency (50%) of the energy volume for the episode is fallen to the range of 0–50 kWh, the episodes' duration (the monitoring's frequency – 40%) – to the range of 0–1 min. In the discharge mode of the electric energy storage unit on the PS 1 the energy volume on the episodes is in the range of 0 - 450 kWh. The energy volume in the discharge mode in 90% of cases doesn't exceed 200 kWh. More than in 90% cases the episodes' duration of the work in the discharge mode doesn't exceed 4 min. The episodes' duration is the range of 0–5 min, herewith the most observed ranges are 0–1 min and 3–4 min with the monitoring's frequency 42% and 35%, respectively. In the charge mode, the energy volume on the episodes of the electric energy storage unit's work is in the range 0–220 kWh, herewith the energy volume up to 160 kWh is observed in 95% cases. The most observed range of the energy volumes' values on the episodes – 20–40 kWh with the monitoring's frequency over 30%, for the range 0–40 kWh the monitoring's frequency is 40%. The cases' duration of the storage unit's work on the PS 1 in the charge mode is in the range of 0–5 min. The most observed range is 0–1 min, the monitoring's frequency for which is over 40%. For number an over 95% case there is observed the episodes' duration no more than 4 min.

The similar energy characteristics were received in the results' processing of the simulation modeling for the storage unit's work on the PS 2. In the Table 2, there are given the results of the energy volume's assessment for the different operating modes by work of the EESU on the buses of the PS. The assessment's results of the episodes' duration of the electric energy storage unit's work on the buses of the PS in the charge and discharge mode are presented in the Table 3.

Table 2. The assessment's results of the electric energy storage unit's energy intensity.

The episode's energy volume	PS 1	PS 2
The storage unit's charge mode	95% – up to 160 kWh	95% – up to 180 kWh
	40% – up to 40 kWh	30% – up to 20 kWh
The storage unit's discharge mode	90% – up to 200 kWh	95% – up to 600 kWh
	50% – up to 50 kWh	40% – up to 100 kWh

The increase of the locomotive's weight leads to the time increase of the electric rolling stock's movement under voltage. The mentioned case explains the falling of the episodes' number of the electric rolling stock in the storage unit's work's mode and the increase of their average duration.

By the trains' circulation of the unitized weight and the heavy freight trains the average duration of the time episode in the charge and discharge mode is comparable and lies in the range of 1.6–3.2 min (Table 4).

Table 3. The assessment's results of the work episodes' duration.

The duration of the work episode	PS 1	PS 2
The storage unit's charge mode	95% – up to 4 min	95% – up to 4 min
	40% – up to 1 min	30% – up to 1 min
The storage unit's discharge mode	90% – up to 4 min	100% – up to 4 min
	40% – up to 1 min	15% – up to 1 min

Table 4. The determination's results of the storage units' work episodes' number.

Object	The episodes' number in day in the work modes, pieces. (the work episode's average duration, min)		
	Discharge	Charge	Expectation
PS 1	40 (2.5)	122 (2.2)	146 (7.3)
PS 2	77 (3.2)	114 (2.4)	157 (5.9)

The difference of the mentioned values of the episodes' number and for the charge and discharge modes makes necessary the assessment's performing of the energy intensity's sufficiency of the electric energy storage unit, which was calculate one way or another, and also of the potential energy deficit in the charge mode or the deficient of the storage unit's energy intensity in the discharge mode for the examined ranges of the device's energy intensity. Let's consider the influence of the setpoints by voltage for the transition to the charge and discharge modes. The dynamics' comparison of the electricity's accumulation during the work of the electric energy storage unit on the PS allows highlighting three possible scenarios of the device's work, such as work with the increase of the energy deficit or excess, the equilibrium consumption. The mentioned variants depend on the setpoints by voltage for the work mode's change. The given case can be illustrated on the example of the electric energy storage unit's work on the buses of the PS 1 at the devise's transition to the charge mode at achievement of the voltage on the buses of the following values – 3600 V, 3900 V and 3880 V (look at the Fig. 5, a, b and c). For the first case the storage of the excess energy is charac-teristically – as you can see from the graph during the day the storing energy volume reaches 4000 kWh. The volume's usage such order on the PS is inexpedient, and the electric energy storage unit's usage with the energy intensity below the achieved brings to the loss of the part of energy. In the second case there is observed a deficit of the storage energy – by results of the day it reaches 250 kWh.

The value's increase of the setpoint's voltage for the inclusion of the above defined level is also inexpedient. The voltage's search, in which within a day the given energy intensity doesn't exceed some value, in the given example it is 400 kWh, and shows, that for the examined case the voltage's value is 3880 V. Subject to the load's cyclicity at the end of the day the accumulated value of the electricity closely to zero.

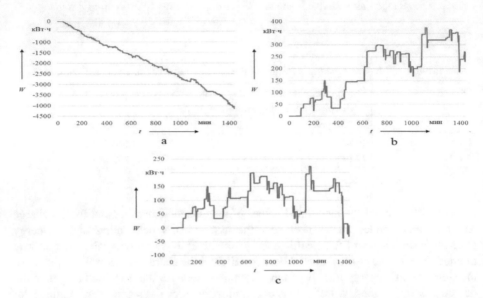

Fig. 5. The change's graph of the accumulated electricity on the PS 1; a – by U = 3600 V, b – by U = 3900 V, c – by U = 3880 V.

The mode's parameters electric energy storage unit's works on the PS 2 differ from the discussed case (PS 1). For the case, when the transition's voltage to the charge mode is 3 600 V, the storage of the excess energy is characteristically – as you can see on the graph (Fig. 6) during the day the stored volume of the energy is over 2000 kWh. By usage of the inclusion's setpoint by voltage at the level of 3900 V, there is observed a deficit of the storing energy– at the end of the day it is 1500 kWh – the electric energy storage unit is in the charge mode short time, and most of time is waiting for the charge. The increase of the value of the setpoint's voltage for the inclusion of the above defined level is also inexpedient. The search of the voltage, for which there is an equilibrium mode of the consumption during the day allow receiving the level of 3790 V. Subject to the load's cyclicity at the end of the day the accumulated value of the electricity is closely to zero, and the required energy intensity is about 400 kWh.

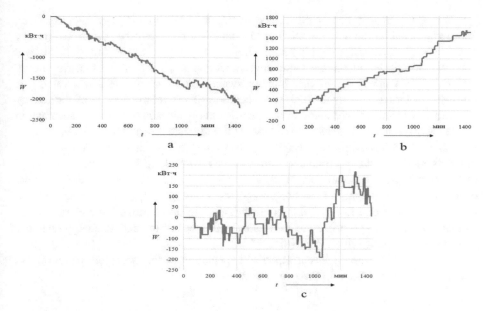

Fig. 6. The changes' graph of the accumulated electricity on the PS 2; a – by U = 3600 V, b – by U = 3900 V, c – by U = 3790 V.

5 The Main Approaches to the Assessment of the Voltage and Energy Intensity

The dynamics' difference of the change of the electricity's volume of the accumulation and the received voltage's levels for the transition to the charge mode of the storage units on the PS is conditioned by the potential values of the recovery energy's receiving and the way's profile on the areas.

The received graphs of the simulation work's load of the electric energy storage units on the PS allow receiving the following work's parameters:

- the potential volume of the receiving and return of the recovery energy;
- the episodes' number by work modes of device;
- the episode's duration of the device's work;
- the necessary device's capacity during the work in the charge/discharge modes;
- the influence of the voltage's setpoints on the device's transition to the charge/discharge mode;
- the influence of the voltage's setpoints on the potential volume of the recovery energy's losses.

By the determination of the energy intensity and capacity of the electric energy storage unit, an approach is used, based on the results of the measurements or modeling. During the measurements or modeling, capacity and energy intensity are setting by the average, maximal or the most observed values.

One of the approaches of the choice of the storage unit's energy intensity is an approach, based on the determination of the statistical probability of the excess energy recovery's appearance for the placement choice. The method of the energy intensity's determination is based on the simulation modeling's results of the electric energy storage unit's work in the electric traction system. The energy intensity's value was proposed to define as a mathematical expectation of the return's value of the energy recovery in the chosen placement of the electric energy storage unit. The determination of the energy intensity's value on the base of the average return's value of the energy recovery has its disadvantages, which are expressing, that in this case the cyclicity of the work modes of the electric energy storage unit isn't taking in account. The noticed assumption brings to the number of the inaccuracies in the calculations. For example, by following successively of the two charges or discharge's episodes, the second and the next episodes due to the full charge or discharge don't move the electric energy storage unit to the appropriate mode. It takes to the loss of the excess energy recovery or to the energy's deficit, respectively.

In the examined case, the mathematical expectation of the stored electric power in the storage unit is defined in the expression:

$$M(W_{exp}) = \int\limits_{0}^{W_{max}} f(W_{exc}) W_{exc} dW_{exc}, \tag{4}$$

where W_{max} – the maximal energy recovery's value at the connection point of the electric energy storage unit to the electric traction system;

$f(W_{exc})$ – the distribution's function of the probabilities' density of the excess energy recovery's appearance at the connection point of the electric energy storage unit.

In the case, if the electric energy storage unit's energy intensity in the electric traction system $W_{eesu} < W_{max}$, the excess electric power's volume W_{exc}, which can't be accepted of the electric energy storage unit, is:

$$M(W_{exp}) = \int\limits_{W_{EESU}}^{W_{max}} f(W_{exp}) W_{exp} dW_{exp}, \tag{5}$$

This way there is define the electric energy storage unit's capacity – as the mathematical expectation of the capacity, which is observed by the electric power's return. In the described method were not discussed the ways of the energy intensity and capacity's determination, appropriated to the storage unit's discharge mode.

There are also known other methods of the energy intensity and capacity's determination of the electric energy storage unit. If we consider the load's graph of an independent area of the power system, it is offering to use the electric energy storage unit for the daily graph's equalization. In the proposed method the electric energy storage unit's capacity is defined in the expression:

$$P_{EESU} = P_{\max} - P_{\text{aver}}, \tag{6}$$

where $P_{\max}$ – the maximal capacity of the power station (consumer); P_{aver} – the average capacity.

Because it is offering to use the electric energy storage unit for the daily graph's equalization of the load, the storage unit's energy intensity is determined out of the diagram of the energy storage's change, which is formed on the basis of the capacity's graph of the electric energy storage unit in the both work modes (charge/discharge). In this case energy intensity is determined as a maximal scope of the energy's values in the charge modes $W_{\max.\ \text{charge}}$ and the discharge $W_{\max.\ \text{dis}}$:

$$W_{EESU} = W_{\max\ \text{ch}} - W_{\max\ \text{dis}}, \tag{7}$$

If we consider the influence on the energy intensity of the different factors' storage unit, then the following should be considered: the electric rolling stock's capacity, the acceleration, the speed of movement, the duration of the acceleration-run-out-braking's cycle, for example, as it was shown in the considering of the urban transport.

Besides the examined approaches of the determination of the electric energy storage unit's capacity and energy intensity there was proposed a method, based on the statistical processing of the modeling's results and the choice of the most observed values of the necessary capacity and energy intensity by the electric energy storage unit's work in the charge and discharge modes [4]. The received results during the simulation modeling of the load's graph of the electric energy storage unit in the charge and discharge mode allow receiving the most observed ranges of the capacity's values, the energy intensity, and the duration of the work episodes in the different modes.

According to the calculation's results for the conditions of the electric energy storage unit's usage in the electric traction system at the movement's organization with the heavy and long trains, we can do the next assessment of the capacity and energy intensity.

The maximal necessary capacity for the PS 1 and PS 2 for the examined examples, calculated by the expression (12), is 9 mW and 15 mW, the average capacity is 0.5 mW and 1.2 mW, respectively. Herewith the most observed capacity's range in the both cases is limited with capacity of 4 mW. In this case the capacity's choice with the maximal values takes to the excess capacity with the low utilization factor, and in the case of the choice with the average values the storage unit's work will be mainly observed with the overload. On this basis the storage unit's capacity should be chosen with the most observed values, which were received as a result of the simulation modeling and the corrected devices, taking into account the overload.

By the determination of the energy intensity with the average values by the expression (10) for the both examined variants there were received the values less than 100 kWh. If we assess the energy intensity by scope of the energy's changes by the expression (13), then the energy intensity should be chosen for the examined variants of the modes' setting by voltage in the range of 250–450 kWh. The most observed range of the energy intensity without the modes' setting by voltage's level is 160–600 kWh.

In the examined cases, the energy volume for the charge and discharge modes corresponds as 1/3. In this regard for the balance of the energy's volumes it is necessary to consider the electric energy storage unit's work in the charge mode with the adaptive to the load's mode of the substations setpoint by voltage, which allows providing the charge in the mode of the low load of the adjacent traction substations.

Thus, we can formulate three main approaches, which were formed in the question of the determination of the electric energy storage unit's capacity and energy intensity – the receiving of the average values, the parameter's determination on the basis of the maximal values and the calculation of the most observed and necessary values. The usage of the average and maximal values without accounting of the work modes' organization of the storage unit in the electric traction system takes to the insufficient or excess capacity and energy intensity, because in the first case it makes the prerequisites for the preemptive storage unit's work with overload, and in the second case it takes to the devices' low load. The determination's task of the storage unit's capacity and energy intensity in the electric traction system by the simulation modeling of its work must be solved in two steps. First the modeling is performing without restrictions by the capacity and energy intensity in the wide range of voltages on the device' buses, further – for the chosen ranges of the work voltages in the charge and discharge modes there are defined the most observed values of the electric energy storage unit's capacity and energy intensity.

Acknowledgement. The researches are performed with the support of the Russian Foundation for fundamental initiatives under the grant №. 17-20-01148 ofi_m_RZD/18.

References

1. Kiepe Electric liefert ersten 24-Meter-Elektro-Bus nach Linz. https://www.eb-info.eu/aktuell/aus-den-unternehmen/12-10-2017-kiepe-electric-liefert-ersten-24-meter-elektro-bus-nach-linz/559221/. Accessed 18 Dec 2017
2. ÖBB-Hybridlok mit Brennstoffzelle: Versuchslokomotive mit alternativem Antriebskonzept getestet. https://www.eb-info.eu/aktuell/forschung-und-entwicklung/11-10-2017-oebb-hybridlok-mit-brennstoffzelle-versuchslokomotive-mit-alternativem-antriebskonzept-getestet/. Accessed 18 Dec 2017
3. Sun, Y., Zhong, J., Li, Z.: Scheduling of battery-based energy storage transportation system with the penetration of wind power. IEEE Trans. Sustain. Energy 8(1), 135–144 (2017). https://doi.org/10.1109/tste.2016.2586025
4. Yap, H.T.: Hybrid energy/power sources for electric vehicle traction systems. In: Yap, H.T., Schofield, N., Bingham, C.M. (eds.) Second International Conference on Power Electronics, Machines and Drives (Conf. Publ. No. 498) (2004)
5. Wang, J., Rakha, H.A.: Electric train energy consumption modeling. Appl. Energy 193(1), 346–355 (2017). https://doi.org/10.1016/j.apenergy.2017.02.058
6. Grechishnikov, V.A.: The energy storage unit's maintenance on the underground. In: Grechishnikov, V.A., Shevlyugin, M.V. (eds.) The World of Transport, T. 11, no. 5(49), pp. 54–58 (2013)

7. Titova, T.S.: The energy efficiency's increase of the locomotives with the energy storage unit. In: Titova, T.S., Evstaf'yev, A.M. (eds.) The News of St. Petersburg Transport University, T. 14, no. 2, pp. 200–210 (2017)

8. Cheremisin, V.T.: The perspective of the electric energy storage unit's usage in the DC electric traction system. In: Nezevak, V.L., Cheremisin, V.T. (eds.) The Bulletin of the Scientific Research's Results, no. 1(14), pp. 76–83 (2015)

9. Petrushin, D.A.: The calculation of the work modes of the electric traction system with the inertial energy storage. In: Petrushin, D.A. (ed.) Vestnik of Rostov State Transport University, no. 1(41), pp. 79–84 (2011). (in Russian)

10. Kryukov, A.V.: The modes' modeling of the electric traction systems, equipped with the energy storage units. In: Kryukov, A.V., Zakaryukin, V.P., Cherepanov, A.V. (eds.) Proceedings of the Bratsk State University, T. 1, pp. 113–120 (2015). (in Russian)

11. Istomin, S.: The analyse of the lokomotive's work 2ES10 on the DC line of the Sverdlovsk railway. In: Istomin, S., Nezevak, V. (eds.) Electric Lokomotives, no. 4, pp. 186–189. Open Company Oldenburg Publishing House, Munich (2015)

12. Cheremisin, V.T.: The organization of control of the maximal modes of the traction network's work in the conditions of the high-speed and heavy traffic. In: Cheremisin, V.T., Kashtanov, A.L., Nezevak, V.L. (eds.) Proceedings of the TRANS-Siberian Railway, no. 1 (29), pp. 83–90 (2017)

13. Rumshinskiy, L.Z.: T mathematical processing of the experiment's results. Science, Moscow (1971)

14. Venikov, V.A.: Similarity theory and modeling (with reference to the electroenergetics' tasks): book for high schools on speciality «Cybernetics of the electric systems». 3rd edition, rewrote and supplemented. High School, Moscow (1984). (in Russian)

15. Teymourfar, R., Asaei, B., Iman-Eini, H., Nejati, R.: Stationary super-capacitor energy storage system to save regenerative braking energy in a metro line. Energy Convers. Manag. **56**, 206–214 (2012). https://doi.org/10.1016/j.enconman.2011.11.019

16. Gao, Z., Fang, J., Zhang, Y., Jiang, L., Sun, D., Guo, W.: Control of urban rail transit equipped with ground-based supercapacitor for energy saving and reduction of power peak demand. Int. J. Electr. Power Energy Syst. **67**, 439–447 (2015). https://doi.org/10.1016/j.ijepes.2014.11.019

17. Frilli, A., Meli, E., Nocciolini, D., Pugi, L., Rindi, A.: Energetic optimization of regenerative braking for high speed railway systems. Energy Convers. Manag. **129**, 200–215 (2016). https://doi.org/10.1016/j.enconman.2016

18. Vil'gel'm, A.S.: The comparative efficiency of the variants of the energy recovery's usage on the DC railways. In: Nezevak, V.L., Vil'gel'm, A.S., Shatokhin, A.P. (eds.) Science and Education for Transport, T. 1, no. 1, pp. 243–247 (2013)

19. Cheremisin, V.T.: The assessment's results of the work modes of the active and passive posts of sectionalization in the electric traction system with the purpose of the electric energy storage units' parameters' choice. In: Cheremisin, V.T., Nezevak, V.L., Erbes, V.V. (eds.) Proceedings of the TRANS-Siberian Railway, vol. 3, no. 31, pp. 132–143. Omsk State Transport University, Omsk (2017)

20. Karabanov, M.A.: The determination of the placement energy intensity of the electric energy storage unit in the DC electric traction system of a double track area. In: Karabanov, M.A., Moskalev, U.V. (eds.) Science and Technology of Transport, vol. 2, pp. 21–28 (2015)

21. Gorte, O.I.: The method of the parameters' choice the electric energy storage unit with a variable load. In: Gorte, O.I, Zyryanov, V.M., Kir'yanova, N.G. (eds.) Proceedings of the VIII International Scientific Technology Conference, pp. 135–138. Samara State Technical University, Samara (2017)
22. Baluev, D.U.: The calculation's method of the main parameters of the electric energy storage unit by experimental load diagrams. In: Baluev, D.U., Zyryanov, V.M., Kir'yanova N.G. (eds.) Vestnik of Irkutsk State Technical University, T. 22, no. 5(136), pp. 105–114 (2018). (in Russian)

Helicity of the Velocity Field in Evaluating the Efficiency of Turbomachines

Nikolay Kortikov[1]([✉]) [iD] and Andrey Nazarenko[2] [iD]

[1] Peter the Great St. Petersburg Polytechnic University, Polytechnicheskaya 29,
195251 St. Petersburg, Russia
n-kortikov@yandex.ru
[2] Ltd "Constanta-XXI", Seleznevsckay ul. 15/2, 127473 Moscow, Russia

Abstract. Improvement of gas turbine engines (GTE) as a complex technical system requires a reliable estimate of the quantities responsible for the performance indicators of engine efficiency units. The dependence was obtained for calculating the efficiency of the turbine stage, taking into account large-scale vortex structures in the inter-path channel of the turbomachine, and validating it. It is noted that the helicity of the velocity field is equal to the decrease of the total enthalpy for adiabatic gas flows. An example is given for calculation of the efficiency of a model turbine stage taking into account the helicity of the velocity field, and its decrease is shown to be 0.15%.

Keywords: Helicity · Efficiency · Vortex

1 Introduction

Improvement of gas turbine engines (GTE) as a complex technical system requires a reliable estimate of the quantities responsible for the performance indicators of engine efficiency units. The fields of the gas parameters in the sections at the inlet and outlet of the elements are uneven and vortical. To determine the efficiency based on the results of measuring the pressure and temperature of the gas flow (or the results of 3D numerical simulation), their mean values are used, which are calculated using a particular averaging method [1].

The aim of the paper is to develop a methodology for averaging uneven vortex flows, which is based on the conditions for the conservation of mass flow, total enthalpy and impulse (or entropy) flows with the addition of a new invariant, the helicity integral of the gas flow, to the averaging algorithm [2]:

$$H = \int_F \vec{\omega} \cdot \vec{u} \, dF, \; \vec{\omega} = \frac{1}{2} rot \, \vec{u} \tag{1}$$

The term "helicity" refers to the integral over the volume (surface or line) from the scalar product of the velocity vector $\vec{u}$ to its vorticity $\vec{\omega}$, or half the speed rotor $rot\,\vec{u}$. In [3], it is proposed to introduce a refinement into this term: the integral in (1) is called the "helicity

© Springer Nature Switzerland AG 2020
V. Murgul and M. Pasetti (Eds.): EMMFT 2018, AISC 982, pp. 539–545, 2020.
https://doi.org/10.1007/978-3-030-19756-8_51

of the velocity field" or the "helicity integral". The helicity density $(\vec{u} \cdot \vec{\omega})$ is a scalar, the local energy characteristic of the stream, and has the acceleration dimension [4].

When considering the flow in the grids of turbomachines, in the interscapular channel (Fig. 1a) formed by the convex and concave surfaces of two adjacent blades and two limiting end surfaces, the primary and secondary gas flows are conventionally separated [5].

The secondary flow is called the vortex flow (Fig. 1b and c) caused by the terminal phenomena (friction with the end confining walls of the lattice and the influence of the radial gap), when summing it with the main one, the real parameters of the spatial flow of a viscous gas in the lattice are obtained.

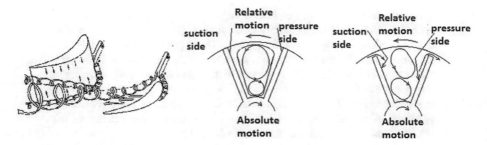

Fig. 1. Physical picture of vortex currents in the inter-path channel (a); a secondary flow pattern in the annular grating channel in the absence of a radial clearance (b); the location of the vortices in the presence of radial clearance (c).

2   The Relationship Between the Helicity Integral and the Parameters of the Process

An analysis of the change in helicity in eddy currents makes it possible to analyze which factors influence the redistribution of the mechanical energy of the flow between the translational and rotational components of the motion, and, consequently, the change in the vortex structure of the flow [6]. Helicity is not zero in systems where vortices do not have a mirror symmetry (for example, two rings linked together).

Helicity closely intertwines with the mathematical theory of nodes and in principle reflects the topology of the system, that is, the relative position of several vortices or parts of the vortex relative to others. The sign of helicity shows the direction of rotation of the vortex with respect to the direction of flow (clockwise or counterclockwise). The behavior of the vorticity in the flows of a compressible gas differs significantly from its behavior in an incompressible fluid. This can be analyzed from the equation of motion of an ideal gas in the form of Gromek - Lamba:

$$\frac{\partial u}{\partial \tau} + \text{grad}\left(\frac{u^2}{2}\right) + (\vec{\omega} \times \vec{u}) = T \cdot \text{grad}(s) - \text{grad}(h) \tag{2}$$

In the case of steady flow, Eq. (2) is simplified and rewritten in the form:

$$(\vec{\omega} \times \vec{u}) = T \cdot \text{grad}(s) - \text{grad}(h^*) \tag{3}$$

where h^* is the specific total enthalpy.

Relation (3) indicates that the vorticity $\vec{\omega}$ is related to the parameters of the gas flow (entropy and enthalpy of the hindered gas). For a plane and axisymmetric case, the solution of Eq. (3) is expressed in terms of the stream function ψ and is written accordingly in the form:

$$\omega = \rho \left(T \frac{\partial s}{\partial \psi} - \frac{\partial h^*}{\partial \psi} \right), \quad \omega = \rho r \left(T \frac{\partial s}{\partial \psi} - \frac{\partial h^*}{\partial \psi} \right) \tag{4}$$

The expression for the helicity integral (for plane and axisymmetric flows) taken along the width of the channel (the lattice spacing) can be written in the form:

$$H = \int_0^l (\omega \cdot u)dl = \int_0^l \left(T \frac{\partial s}{\partial \psi} - \frac{\partial h^*}{\partial \psi} \right)d\psi = \int_0^l (Tds - dh^*) = q_l - \Delta h^* \tag{5}$$

Expression (5) shows that the helicity integral is equal to the difference in the heat influx q_l and the decrease in the enthalpy of inhibition Δh^*. For adiabatic conditions, the presence in the gas flow of large-scale vortex structures affects the change in the value of the total enthalpy, which is equal to the helicity integral:

$$\Delta h^* = -H \tag{6}$$

The value of the isentropic (adiabatic) efficiency of the turbine stage is calculated

$$\eta_{is} = \frac{\overline{T_0^*} - \overline{T_2^*}}{\overline{T_0^*} - \overline{T_{2is}}} = \frac{\left(1 - \overline{T_2^*}/\overline{T_0^*}\right)}{\left[1 - \left(\frac{\overline{p_{2is}}}{p_0^*}\right)^{k-1/k}\right]} \tag{7}$$

where $\overline{p_{2is}}$ is the area-averaged pressure obtained by isentropic expansion in the output section of the stage (the line above is the sign of averaging). The combined use of relations (6) and (7) allowed obtaining an expression for the calculation of efficiency changes $\Delta \eta_{is}$ due to the formation of large-scale vortex structures in the inter-blade channel of the turbomachine:

$$\Delta \eta_{is} = -\frac{H}{c_p \left(T_0^* - \overline{T_{2is}} \right)} \tag{8}$$

In formula (8), the lower indices "0" and "2" correspond to the parameters of the flow at the inlet and outlet of the stage, cp - heat capacity at constant pressure.

3 Numerical Simulation and Helicity of Flow in the Flow Part of the Turbine Stage

Calculation of gas-dynamic processes in the turbine stage was carried out on the basis of Navier—Stokes equations and energy averaged over Reynolds. They are written in an integral form for a moving control volume in a relative coordinate system rotating at the angular velocity of the rotor [7]. The time step is taken to be $5 \cdot 10-6$ s. The mass flow rate at the input and output boundaries was used to control the convergence of the problem. The calculation method is implicit and the Spalart–Allmaras model [10] is used as a turbulence model. As a working medium, air is selected, the thermophysical properties of which were calculated by the model of a perfect gas.

The total pressure at the inlet corresponded to the degree of pressure drop in the stage, equal to 1.4. Reynolds number (Re) for the stator part - $Re_1 = 1.05 \cdot 10^6$, the rotor part - $Re_2 = 2.28 \cdot 10^5$. The temperature at the entrance to the stage - 350 K and $\omega = 5000$ rpm.

For visualization of vortex lines is the normalized helicity (normalized helicity) [8], represents the cosine of the angle between the vectors of velocity and vorticity. In the vicinity of the vortex center, the angle between the velocity and vorticity vectors is small. In the limiting case, when the velocity and vorticity vectors are collinear, the normalized helicity is ± 1, and the current line passing through such a point has zero curvature (straight line).

Visualization of the flow using the method of normalized helicity shows that the vortex structures at the periphery and at the root of the working blade differ [9]. In particular, the center of the separation vortex at the root, which was formed on the trough part of the profile (Fig. 2a), practically overlaps the flow part, causing a dominant role by the amount of losses of vortex structures located on the lower end surface (Fig. 2b).

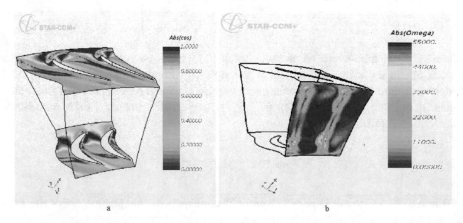

Fig. 2. Visualization of vortex lines in the rotor part of the stage: absolute value of normalized helicity (a); vorticity module in the output section of the computational domain (b).

In turn, the flow pattern at the periphery of the working blade indicates the location of the vortex center line near the Central part of the inter-blade channel (Fig. 2a) and the formation of a single vortex formation (Fig. 2b) at the outlet of the turbine stage.

4 The Method of Averaging the Flow Parameters in Turbomachines

The experience of numerical simulation [11] showed the low suitability of expression (3) for estimating turbine power from the results of numerical simulation. The value of the stagnation temperature at the exit of the stage varies greatly depending on the position of its averaging plane. There are no informed decisions on the placement of these averaging planes.

At the same time, the STAR CCM + provides a direct calculation of the moment developed by the flow on the turbine surfaces. Such an approach makes it possible to determine the useful moment unequivocally and, moreover, it is close to the method used in experimental research of turbines. In this case, the isentropic efficiency will be written as

$$\eta_{is} = \frac{M\pi n}{30 \cdot c_p \cdot \dot{G}\left(T_0^* - \overline{T_{2_{is}}}\right)},\tag{9}$$

where n is the frequency of rotation of the turbine rotor, rpm; M - useful moment developed by the impeller of the turbine without friction of the disk, N m.

The role of internal irreversibility (viscous friction) in estimating the efficiency of a step is more logical to take into account if the adiabatic non-isentropic flow of an ideal gas with a non-zero helicity integral is chosen as the scale. In this case, the internal relative efficiency of the stage will be written as follows [11]:

$$\eta_{non,is} = \frac{M\pi n}{30 \cdot c_p \cdot \dot{G}\left(T_0^* - \overline{T_{2non,is}}\right)}\tag{10}$$

The equality of the cross section momentum for the real (viscous) and canonical flows is used in calculating the averaged parameters of the canonical flow, in particular. Marked equality is written in the form:

$$\frac{k+1}{2k}a_{cr.} \cdot \left(\lambda + \frac{1}{\lambda}\right) \cdot \dot{G} = \frac{k+1}{2k}a_{cr.,H} \cdot \left(\lambda_H + \frac{1}{\lambda_H\left(1 + \frac{H}{c_pT^*}\right)}\right) \cdot \dot{G}_H\tag{11}$$

Here λ, λ_H - the reduced velocities for the averaged three-dimensional and canonical given helicity of the flows, respectively; $a_{cr.}$ - critical speed of sound.

Fulfillment of the conservation conditions for mass flow $\dot{G} = \dot{G}_H$ and the specification of the specific total enthalpy c_pT^* allows us to reduce Eq. (11) to a quadratic equation:

$$\lambda_H^2 - \sqrt{\frac{T^*}{T_H^*}} \cdot \left(\lambda + \frac{1}{\lambda}\right) \cdot \lambda_H + \frac{1}{\left(1 + \frac{H}{c_p T^*}\right)} = 0, \tag{12}$$

where $T_H^* = T^* \left(1 + \frac{H}{c_p T^*}\right)$.

The solution of a quadratic equation is:

$$\lambda_H = \frac{1}{2}\sqrt{\frac{T^*}{T_H^*}} \cdot \left(\lambda + \frac{1}{\lambda}\right) \pm \sqrt{\frac{1}{4}\frac{T^*}{T_H^*} \cdot \left(\lambda + \frac{1}{\lambda}\right)^2 - 1 / \left(1 + \frac{H}{c_p T^*}\right)} \tag{13}$$

which is rewritten (for further efficiency calculations) as follows:

$$\frac{\lambda_H}{\lambda} = \frac{1}{2}\sqrt{\frac{T^*}{T_H^*}} \cdot \left(1 + \frac{1}{\lambda^2}\right) - \sqrt{\frac{1}{4}\frac{T^*}{T_H^*} \cdot \left(1 + \frac{1}{\lambda^2}\right)^2 - \left[\lambda^2 \cdot \left(1 + \frac{H}{c_p T^*}\right)\right]^{-1}} \tag{14}$$

Comparison of expressions (9) and (10) for stage efficiency is based on the use of the following relationship:

$$\Delta\eta = \eta_{is} - \eta_{non,is} = \eta_{is}\left[1 - \frac{(T_0^* - T_{2,is})}{(T_0^* - T_{2non,is})}\right] = \eta_{is}\left[1 - \left(\frac{\lambda}{\lambda_H}\right)^2\right] \tag{15}$$

The averaging of gas flow parameters is based on data on the geometry of the single turbine stage, which are specified in [9]. Using the 3D modeling data and relations (14) and (15), it is possible to obtain refined data regarding the efficiency of the stage. The effect of taking into account the helicity of the velocity field affects the efficiency of the turbine stage, which boils down to its reduction: from 76.2% to 76.05% (a decrease of 0.15%).

5   Conclusions

Thus, the introduction into practice of calculation of dependence (8) and the technique of averaging of parameters of the gas stream considering the contribution of secondary currents to losses of kinetic energy in the form of the Eqs. (9), (14), and (15) allows carrying out an assessment of efficiency of the turbomachine more correctly.

References

1. Derevyanko, V., Zhuravlev, V., Zikeev, V.: The Basics of Designing Turbine Engines. Machinebuilding, Moscow (1988, in Russian)
2. Venediktov, V.: Gas Dynamics of Cooled Turbines. Machinebuilding, Moscow (1990, in Russian)

3. Stepanov, G.: Hydrodynamics of a Lattice of Turbomachines. Science, Moscow (1962, in Russian)
4. Sedov, L.: Methods of similarity and dimension in mechanics. Science, Moscow (1977). (in Russian)
5. Samoilovich, G.: Gas Fluid Dynamics. Machinebuilding, Moscow (1990, in Russian)
6. Augustinovich, V.: Averaging of non-stationary inhomogeneous flows in turbomachines for estimation of their efficiency. Aerosp. Eng. **49**, 63–70 (2017). (in Russian)
7. Alekseenko, S., Kuibin, P., Okulov, V.: Introduction to the Theory of Concentrated Vortices. RAS SB Institute Thermophysics, Novosibirsk (2003). (in Russian)
8. Kurgansky, V.: Helicity in atmospheric dynamic processes. Phys. Atmos. Ocean **53**(2), 147–163 (2017). (in Russian)
9. Mitrofanova, O.: Hydrodynamics and heat transfer of swirling flows in the channels of nuclear power plants. Science, Moscow (2010). (in Russian)
10. Lapshin, V.: Optimization of Flow Parts of Steam and Gas Turbines. St. Petersburg Polytech. University, St. Petersburg (2011). (in Russian)
11. Poludnitsyn, A., Stepanov, R., Frick, P.: Measurement of Helicity of Turbulent Flows by Digital Tracer Methods, vol. 16, pp. 116–123. Perm State University, Perm (2006) (in Russian)
12. Saffman, F.: Dynamics of Vortices. Science World, Moscow (2001). (in Russian)
13. Saburov, E., Karpov, S.: Theory and Practice of Cyclone Separators, Furnaces and Furnaces. Arkhangelsk State Technical University, Arkhangelsk (2001). (in Russian)
14. Scorer, R.: The Aero-and Hydrodynamics of the Environment. World, Moscow (1980). (in Russian)
15. Gladkov, A.: Behavior of vorticity in inhomogeneous flows of compressible gas. Sci. Notes TSAGI **XXX**(1–2), 68–76 (1999)
16. User Guide STAR-CCM + 13.04. CD - adapco. Melville, NY (2017)
17. Volkov, K.: Methods of visualization of vortex flows in computational gas dynamics and their application in solving applied problems. Sci. Tech. J. Inf. Technol. Mech. Optics. **3**(91), 1–10 (2014). (in Russian)
18. Kortikov, N.: Simulation of the joint effect of rotor-stator interaction and circumferential temperature unevenness on losses in the turbine stage. MATEC Web Conf. **245**, 06008 (2018). https://doi.org/10.1051/matecconf/201824506008
19. Bystrov, Y., Isaev, S., Kudryavtsev, N., Leontiev, A.: Numerical Simulation of Vortex Intensification of Heat Exchange in Packages of Tubes. Shipbuilding, St - Petersburg (2005)
20. Smirnov, M., Sebelev, A., Zabelin, N., Kuklina, N.: Effects of hub endwall geometry and rotor leading edge shape on performance of supersonic axial impulse turbine. Part I. In: Proceedings of 12th European Conference on Turbomachinery Fluid Dynamics and Thermodynamics. Stockholm, Sweden (2017). http://doi.org/10.29008/ETC2017-100

Modeling of the Oscillating Mode of the IR-Energy Supply in the Technology of Restoration of Insulating Fingers of Electric Motors of Locomotives

Evgeny Dulskiy[(✉)] [iD], Anatoliy Khudonogov [iD],
Igor Khudonogov [iD], Leonid Astrakhantsev [iD], Pavel Ivanov [iD],
and Albina Tuigunova [iD]

Irkutsk State Transport University, Chernishevsky st. 15, 664074 Irkutsk, Russia
E.Dulskiy@mail.ru

Abstract. This article deals with a completely new technology of restoring the insulating elements of electric motors of traction rolling stock, as illustrated by the insulating fingers of the brush holder brackets. It presents a simulation of the oscillating mode of thermal effects on the insulating element rotating uniformly on a carousel installation, whose principle of operation is also described in the article. The use of the energy balance equation makes it possible to determine the energy losses in the process of drying insulating coatings with IR energy supply. The paper provides formulas for finding losses during the heat of the impregnated IF; loss of heat to the environment through convection; heat losses to the environment through radiation; removal of solvent. The paper shows the geometrical dimensions of the restored surface of the insulating finger, on the basis of which the index of the geometric characteristics of the impregnated insulation is found. Comparative graphs of both continuous and oscillating modes of insulation restoration with an indication of the values controlled during the drying process are clearly demonstrated. A mathematical model for determining the effective duty cycle of the radiating elements and calculating the maximum oscillation cycle duration has been created.

Keywords: Traction motor · Insulating fingers · Oscillating mode ·
IR power supply · Thermal radiation

1 Introduction

The problem of reliability of collector electric motors of traction rolling stock was repeatedly discussed on the pages of various publications. In connection with the re-equipment of railway transport with a fairly new electric locomotive of the Yermak, Donchak and Sinara series, a new problem has emerged connected with the reliability of the insulating fingers (IF) of the brush holder brackets. In the column related to failures of elements of the collector traction motor, there occurs from 5 to 10% of IF failures. Most often, these failures are caused by the overlap and, less commonly, the

© Springer Nature Switzerland AG 2020
V. Murgul and M. Pasetti (Eds.): EMMFT 2018, AISC 982, pp. 546–555, 2020.
https://doi.org/10.1007/978-3-030-19756-8_52

breakdown of the IF [1, 2]. This factor indicates the inefficient technology currently used in the manufacture and restoration of insulating fingers.

2   Materials and Methods

The manufacturing techniques and restoration of IF provides for carrying out several operations on heating of a detail. In factory and depot conditions, such an operation is performed in specialized drying ovens with the convective principle of heating parts before impregnation and after impregnation. Moreover, in depot conditions, a drying oven with a power of 100 kW is placed to heat the stand with only 12 ... 18 fingers, since it is necessary to restore the fingers in the shortest possible time and there is no possibility for their accumulation. The duration of the process of insulation restoration takes more than 12 h in accordance with the technological maps developed to date.

For the purpose of resource saving, employees of the Irkutsk State Transport University offer local technologies in the production and restoration of electric propulsion machines for rolling stock based on the use of thermal (infrared, thermal radiation) incoherent and coherent (laser) radiation. With regard to the production and restoration of IF, several methods and devices are proposed on the basis of coherent and incoherent thermal radiation [3, 4].

One of the most promising installations is the carousel installation, making it possible to restore insulating fingers in the oscillating mode of the infrared power supply. The efficiency of these modes is established for technologies using infrared energy (IR) radiation [5]. A general view of the proposed carousel installation is shown in Fig. 1.

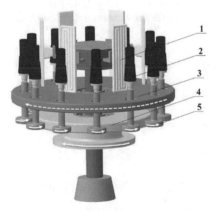

1 – pulse ceramic infrared radiating element; 2 – insulating finger; 3 – carousel;
4 – carousel drive; 5 – IF drive

Fig. 1. General view of the carousel installation.

A set of IFs for a single six-pole traction motor is screwed into the carousel table. The sequence of the recovery technology by applying a polymer to the IF surface, followed by drying, is traditional. Frequency-adjustable carousel drives and IFs are independent, which makes it possible to use the double Fourier transform when developing the theory of hardening of IFs by thermal radiation [6, 7].

3   Results

At the first stage of building a mathematical model, we use the differential equation of energy balance. Preliminary experimental studies have shown that pulsed ceramic radiating elements with a power of 500 W are the most effective for this process [8, 9].

In the case when radiation acts on the polymer from one side, the elementary energy consumed by the radiation source during the time $d\tau$ is equal to

$$dA = P \cdot d\tau \tag{1}$$

where P is the power of the radiation source, W; τ is the process time, s.

Assuming that the intermediate medium between the radiating element and the insulating material does not absorb energy, the elementary energy absorbed by the polymer over time $d\tau$ is defined as

$$dA' = \eta \cdot A_\lambda \cdot P \cdot d\tau \tag{2}$$

where η is the radiating element efficiency; A_λ is the absorbing capacity of the impregnating material (polymer).

The absorbed energy from the radiating elements will be spent on: heating the impregnated PI, heat loss to the environment through convection and radiation, and solvent removal

$$\eta \cdot A_\lambda \cdot P \cdot d\tau = C \cdot dt + P_k \cdot d\tau + P_r \cdot d\tau + P_{\text{req}} \cdot d\tau \tag{3}$$

where C is the heat capacity of the impregnating insulation material, J/K; t is the material temperature, K.

Heat capacity of impregnating insulation material is

$$C = c_{spec} \cdot M \tag{4}$$

where c_{spec} is the specific heat capacity of the polymer, $J/K \cdot kg$; M is the mass of the polymer, kg.

Specific heat of polymers

$$c_{\text{spec}} = \frac{c_{\text{s.is}}(100 - \omega) + c_s \cdot \omega}{100} = \frac{c_{\text{s.is}} \cdot 100 + c_s \cdot u}{100 + u} \tag{5}$$

where $c_{s.is}$, c_s are, respectively, the specific heat capacity of the insulation bonding and solvent, J/K · kg; ω, u are, respectively, the relative and absolute content of solvent in the impregnating material, %.

Heat losses $P_c \cdot d\tau$, resulting from convective heat exchange between the material and the surrounding surfaces, are

$$P_c = \alpha_c \cdot F \cdot (t - t_0) \tag{6}$$

Where α_c is the convective heat transfer coefficient, W/K, m^2; F is the heat exchange area, m^2; t is the material temperature, K; t_0 is the ambient temperature, K.

Thermal losses P_r, resulting from radiant heat transfer between the irradiated material and the surrounding surfaces, are

$$P_r = \alpha_r \cdot F \left[\left(\frac{t}{100} \right)^4 - \left(\frac{t_0}{100} \right)^4 \right] \tag{7}$$

where α_r is the heat exchange coefficient of radiation, W/K^4 m^2.

Power required to remove solvent

$$P_{req} = i \cdot r \tag{8}$$

where i is the removal rate, kg/s; r is the latent heat of solvent removal, J/kg.

Let us introduce the concept of the temperature of the material above the ambient temperature

$$\theta - t - t_0 \tag{9}$$

Equation (3) will become

$$(\eta \cdot A_\lambda \cdot P - ir) \cdot d\tau = C \cdot d\theta + (P_k + P_r) \cdot d\tau \tag{10}$$

The solution of the differential Eq. (10) by the variable separation method, with some assumptions, produces the following result

$$\theta = \theta_{max} \cdot \left(1 - e^{-\tau/T_h} \right) + \theta_{ini} \cdot e^{-\tau/T_h} \tag{11}$$

where θ_{max} is the maximum temperature excess of the IF, K; θ_{ini} is the initial temperature rise of the IF, K; T_h is the time constant of heating of the IF, s; e is the basis of natural logarithms.

The rate of heating of the polymer to the maximum allowable temperature in the process of restoring insulation by infrared heating should not exceed the values obtained by dividing the maximum allowable temperature for this class of insulation by the time constant of its heating

$$V_{max.all.} = \frac{t_{max.all.}}{T_h} \qquad (12)$$

where $V_{max.\,all.}$ is the allowable heating rate for an insulation class, K/s; $t_{max.\,all.}$ is the allowable heating temperature for an insulation class, K; T_h is the time constant of heating of the impregnated IF, s.

The heating time constant is an IF characteristic. It does not depend on the power input and is numerically equal to the ratio of the body heat capacity to its heat transfer.

$$T_h = \frac{c_{spec} \cdot \rho}{\alpha} \cdot \frac{V}{F} \qquad (13)$$

where c_{spec} is the specific heat capacity of the insulation, J/kg K; α is the heat transfer coefficient of impregnated insulation, J/m^2 K s; F is the area of the external surface of the IF, m^2; ρ is the density of impregnated insulation, kg/m^3; V is the volume of impregnated insulation, m^3.

In its essence, σ is a generalized indicator of the geometric characteristics of impregnated insulation. This indicator, as follows from the above, can be determined in the presence of the geometrical dimensions of the material. Figure 2 shows the main geometrical dimensions of IFs of the main series of AC electric locomotives, on the basis of which we obtain the following formula.

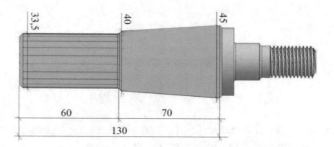

Fig. 2. The geometrical dimensions of the insulating finger of the main series of AC electric locomotives.

$$\sigma = \frac{V}{F} = \frac{\frac{1}{3} \cdot \pi \cdot l \left(\frac{V_{con} + V_{cyl}}{F_{con} + F_{cyl} + F_{end}} r_1^2 + r_1 \cdot r_2 + r_2^2 \right) + \pi \cdot r^2 \cdot l}{\pi \cdot l \cdot (r_1 + r_2) + 2\pi \cdot l \cdot r + \pi \cdot r^2}, \qquad (14)$$

where S_{con}, S_{cyl}, S_t are the area of the conical, cylindrical and end parts, respectively, mm^2; V_{con}, V_{cyl} are volume of the conical and cylindrical part, mm^3; r is the radius of the base of the cylinder, mm; r_1, r_2 are the radii of the bases of the truncated cone, mm; l is the element height, mm.

Due to the fact that the insulating finger rotates around its axis, the drying mode of the impregnated insulation will be oscillating (intermittent). It is known that capillary-porous colloidal bodies, such as impregnated finger insulation, when hardened by IR radiation, have significant gradients of resin and solvent content (up to 90% when impregnated with varnishes) and are heated intensively [10–12].

The rapid increase in the temperature of the impregnated insulation after the critical point leads to a prolonged effect of high temperature on the insulation, which causes a deterioration of its technological properties. A significant temperature gradient, directed opposite to the resin and solvent content gradient, slows the transfer of the solvent from the inner layers of the impregnated insulation to the surface, which also adversely affects the quality of the recovery process.

Hence the need for an intermittent (oscillating) mode of the IR power supply, namely, in the combination of heating the impregnated insulation by the IR radiation and cooling it with air [13]. Let us consider and analyze the infrared power supply in the restoration technology of the insulating fingers of the brush holder brackets (Fig. 3).

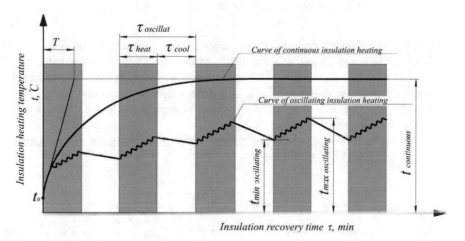

Fig. 3. The schedule of oscillating infrared power supply in the technology of insulating fingers recovery.

During the period of time τ_{heat}, the insulating fingers mounted on the carousel are located opposite the switched on IR radiating elements, thereby heating up with significant evaporation of the liquid in the surface layers. When a uniformly rotating finger is heated, oscillation is also performed due to the fact that the position of the insulating surface of the object changes. This change in position relative to the radiating element has the character of a point moving along a circle, drawing a sine wave, Fig. 4, which has a set of several intervals τ_1 when the point is under the action of the radiating element, and τ_2 when it cools. The set of these time intervals is the total period of heating insulation τ_{heat}.

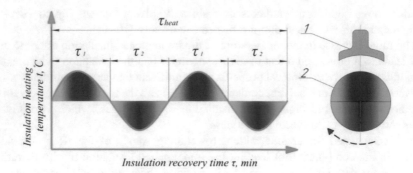

1 – the pulse ceramic infrared radiating element; 2 – the insulating finger.

Fig. 4. The heating of an evenly rotating insulating finger.

During the period of time τ_{cool} the fingers, which are not covered by the IR radiating elements, are cooled as a result of the evaporation of the liquid due to the accumulated heat. The alternation of the segments of the rotating insulating finger that falls within the gaps between τ_{heat} and τ_{cool} occurs with the repetition period of the oscillator $\tau_{oscillat}$. The average value of the power of the electric heater depends on the ratio of the magnitudes of τ_{heat} and $\tau_{oscillat}$.

The ratio of τ_{heat} to $\tau_{oscillat}$ is the coefficient of the relative duration of the IR energy supply and is denoted by the subscript ε.

$$\varepsilon = \frac{\tau_{heat}}{\tau_{oscillat}} = \frac{\tau_{heat}}{\tau_{heat} + \tau_{cool}} = \frac{\tau_{heat}}{\sum \tau_1 + \sum \tau_2 + \sum \tau_{cool}} \tag{15}$$

Therefore, by changing the values of τ_{heat} and $\sigma_{oscillat}$ cycle, it is possible to control the process of drying the insulation. In accordance with the theory of drying, the ratio between the periods of heating τ_{heat} and cooling τ_{cool} of the insulation is determined by the solvent diffusion coefficient. The smaller the diffusion coefficient, the longer pause period you need to have. In the process of drying the insulation, it is necessary to strive for such a ratio τ_{heat}/τ_{cool}, at which the relationship between the solvent content and the heating time is linear [14–16].

4   Discussion

The IF temperature varies along segments of exponential curves and reaches steady-state oscillations with relatively small amplitudes. Due to the cooling of the IF during pauses, the highest temperature $t_{max\ oscillat}$ will be less than the temperature $t_{continuous}$, which would have occurred during prolonged operation. In the case of oscillating infrared power supply, the overheating temperature $t_{max\ oscillat}$ of the oscillator should not exceed the maximum permissible value for this class of insulation.

Thus, the temperature at the end of the heating section τ_{heat} will reach

$$t_{max\,oscillat} = t_{continuous}\left(1-e^{-\tau_{heat}/T_h}\right) + t_{min\,oscillat}e^{-\tau_{heat}/T_h} \tag{16}$$

The temperature at the end of the pause will drop to

$$t_{max\,oscillat} = t_{continuous}e^{-\tau_{cool}/T_c} \tag{17}$$

Substituting in expression (16) the $t_{min\,oscillat}$ value of the oscillator from (17), we get

$$t_{max\,oscillat} = t_{continuous}\left(1-e^{-\tau_{heat}/T_h}\right) + t_{min\,oscillat}e^{-(\tau_{heat}/T_h + \tau_{cool}/T_c)} \tag{18}$$

Solving (18) relative to $t_{continuous}$, we will have

$$t_{max\,oscillat} = t_{continuous}\frac{1-e^{-\tau_{heat}/T_h}}{1 - e^{-(\tau_{heat}/T_h + \tau_{cool}/T_c)}} \tag{19}$$

Since the heat transfer from the surface of the material to the external medium, both during heating and cooling, remains unchanged during oscillating IR energy transfer, we assume that

$$T_h = T_c \tag{20}$$

Taking into account all the above equations, a formula has been obtained for identifying the effective duty cycle of the period of operation of the radiating elements in the insulation restoration technology with an oscillating IR power supply mode

$$t_{max\,oscillat} = t_{continuous}\frac{1-e^{-\tau_{heat}/T_h}}{1 - e^{-\tau_{oscillat}/T_h}} \tag{21}$$

5   Conclusions

Using formula (21) to solve research problems of identifying the effective relative pulse duration of the operation of radiating elements (internal heat and mass transfer) and developing recommendations for calculating the maximum oscillation cycle duration (external heat transfer) will make it possible to create the most modern industrial installations for insulation restoration of IF brush holder brackets of traction motors with oscillating IR power supply mode [17]. For the carousel installation, the oscillating modes of the infrared power supply are provided by moving the fingers along the carousel with the radiating elements always on. This will allow increasing the durability of the radiating element by an order of magnitude, since starting modes with frequent switching on and off will be excluded.

References

1. Shalaev, A.S., Gordienko, A.V.: Possible malfunctions of electric cars. In: Scientific Community of Students of the XXI Century. Engineering Science: Collegiate of Papers on the Materials of the XXV International Student Scientific-Practical Conference, vol. 10, no. 24 (2014)
2. Khudonogov, A.M., Konovalenko, D.V., Sidorov, V.V., Lytkina, E.M.: New method of drying wet insulation of traction motor windings. Development of transport infrastructure is the basis of economic growth in Zabaikalsky region. In: Proceedings of the International Scientific-Practical Conference, pp. 222–230. ZabIZhT (2008)
3. Valishin, A.A., Kartashov, E.M.: Statistical description of thermal motion in polymers. Plastics 7, 36–39 (2006)
4. Lobytsin, I.O., Dulsky, E.Yu., Vasiliev, A.A.: Improving the reliability of the insulating fingers of the brackets of the brush holders of electric machines of the traction rolling stock. Mod. Technol. Syst. Anal. Model. 4(56), 218–224 (2017)
5. Khudonogov, A.M., Konovalenko, D.V., Upyr, R. Yu.: Method of drying the insulation of electrical machines. RF patent 2324278 (2008)
6. Vasiliev, A.A., Alekseev, D.V., Khudonogov, I. A., Lytkina, E.M.: Rational modes of oscillating IR - power supply in the technology of hardening insulation of the frontal parts of the windings of electrical machines. In: Transport Infrastructure of the Siberian Region: Materials of the II Intercollegiate Scientific and Practical Conference, FSBEI HPE "IrGUPS", 16–18 May 2011. IrGUPS, Irkutsk (2011)
7. Lytkina, E.M.: Improving the efficiency of encapsulation of the insulation of the frontal parts of the windings of traction motors of electric locomotives by infrared radiation. Dissertation for the Degree of Candidate of Technical Sciences, Irkutsk (2011)
8. Khudonogov, A.M., Lytkina, E.M., Dulsky, E.Yu.: The criterion of the validity of the choice of impregnating material in the technology of repair of traction electric vehicles of rolling stock. Increase of traction and energy efficiency and reliability of electric rolling stock. In: Interuniversity Subject College of Scientific Papers, pp. 38–43. Omsk State Transport University, Omsk (2013)
9. Makarov, V.V., Smirnov, V.P., Khudonogov, A.M., Yefremov E.V.: Resource-saving principles of drying technology of wet insulation of electrical equipment of EPS. In: Collegiate of Scientific Papers, vol. 1, pp. 32–37. FESTU Publ., Khabarovsk (2001)
10. Khudonogov, A.M., Smirnov, V.P., Konovalenko, D.V., Gamayunov, I.S., Olentsevich, D. A., Sidorov, V.V., Lytkina, E.M., Ilyichev, N.G.: Principles of energy management in the process of removing moisture from the insulation of the windings of traction electrical machines. In: Kryukov, A.V. (ed.) Energy Saving: Technologies, Instruments, Equipment: Collegiate of Scientific Papers, pp. 125–129. IrGUPS Publ., Irkutsk (2009)
11. Smirnov, V.P., Khudonogov, A.M.: Width-intermittent method of drying wet insulation of traction motors. In: Scientific Problems of Transport in Siberia and Far East, vol. 3, pp. 185–192. IrGUPS, Irkutsk (2003)
12. Dulsky, E.Yu.: Improving the technology of restoring the fingers of the brush holders of electric machines for traction rolling stock. Transport infrastructure of the Siberian region. In: Proceedings of the Sixth International Scientific-Practical Conference, vol. 2, pp. 510–513. IrGUPS Publ., Irkutsk (2016)
13. Sidorov, V.V., Lytkina, E.M., Konovalenko, D.V., Khudonogov, A.M., Garev, N.N., Dulsky, E.Yu., Ivanov, P.Yu.: Three-cycle amplitude-width-intermittent method of drying the insulation of electrical machines. RF Patent 2494517 (2011)

14. Dulsky, E.Yu., Khudonogov, A.M., Lytkina, E.M.: Influence of the chemical properties of polymers and IR power supply modes on the strength and plasticity of insulation in local technologies of extending the service life of electric machines of traction rolling stock. News of TransSib **1**(21), 6–11. Omsk State Transport University, Omsk (2015)
15. Lobytsin, I.O., Khudonogov, A.M., Dulsky, E.Yu., Tyumentsev, A.V.: Development and research of the installation for drying the insulating fingers of the brush holder brackets. In: Science and Youth: Collected Papers of the Fourth All-Russian Scientific and Practical Conference of Students, Postgraduates and Young Scientists, pp. 123–127. IrGUPS, Irkutsk (2018)
16. Khudonogov, A.M., Lytkina, E.M., Dulsky, E.Yu.: Innovative technology for improving the reliability and extending the life of electric machines for traction rolling stock. Mod. Technol. Syst. Anal. Model. **4** (36), 102–108. IrGUPS, Irkutsk (2012)
17. Dulsky, E.Yu., Kargapol'tsev, S.K.: Encapsulation of electrical insulation from polymer materials of traction motors when exposed to infrared radiation. In: Moskvichev, V.V. (ed.) Safety and Survivability of Technical Systems: Materials and Reports. V All-Russian Conference, 12–16 October 2015, Krasnoyars, vol. 3, pp. 175–180. Siberian Federal University, Krasnoyarsk (2015)

New Lining with Cushion for Energy Efficient Railway Turnouts

Boris Glusberg[1] ⓘ, Alexander Savin[1] ⓘ, Alexey Loktev[2] ⓘ,
Vadim Korolev[2] ⓘ, Irina Shishkina[2(✉)] ⓘ, Diana Alexandrova[2] ⓘ,
and Daniil Loktev[3] ⓘ

[1] Railway Research Institute, Third Mytishchenskaya 10,
107996 Moscow, Russia
[2] Russian Open Academy of Transport of the Russian University of Transport,
Chasovaya Str. 22/2, 125190 Moscow, Russia
shishkinaira@inbox.ru
[3] Moscow State University of Civil Engineering, Yaroslavskoye Shosse 26,
129337 Moscow, Russia

Abstract. The article deals with the design of pads with a cushion made using
different technologies. The analysis of the stress state and the assessment of the
strength of the lining with a cushion made using welding technology and cast
are made. As a result, further directions to improve the strength and performance
and the optimal design of the lining with a cushion are determined.

Keywords: Turnout · Lining with a cushion · Manufacturing techniques ·
Directions of improvement of linings

1 Introduction

Operation of linings with a cushion is accompanied by the appearance of defects in
them, which include, for example, the weakening of rivet joints, the appearance of
cracks and abrasion of metal under the sole of the rail. However, one of the most
significant defects is the fracture of the lining with a cushion, passing, as a rule, along
the inner edge of the frame rail.

The practice of operation of turnouts shows that the destruction of the pads with a
cushion of different designs as a result of the fracture occurs in the operating range of
20–90 million tons gross.

The breaks of the pads with the cushion have common characteristics. The fracture
line in all cases crosses the heterogeneity of the shape of the lining with a cushion,
which is a concentrator of mechanical stresses and is located along the inner edge of the
frame rail. For the lining with a cushion made using rivet technology, the hubs are rivet
holes. In the lining with a cushion made using hot-stamping technology, an acute angle
of the hole remaining after stamping acts as a concentrator. For the lining with a
cushion made using welding technology, the concentrator is a hole for welding.

Fractures are characteristic of fatigue failure, i.e. failure caused by the gradual
accumulation of damage in the material. This type of destruction is very sensitive to the

© Springer Nature Switzerland AG 2020
V. Murgul and M. Pasetti (Eds.): EMMFT 2018, AISC 982, pp. 556–570, 2020.
https://doi.org/10.1007/978-3-030-19756-8_53

effect of stress concentration. A typical fatigue failure scenario is the accumulation of damage on the stress concentrator, the subsequent generation of a fatigue crack on this concentrator, and the propagation of the crack along the main section of the element until its complete destruction.

2   Research Technique

Further analysis of the stress state and assessment of the strength of the lining with a cushion are made on the example of the lining with a cushion made using welding technology.

The padding Assembly with a cushion made using welding technology consists of a sleeper, rubber padding, padding, cushion and a frame rail and wit resting on them (Fig. 1).

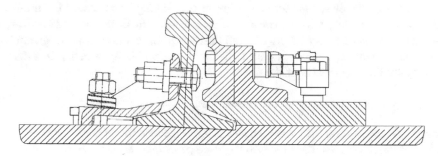

Fig. 1. Pillow-top lining assembly.

From the point of view of mechanics, the cushion lining is a plate on an elastic base. The analytical calculation of such details is difficult, because it requires in the General case the solution of a system of 10 differential equations with 10 unknowns, the main of which is the equation of deflections of a thin plate of the form:

$$\frac{D}{a^2 b^2}\left(\frac{1}{a}\frac{\partial^4 Y}{\partial x^4} + 2\frac{\partial^4 Y}{\partial x^2 \partial y^2} + \frac{a^2}{b^2}\frac{\partial^4 Y}{\partial y^4}\right) = q(x,y) - p(x,y), \tag{1}$$

where a and b are the dimensions of the plate in plan, D is the cylindrical stiffness of the plate, $Y(x, y)$ is the function of plate deflections, $q(x, y)$ is the function of external loads, and $p(x, y)$ is the function of reactive base resistance depending on the deflection function [1].

For the calculation of such elements are used to simplify the design schemes and special approximate tables of values. Given the complex geometry of the pads with a pillow, we can assume that the approximate analytical calculation of the pads with a pillow is impractical, since the cumulative effect of simplifications will not allow describing the behavior of the pads with a pillow with sufficient accuracy. More promising is the use of modern numerical methods of analysis, such as the finite element method, which allows us to obtain a solution that satisfies the equations of mechanics of a deformable solid body, with any predetermined accuracy, without simplifying the geometry and using a small number of other simplifications.

Analyzing the working conditions of the lining with a pillow, we can distinguish the following essential features:

- the presence of an elastic base, significantly affecting the stress-strain (VAT) state of the lining with a pillow;
- the presence of pre-tightening fasteners, resulting in a lining with a pillow there is VAT in the absence of external load;
- cyclic disproportionate loading of the lining with a cushion during the passage of the rolling stock.

The elastic base of the lining with a pillow is due to the presence of a rubber pad between the lining with a pillow and a sleeper. Requirements to the stiffness of this strip are established GOST R 56291-2014 and reach 50–150 MN/m. More precisely, the behavior of the elastic gasket can be described, taking into account the nonlinear stiffness of the gasket, which increases with the load. In GOST R 56291-2014, in particular, results of tests of laying of rail fastenings on compression are given. The load-displacement dependence is essentially nonlinear and is described with sufficient accuracy by the dependence of the form:

$$P = k \cdot w^3, \tag{2}$$

where P - is the load on the gasket (gasket set), k - is the stiffness coefficient, w - is the draft (vertical displacement) of the gasket. More precisely, stiffness can be described in terms of specific pressure p:

$$p = F \cdot k \cdot w^3 \tag{3}$$

$$p = \frac{P}{F}, \tag{4}$$

where F - is the contact surface area of the gasket. The nonlinear stiffness of the gasket, as a function of vertical displacement, can be defined as:

$$K(w) = \frac{\partial p}{\partial w} = 3F \cdot k \cdot w^2 \tag{5}$$

The coefficient k can be determined from the experiment.

Pre-tightening of the mounting bolts is carried out when installing the pad with a pillow in the operating position [2].

In accordance with GOST 16277-93, the lining of the separate bond is made of strips rolled from steel ST4 of various degrees of deoxidation according to GOST 380. It is allowed to produce linings from strips rolled from steel St3 of various degrees of deoxidization with a mass fraction of carbon not less than 0.16% (it is normalized that the total amount of carbon and 1/4 manganese should be not less than 0.28%). Linings make normal and increased accuracy, with heat treatment and without it.

Mechanical characteristics of steels St3 and ST4 are given in Table 1 [3].

Table 1. Mechanical characteristics of steels St3 and ST4.

Steel	$\sigma_{0,2}$, MPa	σ_B, MPa	δ, %	σ_{-1}, MPa
St3sp	195–255	370–530	22–27	195
St3ps	195–245	370–480	23–26	191–213
St3kp	185–235	360–460	24–27	175
St4ps	235–265	410–530	19–24	196–235
St4kp	225–255	400–510	19–25	196–225

Given the working conditions of the lining with a pillow, its strength should be evaluated by two criteria: according to the criterion of strength under static loading and the criterion of cyclic strength. In the calculation it is impossible to take into account all the variety of combinations of loads on the lining with the cushion from trains of different types and weight, since there is no representative statistics of axial loads for turnouts. In this regard, the strength must be justified in the assumption of the most unfavorable combination of forces on the structure at each cycle.

To estimate fatigue life, a fatigue curve is required that specifies the ratio between the number of cycles to failure and the stress amplitude of the symmetric cycle. The curve can be reconstructed from the available literature data using the Weller equation [4], which postulates the linear relation in semi-logarithmic coordinates:

$$\sigma_a = a - b \lg N, \qquad (6)$$

where σ_a is the stress amplitude of the symmetric cycle, N is the number of cycles for failure, a and b are the parameters of the equation. For selection of two parameters of the Weller equation it is necessary to have two points with known relations between stresses and number of cycles.

As the first such point is accepted temporary resistance σB, which obviously causes destruction in one cycle:

$$\sigma_B = a - b \lg 1$$

$$a = \sigma_B \qquad (7)$$

The second point is the endurance limit, which does not cause destruction at the full base of the cycles. 107 cycles are taken as the full base of the cycles in the fatigue test, but the fracture on the fatigue curve occurs already on the basis of $(1 \div 3) \cdot 106$ cycles and it is recommended to take the value of $2 \cdot 106$ cycles as the base of the cycles corresponding to the endurance limit (GOST 25.504-82). Thus, the second point is determined by the ratio:

$$\sigma_{-1} = \sigma_B - b \lg(2 \cdot 10^6)$$

$$b = \frac{\sigma_B - \sigma_{-1}}{6 + \lg 2} \tag{8}$$

Taking from Table 1 the most conservative values of the characteristics $\sigma_V = 360$ MPa and $\sigma_{-1} = 175$ MPa, which correspond to the steel St3kp, we obtain for the parameters of the Weller equation values $a = 360$ MPa and $b = 29.4$ MPa.

To simulate the behavior of the padding with a cushion under load, a mathematical model is developed, which includes the padding, welded pad and rail element. The simulation is performed by the finite element method in ANSYS software package version 19.0 (Fig. 2).

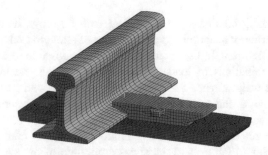

Fig. 2. General view of finite element mathematical model in PC ANSYS preprocessor.

To take into account the uneven elastic base under the pad with a pillow, the base of the pad with a pillow is divided into several separate surfaces on which the boundary conditions of the elastic base are independently assigned. The properties are not applied to all surfaces, but only to those that are in contact with the ribbed rubber gasket, for which the separation of the surfaces is made coinciding with the location of the gasket edges.

The simulation is performed for two productions:

- Task № 1. Pre-tightening of fasteners. Within the framework of the statement, the VAT of the lining with a pillow is determined when tightening the fasteners – two M14 bolts fastening the lining with a pillow to the sleeper. There are no loads on the rail and wit.
- Task № 2. Maximum load on the lining with pillow. As part of this statement, the tightened bolts are fixed in the positions in which they were when they were tightened, and the maximum load is applied to the system, which corresponds to the complete transition of the wheel from the rail to the tip.

To correctly take into account the effects of the elastic Foundation in the first task iteration is determined by the distribution of stiffness of the elastic base. The method of selection is as follows: in the first iteration, all surfaces with elastic base properties are subjected to the same base stiffness of 5 N/mm^3. Then the calculation is made and the total reactive force acting on each surface is determined. The stiffness of each surface is adjusted according to formula (5) and the next iteration of the calculation is performed. Iterations are repeated until all base stiffness in the model stops changing by more than 1%. After the convergence of the solution on the distribution of stiffness is achieved, the stress in the places where the experimental measurement was made is determined and compared with the results of the experiment. If the values do not match, the constant in expression (5) is corrected and the iterative process is repeated.

After achieving convergence on the distribution of stiffness and coincidence with the results of experimental measurements, the second problem is calculated. This uses the constant value in the expression (5), which was determined during the calibration of the first problem, and the iterations necessary to achieve convergence on the distribution of stiffness are repeated. After that, the stress values are also compared with the results of experimental measurements.

As a result of the calibration for the experiment, the value of the constant k' was 1.50 N1/3/MM5/3.

Table 2 shows the results of the calibration and verification of the padding model.

Table 2. Comparison of model calibration results with experimental values.

Settlement case	Tightening of fasteners	Passage of the composition
Model	76 MPa	239 MPa
Experiment	80 MPa	200 MPa

On the basis of the values given in the table, the model of the lining with a cushion on an elastic base can be considered adequate to reality. The stress values in the measurement region are higher than those obtained in the experiment, due to the fact that the calculation takes into account the maximum design load on the pad with a pillow. In further analysis, this case will be used for conservatism.

Figure 3 shows the distribution of von Mises equivalent stresses in the padding with the pad when tightening the joints.

Fig. 3. Distribution of equivalent stresses on the lining with a cushion when tightening the joints.

Figure 4 shows the distribution of von Mises equivalent stresses in the cushion-lined structure.

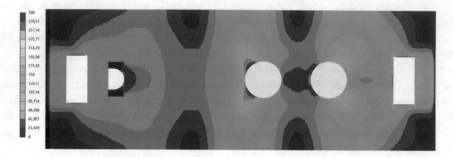

Fig. 4. Distribution of equivalent stresses on the lining with a pillow during the passage of the composition.

As the main loading cycle, a disproportionate cycle is considered, in which one state corresponds to the lining with a pillow loaded only by tightening the fastener, and the other – to the lining with a pillow loaded with a passing composition. That is, one loading cycle corresponds to the transition from VAT problem number 1 to VAT problem number 2.

The estimation of the limiting amplitudes of the cycle is performed on the basis of Goodman's condition, which gives a conservative estimate for carbon steels [5]:

$$\frac{\sigma_a}{\sigma_{-1}} = 1 - \frac{\sigma_m}{\sigma_B} \tag{9}$$

Figure 5 shows the distribution of equivalent (reduced to a symmetric cycle) stress amplitudes.

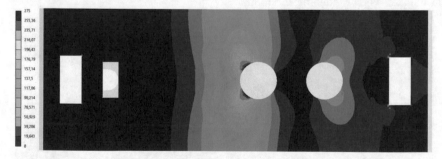

Fig. 5. Distribution of stress amplitudes of the equivalent symmetric cycle in the lining with a cushion of the base structure.

Figure 6 shows the distribution of the expected number of cycles to failure.

Fig. 6. Distribution of the expected number of cycles to failure in the lining with a pillow base design (view of the sole lining with a pillow).

As can be seen from Fig. 6, the dangerous zones of the cushion lining obtained as a result of the simulation completely coincide with the picture of destruction. Thus, the main conclusion of the analysis of the nature of destruction is confirmed – the destruction occurs as a result of the accumulation of fatigue damage to the material on the stress concentrators. The minimum number of cycles to failure is $\sim 10^3$ cycles. In actual operation, the failure occurs at a greater number of cycles, due to the conservative reserves made in the calculation of the load, material properties and methods of calculation.

Zones with the least number of cycles to failure are located on the lower surface of the lining with a pillow, on the contour of the central hole. At this point, there is the origin of the fatigue crack, which extends throughout the section of the lining with a further increase in the number of cycles to its complete separation into parts.

3   Results and Discussion

Optimization of the design of the shoe should be aimed at ensuring the fatigue life of the structure at the load described above. However, it requires compliance with a number of conditions:

- the immutability of the provisions of fasteners $l_{b1x} = l_{b1x0}$, $l_{b2x} = l_{b2x0}$, $l_{b1y} = l_{b1y0}$, $l_{b2y} = l_{b2y0}$;
- invariance of the position of the sole height of the rail $h_r = h_{r0}$;
- invariance of the position of the height of the sole wit $h_o = h_{o0}$;
- immutability of dimensions in terms of $l_x = l_{x0}$, $l_y = l_{y0}$.

In this case, the conditions of strength and reliability must be satisfied:

- providing static strength $\sigma_M \leq [\sigma_s]$;
- ensuring cyclic strength $\sigma_M^{ec} \leq [\sigma_{-1}]$;
- generalized safety factor $\geq$ 20% $[\sigma_s] = \sigma_{0,2}/1, 2$, $[\sigma_{-1}] = \sigma_{-1}/1, 2$.

Here l_{bij} - position of the i-th fastener in the direction j, l_{bij0} - design positions of the same elements; h_r - position of height of a sole of a rail, h_{r0} - its design position; h_o - position of height of a sole of a wit, h_{oo} - its design position; l_j - the overall size of a lining in the direction j, l_{j0} - its design size.

The optimized parameters are a set of geometric dimensions not limited by the above conditions:

$$\vec{X} \in \mathbb{R} \tag{10}$$

And the target optimization function is the reserve coefficient generalized under the conditions of static and cyclic strength:

$$\overrightarrow{X_o} = argmax\, n(\vec{X}), \tag{11}$$

where n - is the generalized factor of safety.

Formally, the optimization problem is written as:

$$\begin{cases} \overrightarrow{X_o} = argmax\, n(\vec{X}) \quad X^{\rightarrow} \in R \\ n = min\left[\frac{\sigma_{0,2}}{\sigma_M}; \frac{\sigma_{-1}}{\sigma_M^{ec}}\right] \\ n \geq 1,2 \\ l_{b1x} = l_{b1x0} \\ l_{b2x} = l_{b2x0} \\ l_{b1y} = l_{b1y0} \\ l_{b2y} = l_{b2y0} \\ h_r = h_{r0} \\ h_o = h_{oo} \\ l_x = l_{x0} \\ l_y = l_{y0} \end{cases} \tag{12}$$

The easiest option is to modify the design of the removal of the lining with a welded cushion holes for welding pillows. Welding pillows in this version is carried out by two longitudinal seams along the long sides of the pillow (Fig. 7).

Fig. 7. Lining with welded cushion without holes in lining.

For the lining of this design, calculations were made similar to the calculation of the lining with the cushion of the original design. Figure 8 shows the distribution of equivalent stress amplitudes. The maximum value of the equivalent amplitude is 147 MPa (in the lining with a pillow of the base structure - 277 MPa) and does not exceed the endurance limit of the material.

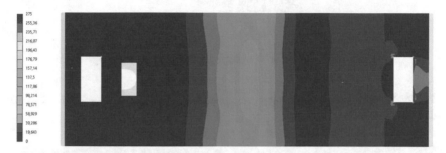

Fig. 8. The distribution of stress amplitudes is equivalent to the symmetric cycle for the lining with a cushion with no holes (sole lining with a cushion).

Figure 9 shows the distribution of the generalized coefficients of stock in the lining of the pillow with the excluded holes. The minimum safety factor for this design is 1.15.

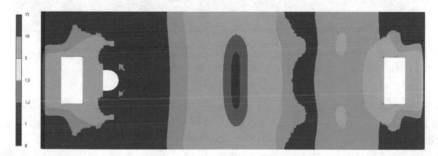

Fig. 9. The generalized distribution of safety factors for the lining with a cushion with no holes (sole lining with a cushion).

The resulting design meets the criteria of strength, but the safety factor in it is relatively small. Analyzing the picture of stresses in Fig. 8, it can be established that the main reason for the small factor of the reserve is the small contact area of the pillow and the lining. The pillow in this design acts on the lining like a stamp and to lower the level of equivalent amplitudes, it is necessary to increase the width of the contact zone between the pillow and the lining (Fig. 10).

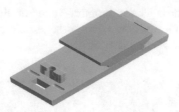

Fig. 10. Lining with welded cushion full-width lining.

Figure 11 shows the distribution of equivalent stress amplitudes for a full-width welded cushion lining.

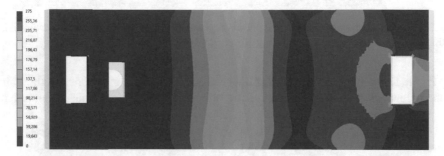

Fig. 11. The distribution of stress amplitudes is equivalent to the symmetric cycle for the lining without holes with a cushion across the full width of lining.

From Fig. 11 it is seen that the maximum value of the equivalent amplitude is 165 MPa, i.e. exceeds a maximum value equivalent to the amplitude for the lining with cushion source design without holes for welding.

Figure 12 shows the distribution of generalized safety factors in the lining without holes with a cushion in the entire width of the lining. The minimum safety factor for this design is 1.05.

Fig. 12. Distribution of generalized safety factors for the lining without holes with cushion in the entire width of the lining (sole lining with cushion).

The value of the safety factor exceeds one (the strength conditions are met), but its minimum value is less than for the lining with a cushion of the original design without holes for welding. The reason for this is the concentration of contact stresses under the boundary of the contact zone on the edges of the lining with the pillow.

To eliminate this disadvantage, the width of the cushion should be reduced to the value when the stress concentration zone under the contact zone boundary is not entirely within the width of the lining (Fig. 13).

Fig. 13. Shoe with welded cushion narrowed in 50 mm.

Figure 14 shows the distribution of equivalent stress amplitudes for the lining without holes with a cushion width of 50 mm already lining width.

Fig. 14. The distribution of stress amplitudes is equivalent to the symmetric cycle for the lining without holes cushion width 50 mm less than the width of the lining.

The maximum equivalent amplitude is 123 MPa, which is less than all other cushion-lined designs.

Figure 15 shows the distribution of generalized safety factors in the lining without holes with a cushion width of 50 mm already lining width. The minimum safety factor for this design is 1.27.

Fig. 15. The generalized distribution of safety factors for liner without holes cushion width 50 mm less than the width of the lining (the lining sole with cushion).

As can be seen from Fig. 15, the value of the safety factor not only exceeds one, but its minimum value exceeds the threshold value of 1.25.

The specified design of a lining with a cushion is optimum among considered, satisfying all requirements of the optimized design.

Another promising way to modify the design is the transition to a solid structure, in which the lining and cushion are one part (Fig. 16).

Fig. 16. Cast lining with small pillow.

Figure 17 shows the distribution of equivalent stress amplitudes in a solid-cast lining with a narrowed cushion. The maximum value of the equivalent amplitude is 148 MPa and does not exceed the endurance limit of the material.

Fig. 17. The distribution of stress amplitudes is equivalent to the symmetric cycle for the one-piece lining with small pillow (sole lining with a cushion).

Figure 18 presents the summary of the reserve factors for the one-piece lining with small pillow. The stock ratios are generally greater than one and the minimum is 1.15.

Fig. 18. The distribution coefficients of the generalized stock for the one-piece lining with small pillow (sole lining with a cushion).

As can be seen from Fig. 18, the most loaded zone in the lining with the cushion of the cast structure is the zone of transition of the pillow into the lining. The minimum value of the safety factor is in this zone. The minimum value of the safety factor is relatively small, this indicates that the solid lining with a narrowed cushion is not the optimal design. On the other hand, the minimum reserve factor on the sole is 1.44, which makes it possible to increase the reliability of the design by increasing the cushion (Fig. 19).

Fig. 19. Cast lining with wide cushion.

Figure 20 shows the distribution of equivalent stress amplitudes for a wide-cushion cast pillow. The maximum value of the equivalent amplitude is 134 MPa and does not exceed the endurance limit of the material.

Fig. 20. The distribution of stress amplitudes is equivalent to the symmetric cycle for the one-piece lining with wide cushion (sole lining with a cushion).

Figure 21 presents a summary of the reserve factors for the one-piece lining with wide pillow. The safety factors, as for a design with a narrowed cushion, exceed one, and the minimum value is 1.22.

Fig. 21. The distribution coefficients of the generalized stock for the one-piece lining with wide cushion (sole lining with a cushion).

4 Conclusions

The design of the solid-cast lining with a wide cushion generally meets the requirements and is optimal among the considered solid-cast structures.

References

1. Gorbunov-Posadov, M.I., Malikova, T.A.: Calculation of Structures on Elastic Base, 2nd edn, Redrafted (in Russian). Stroyizdat, Moscow (1973)
2. Evtukh, E.C.: Effect of Rail Joints on the Contact-Fatigue Strength of Railway Rolling Stock Wheels (in Russian). Bryansk (2014)
3. Reshetov, D.N.: Machine Details: Textbook for Students of Engineering and Mechanical Specialties of Universities, 4th edn, rev. and Extra (in Russian). Mechanical Engineering, Moscow (1989)
4. Sorokin, V.G., Volospikova, A.V., Vyatkin, S.A., et al.: Grade Guide Steels and Alloys (in Russian). Mechanical Engineering, Moscow (1989). Sorokina, V.G. (ed.)
5. Tilkin, M.A.: Handbook of Heat-Treater of the Repair Service (in Russian). Metallurgy, Moscow (1981)

Counter-Rail Special Profile for New Generation Railroad Switch

Boris Glusberg[1] (ID), Alexander Savin[1] (ID), Alexey Loktev[2] (ID),
Vadim Korolev[2] (ID), Irina Shishkina[2(✉)] (ID), Lidia Chernova[2] (ID),
and Daniil Loktev[3] (ID)

[1] Railway Research Institute, Third Mytishchenskaya 10,
107996 Moscow, Russia
[2] Russian Open Academy of Transport of the Russian University of Transport,
Chasovaya str. 22/2, 125190 Moscow, Russia
shishkinaira@inbox.ru
[3] Moscow State University of Civil Engineering, Yaroslavskoye Shosse 26,
129337 Moscow, Russia

Abstract. Studies of the strain-stress distribution of the flange rail assemblies using mathematical modeling methods for the flange rail assemblies of various designs are presented. The characteristics of variants of special profiles are given. The proposal on the expediency of application of the optimized flange rail in the railroad switches construction of modern is formulated.

Keywords: Railway track · Railroad switches · Flange rail special profile · Mathematical modeling · Optimization

1 Introduction

In contrast to the flange rails made from profiles RK or out of the ordinary rails, controls not connected with the running rail run in the path independently of other elements frog site. Dynamic strength tests have shown that the flange rails not connected with the running rails, displacement and additional loads arising from the action of vertical forces transmitted by the railway equipment on the track, in practical assessments may not be taken into account. The stress-strain state of the flange rail and its supporting elements is determined by the horizontal transverse forces acting on its working surface from the passing wheels, transmitted through the rear sides of their ridges.

In dynamic strength tests of flange rails not connected with the running rail, the modern design has been obtained that the greatest tensile loads occur in the upper edge of the head from its non-working side (in places of transition from the bent part to the straight or in the straight part of the flange rail).

The highest values are marked in the middle part of the flange rail. Here, the maximum observed stress values obtained in the tests were 230 MPa. In the bent part of the flange rail, the greatest loads do not exceed 165 MPa. The test results show that

© Springer Nature Switzerland AG 2020
V. Murgul and M. Pasetti (Eds.): EMMFT 2018, AISC 982, pp. 571–587, 2020.
https://doi.org/10.1007/978-3-030-19756-8_54

under unfavorable conditions (widening of the track, narrowing of the gutters) the maximum probable stress values in the bent part can be 270–280 MPa [1].

Design check assemblies with flange rails not tied to the running rails, have the same dimensions ruts and troughs, and nodes and flange rails from specprofile RK, so the conditions of the impact of railway equipment on them similar.

In these circumstances the horizontal shear force acting on the wheels on controls, have the same value.

Based on this, for the case when the counter – rails are not connected to the running rails, made of the same metal as the special profile of the Republic of Kazakhstan, they should be allowed to take the value adopted for special profiles RK-330 MPa.

The method of calculation of the stress-strain state of the flange rail node with a flange rail from the special profile of the RC65, used to optimize it, was developed by prof. Glusberg [2]. In this work, the author used a model of a flange rail in the form of a Kirchhoff–lyave plate with a point fastening in the locations of flange rail bolts. In places of application of transverse forces, a three – dimensional problem of elasticity theory by the Voynovsky-Krieger method was solved.

2   Research Technique

Classical models were used to determine the stress-strain state of flange rail assemblies with flange rails from the brake tire of the slide retarder and to develop an optimized counter-rail profile, and the final calculation was performed by the finite element method [3].

The choice of the design scheme of the flange rail node for the calculation of the stress-strain state of the flange rail depends on the shape of the cross section of the flange rail profile and the method of attaching the flange rail to the supports.

For preliminary calculations, we will use a flange rail model based on classical bending theories. The flange rail is attached to the supports (counter-rail break shoes) with bolts, the axes of which are located horizontally.

This design is an 8-span statically indeterminate continuous beam with consoles at the ends, based on 9 supports, working on a horizontal transverse load in the form of concentrated force transmitted to it by the wheel of the railway equipment.

Since the force may not be located along the axis of symmetry of the flange rail, the flange rail in the track will work on a combined load - bending with torsion. A quasi-static calculation of such a structure can be performed analytically using the force method. Loads and strains due to bending and torsional loads can be determined separately and then summed using the superposition of forces principle.

To test the proposed model of the flange rail node, we calculate the stress state of the flange rail and compare its results with the results of dynamic strength tests [4].

The test results showed that the greatest forces in specific sections of the flange rail are not the same in its length. The maximum forces in the spans decrease as we approach the trapping part of the flange rail from the movement of the railway equipment. At observance of the sizes of a track and troughs on a frog multiple joint, near the catching part of a flange rail, the maximum forces do not exceed 10% from similar sizes in an average part of a flange rail [5].

As mentioned above, the forces acting on the flange rail cause bending and torsion in it.

As a model to calculate bending stresses, revealing the redundancies introduced take cosmipolitan split statically determinate beam with the hinge resting on the supports.

The beam is loaded with a load P, and the torque MK from the impact of the wheels and the moments on the supports M0–M8, compensating for the action of internal forces when replacing the statically indeterminate structure on the split. As a first step in the calculation, consider the application of force and torque in the mean span.

In the matrix form, the system of equations to determine the unknown bending moments will have the form:

$$A \cdot M = P \cdot p \tag{1}$$

where A is a square matrix of 9×9 coefficients with unknown;

M - vector of unknowns;
R – load vector from a single force;
p is the magnitude of the effective force.

To determine the components of the matrix a, we use the equations of three moments:

$$\text{For 0-th support :} \qquad M_0 = 0 \tag{2}$$

$$\text{For 1st support:} \qquad 0.276\,M_1 + 0.5\,M_2 = 0 \tag{3}$$

$$\text{For 2nd support:} \qquad 0.5\,M_1 + 2\,M_2 + 0.5\,M_3 = 0 \tag{4}$$

$$\text{For 3rd support:} \qquad 0.5\,M_2 + 2\,M_3 + 0.5\,M_4 = 0.09375\,R \tag{5}$$

Force and torque are applied in the span between supports 3 and 4

$$\text{For 4th support:} \qquad 0.5\,M_3 + 2\,M_4 + 0.5\,M_5 = -0.09375\,p \tag{6}$$

$$\text{For 5th support:} \qquad 0.5\,M_4 + 2\,M_5 + 0.5\,M_6 = 0 \tag{7}$$

$$\text{For 6th support:} \qquad 0.5\,M_5 + 2\,M_6 + 0.5\,M_7 = 0 \tag{8}$$

$$\text{For 7th support:} \qquad 0.5\,M_6 + 2.076\,M_7 = \tag{9}$$

$$\text{For 8 support:} \qquad M_8 = 0 \tag{10}$$

Bringing the Eqs. 2–10 to the General form, we obtain the components of the matrix A:

$$A = \begin{bmatrix} 0 & 0 & 0 & 0 & 0 & 0 & 0 & 0 & 0 \\ 0 & 4,152 & 1 & 0 & 0 & 0 & 0 & 0 & 0 \\ 0 & 1 & 4 & 1 & 0 & 0 & 0 & 0 & 0 \\ 0 & 0 & 1 & 4 & 1 & 0 & 0 & 0 & 0 \\ 0 & 0 & 0 & 1 & 4 & 1 & 0 & 0 & 0 \\ 0 & 0 & 0 & 0 & 1 & 4 & 1 & 0 & 0 \\ 0 & 0 & 0 & 0 & 0 & 1 & 4 & 1 & 0 \\ 0 & 0 & 0 & 0 & 0 & 0 & 1 & 4,152 & 0 \\ 0 & 0 & 0 & 0 & 0 & 0 & 0 & 0 & 0 \end{bmatrix} \tag{11}$$

The unknown vector M and loads of R:

$$M = \begin{bmatrix} M_0 \\ M_1 \\ M_2 \\ M_3 \\ M_4 \\ M_5 \\ -M_6 \\ M_7 \\ M_8 \end{bmatrix} \tag{12}$$

$$P = \begin{bmatrix} 0 \\ 0 \\ 0 \\ -0.1875 \\ -0.1875 \\ 0 \\ 0 \\ 0 \\ 0 \end{bmatrix} \tag{13}$$

The calculation is feasible for conditions that took place in dynamic strength tests $(P = 110 \text{ kN})$ and for limit loads $(P = 150 \text{ kN})$.

Substituting P and (11); (12) and (13) into Eq. (1), we obtain a system of linear equations of the 9th order.

As a result of its solution, we obtain bending moments acting on the supports of the flange rail.

$$M = \begin{bmatrix} 0 \\ -0.3806 \\ +1.5801 \\ -5.9400 \\ -5.9442 \\ +1.5920 \\ -0.4235 \\ +0.1020 \\ 0 \end{bmatrix} \tag{14}$$

where the components of the vector M are given in [kN m].

The moment in the cross section under the force of Mr is:

$$M_p = M^0 + M_3 + (M_4 - M_3)\frac{x}{l} \tag{15}$$

where M^0 is the moment in a statically determined beam at the point of force application (the beam between supports 3 and 4);

l is the span length of the beam between the supports 3 and 4;

x- the distance from the support 3 to the place of application of force.

Substituting the data, we obtain:

$$M_r = 0.85414 \cdot p,$$

where p is the effective force, or

$$M_r = 128.121 \, kN \, m \; for \; P = 150 \, kN$$

$$M_r = 93.955 \, kN \, m \; for \; P = 110 \, kN$$

The greatest torque acts in the cross section, in which the force of the wheel on the flange rail. With a given loading scheme, and given that the worst possible case is realized when applying force along the upper edge of the flange rail, it will be equal to:

$$M_K = R \cdot, \frac{b}{2} \tag{16}$$

where R is the effective force;

b is the height of the flange rail section.

Substituting the current force, we get:

$$M_K = 3.15 \, \text{kN m for R} = 150 \, \text{kN}$$

$$M_K = 2.31 \, \text{kN m for R} = 110 \, \text{kN}$$

Based on the results of calculations in Fig. 4, diagrams of moments in the flange rail [7] are constructed.

Similarly, bending and torsional moments to the forces in the spans between the other support bearing were calculated. The highest values of the moments obtained in the calculations are given in Table 1.

Table 1. The largest values of bending M_i and torque M_K moments in the spans between the supports of flange rails, kN m.

Case bay numbers	R, kN	0–1	1–2	2–3	3–4
Maximum bending moment, kN m	110	19.309	55.000	82.207	93.955
	150	26.333	75.023	112.100	128.121
Maximum torque, kN m	110	2.310	2.310	2.310	2.310
	150	3.150	3.150	3.150	3.150

Thus, to calculate the greatest stresses in the flange rail and check its strength, the values obtained for loading the span between the supports 3 and 4 should be used.

The greatest bending loads in the flange rails from the "angle of the flange rail SP-850" and "Optimized special profile" arise in the upper edge from the opposite working face of the flange rail, along which its contact with the wheels of the railway equipment occurs.

According to known dependencies:

$$\sigma_i = \frac{M_u}{W_{uy}} \tag{17}$$

where σ_u – the greatest bending loads;
M_i - bending moment;
W_{IU} is the moment of resistance of the flange rail section relative to the y-axis.

Substituting the values, we obtain:

$$\sigma_i = 242 \, \text{MPa for P} = 110 \, \text{kN}$$

$$\sigma_i = 294 \, \text{MPa for P} = 150 \, \text{kN}$$

The greatest shear load from the influence of the torque M_K at the points at which the greatest bending loads occur are determined by the formula:

$$\tau_k = \frac{M_k}{W_k} \tag{18}$$

where τ_k – the maximum shear load from impact torque;
M_K - torque value;
W_k is the torsional resistance moment of the section.

The moment of torsional resistance is determined in the margin of safety:

$$W_k = \beta \cdot b^3; \tag{19}$$

where b is the height of the torsional part of the profile;
β - correction factor depending on the ratio of height and width of the section.

Substituting numerical values, we obtain:

$$W_k = 0.536 \cdot 4.3^3 = 42.62 \, \text{cm}^3$$

$$T_C = 27 \, \text{MPa for P} = 110 \, \text{kN}$$

$$T_C = 37 \, \text{MPa for P} = 150 \, \text{kN}$$

For the class of elements to which the considered flange rail relates, the equivalent stresses are determined by the formula:

$$\sigma_{\text{ecv}} = \sqrt{\sigma_u^2 + 4\tau_k^2} \tag{20}$$

Substituting the numerical values in (20), we obtain:

$$\sigma_{ecv} = 244 \, \text{MPa for P} = 110 \, \text{kN}$$

$$\sigma_{ecv} = 296 \, \text{MPa for P} = 150 \, \text{kN}$$

The permissible loads in the flange rail, taking into account the conditions of its operation, are $[\sigma] = 330$ MPa [8]:

$$\sigma_{ecv} < [\sigma] \tag{21}$$

Thus, the strength of the flange rail of the rolling section is provided.

Comparison of the highest loads obtained in the tests and in the calculation shows that they are 230 MPa and 244 MPa, respectively. That is close enough. This makes it possible to use the considered model and calculation method to optimize the overall design of the flange rail joint.

Horizontal transverse forces transmitted from the flange rail to the stops, with insufficient strength of the stops, can cause crushing of the metal in contact with the flange rail and loss of strength from the effects of bending loads.

The crushing loads are determined by a known formula:

$$\sigma_{cm} = \frac{P}{F}, \tag{22}$$

where σ_{cm} – collapse load;
R is the force acting on the supporting surface;
F-contact area in the places of support of the flange rail on the stops.

Substituting numerical values, we have:

$$F = 2 \cdot (6,3 \cdot 1,6) = 20,16\,cm^2$$

$$\sigma_{cm} = 11000/20,16 = 545\,kg/cm^2 = 54.5\,MPa\ for\ P = 110\,kN$$

$$\sigma_{cm} = 15000/20,16 = 744\,kg/cm^2 = 74.4\,MPa\ for\ P = 150\,kN$$

The value of σ_{cm} does not exceed the yield point, but is close to it, so in operation you should pay increased attention to the work of the support surfaces of the flange rail stops.

To study the load-strain state of the flange rail in order to optimize its geometric dimensions, the finite element method was chosen, since it allows to consider in detail all the features of the cross-section geometries of the flange rails [9].

The finite element method is based on the idea of approximating a continuous function by a discrete model, which is constructed on a set of piecewise continuous functions defined on a finite number of subdomains called finite elements.

The key idea of the method in analyzing the behavior of structures is as follows: the design is modeled by dividing it into regions (finite elements), in each of which the behavior of the medium is described by a separate set of selected functions representing loads and displacements in the specified region. These sets of functions are given in such a form as to satisfy the conditions of continuity of the characteristics described by them in the whole structure.

Boundary conditions for displacements are set by means of restriction of degrees of freedom (by setting zero displacements).

To date, various software systems have been developed that implement the finite element method [10].

In the process of software selection, the characteristics of such software complexes as Pro/Engineer, ABAQUS, MSC NASTRAN and ANSYS were considered.

The ANSYS software package was chosen for the flange rail calculations, as this software package has a wide range of capabilities to perform strength calculations.

The first step was to create a correct geometric model for the study. Half of the counter-rail was modeled. For the calculations, a geometric model was taken, which is a 4-span statically indeterminate continuous beam with a console at the left end (half of the flange rail), based on 5 supports.

The SOLID187 element was selected as the target element type.

The initial data on the geometry of the flange rail special profile for building a computer model of the flange rail were taken from the drawing of the SP-850 flange rail (Fig. 5).

As a method of constructing a grid of nodes and elements, the construction of an ordered grid was used. After generation of the finite element (discrete) model, contact pairs and their properties were determined and boundary conditions were set (Fig. 1).

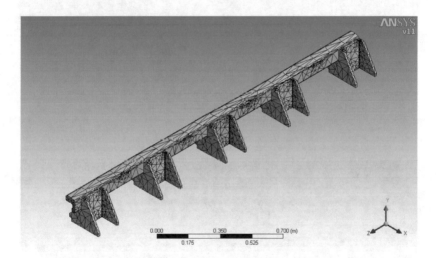

Fig. 1. Model of the angle of the flange rail SP-850, with selected finite elements

Take as boundary conditions that the beam (half of flange rail) is attached to the 5 supports. The origin of the coordinate axes is located at the point of attachment to the 5th (extreme right) support.

The flange rail is rigidly attached to the supports. That is, the movement along the axes x, y, z in the places of attachment of the beam to the supports are taken to be zero.

The load of 0.1; 0.2; 0.6 and 0.9·P is applied to the spans between the supports in order to simulate the operation of the flange rail under train load. Simulation is performed for limit loads (P = 150 kN).

The calculation results for the SP-850 special profile are presented in Figs. 6, 7, 8 and 9 and in Table 2.

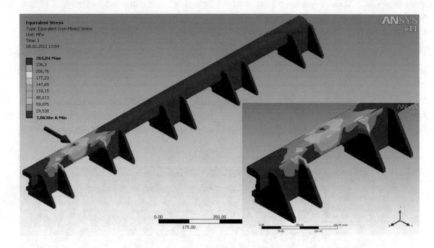

Fig. 2. SP-850 simulation results for load at position 1

Fig. 3. Simulation results of SP-850 for load in position 2

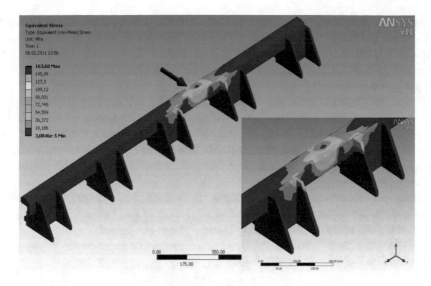

Fig. 4. SP-850 simulation results for load at position 3

Fig. 5. SP-850 simulation results for load at position 4

Table 2. The largest values of allowable loads in the spans between the supports of the SP-850, MPa.

Most loads, MPa, SP-850	Case bay numbers			
	0–1	1–2	2–3	3–4
	265	252	163	56

As can be seen from Figs. 2, 3, 4 and 5 and Table 2, the highest loads for the SP-850 special profile reach a maximum value of 265 MPa, which is lower than the permissible loads in the flange rail, taking into account the conditions of its operation $[\sigma] = 330$ MPa [8]. This suggests that the shape of the SP-850 in terms of excessive strength metal.

In dynamic strength tests, the highest values in the middle part of the flange rail were noted. Here, the maximum observed stress values obtained in the tests were 230 MPa. In the bent part of the flange rail, the greatest loads do not exceed 165 MPa. The test results show that under unfavorable conditions (widening of the track, narrowing of the gutters) the maximum probable load values in the bent part can be 270–280 MPa [9].

Comparing the results, we can conclude: the values obtained are close in value.

Therefore, the task of designing a rational section of the flange rail special profile by removing the metal from its weakly loaded places can be set.

This problem can be solved by the method of successive approximations, by variant calculations on the model described above.

3 Results and Discussion

3.1 Optimization of the Flange Rail Special Profile

The optimization target function will look like this [10]:

$$S = \left\{ min\, S_i | \sigma [\sigma]_{max} | \{\} \right\} \tag{23}$$

where S and S_i mean respectively optimized and i – th cross-sectional area of the flange rail.

The optimized parameters are the dimensions of the cross-section parts of the flange rail.

The constraints are the boundary conditions and dimensions that determine the position of the counter-rail in the track of the frog node.

That is, the position of the top of the flange rail relative to the frog rail, its lateral working face and movement relative to the axes x, y, z in the places of attachment to the supports is regulated.

As a result of the calculations, an optimized version of the flange rail special profile was obtained (Patent certificate for utility model No. 108760 "flange rail corner") [10].

3.2 Result of Calculation

The drawing of the cross section of the final version of flange rail the shape shown in Fig. 6. The results of the analysis of its load state are illustrated in Figs. 7, 8, 9, 10, 11 and 12. The loads values at the characteristic points are given in Table 3.

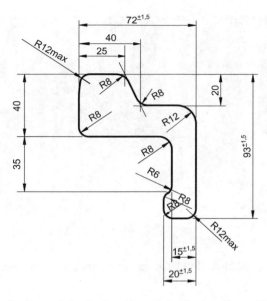

Fig. 6. Sections of the final version of the flange rail special profile

Table 3. Loads values at the characteristic points of the optimized special profile, MPa.

The highest loads, MPa, optimized profile	Case bay numbers			
	0–1	1–2	2–3	3–4
	332	310	214	89

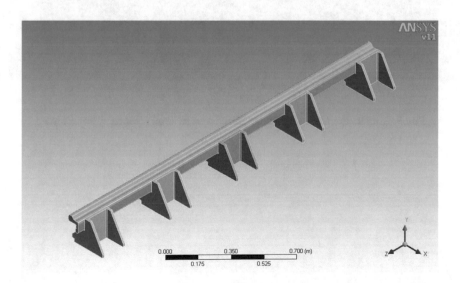

Fig. 7. Model of optimized flange rail special profile

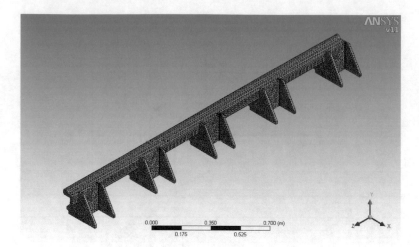

Fig. 8. Model of optimized flange rail special profile with a dedicated finite elements

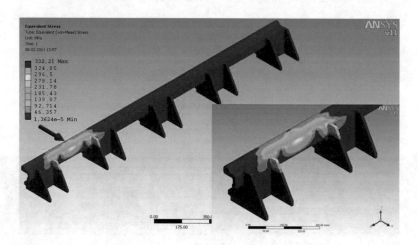

Fig. 9. Simulation results of the optimized special profile for load in position 1

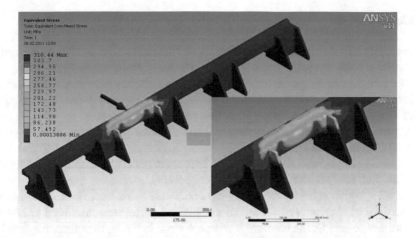

Fig. 10. Simulation results of the optimized special profile the load in the position 2

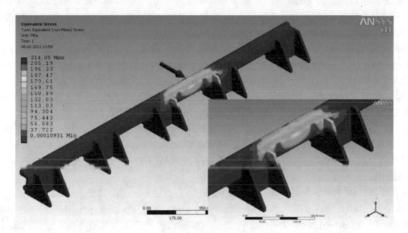

Fig. 11. Simulation results of the optimized special profile for load in position 3

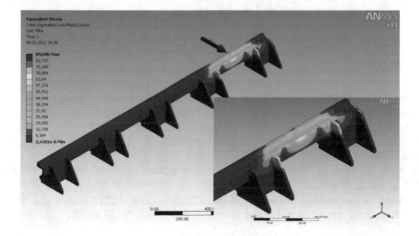

Fig. 12. Simulation results of the optimized shape for load in position 4

4 Conclusions

1. Previously developed models do not allow the calculation of the flange rail, not associated with the rolling rail. To study the load-strain state of the flange rail in order to optimize its geometric dimensions, the finite element method was chosen, since it allows to consider in detail all the features of the cross-section geometries of the flange rails. The ANSYS software package was chosen for the flange rail calculations, as this software package has a wide range of capabilities to perform strength calculations.
2. Conducting loads analysis using ANSYS software package for shape SP-850, it was found that the highest loads for shape SP-850 reaches a maximum value of 265 MPa, which is below the values of permissible loads in centralize subject to the conditions of his work $[\sigma] = 330$ MPa. Obtained in the tests amounted to 230 MPa, so we can conclude: the obtained values are close in value, therefore, the developed mathematical model adequately reflects the loads of the deformed state of the flange rail in the areas of their greatest load. Therefore, the task of designing a rational section of the flange rail special profile by removing the metal from its weakly loaded places can be set.
3. The optimization objective function will minimize the cross-sectional area of the flange rail provided that the strength requirement and the constraint system are met. The constraints are the boundary conditions and dimensions that determine the position of the counter-rail in the track of the frog node. Accordingly, the size of the cross section shape of the SP-850 and variant at the same level, the position of the top of head shape SP-850 and variant flange rail, respectively, the dimensions determining the position of the side of the working face of the flange rail specprofile SP-850 and variant, the horizontal movement of flange rail in the locations of the supports.
4. As a result of the simulation, an optimized version of the flange rail special profile was obtained. The resulting optimized flange rail special profile allows, due to improved geometry of the flange rail, reducing the consumption of metal in the manufacture of flange rails by 22%, which reduces the cost of its manufacture. The service life of flange rails is increased by reducing its rigidity, which will reduce the cost of maintenance of railroad switches.

References

1. Glusberg, B.: Vestnik VNIIZHT, vol. 6, pp. 49–52 (1985)
2. Korolev, V.: The introduction of modern structures and technologies in track facilities. In: Sat. Mat. 1-Oh n-t Conference GETUPS, pp. 77–79 (2008)
3. Korolev, V.: Bulletin of the research Institute of railway transport ("Vestnik VNIIZHT"), vol. 6, pp. 38–39 (2009)
4. Korolev, V.: Bulletin of the research Institute of railway transport ("Vestnik VNIIZHT"), vol. 6, pp. 32–33 (2010)
5. Korolev, V.: Advanced development of science and technology – 2011. In: Sat. Mat. 55th n. - T. Conference Sp. z o.o. Nauka I studia, Przemysl, Poland, pp. 53–55 (2011)

6. Glusberg, B., Korolev, V., Gorbunov, M.: Modern and promising design of the railway track for different operating conditions. In: Works of JSC "VNIIZhT" Intex, Moscow, pp. 82–103 (2013)
7. Korolev, V.: Interuniversity scientific collection. works "Modern problems of perfection of work of railway route". Moscow State Railway University Engineering, pp. 46–48 (2014)
8. Korolev, V., Trykin, A.: Scientific and practical journal "Student of innovation of Russia" №1/2017 Publishing Center Agency of Intellectual Property (LLC "Matess"), pp. 28–30 (2017)
9. Korolev, V.: Popular scientific, production and technical magazine "road and track facilities", vol. 10, pp. 21–24 (2017)
10. Glusberg, B., Korolev, V.: [Bul. No. 27. 2011]. 108760 Russian Federation, IPC E 01 5/18 Area check. The applicant and the patent owner declared. Open joint stock company research Institute of railway transport. No 2011123082/11

Justification of Technical Solutions for Reinforced Concrete Tank Reconstruction

Aleksandr Tarasenko[1] , Petr Chepur[1] , Vadim Krivorotov[2] ,
Evgeniy Tikhanov[2] , and Alesya Gruchenkova[3(✉)]

[1] Industrial University of Tyumen, Volodarskogo Street 38,
625000 Tyumen, Russia
[2] Ural Federal University named after the first President of Russia B. N. Yeltsin,
Mira Street 19, 620002 Ekaterinburg, Russia
[3] Surgut Oil and Gas Institute, Entuziastov Street 38, 628405 Surgut, Russia
alesya2010-11@yandex.ru

Abstract. This article describes the features of the reconstruction of an underground cylindrical reinforced concrete tank with a capacity of 30000 m^3, the service life of which at the time of decommissioning was 43 years. The project proposed new technical solutions: rigid fastening of the columns in the bottom of the tank; pin-edge fixing of roof beams on columns; making the coating sliding along the beams to compensate for shrinkage strains; elimination of expansion joints on the coating, filling of technological expansion joints between the segments with pre-stressed concrete. The necessity of filling the expansion joints with pre-stressed concrete is substantiated to ensure the tightness of the coating, to ensure an acceptable joint operation of the structures of the tank, which is closest to the original typical one. In the proposed scheme, pre-stressed reinforcement, wound on the tank wall plates in the bottom and coating area, creates additional compressive stresses in the bottom and roof, which help to reduce the technological gap. The facility with the proposed new design solutions has successfully passed the state examination.

Keywords: Reinforced concrete tank · Reconstruction ·
Regulatory and technical documentation

1 Introduction

In the early 30s of the last century, it was believed that the construction of reinforced concrete tanks (RCT) for the oil storage is more economically advantageous than metal tanks construction [1–3]. The first typical cylindrical tanks made of reinforced concrete were developed in 1935 and some of them are still in operation. Reinforced concrete underground tanks are still widely used both in the oil industry and for other industrial needs.

The operational requirements for them are extremely diverse, but the methods of building these tanks are generally the same and correspond to the general technologies of building reinforced concrete structures [4–8].

© Springer Nature Switzerland AG 2020
V. Murgul and M. Pasetti (Eds.): EMMFT 2018, AISC 982, pp. 588–594, 2020.
https://doi.org/10.1007/978-3-030-19756-8_55

A variety of standard and individual projects of reinforced concrete tanks has been developed. Depending on the volume of the stored product, its aggressive nature, production technology, seismicity of the construction area, these tanks are divided into the following types: fully buried, semi-buried, rectangular, round, with a stationary roof, with a floating roof, without a roof, with pre-stressed reinforcement, without pre-stressed reinforcement.

Many of these tanks currently in use were built more than 30 years ago. The oil industry, taking into account the peculiarities of continuous production, requires reconstruction of these RCTs, taking into account the modernization of production technology and the elimination of unacceptable defects that are incompatible with the further operation. Technical diagnostics of tanks of the RCT type is carried out only with their service life of over 20 years [9, 10]. At the end of this period, industry regulatory documents regulate the implementation of a partial survey twice a decade and a full one once a decade.

When reconstructing an existing tank, it is important to not only restore or replace building structures with the new ones, but also to restore the joint operation of technological elements, in accordance with the original design model. Given the long service life of such tanks, the lack of executive documentation for construction, including project documentation, the definition of a design model for the joint construction of the tank's building structures is a rather sophisticated problem.

It is also necessary to take into account the fact that many of the tanks described were built on pilot projects. Thus, any project includes elements of research and calculations for new working conditions of the reconstructed elements of a reinforced concrete tank.

2 Materials and Methods

This article will consider the proposed method of repairing an underground cylindrical reinforced concrete tank with a capacity of 30000 m^3, the service life of which at the time of decommissioning was 43 years. In the course of its operation, several capital repairs were carried out related to the reinforcement of supporting structures (Fig. 1), as well as routine repairs of embedded pipes in the coating. The design feature of this tank is the presence of a 200-mm water screen on the surface of the coating (Fig. 2), which prevents its heating and subsequent excessive evaporation of stored oil. Hatch tie-ins in the cover are sealed. A platform is organized above the water screen to access the hatches and branch pipes on the roof. In winter, the water screen is not in use and is emptied through the trays to the relief.

Oil pumping from the tank is carried out by gravity through the inset passages of the dispensing pipelines in the bottom, equipped with a bottom valve. Oil injection is carried out in the same way - through the lower inset of pipelines. In this regard, the laying of pipelines to the tank is made at a considerable depth from the daylight surface (9–12 m) with the location of the main valves in the underground room of the tank control chamber (TCC) and with the serviced postern for laying pipelines from TCC to the tank. The presence of postern allows inspecting the nodes of the passage of the receiving-distributing pipelines in the bottom of the tank [11, 12].

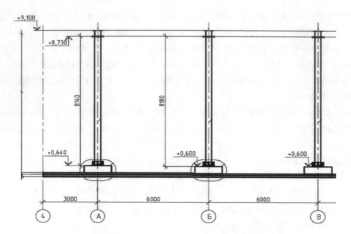

Fig. 1. Bearing elements of RCT-30000 design.

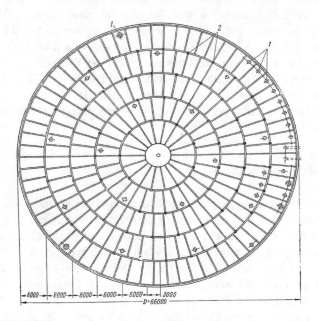

Fig. 2. RCT-30000 coverage plan: 1 - technological hatches; 2 - column locations.

In connection with the change in the process of pumping oil at the station, on the instructions of the customer, it was necessary to install a semi-submersible pump into the tank to change the technology of its pumping to ensure pumping pressure in the process pipeline at higher level. Moreover, the input of the receiving pipeline had to be done through the roof.

Thus, in order to ensure unconditional leak tightness of the bottom, it was supposed to dismantle the tank control chamber, postern, to make plugs of the piping entry points in the bottom, ensuring their unconditional tightness. The task was also assigned to equipping the tank with advanced systems for erosion of bottom sediments and with fire extinguishing systems [13–16].

The results of the complex diagnostics of this tank showed unsuitability for further operation (the "rejected" category) of the following tank elements:

- roof slabs - 60% of the total number are unsuitable for further operation;
- roof beams - 50% of the total number are unsuitable for further operation;
- roof columns - 50% of the total number are unsuitable for further operation;
- weakening of pre-stressed wall reinforcement - 20% of the total number.

3 Experimental Part

Based on the data obtained, a decision was made in the project on the reconstruction of the tank with the implementation of a new monolithic coating instead of the old precast reinforced concrete.

Furthermore, the project provided for ensuring maximum tightness of the wall and bottom.

To this end, it was intended to completely remove the existing sprayed plaster, up to 50 mm thick, saturated with oil, plugging and monolithing of the bottom pipelines inlets and constructing a new reinforced concrete bottom 200 mm thick over the old one. Further, the wall and bottom surfaces should be covered with a composite reinforced coating based on glass mats, 1.5 mm thick [17].

The structure of the monolithic reinforced concrete covering of the tank according to the project is made in the form of a plate with a thickness of 150 mm, supported by reinforced concrete radial beams (Fig. 3).

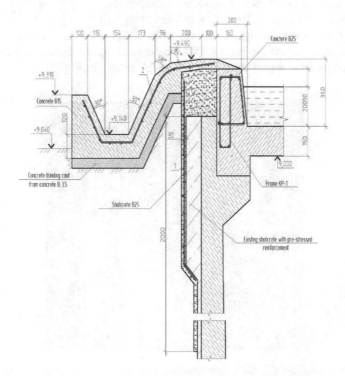

Fig. 3. Structure of a monolithic reinforced concrete belt.

During the design, there were contradictions with the customer on the construction and design scheme of the projected beam.

Thus, the customer's regulatory documents assume a design scheme with installation of columns on the bottom in massive column footings with rigid fixing of beams in the column. The coating is made sliding along the beams (on pergamine gaskets) with the expansion joints between the beam segments, with their fixation with thiokol joint sealant [18].

4   Results and Discussion

However, when performing calculations of this construction, it turned out that the design scheme proposed by the customer had a number of disadvantages, namely:

- the design scheme is not statically determinable, since it allows movement of the column in the friction node of the column footing and the bottom;
- the design scheme is not statically determinable, since it allows displacement of the coating segments under the action of compressive forces transmitted from the wall under the influence of prestressed reinforcement;
- the presence of expansion joints on the coating, taking into account the operating experience of already performed similar objects, causes problems with the tightness of the coating and the leakage of the water shield into the tank.
- Considering the stated disadvantages, the following was proposed:
- rigid fixation of columns in the bottom of the tank;
- hinge fixation of coating beams on columns (Fig. 4);
- making the coating sliding on the beams, to compensate for shrinkage strains;
- elimination of expansion joints on the coating, filling the technological expansion joints between the segments with pre-stressed concrete (Fig. 5).

Fig. 4. Stage of RCT-30000 reconstruction: installation of supporting metal beams.

Fig. 5. Stage of RCT-30000 reconstruction: concreting the outer belt.

Herewith, filling the expansion joints with prestressed reinforced concrete, firstly, allows to ensure the unconditional tightness of the coating, and secondly, to ensure an acceptable joint operation of the tank structures, closest to the original one. A typical tank layout can be simplified as a hollow cylinder with two seized bottoms. In this scheme, the pre-stressed reinforcement wound over the wall plates of the tank in the zone of the bottom and the cover performs the function of jamming and creates additional compressive stresses in the bottom and cover.

The proposed technical solutions in the completed project made it possible to successfully pass the state expertise on this object, as well as to solve the operational problem associated with the loss of coating tightness.

5   Conclusion

The authors have proposed new technical solutions for the reconstruction of an underground reinforced concrete tank, taking into account the results of a comprehensive diagnosis of this tank.

The design scheme for the construction of a monolithic reinforced concrete belt has been developed and substantiated.

The facility with the proposed new design solutions has successfully passed the state examination.

References

1. Slepnev, I.V.: Stress-strain elastic-plastic state of steel vertical cylindrical tanks with inhomogeneous base settlement. Moscow Engineering and Building Institute, Moscow (1988)
2. Tarasenko, A.A., Chepur, P.V., Kuzovnikov, E.V., Tarasenko, D.A.: Calculation of the stress-strain state of the receiving and distributing branch pipe with defect for the purpose of justification of possibility of its further operation. J. Fundam. Study **9**(7), 1471–1476 (2014)

3. Chepur, P.V., Tarasenko, A.A.: Methods of determining the need for repair of the tank in the sediments of the base. J. Fundam. Study **8**(6), 1336–1340 (2014)
4. Nekhaev, G.A.: Design and calculation of steel cylindrical tanks and tanks low pressure. Association of Construction Universities, Moscow (2005)
5. Russian Standard GOST 31385-2008: The vertical cylindrical steel tanks for oil and oil-products. General technical conditions. Standartinform, Moscow (2010)
6. Russian Standard RD-23.020.00-KTN-283-09. The rules of the repair and reconstruction of tanks for oil storage capacity of 1000-50000 cubic meters. OAO AK "Transneft", Moscow (2009)
7. Russian Standard GOST 52910-2008: The vertical cylindrical steel tanks for oil and oil-products. General technical conditions. Standartinform, Moscow (2008)
8. Russian Standard RD-23.020.00-KTN-018-14: Main pipeline transport of oil and oil products. Vertical steel tanks for oil storage capacity of 1000-50000 cubic meters. The design standards. OAO AK "Transneft", Moscow (2014)
9. Tarasenko, A.A., Nikolaev, N.V., Khoperskiy, G.G., Sayapin, M.V.: The stress-strain state of the tank wall at non-uniform settlements of the foundation. J. News Univ. "Oil Gas" **3**, 75–79 (2007)
10. Tarasenko, A.A., Chepur, P.V.: Deformation of large fixed roof tank with nonuniform sediments grounds. J. Fundam. Study **11**(2), 296–300 (2014)
11. Tarasenko, A.A., Nikolaev, N.V., Khoperskiy, G.G., Ovchar, Z.N., Sayapin, M.V.: Investigation of the influence of the transfer nozzles reception on the stress-strain state of the wall of vertical cylindrical tanks. J. News Univ. "Oil Gas" **2**, 59–68 (2005)
12. Tarasenko, A.A., Chepur, P.V.: Stress-strain state of the upper support ring of the tank at non-axisymmetric deformations of the body. J. Fundam. Study **11**(3), 525–529 (2014)
13. Konovalov, P.A., Mangushev, R.A., Sotnikov, S.N., Zemlyansky, A.A.: Foundations steel tanks and deformation of it's bases. The Association of Construction Universities, Moscow (2009)
14. Zemlyansky, A.A.: The design principles and experimental and theoretical studies of large tanks, Balakovo (2006)
15. Vasilev, G.G., Salnikov, A.P.: Analysis of causes of accidents with vertical steel tanks. J. Oil Ind. **2**, 106–108 (2015)
16. Yasin, E.M., Rasshchepkin, K.E.: The upper zones stability of vertical cylindrical tanks for oil-products storage. J. Oil Ind. **3**, 57–59 (2008)
17. Tarasenko, A., Chepur, P., Gruchenkova, A.: Determining deformations of the central part of a vertical steel tank in the presence of the subsoil base inhomogeneity zones. In: AIP Conference Proceedings, vol. 1772, p. 060011 (2016)
18. Korobkov, G.E., Zaripov, R.M., Shammazov, I.A.: Numerical modeling of the stress-strain state and stability of pipelines and tanks in complicated operating conditions. Nedra, St. Petersburg (2009)

Study of the Deformation Process of the Tank Stiffening Ring During Settlement

Petr Chepur[1] (iD), Aleksandr Tarasenko[1] (iD),
and Alesya Gruchenkova[2](✉) (iD)

[1] Industrial University of Tyumen, Volodarskogo Street 38, 625000 Tyumen,
Russia
[2] Surgut Oil and Gas Institute, Entuziastov Street 38, 628405 Surgut, Russia
alesya2010-11@yandex.ru

Abstract. The article deals with the issue of deformation of the upper edge of
the shell during the development of non-uniform tank settlement. Works on the
mathematical representation of the process of changing the stress-strain state of
the various tank elements during non-uniform settlement were analyzed. Models
of RVS-20000 have been created for calculating the SSS using simplified and
extended (full) schemes. In the proposed model, the authors set the task to take
into account the maximum number of geometrical elements of the tank metal-
work. Particular attention was paid to the modeling of the junctions between the
wall and the edge, the wall and the roofing, the roof beams and the stiffness ring.
In this work, the curves of vertical and horizontal wall displacements were
obtained for complete and simplified design schemes. The need to take into
account the size of the protrusion of the annular plate, support stiffness ring,
beam roof structure, and sheet roof structure when calculating the tank SSS in
case of non-uniform settlement was proved; these factors significantly affect the
final rigidity of the entire tank.

Keywords: Tank · Non-uniform settlement · Finite element method

1 Introduction

Non-uniform settlement (Fig. 1) of the outer contour of the bottom of the tank has a
significant impact on the stress-strain state of its structural elements [1–5]. From the
literature it is known that settlement is one of the most dangerous factors, often leading
to damage and destruction of VSTs [6, 7].

The solution can be obtained by analytical methods for non-axisymmetric deformation only by introducing a number of assumptions that significantly reduce the
accuracy of the results [8, 9].

The authors set the task to obtain the dependencies between the values of the
subsiding zone, the values of the resulting stresses and displacements of the VST
elements.

With the development of settlement of the lower contour, the upper edge of the tank
wall also moves in the radial direction, the amount of deformation depends on the
shell's own rigidity.

© Springer Nature Switzerland AG 2020
V. Murgul and M. Pasetti (Eds.): EMMFT 2018, AISC 982, pp. 595–602, 2020.
https://doi.org/10.1007/978-3-030-19756-8_56

Fig. 1. Non-uniform settlement of the outer contour of the bottom of the tank.

Having a similar dependency for each tank size, it is possible to determine, based on geodetic measurements, whether non-uniform settlement of the outer contour of the bottom develops or deformation of the shell is due to defects in the installation of metal structures of the hull and bottom. An approximate solution to the problem using numerical methods was obtained in [1, 5]. The author used the LIRA computing complex in a linear formulation. The assumptions made in the paper led to a lack of accuracy of the solution and reduced the practical significance of the study. The capabilities of the program did not allow the stiffness ring and the roof of the tank to be included in the model, and their effect was replaced by zero axial displacements for the upper edge of the tank wall.

This approach does not reflect the real rigidity of VST structures and significantly distorts the results of calculations. With the advent of new computational packages, it became possible to model and investigate the entire structure, including all the elements of its construction.

It is proposed to compare the results of the tank SSS change using the simplified (without taking into account the roof, stiffness ring and other parameters according to [5]) calculation scheme and the one proposed by the authors for various values of the settlement zone.

2   Materials and Methods

The study uses the modern software product ANSYS Workbench 14.5 based on the finite element analysis [10–12].

In the simplified calculation scheme, the boundary conditions are assumed similarly to [5]: the tank is rigidly fixed at 24 points of the corner weld joint, the upper edge of the wall is not fixed. The geometry does not take into account the stiffness ring, roof, edge, central part of the bottom.

The construction of the RVS-20000 geometry in the ANSYS program is implemented in the DesignModeller module, which allows the use of 2D sketching methods, and the subsequent parametric creation of three-dimensional models. To construct RVS-20000, it is necessary to create a two-dimensional sketch for each structural element.

For geometric modeling of the wall, a sketch is built that includes a circle of a given diameter. Then the "Extrude" command is applied to the sketch, performing the function of "extruding" the circle by a specified amount. In this case, the extrusion value is the height of the first belt of the RVS-20000 wall, equal to 1,490 mm. After creating the shell of the first wall belt, a similar operation is performed for belts 2–8.

Technical and geometric parameters of the RVS-20000 tank are given in Table 1.

Table 1. Main technical and geometric parameters of RVS-20000.

Parameter	Value	Unit
Tank diameter	45600	mm
Tank height	11920	mm
Thickness of the 1st wall belt	13	mm
Thickness of the 2–8th wall belts	11	mm
Design innage height	10880	mm
Stored oil density	875	kg/m^3
Yield strength of 09G2S steel	325	MPa

When modeling non-uniform tank settlement, it is necessary to set the main parameters that will be used in the preparation of the calculation scheme. The main geometric characteristics of non-uniform settlement of the outer contour of the bottom are the length along the arc and an array of points with vertical marks. The non-uniform settlement zone of the tank is modeled by cutting a segment of the foundation ring of size L.

For convenience of presenting the results of calculations, it is proposed to use the dimensionless parameter n adopted in the theory of shells, which takes into account the size of the loading zone (tank settlement):

$$n = \pi R/L \qquad (1)$$

where

L – arc size of the zone of non-uniform settlement, m;
R – tank radius, m.

3 Experimental Part

To obtain reliable results of the deformation of the VST structures with non-uniform settlement, the authors developed a model that takes into account the physical and geometric non-linearity of the processes of deformation of metal structures.

In the design model, the following elements are taken into account: the foundation ring, the wall, the annular plate, the central part of the bottom, the support ring, beams, and roofing sheets. The model consists of beam (BEAM4, BEAM188), shell (SHELL181) and "contact" (CONTA175, TARGE170) finite elements [13, 14]. The foundation and the central part of the bottom have a contact interaction with the subgrade; this problem is solved in an elastic formulation with the bedding ratio of the subgrade which is assumed to be 200 MN/m^3 [15–18].

Wind load is applied to the wall of the tank in accordance with the values adopted for the Ist wind region. Snow load is applied to the surface of the roofing; the value is given for the Vth snow region.

Thus, loading of metal structures is selected for the most unfavorable conditions with maximum indices. The adequacy of the proposed numerical model was verified by solving the axisymmetric problem and comparing with the analytical solution in [2]; the error was no more than 2%.

4   Results and Discussion

Table 2 shows the maximum values of vertical (u) and radial wall displacements (W) and maximum equivalent stresses (according to von Mises, σ_{equiv}), arising in the metal structures of RVS-20000, obtained from the results of numerical calculations performed for the above described models of the VST with the dimensionless parameter n = 1...6.

Table 2. Main technical and geometric characteristics of RVS-20000.

n	Without the stiffness ring and fixed roof			With the fixed roof and support ring		
	W, mm	u, mm	σ_{equiv}, MPa	W, mm	u, mm	σ_{equiv}, MPa
1	629.3	180.6	263.8	8.92	11.24	38.62
2	833	199.8	195	5.67	3.22	28.8
3	878.1	112	138.8	3.51	1.56	23.67
4	439.6	46.3	88.3	2.47	1.06	20.29
5	171.1	8.26	39.5	1.84	0.66	17.12
6	101.5	5.02	26.8	1.7	0.58	16.55

Figure 2 shows the results of calculations of the radial displacements of the RVS-20000 tank wall with the maximum value of the non-uniform settlement zone n = 1 (72 m).

Figure 4 explains the nature of the deformation of the VST structures for n = 1.

Using the values from Table 2, the functional dependencies of the deformations and stresses in the VST metal structures on the size of the non-uniform settlement zone, obtained for the simplified computational model (Fig. 3) and the model with a maximum degree of detail (with a fixed roof and support ring) (Fig. 5).

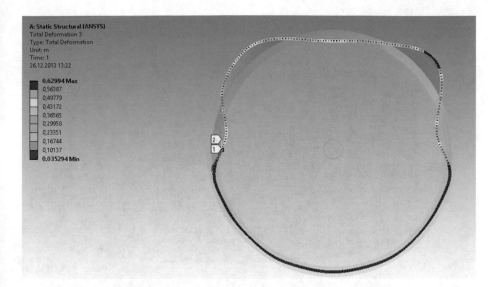

Fig. 2. Radial displacements of the upper edge of the 8th wall belt for $n = 1$.

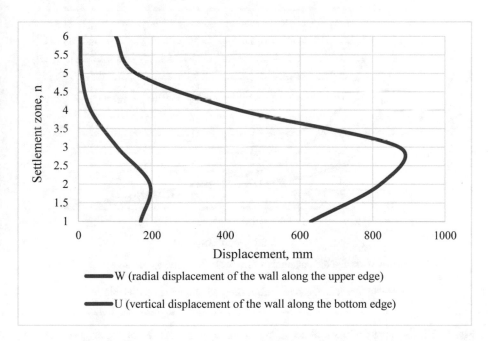

Fig. 3. Displacements of the wall without a roof and support ring.

Fig. 4. Displacements of the VST structures for n = 1.

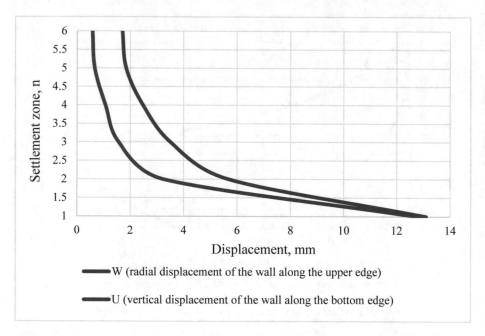

Fig. 5. Displacements of the wall according to the scheme proposed by the authors (ANSYS).

When modeling the non-uniform settlement of RVS-20000, areas with high stress concentrations, the values of which exceed 300 MPa, are established. Such stresses arise in metal mounting plates connecting the supporting roof beams and the support ring.

Under non-axisymmetric loads on the tank body, the support ring and the beam frame are subjected to bending stresses, however, their spatial deformation patterns do not lie in the same plane: the roof structures move along the vertical axis of the VST, and the deformation of the stiffness ring is characterized by torsion relative to the butt joint with the wall on the 8th belt.

Graphs in Figs. 3 and 5 show that with non-axisymmetric loading of the VST, values of the SSS parameters of the structure differ several times for calculation schemes that take into account and those that do not take into account the particularities of the roof geometry.

Analyzing the obtained dependencies, it was established that when calculating the stress-strain state of the tanks, especially in the presence of non-axisymmetric loading factors, the actual rigidity of the VST should be taken into account.

Not taking into account such structural elements as the annular plate, support stiffness ring, beam structure of a fixed roof, and roofing sheets leads to large errors in the calculations, which does not allow them to be used in solving practically important production problems.

The use of geometrically and physically accurate models, as well as modern design packages, allowed us to obtain an adequate picture of the stress-strain state of the VST metal structures suitable for practical use.

5 Conclusion

In the ANSYS software package, a finite element model of the RVS-20000 tank was created, which takes into account the maximum number of structural elements and the physical properties of the steel used, with the development of non-uniform settlement of the bottom outer contour. It was established that when calculating the stress-strain state of tanks, the actual stiffness of the VST should be considered. Not taking into account such structural elements as the annular plate, support stiffness ring, beam structure of a fixed roof, and roofing sheets can lead to errors in the calculations up to 1300%. A relationship between the amount of settlement and the radial displacement of the upper edge of the RVS-20000 wall was obtained.

To obtain functional dependencies that characterize the actual structural rigidity of the tank, when calculating the stress-strain state of the VST in a non-axisymmetric formulation, it is necessary to use the structure model with the maximum degree of detail of the structural geometry. Not taking into account such elements as the annular plate, bottom, support ring and structural elements of the fixed roof of the VST, leads to large errors in the calculation of the SSS. Such an approach will allow forming adequate requirements of the normative technical documentation in terms of limiting the amount of tank settlement with the available stiffness elements for permissible stresses.

The above results of the study were used to develop the theoretical foundations of the methodology and algorithm for determining the need for repair of RVS-20000 when non-uniform settlement of the bottom outer develops.

References

1. Slepnev, I.V.: Stress-strain elastic-plastic state of steel vertical cylindrical tanks with inhomogeneous base settlement. Moscow Engineering and Building Institute, Moscow (1988)
2. Tarasenko, A., Chepur, P., Gruchenkova, A.: Determining deformations of the central part of a vertical steel tank in the presence of the subsoil base inhomogeneity zones. In: AIP Conference Proceedings, vol. 1772, p. 060011 (2016)
3. Vasilev, G.G., Salnikov, A.P.: Analysis of causes of accidents with vertical steel tanks. J. Oil Ind. **2**, 106–108 (2015)
4. Guan, Y., Huang, S., Zhang, R., Tarsenko, A.A., Chepur, P.V.: Influence of laminated rubber bearings parameters on the seismic response of large LNG storage tanks. J. World Inf. Earthq. Eng. **1**, 219–227 (2016)
5. Tarasenko, A., Gruchenkova, A., Chepur, P.: Joint deformation of metal structures in the tank and gas equalizing system while base settlement progressing. J. Procedia Eng. **165**, 1125–1131 (2016)
6. Konovalov, P.A., Mangushev, R.A., Sotnikov, S.N., Zemlyansky, A.A.: Foundations steel tanks and deformation of it's bases. The Association of Construction Universities, Moscow (2009)
7. Zemlyansky, A.A.: The design principles and experimental and theoretical studies of large tanks, Balakovo (2006)
8. Nekhaev, G.A.: Design and calculation of steel cylindrical tanks and tanks low pressure. The Association of Construction Universities, Moscow (2005)
9. Yasin, E.M., Rasshchepkin, K.E.: The upper zones stability of vertical cylindrical tanks for oil-products storage. J. Oil Ind. **3**, 57–59 (2008)
10. Bruyaka, V.A., Fokin, V.G., Soldusova, E.A., Glazunova, N.A., Adeyanov, I.E.: Engineering analysis in ANSYS Workbench. Samara State Technical University, Samara (2010)
11. Beloborodov, A.V.: Evaluation of the finite element model construction quality in ANSYS. Ural State Technical University, Ekaterinburg (2005)
12. Russian Standard RD-23.020.00-KTN-018-14: Main pipeline transport of oil and oil products. Vertical steel tanks for oil storage capacity of 1000-50000 cubic meters. The design standards. OAO AK "Transneft", Moscow (2014)
13. Efemenkov, I.V.: Modelling and optimization of the welding metallurgical process technological parameters software SYSWELD. J. Success Mod. Sci. **9**, 108–111 (2016)
14. Bilenko, G.A.: Analysis of structures residual stresses and deformations after welding working under pressure in the program SYSWELD. J. Metall. Struct. **8**, 32–34 (2012)
15. Russian Standard GOST 31385-2008: The vertical cylindrical steel tanks for oil and oil-products. General technical conditions. Standartinform, Moscow (2010)
16. Russian Standard RD-23.020.00-KTN-283-09: The rules of the repair and reconstruction of tanks for oil storage capacity of 1000-50000 cubic meters. OAO AK "Transneft", Moscow (2009)
17. Russian Standard GOST 52910-2008: The vertical cylindrical steel tanks for oil and oil-products. General technical conditions. Standartinform, Moscow (2008)
18. Russian Standard RD-23.020.00-KTN-170-13: Requirements for the installation of metal structures vertical cylindrical tanks for oil and oil-products storage at new building facilities, technical reequipment and reconstruction. OAO AK "Transneft", Moscow (2013)

Modeling of the Stress-Strain State of Railway Wheel and Rail in Contact

Valerii Prokopev[1]([✉]) [iD], Tatiana Zhdanova[1] [iD],
and Barasbi Kuschov[2] [iD]

[1] Moscow State University of Civil Engineering, Yaroslavskoe shosse 26,
129337 Moscow, Russia
viprokopiev@mail.ru
[2] CJSC "Fodd", Leninsky pr. 148, 119571 Moscow, Russia

Abstract. This article lists the methods implemented in ANSYS Mechanical, which solves contact problems in mechanics. The results of an example of an analysis of the stress-strain state of contact between the wheel and the rail of a railway transport with the geometry maximally close to reality with static lifting using the ANSYS Mechanical program are presented.

Keywords: ANSYS Mechanical · The solution of contact problems · Contact of wheel and rail of railway transport

1 Introduction

Large values of stresses in the contact area of the railway wheel and rail lead to wear of the contacting surfaces [1]. A significant part of the rails (20 to 40% of all seized) is caused by contact-fatigue defects. To prevent them, it is necessary to carry out a whole complex of measures: to improve the cleanliness of rail steel, to strictly observe the technology of manufacturing rails, to optimize their non-destructive testing, to control the intensity of wear of rails and wheels. Solution of contact problem tasks with consideration of the nonlinearities of materials, generally only possible with usage of numerical methods [2]. At present, the finite element method is widely used and implemented in a number of known software complexes [3, 4]. In ANSYS Mechanical software complex, we research stress-strain state of railroad wheel and a rail in contact with maximum approximation to the real geometry in static and dynamic settings.

In contact problems, the contact zone is unknown until the problem is solved. Depending on the loads, material properties, boundary conditions and other factors, the surfaces can come into contact with each other and break contact suddenly and unpredictably. On the other hand, friction of the interacting contacting surfaces in the contact zone must be taken into account, which can affect the convergence process in computer simulation.

© Springer Nature Switzerland AG 2020
V. Murgul and M. Pasetti (Eds.): EMMFT 2018, AISC 982, pp. 603–614, 2020.
https://doi.org/10.1007/978-3-030-19756-8_57

The effect of surface roughness on friction in contacting bodies was first investigated by Bowden and Tabor in the mid-20th century, which led to the expansion of research directions in tribology and the revival of Coulomb's ideas on adhesion as a possible mechanism of friction. Innovative works in this field include the works of Archard (1957), Greenwood and Williamson (1966), Bush (1975) and Persson (2002). One of the most important impulses for the accurate calculation of load conditions in contact was the development of railway transport since stresses in wheel-rail interaction area can reach the maximum permissible load for steel. Research on interacting contacting surfaces made it possible to formulate appropriate mathematical models [5–11]. Computer modeling of contact problems is described in [12].

Following algorithms are implemented in ANSYS Mechanical [13]:

– Lagrange multiplier method (Augmented Lagrangian, KEYOPT(2) = 0);
– Penalty method (Penalty method, KEYOPT(2) = 1);
– Lagrange multiplier method in the direction normal to the contact surface and Penalty method in the direction of the tangent (KEYOPT(2) = 3);
– Lagrange multiplier method in the direction normal to the contact surface and tangent;
– Constraint method in internal points (internal multipoint constraint (MPC), KEYOPT(2) = 2).

The results of numerical simulation are given in Sect. 5. The conclusion in Sect. 4 finalizes the paper.

2   Static Formulation

An example of a wheel and a rail contact calculation with maximum approximation to the real geometry in a static formulation is given in [14].

Let's consider the most common railway wheel A1 - solid rolled with a planoconical disc (GOST 10791-2011) and a rail type P65 (GOST R 51685-2013).

Materials with following mechanical properties were used for the railway wheel and the rail:

– Railway wheel: $E = 2.1 \cdot 105$ MPa, $v = 0.3$, $\sigma_t = 820$ MPa, hardening modulus $Ehrd = 2.1 \cdot 102$ MPa.
– Rail: $E = 2.1 \cdot 105$ MPa, $v = 0.3$, $\sigma_t = 1080$ MPa, hardening modulus $Ehrd = 2.1 \cdot 102$ MPa.

Coefficient of friction $\mu = 0.15$.

The hardening modulus is used in a formulation that takes into account the nonlinear properties of the material, friction, and geometric nonlinearity.

The finite-element model is shown in Figs. 1 and 2.

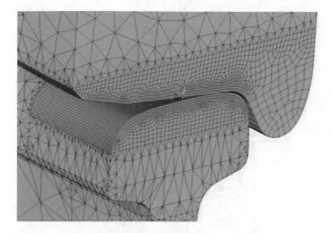

Fig. 1. The finite-element model

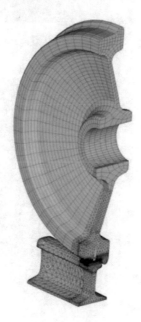

Fig. 2. The finite-element model

An additional calculation for the effect of the kinematic load was made to illustrate isopoles of displacements, von Mises stress, and contact pressure in the wheel area: displacement of 0.25 mm, applied to the wheel hub and directed toward the rail. The results of the calculation are shown in Figs. 3, 4 and 5.

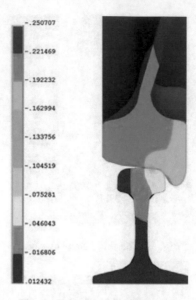

Fig. 3. Isopole of displacements

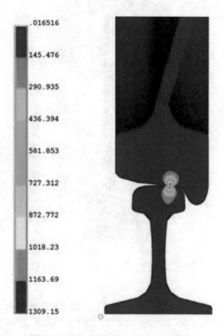

Fig. 4. The Mises stress iso-field

Initially, calculations were made for a linear material. The results of the calculation are shown in Figs. 6 and 7.

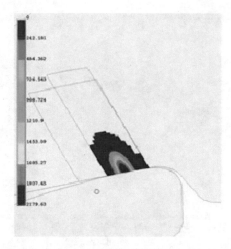

Fig. 5. Isopole of contact pressure in the test calculation

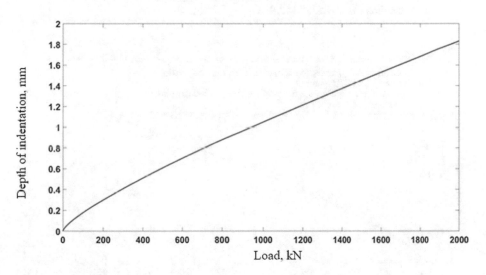

Fig. 6. Dependence of the depth of indentation d on the magnitude of the load F for the numerical solution *(max = 1.83 mm)*

In Figs. 8 and 9, we compare the results of calculations with and without friction, geometric nonlinearity and nonlinear properties of the material (Fig. 10).

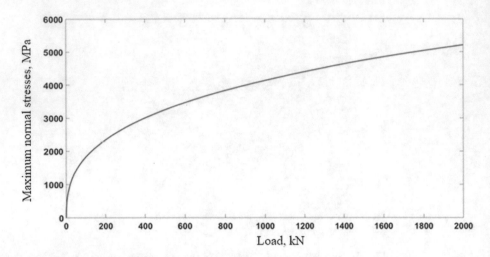

Fig. 7. Dependence of the maximum normal stresses on σ_{max} on the magnitude of the load F for the numerical solution *(max = 5208 MPa)*

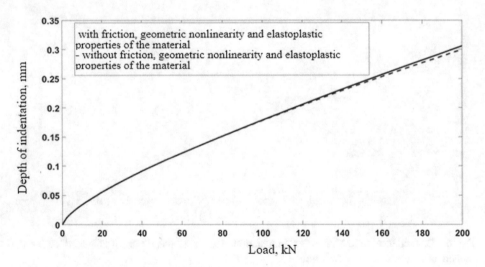

Fig. 8. Dependence of the depth of indentation d on the magnitude of the load F for the numerical solution with *(max = 0.31 mm)* and without *(max = 0.30 mm)* friction, geometric nonlinearity and nonlinear properties of the material

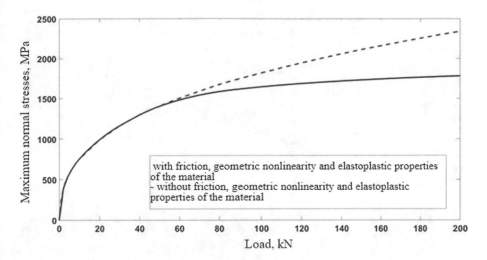

Fig. 9. Dependence of the maximum normal stresses on σ_{max} on the magnitude of the load F for the numerical solution with *(max = 0.31 mm)* and without *(max = 0.30 mm)* friction, geometric nonlinearity and nonlinear properties of the material

Figures 11 and 12 present the results of calculations for various wheel positions on the rail, taking into account friction, geometric nonlinearity and elastoplastic properties of the material.

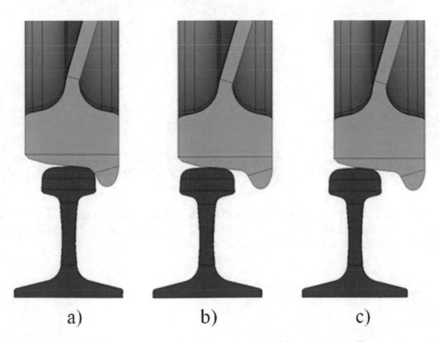

Fig. 10. Different wheel positions on the rail: (a) central position; (b) 2 cm offset; (c) 3 cm offset

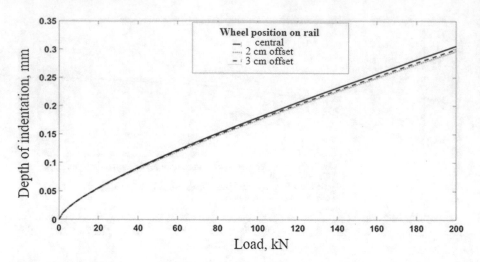

Fig. 11. Dependence of the depth of indentation d on the magnitude of the load F for the numerical solution with friction, geometric nonlinearity and nonlinear properties of the material

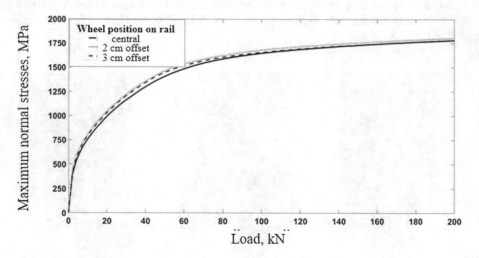

Fig. 12. Dependence of the maximum normal stresses on σ_{max} on the magnitude of the load F for the numerical solution with *(max = 1782* MPa*)* and without *(max = 2340* MPa*)* friction, geometric nonlinearity and nonlinear properties of the material

3 Dynamic Setting

In dynamic setting, rolling of the wheel along the rail is calculated from the action of a vertical load equivalent to a mass of 10 tons per wheel plus the own weight of the wheel.

We model part of the axis to which torque is applied. During the first second, torque evenly reaches *144.5* kNm and remains constant.

The effects of slippage are not taken into account, since the friction coefficient is assumed to be infinite. Practice of calculations shows that this assumption has no significant effect on the normal stresses and the depth of indentation.

It is believed that the deformations are finite, changes in shape are taken into account: areas of cross-sections, thicknesses, etc. Displacements and rotations can be arbitrary.

At large displacements, it is assumed that rotations are finite, but mechanical deformations (which causing stresses) are calculated with linearized expressions. The construction is assumed not to change its form, except for the case of motion in the form of a rigid whole.

Let's look at the rolling a wheel on a rail task.

The finite element model is shown in Fig. 13. Since the position of the contact varies with time along the rail and the rim of the wheel, the mesh is done sufficiently fine in those areas. For the convenience of applying torque, an axis element is modeled in the center.

For the solution of dynamic problem, the Newmark method or HHT (Hilbert-Hughes-Taylor) method is applied with the procedure of stepwise time loading for implicit schemes of integrating the equations of motion.

Fig. 13. Finite-element model of the wheel-rail system for the dynamic problem (the number of finite elements is 547149).

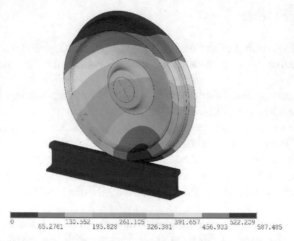

Fig. 14. Final position of the wheel with isopoles of total displacements for the dynamic problem (The number of finite elements is 547149)

At the first step of the calculation, the wheel is secured from displacements along the rails and contacts the rail only under the influence of the vertical load. On the next steps, the fastening along the rails is removed and the wheel begins to roll. The first 5 s after the start of the movement are researched.

The final position of the wheel with the isopoles of total displacements is shown in Fig. 14.

The plot of the movement of the wheel as a whole along the rail is shown in Fig. 15.

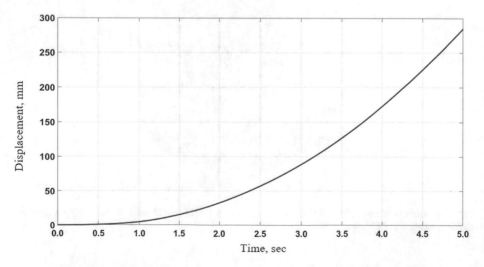

Fig. 15. The plot of the movement of the wheel as a whole along the rail

The change of the indentation and maximum normal stresses values over time is shown in Figs. 16 and 17.

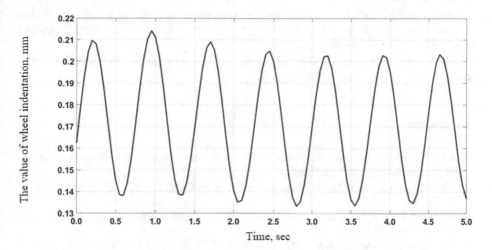

Fig. 16. The value of wheel indentation.

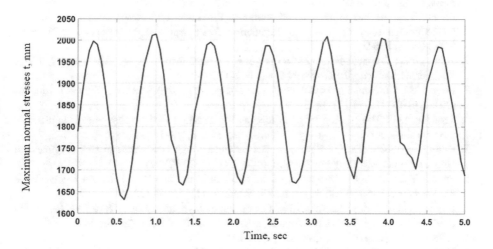

Fig. 17. Maximum normal stresses.

4 Conclusion

Results of the calculations show that consideration of geometric nonlinearity, friction and elastoplastic properties of the material under static loading leads to a significant decrease in stress in the contact area with a slight change in the depth of indentation.

Different positions of the wheel on the rail slightly affect the stresses in the contact area and the depth of penetration (up to 5%).

As a solution of the contact problem of rolling wheel in a dynamic setting, conclusion was made that the desired characteristics (the depth of indentation and the maximum normal stresses) fluctuate around the values obtained in the static solution.

Maximum amplitude of the depth of indentation is 0.038 mm, which is approximately 11% of the average value.

Maximum amplitude of the maximum normal stresses is 195 MPa, which is approximately 11% of the average value.

References

1. Sakalo, V.M., Kossov, V.S.: Kontaktnye zadachi zheleznodorozhnogo transporta. Mashinostroenie, Moskva (2004). (in Russian)
2. Dzhonson, K.L.: Mekhanika kontaktnogo vzaimodejstviya. Mir, Moskva (1989). (in Russian)
3. Morozov, E.M., Zernin, M.V.: Kontaktnye zadachi mekhaniki razrusheniya. Mashinostroenie, Moskva (1999)
4. Kaplun, A.B., Morozov, E.M., Olfer'eva, M.A.: ANSYS v rukah inzhenera: Prakticheskoe rukovodstvo. Knizhnyj dom «LIBROKOM», Moskva (2009). (in Russian)
5. Johnson, K.L.: Contact Mechanics, 6. Nachdruck der 1. Auflage. Cambridge University Press, Cambridge (2001)
6. Popov, V.L.: Kontaktmechanik und Reibung. Ein Lehr- und Anwendungsbuch von der Nanotribologie bis zur numerischen Simulation, 328 p. Springer, Heidelberg (2009). ISBN 978-3-540-88836-9
7. Popov, V.L.: Contact Mechanics and Friction. Physical Principles and Applications, 362 p. Springer, Heidelberg (2010). ISBN 978-3-642-10802-0
8. Popov, V.L.: Mekhanika kontaktnogo vzaimodejstviya i fizika treniya. Fizmatlit, Moskva (2012). ISBN 978-5-9221-1443-1
9. Sneddon, I.N.: The relation between load and penetration in the axisymmetric boussinesq problem for a punch of arbitrary profile. J. Eng. Sci. **3**, 47–57 (1965)
10. Hyun, S., Robbins, M.O.: Elastic contact between rough surfaces: effect of roughness at large and small wavelengths. Tribol. Int. **40**, 1413–1422 (2007)
11. Popov, V.L.: Method of reduction of dimensionality in contact and friction mechanics: a linkage between micro and macro scales. Friction **1**(1), 41–62 (2013)
12. Burago, N.G., Kukudzhanov, V.N.: Obzor kontaktnyh algoritmov. Izvestiya Rossijskoj akademii nauk. Mekhanika tverdogo tela, №1, pp. 44–85 (2005). (in Russian)
13. ANSYS 14.5 Users Guide. Canonsburg (2012)
14. Prokopev, V.I., Zhdanova, T.V., Kuschov, B.S.: Modeling of the stress-strain wheel and rail. Vestnic Chuvash Pedagogicheskogo Universiteta, № 4 (2017)

Free Oscillations of Semi-underground Trunk Thin-Wall Oil Pipelines of Big Diameter

Vladimir Sokolov and Igor Razov[✉]

Industrial University of Tyumen, Volodarskogo Str. 38, 625001 Tyumen, Russia
razovio@tyuiu.ru

Abstract. This article solves the task of determining frequencies and modes of free oscillations of trunk thin-wall oil pipelines of big diameter in case of semi-underground laying. The task will be solved on the basis of geometrically non-linear variant of semi-moment theory of cylindrical shells of moderate flexure. The resolving equations have been acquired, which make it possible to investigate the frequencies of free oscillations of oil pipeline depending on the physical and geometrical parameters in the process of operation. The oil pipeline is exposed to the effect of stationary internal working pressure, parameter of longitudinal compressive force, influence of soil reactive pressure at different geometrical characteristics as well as influence of fluid flow hydrodynamic pressure on the pipe internal wall. The fluid hydrodynamic pressure has been determined on the basis of theory of potential flow of ideal incompressible fluid in cylindrical coordinates with the use of modified Bessel functions of the first genus. The influence of soil surrounding the pipe corresponds to a non-uniform load caused by the bearing pressure of soil; the value of bearing pressure will depend on soils under consideration.

The methods of dynamic calculation of semi-underground oil pipelines corresponding to the real operating conditions have been developed and improved. The obtained solutions can be used for engineering calculations, which make it possible to determine the spectrum of frequencies of free oscillations. The entire oil pipeline system will be constructed on the basis of calculated frequencies in order to prevent a dangerous resonance effect, which contributes to the increased oil pipeline reliability for the whole period of operation.

1 Introduction

Dynamic calculation of thin-wall pipelines of big diameter is one of the actual design tasks. Its main task is to determine dynamic characteristics of natural frequencies in particular. According to SNiP 2.05.06, SP 36.13330.2012, existing methods of dynamic calculation are mainly based on the core theory. The use of this approach is more applicable to the calculation of thick-wall pipes at ratios of wall thickness h to radius R more than 1/20 (h/R > 1/20).

In case of the rod model, the influence of the internal working pressure, which is significant for thin-wall pipes (h/R < 1/20) is not taken into account. As it was proved in works of Ilyin [1] natural frequencies of thin-wall pipes can increase up to 80% with increasing an internal working pressure of 2 to 10 MPa. Therefore, to calculate

© Springer Nature Switzerland AG 2020
V. Murgul and M. Pasetti (Eds.): EMMFT 2018, AISC 982, pp. 615–627, 2020.
https://doi.org/10.1007/978-3-030-19756-8_58

thin-wall pipes h/R < 1/20 (D > 1000 mm), it is required to consider a thin-wall, uniform, isotropic cylindrical finite length shell as a basic design model.

To identify the pipelines resonance zones during operation, it is required to more reliably determine values of natural frequencies corresponding to the real operating conditions. Under conditions of structures resonance due to the oscillation process, a material low-cycle fatigue increases, which significantly reduces the pipeline reliability.

For the problem on determination of pipelines natural frequencies, we should be note out of works the authors: Bereznev [2], Djondjorov [3], Efimov [4], Lilkova–Markova [5], Razov and Sokolov [6–10].

Despite a rather large number of works devoted to determining the pipelines natural frequencies, the problem related to the dynamic calculation of semi-underground pipelines remains open, since the available literature provides no single calculation procedure, which allows taking into account the influence of no uniform soil pressure on the outer pipe surface.

2   Methods

To solve the problem on natural vibrations of main thin-wall oil pipeline of big diameter in case of semi-underground laying (Fig. 1), let's consider the calculation model of the pipeline as an uniform isotropic cylindrical shell with closed cross-section, finite length L radius of the middle surface R, and a wall thickness h (Fig. 2).

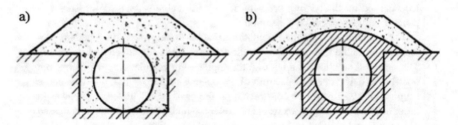

Fig. 1. Semi-underground pipeline laying: (a) – in filling with mineral soil; (b) – in filling with hydrophobizated soil.

The following designations are used in the figure: Ru, Rv, Rw – components of movement in the longitudinal, circumferential and radial directions, x – axial coordinate along the shell axis, θ – vectorial angle of rotation in the circumferential direction, p_0 – internal working pressure, $q_{eqv.}$ – non-uniform effect of elastic soil pressure on the outside perimeter of the oil pipeline.

The shell is subject to the action of stationary internal working pressure p0, longitudinal compression force F, fluid flow rate and the effect of reactive soil pressure on the outside perimeter of the pipe cross-section (Fig. 3).

When studying the soil interaction with the outside pipe perimeter, we can suppose that an apparent soil mass does not depend on pressure intensity, and the decision on its uniform distribution along the circumference perimeter allows to solve the task of closed.

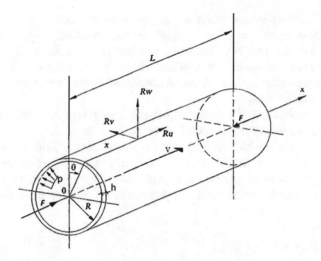

Fig. 2. Calculation model of main thin-wall oil pipeline of big diameter in case of semi-underground laying as an uniform isotropic cylindrical finite length shell.

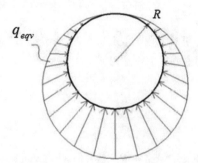

Fig. 3. Distribution of no uniform soil pressure on the outside perimeter of the pipe cross-section

No uniform effect of elastic soil pressure on the pipe cross-section in the first approximation is described by law of (Fig. 2):

$$q_{eqv} = kRw(1 - \alpha \cos \theta) \qquad (1)$$

where q_{eqv} – load due to the action of soil pressure on the external pipe wall; k – stiffness factor of elastic medium, [N/m³], α – factor dependent on soil characteristics determined experimentally for each specific case.

To solve the problem on natural vibrations of semi-underground oil pipeline, the following is required:

– Obtaining the equation of motion of the shell middle surface based on the geometrically nonlinear variant of the semi-membrane theory of cylindrical shells with

a mid-bend taking into account all components of inertia forces, parameter of longitudinal compression force, internal working pressure, geometric characteristics, fluid flow rate and influence of no uniform effect of the elastic soil pressure.
- To take into account the influence of no uniform elastic soil pressure on the outside oil pipeline perimeter in the differential equations of middle surface element motion of thin cylindrical shell;
- To solve the obtained differential equations taking into account assumptions of semi-membrane theory of mid-bend cylindrical shells;
- To obtain the solution to determine natural frequencies of semi-underground thin-wall oil pipeline of big diameter for the case of hinge support of shell ends.
- To analyze the obtained solutions and to propose the recommendations for practical use of the obtained method for determination of oil pipeline basic frequencies.

3   Description of Studies

Based on the geometrically nonlinear variant of the semi-membrane theory of cylindrical shells Ilyin [1]; being deformed, the equation of shell movement under forces is as follows.

$$
\begin{aligned}
&\frac{\partial^2 T_1}{\partial \xi^2} + \frac{\partial}{\partial \xi}\left(\tau \frac{\partial M_2}{\partial \theta}\right) - \frac{\partial^2}{\partial \theta^2}\left(\frac{R_2^*}{R_1^*} T_1\right) - \frac{1}{R^2}\frac{\partial^2}{\partial \theta^2}\left(R_2^* \frac{\partial^2 M_2}{\partial \theta^2}\right) - \\
&- \frac{\partial}{\partial \theta}\left(\frac{1}{R_2^*}\frac{\partial M_2}{\partial \theta}\right) + R\frac{\partial X_1}{\partial \xi} - R\frac{\partial X_2}{\partial \theta} - \frac{\partial^2}{\partial \theta^2}\left(R_2^* X_3\right) = 0,
\end{aligned}
\tag{2}
$$

where:

- $X_1 = -Rh\rho\frac{\partial^2 u}{\partial t^2}$,
- $X_2 = -Rh\rho\frac{\partial^2 v}{\partial t^2}$,
- $X_3 = -Rh\rho\frac{\partial^2 w}{\partial t^2} - \rho_0\Phi_{mn}(R^2\frac{\partial^2 w}{\partial t^2} + V^2\frac{\partial^2 w}{\partial \xi^2}) + p_0 - kRw(1 - \alpha\cos\theta)$,
- $\xi = \frac{x}{R}$ – non-dimensional cylindrical shell coordinate in longitudinal direction;
- ρ- shell material density;
- ρ_0- density of fluid flowing through the pipe;
- R_1^* and R_2^* – deformed shell curvature radius (index 1 is the longitudinal direction, index 2 – circumferential);
- Φ_{mn}- parameter of the modified Bessel function of the first genus;
- V – fluid flow rate:

$$
\frac{1}{R_2^*} = \frac{1}{R}\left(1 - \frac{\partial \vartheta_2}{\partial \theta}\right); \quad \frac{1}{R_2^*} = \frac{1}{R}\left(1 - \frac{\partial \vartheta_2}{\partial \theta}\right);
\tag{3}
$$

- ϑ_2- deflection angle of the tangent to the middle line of the cross-section;
- τ- torsion of the middle surface; T_1- standard longitudinal force; M_2- bending moment.

To pass on to the solution of the problem in displacements, we have the basic relations of the semi-membrane theory of shells:

$$T_1 = Eh\varepsilon_1, M_2 = D\chi_2, \tag{4}$$

Where: E - shell material elastic module; ε_1 – longitudinal deformation presented as two parts:

$$\varepsilon_1 = \varepsilon_0 + \frac{\partial u}{\partial \xi}, \varepsilon_0 = \frac{F}{EA}, \tag{5}$$

Initial deformation ε_0 is determined in the assumption of shell cross-section no deformity, A – cross-section area.

Ratios between deformations and displacements of semi-membrane theory are as follows:

$$\frac{\partial v}{\partial \theta} + w = 0; \frac{\partial v}{\partial \xi} + \frac{\partial u}{\partial \theta} = 0; \vartheta_2 = \frac{\partial w}{\partial \theta} - v; \tau = -\frac{1}{R}\frac{\partial \vartheta_2}{\partial \xi}, \chi_2 = -\frac{1}{R}\frac{\partial \vartheta_2}{\partial \theta}; \tag{6}$$

Substitute ratios (3)–(6) into the Eq. (2), after some conversions and omission of nonlinear terms, we obtain the linearized equation of shell movement in displacements:

$$\frac{\partial^3 u}{\partial \xi^3} + h_v^2 \frac{\partial^3}{\partial \theta^3}(\vartheta_2 + \frac{\partial^2 \vartheta_2}{\partial \theta^2}) + 2\frac{\partial^2}{\partial \theta^2}(\frac{\partial^2 w}{\partial \xi^2}\varepsilon_0) - \frac{R}{Eh}p_0\frac{\partial^3 \vartheta_2}{\partial \theta^4} + \frac{R^2 k}{Eh}\frac{\partial^2 w}{\partial \theta^2} - \frac{R^2 k\alpha}{Eh}\frac{\partial^2 w}{\partial \theta^2}\cos\theta + \frac{2kR^2}{Eh}\frac{\partial w}{\partial \theta}\sin\theta +$$
$$+ \frac{kR^2\alpha w \cos\theta}{Eh} + \rho_0\Phi_{mn}\frac{R}{Eh}(R^2\frac{\partial^2 w}{\partial t^2} + V^2\frac{\partial^2 w}{\partial \xi^2}) - \frac{R^2 \rho}{E}(\frac{\partial^3 u}{\partial \xi \partial t^2} - \frac{\partial^2 v}{\partial \theta \partial t^2} - \frac{\partial^4 w}{\partial \theta^2 \partial t^2}) = 0, \tag{7}$$

Where: u, v, w – components of shell middle surface displacement assigned to radius R; h_v- parameter of relative shell thickness:

$$h_v = \frac{h}{R\sqrt{12(1 - v^2)}} \tag{8}$$

Where: n – Poisson's ratio.

The obtained equation system (7) contains four unknown coordinate functions and time t: u, v, w and ϑ_2. We add the Eq. (7) with assumptions of semi-membrane theory of shells (5)–(6) and decide this system by the method of variables separation, Fourier method. We represent the function $w(\xi, \theta, t)$ for the case of hinged support of the ends of the shell, which satisfies the periodicity condition along the circumferential coordinate θ as follows:

$$w = \sum_m \sum_n f(t) \sin(\tilde{\lambda}_n \xi) \cos(m\theta), \tag{9}$$

Where: $\tilde{\lambda}_n = \frac{n\pi R}{L}$; $m, n = 1, 2\ldots$- wave numbers in the circumferential and longitudinal directions.

Other components of displacements and a deflection angle of the tangent are determined from the ratios of semi-membrane theory of shells (3)–(6):

$$u = -\sum_m \sum_n \frac{\tilde{\lambda}_n}{m^2} f(t) \cos(\lambda_n \xi) \cos(m\theta),$$
$$v = -\sum_m \sum_n \frac{1}{m} f(t) \sin(\tilde{\lambda}_n \xi) \sin(m\theta), \qquad (10)$$
$$\vartheta_2 = -\sum_m \sum_n \frac{m^2-1}{m} f(t) \sin(\tilde{\lambda}_n \xi) \sin(m\theta).$$

Assuming that the shell natural vibrations are harmonic, we have the following:

$$f(t) = \sin \omega_{mn} t, f''(t) = -\omega_{mn}^2 \sin \omega_{mn} t, \qquad (11)$$

Where: $\omega_{mn}-$ is cyclic natural bending vibrations of the shell in forms $m, n = 1, 2, 3....$

Substituting (9)–(10) into the Eq. (7) and making equal the factors at the same trigonometric functions $\cos m\theta$ at $m = 1, 2, 3...$, we obtain a continuous system of homogeneous linear algebraic equations consisting of factors where b_m for the radial displacement component w and factors a_{ij}:

$$\text{at } m= 1 a_{1,1} b_1 + a_{1,2} b_2 = 0$$
$$\text{at } m= 2 a_{2,1} b_1 + a_{2,2} b_2 + a_{2,3} b_3 = 0, \qquad (12)$$
$$\text{at } m= 3 a_{3,2} b_2 + a_{3,3} b_3 + a_{3,4} b_4 = 0$$

The obtained system of homogeneous linear algebraic equations can be written in the compact form:

$$a_{m,m-1} b_{m-1} + a_{m,m} b_m + a_{m,m+1} b_{m+1} = 0, \qquad (13)$$

Where: $m = 1, 2, 3... \ m - 1 > 0$, and factor a_{ij} is determined using the following formulas:

$$a_{m,m} = A_{mn} - B_{mn} \omega_{mn}^2;$$

$$a_{m,m\pm 1} = -\frac{m^2 (m \pm 1)^2}{2} k^* \alpha;$$

$$A_{mn} = \lambda_n^4 + m^4 (m^2 - 1)(m^2 - 1 + p^*) + k^* m^4 - \lambda_n^4 m^4 P/n^2 - \lambda_n^2 \rho_0^* \Phi_{mn} V^2 m^4 h_v; \quad (14)$$

$$B_{mn} = \rho^* R \cdot h(\lambda_n^2 h_v + m^2 + m^4) + \rho_0^* \Phi_{mn} R^2 m^4,$$

Where:

- $p^* = p_0 \frac{R}{Ehh_v^2},$
- $\rho^* = \rho \frac{R}{Ehh_v^2},$
- $\rho_0^* = \rho_0 \frac{R}{Ehh_v^2},$
- $k^* = \frac{R^2 k}{Ehh_v^2}$

Factors of this equations system (8) are non-dimensional at internal working pressure p_0 in MPa, stiffness factor of elastic medium k in N/m^3 and shell material density ρ_0 in $\frac{N}{m^4}c^2$.

For the detailed analysis of the obtained system of homogeneous linear algebraic Eqs. (8), it can be represented in the matrix form:

$$
\begin{pmatrix}
a_{1,1} & a_{1,2} & 0 & 0 & \cdots \\
a_{2,1} & a_{2,2} & a_{2,3} & 0 & \cdots \\
0 & a_{3,2} & a_{3,3} & a_{3,4} & \cdots \\
\cdots & \cdots & \cdots & \cdots & \cdots \\
\cdots & \cdots & \cdots & \cdots & \cdots
\end{pmatrix}
\begin{pmatrix}
b_1 \\
b_1 \\
b_1 \\
\cdots
\end{pmatrix}
= 0. \tag{15}
$$

Since b_m differs from zero $b_{mn} \neq 0$, to determine the frequency spectrum, it is required to equal the system determinant (15) to zero that consists of the system factors $a_{m,m}$, $a_{m,m\pm1}$:

$$
\begin{vmatrix}
a_{1,1} & a_{1,2} & 0 & 0 & \cdots \\
a_{2,1} & a_{2,2} & a_{2,3} & 0 & \cdots \\
0 & a_{3,2} & a_{3,3} & a_{3,4} & \cdots \\
\cdots & \cdots & \cdots & \cdots & \cdots \\
\cdots & \cdots & \cdots & \cdots & \cdots
\end{vmatrix}
= 0. \tag{16}
$$

Therefore, the assigned task for determination of natural frequencies of the specified straight oil pipeline by shell vibration mode reduces to the task for own values of the matrix of homogeneous linear algebraic equation system factors (15).

Let's consider the average system of the homogeneous linear algebraic equation system obtained from (15) at m = 1, 2, 3. In the matrix form this system is as follows:

$$
AB = \begin{pmatrix}
a_{1,1} & a_{1,1} & a_{1,1} \\
a_{1,1} & a_{1,1} & a_{1,1} \\
a_{1,1} & a_{1,1} & a_{1,1}
\end{pmatrix}
\times
\begin{pmatrix}
b_1 \\
b_2 \\
b_3
\end{pmatrix}
= 0. \tag{17}
$$

Reduction of continuous system of homogeneous linear algebraic Eq. (15) will not significantly effect on the accuracy of problem solution as this system is regular. The study performed according to the method Berezin, I. S. (1962) showed that the modules sum of minor terms factors for each line of matrix A divided by a factors module at the main diagonal term is less than 1 at any parameter.

All elements of matrix B differ from zero $(b_m \neq 0)$, therefore, factor matrix determinant A shall be 0:

$$
|A| = \begin{vmatrix}
a_{1,1} & a_{1,1} & a_{1,1} \\
a_{1,1} & a_{1,1} & a_{1,1} \\
a_{1,1} & a_{1,1} & a_{1,1}
\end{vmatrix}
= 0. \tag{18}
$$

The set problem for determination of natural frequencies of the specified pipeline section reduces to the problem of matrix A free values. The determinant (18) is reduced to the form of the characteristic equation of matrix A. For this purpose, this determinant is written by substituting the values of the diagonal elements $a_{m,m}$ by formulas (14):

$$|A| = \begin{vmatrix} A_{1n} - B_{1n}\omega_{1n}^2 & a_{1,2} & a_{1,3} \\ a_{2,1} & A_{2n} - B_{2n}\omega_{2n}^2 & a_{2,3} \\ a_{3,1} & a_{3,2} & A_{3n} - B_{3n}\omega_{3n}^2 \end{vmatrix} = 0.$$

Each line of this determinant is multiplied by $\frac{1}{B_{mn}}$. As a result, under the determinant property, the following can be obtained:

$$\frac{1}{B_{1n}B_{2n}B_{3n}} \times \begin{vmatrix} A_{1n} - B_{1n}\omega_{1n}^2 & a_{1,2} & a_{1,3} \\ a_{2,1} & A_{2n} - B_{2n}\omega_{2n}^2 & a_{2,3} \\ a_{3,1} & a_{3,2} & A_{3n} - B_{3n}\omega_{3n}^2 \end{vmatrix} = 0.$$

Since determinant $\frac{1}{B_{1n}B_{2n}B_{3n}}$ differs from zero, the determinant being the characteristic equation of matrix A is 0:

$$|A - \lambda E| = \begin{vmatrix} d_{1,1} - \lambda & d_{1,2} & d_{1,3} \\ d_{2,1} & d_{2,2} - \lambda & d_{2,3} \\ d_{3,1} & d_{3,2} & d_{3,3} - \lambda \end{vmatrix} = 0, \tag{19}$$

where the following designations are used:

$$\lambda = \omega^2, \qquad d_{m,m} = \frac{A_{mn}}{B_{mn}},$$
$$d_{m,m\pm1} = \frac{a_{m,m\pm1}}{B_{mn}} \tag{20}$$

Opening the determinant of the form (19), we obtain the characteristic equation of degree equal to matrix A order which roots λ_1 called eigen values determine the squares of the natural frequencies of the considered pipeline section.

At the left, the characteristic equation of matrix A of order k contains a k-power polynomial of λ with higher coefficient equal to one, that is, the characteristic equation degree is determined by the matrix A order.

If the first order matrix is taken at the wave number m = 1, then the linear characteristic equation in λ is obtained from (19):

$$d_{1,1} - \lambda = 0. \tag{21}$$

Substituting the corresponding values as per (20), the equality is obtained:

$$\omega_{1n}^2 = \frac{A_{1n}}{B_{1n}},$$

from which at the factors value as per (14) the formula for the square of the natural frequency of the straight oil pipeline section is obtained in case of the semi-underground laying with a non-deformable cross section contour (at m = 1):

$$\omega_{mn}^2 = \frac{\lambda_n^4 + k^* - \lambda_n^4 P/n^2 - \tilde{\lambda}_n^2 \rho_0^* \Phi_{1n} V^2 h_v}{2\rho^* R \cdot h(\lambda_n^2 h_{\tilde{v}}) + \rho_0^* \Phi_{1n} R^2}. \tag{22}$$

Considering the second order matrix at wave numbers m = 1 and 2, we have from (19):

$$\begin{vmatrix} d_{1,1} - \lambda & d_{1,2} \\ d_{2,1} & d_{2,2} - \lambda \end{vmatrix} = 0, \tag{23}$$

from which the square characteristic equation is obtained:

$$\lambda^2 - (d_{1,1} + d_{2,2})\lambda + (d_{1,1}d_{2,2} - d_{1,2}d_{2,1}) = 0, \tag{24}$$

its solution provides two roots λ_1 and λ_2 which determine the squares of cyclic natural bending frequencies of the oil pipeline section ω_{1n}^2 and ω_{2n}^2 as per the corresponding forms of vibrations at m = 1 and 2.

For the third order matrix obtained at wave numbers m = 1, 2, 3, the determinant (19) can be obtained by opening of which the cubic characteristic equation of matrix A can be obtained:

$$\lambda^3 - I_1\lambda^2 + I_2\lambda - I_3 = 0, \tag{25}$$

fixed factors of which are as follows:

$$
\begin{aligned}
I_1 &= d_{1,1} + d_{2,2} + d_{3,3}, \\
I_2 &= d_{1,1}d_{2,2} + d_{2,2}d_{3,3} + d_{3,3}d_{1,1} - d_{1,2}d_{2,1} - d_{2,3}d_{3,2} - d_{3,1}d_{1,3}, \\
I_1 &= d_{1,1}d_{2,2}d_{3,3} + d_{1,2}d_{2,3}d_{3,1} + d_{2,1}d_{3,2}d_{1,3} - d_{1,1}d_{2,3}d_{3,2} - d_{2,2}d_{3,1}d_{1,3} - d_{3,3}d_{2,1}d_{1,2}
\end{aligned}
\tag{26}
$$

When solving the cubic Eq. (30), there are three real roots that is three eigen values $\lambda_1, \lambda_2, \lambda_3$ of matrix A which are the squares of cyclic bending frequencies ω_{mn}^2 of the specified straight pipeline section with flowing fluid as per vibration modes at m = 1, 2, 3. The natural frequencies root of the straight oil pipeline ω_{mn}^2 is determined using the following formula:

$$\omega_{mn}^2 = \frac{\lambda_n^4 + m^4(m^2 - 1)(m^2 - 1 + p^*) + k^*m^4 - \lambda_n^4 m^4 P - \lambda_n^2 \rho_0^* \Phi_{mn} V^2 m^4 h_v}{\rho^* R \cdot h(\lambda_n^2 h_v + m^2 + m^4) + \rho_0^* \Phi_{mn} R^2 m^4}, \tag{27}$$

Where: $\lambda_n = \frac{n\pi R}{L\sqrt{h_v}}$, $p^* = p_0 \frac{R}{Ehh_v^2}$, $\rho^* = \rho \frac{R}{Ehh_v^2}$, $\rho_0^* = \rho_0 \frac{R}{Ehh_v^2}$, $k^* = \frac{R^2 k}{Ehh_v^2}$.

The Eq. (32) is obtained for the oil pipeline section with the hinged end sections.

To determine values of frequencies ω_{mn}, it is also required to determine a Bessel function parameter Φ_{mn} and its derivative using the following formula (28). To determine values:

a. modified Bessel functions of the first genus of m order from action argument $\lambda_n = \frac{n\pi R}{L}$:

$$I_m(\lambda_n) = \left(\frac{\lambda_n}{2}\right)^m \sum_{k=0}^{\infty} \frac{\left(\frac{\lambda_n}{2}\right)^{2z}}{z!G(m+z+1)}, \tag{28}$$

where m – order of Bessel function (wave number) determined by the form shell vibrations ($m = 1, 2, 3...$);

b. gamma-function $G(m + z+1)$ which special values are determined by equation:

$$G(n) = (n-1)! = 1 \cdot 2...(n-1), \quad N = 3, 4, 5... \tag{29}$$

Where: $G(1) = 1$, $G(2) = 1$, $G(3) = 2! = 2$ etc.;

c. Bessel derived functions calculated using the formula

$$I'_m(\lambda_n) = \frac{1}{2}(I_{m-1}(\lambda_n) + I_{m+1}(\lambda_n)). \tag{30}$$

In Fig. 4, using the formulas (28)–(30), diagrams of functions $\Phi_{mn}(\tilde{\lambda}_n) = \frac{I_m(\tilde{\lambda}_n)}{\lambda_n I'_m(\lambda_n)}$ are shown, which are plotted at wave numbers $m = 1, 2, 3$, and within the range λ_n of 0 to 1.0, this method ensures determining the natural frequencies of a surface oil pipeline section using the following formula (27).

4 Results and Discussion

The following initial data were taken to study the solutions obtained: a pipeline with a ratio of wall thickness to radius (1/50), (1/40), (1/30), internal working pressure 5 MPa, length parameter L/R = 10, soil coefficient varies in range k by 1 to 50 MN/m³.

Analysis of the graph in Fig. 1 showed that a change in the soil bed coefficient significantly affects the natural oscillation frequencies of the pipeline increasing them. This effect is more significant on pipelines with the ratio h/r = 1/50, for example, when the bed coefficient k varies from 0.1 to 5 kg/cm3, the increase in frequencies ω21 is 73% for gas pipelines laid by a semi-subterranean method and 29% with the ground method. At a ratio of h/r = 1/30, the frequency increase was 53% for the semi-subterranean method and 20% for the terrestrial method.

The difference in the frequency growth tendencies is due to the higher rigidity of the pipeline at h/r = 1/30, due to the greater wall thickness.

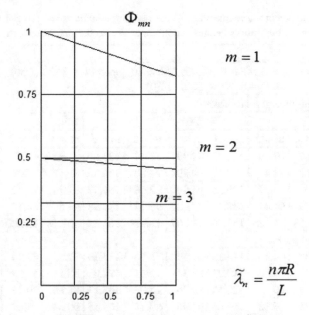

Fig. 4. Diagrams of the functions $\Phi_{mn}(\tilde{\lambda}_n)$.

Differences in the frequency growth trends for different ways of laying are caused by the greater area of contact of the pipeline with the ground with a semi-underground laying (Fig. 5 and Table 1).

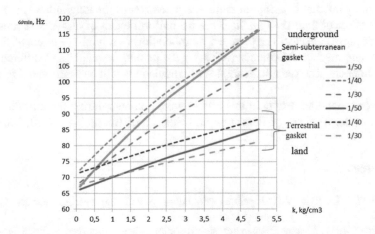

Fig. 5. Dependence of the frequency ω_{21} of the sections of a semi-underground gas pipeline with a length L = 10R from the change in the soil bed coefficient for pipes with different fineness coefficients h/r.

Table 1. Dependence of the frequencies of free bending vibrations ω_{21} (Hz) of gas pipelines on the soil coefficient k, the thin-walledness coefficient h/r and the parameter of length.

k	H/r								
	n = 1; p_0 = 5 MPa; P = 0.05; L/R = 10, pipes 14 × 1420 (1/50), 15 × 1220 (1/40), 20 × 1220 (1/30)								
	Semi-underground gasket								
	1/50			1/40			1/30		
	m = 1	m = 2	m = 3	m = 1	m = 2	m = 3	m = 1	m = 2	m = 3
0.1	75.82	67.19	118.73	87.99	72.41	133.60	87.85	68.43	137.58
0.5	78.67	72.45	122.17	90.29	77.00	136.46	89.58	72.09	139.68
1	82.09	78.54	126.35	93.09	82.38	139.97	91.70	76.42	142.25
2.5	91.59	94.47	138.11	101.02	96.74	149.98	97.80	88.14	149.71
5	105.53	116.25	155.76	113.07	116.81	165.33	107.18	104.80	161.38
	Land gasket								
0.1	75.33	66.26	118.13	87.59	71.60	133.10	87.55	67.79	137.22
0.5	76.25	68.00	119.25	88.34	73.11	134.03	88.11	68.98	137.89
1	77.39	70.12	120.63	89.26	74.95	135.17	88.80	70.45	138.73
2.5	80.71	76.12	124.66	91.96	80.23	138.55	90.84	74.68	141.21
5	85.97	85.18	131.12	96.30	88.33	144.00	94.16	81.25	145.24

5   Conclusion

Based on the geometrically nonlinear version of the semi-membrane theory of cylindrical shells, a method was obtained to determine the natural frequencies of the semi-underground main straight thin-wall oil pipeline of big diameter subjected to the action of a stationary internal working pressure, the parameter of the longitudinal compression force, the effect of fluid flow rate and the no uniform effect of elastic soil pressure on the outside pipe perimeter. The obtained method allows studying the effect of the above geometric and mechanical characteristics for the natural frequencies at different values of wave numbers. The specific examples of calculations will be represented in the next works.

The corresponding results in this article were obtained during the execution of the state task of the Ministry of Education and Science of Russia 7.4794.2017/8.9.

References

1. Akselrad, E.L., Ilyin, V.P.: Pipelines calculation, p. 240. Engineering Industry, Leningrad (1972)
2. Bereznev, A.V.: Natural frequencies and vibration modes of curved sections of steel and polyethylene pipelines with flowing fluid. Civ. Eng. Bull. **3**(4), 20–25 (2005)
3. Djondjorov, P., Vassilev, V., Dzhupanov, V.: Dynamic stability of fluid conveying cantilevered pipes on elastic foundations. J. Sound Vib. **247**(3), 537–546 (2001)

4. Efimov, A.A.: Natural vibrations of sea deep water oil pipeline of big diameter. Civ. Eng. Bull. **4**(17), 26–29 (2008)
5. Lilkova-Markova, S.V.: Vibration of a pipe on elastic foundation. Sadhana **29**(3), 259–262 (2004)
6. Razov, I.O.: Stresses and displacements on contact surface of ground pipeline of bid diameter. Civ. Eng. Bull. **3**(50), 105–108 (2015)
7. Razov, I.O.: Natural vibrations of ground gas pipelines compressed by longitudinal force taking into account elastic soil bed. Civ. Eng. Bull. **1**(36), 29–32 (2013)
8. Razov, I.O.: Study of natural vibrations of ground thin-wall gas pipelines of big diameter. Civ. Eng. Bull. **4**, 100–104 (2013)
9. Sokolov, V.G.: Natural vibrations and static stability of oil pipeline of big diameter taking into account fluid flow, longitudinal compression force and elastic bed. Civ. Eng. Bull. **1**, 49–53 (2014)
10. Sokolov, V.G.: Vibrations, static and dynamic stability of pipelines of big diameter: dis….D. Eng. 23 May 2017, Saint-Petersburg, p. 314 (2011)

Power Contour Diagram of the Device for Supplying Auxiliary Loads Onboard DC Locomotive and Methods of Forming Its Output Voltage

Mikhail Pustovctov[(⊠)] [iD]

Don State Technical University, sq. Gagarin, 1, 344010 Rostov-on-Don, Russia
mgsn2006@rambler.ru

Abstract. Currently, onboard the mainline electric locomotives, three-phase and single-phase alternating current consumers are quite common. One of the tasks solved under the development of devices for auxiliary power supply of a locomotive is selecting a method to convert high voltage of the overhead system into low three-phase alternating voltage. A brief analysis of the semiconductor converter circuits used to supply auxiliary loads onboard the DC locomotives at nominal contact voltage of 3 kV is given. A power circuit diagram of the supply device for auxiliary circuits of a DC locomotive, different from the widespread ones, is proposed. Computer-aided simulation of the device characteristics in a static mode is carried out. The device contains a capacitive DC voltage divider, an autonomous voltage-source inverter, a three-phase transformer with a specific III/Y winding arrangement. A comparative review of the device output performance through different methods of forming a regulated voltage at the inverter output using pulse-width modulation is undertaken. A vertical method of forming the output voltage with a triangular bidirectional reference signal under modulation voltage forms of "meander", "meander with pause" (at the modulating pulse length of 120 electrical degrees per half-period), "sine with overmodulation" and "sine with overmodulation and the third-harmonic premodulation" (the introduced third harmonic is in phase with the first one) is studied. The use of overhead system voltage, efficiency, harmonic composition of currents and voltages, values of total harmonic distortion, power factor were considered. The options for the formation of the output voltage using a sinusoidal modulating signal are recognizes the best. A cost reduction of a set of the converter power switches is expected to be 1.5–2.0 times compared with the known technical solutions in the considered area due to the use of transistors of a lower voltage class.

Keywords: Autonomous voltage-source inverter · Three-phase transformer · DC locomotive · Auxiliary circuit supply · Pulse width modulation · Harmonic composition

© Springer Nature Switzerland AG 2020
V. Murgul and M. Pasetti (Eds.): EMMFT 2018, AISC 982, pp. 628–636, 2020.
https://doi.org/10.1007/978-3-030-19756-8_59

1   Introduction

Currently, onboard the mainline electric locomotives, three-phase and single-phase alternating current consumers are quite common. One of the tasks solved under the development of devices for auxiliary power supply of a locomotive is selecting a method to convert high voltage of the overhead system into low three-phase alternating voltage. Some circuit solutions for DC locomotives based on the two-level and three-level full-bridge autonomous voltage-source inverters (AVI), including devices using three-phase transformers (T), are published by domestic and foreign experts [1–3]. These solutions have a number of drawbacks. The application of a two-level AVI bridge circuit is an option with a minimum number of semiconductor switches (6 pcs) and the highest reliability, but requires, at the voltage of the overhead system of 3 kV, the use of transistors with the operating voltage of 6.5 kV, which are expensive. Features of the formation of the output voltage in a two-level three-phase bridge AVI for the case when the negative terminal of the input DC voltage source characteristic of the electrified rail transport is grounded, are considered in [4]: with an open upper phase transistor at the output of this phase of the AVI relative to the earth, a potential equal to the total constant voltage at the AVI input arises. The use of a three-level full-bridge AVI allows us to restrict ourselves to cheaper power transistors with the operating voltage of 3.3 kV. In this case, their number doubles, the scheme and control algorithms of the AVI get more complicated. This work objective is to propose an option of the power supply de-vice for auxiliary circuits of a DC electric locomotive with a three phase T and AVI, which allows using the minimum number of relatively low-voltage power transistors with a simple connection diagram.

2   Materials and Methods

Due to its compactability and cheapness [5], it is advisable to use not a three-phase group of single-phase T, but T with a single magnetic core, for example, a three-rod one.

The proposed AVI scheme assumes an independent (without electrical connections with each other, according to an open scheme) connection of the primary T winding phases (Fig. 1).

The device in Fig. 1 converts a DC contact voltage into a three-phase AC voltage, the frequency and magnitude of which can be regulated.

Each phase of the primary winding of a three-phase T is connected to a converter cell, which is a single-phase half-bridge AVI. For example, A phase is connected to the cell that includes VT1 and VT2 transistor switches, VD1 and VD2 diodes, C1 and C2 capacitors. VT1 and VT2 transistors open alternately for equal periods of time forming an alternating voltage on A phase of the T primary winding.

The series-connected C7, C8 and C9 capacitors are a capacitive divider sharing the input DC voltage of overhead system into three equal parts (according to the number of T phases and the converter cells). Pairs of transistors in the other two phases operate in the same way as in A phase, but with a time shift of 120° el. and 240° el., forming a three-phase symmetric system of supply voltages on the T windings.

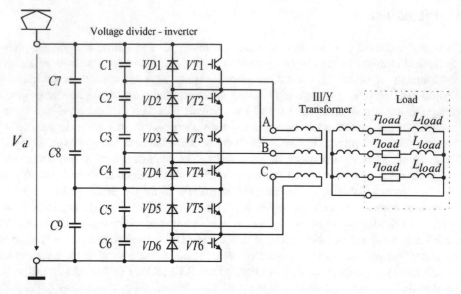

Fig. 1. Basic circuit diagram of the device for supplying auxiliary circuits of DC locomotive.

2.1 Output Voltage of the Inverter Methodology

Let us compare options for the AVI key management using rectangular and sinusoidal methods (terminology according to [6]) of the pulse-width modulation (PWM) under the condition when maximum possible output voltage of the AVI is obtained.

Figure 2 shows methods for the implementation of the PWM phase voltage: (a) rectangular PWM (method 1), (b) rectangular PWM at the waveform of the modulating voltage of "meander with pause" with the pulse width of 120° el. (method 2), (c) sinusoidal PWM (methods 3 and 4). As method 4, we use a sinusoidal PWM with the third harmonic premodulation that has 0.167 amplitude of the first harmonic and overmodulation of the first harmonic by $\pi/2$ factor [7]. The introduced third harmonic is in phase with the first one.

When analysing the operating modes of the power supply unit for the auxiliary circuits of an electric locomotive, we use the mathematical model of a three phase T according to [8, 9]. The computer simulation results through PSpice [10–14] for comparing the control methods of the converter are summarized in Table 1.

The following assumptions were accepted in the simulation: nonlinearity of the T magnetization curve is not considered; idealized voltage-controlled switches are used instead of transistors. In all cases, the T parameters and loads are the same. The load is symmetrical with $\cos\phi = 0.88$. The modulating voltage frequency is 50 Hz, $V_d = 3300$ V. All values in Table 1 are presented in relative units. The instantaneous value of voltage on one transistor did not exceed 1220 V.

For any of the considered methods, the harmonics are small with the orders multiple of three, in the phase and linear voltages of the secondary T winding.

This has a positive effect on the harmonic composition of the secondary current. The potential of the neutral point of the load (Fig. 1) is close to zero.

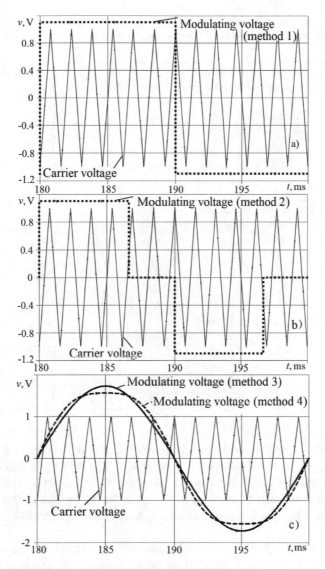

Fig. 2. Methods for implementation of PWM of phase voltage.

From the point of view of the harmonic composition of currents and voltages generated by the converter, it is preferable that an odd number of periods of the carrier frequency voltage fits within the period of the modulating voltage. Otherwise, currents and voltages will contain even harmonics (according to Table 1, this has little effect on the energy characteristics of the electrical system). Method 2 enables to drastically reduce the current of the 3rd harmonic in the primary T winding (up to 11.09% of the 1st harmonic). With method 4: the 3rd current harmonic in the primary T winding is 150.62% of the 1st, which is lower than under method 1, where it is 212.76%.

Table 1. Comparative results of calculating characteristics of device for supplying three-phase and single-phase auxiliary circuits of DC locomotive through various methods of generating AVI output voltage.

Name of characteristics	Phase voltage of 1st harmonic of primary T winding	Active power at T output	Power factor at T input	T efficiency	$\eta_T \cdot \cos\phi_1$	-
Conventions	$\overset{*}{V}_{1ph1}$	$\overset{*}{P}_2$	$\cos\phi_1$	η_T	K_E	$K_E \cdot \overset{*}{V}_{1ph1}$
At carrier voltage frequency of 650 Hz (13 periods of PWM carrier voltage on one modulating period)						
Method 1	1.000 (520.9 V)	1.000 (73.2 kW)	0.845	0.917	0.75	0.775
Method 2	0.859	0.759	0.829	0.970	0.804	0.691
Method 3	0.946	0.877	0.828	0.962	0.797	0.754
Method 4	0.969	0.925	0.833	0.939	0.782	0.758
At carrier voltage frequency of 600 Hz (12 periods of PWM carrier voltage on one modulating period)						
Method 1	1.000 (518.7 V)	1.000 (73 kW)	0.847	0.910	0.771	0.771
Method 2	0.873	0.761	0.834	0.966	0.806	0.704
Method 4	0.966	0.925	0.825	0.961	0.793	0.766

The calculated voltage and current curves are shown in Fig. 3: (a) method 1, (b) method 2, (c) method 4. Curves: 1 is voltage of V_{1ph} primary T winding; 2 is I_{1ph} current of the primary T winding; 3 is I_{2ph} current of the secondary T winding; 4 V_{2ph} is voltage of the secondary T winding; 5 V_{2ph-ph} is linear voltage of the secondary T winding inverted in sign. The curves according to method 3 are close to those shown in Fig. 3(c).

The harmonic composition of the obtained voltage and current curves is given in Table 2, where the values of the total harmonic components THD, % are also presented. The 1st harmonic frequency is 50 Hz. The PWM carrier voltage frequency is 650 Hz.

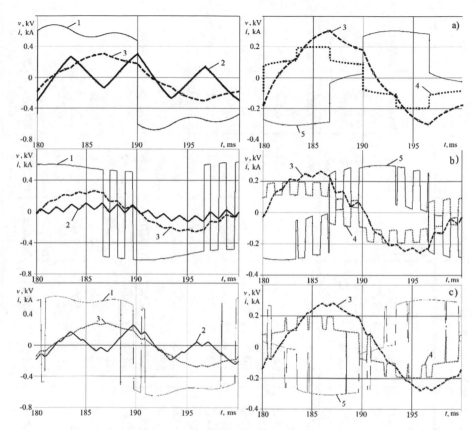

Fig. 3. T voltages and currents under different methods for implementation of PWM of AVI phase voltage.

3   Conclusion

Summing up, it can be stated that the proposed power supply scheme for auxiliary circuits of a DC electric locomotive is based on a three-phase half-bridge AVI and a three-phase T with a III/Y winding circuit, which has the following advantages: the number of power transistors is relatively low (it is acceptable to use transistors with the operating voltage of 2.5 kV, considering the possibility of $V_d = 4$ kV in the overhead system), which reduces the cost of the device.

The cost of a set of semiconductor switches of HVIGBT type can be reduced by 1.5–2.0 times compared with the known technical solutions. The AVI switches management - sinusoidal PWM - is recommended as the best method by energy characteristics, the input voltage use, and aspects of providing electromagnetic compatibility. Methods of forming the output voltage of the proposed device do not differ from those used in the well-known circuits, that is, the proven technologies for constructing a converter control system can be used.

Table 2. Estimated harmonic composition of voltages and currents under different methods of implementation of PWM of AVI phase voltage.

Method	$V_{1ph\,k}$, %		$I_{1ph\,k}$, %		$V_{2ph\,k}$, %		$I_{2ph\,k}$, %		$V_{2ph-ph\,k}$, %	
k harmonic order	1	2	1	2	1	2	1	2	1	2
	2	3	4	5	6	7	8	9	10	11
1	100.00	100.00	100.00	100.00	100.00	100.00	100.00	100.00	100.00	100.00
3	39.766	6.274	212.76	11.090	0.448	5.072	0.190	2.804	0.560	7.510
5	18.508	23.791	5.371	5.818	16.742	21.978	6.585	8.643	16.823	15.231
7	13.485	11.592	2.801	2.070	11.906	10.292	3.499	3.080	11.805	12.455
9	10.629	15.081	18.998	3.509	0.346	12.994	0.026	2.860	0.405	14.424
11	8.451	36.486	1.157	5.733	7.573	31.862	1.414	5.892	7.659	23.834
13	7.224	42.491	0.836	47.743	6.279	5.492	1.026	0.931	6.178	7.264
15	6.283	31.145	6.740	3.420	0.340	27.038	0.011	3.675	0.394	24.257
17	5.481	12.431	0.495	1.449	4.930	10.901	0.596	1.224	5.017	15.037
19	4.939	4.095	0.400	0.562	4.250	3.575	0.481	0.423	4.148	4.977
21	4.475	8.719	3.428	1.433	0.338	7.046	0.007	0.703	0.391	5.319
23	4.060	5.023	0.277	0.750	3.670	4.623	0.326	0.441	3.756	7.155
25	3.757	3.489	0.235	0.282	3.202	2.927	0.276	0.307	3.099	3.785
27	3.481	5.917	2.067	0.739	0.337	4.784	0.007	0.381	0.390	3.647
29	3.229	3.528	0.184	0.554	2.932	2.949	0.205	0.229	3.018	4.249
31	3.034	1.036	0.157	0.194	2.562	1.253	0.181	0.086	2.457	2.058
33	2.852	3.495	1.389	0.216	0.338	2.864	0.005	0.189	0.390	1.771
35	2.683	4.140	0.127	0.251	2.448	3.417	0.141	0.158	2.532	1.875
37	2.548	8.223	0.115	0.305	2.131	6.990	0.127	0.288	2.024	5.848
39	2.418	11.888	0.996	4.037	0.338	2.014	0.004	0.123	0.390	0.924
Effective value of 1st harmonic	V_{1ph1}, V 520.864	447.176	I_{1ph1}, A 66.013	213.853	V_{2ph1}, V 134.816	115.820	I_{2ph1}, A 201.675	173.191	$V_{2ph-ph1}$, V 234.969	208.547
THD, %	50.460	75.136	213.853	50.198	24.729	53.633	7.718	12.335	24.742	47.779

Method	V_1ph1, V (3)	V_1ph1, V (4)	I_1ph1, A (3)	I_1ph1, A (4)	V_2ph1, V (3)	V_2ph1, V (4)	I_2ph1, A (3)	I_2ph1, A (4)	V_2ph-ph1, V (3)	V_2ph-ph1, V (4)
Effective value of 1st harmonic	492.928	504.700	63.956	63.204	126.670	129.690	189.136	193.041	219.677	229.134
THD, %	54.417	50.756	113.094	153.446	35.452	30.576	5.696	4.982	37.423	29.709
k harmonic order	V_{1phk}, %		I_{1phk}, %		V_{2phk}, %		I_{2phk}, %		$V_{2ph-phk}$, %	
1	100.00	100.00	100.00	100.00	100.00	100.00	100.00	100.00	100.00	100.00
3	21.530	27.982	108.05	150.62	1.371	0.608	0.582	0.444	1.392	0.643
5	1.754	6.256	2.553	2.883	1.559	5.941	0.584	2.281	1.572	7.296
7	3.859	5.819	10.252	4.402	2.441	1.213	0.675	0.315	1.674	1.202
9	12.113	10.957	2.496	2.999	11.127	9.802	2.634	2.206	10.852	9.855
11	20.930	15.727	1.851	5.420	19.178	13.847	3.732	2.602	19.923	13.773
	2	3	4	5	6	7	8	9	10	11
13	24.814	18.032	29.989	22.664	0.592	0.402	0.076	0.061	1.496	0.332
15	20.995	16.740	2.602	1.433	18.137	15.225	2.522	2.106	18.785	13.491
17	12.892	13.739	1.876	4.330	11.457	10.832	1.418	1.258	11.034	11.046
19	8.431	11.138	3.465	6.770	5.415	6.481	0.581	0.691	5.516	6.263
21	11.086	9.884	0.663	1.412	10.008	8.368	1.013	0.830	10.253	4.920
23	10.490	7.986	6.166	5.123	1.613	1.400	0.133	0.168	1.547	1.732
25	4.662	4.274	0.896	0.743	3.195	4.105	0.265	0.336	3.802	2.906
27	3.591	2.509	0.532	1.155	3.070	0.870	0.262	0.065	4.077	1.036
29	9.751	5.937	4.898	2.268	2.121	3.952	0.114	0.233	2.759	4.367
31	11.313	7.690	0.668	0.351	10.023	7.248	0.628	0.420	13.390	9.749
33	8.175	7.083	2.588	1.348	5.329	5.263	0.316	0.230	5.713	5.140
35	2.601	5.699	0.533	1.730	1.787	1.228	0.065	0.040	2.315	0.717
37	2.855	5.721	0.479	1.025	1.779	3.133	0.136	0.145	2.701	1.250
39	5.360	6.226	1.472	1.503	1.710	4.679	0.108	0.202	0.709	4.246

References

1. Khomenko, B.I., Kolpakhchyan, G.I., Pekhotskiy, I.V.: Electr. Locomot. Build. **45**, 184–191 (2003)
2. Umezawa, K.: FUJI Electr. Rev. **58**(4), 175–181 (2012)
3. Macan, M., Bahun, I., Jakopovic, Z.: 17th International Conference on electrical drives and power electronics (EDPE 2011), The High Tatras, Slovakia, 28–30 September 2011, pp. 49–54. Stará Lesná (2011)
4. Pustovetov, M.Yu.: J. Transsib Railw. Stud. **4**(12), 116–122 (2012)
5. Voldek, A.I.: Electrical Machines, Leningrad, Energy, p. 840 (2005)
6. Burkov, A.T.: Electronic engineering and converters, Moscow, Transport, p. 464 (2006)
7. Kurochka, A.A., Kabanov, D.A., Lushnikova, L.D.: Vestnik VELNII **1**, 156–163 (2004)
8. Pustovetov, M.Yu.: Russian Electrical Engineering 86(2), 98–101 (2015)
9. Pustovetov, M.: Proceedings of International Forum, 23–25 November 2016, pp. 84–93. Publishing House of Kalashnikov ISTU, Izhevsk (2017)
10. Keown, J.: OrCAD PSpice and Circuit Analysis, 4th (edn.), 609 p. Prentice Hall, Upper Saddle River (2007)
11. Ben-Yaakov, S., Peretz, M.M.: IEEE Power Elec. Soc. Newsletter, Fourth Quat. (2006)
12. Ben-Yaakov, S., Fridman, I.: 23rd IEEE Israel Convention, Tel-Aviv, pp. 342–345 (2004)
13. Şchiop, A., Popescu, V.: Rev. Roum. Sci. Techn. Électrotechn. et Énerg. 52(1), 33–42. Bucarest (2007)
14. Rashid, M.H.: SPICE for power electronics and electric power, 3rd edn, p. 524. CRC Press, Boca Raton (2012)

Materials Systems and Structures

Shear Stiffness of the Steel Roof Panels

Olga Tusnina[(⊠)]

Moscow State University of Civil Engineering,
Yaroslavskoe Shosse 26, 129337 Moscow, Russia
TusninaOA@gic.mgsu.ru

Abstract. In the paper, the results of the analysis of shear stiffness of steel roof panel are represented. Steel panels are used in the roofs of workshops with increased heat, gas and dust emissions. In the paper, there was analyzed the possibility of accounting for steel panels as fastening elements for the upper belt of the trusses, as well as the possibility of taking into account the roof panels in the spatial framework, as elements of the shear disk of the roofing through which efforts are transferred from one frame to another. The shear stiffness of the panel was determined using the FEMAP 11.1.2 software package, when the panel was modeled by solid finite elements. In the software complex Lira-SAPR 2017, the calculation was performed using beam (for beams and edges of the roof panel) and plate (for steel sheet) finite elements. A good agreement of the results obtained with the use of beam and plate finite elements with the results obtained with the use of solid finite elements was found. The optimal plate finite element mesh of the sheet was determined.

Keywords: Shear stiffness · Roof panel · Steel framework

1 Introduction

Steel industrial buildings are often equipped with heavy-duty cranes (groups A7, A8). So, the frame is subjected to significant both vertical and horizontal loads. For a competent assessment of the bearing capacity of structures in this case, it is necessary to analyze the spatial work of the framework [1–4].

The spatial work of the frame of an industrial building is determined by a number of the constructive factors (for example, the construction of walling and roof fencing structures, the schemes and stiffness of horizontal and vertical bracings, the stiffness of column, the height of buildings, the pith of frames, etc.).

Spatial analysis efficiency increases in the following cases:

- with an increase of the height of the building and the rigidity of the longitudinal bracings;
- with a decrease of column stiffness and frame pitch.

The shear stiffness of the roof affects the perception and distribution of horizontal loads between the frames of the building.

The type of roof enclosing structures (trapezoidal sheet, sandwich panels, precast concrete slabs, etc.) [5–9], the type of roof fencing fastening to the bearing structures

© Springer Nature Switzerland AG 2020
V. Murgul and M. Pasetti (Eds.): EMMFT 2018, AISC 982, pp. 639–647, 2020.
https://doi.org/10.1007/978-3-030-19756-8_60

influence on the shear stiffness of roof. Also, the roof fencing can influence on the bearing capacity of the bearing roof structures (beams, trusses). Enclosing roof structures most influencing on the behavior of cold-formed thin-walled purlins. In the roofs of industrial buildings with elevated heat, dust and gas emissions (workshops of ferrous metallurgy facilities - converter workshops, continuous casting of steel, mixing rooms, etc.), steel panels are used as enclosing roof structures. The use of such panels is caused by increased requirements for the durability of the roof, when the roof must be made of a smooth steel sheet.

There is a typical series of such panels, developed in 1977 by the Russian design institute TSNIIPROEKTSTALKONSTRUKTSIYA named after Melnikov.

Various structural solutions of the panels are designed for the perception of a total vertical load of 170, 320, 450 and 760 kg/m^2. The specified load includes panel dead load, the load from snow, taking into account the increase in the places of difference in heights, wind, dust and communications. It is concluded that taking into account in the spatial work of the framework the roof panels reduces the force in the cross-braces of horizontal braces along the upper truss belt by up to 50%. So, the shear stiffness of such a panel itself and the bearing capacity of the joints between panel and roof bearing structure should be analyzed.

2 Materials and Methods

In this paper, the roof panel Sch1-450 (designed on the perception of the distributed load 450 kg/m^2) with dimensions of 3 × 12 m, is considered. The frame of the panel is formed by the longitudinal beams made of rolled channels 40 and transverse ribs with a pitch of 2 m made from angels 100 × 63 × 7. Panel flooring is made of smooth steel sheet of 4 mm thick (Fig. 1). Material of construction of the panel is steel S255.

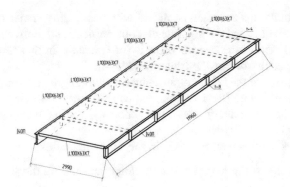

Fig. 1. Steel roof panel Sch1-450.

The 3 m wide panels are leaning on roof truss in its joints, the longitudinal beams of the panels are welded to the belt of the truss by assembly welding. Such panels can be considered as fastening elements of the upper belt of the truss as the elements providing effective length of the upper belt of the truss from its plane of 3 m.

Taking into account the assumption that the panels are taken into account when assigning the effective length of the truss belt from the plane, it is necessary to ensure the perception of fictitious force arising from the loss of stability of the belt, both by the panels themselves and by their fastenings.

Fictitious force in the loss of stability of the belt, is determined by the formula:

$$Q_{fic} = 7.15 \cdot 10^{-6}(2330 - E/R_y)N/\varphi \tag{1}$$

where N – axial force in the belt of the truss; φ – stability factor under central compression.

For example, we define a fictitious force for a truss belt made of paired angles L200 × 20 from steel S255, where a force of 1500 kN acts.

The slenderness of the upper belt from the plane of the truss, taking into account its effective length of 3 m: $\lambda = 300/8.928 = 33.6$.

The relative slenderness of the upper belt of the truss:

$\bar{\lambda} = \lambda\sqrt{R_y/E} = 33.6\sqrt{24/20600} = 1.15$.

For the type of cross-section c coefficient δ calculated (according to Russian standard SP 16.13330.2017 Steel structures):

$$\delta = 9.87(1 - \alpha + \beta\bar{\lambda}) + \bar{\lambda}^2 = 9.87 * (1 - 0.04 + 0.14 * 1.15) + 1.15^2 = 12.4$$

Stability factor:

$$\varphi = 0.5(\delta - \sqrt{\delta^2 - 39.4\bar{\lambda}^2})/\bar{\lambda}^2$$
$$= 0.5 \cdot (12.4 - \sqrt{12.4^2 - 39.4 \cdot 1.15^2})/1.15^2 = 0.876$$

A fictitious force: $Q_{fic} = 7.15 \cdot 10^{-6}(2330 - E/R_y)N/\varphi = 18.2\,\text{kN}$.

Welds should be calculated on the force Q_{fic} in fastening of the panel to the truss. Assembly weld, made by electrode E42 (R_{wf} = 18 kN/cm^2) by manual welding (β_f= 0.7, β_z = 1) with a effective length of 100 mm and a 6 mm leg, carries the most dangerous section in this case on the weld metal:

$$\beta_f k_f l_{wf} R_{wf} \gamma_c = 0.7 \cdot 0.6 \cdot 10 \cdot 18 \cdot 1 = 75.6\,\text{kN}$$

Taking into account that 4 panels lean on one truss node at once, and the load is perceived by four welds, one panel assembly joint takes up a load of 302 kN, which greatly exceeds the fictitious force arising from the loss of stability of the truss belt. Thus, the inclusion of panels as a detachable belt for trusses of elements is quite legitimate. In addition, taking into account the large bearing capacity of the panel fastening unit and sufficient internal shear stiffness of the panels, they are included in the spatial work of the framework and work as a horizontal shear disk in the level of the upper belts of trusses or beams, ensuring the redistribution of horizontal forces between the main supporting structures of the roof.

The presence and accounting of the roof panels allows to take into account the actual reduction of moments from the plane in the beams, taking into account their redistribution from one beam to the neighboring through the roof panels.

Thus, when performing a spatial analysis of the framework of industrial workshops, in the roofs of which steel panels are used, it is necessary to take into account the inclusion of panels in the work of the frame in order to obtain reliable results. Neglect of the roof panels when calculating entails unnecessarily high efforts in the beams, caused by the action of horizontal forces, in particular, bending moments from the plane of the beams.

3 Results

The shear stiffness of the panel is analyzed below.

First of all, it is necessary to determine the optimal finite element mesh of steel sheet. For this we consider a 4 mm thick plate with dimensions 12 × 3 m, fixed in the corners on one side and loaded with load along the free side. The total load value is 1000 kN. The analyses were done with the use of computing program Lira-SAPR 2017. Consider the following schemes with the number of finite elements along the width of the panel 1, 2, 4, 6, 8, 16 and 32 finite elements (Fig. 2).

The graph of the dependence between shear stresses and number of elements along the panel width and the graph of the dependence between horizontal displacement and number of elements along the panel width are shown on Fig. 3 and 4.

The results shown on graphs are also represented in the Table 1.

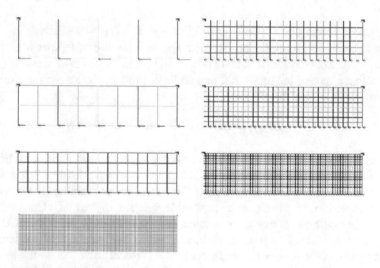

Fig. 2. Considered finite element meshes of the plate with sizes 3 × 12 m (1, 2, 4, 6, 8, 16 and 32 finite elements along the width of plate).

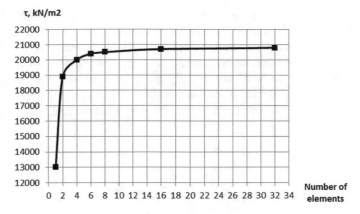

Fig. 3. The graph of the dependence between shear stresses and number of elements along the panel width.

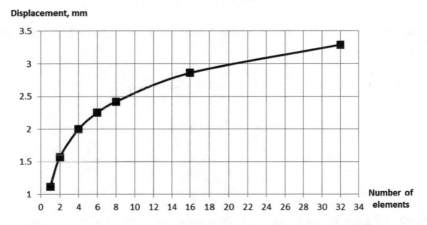

Fig. 4. The graph of the dependence between horizontal displacement and number of elements along the panel width.

Table 1. Comparison of the results obtained with different finite element meshes.

Number of elements	Displacement	Difference, %	Shear stress τ, kN/m²	Difference, %
1	1.12	40.18%	13000	−31.22%
2	1.57	27.39%	18900	−5.50%
4	2	12.50%	20000	−1.96%
6	**2.25**	**7.56%**	**20400**	**−0.49%**
8	2.42	18.18%	20500	−0.97%
16	2.86	15.03%	20700	−0.48%
32	3.29	–	20800	–

Analyzing the graphs and the Table 1, we can conclude that a mesh with 6 finite elements along the width of sheet is sufficient. This mesh should be used to model steel sheet of the panel in the finite element analysis. The shear stiffness of a steel sheet 4 mm thick can be taken into account using the hinge-rod approximation, that is, as a hinge-rod system, the rods of which only work in tension/compression (Fig. 5). Stiffness of rods assigned in accordance with the "Design recommendations on membrane coatings on a rectangular plan for the reconstruction of buildings and structures".

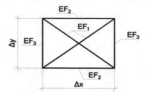

Fig. 5. The scheme of hinge-rod cell.

For a rectangular cell with dimensions Δx, Δy, the areas of the rods are determined as follows:

$$F_1 = \frac{3}{16} \Delta x t \frac{\beta}{\cos^3 \alpha}$$

$$F_2 = \frac{3}{16} \Delta x t \frac{3 - \beta^2}{\beta} \tag{2}$$

$$F_3 = \frac{3}{16} \Delta x t \frac{3\beta^2 - 1}{\beta^2}$$

where t - thickness of the plate; $\beta = ctg\alpha = \Delta x / \Delta y$.

Replace the sections of the sheet between the edges by the hinge-rod cells, the size of 3×2 m, the stiffnesses of the rods are calculated by the above Eqs. (1) in the Table 2.

Table 2. Calculation of the stiffness of the elements of hinge-rod cell.

Parameter	Value
Thickness of sheet, cm	0.4
Size of cell Δx, cm	300
Size of cell Δy, cm	200
Young's Modulus E, kN /cm^2	20600
Axial stiffness EF_1(see Fig. 5), kN	873984
Axial stiffness EF_2 (see Fig. 5), kN	1184500
Axial stiffness EF_3(see Fig. 5), kN	463500

To analyze shear stiffness of the panel, we consider the following problem. In the computing program Lira-SAPR 2017, there were created two finite element models of the considered panel Sch1-450. In the first model (Scheme 1), the steel sheet was modeled by plate finite elements, in the second model (Scheme 2), the sheet was modeled with the use of hinge-rod approximation. The panel was loaded along the free side with a total load of 700 kN (Fig. 6).

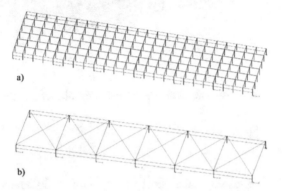

a)

b)

Fig. 6. The finite element models of the panel (a) steel sheet modelled by plate finite elements (Lira-SAPR 2017. Scheme 1); (b) steel sheet modelled by hinge-rod approximation (Scheme 2).

The model of the panel made with the use of solid finite elements was created in the computing program FEMAP 11.1.2. The same scheme of loading and boundary conditions as for models in Lira-SAPR was assigned. The shear stiffness of the panel C_p is the ratio between force and displacement in the direction of force action:

$$C_p = \frac{F}{\Delta} \qquad (3)$$

On the Fig. 7, the contour plots of the displacement in the direction of force action obtained for scheme 3 (with the use of solid finite elements) in program FEMAP 11.1.2 are shown. Displacement of free edge of the panel is 5.03 mm. The displacements obtained for scheme 1 and scheme 2 are represented in the Table 3. The stiffness of the panel was calculated by the Eq. (2) for all of schemes (Table 3)

Table 3. Calculation of the stiffness of the elements of hinge-rod cell.

Scheme	Displacement, mm	Stiffness C_p, kN /m	Difference in stiffness
Lira-SAPR 2017. Scheme 1 (steel sheet modelled by plate finite elements)	5.27	132827	−4.7%
Lira-SAPR 2017. Scheme 2 (steel sheet modelled by hinge-rod approximation)	5.67	123456	−12.7%
FEMAP 11.1.2. Scheme 3 solid finite elements	5.03	139165	−

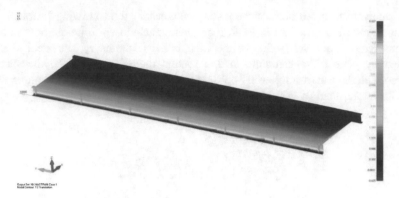

Fig. 7. The horizontal displacements of the steel panel in the direction of the load, mm (FEMAP 11.1.2. Scheme 3).

4 Discussion

The presence of a sufficiently thick steel plating with a thickness of up to 4 mm, supported by transverse stiffening ribs, forms an elastic flat system that can be included in the joint work with the panel longitudinal beams to transfer longitudinal and transverse forces acting in the plane of the roof of the workshop. The possibility of incomplete inclusion of steel sheet in the work of the plate, which is due to shear lag and possible loss of local stability should be taking into account. To correctly account for the operation of panels as parts of the building framework, it is necessary to develop numerical models that reflect the stress-strain state of such structures.

Panels consisting of beams and plating supported by ribs form a disc with significant shear stiffness, which excludes the loss of the overall stability of the truss structure (upper truss belt, beams) at the locations of these panels. At the same time, as the effective length in the presence of the traditional system of braces between trusses should be taken the distance between the longitudinal edges of the panels, provided that the strength of the attachment joints of the panels to the trusses is provided. The strength of the joints should be checked not only for the effect of the fictitious force arising from the loss of stability of the truss structure but also for the longitudinal forces associated with wind, technological (including crane) and temperature effects.

In order to accurately determine the forces at the fastening points of the panels, one should consider the spatial design schemes of the frames with their closest fit to the actual construction (joints, bond arrangements, walling and roof enclosing structures).

The results presented in the article allow concluding about the possible design schemes of panels allowed. It has been established that it is possible with sufficient accuracy for practical calculation to use models of panels made using plate finite elements. The use of hinge-rod approximation is allowed.

5   Conclusions

Based on the results of carried out finite element analysis, the following conclusions can be made:

- welds in the typical bearing joints of panels on truss provides the perception of fictitious force arising from the loss of stability of truss belt, so the panels can be considered as detachments of the upper belt from buckling;
- roof made of steel panels has a significant shear stiffness and the panels should be taken into account analysing the spatial framework;
- the results sufficiently accurate for practical calculations can be obtained by modelling the frame of the panel with beam finite elements, and the steel sheet of the panel with plate finite elements, also hinge-rod approximation for modelling of steel sheet is allowed;
- the sufficient mesh of plate finite elements on the steel sheet of the panel is 6 elements along the panel width;
- the bearing joints of the panels should be calculated on the axial forces acting in the beams of the panels.

References

1. Zolina, T.V., Tusnin, A.R.: Justification for requested accounting side forces generated by crane impact on building frame. Ind. Civ. Eng. **5**, 17–23 (2015)
2. Davydov, E.Yu.: The determination of spatial work of frame of buildings. Struct. Mech. Eng. Constr. Build. **4**, 56–62 (2010)
3. Lewinski, J., Magnuoka Blandzi, E., Szyc, W.: Determination of shear modulus of elasticity for thin-walled trapezoidal corrugated cores of seven-layer sandwich plates. Eng. Trans. **63**(4), 421–437 (2015)
4. Davydov, E.Yu.: The use of profiled sheets in construction of curvilinear extended surfaces. Ind. Civ. Eng. **10**, 33–35 (2010)
5. Avci, O., Lutrell, L.D., Mattingly, J., Samuel, W.: Eastering diaphragm shear strength and stiffness of aluminum roof panel assemblies. Thin-Walled Struct. **106**, 51–60 (2016)
6. Todd, A., Helwig, J.A.Y.: Shear diaphragm bracing of beams. I: stiffness and strength behavior. J. Struct. Eng. **134**(3), 356–384 (2008)
7. Helwig, T.A., Jura, J.A.: Shear diaphragm bracing of beam. II: design requirements. J. Struct. Eng. **134**(3), 360–361 (2008)
8. Korcz-Konkol, N., Urbanska-Galewska, E.: Influence of sheet/purlin fasteners spacing on shear flexibility of the diaphragm. In: MATEC Web Conference, vol. 219, p. 02007 (2018)
9. Balazs, I., Melcher, J., et al.: Stabilization of beams by trapezoidal sheeting: parametric study. In: Proceedings of the 3rd European Conference of Civil Engineering, Paris, pp. 223–227 (2012)

Mechanical Properties of Building Mortar Containing Pumice and Coconut-Fiber

Van Lam Tang[1,2](✉) ⓘ, Kim Dien Vu[2] ⓘ, Van Phi Dang[1] ⓘ,
Tai Nang Luong Nguyen[3] ⓘ, and Dinh Trinh Nguyen[4] ⓘ

[1] Hanoi University of Mining and Geology, 18 Pho Yen, Duc Thang,
Bac Tu Liem, Hanoi, Vietnam
lamvantang@gmail.com
[2] Moscow State University of Civil Engineering, Yaroslavskoe Shosse 26,
Moscow 129337, Russia
[3] University of Fire Fighting and Prevention,
243 Khuat Duy Tien, Thanh Xuan, Hanoi, Vietnam
[4] Thuyloi University, 175 Tay Son, Dong Da District, Hanoi, Vietnam

Abstract. Coconut-fiber is a natural fiber extracted from the outer shell of the coconut fruit. In this study, the length of the coconut-fiber varies from 0.3 to 0.8 mm, but the average length range from 10 to 30 mm is used in mortar mixtures.

The purpose of this work was to study the mechanical properties of building mortar with varying contents of coconut-fiber and pumice. The Russian standard GOST 30744-2001 and absolute volume method were used to determine the compositions of the mortar. In addition, the compressive strength, tensile strength, modulus of elasticity of the mortar-specimens were determined. The results of this study show that due to the different diameter and length property of the coconut-fiber and in combination with the large surface area of pumice, the workability of the mortar mixture was decreased with an increase in the content of coconut-fiber and pumice. Furthermore, by increasing the content of these materials, both the compressive and flexural strength of tested mortar samples were decreased, but only to increase the ratio of compressive to flexural strength of these samples. In the future research, superplasticizers can be used to improve the flow characteristics of this mortar with high contents of coconut-fiber and pumice.

Keywords: Coconut-fiber · Pumice · Building mortar · Compressive strength · Flexural strength · Modulus of elasticity

1 Introduction

Agricultural production, which is the basis of Vietnam's main development strategies, varied significantly from year to year after the country's reunification in 1975. In human life, agricultural materials played an important role, especially as food resources. However, the waste product from agriculture industries has not been managed properly. The results of previous researches show that there are about 0.65 billion tons of agricultural waste every year in China [1, 2] and India produces more than 400

© Springer Nature Switzerland AG 2020
V. Murgul and M. Pasetti (Eds.): EMMFT 2018, AISC 982, pp. 648–659, 2020.
https://doi.org/10.1007/978-3-030-19756-8_61

million tons of agricultural waste [3, 4]. In addition, in the past years, Malaysia had recorded 998 billion tons of agricultural waste. Furthermore, the results of the studies [5, 6] the total amount of agricultural waste in Vietnam is about 80 million tons, including the waste from coconut, tree's coconut in the Southern and rice straw and husks in the North [7, 8]. These wastes were often thrown at landfills, and only a small amount of them was recycled. This has contributed to the environmental pollution that keeps occurring in the territory of Vietnam. In Vietnam, in recent years, two methods are practiced in managing the waste from the coconut tree and husks of coconut; either disposed to landfills or combustion process of the coconut-fiber. These practices produce more disadvantages rather than advantages. Generally, a farmer in Vietnam makes an easy decision with burning the coconut-fiber. This activity resulted with the release of gas emission to the atmosphere, thus contribute to pollution. The previous researches have used plant fiber as an alternative source of steel and artificial fibers to be used in composites such as cement paste and foamed concrete to increase its strength properties [9–11]. Some of the plant fiber are coir, pineapple leaf, bamboo, palm, banana, cotton, and sugarcane which have found that the use of these plant, as additive in foamed concrete mixture, plus it is economical for increasing the tensile strength, shear strength, flexure strength, and lightweight from the current concrete mixture [12, 13].

Globally, 19.6 Mt of volcanic pumice is mined every year with Turkey remaining the dominant producer with a production of 4.2 Mt per year [14, 15]. Pumice is normally used as an aggregate in lightweight building blocks, concrete and assorted building products [16]. The use of pumice as a coating for the reinforcement of steel against corrosion, and concrete abrasions and their suitability of using in load-bearing and enclosure structures was studied [17]. Whereas, in Vietnam, the use of pumice as additives and aggregate of cement pastes, mortar and concrete are still limited. The objective of the current study was to investigate the mechanical properties of building mortar containing contents (0 ÷ 3)% of coconut-fiber and (0 ÷ 30)% of pumice by mass Portland cement.

2 Experimental Investigation

2.1 Materials

1. Quartz sand (QS) sourced original from "Lo river" in the North of Vietnam was used as fine aggregate in mortar mixtures. The physical properties of the fine aggregate are presented in Table 1. In addition, the particle size distributions details of this fine aggregates are shown in Fig. 1.

Table 1. The physical properties of fine aggregate.

Aggregate size (mm)	Loose density (kg/m³)	Dry density (g/cm³)	Saturated density (kg/m³)	Water absorption (%)	Fineness modulus
0.15 ÷ 5	1490	2.65	2682	0.50	3.1

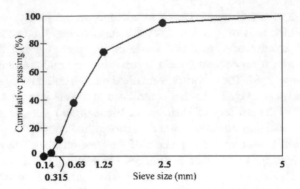

Fig. 1. Sieve analysis of fine aggregate

2. The cement used was locally Grade 40-Portland cement (PC), which manufactured at "But Son" factory in Vietnam with specific a weight of 3.15 g/cm^3. The experimental results of physical properties, mechanical properties, and Portland cement's mineralogical composition are presented in Tables 2 and 4, respectively.

Table 2. Physical and mechanical properties of "But Son" Portland cement.

Specific weight (g/cm^3)	Surface area (cm^2/g)	Passing sieve 10 μm (%)	Time of setting (min)		Compressive strength (MPa)			Standard consistency %
			Initial	Final	3 days	7 days	28 days	
3.15	3685	18	123	248	30.8	42.5	51.3	29.5

3. Pumice (PS) used in this investigation was obtained from mines in the North of Vietnam. They are dried in Drying Oven Construction Lab Equipment (Model ECDO-3 Lab, Zhejiang, China) at a temperature of 100 ± 10 °C within 2 h. After that, the pumice was cooled in one day to the temperature of an environment. Next, the pumice was grounded by a ball mill in 1 h in order to obtain fine powder similar to Portland cement (Fig. 2). The physical and chemical properties of pumice are shown in Tables 3 and 4, respectively.

Table 3. The physical properties of pumice in Vietnam.

Specific weight (g/cm^3)	Unit weight of natural porous state (kg/cm^3)	Surface area (cm^2/g)	Passing sieve 10 μm (%)
2.48	850	3850	15

Table 4. The chemical properties of Portland cement and pumice.

Materials	SiO$_2$	Al$_2$O$_3$	Fe$_2$O$_3$	SO$_3$	K$_2$O	Na$_2$O	MgO	CaO	P$_2$O$_5$	LOI Loss on ignition	other
Portland cement	21.5	4.2	5.6	3.2	–	–	1.8	60.8	–	2.4	0.5
Pumice	66.94	8.72	11.76	0.03	–	–	–	2.5	1.04	3.71	5.3

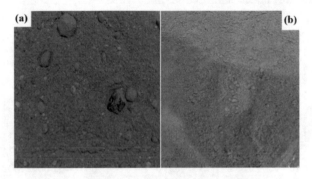

Fig. 2. (a) - The natural appearance of pumice and (b) - After drying and fine grinding of pumice.

4. The coconut-fiber (CF) is extracted from the outer shell of a coconut, which sourced from Southern Vietnam was used as the fibers in mortar mixtures. The manufacturing process of coconut-fiber is illustrated in Fig. 3.

Fig. 3. Technological scheme for obtaining coconut-fiber from the coconut husks.

Table 5 Showed the physical properties of this coconut-fiber.

Table 5. The physical properties of coconut-fiber.

Length (mm)	Diameter (mm)	Density (g/cm^3)	Tensile strength (MPa)	Swelling in water (diameter) (%)
10 ÷ 30	0.3 ÷ 0.7	1.44	340 ÷ 385	4 ÷ 5

5. Local tap water (W) was used for both mixing concrete and curing of the test patterns in this study.

2.2   Mixture Proportions and Samples Preparation

In this study, the proportions of the building mortar were designed following the guidelines of GOST 30744 - 2001 standard (Russian) at the ratios of raw materials: $\frac{QS}{BID} = 3$, $\frac{W}{BID} = 0.5,0$ $\frac{PS}{PC} = 0 \div 0.3$ and $\frac{CF}{PC} = 0 \div 0.03$ with   BID – bind and BID = PC + PS.

The ratios of raw materials and ingredient proportions for the mortar containing varying amounts of pumice and coconut-fiber are shown in Table 6.

Table 6. Mix compositions of the mortar mixture for a batch.

Sample ID.	Compositions	Ratios of raw materials				Compositions of mortar mixture (kg/m³)				
		$\frac{PS}{PC}$	$\frac{CF}{PC}$	$\frac{QS}{BID}$	$\frac{W}{BID}$	PC	PS	CF	QS	W
ID.1	0%PS + 0%CF	0	0	3	0.5	513	0	0	1539	256
ID.2	10%PS + 1%CF	0.1	0.01	3	0.5	463	46.30	4.63	1528	255
ID.3	10%PS + 2%CF	0.1	0.02	3	0.5	461	46.15	9.23	1523	254
ID.4	10%PS + 3%CF	0.1	0.03	3	0.5	460	46.00	13.80	1518	253
ID.5	20%PS + 1%CF	0.2	0.01	3	0.5	423	84.62	4.23	1523	254
ID.6	20%PS + 2%CF	0.2	0.02	3	0.5	422	84.37	8.44	1519	253
ID.7	20%PS + 3%CF	0.2	0.03	3	0.5	421	84.12	12.62	1514	252
ID.8	30%PS + 1%CF	0.3	0.01	3	0.5	390	116.86	3.90	1519	253
ID.9	30%PS + 2%CF	0.3	0.02	3	0.5	388	116.55	7.77	1515	253
ID.10	30%PS + 3%CF	0.3	0.03	3	0.5	387	116.24	11.62	1511	252

2.3   Test Methods

The workability of the mortar mixture was measured using a standard slump cone with dimensions of $100 \times 200 \times 300$ mm as described by ASTM C143 standard.

The compressive strength (R_{cs}) and flexural strength (R_{fs}) of experimental mortar is determined by a $40 \times 40 \times 160$ mm specimen by Russian standard GOST 10180-2012 at the ages of 3, 7, 14, and 28 days (Figs. 4 and 5). These mortar-samples in this investigation are demolded after 24 h later casting and placed in a 25 ± 2 °C water curing tank until ages of the experimental plan.

Fig. 4. Experimental mortar beam-specimens

Fig. 5. Destruction of mortar-specimens tested

The modulus of elasticity of the experimental mortar was measured in cylindrical specimens sized of diameter × height = 50 × 100 mm according to ASTM C469-02 Standard at age 28-day [18].

In this study, uniaxial compressive tests on mortar samples were performed with a constant loading rate of 500 N/s on system Controls Advantest 9 (Figs. 6 and 7).

Fig. 6. Flexural strength test of mortar specimens

Fig. 7. Compressive strength test of mortar specimens

3 Results and Discussion

3.1 Effect of Coconut-Fiber and Pumice on Properties of Fresh Mortar Mixtures and the Physical Properties of Mortar

Table 7 shows properties of fresh mortar mixtures and the basic physical properties of mortar containing different contents of coconut-fiber and pumice.

- For fresh mortar mixtures: The fresh unit weight and flow slump values were in the range of 2278 ÷ 2308 kg/m^3 and 490 ÷ 572 mm, respectively.
- For mortar at age 28-day: The average density values and water absorption were in the range of 2219 ÷ 2268 kg/m^3 and 8.3 ÷ 13.4%, respectively.

By increasing the contents of pumice and coconut-fiber, both the flow slump values and the average density of the fresh mortar mixtures were decreased. These were mainly due to the shrink-resistance of coconut-fiber and the surface area of pumice (3850 cm^2/g) high compared with Portland cement (3685 cm^2/g). These results are similar to the results presented in published studies [9, 10, 14].

The relation between dry density and water absorption of mortar at age 28-day containing different contents of coconut-fiber and pumice is illustrated in Fig. 8 and shown in formula (1).

$$Y = 2314 - 7.186 \cdot x \left(kg/m^3 \right) \, with \, R^2 = 0.842 \qquad (1)$$

The correlation coefficient value R^2 = 0.842 of Eq. (1) represents a very strong negative correlation between the dry density and water absorption of mortar incorporating different amounts of coconut-fiber and pumice.

3.2   The Combined Effect of Coconut-Fiber and Pumice on Mechanical Properties of Mortar

Table 7. Properties of fresh mortar mixtures and the basic physical properties of mortar.

Sample ID.	Compositions	Fresh mortar mixture		Mortar at age 28-day	
		Fresh unit weight (kg/m³)	Flow slump value (mm)	Dry density (kg/m³)	Water absorption (%)
ID.1	0%PS + 0%CF	2308	572	2268	8.3
ID.2	10%PS + 1%CF	2296	565	2252	8.5
ID.3	10%PS + 2%CF	2294	550	2245	8.8
ID.4	10%PS + 3%CF	2291	548	2239	9.9
ID.5	20%PS + 1%CF	2289	523	2242	10.2
ID.6	20%PS + 2%CF	2286	522	2235	10.4
ID.7	20%PS + 3%CF	2284	518	2229	11.9
ID.8	30%PS + 1%CF	2283	508	2232	12.2
ID.9	30%PS + 2%CF	2280	502	2226	12.5
ID.10	30%PS + 3%CF	2278	490	2219	13.4

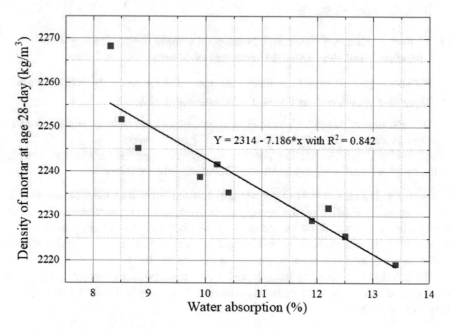

Fig. 8. The relationship between dry density and water absorption of mortar.

The data in Table 8 show the mechanical properties of experimental mortar containing varying contents of PS and CF at different curing times. As the results of this study, the compressive strength of the mortar-sample was in the range of (44.0 ÷ 54.2) MPa at 28 days curing time. Whereas, the flexural strength and modulus of elasticity values were, respectively, in the range from 6.47 to 7.20 MPa and from 16494 to 18626 MPa at the same age.

For clear observation, the speed of development in compressive and flexural strengths of the mortar samples was plotted in Figs. 9 and 10.

It can be seen from Table 7 that the mortar mixture compositions calculation with varying contents of coconut-fiber (from 0% to 3%) and pumice (from 0% to 30%), as represented graphically above, allows obtaining quickly high early-strength mortar. In

Table 8. Mechanical properties of mortar at different curing ages.

Sample ID.	Compositions	Compressive strength (R_{cs}) of mortar (MPa)				Flexural strength (R_{fs}) of mortar (MPa)				$\dfrac{R_{CS}}{R_{fs}}$	Modulus of elasticity at 28 days (MPa)
		3 days	7 days	14 days	28 days	3 days	7 days	14 days	28 days	28 days	
ID.1	0%PS + 0%CF	21.4	35.4	47.7	54.2	3.18	5.45	7.54	8.38	6.47	18626
ID.2	10%PS + 1%CF	19.4	33.1	45.7	52.5	3.14	5.24	7.19	7.94	6.61	18332
ID.3	10%PS + 2%CF	20.0	33.8	46.5	52.8	3.04	5.11	7.04	7.86	6.72	18384
ID.4	10%PS + 3%CF	21.0	34.5	47.6	53.1	2.99	4.78	6.66	7.47	7.11	18436
ID.5	20%PS + 1%CF	17.3	29.9	41.9	46.0	2.65	4.51	6.18	6.94	6.63	17159
ID.6	20%PS + 2%CF	18.5	31.6	43.0	48.3	2.72	4.58	6.35	7.16	6.75	17583
ID.7	20%PS + 3%CF	19.5	32.5	44.2	48.7	2.68	4.29	6.12	6.81	7.15	17656
ID.8	30%PS + 1%CF	16.0	27.7	38.8	42.5	2.58	4.17	5.74	6.38	6.66	16494
ID.9	30%PS + 2%CF	16.5	28.3	39.2	43.1	2.48	4.11	5.67	6.35	6.79	16610
ID.10	30%PS + 3%CF	17.3	29.7	39.4	44.0	2.43	3.98	5.49	6.11	7.20	16782

details, the compressive and flexural strength at the age of 3 days are in the range from 37% to 40.5% in comparison to 28 days period. From the graph that represents the

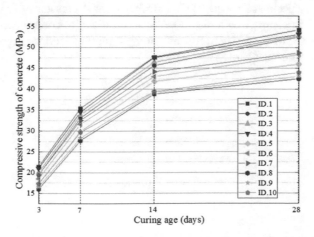

Fig. 9. The development in compressive strength of mortar samples at different ages.

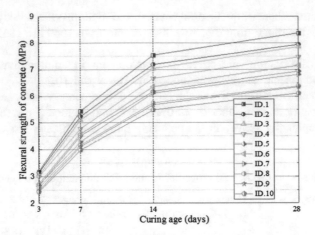

Fig. 10. The development in flexural strength of mortar samples at different ages.

variation in the compressive and flexural strength of mortar-samples at different curing times (Figs. 4 and 5), it could be seen that the strengths of mortar samples were decreased with an increase in the contents of coconut-fiber and pumice of the mortar mixtures. The main purpose of including coconut-fiber in mortar is to increase the flexural strength of mortar-sample which makes the concrete work more efficiently as a flexural member, similar to the results presented in published studies [11, 12]. However, experimental results of this study show that the combined effect of coconut-fiber and pumice was only to increase the compressive to flexural strength ratio of these samples.

The relation between compressive to flexural strength ratio $(\frac{R_{cs}}{R_{fs}})$ and coconut-fiber content of mortar-samples is shown in Fig. 11 and shown in formulas (2–4) (Table 9).

From Fig. 12, it is found that the modulus of elasticity for a control specimen (ID.1) is 18626 MPa, which was the high value among the mortar-specimens tested. Whereas, with increasing the contents of pumice and coconut-fiber the average value of modulus of elasticity for specimens tested (ID.2 - ID.10) were decreased in the range 1.02% from to 11.45% in comparison to the modulus of elasticity of control specimen (ID.1) .

Where: y is the compressive to flexural strength ratio of mortar at age 28-day and x are contents of coconut-fiber.

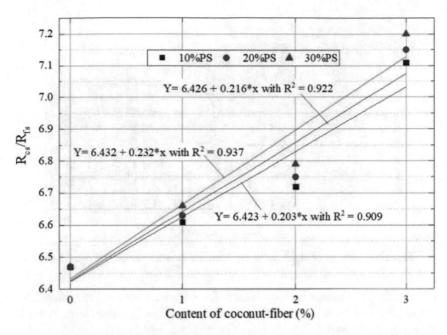

Fig. 11. The relationship between compressive to flexural strength ratio and coconut-fiber content of mortar sample

Table 9. Content of Fig. 11.

Content of pumice	Regression equation	R^2	No
10% PS	y = 6.423 + 0.203*x	0.909	(2)
20% PS	y = 6.426 + 0.216*x	0.922	(3)
30% PS	y = 6.432 + 0.232*x	0.937	(4)

4 Conclusions and Future Work

Based on the results of the present study, the following conclusions were made:

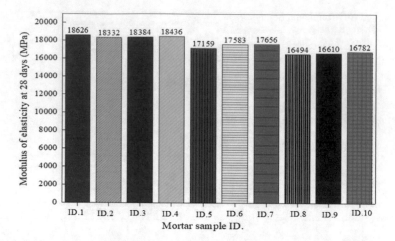

Fig. 12. Modulus of elasticity of experimental mortar samples at age 28-day.

1. It is possible to produce building mortar from Vietnamese local raw materials with contents (0 ÷ 3)% of coconut-fiber and (0 ÷ 30)% of pumice, with relate to the mortar mixture workability by flow slump from 490 to 572 mm standard cone, (44.0 ÷ 54.2) MPa compressive strength, (6.47 ÷ 7.20) MPa tensile strength and (16494 ÷18626) MPa of modulus of elasticity at the age of 28 days with normal hardening.

2. Due to the different diameter and length property of the coconut-fiber and in combination with the large surface area of pumice, the workability of the mortar mixture was decreased with an increase in the content of coconut-fiber and pumice.

3. By increasing the contents of pumice and coconut-fiber, both the compressive and flexural strength of tested mortar samples were decreased, but only to increase the compressive to flexural strength ratio of these samples.

4. The results of this study show that the workability of the mortar mixture was decreased with an increase in content of coconut-fiber and pumice. In the future work, certain admixtures such as air entraining agents and superplasticizers can be used to improve the flow characteristics of this mortar.

References

1. Fanfei, M., Mingxu, Z., Yu, Z., Yan, C., Wei-Ping, P.: An experimental investigation into the gasification reactivity and structure of agricultural waste chars. J. Anal. Appl. Pyrol. **92**(1), 250–257 (2011). https://doi.org/10.1016/j.jaap.2011.06.005
2. Min, F., Zhan, M., Chen, Q., Chen, M.: Study on pyrolysis gasification semicoke characteristics of fresh biomass by FTIR. J. Anhui Univ. Sci. Technol. (Nat. Sci.) **28**, 55–58 (2008)
3. Prithivirajan, R., et al.: Bio-based composites from waste agricultural residues: mechanical and morphological properties. Cellul. Chem. Technol. **49**(1), 65–68 (2015)
4. Noridah, M., Muhamad, A.I., et al.: Mechanical properties and flexure behaviour of lightweight foamed concrete incorporating coir fibre 2017. In: IOP Conference on Series:

Earth and Environmental Science, vol. 140, p. 012140 (2013). https://doi.org/10.1088/1755-1315/140/1/012140

5. Nguyen, T.S., Duong, T.L., et al.: Online catalytic deoxygenation of vapour from fast pyrolysis of Vietnamese sugarcane bagasse over sodium-based catalysts. J. Anal. Appl. Pyrol. **127**, 436–443 (2017). https://doi.org/10.1016/j.jaap.2017.07.006

6. Son, P.T.: Forecasted trends of green energy development in Vietnam. Energy Syst. Res. **1** (2), 51–56 (2018)

7. Lam, V.T., et al.: Effect of rice husk ash and fly ash on the compressive strength of high performance concrete. In: E3S, vol. 33, p. 02030 (2018). https://doi.org/10.1051/e3sconf/20183302030

8. Lam, V.T., Bulgakov, B.I., Aleksandrova, O., Pham, N.A., Yuri, M.: Effect of rice husk ash on hydrotechnical concrete behavior. In: IOP Conference on Series: Materials Science and Engineering, vol. 365, p. 032007 (2018). https://doi.org/10.1088/1757-899x/365/3/032007

9. Majid, A.: Natural fibres as construction materials. J. Civ. Eng. Constr. Technol. **3**(3), 80–89 (2012). https://doi.org/10.5897/JCECT11.100

10. Majid, A.: Coconut fibre: a versatile material and its applications in engineering. J. Civ. Eng. Constr. Technol. **2**(9), 189–197 (2011)

11. Noridah, M., Muhamad, A.I., et al.: Mechanical properties and flexure behaviour of lightweight foamed concrete incorporating coir fibre. Earth Environ. Sci. **140** (2018). https://doi.org/10.1088/1755-1315/140/1/012140

12. Mahyuddin, R., Wai, H.K., Noor, F.A.: Strength and durability of coconut-fiber-reinforced concrete in aggressive environments. Constr. Build. Mater. **38**, 554–566 (2013). https://doi.org/10.1016/j.conbuildmat.2012.09.002

13. Javad, T., Alireza, A., Momtazi, A.S.: Using wood fiber waste, rice husk ash, and limestone powder waste as cement replacement materials for lightweight concrete blocks. Constr. Build. Mater. 50, 432–436 (2014). https://doi.org/10.1016/j.conbuildmat.2013.09.044

14 Crangle, R.D.: Pumice and pumicite. US Geol. Surv. Miner. Yearb. Mineral Commod. Summ. **10**, 124–125 (2010)

15. Hossain, K.M.A., Ahmed, S., Lachemi, M.: Lightweight concrete incorporating pumice based blended cement and aggregate: mechanical and durability characteristics. Constr. Build. Mater. **25**, 1186–1195 (2011). https://doi.org/10.1016/j.conbuildmat.2010.09.036

16. Sari, D., Pasamehmetoglu, A.G.: The effects of gradation and admixture on the pumice lightweight aggregate concrete. Cement Concr. Res. **35**, 936–942 (2005). https://doi.org/10.1016/j.cemconres.2004.04.020

17. Hanifi, B., Huseyin, Z., Gulay, Z., Fatma, Y.: The use of pumice as a coating for the reinforcement of steel against corrosion and concrete abrasions. Corros. Sci. **50**, 2140–2148 (2008). https://doi.org/10.1016/j.corsci.2008.03.018

18. ASTM C469 – 02: Standard Test Method for Static Modulus of Elasticity and Poisson's Ratio of Concrete in Compression. ASTM International, West Conshohocken, PA, 87 p. (2002)

Effects of High Temperature on High-Performance Fine-Grained Concrete Properties

Van Lam Tang[1](✉) iD, Kim Dien Vu[2] iD, Dinh Tho Vu[2] iD,
Boris Bulgakov[2] iD, Sophia Bazhenova[2] iD,
and Tai Nang Luong Nguyen[3] iD

[1] Hanoi University of Mining and Geology, 18 Pho Yen, Duc Thang,
Bac Tu Liem, Hanoi, Vietnam
lamvantang@gmail.com
[2] Moscow State University of Civil Engineering, Yaroslavskoe Shosse 26,
129337 Moscow, Russia
[3] University of Fire Fighting and Prevention,
243 Khuat Duy Tien, Thanh Xuan, Hanoi, Vietnam

Abstract. The dense development of high-rise construction in urban areas in Vietnam requires the creation of new concretes with essential properties for fire safety solutions and for high temperature. In this study, the effects of high temperature on high performance fine-grained concrete properties were investigated. Concrete samples were exposed to high temperatures at 300 °C at 4, 5, and 6 h, then cooled to ambient temperature before tests. Two mixtures of Normal Fine-grained Concrete (NFC) and High-Performance Fine-grained Concrete (HPFC) containing 10% silica fume (SF) and 50% bottom ash (BA) were designed in accordance with an absolute volume method. Mass loss, residual compressive strength, and X-ray analysis were performed to investigate the effect of high temperature at different times on the performance of NFC and HPFC. The results of this study showed that the compressive strength of HPFC mixture containing SF and BA obtained is significantly greater than that of the NFC at 300 °C for different curing ages. This can be explained by enhanced reactivity of SiO_2 amorphous in the SF and BA contents in HPFC, which binds more calcium hydroxide at higher temperatures, a percentage of calcium silicate hydrates increases, and the presence of Tobermorite and Xonotlite secondary particles are confirmed through XRD analysis.

Keywords: High temperature · Curing regime · Xonotlite · Tobermorite · Silica fume · Bottom ash · High-performance fine-grained concrete

1 Introduction

The structural stability of concrete and reinforced concrete of high-rise buildings exposed to fire is gaining a significant role in the design process, as users and authorities are increasingly demanding for fire safety solutions [1]. Recent years in Vietnam, with the recent rapid developments of high-rise buildings in urban areas,

© Springer Nature Switzerland AG 2020
V. Murgul and M. Pasetti (Eds.): EMMFT 2018, AISC 982, pp. 660–672, 2020.
https://doi.org/10.1007/978-3-030-19756-8_62

where fire safety measures are difficult to implement [2]. In recent times, fires have occurred in high-rise buildings and houses, causing a great deal of damage to property, health, and lives of people. Only in 2017 in this country occurred 4100 fires and is particularly serious fire at Carina Plaza apartment building Ho Chi Minh city in March 2018 [3]. As we know, concrete is a non-combustible material, and, as such, it does not increase the fire load and constitutes a natural barrier preventing the spread of fire [4, 5]. When exposed to fire temperatures concrete does not release any toxic gases or smoke. However, in high-temperature conditions of fires, its internal structure undergoes several physical transformations accompanied with chemical reactions, high result in irrecoverable changes affecting the performance and in the worst case leading to total destruction of the material. Currently, in Vietnam, there is no standardized test methods and is difficult to definitively determine the effect of high-temperature conditions on concrete. The studies [6–8] showed that the fire resistance of concrete is increased with the replacement of brick and steel industry waste such as steel slag, bottom ash, crushed bricks and crushed tiles used as aggregate in concrete. The results of the studies [9–12] show the effects of fly ash, crushed quartz, ground-granulated blast-furnace slag, and other mineral admixture on the mechanical properties of concrete at high temperature. Besides that, some studies [13, 14] the annual amount of industrial waste is more than 150 million tons in Vietnam. In which, metallurgical slag is about $45 \div 55$ million tons, ash and slag TPP is nearly $50 \div 60$ million tons. In 2016, TPP "Vung Ang" produces about 3000 tons of ash and slag waste daily. In addition, an enormous number of gaseous substances and solid particles formed as a result of solid fuel combustion enter the atmosphere through the smokestacks of this power plants, which have caused serious environmental pollution in Vietnam central provinces [15]. The objective of the current study was to investigate the combined effects of high temperature on properties of high-performance fine-grained concrete containing silica fume and bottom ash, which are intended for concrete blocks in the High-Rise Construction.

2 Experimental Details

2.1 Materials

1. The cement used was ordinary Portland cement (OPC) (40 Grade) manufactured at "Tam Diep" factory (Vietnam), the specific weight of 3.15 g/cm^3. The experimental results of physical and mechanical properties of cement are presented in Table 1 and the results of the chemical compositions are presented in Table 2.

Table 1. Mineralogical composition, physical and mechanical properties of "Tam Diep" Portland cement.

Mineral composition (%)					Time of setting (min)		Compressive strength (MPa)			Standard consistency (%)
C$_3$S	C$_2$S	C$_3$A	C$_4$AF	Other	Initial	Final	3 days	7 days	28 days	
56.15	22.47	5.14	12.25	3.99	142	235	35.1	40.4	52.3	29.5

2. Good quality river sand was used as a fine aggregate, which produced from the quartz sand (QS) of "Lo River" (Vietnam). The fineness modulus MK = 3.1, specific gravity and dry density are 2.65 g/cm^3 and 1650 kg/m^3. The particle size distributions details of fine aggregates are shown in Fig. 1.

3. Bottom Ash (BA) TPP "Vung Ang" (Vietnam) class F and Silica Fume SF-90 (SF90) (Vina Pacific). The chemical composition and physical properties of the BA TPP "Vung Ang" and silica Fume SF-90 are presented in Table 2 and their particle size distribution are presented in Fig. 1.

Table 2. Chemical compositions and physical properties of Portland cement, BA TPP "Vung Ang" and Silica fume SF-90.

Chemical components (wt.%)	BA TPP "Vung Ang"	Silica Fume SF-90	Portland cement
SiO_2	61.22	91.65	20.4
Al_2O_3	21.17	2.25	4.4
Fe_2O_3	5.85	2.47	5.4
SO_3	2.42	–	3.4
K_2O	1.25	–	1.2
Na_2O	1.23	0.55	0.3
MgO	0.57	–	2.5
CaO	1.12	0.51	60.2
P_2O_5	1.03	0.03	–
LOI	4.14	2.54	2.2
Average particle size (μm)	6.15	0.243	8.365
Specific gravity (g/cm^3)	2.35	2.15	3.15
Dry density (kg/m^3)	575	760	1250
Surface area (m^2/g)	5.82	14.45	0.365

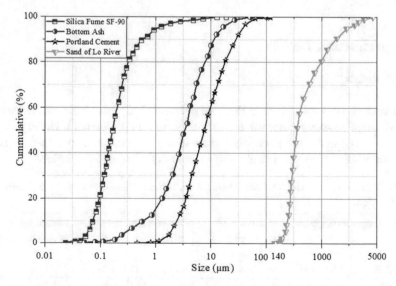

Fig. 1. Sieving analysis of Silica Fume, Portland cement, Bottom Ash and sand of Lo River.

4. Superplasticizer SR 5000F "SilkRoad" (SR5000) (Korea). It is a new generation of chemical additives based on polycarboxylate ethers with a specific weight of 1.1 g/cm^3 at 25 ± 5 °C.
5. Ordinary clean tap water (W) was used for both mixing concrete and curing of test specimens.

2.2   Mixture Proportions and Samples Preparation

Absolute Volume Method. The Normal Fine-grained Concrete (NFC) and High-Performance Fine-grained Concrete (HPFC) mixtures were designed in accordance with the absolute volume method and NSC as the control mixture of concrete in this experimental study.

Mix compositions of fine-grained concrete. In the case of this study, the initial ratios of raw materials by weight in concrete mixtures for the production of NFC and HPFC are given in Table 3.

Table 3. Ratios of raw materials used in the preliminary composition.

Ratios value	$\frac{QS}{OPC}$	$\frac{SF}{OPC}$	$\frac{BA}{OPC}$	$\frac{SR5000}{OPC}$	$\frac{W}{OPC}$	The volume of air in concrete
NFC	1.2	0	0	0	0.4	2%
HPFC	1	0.5	0.1	0.02	0.28	2%

Based on the ratios of raw materials in Table 3 and combined with absolute volume method, the preliminary of the material compositions in the dry state of the NFC (control mixture) and HPFC mixtures are shown in Table 4.

Table 4. Mix compositions and properties of fresh fine-grained concrete.

Mixture	Compositions of the concrete mixture (kg/m^3)						Slump flow (cm)	Slump (cm)	Average density (kg/cm^3)
	OPC	SF	BA	SR5000	QS	W			
NFC	826	0	0	0	991	330	–	15	2082.1
HPFC	629	62.9	315	6.29	1006	214	65	–	2170.2

2.3   Test Methods

Workability of Concrete Mixtures. Properties of the fresh concrete, including slump, slump flow, which are determined by standard slump cone with dimensions of $100 \times 200 \times 300$ mm according to ASTM C143 and average density were measured right after mixing and its results are shown in Table 4.

Compressive strength. The compressive strength of NFC and HPFC test was performed after 3, 7, 14, and 28 days on cubic samples of $70 \times 70 \times 70$ mm by Russian standard GOST 10180-2012. These cube samples are demolded after 24 h later casting and placed in a 25 ± 2 °C water curing tank until ages of the experimental plan.

High-temperature condition. Fine-grained Concrete cubes were cured under hot air condition as shown in Fig. 3. The effect of high temperature at 300 °C in 4 h, 5 h and 6 h at different ages were studied for normal fine-grained concrete without mineral additives at the age 28 days. In addition, in this paper are studied the effect of normal and high temperatures, respectively, 25 °C and 300 °C on the properties of NFC and HPFC, which contains 10% silica fume and 50% bottom ash by mass Portland cement at the ages of 3, 7, 14, and 28 days.

Test procedures. In this work, uniaxial compressive tests on NFC and HPFC samples (for each concrete sample) were performed with a constant loading rate of 3000 N/s on system Controls Advantest 9. The reason of choosing 3000 N/s is to keep the loading rate to a minimum in the comparison of test NFC and HPFC results (Fig. 2).

a   b

Fig. 2. a - Bottom ash TPP «Vung Ang», b - Determine the weight of concrete specimens.

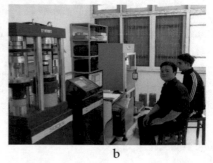

a   b

Fig. 3. a - Hot air curing, b - Compressive strength test of Fine-grained concrete.

Characterization of X-ray and of laser granularity. X-ray diffraction was performed with an XRD "Model XDA-D8 Advance" diffractometer of a company "Bruker" sensing with □□□□ configuration and Cu k□ radiation (□ = 1.54 Å). The angular range from 0 to 90° was performed, to know the different crystalline phases of C-S-H family members developed under high temperature. Laser (Model BT-9300Z, China) granularity analyzer was used to test the granularity distribution of silica fume, bottom ash, and Portland cement.

3   Results and Discussion

3.1   Effect of the High-Temperature Condition in Different Times on Weight and Compressive Strength of Fine-Grained Concrete at Age 28 Days

The temperature of curing regime has a vital role in the development of concrete properties. Weight and compressive strength of NFC and HPFC at age 28 of days obtained at 300 °C for 4, 5, and 6 h is shown in Table 5.

Table 5. Weight and compressive strength of NFC and HPFC at age 28-day at 300 °C in different times.

Type of concrete	The weight of NFC at age 28 days (g)				Compressive strength at age 28 days (MPa)			
	25 °C	At 300 °C for 4 h	At 300 °C for 5 h	At 300 °C for 6 h	25 °C	At 300 °C for 4 h	At 300 °C for 5 h	At 300 °C for 6 h
NFC	777.06	686.61	675.50	668.66	54.40	61.74	67.32	68.85
HPFC	789.15	727.99	708.26	699.27	89.50	09.59	111.98	115.23

Weight Loss of Concrete Specimens. For weight loss assessment, the weights of the concrete cubes were measured before (at 25 °C) and after the exposure to elevated temperatures at 300 °C. The impact of high temperature on the mass loss of both NFC and HPFC containing 50% of bottom ash are shown in Fig. 4.

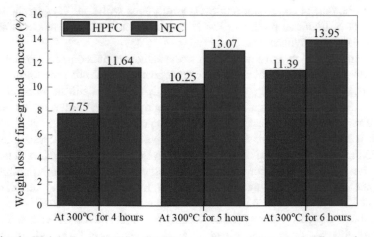

Fig. 4. Weight loss of NFC and HPFC specimens at 300 °C in different times.

The mass loss of all the investigated NFC and HPFC specimens is expressed as a percentage of the original mass at the ambient temperature (25 °C) to the mass after exposure to temperature at 300 °C. Figure 4 further displays that at 300 °C in different times, the weight loss of NFC was the tendency to increase, when compared to HPFC containing bottom ash and silica fume.

In theory, the mass loss in the concrete samples at high temperatures could be attributed to the extra amount of free water, the release of both gel and capillary water, as well as the decomposition of calcareous aggregates, a liberation of carbon dioxide (CO_2) and sloughing off of the concrete surface, which therefore altered the mechanical properties of the concrete [16, 17] (see in Fig. 5).

A - The decomposition of calcareous aggregates and the liberation of CO_2.
B - The sloughing off of the concrete surface.
C - The release of both gel and capillary water.

Fig. 5. The fine-grained concrete surface at 300 °C for 6 h.

Compressive strength development. Data from Table 5 show that as hot air curing at 300 °C for 4, 5 and 6 h, compressive strength is observed to increase, respectively, to 13.49%, 23.75% and 26.56% for NFC, 22.45%, 25.12% and 28.75% for HPFC in comparison with strength at 28 days that obtained at 25 °C. Heat treatment essentially accelerates the pozzolanic reaction of SiO_2 amorphous with $Ca(OH)_2$, facilitating large development of hydrated products. The strength gain is attributed to well these products under heating environment.

According to [18] reactive powder concrete samples when cured in the temperature range of 100 °C ÷ 300 °C have shown the presence of needle-like the fibrous structure of Tobermorite and Xonotlite, which is a hydrated calcium silicate mineral ($Ca_5.Si_6.O_{16}(OH)_2.4H_2O$) and ($Ca_6.Si_6.O_{17}(OH)_2$), respectively. The results of the study [19] also have reported the presence of Tobermorite and Xonotlite at this temperature. It is known that compressive strength, porosity, and permeability of the crystalline calcium silicate hydrate matrix formed at temperatures from 150 °C to 200 °C is dependent in part on the phase formed, for example, Truscottite, Tobermorite, Xonotlite. These crystals become stable and continue their growth, leading to the development of higher strengths of fine-grained concrete, when the heat curing is continued. The relationship between the compressive strength and weight of NFC and HPCF at age 28 days at 300 °C in different times. The given data in Table 5 was observed that the cube compressive strength values could be correlated with the corresponding mass of concrete. Figure 6 displays a good relationship amongst the ratio (%) of residual compressive strength at 300 °C to original compressive strength at 25 °C and ratio (%) of the weight of concrete at 300 °C to original weight at 25 °C.

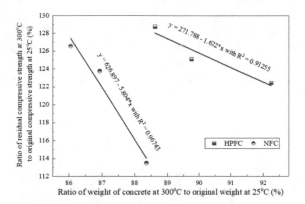

Fig. 6. The relationship between relative compressive strength and weight of NFC and HPCF at age 28-day at 300 °C in different times.

To correlate the experimental data, a linear regression method was applied, resulting in Eqs. (1) and (2) with a correlation coefficient values (R^2) of between approximately 0.91255 and 0.96743 for all samples of NFC and HPFC, which signified good confidence for the relationships. They are as the following:

$$+ \text{ For NFC: } y = 626.897 - 5.804 \cdot x \text{ (with } R^2 = 0.96743) \tag{1}$$

$$+ \text{ ForHPFC : } y = 271.788 - 1.622 \cdot x \text{ (with } R^2 = 0.91255) \tag{2}$$

Where: y is the ratio of residual compressive strength at 300 °C to original strength at 25 °C (%) and x signifies the ratio of the weight of concrete at 300 °C to original weight at 25 °C (%).

3.2 Effect of High-Temperature Condition on Properties of NFC and HPFC at Different Ages

The compressive strength of NFC and HPFC obtained at 25 °C and 300 °C for 6 h and at different curing, ages are shown in Table 6.

Table 6. Compressive strength of NFC and HPFC at different ages at 25 °C and 300 °C for 6 h.

Curing age (days)	The compressive strength of NFC (MPa)			The Compressive strength of HPFC (MPa)		
	At 25 °C	At 300 °C for 6 h	% at 300 °C with at 25 °C	At 25 °C	At 300 °C for 6 h	% at 300 °C with at 25 °C
3	25.7	40.6	157.98	42.2	85.3	202.13
7	41.2	55.3	134.22	71.6	103.3	144.27
14	49.6	63.2	127.42	84.1	111.2	132.22
28	54.4	68.85	126.56	89.5	115.23	128.75

From Table 6, it can be seen that as the temperature of 300 °C for 6 h, the strength of NFC and HPFC observed to quickly increase at early age (3-day) and to slowly increase at late ages (14 and 28 days) in comparison with the strength of concrete that obtained at ambient temperature curing (at 25 °C). For 3 days of hot air curing at 300 °C for 6 h, the HPFC compressive strength was 102.13% higher, when compared to 3 days strength obtained at 25 °C, but for NFC only of 57.98%. However, at 28-day age, compressive strength for HPFC was 28.75% and for NFC was 26.56% higher than those of ambient temperature curing regime.

The relationship existing between the compressive strength and curing age for NFC and HPFC at 25 °C and 300 °C is shown in Fig. 7.

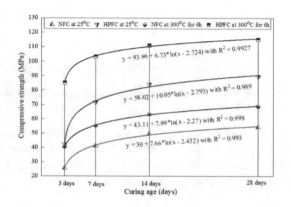

Fig. 7. Compressive strength variation of NFC and HPFC for different curing ages.

The cement concrete strength level and rate of gain, they are a complex function of many different factors. Hydration rate and percentage are two factors related to the used Portland cement [20]. Besides mixture compositions, aggregate type and properties, curing time, different curing regime and method are some factors among the factors affecting both strength level and the gain rate at different ages of concrete [21]. According to data experimental results, for all NFC and HPFC specimens, the relationship between curing age and the compressive strength given by the following formulas:

$$+ \text{For NFC at } 25°\text{C}: \ y = 30 + 7.66 \cdot \ln(x - 2.432) \text{ with } R^2 = 0.993 \qquad (3)$$

$$+ \text{For HPFC at } 25°\text{C}: y = 58.02 + 10.05 \cdot \ln(x - 2.793) \text{ with } R^2 = 0.989 \qquad (4)$$

$$+ \text{ For NFC at } 300°\text{C in 6 hours}: y = 43.11 + 7.99 \cdot \ln(x - 2.27) \text{ with } R^2 = 0.998 \qquad (5)$$

$$+ \text{For HPFC at } 300°\text{C in 6 hours}: y = 93.96 + 6.73 \cdot \ln(x - 2.724) \text{ with } R^2 = 0.9927 \qquad (6)$$

Where: y is the compressive strength of concrete (MPa) and x is curing age (days). These various equations for prediction of compressive strength derived from curing age also have been proposed by published study [22].

Based on the results of this study, the correlation between curing times and the ratio of residual compressive strength at 300 °C to original compressive strength at 25 °C of NFC and HPFC at different curing times was illustrated in Fig. 8 and shown by Eqs. (7) and (8).

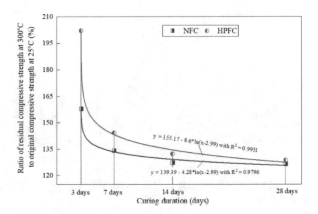

Fig. 8. Effect of high temperatures on compressive strength variation of NFC and HPFC at different curing ages.

$$+ \text{For NFC} : y = 139.39 - 4.28 \cdot \ln(x - 2.99) \text{ with } R^2 = 0.9796 \qquad (7)$$

$$\mid \text{For HPFC}: y = 155.17 - 8.6 \cdot \ln(x - 2.99) \text{ with } R^2 = 0.9931 \qquad (8)$$

In Eqs. (7) and (8), x is curing age (days) and y is the ratio (%) of residual compressive strength at 300 °C to original compressive strength at 25 °C of NFC and HPFC. The values R^2 of Eqs. (7) and (8), respectively 0.9796 and 0.9931 represents a very strong negative correlation between the two compared parameters of the curing age and compressive strength variation for the NFC and HPFC, which included amounts of SF-90 and BA.

3.3   X-Ray Diffraction (XRD)

The results of XRD of the NFC and HPFC exposed to high temperatures of 300 °C for 6 h are illustrated in Fig. 9. The XRD analysis shows that at the high temperatures the compositions of Normal Fine-grained Concrete and High-Performance Fine-grained were significantly affected by hot air curing.

From the data given in Fig. 9, it can be seen at 3 days at 300 °C, the presence of calcium hydroxide $(Ca(OH)_2)$ in both NFC and HPFC specimens, but the content of calcium hydroxide in NFC was higher than of HPFC. This can be explained by enhanced reactivity of SiO_2 amorphous in the bottom ash and silica fume contents in HPFC, which binds more calcium hydroxide at higher temperatures.

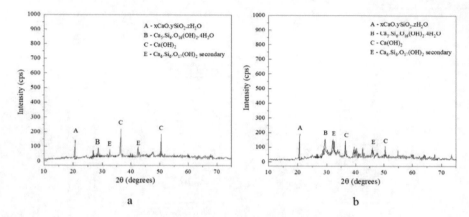

a   b

Fig. 9. a - XRD analysis of NFC samples at 3 days at 300 °C, b - XRD analysis of HPFC samples at 3 days at 300 °C.

The reaction of SiO_2 amorphous is consistent with portlandite decreases, the percentage of secondary calcium silicate hydrates increases and the presence of small sponge ball shape of Tobermorite and Xonotlite secondary particles were observed in 3 days at 300 °C of Fine-grained Concrete samples, which are compressive strength enhancing compound in HPFC higher than those of NFC.

4   Conclusions and Future Work

The experimental study was carried out in the laboratory of Hanoi University of Mining and Geology to find out the effect of high temperatures on high performance fine-grained concrete properties. The results of the present study support the following conclusions:

1. Under hot air curing at 300 °C in 4, 5 and 6 h, compressive strength is observed to increase, respectively, to 13.49%, 23.75% and 26.56% for NFC, 22.45%, 25.12% and 28.75% for HPFC in comparison with strength at 28 days that obtained at 25 °C. Still, in that curing regime, weight loss of specimens was also increased, respectively, to 7.75%, 10.25% and 11.39% for HPFC, 11.64%, 13.07% and 13.95% for NFC.

2. At 300 °C for 6 h, the strength of NFC and HPFC observed to quickly increase at early age and to slowly increase at late ages in comparison with the strength of concrete that obtained ambient temperature curing. For 3 days of hot air curing at

300 °C for 6 h, the HPCF compressive strength was 102.13% higher, when compared to 3 days strength obtained at 25 °C, but for NFC only of 57.98%. However, at 28-day age, strength for HPFC was 28.75% and for NFC was 26.56% higher than those of ambient temperature curing regime.

3. The results of this study showed that during the increase in curing temperature from 25 °C to 300 °C at different curing times, the strength of NFC and HPFC is also increasing, which shows a promising positive influence of heat treatment to improve microstructure of Fine-grained Concrete in the future work. In addition, in the microstructure of NFC and HPFC, the presence of secondary hydrated products such as Tobermorite and Xonotlite secondary particles are confirmed through XRD analysis.

Acknowledgments. The authors greatly appreciate the valuable assistance provided by the fellows of the Department "Technology of Binders and Concretes" at the National Research Moscow State University of Civil Engineering (Russian Federation) and the Faculty of Civil Engineering's laboratory of Hanoi University of Mining and Geology (Vietnam) during this investigation.

References

1. Lam, V.T., Boris, B., Sofia, B., Olga, A., Anh, N.P., Tho, D.Vu.: Effect of rice husk ash and fly ash on the workability of concrete mixture in the high-rise construction. In: E3S Web of Conferences, vol. 33, p. 02029 (2018). https://doi.org/10.1051/e3sconf/20183302029
2. Lam, T.V., Luong, N.T.N.: The scientific basis of the use of mineral additives for manufacture concrete and mortar. Fire Prev. Fire Eng. J. **108**, 36–37 (2018)
3. Hoa, B.T., Duc, V.M., Hoa, N.N.: Study on the production of manufacture adhesives from Portland cement mixture. J. Constr. Sci. **11**, 87–92 (2012)
4. Tomasz, D., Wioletta, J., Mariusz, T., Artur, K., Jerzy, G., Ritoldas, Š.: Effects of high temperature on the properties of high performance concrete. Proc. Eng. **172**, 256–263 (2017). https://doi.org/10.1016/j.proeng.2017.02.108
5. Venkatesh, K.: Properties of concrete at elevated temperatures. ISRN Civil Eng. Article ID 468510, 15 p. (2014). https://doi.org/10.1155/2014/468510
6. Masood, A., Shariq, M., Masroor, Alam, M., Ahmad, T., Beg, A.: Effect of elevated temperature on the residual properties of quartzite, granite and basalt aggregate concrete. J. Inst. Eng. (India): Ser. A **12**, 13 p. (2018). https://doi.org/10.1007/s40030-018-0307-6
7. Netinger, I., Kesegic, I., Guljas, I.: The effect of high temperatures on the mechanical properties of concrete made with different types of aggregates. Fire Safe. J. **46**, 425–430 (2011). https://doi.org/10.1016/j.firesaf.2011.07.002
8. Hachemi, S., Ounis, A.: Performance of concrete containing cru-shed brick aggregate exposed to different fire temperatures. Eur. J. Environ. Civ. Eng. **19**, 805–824 (2015). https://doi.org/10.1080/19648189.2014.973535
9. Subekti, S., Bayuaji, R., Darmawan, M.S., Husin, N.A., Wibowo, B., Anugraha, B., Irawan, S., Dibiantara, D.: Review: potential strength of fly ash-based geopolymer paste with substitution of local waste materials with high-temperature effect. In: IOP Conference Series: Materials Science and Engineering, vol. 267, p. 012001 (2017). https://doi.org/10.1088/1757-99x/267/1/012001

10. Yazici, H., Yardimci, M.Y., Aydin, S., Karabulut, A.S.: Mechanical properties of reactive powder concrete containing mineral admixtures under different curing regimes. Constr. Build. Mater. **23**, 1223–1231 (2009). https://doi.org/10.1016/j.conbuildmat.2008.08.003

11. Courtial, M., De Noirfontaine, M.N., Dunstetter, F., Signes-Frehel, M., Mounanga, P., Cherkaoui, K., Khelidj, A.: Effect of polycarboxylate and crushed quartz in UHPC: microstructural investigation. Constr. Build. Mater. **44**, 699–705 (2013). https://doi.org/10.1016/j.conbuildmat.2013.03.077

12. Liu, Y., Presuel, M.F.: Effect of elevated temperature curing on compressive strength and electrical resistivity of concrete with fly ash and ground-granulated blast furnace slag. ACI Mater. J. Title No. 111-M47, 531–541 (2014). https://doi.org/10.14359/51686913

13. Lam, T.V., Hung, N.S., Bulgakov, B.I., Aleksandrova, O.V,. Larsen, O.A., Orekhova, A. Yu., Tyurina, A.A.: Ispol'zovanie zoloshlakovyh othodov v kachestve dopolnitel'-nogo cementiruyushchego materiala (in Russian). Nauchno-teoreticheskij zhurnal « Vestnik BGTU im. V.G. SHuhova » . №. 8, pp. 10–18 (2018). https://doi.org/10.12737/article_5b6d58455b5832.12667511

14. Lam, T.V., Boris, B., Olga, A., Oksana, L., Anh, P.N.: Effect of rice husk ash and fly ash on the compressive strength of high performance concrete. In: E3S Web of Conferences, vol. 33, p. 02030 (2018). https://doi.org/10.1051/e3sconf/20183302030

15. Lam, T.V., Bulgakov, B.I., Aleksandrova, O.V., Larsen, O.A.: Vozmozhnost' ispol'zo-vaniya zol'nyh ostatkov dlya proizvodstva materialov stroitel'nogo naznacheniya vo V'etname (in Russian). Nauchno-teoreticheskij zhurnal « Vestnik BGTU im. V.G. SHuhova » , №. 6, pp. 06–12 (2017). https://doi.org/10.12737/article_5926a059214ca0.89600468

16. Düğenci, O., Haktanir, T., Altun, F.: Experimental research for the effect of high temperature on the mechanical properties of steel fibre-reinforced concrete. Constr. Build. Mater. **75**, 82–88 (2015). https://doi.org/10.1016/j.conbuildmat.2014.11.005

17. Ma, Q., Guo, R., Zhao, Z., Lin, Z., He, K.: Mechanical properties of concrete at high temperature: a review. Constr. Build. Mater. **93**, 371–383 (2015). https://doi.org/10.1016/j.conbuildmat.2015.05.131

18. Parameshwar, N.H., Subhash, C.Y.: Effect of different curing regimes and durations on early strength development of reactive powder concrete. Constr. Build. Mater. **154**, 72–87 (2017). https://doi.org/10.1016/j.conbuildmat.2017.07.181

19. Tam, C., Tam, V.W.: Microstructural behaviour of reactive powder concrete under different heating regimes. Mag. Concr. Res. **64**, 259–267 (2012). https://doi.org/10.1680/macr.2012.64.3.259

20. Abdelaty, M.: Compressive strength prediction of Portland cement concrete with age using a new model. HBRC J. **10**(2), 145–155 (2014). https://doi.org/10.1016/j.hbrcj.2013.09.005

21. Hewlett, P.C.: Lea's Chemistry of Cement and Concrete, 4th edn., 1092 p. Elsevier Butterworth – Heinemann, New York (2006)

22. Colak, A.: A new model for the estimation of compressive strength of Portland cement concrete. Cem. Concr. Res. **36**(7), 1409–1413 (2006). https://doi.org/10.1016/j.cemconres.2006.03.002

Fire Retardant Coating for Wood Using Resource-Saving Technologies

Svetlana Belykh$^{(\boxtimes)}$ ⓘ, Julija Novoselova ⓘ, and Denis Novoselov ⓘ

Bratsk State University, st. Makarenko, 40, Bratsk, Russia
sveta.belyh@mail.ru

Abstract. Compositions and technology for producing a fire retardant coating for wood are developed. The fire retardant compositions include sodium liquid glass with the addition of a surfactant and a filler (black shale). The results of the study of the influence of the composition of liquid glass composites on the properties of fire retardant coatings obtained from them are presented. The advantages of using by-products of industrial production in the technology of producing fire retardant coatings for wood are shown. A method is proposed for improving the process of preparing liquid glass from microsilica in relation to fire retardant coatings. The advantages of the developed material for the protection of wood from fire: simplification of the composition of the liquid glass composite; reduction of cost of the liquid composite; increased adhesion and fire retardant efficiency; ensuring environmental safety.

Keywords: Wood · Fire retardant coating · Liquid glass ·
Liquid glass composite · Microsilica · Black shale · Adhesion ·
Fire retardant efficiency

1 Introduction

Fire protection of buildings, structures, and constructions made of wood and materials based on it is the most important activity of builders, designers, and government agencies that carry out fire-prevention supervision. At the design stage of wooden structures, it is necessary to provide for such engineering and technical solutions, with the help of which, in case of a fire, the spread of fire can be prevented, and people can be evacuated without threatening their life and health [1–3].

The construction industry needs constant modernization of the technologies used, as well as an increase in the efficiency of the used building materials. Based on an analysis of the situation on the modern market, it has been established that the Russian economy is developing according to laws similar to global ones. Nowadays, many domestic enterprises cannot compete with foreign ones [4–6] due to the imperfection of the used technologies and the lack of production efficiency. This especially applies to resource saving. It is known that in the domestic market of Russia, energy prices are growing faster and faster up to the level formed by the international market, which already poses a problem for Russian enterprises - the development and use of energy- and resource-saving production technologies that will significantly reduce the prime cost and, accordingly, the price of the products.

© Springer Nature Switzerland AG 2020
V. Murgul and M. Pasetti (Eds.): EMMFT 2018, AISC 982, pp. 673–681, 2020.
https://doi.org/10.1007/978-3-030-19756-8_63

Based on the analysis of the sci-tech and patent literature, it has been established that fire retardant coatings for wood by the method of applying to the surface (varnishes, paints, coatings, intumescent compositions) are most effective both in terms of fire retardant efficiency and the use of energy-saving technologies, in contrast to expensive impregnation of wood and products based on it with salt solutions (fire retardants). The existing range of fire retardant compositions for wood is represented either by expensive imported materials or by compositions using polymeric binders with organic solvents, which is unsafe from both technological and environmental points of view [7]. There are methods for producing fire retardant coatings based on liquid glass, which is the most affordable and promising binder in the field of creating fire retardant compositions. The advantage of aqueous alkaline solutions of silicates is their ability to maintain a viscoplastic state at the stage of preparation of the raw mix. This feature increases the efficiency and manufacturability of the production of materials based on them, and also leads to the widespread use of liquid glass in construction, in particular, in obtaining fire retardant compositions. However, the existing level of properties of fire retardant compositions based on liquid glass does not always satisfy consumers in terms of durability and fire retardant efficiency for thin-film coatings and need significant improvement. Analyzing the compositions of fire retardant compositions based on liquid glass, one can also point out the existing problems of adhesion of liquid glass compositions to wooden surfaces, which depend on the type of wood, the quality of surface pretreatment, and the viscosity of liquid glass compositions.

The availability and low cost of raw materials, as well as the simplicity of the used production technologies increase the competitiveness of building materials from secondary mineral resources and by-products of industrial production. There are industrial enterprises in the city of Bratsk and the region of Eastern Siberia, whose by-products, as well as secondary mineral resources, have stable chemical composition and physical properties that determine the main directions of their use in the construction industry.

Microsilica is a large-tonnage fine by-product in the production of ferrosilicon at the company "Bratsk Ferroalloy Plant" LLC. Microsilica has a stable chemical composition with a predominant content of aluminosilicate component. These properties determined its usage as a basis for the production of liquid glass according to the low-power production method, which was developed and improved by scientists of the Department of Building Materials Science and Technology of FSBEI HE "BrSU" (BrGU).

It is known that to increase the performance properties of liquid glass compositions, finely ground porous fillers, fibrous materials, hardeners, and additives are introduced into their composition. The chemical composition of the fillers is represented mostly by the aluminosilicate component.

Black shale are dispersed by-products obtained in the processing during the extraction of gold by gold mining enterprises in the Bodaybinsky district of the Irkutsk region and are mainly represented by alumina silicate composition. The volume of processed black shale rocks by the enterprises of the Bodaybinsky district is about 20 million m^3 per year. So far, only gold is of industrial interest in these materials, and their implementation after processing is not conducted at all, black shale is being stored in stock piles and storages. The size of black shales is close to 0.01 mm. The condition for the preparation of liquid glass compositions is the dispersity of the filler, so black shale can be considered as a raw material of high degree of technological readiness. The use of local raw materials will expand the range of liquid glass compositions, reduce their cost, and eliminate scarce materials.

The purpose of this work is to develop and obtain fire retardant coatings for wood using resource-saving technologies. To achieve the goal, of the following tasks are set:

- assessment of the use of local raw materials;
- study of black shale as a filler in the composition of the liquid glass composition;
- development and optimization of compositions for the preparation of a liquid glass composition;
- development of technology for producing liquid glass composition;
- study of the rheological properties of the raw material mixture, fire retardant ability, adhesive strength, and overall efficiency of a fire retardant coating for wood.

2   Materials and Methods

Experimental samples of wood of pine and larch with a size of $30 \times 60 \times 150$ mm, which were previously dried to constant weight, were used for the study.

When developing liquid glass compositions and obtaining fire retardant coatings for wood, sodium liquid glass was used as a binder, obtained with the participation of the authors from microsilica, a by-product formed during the production of ferrosilicon at the company "Bratsk Ferroalloy Plant" LLC, and also from silicate lump (GOST 13078-81(Russian State standard)). The authors established that the best fire retardant properties, as well as adhesion to a wooden surface, were achieved using liquid glass from microsilica with a silicate module n = 3 and density ρ = 1.25 g/cm³. The authors proposed a method for improving the process of preparing liquid glass from microsilica in relation to fire retardant coatings. To improve the adhesiveness of liquid glass, the additive surfactant was added in its composition. When calculating the composition of liquid glass, 10% of microsilica was replaced with fine ground quartz sand corresponding to GOST 8736-2014, which made it possible to improve the quality of liquid glass from microsilica. Table 1 presents the main technical characteristics of microsilica liquid glass. Table 2 presents the technical characteristics of sodium liquid glass from silicate lumps.

Table 1. Technical characteristics of sodium liquid glass from microsilica

Liquid glass properties		Compositions, wt.%		
Silicate module	Density, g/cm³	SiO_2	Na_2O	H_2O
3	1.25	25.05	8.35	66.6

Table 2. Technical characteristics of sodium liquid glass from silicate lumps (GOST 13078-81)

Density, g/cm^3	Viscosity V 3–4, s	Mass fraction of oxides, %					Silicate module	Solid residue
		SiO$_2$	R$_2$O$_3$	CaO	SO$_3$	Na$_2$O		
1.36	25–30	74.18	0.25	0.05	0.06	25.46	2.93	50

As a dispersed filler, in developing compositions of liquid glass compositions, the host valuable components were used - black shales crushed as a result of processing in gold mining at gold mining enterprises in the Bodaybinsky district of the Irkutsk region. Table 3 presents the chemical composition of black shale, averaged over samples. There is a small amount of impregnated pyrite emissions (0.63–2.3 wt.%) in samples of black shale.

Table 3. Chemical composition of black shale (wt.%)

SiO$_2$	Al$_2$O$_3$	FeO	MgO	Fe$_2$O$_3$	K$_2$O	CO$_2$	CaO	Na$_2$O
59.1	16.55	4.6	3.15	2.75	2.6	2.6	1.83	1.45

As an additive surfactant to provide the necessary conditions for wetting wooden surfaces, synthetic foaming agent PO-6 [TU 0258-148-05744685-98 (Russian Technical Regulations)] was introduced into the composition of liquid glass compositions, which is the composition of an aqueous solution of triethanolamine salts of primary alkyl sulfates with stabilizing additives.

The liquid glass composition was prepared by introducing an additive of surfactant into the liquid glass and the subsequent introduction of a fine-dispersion filler with constant stirring using a high-speed mixer. The resulting composition was applied to the surface of wooden samples with a paint brush in three layers.

Conditional viscosity of the obtained raw mixture was determined in laboratory conditions, using a viscometer to determine the viscosity of coating compositions VZ-4 (GOST 9070-75 N).

The adhesion of the fire retardant liquid glass composition to wooden surfaces was determined experimentally, according to the cross-cut test method described in GOST 15140. The essence of this method is to apply cross-cuts to the finished coating and visual assessment of the condition of the coating using a four-point system.

The flame retardant efficiency of the developed fire retardant coatings was assessed using an experimental laboratory setup using the "fire tube" method described in GOST 16363. To do this, the coated wood sample was exposed to fire for 2 min at a gas burner flame height of 23–25 cm. After fire tests, the weight loss of the test samples was calculated in percentage. The optimization of the compositions of fire retardant liquid glass compositions was carried out using mathematical modeling methods. The "MODEL_NR" program was used for mathematical processing of the results.

In the paper, modern methods of physicochemical studies are used: X-ray phase, synchronous thermal, electron microscopic.

3 Results

Based on scientific research and laboratory experiments, the optimum viscosity of the raw mixture was established, which is 25–30 s by the VZ-4 viscometer. It is known that the properties of fire retardant compositions depend on their composition and structure. Viscosity is the most important physico-chemical characteristic of liquid glass compositions, which allows obtaining high-quality coatings of painting consistency. It is the composition viscosity that determines the uniform application and coating of various wooden surfaces without prior treatment (both smooth and rough, with all the bulges and bumps), which, in turn, determines the adhesion and the required fire retardant efficiency of the material, as well as ensures optimal consumption of liquid glass compositions. The optimal viscosity of the liquid compositions is provided when the content of the filler in an amount of from 10 to 25 wt.%. In this case, the yield tensile strength, which characterizes the shear resistance and indicates the amount of the liquid glass composition that does not flow under its own weight on a vertically oriented wooden surface, was from 10 to 15 g/100 cm^2.

According to the results of the study, the optimal limits for the content of raw materials in the compositions of fire retardant compositions, depending on the liquid glass used, were established. The authors considered the process of forming the structure of liquid glass compositions. The introduction of black shale as a quartz-containing filler (SiO2-59.1%) will increase the silicate modulus of the composition and accelerate the process of structure formation. Liquid glass from microsilica is more viscous, contains a certain amount of undissolved microsilica particles. Therefore, for liquid glass from microsilica, a smaller amount of filler and a larger amount of surfactant additive are required to provide the necessary wetting condition for the wood. Figures 1 and 2 show the effect of the amount of black shale on the fire retardant efficiency (Fig. 1) and adhesion (Fig. 2) depending on the binder used.

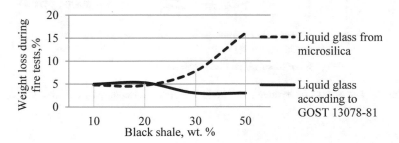

Fig. 1. The effect of black shale on the fire retardant efficiency of coatings

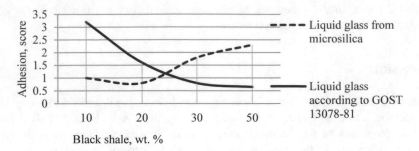

Fig. 2. The effect of black shale on the adhesion of fire retardant coatings

On the basis of the conducted studiy, it was established that to obtain fire retardant coatings based on liquid glass from microsilica, the optimum content of black shales is of 10–25 wt.%. Fire retardant coating based on liquid glass according to GOST 13078-81 is characterized by the best indicators of adhesion and fire retardant efficiency when the content of black shale is of 20–35 wt.%.

Figures 3 and 4 show the effect of surfactant on fire retardant efficiency (Fig. 3) and adhesion (Fig. 4) of fire retardant coatings for wood depending on the binder used. The best indicators of fire retardant efficiency and adhesion of compositions based on liquid glass from microsilica are achieved by using a surfactant in the amount of 1–2 wt.%. For compositions based on liquid glass according to GOST 13078-81, the content of surfactant additive in the amount of 1% will be optimal.

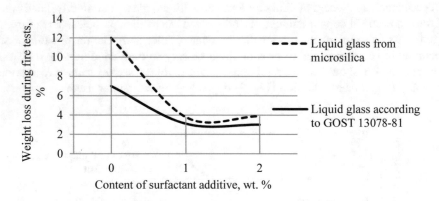

Fig. 3. The effect of surfactants on the fire retardant efficiency of coatings for wood

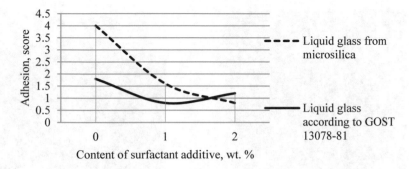

Fig. 4. The effect of surfactants on the adhesion of fire retardant coatings for wood

Using the results of X-ray phase analysis, it was established that the products of hardening of liquid glass compositions are aluminosilicate compounds of the frame-work structure similar to natural zeolites and micas: quartz (SiO_2), muscovite ($KAl_3Si_3O_{10}(OH)_2$), corundum (Al_2O_3), hematite (Fe_2O_3) (Table 4). For fire retardant compositions based on liquid glass from microsilica and based on liquid glass according to GOST 13078, an identical phase composition is established with a small discrepancy in quantitative ratios. All identified minerals are temperature resistant and are used in the refractory industry. This study confirmed the efficiency of liquid glass from microsilica as a binder in the composition of fire retardant compositions.

Table 4. Phase composition of flame retardant compositions at the age of 12 months, depending on the binder used

Revealed new formations	Fire retardant composition on the basis of liquid glass according to GOST 13078	Fire retardant composition based on liquid glass from microsilica
Quartz (SiO_2), %	80.86	79.51
Muscovite ($KAl_3Si_3O_{10}(OH)_2$), %	13.92	16.99
Corundum (Al_2O_3), %	3.43	1.48
Hematite (Fe_2O_3), %	1.79	2.02

The results of electron microscopic studies (see Fig. 5) established the penetration of the fire retardant composition into wood fibers, which is evidence of good adhesion of the fire retardant coating to the wooden surface. The thickness of the fire retardant coating on the surface of the wood was about 200 µm (0.2 mm).

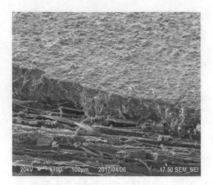 

Fig. 5. Electronic microscopic images of a wood sample with a fire retardant coating (in section). Zooming 100

According to the results of thermal analysis, the effect of the fire retardant action of the developed liquid glass compositions was determined, which is explained by the formation of pores upon swelling and filling them with water vapor and gases that do not support combustion.

4 Discussion

As a result of scientific research, the compositions and method for producing liquid glass compositions for protecting wood from fire using liquid glass according to GOST 13078-81 and liquid glass from microsilica made by low energy-intensive technology as a binder have been developed. The technical and operational performance of fire retardant coatings based on liquid glass and black shale with the addition of PO-6 surfactant has been studied.

5 Conclusions

– It was established that during the structure formation of the developed liquid glass compositions with black shales and the addition of the PO-6 surfactant, the compounds are formed: quartz (SiO_2), muscovite ($KAl_3Si_3O_{10}(OH)_2$), corundum (Al_2O_3), hematite (Fe_2O_3), having high temperature resistant.
– The optimum viscosity of the raw material mixture 25-30 s according to VZ-4 viscometer was established, which ensures uniform application and high-quality coating, which was achieved when from 25 to 40% of black shale was contained in liquid glass according to GOST 13078 and from 10 to 25% black shale was contained in the composition of liquid glass from microsilica. When the additive surfactant PO-6 in the amount of 1% was used in the composition, the penetration of the liquid glass composition into the wood fibers was established, which made it possible to ensure high adhesion of the developed fire retardant materials to wooden surfaces.

- In case of high-temperature exposure, pores are formed during swelling, filled with gases that do not support combustion. It is established that a three-layer liquid composition with black shales and a foaming additive PO-6, about 200 μm thick (0.2 mm) on the surface of wood, corresponds to the 1st group of fire retardant efficiency according to GOST 16363-98.

The economic efficiency and manufacturability of the liquid glass composition are due to the small number of components, the use of available raw materials of a high degree of technological readiness.

Involvement in the production of secondary mineral resources provides economic efficiency, environmental stability and safety. The use of liquid glass from microsilica as a binder to create a fire retardant composition allows expanding the range of building materials, reducing energy costs, and simplifying the technology for producing fire retardant compositions.

The composition of the raw material mixture for producing a fire retardant coating for wood is confirmed by the RF patent No. 2613515 (the date of State registration in the State Register of Inventions of the Russian Federation is March 16, 2017).

Safety is confirmed by the radiation control protocol No. 7763 dated September 25, 2018 of the accredited testing laboratory center of the East-Siberian Railway Branch of the Federal Budgetary Healthcare Institution "Center for Hygiene and Epidemiology for Railway Transport".

References

1. Spear, M.J.: Preservation, protection and modification of wood composites. In: Wood Compo-sites, pp. 253–310 (2015)
2. Cueff, G., Mindeguia, J.-C., Dréan, V., Breysse, D., Gildas, A.: Experimental and numerical study of the thermomechanical behaviour of wood-based panels exposed to fire. Constr. Build. Mater. **160**, 668–678 (2018)
3. Makovická, F.K., Osvaldová, L., et al.: The effect of synthetic and natural fire-retardants on burning and chemical characteristics of thermally modified teak wood. Constr. Build. Mat. **200**, 551–558 (2019). https://doi.org/10.1016/j.conbuildmat.2018.12.106
4. Carosio, F., Cuttica, F., Medina, L., Berglund, L.A.: Clay nanopaper as multifunctional brick and mortar fire protection coating—wood case study. Mat. Des. **93**, 357–363 (2016)
5. Olawoyin, R.: Nanotechnology: the future of fire safety. Saf. Sci. Part A **110**, 214–221 (2018)
6. Kozlowski, R., Muzyczek, M.: Smart environmentally friendly composite coatings for wood protection. In: Smart Composite Coatings and Membranes Transport, Structural, Environmental and Energy Applications Woodhead Publishing Series in Composites Science and Engineering, pp. 293–325 (2016)
7. Vakhitova, L.: Fire retardant nanocoating for wood protection. In: Nanotechnology in Eco-efficient Construction. Materials, Processes and Applications Woodhead Publishing Series in Civil and Structural Engineering, 2nd edn., pp. 361–391 (2019)

Parametric Optimization of Steel Frames Using the Job Search Inspired Strategy

Igor Serpik[(⊠)] [iD]

Bryansk State Engineering Technological University,
Stanke Dimitrov av. 3, 241037 Bryansk, Russia
inserpik@gmail.com

Abstract. An algorithm for parametric optimization of steel plane frames is proposed, which is based on a meta-heuristic job search inspired strategy and genetic algorithms using the main and elite populations. A feature of this computational scheme is the ability to effectively solve an optimum problem without introducing penalty functions. This allows us to strictly consider the constraints in any algorithm run. Tension-compression deformations, bending and pure torsion of the rods are taken into account. The goal is to minimize the mass of the frame rods, taking into account active constraints on stresses, displacements and overall stability, including the stability of individual rods. The cross sections of rods vary on discrete sets of admissible options. Analyzing the considered structure's deformations is performed using the finite-element method in the form of the displacement approach. The constraints on strength and stiffness are checked by iterative calculation of the stress-strain frame state, using a tangential stiffness matrix, which is formed considering the influence of normal forces on the bending of the rods. Information about the convergence of this computational process is required for the stability assessment of the structure. At the elite population formation stage, additional control is carried out on the stability of the frame's options on the basis of checking the positive definiteness of the tangent stiffness matrix. The efficiency of the proposed approach to the optimal design of frame structures is illustrated by the example of a rod system made of round tubes.

Keywords: Steel frames · Optimization · Meta-heuristic algorithms ·
Job search inspired strategy · Overall stability

1 Introduction

At the present time, the proportion of buildings and constructions with steel load-bearing structures is increasing. At the same time frame systems are widely used. Designing such objects is often associated with the choosing the cross sections of rods on discrete sets of feasible options. The approach of meta-heuristic algorithms is among the most effective ones in optimal design. In order to optimize the load-bearing systems, the following meta-heuristic approaches have shown themselves quite effectively: genetic algorithms [1], particle swarm approach [2], harmony search [3], imperialist competitive algorithm [4], dolphin echolocation [5], simulated annealing [6], ray algorithm [7], colliding bodies optimization [8], mine blast algorithm [9], firefly

© Springer Nature Switzerland AG 2020
V. Murgul and M. Pasetti (Eds.): EMMFT 2018, AISC 982, pp. 682–691, 2020.
https://doi.org/10.1007/978-3-030-19756-8_64

algorithm [10], teaching-learning-based optimization [11] and others. The problem of optimizing the steel frame structures using meta-heuristic algorithm is discussed in numerous researches [12–14]. The complicated combination of constraints has to be taken into account in the optimal design of structural systems, such as conditions for ensuring their strength, rigidity and stability. Constraints are usually taken into account on the basis of penalty functions when meta-heuristic iterative algorithms are used for deformable objects [15]. In many cases this leads to a distortion of the problem conditions and significant instability of the solutions obtained. According to the research [16], a meta-heuristic algorithm for parametric optimization of plane girders, based on job search inspired (JSI) strategy, was proposed. An optimal search algorithm with constraints in this strategy is placed in accordance with the possible behavior of the person seeking employment with the highest salary, while satisfying his preferences for working conditions and the requirements of employers. This strategy allows to effectively take constraints into account without penalty functions. The meta-heuristic algorithm using the JSI strategy to optimize steel frames is developed. At the same time, genetic operations of mutation, selection and crossing-over are performed in order to implement the number of steps in this strategy. A farm project is interpreted both as a job vacancy and as an individual in a population. The constraints on the overall stability of the structure are verified based on a combination of a simplified scheme. Such scheme involves analyzing the convergence of the iterative calculating process for structure using the finite-element method and a tangential stiffness matrix [14]. The positive definiteness of the matrix, when an individual is included in the elite population, is controlled.

2 Setting of the Optimization Problem

It is believed that the steel frame is made of straight rods with cross-sections that are constant in length. After minimizing the weight W of all rods in the structure:

$$W(S_1, S_2, \ldots, S_N) \rightarrow min \tag{1}$$

where S_n is the set of permissible cross-sections for rods of a group $n(n = 1, .2, \ldots, N)$, N is the total number of such groups.

It is believed that in the general case the following limitations can be taken into account:

1. Strength Condition:

$$\Phi_\sigma = \max_{i=1,\ldots,i_o} \frac{\sigma_{mi}}{R_y} \leq 1 \tag{2}$$

where Φ_σ is the indicator characterizing the satisfaction of the stress limit for the frame as a whole, σ_{mi} is the highest von Mises stress in bar i, i_o is the total number of bars, R_y is the design steel resistance assigned by yield point (SP 16.13330.2017 SNiP II-23-81*, in Russian).

2. Stiffness requirements. For each node j of the discretized structure in the direction of the OX_k axis of the Cartesian coordinate system, the following condition needs to be fulfilled:

$$\Phi_\delta = \max_{\substack{j=1,\,\ldots,\,j_o \\ k=1,\,2,\,3}} \frac{|\delta_{jk}|}{[\delta_{jk}]} \leq 1 \qquad (3)$$

where Φ_δ is the indicator characterizing the satisfaction of the stiffness limit for the structural system, δ_{jk} is the projection of the node j displacement vector of the on the OX_k axis, $[\delta_{jk}]$ is the modulus of this value, j_o is the total number of nodes.

3. The overall stability of the frame structure, including the stability of individual rods.
4. Providing local structural strength.
5. Local stability of thin-walled rod elements.
6. The stability of the plane rod bending form.

Constraints 1–3 are active and directly taken into account during the optimization process. The remaining constraints are taken into account when selecting the initial prerequisites for optimal design and are monitored after the search is over.

When controlling the fulfillment of active constraints, it is assumed, that the material of the structure works in the conditions of linear elasticity. In the general case, deformations of tension-compression, bending and pure torsion of the rods are considered. The stress-strain state calculation of an object is performed using the finite-element method in the form of the displacement approach [17]. The iterative process is carried out, the calculation of the first iteration is performed using ordinary stiffness matrices of the finite-element method, and the following system of linear algebraic equations is solved in each subsequent iteration r:

$$[K]_\tau^{(r)} \{\delta\}^{(r)} = \{Q\} \quad (r = 2, 3, ..) \qquad (4)$$

where $[K]_\tau^{(r)}$ is the tangential stiffness matrix [18, 19] of the finite-element model, determined by the following dependence

$$[K]_\tau^{(r)} = [K] + \left[K_G \left(\{N\}^{(r-1)} \right) \right] \qquad (5)$$

$\{\delta\}^{(r)}$ is a vector of nodal displacements calculated in iteration r, $\{Q\}$ is an equivalent nodal force vector, $[K]$ is a casual stiffness matrix of a finite-element model, $\left[K_G \left(\{N\}^{(r-1)} \right) \right]$ is a geometric matrix of a finite elements system expressed in terms of normal forces grouped into a vector $\{N\}^{(r-1)}$, obtained in iteration $r-1$.

In order to check the condition 3, a solution to the eigenvalue problem for matrices is not provided. According to the rationale of the research [14], it is obtained that the following requirement should be fulfilled for a given iteration number $r_o \geq 3$.

$$\Phi_b = 1 + \left| \frac{U^{(r_o)} - U^{(r_o-1)}}{U^{(r_o-1)}} \right| \leq \alpha \tag{6}$$

where Φ_b is the indicator characterizing the fulfillment of this requirement, $U^{(r_o)}$, $U^{(r_o-1)}$ is the deformation energy of the discretized object, obtained in iterations r_o and $r_o - 1$, respectively, α is a given positive number.

As shown by the calculations, a sufficiently effective verification of this type is provided for $r_o = 5$, $\alpha = 1.001$. If all the regulatory requirements are fulfilled, the iterative process (4) almost converges in 3–5 iterations.

Satisfying the positive definiteness condition of the tangent stiffness matrix of the system for every $r \geq 2$ iteration was considered in addition to testing general stability. This check is carried out within the framework of the decomposition procedure of the matrix into triangular factors.

3   Algorithm of Optimal Search

Within the JSI strategy framework it is believed that the job applicant is aimed at getting the job with the highest salary $F = 1/W$. The set of vacancies V, each of which is a set of parameter values of an applicant, serves as a discrete set where the search is performed. Relatively fast search actions (analyzing vacancies, sending CVs, making phone calls, etc.) and much more time-consuming actions in an interview format (with possible exams) are considered. It is believed that the applicant receives information on the values of F for vacancies within the framework of "quick" actions, as well as information on the partial fulfillment of their own conditions and employers' requirements (optimization constraints T_1). Testing for a number of more labor-intensive constraints, which will be denoted as T_2, is carried out at the interview. In addition, the criterion T_3 that can be used for additional assessments of the best jobs is introduced. Constraints T_1 are taken into account when defining discrete sets of valid rod sections and ensuring the condition $F > F_A$. Constraints T_2 are associated with satisfying inequalities (2), (3), (6). Checking the provision of positive definiteness of tangent matrix was taken as an indicator T_3.

Let us highlight the set of vacancies $V_1 \subset V$, for which T_1 constraints are fulfilled. Let the applicant to choose a number of vacancies v_i at the initial stage in some way from a set $V_{1A} \subset V_1$, that satisfies the salary condition $F > F_A$, where F_A is a given value, which may later change.

Let us operate with the main population Π with the size N_Π and an auxiliary elite population Ψ, the size of which depends on the results of the iterative process, but does not exceed the value N_Ψ. Initially $F_A = 0$ is set, the population Π is formed from the sections with maximum serial numbers within the sets of permissible rod cross sections, and the population Ψ is left empty at this stage. For each $s \geq 1$ iteration of the JSI strategy, the following sequence of steps will be considered:

Step 1. Part replacement from vacancies v_i is carried out using random variations, in compliance with the requirement $v_i \in V_{1A}$ for new vacancies. A mutation is performed for individuals of the population Π using the work algorithms [20, 21]. If the iteration number is greater than a certain number s_1, then this procedure is implemented for randomly selected parameters $n_1 = \max(1, \lfloor \lambda n_o \rfloor)$ in each individual of the population, where λ is the specified value on the $(0 \le \lambda < 1)$ segment, n_o is the total number of parameters. When $s \le s_1$, the number n_1 can be obtained by multiplying λ by the number d $(1 < d \le \lfloor n_o/n_1 \rfloor)$. For each parameter to be changed, the value p_a of the interval [0, 1] is selected using a random number generator with a uniform distribution law and is compared with the value $m_a(0 < m_a < 1)$. If $p_a > m_a$, any of the valid parameter values can be selected with equal probability, otherwise the position number of the parameter in the set of valid options is assigned by randomly changing its current value by 1–2 units. A mutation operation for an individual can be performed multiple times until the condition $F_i \ge F_A$ is satisfied.

Step 2. The condition T_2 is checked for the considered vacancy group. If these conditions are met for any vacancy v_i, $F_A = F_i$ is assumed, where F_i is the salary value of this vacancy. In this case, the following actions are implemented:

   2a. Satisfaction of constraints T_2 for population Π individuals is checked. Thus, the proficiency ratio value of each i-th individual is determined from the point of view of the JSI strategy:

$$k_p = \frac{1}{\max(\Phi_\sigma, \Phi_\delta, \Phi_b)} \tag{7}$$

   Fulfilling the conditions $k_p \ge 1$ and $F_i > F_A$ leads to the appointment of a new value $F_A = F_i$.

   2b. Each of the individuals of the population Ψ is sequentially added to the population Π, which is larger in terms of the k_p value for the worst individual of the population Ψ and satisfies the constraint T_3. At the same time, the set of gene values of an individual i has to be absent in the elite population. In case the Ψ population size becomes equal $N_\psi + 1$, the variant of the project with the smallest k_p value is excluded from it.

   2c. Checking the individuals of the population Ψ for the fulfillment of the condition $F_i \ge F_A$ was carried out. If this condition is not met, then the individual is removed from the database. If the value F_A was changed at stage 2a, then at stage 2c only the individuals that satisfy the condition $F_i = F_A$ can remain in the database.

Step 3. Based on results of checking the conditions T_2, assessing the proficiency ratio of the applicant is performed in relation to vacancies v_i. On the basis of these results, a group of vacancies v_i^c is selected from the set V_{1A}. The group is close to vacancies v_i, for which proficiency ratio will be greatest. At the same time, selection and single-point crossingover operations are performed. Individuals with a greater value of k_p ratio are considered more adapted. The roulette method, which defines the length of the cut on the segment [0, 1] for the individual i of is used as follows when selecting pairs of individuals:

$$\Delta_i = t_i / \sum_{j=1}^{N_\Pi} t_j \tag{8}$$

where

$$t_j = k_{pj}^{\beta} \tag{9}$$

k_{pj} is value k_p for the individual j, β is a given constant.

Step 4

4a. The operations of step 2 are implemented on the basis of the population Π, obtained according to the crossover results. In this case, when replenishing the database Ψ from the database Π, additional check of the $F_i \geq F_A$ condition for the individual is performed, since crossingover can cause its violation.

4b. The $F_i \geq F_A$ condition is checked for all individuals of the population Π. If this requirement is violated for the considered individual, it is replaced by the best individual placed in the population Ψ, provided it is not present in the population Π. If there is no such individual, then it is replaced by an individual, which is specified by randomly chosen values of the variable parameters. The search result for the current iteration is the individual corresponding to the last F_A value found at steps 2 or 4.

Studying the performance of the proposed iterative procedure showed that it is advisable to adopt $N_{\Pi} = 15...25$, $N_{\Psi} = 10...20$, $\beta = 100...150$, $m_a = 0.9$, $\lambda = 0.1...0.2$, $s_1 = \theta n n_o$, d = 3...6, where $\theta = 0.2...0.5$.

4 Results

Let us consider the capability of the developed computational scheme on the example of optimizing the plane framework (Fig. 1), which is subjected to deformations both in and out of its plane. It was assumed that the rods are made of round pipes according to GOST 32931-2015 "Truby` stal`ny`e profil`ny`e dlya metallokonstrukcij. Texnicheskie uslo-viya". S245 steel was taken as the pipe material (SP 16.13330.2017 SNiP II-23-81*, in Russian). The elasticity modulus of the material $E = 2.06 \cdot 10^5$ MPa and specific gravity $\chi = 77$ kN/m^3 were taken into account. The frame has supports in the nodes A, B in the form of hinged points and is fixed in the nodes D by displacements in the direction of the axis OZ. Concentrated forces $P_1 = 295$ kN, $P_2 = 100$ kN, $P_3 = 50$ kN, and distributed load $q = 28.5$ kN/m were taken into account, as well as the weight of the rods, which was adjusted during the optimization process depending on the values of the variable parameters. $R_y = 240$ MPa was taken into account. 215 rod finite elements were used for discretization of the structure.

The rods were combined into 16 groups (Table. 1) For each of the groups, any of the following 19 combinations of D × s cross section pipe sizes was allowed: 73 × 3.5, 89 × 4, 102 × 4, 140 × 3, 140 × 3.5, 152 × 3.5, 152 × 4, 152 × 5, 152 × 5.5, 168 × 6, 168 × 7, 177.8 × 8, 219 × 10, 219 × 11, 219 × 13, 219 × 16, 219 × 18, 219 × 20, 273 × 18 (mm), where D is the outer diameter, s is the thickness. The iteration process was carried out under the following conditions: $N_{\Pi} = N_{\Psi} = 20$, $\beta = 120$, $n_1 = 1$, $s_1 = 300$, d = 5. Ten runs of the optimization process were performed with implementing the 5000 iterations of the optimal search. As a result, structure types with the values of W, which fit into a rather short numerical segment [72.1...72.9] (kN) were obtained for 9 launches. In one run it turned out $W = 80.5$ kN. In this case, the smallest of the W values was achieved in 5 runs, which corresponds to high stability of the results for algorithms of this type.

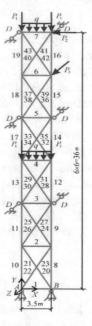

Fig. 1. Framework structure

Data on selecting the cross sections for the best solutions are given in the Table. 1. Post-processor checks for this variant of the load-bearing system confirmed satisfying all the constraints set. The convergence graphs of the optimal search procedure associated with achieving the best and worst results by the value of the objective function are represented in Fig. 2.

Table 1. Combining rods into groups and the best optimization result.

Group number	Rod number	$D \times s$ (mm)
1	1–3	73 × 3.5
2	4	177.8 × 8
3	5, 6	73 × 3.5
4	7	168 × 6
5	8, 9	219 × 10
6	10, 11	273 × 18
7	12	73 × 3.5

(*continued*)

Table 1. (*continued*)

Group number	Rod number	$D \times s$ (mm)
8	13	273 × 18
9	14–16	273 × 18
10	17–19	219 × 13
11	20–23	152 × 3.5
12	24–27	152 × 3.5
13	28–31	140 × 3.5
14	32–35	152 × 4
15	36–39	152 × 4
16	40–43	152 × 4

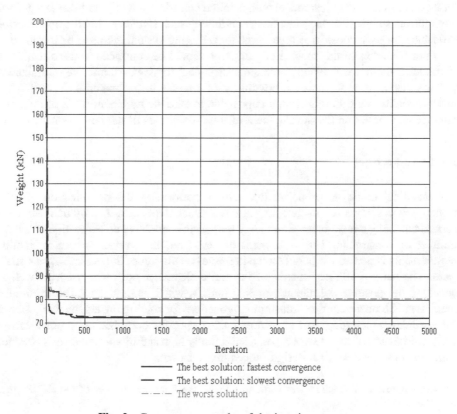

——— The best solution: fastest convergence
— — The best solution: slowest convergence
— · — · — The worst solution

Fig. 2. Convergence graphs of the iterative process.

5 Discussion

It should be noted that, compared with the algorithms for the optimal design of frame structures considered in the researches [12–14], the JSI strategy eliminates the difficulties associated with selecting the conditions for introducing penalty functions. In contrast to the procedures studied in the researches [12, 13], the proposed approach ensures that both the rod stability and the overall stability of the load-bearing system are monitored. Compared to the research [14], checking the overall stability is not limited to analyzing the convergence of the iterative process of calculating the structural stress-strain state based on using the tangent stiffness matrix. During the selection of elite specimens, a direct check of the positive definiteness of these matrices is carried out, which makes it possible to guarantee the accuracy of final decisions on ensuring overall stability. This additional check practically does not increase the labour content of the calculations, since it is realized within the framework of the tangent matrix decomposition into the product of triangular matrices.

In any case, the designer has the opportunity to perform a refined assessment of the structure example working under load, which is a result of an optimum search process. In particular, it is possible to calculate this object in a geometrically nonlinear formulation based on the shell analytical model with detailed consideration of both local and general strength and stability. At the same time, conducting such studies within the optimization procedure for each considered type of the load-bearing system seems to be irrational.

Further studies of the considered problems should be continued in the direction of introducing the overall stability assessment for steel frameworks into the optimization process, taking into account the possibility of stresses in the material exceeding the proportionality limit. In this case, it is possible to take the nonlinear work of materials into account, based on the development of the possibilities of the iterative scheme (4).

6   Conclusions

An analytical model is proposed that allows performing the optimization of steel framework structures on discrete sets of cross sections of rods. The optimal search is carried out using a meta-heuristic strategy that ensures effective consideration of a large number of constraints. The assessment of satisfying the overall structural stability requirements is performed based on the iterative calculation of the object's stress-strain state. The tangent stiffness matrix of the finite element model was used, as well as checking the positive definiteness of this matrix, which does not require the implementation of a relatively time-consuming procedure for solving the eigenvalue problem for matrices. According to the example of a specific optimization problem for a framework made of round steel pipes, a sufficiently high stability of the results obtained using the presented optimal design procedure is shown.

Acknowledgment. The study was carried out with the financial support of the RFBR grant No. 18-08-00567.

References

1. McCall, J.: Genetic algorithms for modelling and optimization. J. Comput. Appl. Math. **184** (1), 205–222 (2005)
2. Perez, R.E., Behdinan, K.: Particle swarm approach for structural design optimization. Comput. Struct. **85**, 1579–1588 (2007)

3. Lee, K.S., Geem, Z.W., Lee, S.-H., Bae, K.-W.: The harmony search heuristic algorithm for discrete structural optimization. Eng. Optim. **37**(7), 663–684 (2005)
4. Kaveh, A., Talatahari, S.: Optimum design of skeletal structures using imperialist competitive algorithm. Comput. Struct. **88**(21–22), 1220–1229 (2010)
5. Kaveh, A., Farhoudi, N.: A new optimization method: Dolphin echolocation. Adv. Eng. Softw. **59**, 53–70 (2013)
6. Lamberti, L.: An efficient simulated annealing algorithm for design optimization of truss structures. Comput. Struct. **86**(19-20), 1936–1953 (2008)
7. Kaveh, A., Khayatazad, M.: A new meta-metaheuristic method: ray optimization. Comput. Struct. **112–113**, 283–294 (2012)
8. Kaveh, A., Mahdavi, V.R.: Colliding bodies optimization method for optimum design of truss structures with continuous variables. Adv. Eng. Softw. **70**, 1–12 (2014)
9. Sadollaha, A., Bahreininejada, A., Eskandarb, H., Hamdia, M.: Mine blast algorithm for optimization of truss structures with discrete variables. Comput. Struct. **102–103**, 49–63 (2012)
10. Miguel, L.F.F., Lopez, R.H., Miguel, L.F.F.: Multimodal size, shape and topology optimization of truss structures using the firefly algorithm. Adv. Eng. Softw. **56**, 23–37 (2013)
11. Degertekin, S.O., Hayalioglu, M.S.: Sizing truss structures using teaching-learning-based optimization. Comput. Struct. **119**, 177–188 (2013)
12. Alberdi, R., Khandelwal, K.: Comparison of robustness of metaheuristic algorithms for steel frame optimization. Eng. Struct. **102**, 40–60 (2015)
13. Tejani, G.G., Bhensdadia, V.H., Bureerat, S.: Examination of three meta-heuristic algorithms for optimal design of planar steel frames. Adv. Comput. Des. **1**(1), 79–86 (2016)
14. Serpik, I.N., Alekseytsev, A.V., Balabin, P.Y., Kurchenko, N.S.: Flat rod systems: optimization with overall stability control. Mag. Civil Eng. **76**(8), 181–192 (2017)
15. Stolpe, M.: Truss optimization with discrete design variables: a critical review. Struct. Multi. Optim. **53**(2), 349–374 (2016)
16. Serpik, I.N.: A metaheuristic job search inspired strategy for optimization of bearing structures (in Russian). In: Proceedings of the 7th Scientific and Practical Internet Conference on Interdisciplinary Research in Mathematical Modelling and Informatics, Tolyatti, Russian Federation, pp. 40–43 (2016). https://elibrary.ru/item.asp?id=25678372
17. Zienkiewicz, O.C., Taylor, R.L., Fox, D.: The Finite Element Method for Solid and Structural Mechanics. Elsevier, Oxford (2014)
18. Wriggers, P.: Nonlinear Finite Element Methods. Springer, Heidelberg (2008)
19. Sergeyev, O.A., Kiselev, V.G., Sergeyeva, S.A.: Overall instability and optimization of bar structures with random defects in case of constraints on faultless operation probability (in Russian). Mag. Civil Eng. **44**(9), 30–41 (2013)
20. Serpik, I.N., Alekseytsev, A.V., Balabin, P.Y.: Mixed approaches to handle limitations and execute mutation in the genetic algorithm for truss size, shape and topology optimization. Period. Polytech. Civil Eng. **61**(3), 471–482 (2017)
21. Serpik, I.N., Mironenko, I.V., Averchenkov, V.I.: Algorithm for evolutionary optimization of reinforced concrete frames subject to nonlinear material deformation. Proc. Eng. **150**, 1311–1316 (2016)

Stress-Strain State Generation Within a Stress Concentration Zone Using the Photoelasticity Method

Lyudmila Frishter[✉] [iD]

Moscow State University of Civil Engineering, Yaroslavskoye Shosse 26,
129337 Moscow, Russia
lfrishter@mail.ru

Abstract. Combined stress-strain state of composite assemblies and structures occurs within a stress concentration zone due to both the boundary form or "geometric factor" and the finite discontinuity of specified forced deformations, stress-strain characteristics appearing on the contact surface of structural elements. Analysis of the local stress-strain state within a composite range of essential structural non-homogeneity is valid for calculation and design of structural elements of nuclear power plants, power-generation facilities, waterworks in case of discontinuities of the structural boundary form, e.g. angular or stepped form. The article briefly describes analysis methods for specific solutions of the elasticity problem caused by the boundary form or "geometric factor". The experimental photoelasticity method, which is a continual method, and the forced deformation unfreezing method enable the generation of the stress-strain state within the stress concentration zone on the contact surface of composite structures with a forced deformation jump on models made of an optically responsive material. Brief summary is provided regarding capabilities of the experimental method of deformation unfreezing for solution of forced deformation problems. The experimental solution was obtained on a model with an angular cut-out of the boundary, where the apex encloses the forced (thermal) deformation discontinuity. The fringe pattern, obtained on the model using the methods of photoelasticity and deformation unfreezing in the stress concentration zone – within the angular cut-out and forced deformation jump areas – is characterized by considerable gradients of experimental data: the fringe order (isochrome), parameters of isoclines, stresses and strains. Modern techniques of experimental data visualization based on digital shooting enhance capabilities of the photoelasticity method. The purpose of this work is to assess the capabilities for generation of stress state deformations within an area, which is as close as possible to the stress concentration area during digital shooting and data processing using standard stress separation methods.

Keywords: Tress-strain state · Stress concentration · Forced deformations ·
Area fragmentation · Data processing · Energy structures

© Springer Nature Switzerland AG 2020
V. Murgul and M. Pasetti (Eds.): EMMFT 2018, AISC 982, pp. 692–700, 2020.
https://doi.org/10.1007/978-3-030-19756-8_65

1 Introduction

Studies of the local stress-strain state (SSS) within a complex area of essential structural non-homogeneity is valid for calculation and design of structural elements of nuclear power plants, power-generation facilities, waterworks in case of discontinuities of the structural boundary form, e.g. angular or stepped form. The stress-strain state (SSS) of composite assemblies and structures is characterized by a considerable stress concentration at interfaces between elements with different boundary designs: singular lines, points, e.g. re-entrant angle, etc. Combined SSS is generated within a stress concentration zone due to both the boundary form or "geometric factor" and the finite discontinuity of specified forced deformations, stress-strain characteristics entering the irregular point of the area boundary.

The experimental photoelasticity method, which is a continual method, and the forced deformation unfreezing method enable the SSS generation within the stress concentration zone on the contact surface of composite structures with a forced deformation jump on models made of an optically responsive material [1–13].

Numerical calculation techniques in the stress concentration zone, based on computational region discretization, require additional assessments for SSS within the area of the elastic body's singular point for verification of solution results.

The behavior of solutions of Laplace, Poisson and elliptical equations for domains with non-smooth boundaries is addressed by works of G.L. Hesin, M.M. Froch, A. Durelli, W. Riley, A. Ya Aleksandrov, M.Kh. Akhmetzyanov, A. Kobayashi, V. A. Kondratyev, V. V. Fufayev, M. L. Williams, Ya. S. Uflyand, A. I. Kalandiya, G. P. Cherepanov, D. B. Bodzhi, O. K. Aksentyan, K. S. Chobanyan, I. T. Denisyuk, I.A. Razumovsky [1], Koshelenko, Poznyak [2], Kuliyev [3] and others.

The fundamental work of V. A. Kondratyev proves that the solution of the general elliptic boundary condition in the neighborhood of irregular points of the domain boundary is represented by an asymptotical series and an infinitely differentiable function. Members of this series contain solutions of homogeneous boundary problems for model domains: a wedge or a cone. These solutions depend on local characteristics – the measure of the solid or plane angle and the boundary condition type, mechanical behavior of piecewise homogeneous bodies. Values of solution expansion coefficients in the neighborhood of the singular point are unknown and depend on the problem in general. Methods for determination of the said expansion coefficients are complicated and hardly accessible for practical determination of stresses in composite structures with a complex boundary form.

Works of I. T. Denisyuk provide elastic solution asymptotics for the plane complex domain with corner points on the boundary line. Advantages of the fundamental study of the singular solution of the boundary problem in the work of Kuliyev [3] include the possibility of application of its results for problems of residual stresses and crack propagation across the interface of heterogeneous materials [1, 3–12]. It was shown [1, 7, 12–17] that the stress singularity series depends on the degree of approximation of the apex of a crack to the interface area of materials with different modules.

Works of M. L. Williams, referred to by many authors, state that stresses, deformations, Airy's stress function near the apex of a sector with straight-line sides have a power series solution. In works of A. I. Kalandiya, K. S. Chobanyan, L. A. Bagirov, O. K. Aksentyan, G. P. Cherepanov, V. P. Netrebko, V. Z. Parton and many others, the solution of a homogeneous boundary problem in the neighborhood of an irregular boundary point is regarded as a power series.

In works Aksentyan, Netrebko [7] for SSS study in the neighborhood of an irregular point on a singular line of the elastic body boundary a local curvilinear coordinate system is introduced to be used for Lame's equations. At the approach to the irregular boundary point from within the domain, the elastic problem solution is reduced to solution of two homogeneous elastic problems: plane deformation and out-of-plane deformation or antiplane shear.

Presentation of the elastic problem solution in the neighborhood of an irregular point on the singular boundary line as two homogeneous plane elastic problems is possible, if: (a) specified forced deformations, volume forces are persistent across the elastic body domain; (b) specified forced deformations, volume forces are piecewise continuous functions, while the jump of values of forced deformations, volume forces across the internal domain interface reaches the singular line of the body boundary; (c) the interface of domains V_1 and V_2 of the elastic complex body V, having different stress-strain behaviors $E_i, v_i, i = 1, 2$ respectively, reaches the singular line of the body boundary. In case of (c), the elastic problem solution in the neighborhood of an irregular boundary point is reduced to two plane elastic problems for a complex body.

In works [15–17], the similarity theory is used in studies of the elastic problem solution for movements in the neighborhood of an irregular point on a singular line of the domain boundary.

They cover a small neighborhood of an irregular point on the singular line – line of discontinuity, e.g. boundary conditions or first-order derivatives of the domain surface function. For the small neighborhood of an irregular point of the surface boundary, the following similarity group is applied: $x_1 = tx;$ $y_1 = ty;$ $z_1 = z; t > 0,$ where t – group parameter [12–14]. While writing the Lame's equation in the small neighborhood of an irregular point on the singular line and proceeding to the limit at $t \to +\infty$, the elastic problems solution is reduced to solution of two homogeneous plane elastic problems: plane deformation and out-of-plane deformation.

The works G. P. Cherepanov introduces the concept of a canonical singular problem that characterizes the SSS singularity in the neighborhood of an irregular boundary point, for which two following theorems are provided.

Any canonical singular problem is corresponded by a transcendental equation, where each root corresponds to a homogeneous solution, and the number of arbitrary real constants in this solution is equal to the root multiplicity. In the infinitesimal neighborhood of a singular point, the solution of a correct boundary problem in the elasticity theory behaves as the asymptomatically highest proper function of a corresponding canonical singular problem.

Analytical solutions of the elasticity problem in the area of an irregular point of the domain boundary line are characterized by the solution singularity, resulted from idealization of the boundary form.

Theoretical and experimental studies of the stress concentration resulted from the boundary form are described in works of G. Neyber, R. Peterson, N. G. Savin and V. I. Tulchy, B. N. Ushakov, I. P. Fomin, N. A. Makhutov, V. V. Vasilyev, V. P. Netrebko and many others.

Freezing of workpiece deformations and consequent annealing (unfreezing) of a composite polymer model with the use of the photoelasticity method enable the SSS generation in the model corresponding to the desired result [1–7]. The advantage of the freezing method is that reproduction of the temperature pattern in the model is not required, while standard experimental equipment is used and SSS modeling for structures with complex boundary forms (geometrical concentration) is carried out.

The freezing method is virtually universal and recommended for solution of the plane elasticity problem, as well as for some cases of volume problem characterized by the presence of temperature effect limitations. For particular cases, when thermal deformations in one direction do not cause any stress, many engineering problems are solved: studies of the thermal stress state of structural elements of nuclear reactors, energy equipment, in weld and threaded joints, structures with concentration zones, in machine engineering, hydraulic engineering, etc. [1, 2, 4–7, 15–17].

Modeling of problems with standard forced deformations, not satisfying the conditions of compatibility of strain components, is described in works of G. S. Vardanyan, N. I. Prigorovsky, S. E. Bugayenko, M. N. Dveres, B. N. Evstratov, I.A. Razumovsky, V. N. Bronov, V. N. Savostyanov, L. Yu. Frishter and others.

The disadvantage of the unfreezing method, associated with the impossibility of volume deformation modeling on models due to material incompressibility in the high-elastic state, has been overcome in works of S. E. Bugayenko, N. I. Prigorovsky, M. N. Dveres, B. N. Evstratov, V. N. Savostyanov, D. I. Omelchenko and others.

The fringe pattern, obtained on the model using the methods of photoelasticity and deformation unfreezing in the stress concentration zone is characterized by considerable gradients of experimental data: the fringe order (isochrome), parameters of isoclines, stresses and strains. Modern techniques of experimental data visualization based on digital shooting enhance capabilities of the photoelasticity method. Therefore, it becomes necessary to assess the capabilities for generation of stress state deformations within an area, which is as close as possible to the stress concentration area during digital shooting and data processing using standard stress separation methods.

The purpose of this work is to analyze the capabilities for generation of stress state deformations using the unfreezing method within an area of a boundary angular cut-out under the influence of rupturing forced (thermal) deformations with the use of experimental data visualization during digital shooting and data processing using standard stress separation techniques based on the photoelasticity method.

2 Materials and Methods

The experimental solution of the elasticity problem with forced deformations for a plane domain, using a composite beam 180 mm long and 24 mm wide as an example model, involves the use of the deformation unfreezing method [1, 2, 15–17].

In one of the beam areas Ω_2 thermal deformations $\alpha T \delta_{ij}$ are created, and the other area Ω_1 is not stressed. The thermal deformation jump across the interface reaches the irregular boundary point O(0, 0) – the apex of the domain cut-out.

Experimental solutions are analyzed in the neighborhood of the boundary cut-out apex for the beams with the ends having different apex angles: (a) squared end $2\alpha = 180^0$; (b) "skew" squared end $\alpha + \beta = 180^0$, $\alpha = 105^0$, $\beta = 75^0$; (c) symmetrical cut-out angle end $2\alpha = 260^0$.

The use of detailed experimental data from digital shooting enables fragmentation of the beam end area in such a way that standard stress separation techniques of the photoelasticity method [1, 2, 13, 14] become applicable: graphical method and method of shear stress difference in the area, which is as much close to the stress concentration zone as possible.

By selecting different fragments of the beam area, both isostatic patterns for the beam end in general and isostatic patterns for the area adjacent to the stress concentration zone have been obtained.

As an example, Fig. 1 provides a catalog of experimental data for a selected fragment of area Ω_2 of the beam end with the apex angle $2\alpha = 260^0$.

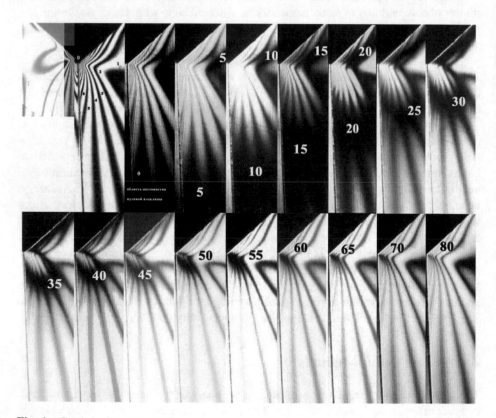

Fig. 1. Catalog of experimental data for a selected fragment of the beam end area with the apex angle.

For the selected fragment of the beam end area, Fig. 2 provides experimental data: (a) fringe pattern; (b) combined fringe and isocline pattern.

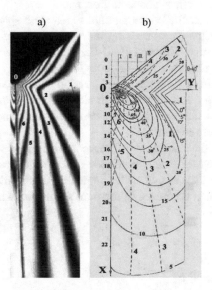

Fig. 2. Input data for the beam end area with the apex angle: (a) isochrome pattern; (b) combined isochrome and isocline pattern with a pitch.

According to Fig. 2, isostatic lines have been constructed for the selected area fragment using the graphical method, as shown in Fig. 3.

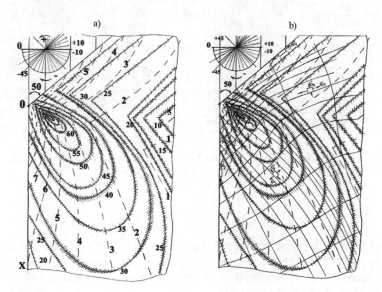

Fig. 3. Graphical method of stress separation in the beam end area with the apex angle: (a) isochrome and isocline pattern with indication of primary stress directions; (b) combined isostath, isochrome and isocline pattern.

According to experimental data in Fig. 2, with the help of the shear stress difference method stress distribution diagrams have been constructed in several vertical beam sections quite close to the stress concentration zone.

Auxiliary sections I, II, III, IV are parallel to axis and space from this axis by the distance, equal to respectively, mm. Using the shear stress difference method, stresses have been obtained in midsections I-II, II-III, III-IV, spaced from axis OX by the distance equal to mm, mm, mm, mm respectively.

Figure 4 provides stress diagrams in midsection of the beam end.

The stress separation result shows that in all reviewed beam sections with different boundary cut-out angles there are points, where, and areas inclined at to the primary area are in the pure shear conditions.

Points, where pure shear areas are observed, are located in the summits of acute angles of isochromes and their neighborhoods. The stress separation result shows that the line connecting the summits of acute angles of isochromes, entering the cut-out apex O(0, 0) at the domain boundary, is the pure shear line, where pure shear areas are observed in every point.

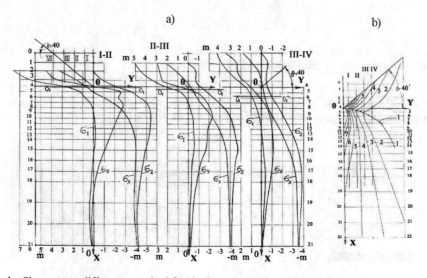

Fig. 4. Shear stress difference method for the beam end area with the apex angle: (a) stresses in midsections I-II, II-III, III-IV of the beam end; (b) pure shear line.

3   Results

Using plane models of the photoelasticity and deformation unfreezing method, the stress state has been generated in the area of the angular boundary cut-out, the apex of which includes the forced (thermal) deformation rupture. The obtained results of stress separation show that modern capabilities of experimental data visualization during digital shooting and data processing, application of standard stress separation methods enable the generation of the stress state in the neighborhood adjacent to the stress concentration zone, which enhances the capabilities of the photoelasticity method for experimental solution analysis. For the domain of a singular elastic solution (stress concentration), where fringe patterns are unreadable at any magnification of the domain fragment, it is necessary to develop a method, which enables extrapolation of reliable experimental data on the stress concentration zone.

4 Discussion

Studies of the stress-strain state of a structure in the area of the angular boundary cut-out, the apex of which includes the forced (thermal) deformation rupture, have multiple approaches. Analytical solution in the area of an irregular boundary point – the angular cut-out apex – is characterized by the solution singularity, resulted from idealization of the boundary form. Numerical calculation techniques in the stress concentration zone, based on computational region discretization, require additional assessments for SSS within the area of the elastic body's singular point for verification of solution results. The photoelasticity and deformation unfreezing methods enable the generation of the stress state within an area, which is as close as possible to the stress concentration area during digital shooting and data processing using standard stress separation methods. The advantage of the freezing method is that reproduction of the temperature pattern in the model is not required, while standard experimental equipment is used and SSS modeling for structures with complex boundary forms (geometrical concentration) is carried out. The unfreezing method is virtually universal and recommended for solution of the plane elasticity problem. For the domain of a singular elastic solution (apex of the angular boundary cut-out), fringe patterns are unreadable at any magnification of the domain fragment. Taking into account advantages of one or another approach to SSS studies in the structural heterogeneity zone under the influence rupturing forced deformations, it is advisable to apply an integrated approach: experimental, analytical or numerical for data evaluation.

5 Conclusions

Modern techniques of experimental data visualization based on digital shooting and data processing, application of standard stress separation methods enable the stress state generation in the neighborhood adjacent to the stress concentration zone, thus enhancing the capabilities of the photoelasticity method for experimental solution analysis.

References

1. Razumovsky, I.A.: Interference-Optical Methods of Deformable Solid Mechanics. Publisher MGTU named after N. Bauman, Moscow (2007). (in Russian)
2. Koshelenko, A.S., Poznyak, G.G.: Theoretical foundations and practice of photomechanics in mechanical engineering, Granitsa, Moscow (2004). (in Russian)

3. Kuliev, V.D.: Singular Boundary Value Problems. Nauka, Moscow (2005). (in Russian)
4. Pestrenin, V.M., Pestrenina, I.V., Landik, L.V.: The stress state near a singular point of flat composite design. Vestnik TGU Math. Mech. **4**, 80–87 (2013). (in Russian)
5. Vardanjan, G.S., Savostyanov, V.N., Frishter. L.Ju.: The development of methods of experimental mechanics. In: Makhutova, O.N., et al. (eds.) IMASH RAS, Moscow, pp. 60–68 (2003). (in Russian)
6. Albaut, G.N., Kharinova, N.V., Sadovnichy, V.P., Semenova, Zh.I., Fedin, S.: Nonlinear problems of mechanics of destruction. Bull. Lobachevsky Univ. Nizhni Novgorod **4**, 1344–1348 (2011). (in Russian)
7. Netrebko, V.P.: Photoelasticity method-based study of stress intensity factors near inclined cracks in orthotropic plates, pp. 69–77. Mechanical Engineering Research Institute of the Russian Academy of Sciences, Moscow (2003). (in Russian)
8. Xu, L.R., Kuai, H., Sengupta, S.: Exp. Mech. **44**, 608–615 (2004)
9. Xu, L.R., Kuai, H., Sengupta, S.: Exp. Mech. **44**, 616–621 (2004)
10. Yao, X.F., Yeh, H.Y., Xu, W.: Int. J. Solid Struct. **43**, 1189–1200 (2006)
11. Frishter, L.: The analysis of the methods of research the influence of nonuniform local stress on a crack resistance of protective coating. In: MATEC Web of Conferences, vol. 86, p. 04022 (2016)
12. Frishter, L.: Evaluations of the solution to the homogeneous plane problem of the theory of elasticity in the neighborhood of an irregular boundary point. In: MATEC Web of Conferences, vol. 117, p. 00047 (2017)
13. Vardanyan, G.S., Frishter, L.Yu.: Int. J. Comput. Civ. Struct. Eng. **3**, 75–81 (2007)
14. Frishter, L.Yu.: Photoelasticity-based study of stress-strain state in the area of the plain domain boundary cut-out area vertex. In: Advances in Intelligent Systems and Computing, vol. 692, pp. 836–844 (2017). https://doi.org/10.1007/978-3-319-70987-1_89
15. Matveenko, V.P., Nakarjakova, T.O., Sevodina, N.V., Shardakov, I.N.: Stress singularity at the top of homogeneous and composite cones with different boundary conditions. J. Math. Mech. **72**, 477–484 (2008). (in Russian)
16. Albaut, G.N., Tabanyukhova, M.V., Kharinova, N.V.: Determination of the first stress intensity factor in elements with angled cut-out. In: Experimental Mechanics and Calculation of Structures (Kostinsky Readings), pp. 166–175. Moscow State University Press, Moscow (2004). (in Russian)
17. Frishter, L.: Stress-deformed state in the plane domain boundary angle cutout zone. In: MATEC Web of Conferences, vol. 196, p. 01036 (2018)

Scanning the Layered Composites Using Subminiature Eddy-Current Transducers

Sergey Dmitriev[1] , Alexey Ishkov[2] , Alexey Grigorev[1] ,
Lilia Shevtsova[3] , and Vladimir Malikov[1(✉)]

[1] Altai State University, Lenina ave. 61, 656038 Barnaul, Russia
osys@me.com
[2] Altai State Agricultural University, Krasnoarmeyskiy ave. 98,
656049 Barnaul, Russia
[3] Novosibirsk State Technical University, Karla Marksa ave. 20,
630073 Novosibirsk, Russia

Abstract. On the basis of superminiature eddy current transducer, a measuring system is designed that allows scanning of composite materials for detection of small defects. The results of the study of discontinuity defects simulated in metal-polymer layered composites of aluminum-polyethylene-aluminum systems up to 10-layer composite are described. Composites were made by alternating layers of aluminum or copper foil, 20…100 μm thick, with layers of a film, 20 μm thick. Defects in the structure of the material were modeled by skipping or increasing the number of separate layers, and defects in continuity and bridges were modeled by cutting a round or rectangular hole in the foil or dielectric layer. Visual images of model defects are obtained by Fourier-transformer of the IENM-5FA device. The dependence of the eddy current transducer signal on the presence of a defect in the layered structure is established.

Keywords: Eddy-current transducer · Composite materials ·
Electrical conductivity · Defect

1 Introduction

The eddy-current method (ECM), together with ultrasonic scanning and fluoroscopy, is one of the main methods to inspect defects for the purpose to control various materials and products in modern technology. The method is based on the analysis of an interaction of the electromagnetic field of a special sensor—the eddy-current transducer (ECT)—and the object being investigated. The properties of the investigated object, the nature and topology of the defects detected by ECM, design of such transducers and the principles taken as a basis for defects examination change depending on the goals.

For example, to determine the thickness of conductive and dielectric coatings applied on various structures, as well as to study the properties of conductive non-ferromagnetic materials correlating with conductivity, ECTs are often made according to a scheme of attachable differential transformer transducer with at least two coils: transmitting or energizing that induces magnetic field in the sample under investigation

© Springer Nature Switzerland AG 2020
V. Murgul and M. Pasetti (Eds.): EMMFT 2018, AISC 982, pp. 701–708, 2020.
https://doi.org/10.1007/978-3-030-19756-8_66

and energizes eddy-currents in it, and pickup or measuring coil designed to measure the characteristics of the eddy-current field.

At present, metal-polymer laminated composites (MPLC) hold a specific place among objects subject to non-destructive testing. The simplest MPLC is, for example, sheet materials of facing panels, honeycomb panels, sound and heat insulation panels for aircrafts, coatings reflecting the radiation on the equipment and instrumentation, decorative metal-plastic panels, materials for the production of printed circuit boards for radio-electronic devices, and other similar composites with one or two and more metal layers separated by dielectric interlayers. MPLC includes metal-plastic pipes widely used in everyday life where a metallic, aluminum layer of the composite is located between layers of cross-linked high-density polyethylene and low-density polyethylene (HD polyethylene, LD polyethylene) of different thickness.

Defects inspection of MPLC is performed to determine the following standard defects: defect of the metallic and (or) polymer layer uniformity, the number and thickness of the layers, conducting and non-conducting strips between layers, deformation of the metal layer surface, change in the state of boundary between the metal and polymer layers of the MPLC. In this case, one of the universal parameters sensitive to all the listed defects of MPLC is the local electrical conductivity of the material and its distribution over the surface, associated with the topology of the electromagnetic field in the material interacting with these defects.

According to the literature review, there are many types of inspections to evaluate composites and there are many proposed methods by researchers for each of which. For damage identification in aircraft composite structures, aircraft composites assessment, and health monitoring of aerospace composite structures the suggested methods in the literature is ultrasonic testing [1, 2], thermographic testing [3, 4], vibration methods [5, 6], infrared thermography [7], shearography [8], and XCT [9].

Ultrasonic testing is the most applied method in health monitoring of a composite wing-box structure [10], damage identification in aircraft composite structures [11], aircraft composites assessment [8], health monitoring of aerospace composite structures [5], and structural health monitoring (SHM) [2].

Pulse echo ultrasonic method can readily locate defects in homogeneous materials. In this method, the operator more concerns about the transit time of the wave and the energy loss due to attenuation and wave scattering on flaws. It helps to locate inconsistency in a material whether it is homogeneous or heterogeneous [12]. For large defect detection, location, and imaging purposes, and quality control, ultrasonic pulse velocity measurements are quite suitable [13].

Electromagnetic Testing (ET) methods use magnetism and electricity to detect and evaluate fractures, faults, corrosion or other conditions of materials. ET induces electric currents, magnetic fields, or both inside a test object and observes the electromagnetic response. Electromagnetic (EM) methods include Eddy Current Testing (EC) [14], Remote Field Testing, Magnetic Flux Leakage and Alternating Current Field Measurement. In each of these techniques, the underlying physics is fundamentally different as the fields described by different classes of partial differential equations [15].

Earlier, we reported on the successful development and testing of a virtualized measuring instrument - a conductivity meter for non-ferromagnetic materials—CMNFM-5FA [14, 15].

The subminiature ECT (SMECT) [18, 19] of the original design is used as a sensor in this device, it is made according to a differential scheme of switching on of the coils of a transformer ECT and allowing to localize the control area up to 50 μm^2. Measurement of local electrical conductivity on the surface of a sample under investigation signal of the measuring coil of SMECT in real-time mode makes it possible to use this device as the defectoscope, comparing the data obtained at the object under investigation with the image of the defect previously obtained on the material model.

The purpose of this work was to investigate the possibility to detect defects such as discontinuity of MPLC based on aluminum and LDPE type A1-(LDPE-Me)p-A1, as well as visualization of such defects by means of CMNFM-5FA.

2 Materials and Methods

To conduct the research, the CMNFM-5FA was used, equipped with a wireless hand-held SMECT with a locality of 0.1–0.5 mm^2.

The software of the CMNFM-5FA was installed on a PC. The energizing parameters of the SMECT: current 30–35 mA, voltage 3.5–4.5 V, operating frequency 0.3–300 kHz. Absolute calibration of the device was carried out was provided according to a standard copper sample with a certified electrical conductivity of 415 ± 2 mS/m.

MPLC and models of their defects were made by interchanging layers of aluminum or copper foil with thickness of 20–100 μm and with layers of polyethylene film with thickness of 20 μm or with layers of paraffined paper with thickness of 50–100 μm. Samples were cut out of these materials with a size of 20 × 50 mm, a package of the given structure was formed, a foil layer or dielectric with a defect was inserted and pressed at a temperature of 110–120 °C and pressure of 5–10 MPa within 3.5 min. Defects in the MPLC structure were simulated by skipping or increasing the number of separate layers, and defects of continuity and strips were simulated by cutting out of a foil or dielectric layer in a form of a circular or rectangular hole.

3 Results and Discussion

The main characteristic of MPLC sensitive to investigated defects is the electrical conductivity of the material of the metal layer.

As it is shown in [17], it enters the nonhomogeneous Helmholtz equation for the vector potential of the eddy-current field of a multilayer medium along with the magnetic permeability of the material and approaches the value of the electrical conductivity of a solid bulk material as the number of layers in the MPLC increases and the thickness of the dielectric layers decreases. At the same time, the contribution to the value of the voltage introduced into the measuring winding of the SMECT with high locality and the ratio of the radii of the measuring and transmitting coils is not less than 0.2–0.4 from each new layer will be from 10 to 25% which is enough to record it by any measuring instrument with an absolute permissible error not exceeding 3.5%.

Increase in the thickness or the number of dielectric layers in the MPLC is resulted in approximately the same values of the input to the value of the applied voltage of the

SMECT. However, in order to make a decision on the suitability or unavailability of a sample of the MPLC under investigation only by the value of this parameter it is required to adjust the defectoscope for the gap between the material surface and the SMECT, since in the standard design it defines the measured parameter of the sensor like the values given above. For this purpose, the dependence of the readings at CMNFM-5FA on the distance between the surface of the MPCL and the SMECT was obtained.

At the same time, the following method is known to inspect defects. In order to ensure that the readings at the device are adjusted depending on the gap between the sensor and the sample surface within wide limits and increase the sensitivity, the material under investigation is placed on the surface of a more massive known and defect-free material.

The dependence of the readings at CMNFM-5FA is shown in Fig. 1 when removing its sensor from the surface of the MPLC A1- LDPE-A1, located on a copper base with 5 mm thickness.

MPLC used in this experiment was obtained by interchanging two layers of aluminum foil (20 μm) with one layer of high-density polyethylene (20 μm), measurements were made at exciting frequency of 35 kHz, and the base material was copper.

As it follows from Fig. 1, a sample of this MPLC acts as a screen with respect to the copper base, this screen is inserted into the gap between the sensor of the device and the copper base and it gradually weaken the measured signal. At a distance of 100–250 μm from the MPLC surface, there is an area where the readings remain constant. It is obvious that the topology of the eddy-current field of the sample under investigation at such a distance from the conductive copper base is received by the sensor of the device as a field from a homogeneous medium and the appearance of any defect in the material will cause deviation of its readings from the mean values measured at a distance of 100–250 μm.

In this case, the defects inspection of MPLC is possible when setting up the device to zero on the defect-free area (with a constant value of reference electrical conductivity) and using a hollow probe tip that provides the constant distance between the material surface and the SMECT in a range of 100–250 μm.

The measurements carried out by us for model defects of the type of violation of the number and order of layers for multilayer MPLC systems A1 (Cu)-LDPE-A1 (Cu) showed a change in the reference electrical conductivity up to 3–7 mV in the defect area, with an absolute error of 0,1-0.1-0.3 conventional units, i.e. 20–30-time excess of the defect signal over the noise. The appearance of any of these defects in the area under investigation causes a sudden change in the topology of the field and the relative readings of the device.

The defect may also be represented by amplitude-time dependence of the readings, that has been already implemented in CMNFM-5FA, with manual or automatic scanning of the object's surface. Diagnostics of MPLC continuity defects, interlayer closure, strips, etc. is possible with direct contact scanning of the sample surface by a sensor. Therein, time is converted into the position of the sensor relative to the starting point of the movement, taking into account the speed of the sensor.

In this case, the dependence of the amplitude of the sensor signal on the sensor position in real time, displayed by CMNFM-5FA, can serve as a defect image.

In this case, the main frequency and amplitude of the signal are displayed on the left screen, and the transducer coordinate—along the X axis.

Figures 2 and 3 show typical images of some MPLC model defects of layer discontinuity type, obtained with a device, when a sample of MPLC was placed on a dielectric base.

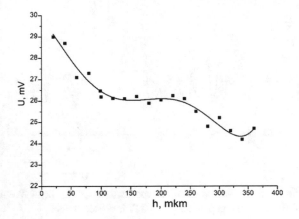

Fig. 1. Dependence of the reference electrical conductivity of a 3-layer MPLC L1 - LDPE-L1 placed on a copper base on the gap to the SMECT.

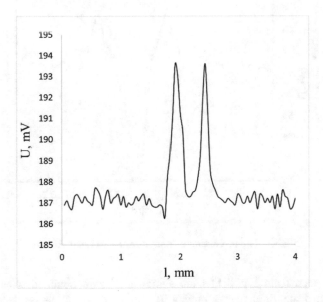

Fig. 2. Dependence of the reference electrical conductivity of a 3-layer MPLC L1 - LDPE-L1 placed on a copper base on the gap to the SMECT. Visible defect.

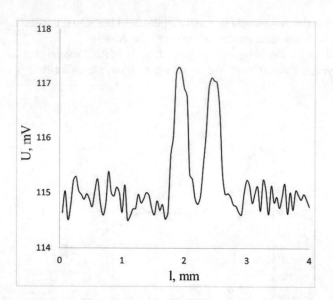

Fig. 3. Dependence of the reference electrical conductivity of a 3-layer MPLC L1 - LDPE-L1 placed on a copper base on the gap to the SMECT. Invisible defect.

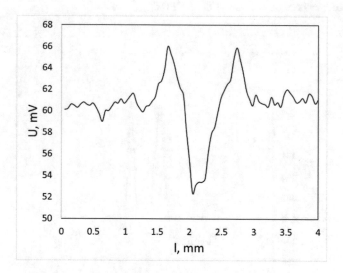

Fig. 4. The discontinuity image in the isolated Al-layer

It can be seen from Fig. 3 that the signal amplitude obtained when scanning a defect of layer discontinuity type, hidden in a 2-layer MPLC, as per selected control mode is in general similar to the signal amplitude obtained from the visible defect (Fig. 2).

When obtaining a signal from such a defect, the device displays the dependence of the amplitude of its signal on the transducer position. This allows to perform initial adjustment of the device and provides the operator's training in the diagnosis of defects directly from their visual models, hidden between the layers of MPLC.

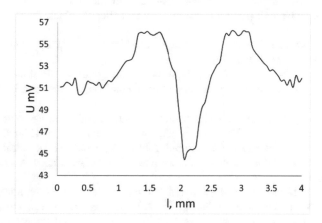

Fig. 5. The discontinuity image in the 2-layer Al-LDPE composite, scanned from the dielectric side

With an increase in the number of composite layers over the defect, the image of the defect significantly changes on the screen (Figs. 4 and 5).

As it can be seen from Fig. 4, there is an obvious signal omission during the continuous scanning of MPLC in the defect image, however, an area of significant increase in the signal amplitude appears in the area of defect at main excitation frequency, that is symmetrically bounded on the left and on the right by less intense signals.

If the material structure becomes more complicated and the defect is removed, this area gradually expands (Fig. 5), that provides its more correct identification both in the lower, inaccessible layer, and in the inner layers. At the same time, the dependence of the frequency of the SMECT signal on the level of its amplitude makes it possible to easily reconstruct the eddy-current field topology near the defect and along the MPLC layers.

4   Conclusions

1. The readings of CMNFM-5FA when measuring the electrical conductivity of MPLC type A1 (Cu) - (LDPE-A1 (Cu)) n-A1 (Cu) using SMECT with locality up to 50 μm^2 do not depend on the gap in the range from 100 to 250 μm.
2. High locality of the sensor and the determined dependence of electrical conductivity of MPLC on model defects make it possible to carry out defect inspection of materials directly as per CMNFM-5FA readings.
3. The type of MPLC defect, its localization in the layers, and the field topology near the defect can be identified by its image obtained using the device.

References

1. Katunin, A., Danczak, M.: Automated identification and classification of internal defects in composite structures using computed tomography and 3D wavelet analysis. Arch. Civ. Mech. Eng. **15**(2), 436–448 (2015)
2. Staszewski, W.J., Mahzan, S., Traynor, R.: Health monitoring of aerospace composite structures – active and passive approach. Comput. Sci. Technol. **69**(11–12), 1678–1685 (2009)
3. Katunin, A., Dziendzikowski, M.: Damage identification in aircraft composite structures. Compos. Struct. **127**, 1–9 (2015)
4. Maierhofer, C., Steinfurth, H.: Characterizing damage in CFRP structures using flash thermography in reflection and transmission configurations. Comput. Part B: Eng. **57**, 35–46 (2014)
5. Panopoulou, A., Roulias, D.: Intelligent health monitoring of aerospace composite structures based on dynamic strain measurements. Expert Syst. Appl. **39**(9), 8412–8422 (2012)
6. Fassois, S.D., Marioli-Riga, Z.P.: Vibration-based skin damage statistical detection and restoration assessment in a stiffened aircraft panel. Mech. Syst. Signal Process. **22**(2), 315–337 (2008)
7. Meola, C., Carlomagno, G.M.: Infrared thermography to evaluate impact damage in glass/epoxy with manufacturing defects. Int. J. Impact Eng. **67**, 1–11 (2014)
8. Růžek, R., Jironč, J.: Ultrasonic C-Scan and shearography NDI techniques evaluation of impact defects identification. NDT and E Int. **39**(2), 132–142 (2006)
9. Spearing, S.M.: Comput. Part A: Appl. Sci. Manuf. **52**(1), 62–69 (2013)
10. Grondel, S., Delebarre, C.: Health monitoring of a composite wingbox structure. Ultrasonics **42**(1–9), 819–824 (2004)
11. Polimeno, U., Meo, M.: Detecting barely visible impact damage detection on aircraft composites structures. Comput. Struct. **91**(4), 398–402 (2009)
12. Warnemuende, K.: Amplitude modulated acousto-ultrasonic non-destructive testing: damage evaluation in concrete. Ph.D. Wayne State University, Ann Arbor (2006)
13. Oguma, I.: Ultrasonic inspection of an internal flaw in a ferromagnetic specimen using angle beam EMATs. Przeglad Elektrotechniczny **88**(7B), 78–81 (2012)
14. Hoshikawa, H., Kojima, G.: Eddy current nondestructive testing for carbon fiber-reinforced composites. J. Pressure Vessel Technol. **135**(4), 041501 (2013)
15. Yang, S.-H., Kim, K.-B.: Non-contact detection of impact damage in CFRP. NDT & E Int. **57**(1), 45–51 (2013)
16. Dmitriev, S.F., Malikov, V.N.: Subminiature eddy-current transducers for conductive materials and layered composites research. In: Advances in Intelligent Systems and Computing, vol. 692, pp. 655–665 (2017)
17. Dmitriev, S., Ishkov, A., Malikov, V., Sagalakov, A., Katasonov, A.: Non-destructive testing of the metal-insulator-metal using miniature eddy current transducers. In: IOP Conference Series: Materials Science and Engineering, vol. 71, pp. 1–6 (2015)
18. Dmitriev, S., Malikov, V., et al.: Subminiature eddy current transducers for studying semiconductor material. J. Phys.: Conf. Ser. **643**, 1–7 (2015)
19. Malikov, V., Davydchenko, M., Dmitriev, S., Sagalakov, A.: Subminiature eddy-current transducers for conductive materials research. In: International Conference on Mechanical Engineering, Automation and Control Systems, Tomsk, pp. 1–4 (2015)

Method for Studying Deformation of Non-woven Heat-Insulating Building Materials

Liubov Lisienkova[1]([⊠]) [iD], Lyudmila Nosova[2] [iD],
Ekaterina Volkova[3] [iD], and Ekaterina Baranova[3] [iD]

[1] Moscow State University of Civil Engineering,
Yaroslavskoe shosse, 26, Moscow 129337, Russia
lisienkovaln@mail.ru
[2] South Ural State Humanitarian and Pedagogical University,
Chelyabinsk, Russia
[3] South Ural State University (National Research University),
Zlatoust Branch, Chelyabinsk, Russia

Abstract. The aim of the study was to develop a method and means of the research of the deformation of non-woven heat-insulating building materials. The characteristics of materials deformation under various conditions of cyclic compression have been studied, the parameters and test modes have been determined. The design and principle of operation of a device for cyclic compression have been shown. The method of research of the materials deformation with different compression parameters has been presented, and the results of experimental tests of samples of non-woven heat-insulating materials have been presented. It has been shown that the method of cyclic compression allows predicting the behavior of non-woven heat-insulating materials in the processes of installation and operation of products. A promising application of the method is to study the patterns of change in the thermal conductivity of materials under cyclic compression conditions.

Keywords: Heat-insulation · Energy saving building structures ·
Non-woven heat-insulating materials · Deformation · Cyclic compression

1 Review of Construction Heat-Insulating Materials

The large-scale construction carried out in Russia requires a huge amount of building materials, among which thermal insulation materials produced in different ways and from different types of raw materials are very important. The role of thermal insulation is constantly increasing due to the general trend of reducing heat losses and saving energy. The range of existing building heat-insulation materials is quite wide and covers light (soft) insulation, as well as heat and windproof, withstand increased load.

Nowadays, traditional heat-insulating building materials based on mineral fibers (Russian standard specification /GOST (RSS) 31309–2005), mineral wool boards with a synthetic binder (RSS 9573–96), heat-insulating products made of glass staple fibers (RSS 10499–95), boards and products heat-insulating polyfoam based on

© Springer Nature Switzerland AG 2020
V. Murgul and M. Pasetti (Eds.): EMMFT 2018, AISC 982, pp. 709–722, 2020.
https://doi.org/10.1007/978-3-030-19756-8_67

phenol-formaldehyde resins (RSS 20916–87, RSS 22546–77), mats of mineral wool heat-insulating (RSS 21880–94), heat-insulating materials foam (RSS 22 546–77), thermally insulated mineral wool mats vertically layered (RSS 23307–78), mineral wool (RSS 4640–93), thermal insulation boards made of mineral wool on a bitumen binder, etc. are widely known on the Russian building materials market. Fiberglass insulation materials under the ISOVER brand are also known: ISOVER construction insulation, ISOTEC technical insulation - for insulation of pipelines, air ducts, industrial tanks, chimneys, various technological equipment. Traditional heat-insulation materials (mats, boards, mineral wool, etc.) are used as (RSS 31309–2009):

- non-loaded thermal insulation (for horizontal surfaces of mansard and attic rooms, etc.) with a bulk density of not more than 75 kg/m3;
- loadable thermal insulation in terms of the impact of the following types of load: compression, tension and separation of the layers.

Thermal insulation materials based on mineral fibers have significant drawbacks: during operation, fine dust of basalt and glass fibers and harmful substances (hydrocarbon vapors) can be released. They have a negative effect on the respiratory system and human skin [1, 2]. Organic or inorganic binders, as well as components of dedusting organic additives in materials based on mineral fibers, also have a negative effect on human health and the environment. To strengthen the structure of insulating materials made of glass or basalt fibers, phenol-formaldehyde resins are used. They are destroyed during long-term operation. Formaldehyde is released into the environment and atmosphere, which is also harmful to human health.

2 Prospects and Problems of Using Non-woven Heat-Insulating Materials in Construction

In recent years, non-woven heat-insulating materials produced from synthetic fibers (mainly polyester) have been used in construction. The scope of their application is constantly expanding: from the thermal insulation of building enclosure for buildings, structures, pipelines and equipment to sound-absorbing and soundproofing structures [3–5]. It is promoted by a variety of production methods and the formation of a fibrous layer, the possibility of creating porous structure materials, as well as their environmental safety [6].

The largest Russian producers of non-woven heat-insulating materials: Thermopol, LLC (Moscow, Russia), Ves Mir Non-Woven Fabric Factory, LLC (Podolsk, Russia), Comitex, OJSC (Syktyvkar, Russia), Sibur-Geotekstil, LLC (Surgut, Russia), Nomateks, LLC (Ulyanovsk Region, Russia), Holteks-Avto, CJSC (Moscow, Russia), Inzensk Non-woven Fabrics Factory, LLC (Ulyanovsk Region Russia), Dmitrovsk Artificial Fur, OJSC (Moscow Region, Russia), Tuymazyteks, CJSC (Republic of Bashkortostan), Nimprotex, OJSC (Zheleznogorsk, Kursk region, Russia).. Various types of synthetic insulators with different trade names are widely represented on the Russian market: sintepon, Hollofiber, Tinsuleyt, Faybertek, as well as Shelter, Izosoft, Sherstippon [2].

Important indicators of the quality of heat-insulating materials is the ability of materials during installation and operation to quickly restore its original shape, maintaining a constant thickness, which is important for thermal insulation.

The coefficient of thermal conductivity of materials at a temperature of 25 °C is the most important characteristic in construction. Effective are materials with a value of thermal conductivity lower than 0.04 W/(m*K) for operating conditions in a loaded enclosing structure. In particular, for bricks, this figure is 0.5 W/(m*K).

Depending on the operating conditions of non-woven heat-insulating materials in building structures, the following indicators of material properties (RSS 31309–2005) are defined: a compressibility at the specific load of 2000 PA, an elasticity (the ability of a material to restore shape after removal of a load), a compressive strength at 10% deformation, an ultimate tensile strength, a layer tearing strength, a vapor permeability, water resistance, a thermal conductivity of the products at temperatures of 25 °C and 100 °C. Requirements for thermal conductivity of the products under different operating conditions are given in SNiP 23-02-2003 «Thermal insulation of buildings» , and fire-technical characteristics (combustibility group, flammability group, smoke-generating group) - in the document NPB 244-97 «Fire safety standards» .

The properties of non-woven fabrics Hollofiber brands 500 and 1000 were investigated in [2]. The values of thermal conductivity are 0.039–0.041 W/(m*K). It is important to note that the polyester fibers used in non-woven insulation materials do not spread the flame. When open, they melt, and at removal of an open flame self-extinguish. Polyester fibers soften at a temperature of 180–300 °C, and ignite at 450–485 °C. Non-woven heat-insulating materials have promising development and application in cottage and hangar construction. Materials are recommended for use in frame enclosing structures and as thermal insulation of vertical, horizontal and inclined surfaces, for insulation of floors when laying material from above. Non-woven fabrics can be produced in various ways and have a different structure and properties [6]. Elasticity and compressibility are important indicators of the mechanical properties of non-woven thermal insulation fabrics and are associated with the characteristics of their structure [7]. A change in the deformation of materials during the operation of building structures leads to a change in geometrical, mechanical, and thermophysical properties. The main type of deformation of non-woven thermal insulation materials during operation is compression deformation. However, the properties of non-wovens have not been sufficiently studied. For example, there is no information about the molding ability of such materials. There is no data on the patterns of deformation of non-woven thermal insulation materials under the action of moisture, temperature and pressure. There is no theoretical justification and recommendations for a rational choice of modern non-woven heat-insulating materials from synthetic fibers, taking into account the specific operating conditions.

The problem of the effective use of modern non-woven heat-insulating materials in building structures is due to the lack of objective research methods and tools patterns of changes in the thermal, structural, physic and mechanical characteristics of materials when exposed to factors of installation and operation of products.

The purpose of the study is the development of research methods and tools for studying the deformation of non-woven heat-insulating materials under cyclic compression. The objectives of the study to develop methods and a tool for evaluating the deformation of non-woven materials under compression, experimental studies of the deformation of non-woven fabrics of various production methods and structures under cyclic compression, the development of practical recommendations for applying the obtained results.

3   Justification of Research Methods

3.1   Analysis of Factors Affecting the Structure and Properties of Non-woven Thermal Insulation Materials

During the life cycle, non-woven materials experience various external influences, the effect of which leads to a change in the initial properties of materials (see Fig. 1).

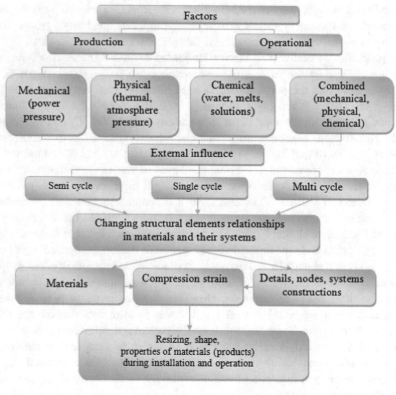

Fig. 1. Compressive volumetric strain factors of non-woven materials during installation and operation of structures

The main factors affecting materials during installation and operation of building structures are time, mechanical and thermal effects, moisture. Deformation of compression of materials can lead to changes in the structure of materials due to the displacement of structural elements. As a result, the geometrical (thickness, density), physical (heat-shielding properties) and mechanical (rigidity, elasticity) properties will change. The main operational factors affecting the materials are: mechanical, climatic, physical and chemical. During the operation of products, there is also a possibility of exposure of external factors that are not provided by the operating conditions (random). The force acting on the material during installation and operation of the structure may be static or dynamic. Dynamic actions include vibration and shock. The impact of compressive forces may be cyclical when the load alternates with the rest of the materials. The value of force pressure, as a rule, does not exceed or it is significantly less than the limit values of the material strength (breaking load).

Repeated cyclic loads lead to a change in the structure, which in turn affects the properties of materials and their systems. The resulting deformation of the compression elements of the structure of the non-woven material will cause a change in the geometric, mechanical and thermal characteristics (thickness, density, stiffness). The nature of the operational loads on non-woven material depends on the type of building structure. The most susceptible to structural changes in compression are non-woven bulk materials, which are used as heat-insulating materials. Compression refers to the main types of deformation of heat and shock insulation materials. Changing the thickness of the bulk non-woven materials leads to changes in ergonomic indicators (thermal properties) of products. The quality of non-wovens depends on changes in thickness and elastic properties in compression during operations. The pressures at which the mechanical properties of textile fibers start to deteriorate during mass compression: for cotton fiber is $981*10^2$ MPa, for polyester fibers is $1962*10^2$ MPa, for viscose fibers is $14715*10$ MPa, for polyamide fibers is $14715*10^2$ MPa.

Under compressive forces, changes in external (thickness) and internal bonds (the conformation of macromolecules and supramolecular formations) in bulk non-woven materials are possible. The changes in structure and properties are affected by the conditions of compression (see Fig. 2). There is an increase in cross-sectional area and volume reduction under conditions of free compression in bulk non-woven materials. It is associated with convergence, straightening, wrinkling of fibers. Reducing the volume of the material during compression leads to a decrease in porosity and an increase in the average density of the material.

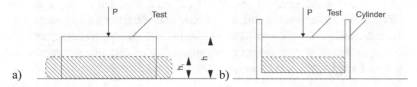

Fig. 2. The materials deformation scheme in various conditions of compression: a - free, b - constrained

Non-wovens are subjected to cyclic compressive forces at all stages of the life cycle: storage, transportation, installation, operation and maintenance. The value of these efforts does not exceed 5–20% of the ultimate strength of the material.

The most important role for non-woven materials are physical properties (heat-shielding properties, heat capacity), mechanical (dimensional stability) properties. Thermophysical properties of nonwoven materials largely depend on thickness, bulk density, which change under compressive forces. Bulk non-woven fabric, compressed even under the action of small loads. The most intensive change in thickness, and, accordingly, porosity in high-volume fibrous materials, occurs in the range of values from 0.1 to 1.0 kPa. In the scientific and technical literature there is practically no information on the studies of the properties of non-woven materials under the action of compressive forces. The results of the study allow us to conclude that the force applied to the material leads to a deterioration of its heat-shielding properties. The degree of reduction of thermal resistance of insulation depends on the pressure and the initial density of the material. The underestimation of the influence of changes in the thickness of materials on thermal properties can lead to a distortion of the actual values of thermal resistance of non-woven heat- insulating materials during operation. If the design of the structure to use the thermal resistance of non-woven materials in the original (uncompressed) form, during the operation of the product the actual value of thermal resistance of the building system can be several times lower than the required one. Therefore, when designing structures, it is necessary to take into account changes in the structure and properties of non-woven heat-insulating materials during compression in operation.

3.2   Justification of Indicators and Methods of Evaluation of Deformation of Materials Under Compression

Currently, standard and original methods, devices and installations are used to determine the properties of materials in compression. The disadvantages of the known methods and of tools the deformation of materials under compression include: single-cycle effect, the test conditions do not simulate real ones; the applicability of methods and instruments for a particular type of material; the complexity and bulkiness of the test equipment; no versatility (can not simulate the conditions of production and operation). Semi-cycle characteristics of the properties of materials under compression (tensile strength, fracture deformation, modulus of elasticity, compression stiffness) are mainly determined when samples are tested on machines, with sample destruction. These methods are applicable to solid, monolithic materials, but they cannot adequately assess the deformation of bulk non-woven materials. Single-cycle characteristics of the materials properties allow us to estimate the materials elasticity under compression. For semi-cycle and single-cycle compression tests, various cross-beams are used, installed on the upper and lower rods of the breaking machine, or other mechanisms equipped with measuring devices.

The thickness of fibrous materials is usually measured on a thickness gauge with a compressive force $F = 0.2...100$ cH. Therefore, the experimentally determined thickness h, mm of the material will differ from the actual thickness in the initial uncompressed state by value Δh. The Δh value depends on the material properties and compressive force.

One of the characteristics of the properties of materials under compression is hardness. Hardness H, Pa, is a characteristic of the material, reflecting its strength and plasticity, determined by relative and absolute methods according to RSS 263 and RSS 20403. The methods for determining hardness are suitable for solid materials with a monolithic structure with a thickness more than 6 mm.

The disadvantage of which is the realization of only single-cycle constrained compression. The device PRS-1 to assess the deformation of the fibers has a complex structure and can only be used for specific types of non-woven materials. The main disadvantage of the methods used is that they do not take into account the influence of external factors of production and operation on the materials properties. Methods and tools are suitable for solid materials with a monolithic structure, and can not objectively assess the deformation of non-woven insulation materials. The indicators are- the elastic and residual parts of the total deformation. They allow us to objectively evaluate the effect of compression deformation on the properties of materials. Therefore, methods and tools for studying of the materials deformation for clothing under cyclic compression conditions are promising and allowing us to study the dynamics of changes in the components' deformation simulating the external factors effects.

4   Characteristics of Objects and Method of Study of Materials Deformation Under Compression

Non-woven insulating and warming materials were investigated as objects. They differed in composition, production method and purpose, made in accordance with RSS 19008, RSS 14253, RSS 6418. Characteristics of the objects of study are presented in Table 1.

Table 1. Materials characteristics

No.	Material, article	RSS, TU	Method of obtaining	Thickness, mm	Fibrous composition, %	Surface density Ps, g/m2
1	Non-woven canvas, code 927622	RSS 19008	Sewing	4.8	Wool – 85, Polyester – 15	215
2	Non-woven canvas, code 917618	RSS 14253	Sewing	2.8	Cotton – 100	190
3	Felt	RSS 6418	Felting	6.8	Wool – 100	200
4	Sherstipon	–	Thermobonded	24.4	Polyester – 40 Wool – 60	300
5	Sherstipon	–	Thermobonded	14.9	Polyester – 40 Wool – 60	300
6	Tinsulate	Modification R 150	Thermobonded	15.8	Polyester – 100	100
7	Sintepon	SK150/300	Thermobonded	7.8	Polyester – 100	140
8	Holofiber	Thermopol, LLC (Russia)	Thermobonded	12.0	Polyester – 100	130

The selection and preparation of samples of materials for testing was carried out in accordance with RSS 13587 "Non-woven canvas and non-woven products rules of acceptance and sampling methods", RSS R ISO 15902.2 "Non-woven canvas. Methods for determining the structural characteristics", RSS R ISO 12023 "Textile materials. Fibers. Method for determining the thickness". Before testing, samples were kept for 12 h under normal atmospheric conditions (temperature $20 \pm 30C$ and relative humidity $65 \pm 5\%$ by psychrometer) in accordance with RSS 10681 "Textile materials. Climatic conditions for the conditioning of samples and methods for their determination", tests were conducted under the same conditions. The objects were tested under cyclic compression. To simulate the factors of production and operation, we determined the deformation of compression of materials in the conditioned state, after moistening (Wf = 40%), after wet-heat treatments. Compression deformation was determined by measuring the results of testing materials.

The total compression strain:

$$L_{total} = h_{max} = h_0 - \Delta h \qquad (1)$$

where h_{max} - the value of the maximum sample forcing, mm;
h_0 - the value of the maximum forcing on the scale, mm;
Δh - the zero level of the sample surface, mm.
Reversible deformation:

$$\Delta \varepsilon_{re} = (h_{max} - h_i)/h_{max} \qquad (2)$$

where h_i is the last measurement value of the reader, mm.
Irreversible (residual) deformation:

$$\Delta \varepsilon_{irr} = h_i/h_{max} \qquad (3)$$

When creating an installation, the main task was to develop a device that allows you to implement various compression conditions (constrained, free) with high accuracy of strain measurement. The use of differential photosensors and the absence of pressure from the meter ensures objectivity of the test results. The use of photosensors provides a constant absolute instrument error of the measuring device of 10–6 m.

The device combines the test zones of the sample and its measurements, which makes it possible to measure the strain during the load and rest periods.

The device allows us to study both the technological properties of materials (the ability to deform) and the most important operational characteristics (elasticity). At the same time, various climatic conditions may be created during testing, including a humid (or other) environment. The compression device is a steel cylinder with diameter Dc = 35 mm and height H = 20 mm; steel, removable indenter with a diameter of D = 30 mm and a thickness of 10 mm (see Fig. 3). The indenter due to the design features provides a uniform force pressure on the material with different relief and uneven thickness. The papers [8, 9] present the description and the principle of device operation.

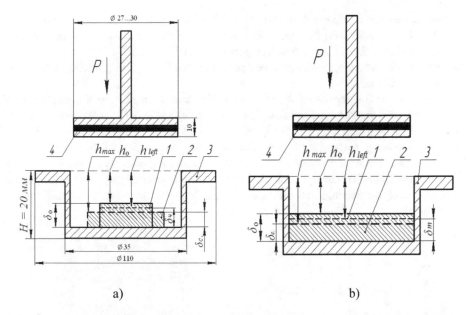

a)   b)

Fig. 3. Materials compression scheme: a - free compression; b - constrained compression: 1 - test to load; 2 - test under load; 3 - cylinder; 4 - indenter

5 Experimental Studies of the Deformation of Nonwovens Under Cyclic Compression

Preliminary experiments were carried out to select the optimal test parameters. The compression parameters of the materials are presented in Table 2. The relative error of the results of measuring the materials deformation with a thickness of 0.1–30 mm by cyclic compression was 3.1–12.3%, which indicates the reliability of the experimental data.

The value of the relative error does not exceed 5% with a confidence of 0.95, and the coefficient of variation is 10%.

Table 2. Compression parameters of nonwoven materials

Pressure, kPa	Compression Conditions	Sample diameter, mm	Compressive load Rc, daN							
			0.5	1.0	1.5	2.0	2.5	3.0	3.5	4.0
	Constrained	30.0	0.07	0.14	0.21	0.28	0.35	0.43	0.50	0.57
	Free	20.0	0.16	0.32	0.48	0.64	0.80	0.96	1.12	1.27
		15.0	0.28	0.56	0.85	1.13	1.41	1.69	1.98	2.26

Test parameters: indenter diameters D = 30 mm, samples of materials d1 = 10 ...
25 mm, d2 = 27...30 mm; the load value in the cycle Rc = 1.5 daN, the compression
time tn = 5 s and the rest to = 5 s; test period n = 100 cycles. For given parameters,
the pressure on the sample with free compression is 0.5 kPa, at the constrained compression it is 0.2 kPa.

According to the results of experimental data, the main statistical characteristics
were determined by known formulas. The value of the relative standart error did not
exceed 10% with a confidence of 0.95, the coefficient of variation did not exceed 10%.
Table 3 presents a fragment of experimental data, the results of the experiments.
Figure 4 shows the experimental curves of changes in the thickness of materials under
cyclic compression. The analysis of the experimental results allows us to estimate the
reversible and irreversible deformation of the samples in various conditions of
compression.

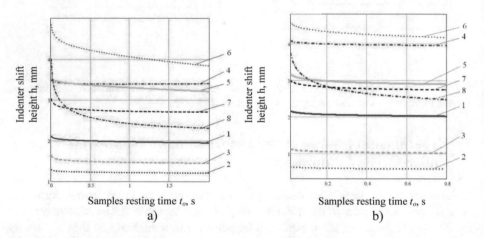

Fig. 4. Experimental graphs of intender shift in the materials thickness (Table 1) under cyclic
compression n = 100 cycles, tn/to = 5/5 s: a - constrained compression; b - free compression

The optimum test conditions were experimentally established: the load and rest
time of the sample material in a cycle of 5 s, the force pressure parameter Rc = 1.5
daN, the loading period 1–100 cycles. The selected test modes allow you to explore the
deformation of materials of different composition and structure.

Table 3. Experimental measurements of the materials thickness and deformation after 100 compression cycles

N	Samples	Compression	Pre. processing	Sample thickness, mm			Deformation after compression		
				before compression δ_0	in compression δ_c	after the rest δ_m	full, % ($\times 102$)	irreversible, parts.	
				The number of compression cycles					
				–	1	100	1	100	100
1	Non-woven canvas, sewing, code 927622	Constrained	–	4.8	4.11	4.0	0.29	0.58	0.31
			wet	5.7	3.5	2.97	0.39	0.63	0.10
		Free	–	4.7	4.3	4.1	0.26	0.44	0.14
			wet	5.8	4.07	3.97	0.48	0.62	0.31
2	Non-woven canvas, sewing, code 917618	Constrained	–	2.8	2.5	1.93	0.11	0.21	0.10
			wet	3.0	2.93	1.43	0.67	0.93	0.19
		Free	–	2.6	2.51	1.97	0.39	0.46	0.17
			wet	2.9	2.12	1.27	0.5	0.81	0.32
3	Felt	Constrained	–	6.0	5.91	5.76	0.04	0.20	0.15
			wct	6.6	6.13	5.82	0.04	0.31	0.16
		Free	–	5.9	5.53	5.0	0.03	0.27	0.11
			wet	6.5	6.3	5.63	0.03	0.34	0.19
4	Sherstipon Polyester – 40 Camel's wool – 60	Constrained	–	24.4	23.87	23.46	0.25	0.24	0.13
			wet	18.6	18.4	17.9	0.33	0.30	0.19
		Free	–	24.2	23.83	23.2	0.23	0.22	0.13
			wet	19.3	18.31	18.8	0.31	0.31	0.13
5	Sherstipon Polyester – 40 Sheep's wool – 60	Constrained	–	14.9	13.93	12.9	0.13	0.19	0.24
			wet	16.2	15.62	14.5	0.38	0.34	0.10
		Free	–	14.7	13.58	12.5	0.23	0.21	0.12
			wet	17.0	16.28	15.01	0.09	0.26	0.23
6	Tinsulate (Russia) R 150	Constrained	–	15.8	14.81	13.72	0.28	0.25	0.13
			wet	15.9	14.21	13.9	0.18	0.36	0.05
		Free	–	15.6	14.64	13.53	0.34	0.34	0.26
			wet	16.1	15.63	14.2	0.27	0.33	0.11
7	Sintepon, code SK150/300	Constrained	–	7.80	7.34	6.49	0.18	0.18	0.16
			wet	12.0	8.44	9.23	0.13	0.77	0.13
		Free	–	7.5	6.43	6.27	0.40	0.40	0.15
			wet	12.6	10.88	9.75	0.16	0.66	0.16
8	Holofiber	Constrained	–	11.83	10.52	9.26	0.10	0.21	0.13
			wet	19.68	15.74	11.4	0.19	0.42	0.16
		Free	–	12.58	10.99	8.52	0.12	0.32	0.11
			wet	17.60	14.59	8.23	0.17	0.53	0.16

6   Analysis of the Results of Experimental Studies of Non-woven Deformation

The results of materials deformation after 100 compression cycles are presented in Fig. 3. Analysis of the results showed that the irreversible deformation of samples from synthetic fibers (No. 6, 7, 8) is 15–20% less than that of samples made of natural fibers (No. 1–5). This is due to the elastic properties of synthetic fibers. But under free compression, the irreversible deformation of sample No. 6 increases by 29%, which is associated with the best material recoverability under conditions of constrained compression.

The highest values of irreversible deformation after 100 compression cycles under constrained compression in non-woven canvas with an investment of wool fibers (samples No. 1, 4, 5), due to the morphology of the structure of wool fibers. Samples with an investment of synthetic fibers (samples No. 6–8) are better restored in conditions of constrained compression. In the free state, the irreversible deformation increases by 15–25%.

The results of the evaluation of the materials deformation in the conditioned and wet conditions at an air temperature of 20 °C are presented in Fig. 5.

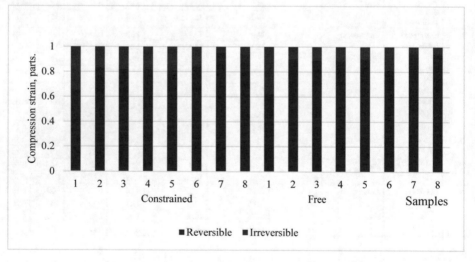

Fig. 5. The proportion of residual strain after 100 compression cycles (P = 1.5 daN, tagr / totd = 5/5). Samples: non-woven canvas, sewing: 1 - half-wool canvas, 2 - cotton; 3 - felt; Shestipon non-woven fabric: 4 - camel wool, 5 - sheep wool; 6 - Tinsulate; 7 - Sintepon; 8 - Holofiber (Table 1)

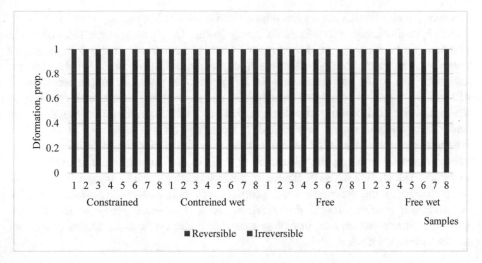

Fig. 6. Residual deformation of non-woven canvas after 100 compression cycles (P = 1.5 daN, tagr /totd = 5/5) when exposed to wetting. Samples: canvas, sewing: 1 - wool, 2 - cotton; 3 - felt; Shestippon: 4 - camel wool, 5 - sheep wool; 6 - Tinsulate; 7 - Sintepon; 8 - Holofiber (Table 1)

Analysis of the results in Fig. 6 shows that after 100 compression cycles in cramped conditions, wetting significantly affects the deformation of the objects under study. The proportion of irreversible deformation in non-wovens (sample No 1, 2) increases by 35–39% when wetted. For non-woven materials made of synthetic fibers (samples No. 6–8), the proportion of irreversible deformation increases by 25–30%. Insignificant changes in irreversible deformation were observed only in sample No. 3 upon wetting. Obviously, this is due to the more dense structure of this material compared to other objects. Under conditions of free compression of materials in the wet state: for samples 1–5, irreversible deformation is less by 15–29% than in the constrained state. In general, after 100 cycles of constrained and free compression and wetting, the residual strain increases. Therefore, in the operation of products in areas with high humidity or heavy precipitation, we should carefully approach the choice of thermal insulation materials in building structures.

7 Conclusion

The study obtained the following results. Modern non-woven heat-insulating materials are used in thermal insulation of enclosure for buildings, structures, pipelines and equipment, as well as in sound-absorbing and sound-insulating structures. The main type of deformation of non-woven insulation materials during operation is compression deformation. Elasticity and compressibility are important indicators of the mechanical properties of non-woven insulating materials. The problem of effective use of non-woven heat-insulating materials is due to the lack of objective methods and tools of studying the patterns of change in the properties of materials during the life cycle.

A method and a device have been developed for evaluating the deformation of non-woven heat-insulating materials under compression. The optimal parameters of cyclic compression are established, providing an objective assessment of the deformation of non-woven materials during production and operation. The advantage of the method: automatic test mode and measurement using photo sensors, the implementation of different compression conditions (constrained or free, climatic conditions). Analysis of the results of experimental studies showed that the deformation of non-woven materials under compression varies ambiguously, depending on the structure of materials, compression parameters. It has been experimentally shown that the deformation of non-woven materials under cyclic compression leads to a change in the thickness of the materials. The change in the thickness of materials during cyclic compression can lead to a change in the thermophysical properties of non-woven insulating materials. A promising continuation of this work is the experimental study of the thermal resistance of non-woven heat-insulating materials under cyclic compression conditions on the developed device.

References

1. Smirnov, T.: Building non-woven thermal insulation materials based on mineral fibers. Non-woven Mater. **1**(6), 1–6 (2009)
2. Mukhamedzhanov, G.: Non-woven insulation materials and products. Polymeric Materials, pp. 26–29, March 2013
3. Rumyantcev, B., Zhukov, A.: Principles of creation of new building materials. Internet-vestnik VolgGASU. Ser. Politematicheskaya **3**(23), 19 (2012). www.vestnik.vgasu.ru. Accessed 21 Jan 2019
4. Hanisch, V., Kolkmann A., Roye A., Gries T.: Influence of machine settings on mechanical performance of yarn and textile structures, In: ICTRC 2006 1st International RILEM Symposium on Textile Reinforced Concrete (2006)
5. Hanisch, V., Kolkmann, A., Roye, A., Gries, T.: Yarn and textile structures for concrete reinforcements. In: Proceedings FERRO8, Bangkok (2006)
6. Treshalin, M., Mukhamedzhanov, G., Telitsyn, A., Mandron, V., Treshchalina, A.: Production and test methods for non-wovens, Moscow (2008)
7. Serebryakova, I.: The study of the mechanical properties of non-woven fabrics. Izvestiya Vysshikh Uchebnykh Zavedenii, Seriya Teknologiya Tekstil'noi Promyshlennosti **3**, 8–11 (2012)
8. Lisienkova, L., Deryabina, A.: Izvestiya Vysshikh Uchebnykh Zavedenii, Seriya Teknologiya Tekstil'noi Promyshlennosti, 32–36 (2013)
9 Lisienkova, L., Deryabina, A., Tarasova, O.: Izvestiya Vysshikh Uchebnykh Zavedenii, Seriya Teknologiya Tekstil'noi Promyshlennosti, 29–34 (2015)

Influence of Orientation of Cement Stone Loaded at Early Age on Strength Properties

Yuriy Galkin$^{(\boxtimes)}$ and Sergey Udodov

Kuban State Technological University, Moskovskaya 2,
350072 Krasnodar, Russia
tcmii@mail.ru, udodov-tec@mail.ru

Abstract. The main purpose of the work was to study the tensile bending strength of samples preloaded at early age and to check the influence of orientation factor of the preloaded sample. Besides, it was expected to consider the effect of active mineral additives on this parameter. The objects of the study are samples of cement-sand grout with Portland cement produced by various manufacturers. By changing the orientation of the preloaded samples under testing, the difference in bending strength characteristics of the samples was established. The maximum increase was observed in a plane perpendicular to the direction of compression stress. It has been established that adding of mineral additives in order to replace cement in a certain proportion leads to loss of the observed effect. The study carried out by using statistical methods made it possible to establish effect of orientation of the sample on tensile bending strength, as well as on compression strength slightly.

Keywords: Early loading · Cement stone · Microsilica ·
Tensile breaking strength · Anisotropic properties

1 Introduction

The continued acceleration of construction leads to greater demands on construction materials, concrete in particular, not only in terms of early age strength but also concerning the reduction of time period from the moment of molding to application of process impact and construction loads at the earliest stages.

Early loading as a process method accelerating concrete strength development of monolithic structures at temperatures below zero has been considered in the early writings of Bayburin [1]. Transforming the strength formula of reinforced concrete applied for the design purposes, he introduces the concept of the ratio of permissible loading intensity β:

$$R_b = \frac{0.5 \times (R_s \times A_s)^e}{(R_s \times A_s \times h_0 - \beta \times Mc) \times b}, \tag{1}$$

where $R_s \times A_s$ – design strength and area of reinforcement; h_0, b – effective height and cross-sectional width; M_c – sectional moment induced by process loading; β – ratio of permissible loading intensity.

© Springer Nature Switzerland AG 2020
V. Murgul and M. Pasetti (Eds.): EMMFT 2018, AISC 982, pp. 723–730, 2020.
https://doi.org/10.1007/978-3-030-19756-8_68

Ratio of permissible loading intensity is determined experimentally, i.e. it depends both on compounding-and-process factors and loading parameters.

As it has been proven by experimentation of the author, samples with strength $(0.2...0.8)$ R_{28} loaded with intensity 0.3–0.75 of breaking load 90–180 days later showed increase in strength by 22–28% and elastic module by 15% and 19% for direct compression and eccentric compression in comparison with samples that were not subjected to loading.

The author proposed to apply loading of the same intensity for early loading of reinforced concrete beams and walls that are hardening under conditions of low temperatures. At $R_0 = (0.35–0.65)$ R_{28}, increase of deformation within the compressed area and bending of the beams stops in 5–7 days while their strength and crack resistance increase.

Early loading of walls under real conditions led to increase in strength by 16–23%. It helped to reduce the temperature of isothermal aging and to decrease the rate of temperature rise and cooling without increasing the total duration of heat treatment [1].

The subject has been researched extensively in the works of Golovnev and Berkovich [2, 3], as well as in works of Nikonorov and Tarasova [4]. Through the research of monolithic frame of a 24-storey building, the latter proposed method providing early removal of floor formwork in case of application of beam-and-column formwork by using additional support columns. After curing the concrete slabs up to the required strength degree, the formwork was removed, and temporary support columns were installed to transfer the load from the stripped flooring slabs to the underlying slabs. This approach reduced the curing period of concrete from 3 to 5 days, as well as the cost of electricity for heating concrete to about 50 kW per 1 cu.m. [4].

Analysis of the existing experience makes it obvious that concrete is often subjected to early loading. As A.V. Satalkin noted, the effectiveness of such loading may be attributed to plastic flow at an early age (early-age creep), the value of which serves as the criterion for compaction and hardening of the structure caused by the flow. According to his observations (later it was also noted by A. V. Kosolapov, Yu. A. Mamontov and I. I. Shukenov) the crack resistance of the structure subjected to short-term pressing at an early age grows as well.

Considering the particular relevance of the issue of improving the bending resistance of concrete, as well as the available results of the experimental work of the author [5, 6], the purpose of the work was to study the effect of spatial orientation of the samples during their early loading on their strength characteristics under tension.

In this regard, the following tasks have been formulated:

1. obtain data on tensile strength when splitting for two options: splitting along the impact direction and splitting perpendicular to it;
2. check the effect of the orientation factor on compression strength;
3. assess the effect of the cement paste content (by replacing part of the cement with mineral additives) on the tensile breaking strength after early short-term loading.

2   Materials and Methods

The samples have been made from the cement stone grout with the following content C:S:W = 1:1:0.4. Portland cement Cem I 42.5 H of "Verhnebakanskiy" and "Novoroscement" plants was used as a binder; sand of "Cube" pit (Krasnodar region) with fineness ratio 1.9–2.0 and rest on sieve 0.63 to 0.02% was used as a fine aggregate.

Introducing of inert aggregate to the grout stems from the desire to reduce the variation in strength properties due to shrinkage loss as well as to thermal stresses during the hydration of cement stone, which may occur when pure cement paste is used. In order to assess the reduction of cement binder content, condensed microsilicasuspension MKU - 85 by "Kuznetskiye Ferrosplavy" OJSC produced as per TU 5743-048-02495332-96 and GOST 56178-2014. Condensed microsilicasuspension is an ultrafine material consisting of spherical particles obtained in the course of the gas cleaning of furnaces while producing of silicon-containing alloys. The main component of the material is amorphous silicon dioxide.

After 24 h from the moment of gauging, the samples of 70 × 70 × 70 mm were relieved of moulds. Then the strength of the composition at the current age was determined by compression testing a series of three control samples. Loading was performed in stands in lots of 12 pieces. The load intensity was equal to 10% of the breaking load at the current age ($0.1 \times P_{brek.\ 1\ day}$). The control samples were located next to the pre-loaded ones under the same temperature and humidity conditions (t = 21–22 °C, φ = 55–60%). The holding room had no windows. It excluded the influence of draughts in compliance with recommendations. On each section of the stand, dynamometers of standard compression DOSM 3–5 were installed to ensure the constancy of the load value applied to the samples. The difference in readings among the 6 samples did not exceed 3–4%. After 24 h of exposure, the loaded samples were taken away and subjected to splitting tension tests together with the control samples. Randomly selected samples were split by the impact of pressing cylinders along the axis of the pre-applied load. The other samples of the lot were turned so that the plane compressed during the early loading became perpendicular to the splitting impact. The obtained data were processed by the methods of mathematical statistics. Besides, estimation of gross error was made in compliance with the criterion $\tau_i \leq \tau_{tab}$. At the selected probability values and significance level it was equal to [7] 1.98.

3   Results and Discussion

3.1   Samples Containing Portland Cement as a Binder

The results of strength tests of samples prepared after an early short-term loading are shown in Table 1.

According to the data of Table 1, pre-loading leads to a difference in the strength characteristics in a certain direction. In this particular case, strength characteristics of plane perpendicular to the plane of loading have been changed. At the same time, the maximum homogeneity of values was achieved, which is confirmed by a low variation

Table 1. Tensile splitting strength of the samples (cement "Verhnebakanskiy").

Direction of load applied	Orientation of the sample under tensile testing	Splitting force, kN	$R_{tt} = \gamma \dfrac{2F}{\pi A}$, $\gamma = 0.87$	Average Rtt, MPa
		17.1	1.934	
		16.6	1.877	
		17.4	1.968	1.924
		16.9	1.911	
		17.6	1.990	
		16.5	1.866	
		13.0	1.470	
		12.8	1.448	
		13.7	1.549	1.530
		14.3	1.617	
		13.8	1.561	
		13.6	1.538	
		14.8	1.674	
		15.1	1.708	
	Control samples (not preloaded)	16.1	1.821	1.723
		15.0	1.696	
		16.2	1.832	
		14.2	1.606	

ratio equal to 2.6% (see Table 1). The average force applied when splitting a turned sample has increased by 25% relative to the value for a series of samples tested in a position similar to the one during pre-loading. It also exceeded the strength of the control sample by 11.7% (see Table 1).

The samples contain Portland Cem I 42.5 N Novoroscement. The same content was adopted: C:S:W = 1:1:0.4. Due to the fact that the normal density of the cement paste for the selected binder is less than that of Portland cement produced by Verkhnebakanskiy plant, the grout flow, measured by using Suttord viscometer, was 120–130 mm. The size and number of samples for testing, the age of loading and load duration were the same as for the samples of "Verkhnebakanskiy" Portland Cement. The test results, as well as the statistical processing are shown in Table 2.

As can be seen from Table 2, the strength values were sufficiently homogeneous and stable. Variation ratio did not exceed 5.7%, and there were also no gross errors. As in the first case, the samples split after 900 rotation had the highest strength. The difference between strength of the turned samples and "still" samples was about 19%, and the excess over the strength of control samples was 16%. The lowest average value was obtained while testing "still" samples (in a position similar to the one during pre-loading).

The observed anisotropy of bending resistance is probably related to the ongoing physicochemical processes that, under conditions of compression and creeping deformations of cement stone, can lead to local compaction of its components [8],

Table 2. Tensile splitting strength of the samples (cement "Novoroscement").

Direction of load applied	Orientation of the sample under tensile testing	Splitting force, kN	$R_{tt} = \gamma \dfrac{2F}{\pi A}$, $\gamma = 0.87$	Average Rtt, MPa
		21.9	2.477	
		19.3	2.183	
		20.6	2.330	2.283
		19.6	2.217	
		20.9	2.364	
		18.8	2.126	
		16.7	1,889	
		17.0	1.923	
		16.2	1.832	1.911
		18.1	2.047	
		15.8	1.787	
		17.6	1.990	
		17.3	1.956	
		16.7	1.889	
	Control samples (not preloaded)	18.9	2.137	
		16.8	1.900	1.962
		16.5	1.866	
		17.9	2.024	

reduction of porosity, redistribution of forces [9], convergence of layers of the calcium hydrosilicate gel (CSH), and the diffusion of moisture [10].

In this particular case, the issue of strengthening should be studied from the standpoint of change in the proportion of amorphous and crystalline structure under load and without it, as well as the morphology of hydrated products. Of particular importance is the search and determination of "structural features" for a series of samples with increased strength and clarifying of possible influence of compounding-and-process factors on the increase in tensile strength for cement samples subjected to early loading.

Obtaining of data on change in compression strength of pre-loaded samples depending on their position under test is of particular interest as well. This factor was studied through a cement-sand grout of the following composition: C:S:W = 1:1:0.4. This cement composition is applied, for example, for fixing anchor bolts in concrete foundations when installing equipment and may be exposed to early loading at an early age. Portland cement used is Portland cement Cem I 42.5 H of Verkhnebakanskiy plant. The methods of mixing, loading and exposing of the samples are the same that were used for samples tested for tensile splitting strength. Compression strength was determined as per GOST 10180 with a scaling factor of 0.85. The test results and processing methods of mathematical statistics are shown in Table 3.

The analysis showed that loading for 1 day with a force equal to 10% of the breaking stress at the current age, led to a decrease in the compression strength of samples in comparison to unloaded ones (control samples). The average strength of 6

Table 3. Strength of pre-loaded samples (cement "Verkhnebakanskiy").

Direction of load applied	Orientation of the sample under tensile testing	Splitting force, kN	$R = \alpha \dfrac{F}{A}$, $\alpha = 0.85$ MPa	Average R, MPa
		102.5	11.32	
		97.3	10.75	
		95.3	10.53	10.869
		99.6	11.00	
		103.0	11.38	
		92.4	10.20	
		111.5	12.32	
		101.6	11.22	
		109.1	12.05	11.867
		97.8	10.81	
		110.2	12.17	
		115.3	12.74	
		117.7	13.00	
		109.6	12.11	
	Control samples (not preloaded)	110.4	12.19	12.437
		99.4	10.98	
		118.8	13.12	
		116.6	12.88	

samples of the series is less by almost 4.6%. According to the test, turning of cubes with respect to compression direction increased the difference between the strength of pre-loaded samples and control samples by 12.5%. All the data obtained were included in the confidence interval and did not contain gross errors (Table 3).

Therefore, position of the sample subjected even to short-term (24-h) loading at early age affects the properties of the material. A significant increase in bending resistance in case of early loading of silicate materials (such as Portland cement) was also noted by the author in [5, 6]. Thus, as it has been proven in the experiments with Portland cements of different manufacturers, it is possible to increase the bending resistance of cement composites from 16 to 25% (see Tables 1 and 2) by changing the position of the structure under load. Speaking of compression strength, the differences between pre-loaded samples and control samples are less significant (4.6%), while turning of the samples led to decrease in R_{comp} by 12.5%.

3.2 Samples Containing Mix of Binders (Portland Cement + Microsilica)

In order to assess the effect of mineral additives on tensile strength, the share of which in cement-concrete composites increases every year, 10%, 15% and 20% of cement in cement-sand compound C:S:W = 1:1:0.4 was replaced by microsilicasuspension MKU - 85. Samples—cubes having edge length 70 mm a day after molding were loaded with stress equal to 0.1 R_{comp} (R_{comp} was determined on the basis of the test results of three samples) and kept under load for 24 h. Next, tensile splitting tests were performed as per GOST 10180 for pre-loaded samples in different positions (along and perpendicular to the direction of short-term load). The test results, as well as the plotted approximation curves, are shown in Fig. 1.

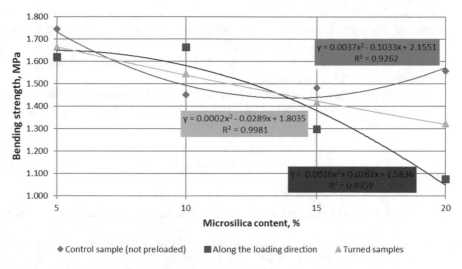

Fig. 1. Bending strength variation depending on microsilica content.

Decrease in strength occurs with an increase in the mineral additive rate in the preloaded samples. The samples split along the direction of the pre-compression stress turned out to be the most sensitive to it. When microsilicasuspension content equals to 20% by weight of the binder, the bending strength drops by 31% comparing with the strength of the control samples. The loss of strength in the samples split after 900 rotation was less ($\Delta R = 15.1\%$), which can be due to the difference in strength caused by change in the position of the sample under test (Tables 1 and 2).

The decrease in strength gain in loaded samples can be attributed to a decrease in the "reactive binder" (Portland cement), which is likely to be deformed under load while strengthening. The strength curve of the control sample declines in the range from 10 to 15%, and then starts to increase again, which may be due to a denser filling of the cement matrix and an increase in strength. Loading at early age leads to creep deformation, which can cause transverse strains during the convergence of CSH gel layers [11], shifts of particles, as well as changes in porosity and diffusion of moisture [12, 13], especially during the early stages hardening. A decrease in strength gain with a decrease in the binder content may indicate a proportional relationship between ΔR and cement content. This compounding factor should be studied further.

4 Conclusions

- The effect of orientation of preloaded samples on tensile splitting strength has been established;
- An increase of R_{bend} in the plane perpendicular to the current load is noted;
- Value of the pedal compressive strength changed slightly (within 4.6%) with a change in the orientation of the samples;
- Change of orientation of the sample leads to a slight variation of compression resistance (4.6% at the extreme);
- When replacing a part of cement with an active mineral additive, the effect of the strength gain decreases, while the factor of the orientation of the sample becomes less significant.

References

1. Baiburin, A.Kh.: Technology of the early age concrete loading. Proc. Eng. **150**, 2157–2162 (2016)
2. Golovnev, S.G., Berkovich, L.A.: Technology of rapid construction of high-rise buildings of reinforced concrete. Vestnik UralNIIProyekt **1**, 30–32 (2009)
3. Golovnev, S.G., Bayburin, A.Kh., Berkovich, A.: New method of construction of monolithic buildings in winter time. Vestnik YuUrGU **15**, 56–58 (2010)
4. Nikonorov, S.V., Tarasova, O.A.: Technology of early loading of monolithic slabs using beam – and – frame formwork. Inzhenerno – stroitelnyy zhurnal **4**, 17–20 (2014)
5. Galkin, Yu., Udodov, S.: Bending strength and phase composition of cement stone, subjected to loading at an early stage. In: MATEC Web of Conferences, vol. 193 (2018). https://doi.org/10.1051/matecconf/201819303031
6. Galkin, Yu., Udodov, S.A., Vasil'eva, L.V.: The phase composition and properties of aluminate cements after early loading. Mag. Civ. Eng. **7**, 114–122 (2017). https://doi.org/10.18720/mce.75.11
7. Usherov-Marshak, A.V.: Market concrete - the theme of concrete and the problem of concrete technology. Stroitelnyye materialy **3**, 5–8 (2008)
8. Pukharenko, Yu.V., Khrenov, G.M.: The problem of technological mechanics in the development of methods of cold forming. Vestnik grazhdanskikh inzhenerov **6**(65), 152–157 (2017)
9. Maksimova, I.N., Akchurin, T.K., Makridin, N.I., Tarakanov, O.V, Tambovtseva, Ye.A.: To the question of correlation of shrinkage and strength of concrete. Internet – vestnik VolGASU **3**(39) (2015)
10. Goncharov, Ye.: Simulation of concrete creep in differential form using rheological models. Tekhnologii betonov **11**, 53–55 (2014)
11. Pignatelli, I., Kumar, A., Alizabeh, R., Pape, Y., Bauchy, M., Sant, G.A.: Dissolution – precipitation mechanism is at the origin of concrete creep in moist environments. J. Chem. Phys. **145**, 231–242 (2016)
12. Kubaneyshvili, A.S., Piradov, A.B., Yuryatin, A.M.: Physico-mechanical properties of concrete hardening under pressure in a confined space. Beton i zhelezobeton **5**, 11–13 (2004)
13. Koval, S.B., Molodtsov, M.V.: Early loading of concrete under different humidity. Vestnik YuUrGU **16**, 15–17 (2011)

Development and Automation of the Device for Determination of Thermophysical Properties of Polymers and Composites

Denis Bakanin[1] , Vladimir Bychkovsky[1] , Nikolai Filippenko[2] ,
Denis Butorin[2(✉)] , and Aleksei Kuraitis[2]

[1] JSC Irkutsk Relay Plant, Baikalskaya st., 239, 664075 Irkutsk, Russia
denis.bakan@mail.ru
[2] Irkutsk State Transport University, Chernyshevskogo st., 15,
664074 Irkutsk, Russia
den_butorin@mail.ru

Abstract. This work is devoted to the development and automation of the device to control phase transformations and determine the thermal properties of polymer and composite materials. The developed device allows you to simultaneously examine two prototypes: one on the thermal properties (melting point, thermal expansion), the other is tested by thermal degradation (heat resistance, the intensity of the thermal phase change). A significant part of the work is devoted to the automation of the developed device, while the following tasks were solved: automation of control of the process of linear uniform heating of the prototype; automation of control of linear thermal expansion of the heated prototype; automation of control of the dynamics of gas evolution during thermal decomposition of the polymer, by automated measurement of the liquid level in the manometer; automation of the process of determining the optical density of the released gas and the indicator during the destruction of the prototype. In the course of the work, the algorithmic support of the processes of control, measurement and processing of experimental data, allowing automation of the control of the experiment, improve the efficiency and accuracy of measurements, the algorithm of the automated control system, implemented in the form of software. The results of experiments conducted on an automated device are presented.

Keywords: Automation · Polymers · Composites ·
Thermophysical properties · Phase transformations

1 Introduction

The current continuously increasing volume of production of polymers and composites is characterized by the need to introduce new advanced materials and products from them, capable of stable and reliable operation under any conditions under the required loads [1–4]. At the same time, the development of new materials is based on the improvement of existing and the creation of new technological processes for their production and processing [1, 2]. And this, in turn, requires a clear representation of the

© Springer Nature Switzerland AG 2020
V. Murgul and M. Pasetti (Eds.): EMMFT 2018, AISC 982, pp. 731–740, 2020.
https://doi.org/10.1007/978-3-030-19756-8_69

processes occurring in the material under various kinds of influence on it, for example, thermal or electrical [5]. Moreover, most of the existing technological processes associated with the production, processing or processing, both the materials themselves and the products from them, involves precisely thermal effects [6]. In this regard, when working with a new or existing material, of paramount importance is the study of the influence of thermal effects on the thermal properties of this material.

A large number of works [7–10] are devoted to the issues of thermal studies of polymeric materials. In the world practice, there are a number of methods and devices for determining the thermal properties of polymers and composites [11–15], but most of them are complex in performance and highly specialized. In addition, it should be noted that the existing range of devices is mainly not automated, and one of the most important tasks of technological progress is the automation of scientific research [1], which will significantly increase the productivity of researchers, reduce the time of experiments and processing of experimental information, improve the reliability and quality of the results. Hence there is a need to develop an automated device capable of determining the complex of thermophysical parameters of polymers and composites at their sequential heating. And also to control the phase transformations occurring in the materials. This was the main purpose of this research.

2   Device Development and Automation

2.1   The Design of the Device

For a comprehensive study of the thermophysical properties of polymer and composite materials, a corresponding device was developed, presented in Figs. 1 and 2. The developed device was based on the device of Vogel V.O. and Alekseev P.G. to determine the thermophysical characteristics of materials.

Fig. 1. Three-dimensional model of the experimental device.

The principle of operation of the device is as follows. After laying a specially prepared polymer or composite sample 16 on the heating element 3, it is installed on the base 1, on top of the surface of the prototype 16, the rod 2 is lowered. Indicator micrometer 15 electronic type, connected to the rod indenter 2, set to zero. At the same time, the second prototype 17 made of a polymer or composite material is placed in a flask 11. The flask 11, in turn, is installed on the thermal spacer 9, a thermal paste is placed in it for tight contact of the base of the flask with the thermal spacer, for better heat transfer to the sample 17 located in the flask 11. The connection of the flask 11 with the liquid pressure gauge 12 is carried out by a heat-resistant silicone tube 13.

The liquid pressure gauge is filled with a diluted indicator (methyl orange). The scale, spread out on the wall of the pressure gauge, allows you to fix the liquid level drops. It should also be noted that the second end of the pressure gauge 12 communicates with the atmosphere.

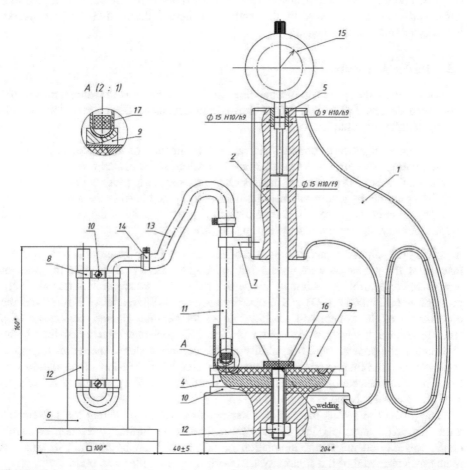

Fig. 2. Scheme of the experimental device for the study of thermophysical properties of polymer and composite materials, where 1-base; 2-rod; 3-heating element; 4-heat insulation plate; 5-sleeve; 6-stand; 7-fasteners; 8-mounting bracket; 9-thermal spacer; 10-basalt wool; 11-bulb; 12-pressure gauge; 13-silicone tube; 14-clamp; 15-micrometer indicator; 16-prototype №1; 17-prototype №2.

Then the heating element 3 is switched on, which provides heating of the sample 16 installed under the rod 2 and the sample 17 installed in the flask 11. Thermal expansion of the prototype, through the rod 2 will have an impact on the indenter of the micrometer 15 and give the data on the linear thermal expansion. The heating in the flask 11 registers the pressure gauge 12 by changing the liquid level, that is, the pressure change in the flask, the dynamics of the liquid level change will be noticeably increased when the material in the flask is destroyed, when a large amount of gas is released from it. And in the flask can be located indicator litmus paper, which will determine the alkaline or acidic composition of the gas from the polymer or composite material.

Thus, the experimental setup presented in the figure above makes it possible to study two prototypes simultaneously: one for thermal properties (melting point, thermal expansion), the other is tested by the method of thermal destruction (heat resistance, intensity of thermal phase change).

2.2   Device Automation

As noted above, the guarantee of obtaining quality indicators is the automation of the developed device. The task of automation of the developed device included the following number of subtasks:

- automation of process control of linear uniform heating of the prototype;
- automation of control of linear thermal expansion of the heated prototype;
- automation of control of the dynamics of gas release during thermal decomposition of the polymer by automated measurement of the liquid level in the pressure gauge;
- automation of the process of determining the optical density of the evolved gas and the indicator during the destruction of the prototype.

In solving the subtask of automation of process control linear (constant speed) uniform heating of the prototype were considered existing in the world practice the laws of temperature control (positional, proportional (P), proportional-integral (PI), proportional-differential (PD) and proportional-integral-differential (PID)). Given the high inertia of the system and the low rate of temperature change, the choice of a suitable control law was carried out with the PID controller, but when tested, the differential component caused instability of the control system. Therefore, to implement the required linear uniform heating of the samples, the PI-law of regulation was experimentally selected, fully satisfying the condition of the problem to be solved. The PID control scheme is shown in Fig. 3.

The process of setting the PI controller consists mainly of setting the set temperature (set point) and the values of the PI coefficients. Preliminary selection of coefficients was carried out in the simulation environment Matlab/Simulink, in which a mathematical model of the control system with a PI controller was built.

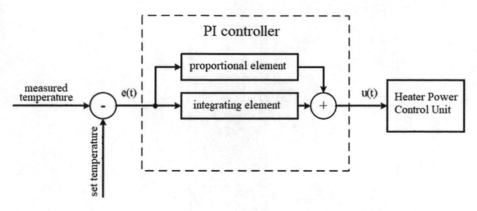

Fig. 3. PI control circuit, where e(t) – misalignment signal, control error; u(t) – control action.

The main task of automation is to ensure uniform heating at a constant speed [1, 5, 21]. To solve this problem, the heating control function was derived, which establishes the dependence of the set temperature (set point) of the Tset point on the heating time τ according to the formula (1):

$$T_{yc} = \frac{\tau}{12}. \tag{1}$$

The obtained dependence (1) provides a constant heating rate equal to 5 °C/min, which ensures uniform heating of the test sample and intensive gas evolution during its destruction.

Control of the heating power is organized by pulse width modulation (PWM), the width of which is regulated by the action of the PI controller. The PWM of the developed system of uniform heating is shown in Fig. 4, PWM period equal to 500 ms was determined experimentally.

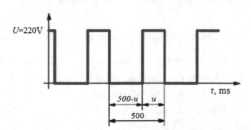

Fig. 4. PWM control of the heating element, where u is the control action of the PI controller.

To control the dynamics of gas release during thermal decomposition of the polymer, it is necessary to measure in real time the liquid level in the pressure gauge. To do this, the design of the "float", shown in Fig. 5.

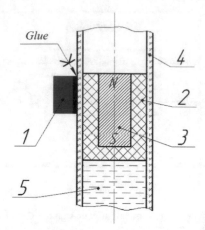

Fig. 5. The design of the "float", where 1 – Hall sensor; 2 – float; 3 – magnet; 4 – liquid pressure gauge; 5 – liquid.

The principle of operation of the design "float": inside the pressure gauge 4 omitted cylindrical float 2, of a low density material in which the magnet is mounted 3; outside to the tube of the pressure gauge relative to the float installed Hall sensor 1. As a result, when the liquid level 5 in the pressure gauge increases, the float will start to rise up, changing the magnetic field, thereby affecting the Hall sensor.

To determine the optical density of the released gas and the indicator during the destruction of the prototype, optical density sensors were selected, the output voltage of which is proportional to the light intensity. When selecting the sensors of optical density of the released gas in the glass bulb, it was taken into account that the glass bulb transmits certain wavelengths of electromagnetic radiation: near ultraviolet light – 315…400 nm; near infrared radiation – 0.76…2 microns. In addition, given the versatility of the developed installation and a wide range of non-metallic materials with different chemical structures, it was decided to use ultraviolet (UV) and infrared (IR) sensors to determine the optical density of the gas.

2.3   Block Diagram of the Developed Automated Control System

Figure 6 shows the block diagram of the developed automated control system (ACS) experimental installation to determine the thermal properties and control phase transformations in polymer and composite materials. At the first stage, the prototypes are loaded into the device. The microcontroller reads the temperature from the t1 thermocouple and sends it to the PI controller. The PI controller detects and outputs the regulating action through the PWM to a triac power circuit that commutes the high voltage load of the heating element. The heater heats evenly at a constant speed. Also at this time, the microcontroller takes readings from the t2 thermocouple through the amplifier, and readings from the Hall Sensor ΔH – the intensity of gas release during polymer destruction. Then the data is displayed on the computer. Also transmitted to the computer: the value of the linear thermal expansion of the sample with an electronic micrometer L; indicators of the optical density of the gas emitted from the IR sensor and UV sensor; indicators of the optical density of the liquid when indicating methylorange from the RGB color sensor.

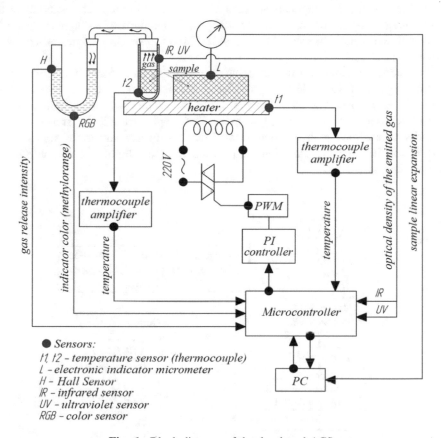

Fig. 6. Block diagram of the developed ACS

At the next stage, the algorithm of the automated device was developed and debugged.

2.4   Algorithm of ACS Work

Figure 7 shows the algorithm of ACS experimental setup to determine the thermal properties and control phase transformations in polymer and composite materials. The ACS algorithm is built in such a way that the operator has the ability to change the parameters of the system depending on the material under study, namely, the heating rate ($v_{heating}$) and the maximum heating temperature (T_{max}). The developed algorithmic software for the measurement and processing of experimental data, allowed to automate the control of the experiment and thereby improve the efficiency and accuracy of measurements.

3 Results and Discussion

With the use of an automated device, studies were conducted to study the effect of temperature on the thermal characteristics of polymer and composite materials in the solid aggregate state. The results of the study on the example of a polymer composite material of the brand armamide PA SV 30-1 ATM are shown in the form of a graph in Fig. 8.

Polymers, depending on the temperature, can be in three States that differ in the nature of thermal motion: glassy, highly elastic and viscous. Each of the physical States is associated with a certain set of properties, and each state corresponds to its own area of technical and technological application. The graph (see Fig. 8) shows that the transition temperature of the polymer studied from the glassy state to the highly elastic state is 150 °C, and the transition temperature from the highly elastic state to the viscous-flow state is 200 °C.

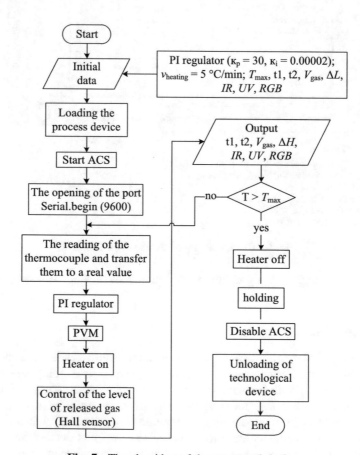

Fig. 7. The algorithm of the automated device.

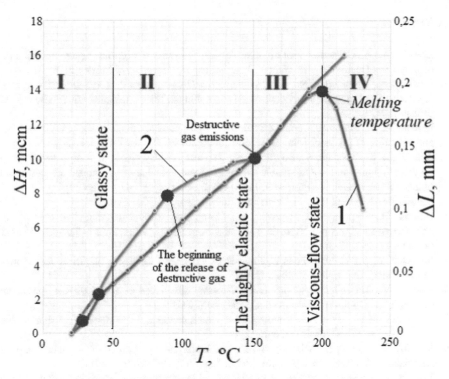

Fig. 8. Experimental curves ΔL = f(T) (1) and ΔH = f(T) (2) obtained in the study of polymer composite material of brand armamide PA SV 30-1 ATM.

4   Conclusion

In the course of this work, the following has been accomplished:

- developed and automated device for control of phase transformations and determination of thermophysical properties of polymer and composite materials;
- developed algorithmic software for control, measurement and processing of experimental data, allowing to automate the control of the experiment, improve the efficiency and accuracy of measurements, the algorithm of the automated control system, implemented as a software package;
- the developed automated installation was tested on polymer and composite materials of the following brands: polyamide PA6, polyurethane SKU-7L, polyvinyl chloride plasticate OMB-60 and armamide PA SV 30-1 ATM. The paper presents the results of a study of polymer composite material brand armamide PA SV 30-1 ATM.

As a result, a working installation with automated control of the process of experiments for the study of thermal properties and phase transformations in polymeric and composite materials was created.

References

1. Livshits, A.: Automated control of technological processes of high-frequency electrothermy of polymers. Doctoral thesis. Irkutsk State Transport University, Irkutsk (2016)
2. Livshits, A., Filippenko, N., Kargapoltsev, S.: High-frequency processing of polymeric materials. The organization of control systems. Irkutsk State Transport University (ISTU), Irkutsk (2013)
3. Larchenko, A.: Automated control system of high-frequency diagnostics in the production and operation of products from polymeric materials. Ph.D. thesis, Irkutsk (2014)
4. Aslamazov, T., Kotenev, V., Sokolova, N., Tsivadze, A.: Synthesis of composites based on polymer binders and water-soluble strongly polar phthalocyanine aimed at developing functional nanomaterials. Prot. Met. Phys. Chem. Surf. **4**(46), 474–478 (2010)
5. Butorin, D.: MATEC Web of Conferences, vol. 216, p. 02003 (2018)
6. Butorin, D., Filippenko, N., Livshits, A.: Complexed method of automated control high phase transformations in polymeric materials. Dev. Syst. Manag. Control Diagn. **10**, 10–18 (2016)
7. Shevchenko, V.: Fundamentals of Physics of Polymer Composite Materials. Lomonosov Moscow state University, Moscow (2010)
8. Volkov, V.: Thermophysical properties of composite materials with polymer matrix and solid solutions. Science of Education, Moscow (2011)
9. Pokhodun, A., Sharkov, A.: Experimental research methods. Measurements of thermophysical quantities. ITMO, St. Petersburg (2006)
10. Volkov, D., Egorov, A., Mironenko, M.: Thermophysical properties of polymer composite materials. J. Sci. Tech. Inf. Technol. Mech. Opt. **2**(108), 287–293 (2017)
11. Temnikova, S.: Thermophysical properties of polymer materials of modified structure based on Pentaplast. Ph.D. thesis, Kazan (2009)
12. Kuchmenova, L.: Thermal properties of polymer-polymer composites based on polypropylene. Ph.D. thesis, Nalchik (2014)
13. Vogel, V., Alekseev, P.: Device for determination of thermophysical characteristics of materials. In: Patent for Invention, № 142058 (1961)
14. Butorin, D., Filippenko, N., Filatova, S., Livshits, A., Kargapoltsev, S.: Development of methods for determining structural transformations in polymeric materials. Modern Technol. Syst. Anal. Model. **4**(48), 80–86 (2015)
15. Volovich, G.: Integral hall sensors. Modern Electron. **12**, 26–31 (2004)

Reinforcement and Strength Analysis of Reinforced Concrete Slabs in Computer Program PRINS

Vladimir Agapov[(✉)] [ID]

Moscow State University of Civil Engineering, Yaroslavskoe shosse, 26,
129337 Moscow, Russia
agapovpb@mail.ru

Abstract. The technique of reinforcement and strength analysis of reinforced concrete slabs implemented in the PRINS computer program developed by the author is described. Reinforcement is carried out on the limiting states of the first and second groups in accordance with normative documents. In this case, at the first stage, a linear static analysis of the structure without taking into account the reinforcement is carried out and internal forces are determined. At the second stage, the required amount of reinforcement in the structure elements is determined by the criteria of strength and crack resistance. The formulas for determining the amount of reinforcement given in the normative documents are approximate, but due to the use of reduced strength characteristics of the materials and various empirical correction factors, they provide an adequate margin of safety. However, the actual safety factor of the structure remains uncertain. In some cases, found by the recommendations of the normative documents, the amount of reinforcement may be insufficient, and in some - excessive. Therefore, to determine the real safety factor in the PRINS program, a non-linear calculation of the reinforced structure is carried out, and, based on the calculation results, the amount of reinforcement is adjusted. An example of reinforcement and calculation of reinforced concrete slabs using the proposed method is given.

Keywords: Reinforced concrete slabs · Reinforcement ·
Finite element method · Physical nonlinearity

1 Introduction

The technique of the reinforcement of concrete slabs, developed on the basis of domestic experts studies [1–3], systematized in a number of modern publications [4, 5] and described in building codes and rules [6, 7, 8] as a guide for designers, for many years was successfully applied in practice for the design of reinforced concrete structures. With the use of the design resistance values less than the characteristics values, as well as various correction factors, this technique provides an adequate safety factor for the designed elements. However, the real values of safety factors remain unknown, and, therefore, the accepted reinforcement cannot be considered as optimal. In addition, studies of a number of authors [9, 10] have shown that, in some cases, the

© Springer Nature Switzerland AG 2020
V. Murgul and M. Pasetti (Eds.): EMMFT 2018, AISC 982, pp. 741–748, 2020.
https://doi.org/10.1007/978-3-030-19756-8_70

reinforcement found on the recommendations of building rules is not enough. Therefore, the latest editions of the building acts [7, 8] recommend that designers, after
determining the reinforcement ratio, carry out calculations in a nonlinear formulation.
The recommendations are general in nature, although some emphasis is placed on the
application of the deformation theory, which is explained by the traditional preference
of this theory by many well-known scientists [1–3]. However, the deformation theory
is valid only for simple loading, whereas in many cases loading is a complex one.
Moreover, when using the deformation theory, it becomes problematic take into
account physical and geometric nonlinearity simultaneously. That is why many experts
on the finite element method use the theory of flow [11-16] when developing the
computer programs. This approach was also used by the author of the article when
developing the PRINS program [9]. Using this program in practice will give the
additional opportunities to designers to analyze the strength of reinforced concrete
structures, which should increase the reliability of the results.

2 Methods

The calculation algorithm adopted in the PRINS program is as follows.

At the first stage, a structure without reinforcement is calculated by the finite
element method and the forces in its elements are found. The formulas recommended
by the building rules [6, 7] establish a direct dependence of the area of the reinforcement on the found forces according to the strength condition. This relationship is
used to determine the area of the reinforcement in the first approximation. Then, when
the reinforcement is found, the crack opening width a_{crc} is determined. If the crack
opening width is less than the limit value $a_{crc,ult}$, then the area of reinforcement found is
taken as final. If the crack opening width is greater than the limit value, then the area of
reinforcement found from the strength condition is corrected. Since the direct dependence of the area of the reinforcement on the given value of the crack opening width
and the found forces does not exist, the correction carried out in an iterative way. In this
case, the value of a_{crc} found from the recommendations of building codes is compared
with the limit value $a_{crc,ult}$, the coefficient $k = a_{crc}/a_{crc,ult}$ is found and the area of
reinforcement from the previous iteration is multiplied by this coefficient. Iterations
continue until a given accuracy is reached.

At the second stage, the design calculation scheme is rebuilt with regard to the
reinforcement found and a non-linear static calculation for the given loads is carried
out. The nonlinear calculation technique implemented in the PRINS program is
described in the author's previous works (see, for example, [13, 14]). In this case,
special multilayered finite elements are used, in which the reinforcement is specified by
a special layer of equivalent thickness with unidirectional properties. Based on the
results of this calculation, a final conclusion is made about the sufficiency or insuffi-
ciency of the reinforcement area found at the first stage.

3   Results

We illustrate the above described algorithm with an example of calculating a single slab. A rectangular floor slab shown in Fig. 1 was calculated. The slab was simply supported along the short sides. It was loaded by own weight and uniformly distributed pressure. The total value of applied load was equal to 20 kPa. The material of slab: heavy concrete of B25 class and reinforcement of A400 class. The finite element calculation scheme with the numbering of elements of the slab is shown in Fig. 2.

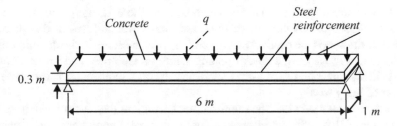

Fig. 1. Reinforced concrete slab

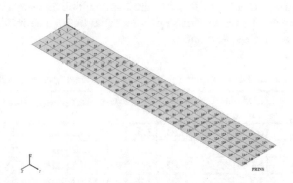

Fig. 2. Calculation scheme of the slab with the numeration of the elements

Fig. 3. Bending moments My in the homogeneous concrete slab

For the calculation of slab reinforcement, the PRINS program uses single-layer finite elements identified as Shell36. These elements can have both quadrangular and triangular shapes. The elements Shell34, Shell26, Shell29 and Shell33 are used in the PRINS program for the non-linear analysis of reinforced concrete slabs. The elements Shell34 and Shell29 have a triangular shape, and the elements Shell26 and Shell33 has the shape of an arbitrary quadrilateral. The Shell34 and Shell26 elements are used to calculate plates of constant thickness, whereas the Shell29 and Shell33 elements can be used to calculate plates of both constant and variable thickness.

The bending moments in the slab at the load of 20 kPa are shown in Fig. 3, and the results of the reinforcement obtained with the aid of PRINS program are shown in Table 1. The following designations are taken in Table 1: $A_{s,x1}$ and $A_{s,x2}$ are the reinforcing steel areas on the side of the upper and lower surfaces of the plate in the direction of the x-axis, respectively; $A_{s,y1}$ and $A_{s,y2}$ is the same in the direction of the y axis; $A_{cr,x1}$ and $A_{cr,x2}$ are the crack opening width on the upper and lower surfaces of the plate in the direction of the x-axis, respectively; $A_{cr,y1}$ and $A_{cr,y2}$ is the same in the direction of the y axis.

In order to make sure that the reinforcement found on the recommendations of the building rules [7] is sufficient or insufficient, a physically non-linear static analysis of the slab for the given loads was carried out. The multilayered finite elements Shell26 were used in this analysis. The same finite element mesh as in the previous analysis was used. The base load was equal to 10 kPa. The load multipliers were assumed to be 0.1 for all loading steps. The number of steps was equal to 20. The results of non-linear analysis are shown in Figs. 4, 5 and 6.

Table 1. The reinforcing steel areas and cracks opening width

Element group number	Element number	$A_{s,x1}$	$A_{s,x2}$	$A_{s,y1}$	$A_{s,y2}$	$A_{crc,x1}$	$A_{crc,x2}$	$A_{crc,y1}$	$A_{crc,y2}$
1	67	0.0	1.3	0.0	15.2	0.0	0.0	0.0	0.302
1	68	0.0	1.3	0.0	15.2	0.0	0.0	0.0	0.302
1	69	0.0	1.3	0.0	15.2	0.0	0.0	0.0	0.302
1	70	0.0	1.3	0.0	15.2	0.0	0.0	0.0	0.302
1	71	0.0	1.3	0.0	15.4	0.0	0.0	0.0	0.302
1	72	0.0	1.3	0.0	15.3	0.0	0.0	0.0	0.302
1	73	0.0	1.3	0.0	15.3	0.0	0.0	0.0	0.302
1	74	0.0	1.3	0.0	15.3	0.0	0.0	0.0	0.302
1	75	0.0	1.3	0.0	15.4	0.0	0.0	0.0	0.302
1	76	0.0	1.3	0.0	15.4	0.0	0.0	0.0	0.302
1	77	0.0	1.3	0.0	15.3	0.0	0.0	0.0	0.302
1	78	0.0	1.3	0.0	15.3	0.0	0.0	0.0	0.302
1	79	0.0	1.3	0.0	15.3	0.0	0.0	0.0	0.302
1	80	0.0	1.3	0.0	15.4	0.0	0.0	0.0	0.302
1	81	0.0	1.3	0.0	15.2	0.0	0.0	0.0	0.302
1	82	0.0	1.3	0.0	15.2	0.0	0.0	0.0	0.302

The curve "load parameter – displacement in the middle of the span" in presented in Fig. 4. The displacements field at the load of 20 kPa is shown in Fig. 5 and the moments in the slab at the same load are shown in Fig. 6.

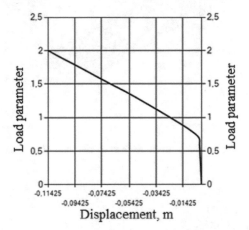

Fig. 4. The curve "load parameter – displacement in the middle of the span"

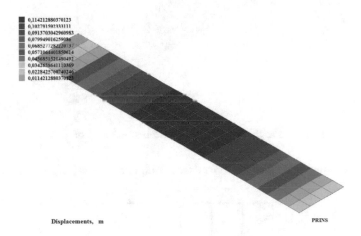

Fig. 5. The displacements in the slab at the load of 20 kPa

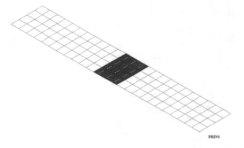

Fig. 6. Cracks on the bottom surface of the slab at the load of 5 kPa

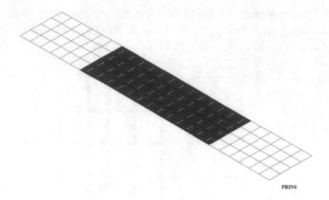

Fig. 7. Cracks on the bottom surface of the slab at the load of 7 kPa

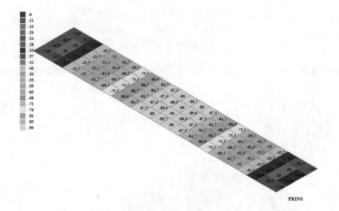

Fig. 8. The moments My in the slab at the load of 20 kPa

Further loading of the slab led to its destruction under a load of 38 kPa (see Fig. 9).

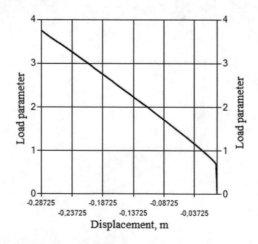

Fig. 9. Load-displacement curve down to failure.

4 Discussion

The results of the non-linear analysis of the slab showed that the reinforcement found according to the building rules [7] using the strength and cracking criteria does not meet the requirements for deformations. It is seen from Figs. 4 and 5 that the maximum deflection at a load of 20 kPa is 11.4 cm, while the allowable value of the deflection is $l/200 = 3$ cm. A large amount of deflection is due to the fact that with a load of 5 kPa the first cracks arise (see Fig. 6), and at a load of 7 kPa the crack formation area occupies a significant part of the lower surface of the slab (see Fig. 7).

Note that the reliability of the nonlinear calculations carried out is confirmed by the observance of the equilibrium conditions of the slab. This follows from Fig. 8.

Thus, we come to the conclusion that it is necessary to increase the area of reinforcement for the slab under consideration. Since it is not possible to establish the analytical dependence of the plate deflection on the load in a nonlinear calculation, this problem has to be solved by the way of selection of the reinforcement area.

By setting the reinforcing steel area $A_s = 36$ cm^2/m, we got the maximum displacement w_{max} (see Fig. 10a). And by reducing the area to the value $A_s = 32$ cm^2/m, we got w_{max} (see Fig. 10b).

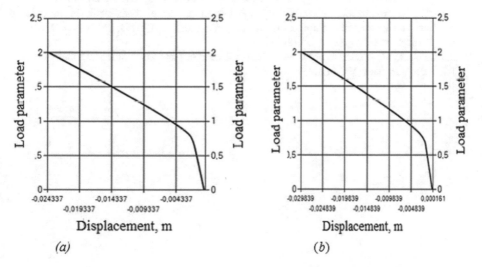

(a)   (b)

Fig. 10. The curve load-displacement at: (a) $A_s = 36$ cm^2/m; (b) $A_s = 32$ cm^2/m

Thus, using the capabilities of the PRINS program, we found the reinforcing steel area $A_s = 32$ cm^2/m, which satisfies all three conditions - strength, crack resistance and deformation for given slab.

The analysis of the slab according to the PRINS program until complete destruction at the final value of the reinforcement area $A_s = 32$ cm^2/m showed that the load-bearing capacity of the slab increased to 66 kPa.

5 Conclusion

One of the main tasks in the designing of reinforced concrete slabs is the selection of the steel reinforcement, ensuring reliable operation of the structure. To solve this problem, designers need the calculation tools for the adequate determining of the stress-strain state of structure both before and after reinforcement. One of such tools is the PRINS program. The implementation of this program in the design practice will allow designers to make more informed decisions when assigning reinforcement for reinforced concrete slabs.

References

1. Malahova, A.: Armirovanie zhelezobetonnyh konstrukcij. MGSU, Moskva (2015)
2. Kuznecov, V.: Zhelezobetonnye i kamennye konstrukcii. Izdatel'stvo ASV, Moskva (2016)
3. Nikolaev, V.: Prochnost' zhelezobetonnyh konstrukcij GES i AES. Gidro-tekhnicheskoe stroitel'stvo 10, 64–69 (2018)
4. Hinton, E., Owen, D.: Finite Element Software for Plates and Shells. Pinerige Press, Swansea (1984)
5. Chen, W.: Plasticity in Reinforced Concrete. J. Ross Publishing, Richmond (2006)
6. Agapov, V.: Realization of reinforced concrete structures analysis with the account of physical and geometrical nonlinearity in computer program PRINS. In: MATEC Web of Conferences, vol. 251, p. 04035 (2018)
7. Zienkiewicz, O., Taylor, R.: The Finite Element for Solid and Structural Mechanics, 6th edn. McGraw-Hill, New York (2005)

Valuation of Glulam Applicability in Public Buildings Architecture

Mikhail Zhuravlev$^{(\boxtimes)}$ and Tatyana Bogdanova

Samara State Technical University, Molodogvardeyskaya st., 194,
443001 Samara, Russia
mihail_zhuravlev@inbox.ru

Abstract. The article contains the research on physical and space forming abilities of glulam (glued laminated timber) and its economic values. The public buildings space forming research is based on followings techniques: walls with protrusion vertical bearing elements; walls with standing out vertical bearing elements; combination of vertical bearing elements; surface formed by crossed girders. All methods used in the article consider world design experience, physical and economic research is based on current normative values and documents and practical test results. The research in the article compares steel, glulam and concrete constructions considering span size, physical characteristics, fire and chemical resistance abilities, lasting quality, environmental protection and reveals the regularities in efficiency of using these materials. One of the most important roles in efficiency calculating plays strength-density ratio (SDR). The comparison results of the material characteristics are presented in the summary table. Combining of glulam space forming properties and constructive characteristics allow receiving general valuation of glulam applicability in public buildings architecture.

Keywords: Glulam constructions · Public buildings architecture ·
Strength-density ratio · Space forming methods · Wood

1 Introduction

Due to technical progress concrete, steel and different synthetic materials are widely used in constructing nowadays. They are economically effective and can be produced in great amounts. Trying to make constructing cheap leaves the construction only one purpose – to be as much durable as possible leaving the aesthetic behind. Thus different decorations are used to hide poor appearance of a construction and to imitate the use of some other materials. But if we look to the past and consider some acknowledged architectural masterpieces, we will see that framing is able to give the building not only the necessary durability but also the good and pleasant aesthetic view of the whole building. In spite of undeniable advantages of modern synthetic materials, we still need natural ones because humanity always wishes natural beauty. And we come to wood, renewable resource which has necessary aesthetic and constructive properties. Glulam technology has eliminated most disadvantages of wood as constructive material and still it is not preferred as framing. For example, in Europe and the USA, the use of

© Springer Nature Switzerland AG 2020
V. Murgul and M. Pasetti (Eds.): EMMFT 2018, AISC 982, pp. 749–760, 2020.
https://doi.org/10.1007/978-3-030-19756-8_71

glulam as basic structure is becoming more and more popular within the last several decades. Different sports complexes, malls and other buildings which have great importance for city life are made of glulam and the amount of such buildings is only increasing. That is why it is necessary to reconsider the valuation of glulam in general to find out the rationality of its use. The purpose of the research is the general valuation of glulam applicability considering its physical and space forming abilities and economic characteristics.

2   Materials and Methods

Our general valuation is based on two aspects:

- glulam as space forming material;
- glulam as constructive material.

2.1   Methods of Glulam General Valuation as Space Forming Material

Several buildings with special requirements were chosen to evaluate glulam as space forming material. These requirements are listed below:

- the building should be public because public buildings express the main tendencies in modern architecture;
- the basic structure of the building should be made of glulam;
- the building is of great cultural or functional importance for the city.
- Several architectural methods for space forming were used to evaluate the glulam space forming abilities:
- walls with protrusion vertical bearing elements;
- walls with standing out vertical bearing elements;
- combination of vertical bearing elements;
- surface formed by crossed girders.

We furthermore analyze the abilities of glulam to meet the requirements listed above. It should be noted that a very important thing here is the influence of the building on the aesthetic perception of the whole building space.

2.2   Methods of Glulam General Valuation as Constructive Material

The general valuation also considers glulam constructive abilities compared with steel and concrete, because these materials are widely used in constructing almost all kinds of buildings nowadays. To compare these materials, we used several very important characteristics, namely:

- technical and economic abilities depending on the span size;
- physical characteristics;
- chemical resistance;
- environmental protection;
- fire resistance;
- lasting quality;

- constructional abilities;
- strength-density ratio (SDR).

Calculations on glulam bearing construction for spans of 6, 12 and 24 m long and its weight in subsection "technical and economic indexes depending on span size" are performed using methods from normative documents. Data on 48 m long spans is based on the experience of large-span construction design with use of glulam [1].

Chemical resistance ability is evaluated on the basis of potassium salt and reagent storehouse design experience with use of large-span glulam constructions [2, 3] and current normative documents. Data on fire resistance limits is based on true events of fire incident with buildings using some glulam constructions. The good example of the safety of glulam is the fire accident happened on the potassium mill storehouse. Its basic constructions were glulam arches. The fire lasted for 2 h but the cross section of the arch showed the decrement of only 20%. Further investigation concluded that decrease of construction bearing ability was negligible [4]. The other fire accident took place during the finishing jobs of an aquapark. The first truss crashed only in 2,5 h of unceasing fire [5]. Mjostarnet, 81 m high building, was started in 2017 in Norway not far from Oslo. Its structural support is made of glulam. The results of fire resistance tests for this building are used in current research [6]. Environmental protection characteristics in producing materials and strength properties are based on normative values with support of current test results and constructing experience [7, 8]. Some normative values are not up-to-date. In this case the actual values are used.

3   Results

3.1   Results of Glulam General Valuation as Space Forming Material

The valuation of glulam as a space forming material is based on the architectural space forming methods listed above found in public architecture.

The structural support elements from the first architectural method are used as part of the whole construction composition that influence the space perception. Standing out vertical ribs set surface plasticity and form rhythmic sections. Examples of using this method by means of glulam are given below. Bournville College, Birmingham, UK [9], Gosta Serlachius Museum, Mantta, Finland [10], St Henry's Ecumenical Art Chapel, Turku, Finland [11] (see Fig. 1).

Bournville College	Gosta Serlachius Museum	St Henry's Ecumenical Art Chapel
Broadway Malyan 2011 Birmingham, UK	MX_SI 2003 Mantta, Finland	Sanaksenaho Architcets 2005 Turku, Finland

Fig. 1. Architectural space forming method "walls with prominent vertical bearing elements"; original work.

The second method allows visually splitting the space into several functional areas. While used in interiors with large halls this method gives an opportunity to create areas of a more comfortable visual size for a human. And as for exterior, this method allows visually creating some transitional area between inner and outer space. A good example of this method is the building of Washington fruit and produce company, Yakima, USA [12], Bournville College, Birmingham, UK [9], Links National Golf Club, Moscow, Russia [13] (see Fig. 2).

Washington fruit & produce co.
Graham baba architects
2016
Yakima, USA

Bournville College
Broadway Malyan
2011
Birmingham, UK

Links National Golf Club
Tammvis
2015
Moscow region, Russia

Fig. 2. Architectural space forming method "walls with flaring vertical bearing elements"; original work.

Splitting the space into some areas is also achieved by using pillars in it. In this case, visual integrality is not infracted. This method is used in the building of Sunbeams Music Centre, Penrith, United Kingdom [14], Agri Chapel, Nagasaki, Japan [15], Wood Innovation and Design Centre, Prince George, Canada [16] (see Fig. 3).

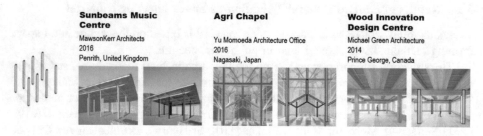

Sunbeams Music Centre
MawsonKerr Architects
2016
Penrith, United Kingdom

Agri Chapel
Yu Momoeda Architecture Office
2016
Nagasaki, Japan

Wood Innovation Design Centre
Michael Green Architecture
2014
Prince George, Canada

Fig. 3. Architectural space forming method "combination of vertical bearing elements"; original work.

The next method, the most up-to-date one, proposes the crossing of two or more rod elements. This method allows to create penetrable flat surface. It is represented in the building of Soma City Home for all, Soma, Japan [17], UC Architecture School Building, Santiago, Chile [18], France's pavilion, Milan, Italy [19] (see Fig. 4).

Soma City Home for all
Klein Dytham architecture
2015
Soma, Japan

UC Architecture School Building
Gonzalo Claro
2016
Santiago, Chile

France's pavilion
XTU Architects
2015
Milan, Italy

Fig. 4. Architectural space forming method "surface formed by crossed girders"; original work.

Constructive system of these buildings is made of glulam and support elements are used as part of interior and exterior that tells us of great aesthetic values of this material.

3.2 Results of Glulam General Valuation as Physical Abilities and Economic Characteristics

Constructive product technical and economic values considering span size

Table 1. Beam efficiency, 6 m

	Glulam	Steel	Concrete
Weight, kg	96	126	945
Max construction height, mm	250	200	300
Price, thousand rub.	24	26	8

Table 2. Beam efficiency, 12 m

	Glulam	Steel	Concrete
Weight, kg	825	506	4600
Max construction height, mm	800	330	890
Price, thousand rub.	50	50	48

Table 3. Truss efficiency, 12 m

	Glulam	Steel	Concrete
Weight, kg	615	615	8350
Max construction height, mm	2400	3000	1800
Price, thousand rub.	82	61	110

Table 4. Beam efficiency, 24 m

	Glulam	Steel	Concrete
Weight, kg	1548	4176	5500
Max construction height, mm	1000	500	1500
Price, thousand rub.	110	127	85

Table 5. Truss efficiency, 24 m

	Glulam	Steel	Concrete
Weight, kg	1430	2400	11200
Max construction height, mm	3000	4150	3240
Price, thousand rub.	133	155	185

Table 6. Beam efficiency, 48 m

	Glulam	Steel	Concrete
Weight, kg	4590	–	–
Max construction height, mm	1100	–	–
Price, thousand rub	587	–	–

Table 7. Truss efficiency, 48 m

	Glulam	Steel	Concrete
Weight, kg	9570	14400	–
Max construction height, mm	6000	4700	–
Price, thousand rub	600	312	–

3.3 Physical Characteristics

The main physical characteristics of glulam in comparison with steel and concrete. Glulam has the lowest density of involved materials. Such characteristics as specific gravity and thermal conductivity directly depend on this characteristic. Thus glulam construction in contrast to steel and concrete will weigh 5 and 1,6 times less respectively.

Low thermal conductivity of glulam is determined by internal constitution of the material. Steel and concrete constructions require additional efforts to achieve necessary heat transfer resistance. The fibrous structure of glulam also provides comfortable acoustic environment due to noise absorption (see Table 8).

Table 8. Physical characteristics

	Glulam	Steel	Concrete
Density, kg/m^3	335–430	7800–7850	2500
Specific gravity, kg/m^3	1.54	7.85	2.5
Thermal conductivity, W/m * °C	0.15	53	1.7
Thermal capacity rate, J/kg * °C	1.6	0.5	0.84
Thermal expansion, 1/°C	0.000005	0.000012	0.000012

Thermal expansion tells us about the linear changing of the product depending on temperature changing. Glulam thermal expansion lengthwise fibre is 2-3 times lower then steel thermal expansion. Thus there is no need to make temperature junctions which is especially important in constructing large-span buildings because in this case there is no need to make special geometry parameter changes in constructive elements.

Thermal capacity rate stands for the quantity of heat necessary to increase the temperature of the material surface by $1°$. The high value of this characteristic means that the surface temperature of a product is comfortable for a human.

Chemical resistance. The experience of reagents storehouse design using glulam constructions confirms glulam high chemical resistance ability (see Table 9).

Steel constructions are not usually used in chemically aggressive environments because of high corrosion. Special constructive and protective modifications are required to achieve necessary chemical resistance value, which increase the price of a construction.

Table 9. Chemical resistance

	Glulam	Steel	Concrete
Chemical resistance	Preserves the integrity in acidic and alkaline medium	Exposed to chemically aggressive environment	Exposed to acidic and alkaline environment

Concrete constructions without special modifications are exposed to acidic and alkaline mediums. Porous concrete is more vulnerable because aggressive environment destroys it from both inner and outer sides. In this case, it is very likely that the reinforcement will be destroyed by the corrosion and the whole construction will lose its durability.

Environmental security. Glulam is a composite material and the most important part in glulam producing technology is the use of special glues and impregnations.

It should be noted that glulam production on all the stages needs many times less energy and resources then the production of steel or concrete items. Wood is a renewable resource and if it is well-time renewed and effective latter-day producing technologies are used then the damage to the environment will be negligible (see Table 10).

Table 10. Environmental security

	Glulam	Steel	Concrete
Environmental security	Production based on renewable resource	Present day volumes of productions considerably pollute the environment	Production causes emissions into the atmosphere

It is necessary to understand the great influence of production process on the environment. Producing steel constructions is inseparably connected with extractive industry that makes a huge contribution into polluting the environment. Concrete production in turn is a very energy-intensive process with emissions of CO_2 into the atmosphere.

Fire resistance. Wood is an inflammable material in contradistinction to steel or concrete and that is why it is preferred not to use it in public buildings architecture. And here we need to consider the changing of physical conditions of glulam, concrete and steel in fire.

Fire resistance limit of wooden constructions depends on 2 factors: time between starting fire and construction ignition and the time between ignition and full loss of constructive durability. For non-modified wood the first time interval is 4 min. Modifying of the external party of glulam elements with coating and dyes increases this time interval. After the ignition, the burning takes place. The cross section of glulam elements during this process changes to 0,6–0,7 mm/min and that is why it takes a long time for a construction to lose its constructive durability, a time enough for people to evacuate and to eliminate the fire (see Table 11).

Table 11. Fire resistance

	Glulam	Steel	Concrete
Time between the fire starting and loss of constructive durability, min	<150	10–15	45–150

For an opened unmodified steel construction fire resistance limit is R10–15. The whole steel construction is getting heated very quickly during burning process and when the critical value is reached the construction abruptly loses constructive durability that leads to crushing of the whole construction.

The fire resistance limit for most frequently used concrete items is R90–150 (for compressed items) and R45–90 (for bending elements). Bending elements have such small values because reinforcement in the stretched zone is closer to the surface of a concrete item.

Lasting quality. There are two factors which mostly influence the glulam lasting quality. The first one is based on the durability of composite material. The basics for glulam is natural fibrous material that tends to change response on some external exposure (strain for example) in a long period of time. Solid wood has non-elastic deformations that greatly decreases the lasting quality of a construction. But this draw is minimized in glulam due to additives and different kinds of glues. In spite of the fact that normative lasting quality value for glulam constructions is 15–20 years, practically it is much higher. Some storehouses and large-span sports facilities built in the middle of the past century are still used in our country [20].

The second factor which influences the lasting quality of glulam construction is the presence of small inner defects which tend to progress if the building is not properly exploited. With proper temperature and moisture content conditions the lasting quality of glulam constructions can exceed 100 years (see Table 12).

Table 12. Lasting quality.

	Glulam	Steel	Concrete
Quality, years	<100	50	150

Lasting quality of steel constructions also depends on treatment conditions because its surface is corrosive. For example, the lifetime of some steel constructions was less the appointed one in spite of plenty of regulations (less than 50 years).

Concrete has the longest lasting quality value of all materials under this review - 150 years. But it should be noted that without special additives into concrete mix the concrete is affected by aggressive environment.

Strength properties. Due to world experience of glulam construction design we can notice that actual strength properties values are many times higher than designed ones. Thus if we use real experience of glulam design the use of such constructions can be rationally justified (see Table 13).

Table 13. Strength properties.

	Glulam	Steel	Concrete
Tensile strength	65–216 MPa	100–570 MPa	240–500 MPa
Compressive strength	40–60 MPa	210 MPa	215–360 MPa
Flexural strength	36–150 MPa	350–470 MPa	5–104 MPa
Weight strength	0.2	0.07	0.2

Specific strength, the ratio of the ultimate strength along the fibers to the density of the material, is a very important indicator in this analysis. This indicator shows the durability of the material depending on its mass. High value of specific strength is mostly important in large-span constructions design because especially in this case it is important to minimize the load of framing on the foundation and at the same time preserve the necessary durability of constructive elements. Comparison of this value leaves steel far behind.

Strength-density ratio (SDR). Strength-density ratio shows the ratio of mechanical property index of the material to its density. This parameter is a summary measure of some internal characteristics of the material and objectively demonstrates their relation. This factor plays the most important role in large-span construction design because in this case it is very important to minimize the amount of used materials and at the same time to use all these materials most efficiently (see Table 14).

Table 14. Strength-density ratio

	Glulam	Steel	Concrete
SDR	2,2	0,55	0,06

Glulam has the highest value of SDR, then comes steel. Concrete has the lowest value. Thus we can conclude that glulam is very efficient in constructing.

4 Discussion

4.1 Glulam is Efficient Space Forming Material

The research showed that glulam is efficient to use as framing material when using different ways of space forming. It allows applying architectural methods such as:

- walls with prominent vertical bearing elements;
- walls with standing out vertical bearing elements;
- combination of vertical bearing elements;
- surface formed by crossed girders.

Glulam is a ductile material and it makes it possible to create complex forms and surfaces.

4.2 Aesthetic Properties of Glulam

All constructive elements of the buildings reviewed above are presented both in exterior and interior that show that the aesthetic properties of glulam are very high. Color, texture, and surface temperature allow minimizing the use of decoration materials, the element is in direct human contact and doesn't need any additional decorations for visual comfort. Thus glulam is a self-sufficient material from the aesthetic point of view.

4.3 General Valuation of Glulam Physical Characteristics

According to the Tables 1, 2, 3, 4, 5, 6 and 7, the most indicative parameter is the weight of a glulam construction. Glulam has the lowest weight in 6, 24 and 48 m long beams and trusses of all compared materials. This indicates a more rational load distribution over the whole length of an item.

The use of steel or concrete beams with 48 m long spans is irrational, in this case glulam beam is the only possible variant.

Thus according to the Tables 1, 2, 3, 4, 5, 6 and 7 the efficiency of using glulam beams and trusses is increasing with the length of the span.

The advantages of compared materials (green mark) are listed below in a summary table based on the Tables 1, 2, 3, 4, 5, 6 and 7, 8, 9, 10, 11, 12, 13 and 14. (see Table 15).

Table 15. Summary of physical characteristics

	Glulam	Steel	Concrete
Span 6 m	○	○	●
Span 12 m	○	○	●
Span 24 m	○	○	●
Span 48 m	○	●	●
Density	○	●	○
Specific gravity	○	●	○
Thermal conductivity	○	●	○
Thermal capacity rate	○	●	○
Thermal expansion	○	○	○
Chemical resistance	○	●	○
Environmental security	○	●	●
Fire resistance limits	○	●	○
Lasting quality	●	○	○
Tensile strength	●	○	○
Compressive strength	●	○	○
Flexural strength	○	○	●
Weight strength	○	●	○
SDR	○	○	●

5 Conclusion

General valuation of glulam applicability in municipal constructions showed that many characteristics of glulam are not lower than the same characteristics of concrete or steel. The summary table shows the advantages of this or that material depending on the characteristics we want to use most.

Self-sufficiency as means of expression is the advantage of glulam constructions.

This research gives an opportunity to make a proper choice of a constructing material in designing based on lots of factors.

Strength-density ratio takes in account the ratio of mechanical characteristics of the material and its density. Glulam has the highest value of this parameter. In addition to unique aesthetic properties and high space forming abilities, glulam appears to be the most efficient material for municipal constructions.

References

1. Turkovskiy, S.B., Pogoreltsev, A.A., Preobrazhenskaya, I.P.: Glulam in modern construction. Part 6. Sport facilities. LesPromInform 4(102), 132–141 (2014)
2. Schmidt, M., Glos, P., Wegener, G.: Gluing of European beech wood for load bearing timber structures. HolzAlsRohund Werkstoff 1(68), 43–57 (2010)
3. Turkovskiy, S.B., Pogoreltsev, A.A., et al.: Glulam in modern construction. Part 8: commercial facilities and storage facilities. LesPromInform 5(103), 146–160 (2014)
4. Merritt, F.S.: Building Design and Construction Handbook, 6th edn. McGraw-Hill, New York
5. Turkovskiy, S.B., Pogoreltsev, A.A., Preobrazhenskaya, I.P.: Glulam in modern construction. Part 4. LesPromInform 1(99), 110–116 (2014)
6. Rune, A.: Mjøstårnet-construction of an 81 m tall timber building. In: Internationales Holzbau-Forum IHF, FORUM 2017, vol. 112, pp. 3–13 (2017)
7. Chubinsky, A.N., Medov, V.S., Slavik, Y.Y.: On the question of laminated wood constructions' durability. Izvestia Sankt-Peterburgskoj lesotehniceskoj Akademii 203, 128–134 (2013)
8. Dunn, A.: Final report for commercial building costing cases studies – traditional design versus timber project. Forest & Wood Products Australia PN308-1213, pp. 1–17 (2015)
9. BroadwayMalyan. https://www.broadwaymalyan.com. Accessed 02 Dec 2018
10. BAX studio homepage. http://baxstudio.com. Accessed 21 Dec 2018
11. Sanaksenaho Architects. http://www.kolumbus.fi/sanaksenaho/. Accessed 17 Nov 2018
12. Graham Baba Architects. http://grahambabaarchitects.com/home. Accessed 05 Dec 2018
13. Tammvis Homepage. http://www.tammvis.ru/proekty. Accessed 03 Dec 2018
14. Mawsonkerr. https://www.mawsonkerr.co.uk/projects/sunbeamsmusiccentre/. Accessed 06 Dec 2018
15. Yu Momoeda Architecture Office. https://www.yumomoeda.com. Accessed 06 Dec 2018
16. Michael Green Architecture. http://mg-architecture.ca. Accessed 08 Dec 2018
17. Toyo Ito & Associates, Architects. http://toyo-ito.co.jp. Accessed 08 Dec 2018
18. Architizer Homepage. https://architizer.com/projects/uc-architecture-school-building/. Accessed 12 Dec 2018
19. https://www.xtuarchitects.com. Accessed 03 Dec 2018
20. Serov, E.N., Labudin, B.V.: Glued Timbering: Present State and Development Problems. LesnoyZhurnal 2(332), 137–146 (2013)

Structure and Properties of Decorative Concrete Impregnated with Vegetable Oil

Viktor Voronin⬭, Oksana Larsen(✉)⬭, Dmitry Zamelin⬭,
and Nikolay Mikhailov⬭

National Research Moscow State University of Civil Engineering,
Moscow 129337, Russia
larsen.oksana@mail.ru

Abstract. The performance of decorative concrete can be improved significantly with the modification by organic oil. The aim of the present study was to
investigate the structure and performance properties of decorative concrete
modified with technical vegetable oil. The studied materials were Portland
cement, cement with low water demand and cement with superplasticizer S-3
addition. Organic drying oil was used as an impregnation mixture. Impact of
superplasticizer S-3 was tested by two ways. Self-vacuuming method was used
to accelerate the impregnation. Pigments were used to improve the appearance
of the concrete surface. Linen oil and concrete samples were heated before
impregnation. Impregnation was made by immersion of the heated samples in
the heated oil. Mechanical strength, frost resistance, porosity, color properties,
water and oil absorption were tested. Durability of decorative concrete has been
tested according to strength and frost resistance Results revealed correlations
between structure of modified concrete and its performance. In this study,
comparison between the various cements was showed, and the most effective
binder for decorative modified concrete was determined.

Keywords: Impregnated concrete · Decorative concrete · Vegetable oil ·
Concrete modification · Superplasticizer

1 Introduction

Concrete is the most widely used construction material in the world. Concrete is used
not only as load bearing material but also an architectural element. There are different
ways to improve the appearance of concrete surface, such as forms and form liners,
white Portland cement and colored cements, decorative concrete. The decorative
concrete is considered a special concrete, which simulates natural rocks. The use of
decorative concretes, simulating rocks, can significantly enhance the appearance of
concrete buildings. Decorative concrete doesn't have strong durability. To increase
longevity, concrete is impregnated by polymers [1–5]. Despite the advantages of this
method, there are significant complexity of the manufacturing process and high cost of
polymer impregnation mixes. To obtain high-quality, decorative concrete can be
impregnated with vegetable oil can be used [6].

© Springer Nature Switzerland AG 2020
V. Murgul and M. Pasetti (Eds.): EMMFT 2018, AISC 982, pp. 761–768, 2020.
https://doi.org/10.1007/978-3-030-19756-8_72

In order to increase durability of decorative concrete and reinforced concrete elements, density and strength of surface layers of concrete should be increased [6]. These layers have increased porosity due to sedimentation and water separation when compaction of concrete mixture. Methods of modification such as coating with a protective layer, hydrophobization and impregnation with various substances are used to increase durability. Effective impregnation mix for concrete should have the following criteria: low price, market availability, melting point within 100–400 °C, low viscosity before solidification, ability to wet the surface of concrete, high mechanical strength and stiffness after solidification, chemical inertness, low vapor pressure above solid phase and fire-resistance.

Vegetable oil meet the requirements for the impregnation mixes. However, strength of solid vegetable oil is lower than polymers. Therefore, then applying vegetable oil as impregnation mixtures should be envisaged a set of measures to improve strength of concrete matrix. Introduction of surfactants contributes to improving strength of concrete especially when combined grinding with cement. Pigments are used in order to achieve artistic expression. Introduction of pigment when combined grinding with surfactants and Portland cement contributes to its more even spread the volume of concrete, saving and milling of pigment. It can be seen from the review of literature above that we need to better understand how to ensure the performance of decorative concrete.

2 Materials and Methods

According to requirements for decorative, sculptural and monumental art, in order to ensure more even spread of filler particles the volume of element and particularly in surface layer, filler particles size should not exceed 20 mm with the main content of granules 5–10 and 2.5–5 mm. Through selection of proportions of decorative concrete revealed filler granules with size less than 1.25 mm change filler color attributes of dark-colored rocks. Therefore, granules with size less than 1.25 mm content in that case should be limited. For light-colored rocks such limitation may not be introduced.

Through varying the proportion between different fraction particles, granulometric composition proportions of fillers for different decorative concrete mixtures were found, as it's shown below in Table 1.

Table 1. Optimal granulometric composition proportions for decorative concrete.

Filler	Fractional composition content (%)				
Granite gray	10	30	45	9	6
Granite red					
Gabbro					
Limestone	6	18	42	20	14
Marble	10	16	45	15	14

Selection of proportions of decorative concrete was made by varying the cement consumption from 300 to 900 kg/m3 with different flowability (slump 2–3 cm; 8–9 cm; 50–55 s). Impact of superplasticizer S-3 was tested by two ways: through mixing with water and by combined grinding S-3 with cement (cement with low water demand).

Based on suggested, concrete proportions with strength classes from 7,5 to 40 were obtained, differ significantly between porosity volume and nature of porosity space (Tables 2, 3, 4) to increase quality of decorative concrete, mixes with superplasticizer S-3 were used to reduce the volume of water and to increase density and mechanical strength. Table 3 shows the proportions of concrete mixtures with superplasticizer S-3, imitating gabbro.

Table 2. Proportions of the materials in Portland cement based mixtures.

Strength class		Proportions of the materials (kg/m^3)			Water/Cement	Slump
Cement	Concrete	Cement*	Water	Filler		
22.5	22,5	297	172	1982	0.58	50 s
32.5	15	300	193	1860	0.64	3 cm
32.5	20	314	188	1838	0.55	2 cm
32.5	20	385	213	1737	0.55	9 cm
42.5	30	409	193	1873	0.47	55 s
42.5	30	546	252	1498	0.46	2 cm
42.5	40	550	212	1602	0.39	50 s
52.5	40	725	284	1265	0.39	2 cm
52.5	40	928	390	1040	0.42	8 cm

*Pigment proportion (iron minimum) - 5% expressed in cement weight.

Table 3. Proportions of the materials in mixtures with S-3.

Strength class		Proportions of the materials (kg/m^3)			Water/Cement	Slump
Cement	Concrete	Cement*	Water	Filler		
22.5	22.5	238	138	2058	0.58	50 s
32.5	15	240	154	2014	0.64	3 cm
32.5	20	275	151	1992	0.55	2 cm
32.5	20	309	170	1914	0.55	9 cm
42.5	30	330	154	1935	0.47	55 s
42.5	30	439	202	1720	0.46	2 cm
42.5	40	436	170	1806	0.39	50 s
52.5	40	582	227	1534	0.39	2 cm
52.5	40	743	312	1175	0.42	8 cm

*Pigment proportion (soot) - 2% expressed in cement weight,
S-3 proportion - 0,8% expressed in cement weight.

Introduction of pigment with large specific surface area, strongly increases water absorption of decorative concrete. Since cement with low water demand based concrete has high strength, binder consumption in decorative concrete is low. This contributes to the perfection of decorative effect and expressiveness of artificial rocks, because a layer of hardened cement between filler particles becomes thinner. Proportions of cement with low water demand based on concrete mixtures are shown in Table 4.

Table 4. Proportions of the materials in mixtures based on cement with low water demand.

Concrete strength class	Pigment	Filler	Proportions of the materials (kg/m^3)			Water/Cement	Slump
			Cement*	Water	Filler		
40	Iron minimum	Granite	270	123	2048	0.46	2.5
40	Soot	Gabbro	270	125	2200	0.46	2.5
40	Chalk	Limestone	270	136	1800	0.5	2.5

For more even distribution of pigment in concrete mixture and to reduce water contents, grinding of Portland cement with S-3 and pigment (cement with low water demand) were used. This cement-based binder has minimum water content among other mineral binders. Consumption of S-3 is 2–3% expressed in cement weight. Specific surface area of cement with low water demand is 4800–5200 cm^2/g. Frost resistance of concrete samples based on cement with low water demand is more than 500 freezing and thawing cycles. Mechanical strength, generally, is 1½–2 times greater than base cement and is heavily dependent on the amount of mineral addition. Technical drying vegetable oil with no additives was used an impregnation composition. Linen oil has a high mechanical strength of oil film, with no softening and melting.

To accelerate the impregnation, self-vacuuming method was used. Oil was heated to a temperature 130–140 °C. Concrete samples were heated to a temperature 200 °C. The amount of absorbed oil while self-vacuuming is increased by 20–25% compared to previous investigations with self-vacuuming. Formed samples with dimensions 4 × 4 × 16 cm were stored under standard conditions for a period of 28 days.

3   Results and Discussion

The amount of absorbed oil was tested during the research. Modified samples were tested for 28-day compressive strength and water absorption. The same properties were received for base concrete samples. Revealed results are shown in Tables 5, 6, and 7.

Table 5. Properties of decorative concrete based on Portland cement.

Porosity (%)	28-day compressive strength (MPa)		Oil absorption (%)		Water absorption expressed in weight	
	Without impregnation	With impregnation	Ow*	Ov*	Without impregnation	With impregnation
14.3	35.1	48.4	5.6	6.15	8.58	0.77
15.2	26.2	35.9	6.0	6.5	9.1	0.81
16.5	24.8	33.5	6.9	7.1	10.23	0.8
15.6	22.1	29.8	6.1	6.7	9.2	0.78
15.4	44.4	62.2	6.2	6.6	9.3	0.82
18.2	42.2	58.7	7.7	8.2	11.2	0.83
16.1	51.5	73.1	6.4	6.9	9.6	0.85
19.3	51.3	73.9	8.5	8.7	12.3	0.87
24.3	50.9	73.8	10.7	10.95	15.0	0.87

*Ow – oil expressed in weight; Ov – oil expressed in volume.

Table 6. Properties of decorative concrete with S-3.

Porosity (%)	28-day compressive strength (MPa)		Oil absorption (%)		Water absorption expressed in weight	
	Without impregnation	With impregnation	Ow*	Ov*	Without impregnation	With impregnation
10.8	35.6	49.1	3.84	4.64	6.4	0.87
12.4	26.6	36.0	4.64	5.3	7.4	0.83
11.6	25.2	33.1	4.32	4.99	7.0	0.87
13.1	22.0	30.2	5.12	5.63	8.0	0.89
11.3	44.1	61.9	4.0	4.86	6.6	0.85
14.7	42.0	59.1	6.0	6.32	9.1	0.81
11.5	51.3	72.5	4.16	4.9	6.8	0.81
15.4	51.5	73.0	6.48	6.6	9.7	0.75
21.8	51.3	73.3	9.68	9.87	13.7	0.75

*Ow – oil expressed in weight; Ov – oil expressed in volume.

By comparing the results from Tables 5 and 6, we can make the conclusion that with the same mechanical strength and consistency binder and water consumption in mixes with S-3 has been reduced by 20%. Porosity of concrete with S-3, and therefore, the amount of absorbed oil is less in comparison with the samples without mineral addition. However, despite the decrease of amount of oil the strength performance of both mixtures remains equal. It shows that in concrete mixtures with less porosity, the use of impregnation mix is more efficient.

Table 7. Properties of concrete samples based on cement with low water demand.

Porosity (%)	28-day compressive strength (MPa)		Oil absorption (%)		Water absorption expressed in weight	
	Without impregnation	With impregnation	Ow*	Ov*	Without impregnation	With impregnation
7.9	35.7	51.1	2.16	3.39	4.3	0.7
8.1	34.0	47.6	2.4	3.48	4.6	0.78
11.6	30.6	42.2	4.16	4.99	6.8	0.81

*Ow – oil expressed in weight; Ov – oil expressed in volume.

It can be clearly seen that cement with low water demand concrete with lower oil content has no impact on concrete strength performance. Therefore, cement with low water demand concrete is the most effective among evaluated concrete mixtures.

For samples impregnated with vegetable oil, hue λ and saturation P were determined. Experimental results are shown in Table 8.

Figure 1 shows the relationship between strength of modified decorative concrete and the amount of absorbed oil. By analyzing of the results, we can make the conclusion that the use of oil is more efficient in impregnation of cement with low water demand based concrete.

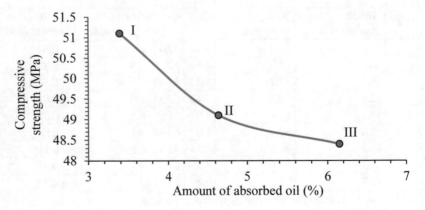

Fig. 1. Relationship between strength of modified decorative concrete and the amount of absorbed oil. I – low water demand cement concrete, II – concrete with S-3, III – Portland cement concrete.

Table 8. Color properties of decorative concrete, impregnated with vegetable oil.

Rocks	Color properties		
	λ (nm)	P (%)	Color
Red granite	601	30	Red
Granite imitation	598	28	Red
Native Gabbro	503	3	Black and red
Gabbro imitation	506	3	Black and red

Figure 2 shows that amount of absorbed oil is in linear relation with respect to the concrete porosity.

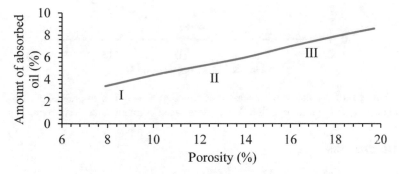

Fig. 2. Amount of absorbed oil as function of porosity. I – low water demand cement concrete, II – concrete with S-3, III – Portland cement concrete.

Figure 3 shows the relationship between frost resistance of decorative oil concretes and depth of impregnation. It can be seen that frost resistance of modified concrete strongly increases. At depth value about 1 cm and more, it only slightly changes.

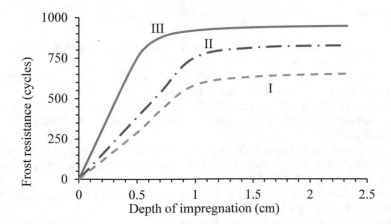

Fig. 3. Frost resistance of decorative oil concretes as function of depth of impregnation. I – Portland cement concrete, II – concrete with S-3, III – low water demand cement concrete.

Figure 4 shows that the increase of strength begins at depth of impregnation near 0.8–1.0 mm. At a higher depth of impregnated layer, strength increases intensively. After depth of impregnation value reaches about 1.5 cm growth of strength is slowing down because of intense compaction of central area of concrete sample.

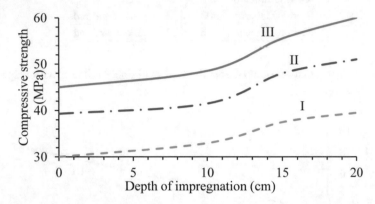

Fig. 4. Strength of decorative oil concretes as function of depth of impregnation. I – Portland cement concrete, II – concrete with S-3, III – low water demand cement concrete.

4 Conclusions

It has been found in this study that durability of decorative impregnated concrete with linen oil depends on structure of base concrete and depth of impregnation. Results of investigation show that strength of concrete matrix and amount of absorbed oil has a crucial influence on mechanical strength performance of modified decorative concrete properties. Results also revealed that concrete based on low water demand cement is the most effective in the production of decorative concrete impregnated with linen oil.

References

1. Bazhenov, Y.M., Alimov, L.A., Voronin, V.V.: Modified concrete with double structure formation. ASV, Moscow (2017)
2. Lollini, F., Carsana, M., Bertolini, L.: A study on the cement-based decorative materials in the San Fedele Church in Milan. Case Stud. Constr. Mater. **7**, 36–44 (2017)
3. Bhutta, M.A.R., Maruya, T., Tsuruta, K.: Use of polymer-impregnated concrete permanent form in marine environment: 10-year outdoor exposure in Saudi Arabia. Constr. Build. Mater. **43**, 50–57 (2013)
4. Baltazar, L., Santana, J., Lopes, B., Rodrigues, M.P., Correia, J.R.: Surface skin protection of concrete with silicate-based impregnations: influence of the substrate roughness and moisture. Constr. Build. Mater. **70**, 191–200 (2014)
5. Kou, S.-C., Poon, C.-S.: Properties of concrete prepared with PVA-impregnated recycled concrete aggregates. Cem. Concr. Compos. **32**(8), 649–654 (2010)
6. Franzoni, E., Varum, H., Natali, M.E., Bignozzi, M.C., Pereira, E.: Improvement of historic reinforced concrete/mortars by impregnation and electrochemical methods. Cem. Concr. Compos. **49**, 50–58 (2014)

Investigation of Concrete Properties with the Use of Recycled Coarse Aggregate

Nikita Dmitriev⬛, Oksana Larsen⁽✉⁾⬛, Vitaly Naruts⬛, and Egor Vorobev⬛

Moscow State University of Civil Engineering, Yaroslavskoe shosse, 26, Moscow 129337, Russia
larsen.oksana@mail.ru

Abstract. This paper presents experimental results of two-stage mixing approach on properties of concrete made with recycled concrete aggregates. Concretes with recycled concrete coarse aggregates have poorer workability and lower mechanical strength than concretes with natural coarse aggregate. Concrete mixtures lose workability in early time due to their increased capillary porosity and presence of interfacial transition zone (ITZ). It consists of ettringite and calcium hydroxide. Due to the existence of two weaknesses in the ITZ, the performance of recycled aggregate concrete is much lower than that of natural aggregate concrete. When formation time of cement-based materials in concrete with recycled concrete aggregate, new appear in addition to the old ones - between recycled concrete aggregate and a new mortar part. Interfacial transition zone is less durable than the interfacial transition zone in recycled concrete aggregate itself. The article describes the method of using two-stage mixing approach (TSMA) with research of the time of formation of concrete made with recycled concrete aggregates. Concrete mixture prepared by two-stage mixing approach showed difference in time formation in early stages, without a sharp loss of workability in contrast to the concrete mixtures with normal mixing. The samples made by TSMA has higher strength than the samples made by normal mixing. The compressive strength in 3, 7, and 28 days increased by 16%, 8%, and 16%, respectively, compared with the samples made by normal mixing. The obtained data showed the positive influence of two-stage mixing on loss of workability in early time and strength of the concrete.

Keywords: Crashed concrete aggregate · Two-stage mixing approach · Time of formation · Coarse aggregate · Fly ash

1 Introduction

Worldwide, the demolition of obsolete for various reasons buildings and houses, resulting in a large amount of construction waste, is occurring everywhere. Its main part is crashed concrete and reinforced concrete structures. In particular, a renovation program has been launched in Moscow, involving the demolition of five-story houses and construction of new houses in their place. These are huge volumes of crashed concrete, which must be at least partially used in new construction. Recycled concrete aggregate is produced from concrete waste by crushing in several stages. It differs in

© Springer Nature Switzerland AG 2020
V. Murgul and M. Pasetti (Eds.): EMMFT 2018, AISC 982, pp. 769–777, 2020.
https://doi.org/10.1007/978-3-030-19756-8_73

properties from aggregates from quarries in that its grains have adhered cement stone, the amount of which in grains of aggregates depends on the technology for producing recycled concrete aggregate at crushing stages.

The period of formation of cement-based materials (PFCM) is the time from the beginning of mixing to the moment of a sharp increase in strength. The presence of aggregates in the cement paste reduces the PFCM, because the aggregates divert some of the mixing water and then the cement paste hardens with effective water-cement ratio ($W/C_{effective}$) (formula 1) - W/C, which would have cement paste if it would have the same properties as the concrete mix (mobility, setting time). By determining the PFCM of the cement paste for at least two W/C, you can determine the effective W/C for concrete mixes. And the difference between these values is the amount of water diverted by the aggregates.

$$W/C_{effective} = \frac{W - k_s \cdot S - k_{CA} \cdot CA}{C} \qquad (1)$$

where $W/C_{effective}$ – effective water-cement ratio; W, S, CA, C is the amount of water, sand, crushed stone (coarse aggregate) and cement, respectively, per 1 m³ of concrete, kg; k_s, k_{CA} – water demand of sand and coarse aggregate, respectively. Another structural characteristic is the volume concentration (VC) of cement paste in the concrete mix:

$$VC = \frac{C}{1000} \left(\frac{1}{\rho_c} + W/C_{effective} \right) \qquad (2)$$

where VC - volume concentration of cement paste in the concrete mix; C - cement consumption, kg; $W/C_{effective}$ is the effective water-cement ratio of the cement paste; ρ_c - cement density, t/m³.

Figure 1 shows that the aggregate reduces the period of formation of cement-based materials compared to cement paste with the same W/C mixing.

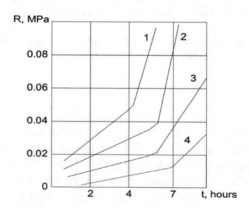

Fig. 1. Influence of sand content on the period of formation of cement-based materials: 1 - cement-sand mortar with W/C = 0.3; 2 - cement-sand mortar with W/C = 0.4; 3 - cement paste with W/C = 0.3; 4 - cement paste with W/C = 0.4.

It is possible to improve the characteristics of secondary crushed stone when crushing according to the regimes in which the weakest link is destroyed - cement stone [1]. It needs either special equipment (vibratory jaw crushers, cone inertial crushers), or crushing at more than one stage in conventional jaw crushers, but in the "block" mode - maximum filling of the working space of the device. In this "soft" mode, the concrete waste is crushed by direct interaction with the material being crushed, and not by strikes against the moving cheek.

Multistage crushing allows improving the properties for sand from concrete waste [2].

Concretes on recycled and concretes on natural aggregates have different adhesion to reinforcement. Adhesion to reinforcement of concrete with recycled concrete aggregate was 35% lower than that of concrete on natural aggregate [3].

There are other ways to improve the properties of recycled aggregate, for example, treatment with polyvinyl alcohol. For this, the recycled aggregate is saturated with polyvinyl alcohol in a desiccator under vacuum pressure. At the age of 90 days, samples of concrete with recycled aggregate, saturated with 10% polyvinyl alcohol solution and dried in air, showed strength similar to the strength of concrete samples on a natural aggregate [4]. Replacing 50% of coarse and 20% of fine aggregate with recycled one does not reduce the strength of self-compacting concrete [5].

Slump of concrete mixes with recycled sand and recycled coarse aggregate remained almost constant, provided that the replacement of natural coarse aggregate with recycled was no higher than 30%. When this value is exceeded, the slump significantly decreased [6]. For concrete hardening in marine conditions, 20% replacement of natural coarse aggregate with recycled is acceptable and will not cause significant changes in strength and durability [7]. Not always, an increase in the number of crushing stages allows improving the quality of recycled concrete aggregate. An increase in the number of crushing stages leads to a decrease in the strength of concrete on a recycled coarse aggregate. It is possible that crushing modes were used in which the destruction of the aggregate grains, and not the separation of adhered cement slurry [8].

Processing recycled concrete aggregate carbon dioxide can significantly improve the properties of concrete with this aggregate. The increase in compressive strength is 22.6% (from 38.6 to 47.3 MPa) compared with the untreated recycled aggregate [9].

Replacement of 25% of natural coarse aggregate with recycled concrete aggregate does not significantly change the strength properties of concrete [10].

Due to the presence of cement mortar with a developed net of capillary pores on such aggregate, it has an increased water absorption. In addition, for concrete mixes on recycled aggregate, mobility is noticeably reduced in the first few minutes after their manufacture. In the hardened concrete, there are two types of interfacial transition zones (ITZ) between the coarse aggregate and the mortar part: the old, which are already in the recycled concrete aggregate, and the new, formed during the formation of cement-based materials of the concrete. Interfacial transition zones consist mainly of fragile minerals - ettringite and calcium hydroxide. In concretes on recycled concrete aggregates, such a zone can be located between the new mortar part and the old mortar part (from concrete waste), which makes this place even more vulnerable from the point of view of strength (Fig. 2).

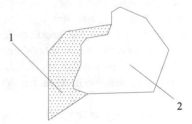

Fig. 2. Recycled coarse aggregate: 1- cement mortar with capillary pores; 2 - grain of natural coarse aggregate.

There are two methods of two-stage mixing [11], which should strengthen the old ITZ and new ITZ due to clogging of capillary pores in the concrete waste and reduce the water absorption of recycled concrete aggregate:

- two-stage mixing approach with addition of silica fume (TSMA$_{(s)}$), which consists in adding a few percent of silica fume in the first stage of mixing;
- two-stage mixing approach with addition of silica fume and cement (TSMA$_{(sc)}$), which consists in adding a few percent of silica fume and cement in the first stage of mixing;

Two-stage mixing consists of several stages:

(1) Mixing the mixture of all aggregates and recycled concrete aggregates with silica fume (and in the case of TSMA$_{(sc)}$ with a small part of cement) - 60 s;
(2) Adding half the mixing water and mixing - 60 s;
(3) Adding total cement and mixing - 30 s.
(4) Adding the second half of the water and mixing - 120 s.

The clogging of capillary pores is achieved by wrapping the recycled concrete aggregate in the second stage with cement paste, and due to the presence of silica fume, calcium hydroxide is bound, resulting in the formation of low-basic hydrocalcium silicates, which even more clogs the pores.

2   Materials

The materials were used:

- ordinary Portland cement CEM I 42.5 R with standard consistency of 28.25% in accordance with the Russian Standard;
- sand as fine aggregate with fineness modulus of 2.3 and true density of 2560 g/cm^3;
- natural and recycled coarse aggregate. As natural coarse aggregate was used gravel with bulk density and true density of 1400 kg/m^3 and 2550 kg/m^3 respectively. Recycled coarse aggregate was obtained by crushing old concrete from demolishing of reinforced concrete with bulk density and true density of 1300 kg/m^3 and 2400 kg/m^3 respectively.

−   fly ash from "Cherepetskaya hydroelectric power plant" with specific surface area
    more than 2500 cm^2/g. Chemical composition of fly ash is given in Table 1.

Table 1. Chemical composition of fly ash.

Oxide, substance	Content of mass fraction, %
Na_2O	0.34
MgO	1.74
Al_2O_3	20.7
SiO_2	59.2
K_2O	2.46
CaO	3.73
TiO_2	0.91
MnO	<0.10
Fe_2O_3	5.64
CO_2	<0.10
S	<0.10
loi	5.26

3   Results and Discussion

To plot the period of formation of cement-based materials versus cement-cement ratio
(W) for three cement pastes (with the amount of water equal to 0.876, 1 and 1.65 of
standard consistency), PFCM were determined by ultrasound method (Fig. 3).

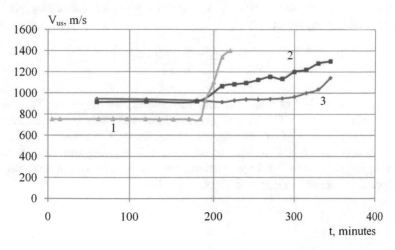

Fig. 3. Graph of the passage of ultrasound through the cement paste from the time of mixing:
1 - W/C = 0.247; 2 - W/C = 0.2825; 3 - W/C = 0.466.

The periods of formation of cement-based materials for each of the cement pastes are: PFCM 1 = 187 min for W/C = 0.247, PFCM 2 = 210 min for W/C = 0.2825, PFCM 3 = 320 min for W/C = 0.466. According to these data, a graph of the period of formation of cement-based materials of cement paste from the water-cement ratio of the cement paste was built (Fig. 4).

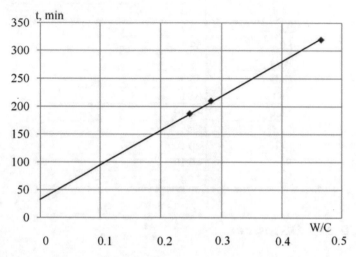

Fig. 4. Relationship between structure formation period and W/C of the cement paste

Then, according to the method for determining the water demand of the aggregate in the concrete mixture [12], the coefficient of water demand (k_{agg}) for recycled concrete was determined, which was 7.5%. The formula for calculating the water requirement of the aggregate:

$$k_{agg} = \frac{W_{mix} - C \cdot W/C_{true}}{P_{agg}} \qquad (3)$$

Where, k_{agg} – coefficient of water demand for investigated aggregate; W_{mix} – is the amount of mixing water, l; C - cement consumption, kg; $W/C_{effective}$ – effective water-cement ratio; P_{agg} – mass of the investigated aggregate, kg. Fly ash was used instead of silica in an amount of 10% by weight of cement. Replacing gravel with recycled concrete aggregate – 30%. Next was selected the composition of the concrete, which was prepared by normal mixing and TSMA$_{(sc)}$ (Table 2).

Table 2. Composition of concrete

Component	Consumption per 1 m³, kg	Consumption per 4.5 dm³, kg
Cement	398	1.791
Sand	717	3.23
Gravel	752.5	3.39
Recycled concrete aggregate (30% replacement)	322.5	1.45
Water	210	0.945
Fly ash	39.8	0.179

6 samples of $10 \times 10 \times 10$ cm were molded, which were then sealed on a vibrating table. For samples of types 1 and 2, the strength was fixed in time using an ultrasonic device Pulsar-2.2. Figure 5 shows that the period of formation of cement-based materials was 160 min for NMA and $\text{TSMA}_{(sc)}$, which corresponds to $\text{w/c}_{effective} = 0.2$.

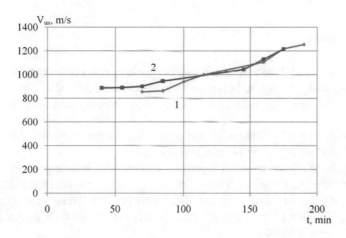

Fig. 5. Graph of the passage of ultrasound through concrete mixes on the time from the moment of mixing: 1 - NMA concrete; 2 - $\text{TSMA}_{(sc)}$ concrete.

Figure 4 shows that in the first 100 min there is a difference in the formation of cement-based materials of concrete. Concrete mixture of normal mixing forms cement-based materials with a lower speed than the concrete mixture of two-stage mixing. This is probably due to the fact that capillary pores of concrete waste are clogged in a two-stage mixing concrete mixture, due to which water enters them in very small amounts or does not flow at all, but instead immediately goes to the cement and the cement-based materials is formed faster. In the concrete mix of normal mixing, these capillary pores do not clog, therefore, processes similar to those of concrete mixes on a light-weight aggregate take place, when the aggregate draws a significant amount of water to itself and then gradually gives it away during the formation of cement-based materials.

Based on PFCM = 160 min, the cement accounts for 79.6 L of water, and the aggregates – 130.4 L, from which the water requirement of the mixture of aggregates – 7.1%. All specimens were tested for compressive strength (Table 3).

Table 3. The test results on the strength of the samples.

	3 day	7 day	28 day
Strength, MPa			
Normal mixing	28.1	33.7	46.7
TSMA$_{(sc)}$	32.6	36.4	54.1
Density, kg/m^3			
Normal mixing	2325	2320	2320
TSMA$_{(sc)}$	2340	2320	2323

The 3, 7 and 28 days compressive strengths increased by 16%, 8% and 16% respectively, compared to the NMA.

4   Conclusion

The main objective of this study is to prove the positive effect of TSMA(sc) on concrete mixture properties. The use of two-stage mixing with addition of mineral additive and a portion of cement at the preliminary mixing stage gives the increasing of strength up to 15%.

In the early stages of TSMA(sc) concrete mixtures the workability is maintained longer than workability of NMA concrete mixtures.

In the case of use of recycled aggregate in concrete, it is important to minimize water absorption with the adhered mortar of the aggregate and with old ITZ. TSMA(sc) allows blocking the access of water to the capillary pores of the adhered mortar and the pores of old ITZ. Due to this method, concrete mixtures retain their workability longer and their strength is higher.

References

1. Leemann, A., Loser, R.: Carbonation resistance of recycled aggregate concrete. Constr. Build. Mater. **204**, 335–341 (2019)
2. Pacheco, J., de Brito, J., et al.: Experimental investigation on the variability of the main mechanical properties of concrete. Constr. Build. Mater. **201**, 110–120 (2019)
3. Florea, M.V.A.: Properties of various size fractions of crushed concrete related to process conditions and re-use. Cem. Concr. Res. **52**, 11–21 (2013)
4. Surya, M., Kanta Rao, V.V.L., Lakshmy, P.: Recycled aggregate concrete for transportation infrastructure. Proc. - Soc. Behav. Sci. **104**, 1158–1167 (2013)
5. Kou, S.C., Poon, C.S.: Properties of concrete prepared with PVA-impregnated recycled concrete aggregates. Cem. Concr. Compos. **32**, 649–654 (2010)

6. Señas, L., Priano, C., Marfil, S.: Influence of recycled aggregates on properties of self-consolidating concretes. Constr. Build. Mater. **113**, 498–505 (2016)
7. Tahar, Z.A., Ngo, T.T., et al.: Effect of cement and admixture on the utilization of recycled aggregates in concrete. Constr. Build. Mater. **149**, 91–102 (2017)
8. Thomas, C., Setién, J., Polanco, J.A., Cimentada, A.I.: Influence of curing conditions on recycled aggregate concrete. Constr. Build. Mater. **172**, 618–625 (2018)
9. Abreu, V., et al.: The effect of multi-recycling on the mechanical performance of coarse recycled aggregates concrete. Constr. Build. Mater. **188**, 480–489 (2018)
10. Xuan, D., Zhan, B., et al.: Durability of recycled aggregate concrete prepared with carbonated recycled concrete aggregates. Cem. Concr. Compos. **84**, 214–221 (2017)
11. Thomas, J., Thaickavil, N.N., Wilson, P.M.: Strength and durability of concrete containing recycled concrete aggregates. J. Build. Eng. **19**, 349–365 (2018)
12. Tam, V.W.Y., Tam, C.M.: Diversifying two-stage mixing approach (TSMA) for recycled aggregate concrete: TSMAs and TSMAsc. Constr. Build. Mater. **22**, 2068–2077 (2008)

Justification of Energy-Saving Technology of Prefabricated Monolithic Slabs of Limestone Blocks

Kirill Leonenko$^{(\boxtimes)}$ ⓘ and Vasiliy Shalenny ⓘ

V.I. Vernadsky Crimean Federal University, Prospekt Vernadskogo 4,
295007 Simferopol, Russia
Leonenkoka@gmail.com

Abstract. The article covers available solutions for arranging horizontal structures: floor slabs for low-rise buildings of the Republic of Crimea. During the arrangement of monolithic and prefabricated monolithic slabs on the specific site – a manor house in the village of Pionerskoie, Republic of Crimea, testing was carried out. At the same time, two prefabricated monolithic constructive technological systems were designed and implemented in practice on the same slab of the basement: monolithic on steel shaped flooring and prefabricated monolithic with filling with blocks of local building material – limestone blocks (shell rock). A comparative analysis was carried out taking into account the criterion of ergonomic indicators. On this basis, a more rational, in terms of the severity of labor of construction workers, technology was proposed in comparison with gas concrete in the MARCO system, which require high energy consumption in autoclave hardening under high temperature and pressure, the use of local raw materials almost excludes them. As a result, it was possible to achieve an increase in their working capacity and a reduction in the level of industrial injuries by one third. This result was obtained by processing the data of filming, timing and physiological state of construction workers, which confirms the feasibility of using the proposed technology and methodology for assessing the severity of their labor in comparison with the arrangement of monolithic reinforced concrete slabs as the most common and currently available technology.

Keywords: Energy efficiency · The severity of labor ·
Method of energy and labor costs count · Ergonomics ·
Prefabricated monolithic slab

1 Introduction

On the territory of the Republic of Crimea, due to high seismicity of the construction area, monolithic reinforced concrete frame structural systems are most prevalent due to their increased rigidity and ease of performance of reinforced concrete works in dismountable and adjustable industrial formwork systems [1] directly at the construction site. At the same time, such technologies have a number of significant drawbacks associated primarily with their increased labor-intensiveness and significantly longer turnaround time. Therefore, both in Russia and abroad, compromise prefabricated

© Springer Nature Switzerland AG 2020
V. Murgul and M. Pasetti (Eds.): EMMFT 2018, AISC 982, pp. 778–786, 2020.
https://doi.org/10.1007/978-3-030-19756-8_74

monolithic constructive-technological systems of low-rise civil engineering are becoming increasingly common [2]. The share of prefabricated monolithic structures for slabs in the EU countries, according to various estimates, ranges from 20% to 35%. In Russia, until 2008, such slabs were not applied at all, though such constructions have a lower own weight, higher indicators of heat protection and sound insulation, and they do not require the use of powerful load-lifting equipment and other related labor costs [3]. Focusing on the well-known foreign prefabricated monolithic systems Porotherm, Teriva, Ytong, Rectolight, in the Russian Federation, the domestic innovative technology SMP MARKO was patented and developed [4]. Only the absence of closely located plants of production of prefabricated elements of this system hinders its distribution in Crimea.

The use of such prefabricated monolithic constructive-technological systems, as compared with the most common monolithic slabs, leads to both a decrease in construction labor intensity and the severity of construction workers as an ergonomic indicator of modern construction processes [5]. Having put forward such a working hypothesis, in the present scientific and applied work the tasks are set and a comparative assessment of the workload of construction workers is made to substantiate the expediency of choosing and improving the technology of assembling prefabricated monolithic slabs of low-rise buildings of Crimea in order to reduce this ergonomic indicator.

2 Materials and Methods

2.1 Object of Research

After analyzing the available literary sources and scientific and production experience of their use, we concluded that an integral scoring of the severity of labor (ergonomic scoring) would be the most reliable for assessing the severity of labor in construction, and the assessment method which is based on the actual heart rate and is convertible into the severity of labor according to the formula of Professor Travin could be duplicative [6]. We decided to use such research methods for a comparative assessment of the severity of labor of construction workers while arranging monolithic and prefabricated monolithic slabs on the specific site – manor house in the village of Pionerskoie, Republic of Crimea. At the same time, two prefabricated monolithic constructive technological systems were designed and implemented in practice on the same slab of the basement: monolithic on steel shaped flooring and prefabricated monolithic with filling with blocks of local building material – limestone blocks (shell rock) (Fig. 1).

Since in terms of cost, the practical application of slabs in the non-removable formwork from galvanized profiled flooring with edges of 60 mm in height were clearly uncompetitive, this technology has not been further investigated according to the selected ergonomic indicators. While the constructive system presented on the right in Fig. 1 is the MARCO system which is more adapted to the conditions of Crimea. Its modification consists in replacing the prefabricated forming-and-reinforcing system with two elements manufactured in construction conditions - a reinforcing spatial frame and a formwork board-base on which, with a gap providing the required thickness of the concrete protective layer, the mentioned reinforcement frame was mounted (Fig. 2). This modification of the well-known innovation system MARCO allows completely eliminate transportation costs for the delivery of prefabricated elements which for the conditions of Crimea are of decisive economic importance due to the range of factory production of such prefabricated elements and complex logistics.

Fig. 1. General view of the slabs prepared for concreting: on steel shaped flooring (left) and with filling limestone blocks (right).

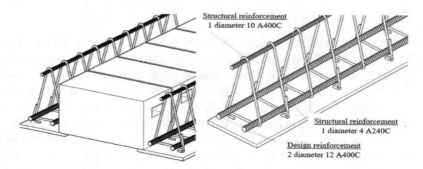

Fig. 2. Elements of the prefabricated system and the obtained cross-sections of the spatial reinforcement frame of the monolithic supporting beam of the projected slab – the object of research.

2.2   Research Methods

The severity of labor is determined based on the integral indicator [7, 8]. It also allows to determine the degree of fatigue and level of working efficiency, increase in labor productivity and the level of occupational injuries. Additional assessment of the severity of labor in the implementation of the considered technologies for arranging prefabricated monolithic slabs was made on the basis of data on the heart rate of workers based on calculations of their energy consumption using the formula of Professor Travin:

$$E = 0.014 \cdot G \cdot t \cdot (0.12 \cdot f - 7),\tag{1}$$

where E is the energy consumption, kcal; G – body weight, kg; t – the duration of physical activity, min.; f is the number of heartbeats (HR), min^{-1}. After that, on the basis of clause 12 of annex 1 to GOST (All-Union State Standard) 12.1.005-88, it is also possible to determine the category of the severity of labor of construction workers. The technological order of works was built according to the Sechenov criterion, the essence of which is that when a type of activity changes, other nerve cells come to excitement, while the excitement of the cells which participated in the former activity is replaced by slowdown which ensures the effect of active rest [8] taking into account the dynamics of the performance of construction workers [9].

2.3   Hardware of the Production Experiment

To obtain quantitative information for calculations using the presented methods, video recording was taken by two cameras: one of the cameras, the Sony Handycam CX625, was mounted on a tripod and recorded general technological process; based on its data, the values of the total dynamic load and the workers moving time were calculated and other similar ones; the second camera, GoPro Hero 7, was attached directly to the worker's chest performing technological works and, based on its data, the values of regional dynamic loads, the monotonicity of works and other similar ones were calculated. The heart rate of the workers was recorded using a wrist sensor (Garmin Vivomove HR) and a chest sensor (Garmin HRM Run) with transmission, registration and processing of data on a personal computer. In addition, the technological process was divided into operations chronometrically: the segments were recorded with a stopwatch. To obtain statistical data of the experiment, 5 construction workers of different qualifications and physical parameters were selected:

(1) Worker of the 2nd category, age 25, height 183 cm, weight 78 kg;
(2) Worker of the 4th category, age 32, height 176 cm, weight 75 kg;
(3) Worker of the 4th category, age 46, height 174 cm, weight 63 kg;
(4) Construction superintendent, age 42, height 185 cm, weight 88 kg;
(5) Head of the site, age 56, height 179 cm, weight 70 kg.

3   Results

Processing the information obtained from the video recordings and the physiological state of the workers who implemented the presented constructive-technological solutions of prefabricated monolithic reinforced concrete slabs on the site made it possible to study the changes in the pulse of the workers (Fig. 3), as well as the temporal and power parameters of the work stages performed during this process. They became the basis for drawing up protocols of the severity of implementation of certain types of work (Table 1) which most objectively characterize the technologies being studied.

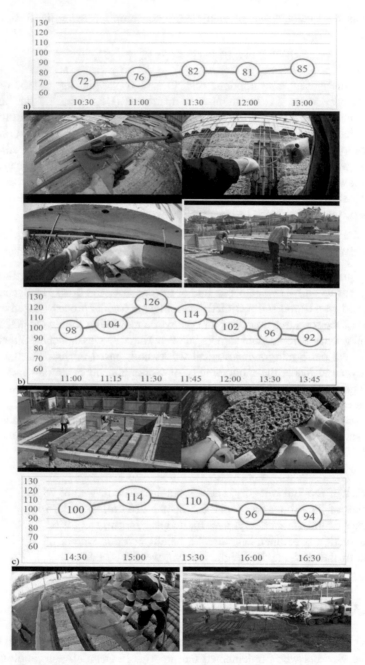

Fig. 3. Diagrams of changes in the pulse of the workers over time.

Diagrams of changes in the pulse of the workers over time to study the complex of technological works: (a) arrangement of an element of a prefabricated monolithic slab, fixing the upper reinforcing mesh to the elements of the slab; (b) internal transport: a device for filling a prefabricated monolithic slab; (c) arrangement of the monolithic part of the plate.

Table 1. Protocol of the assessment of the severity of labor of the workers during the arrangement of prefabricated monolithic slabs of the proposed design.

No.	Working environment factors	Value indicator	Factor score	Factor duration	Specific weight in a shift	Specific severity assessment
1	Physical dynamic load, J: total * 10^5	13.42	4	480	1.00	4
2	Physical dynamic load, J: regional * 10^5	6.80	4	480	1.00	4
3	Physical static load, N·s on two hands * 10^4	238.37	5	90	0.19	0.94
4	Workplace (WP), posture and movement in space	Non-free posture, in an inclined position to 30° up to 25% of time	3	480	1.00	3
5	Shift system	Two shifts	2	480	1.00	2
6	Duration of continuous work during a day, hours.	Less than 8 h	2	480	1.00	2
7	Monotonicity: the number of functions in one operation	5 functions	3	435	0.40	1.20
8	Monotonicity: duration of repetitive functions, seconds	30,00	3	435	0.40	1.20
9	Schedule of work and rest	Reasonable	2	480	1.00	2
10	Neuro-emotional stress	Complicated actions on a given plan with the possibility of correction	3	480	1.00	3
11	Pace (number of movements per hour): Large movements (hands)	320	4	400	0.83	3.33
12	Duration of the concentrated observation, % of a shift	50% of a shift	4	480	1.00	4

From the analysis of time-keeping observations and technological charts on the arrangement of monolithic reinforced concrete slabs it follows that the minimum category of the severity of labor is 4–5 due to the long time of reinforcement lashing during which the worker is in non-free posture, in an inclined position more than 50% of a work shift, and at the same time makes up to 300 inclinations to 60° during a work shift. Thus, the load on the lower back will be above the normative. In this case, the minimum integral score index of labor is 5, and the actual one will be about 5.3, which is at the junction of categories 4 and 5 of the severity of labor from 6. Thus, from the point of view of ergonomics, works on the arrangement of monolithic reinforced concrete slabs are classified as heavy.

Thus, the works on the arrangement of the proposed prefabricated monolithic slabs are characterized by the following integral index of the severity of labor:

$$U_T = 10\left[X_{max} + \frac{\sum_{i=1}^{n} X_i}{n-1} \cdot \frac{6-X_{max}}{6}\right] = 10\left[5 + \frac{30.67}{12-1} \cdot \frac{6-5}{6} - 1\right] = 4.47 \cdot 10 = 44.7. \tag{2}$$

The results obtained correspond to the 3rd category of the severity of labor (3 of 6), but at the junction with 4. In this case, the fatigue index is:

$$Y = \frac{U_T - 15.6}{0.64} = \frac{44.7 - 15.6}{0.64} = 45.47. \tag{3}$$

Then the working capacity will be:

$$R = 100 - Y = 100 - 44.06 = 54.53. \tag{4}$$

Increase of the productivity of labor due to the modernization of the technology of performance of the work is:

$$\Pi_{\Pi T} = \left(\frac{R_2}{R_1} - 1\right) \cdot 100 \cdot 0,2 = \left(\frac{54.53}{41.56} - 1\right) \cdot 100 \cdot 0.2 = 6.24\%. \tag{5}$$

And the predicted level of industrial injuries will be:

$$K = \frac{1}{1.3 - 0.0185 \cdot U_T} = \frac{1}{1.3 - 0.0185 \cdot 44,7} = 2.11 \text{ times} \tag{6}$$

In the case of the arrangement of monolithic slabs, the level of industrial injuries ($U_T = 53$) is 3.13 times. Thus, using the technology of prefabricated monolithic slabs, it became possible to achieve a reduction in the level of industrial injuries by a third.

The energy consumption of the workers during the arrangement of prefabricated monolithic slabs calculated by the formula of Professor Travin is:

$$E = 0.014 \cdot G \cdot t \cdot (0.12 \cdot f - 7) = 0.014 \cdot 70 \cdot 186 \cdot (0.12 \cdot 84 - 7) = 561.42 \, (kcal). \tag{7}$$

Or, on average, 184.07 kcal/hour. On the basis of clause 12 of annex 1 to GOST (All-Union State Standard) 12.1.005-88 the category of the severity of labor is 2a (moderately severe of category I) (3 of 5).

4 Discussion

Due to the considerable specific weight of reinforcement works during the arrangement of monolithic reinforced concrete slabs, all works on their production should be attributed to the 5th category of the severity according to the method of assessment of the integral index. One of the ways to reduce the level of the severity of works on the arrangement of monolithic slabs is the proposed use of tested in practice remote tools for reinforcement lashing [5], however, we consider a rejection of monolithic structures in favor of prefabricated monolithic reinforced concrete slabs as a more effective solution.

At the same time, the integral severity index (4.47) and the severity index calculated by the method of energy consumption based on the workers heart rate (184.07 kcal/hour) turned out to be the best, and the work as a whole should be attributed to work of moderate severity. Only on the basis of improving the working conditions of the performers, the proposed technology for arranging prefabricated monolithic slabs using the filling from local materials for low-rise construction in Crimea can be recommended for widespread use. Moreover, reducing costs, reducing construction time, increasing labor productivity and reducing the level of industrial injuries by a third should be recognized as other concomitant positive factors. Therefore, the proposed innovative technology can have a significantly greater, super cumulative synergistic socio-economic effect the evaluation of which is expected in our further work.

5 Conclusions

(1) Technology of arrangement of reinforced concrete slabs have recently been developed in the direction of increasing the specific weight of prefabricated monolithic constructive and technological systems. In Russia, the national innovative prefabricated monolithic system MARCO has emerged and is developing which we have taken as the basis for further improvement.

(2) Considering the unfavorable conditions for the delivery of prefabricated elements of the MARCO system to our region, as well as the possibilities of using local materials to fill the inter-block space, an improved construction of prefabricated monolithic slab has been proposed and presented in the work.

(3) The design and experimental production of the proposed structural and technological prefabricated monolithic system in the conditions of low-rise suburban development of the central part of Crimea has been made.

(4) As a result of the processing of the data of the camera recording, timing and physiological condition of the construction workers, the expediency of using the proposed technology based on an assessment of the indicators of the severity of their labor in comparison with the arrangement of monolithic reinforced concrete slabs has been shown.

References

1. Kapshuk, O.A., Shalenny, V.T.: Constructability of varieties of modern disassembled and adjustable formwork systems. Eng. Constr. J. **7**, 80–88 (2014)
2. Shalenny, V.T.: Prefabricated monolithic house building. In: Shalenny, V.T., Balakchina, O.L. (eds.) IPR Media, Saratov (2018). (in Russian)
3. Teplova, Zh.S., Vinogradova, N.A.: Prefabricated monolithic slabs of the 'MARKO' system. Constr. Unique Build. Struct. **8**(35), 48–59 (2015)
4. Nedviga, E.S., Vinogradova, N.A.: Systems of prefabricated monolithic slabs. Constr. Unique Build. Struct. **4**(43), 87–102 (2016)
5. Ershov, M.N.: Ergonomics of building processes, affordable solutions. In: Ershov, M.N. (ed.) ASV Publishing House, Moscow (2010). (in Russian)
6. Shalenny, V.T., Leonenko, K.A.: Comparative analysis and justification of the target-specific methodology for assessing the severity of labor of construction workers to improve the ergonomic performance of masonry installation works. Biosph. Compat.: Hum. Reg. Technol. **4**(20), 80–85 (2017)
7. Voronova, V.M., Egel, A.E.: Definition of the category of severity of labor: methodology guidelines for a thesis preparation. Orenburg State Educational Institution, Orenburg State University, Orenburg (2004)
8. Krushelnitskaia, Ya.V.: Measures to improve the working capacity of workers. In: Physiology and Psychology of Labor. Finance and Statistics, Moscow (2003)
9. Human physiology, vol.3. Transl. from English. Ed. by R. Schmidt and G. Tevs, 3rd edn. Mir, Moscow (2005)

Equation of Limit Condition of the Three-Parameter Mohr-Coulomb Criterion

Anatoly Aleksandrov[ID], Gennadiy Dolgih[ID],
and Aleksander Kalinin[✉][ID]

Siberian State Automobile and Highway Academy, Mira 5,
644080 Omsk, Russia
alexsandr55ne@mail.ru

Abstract. The analysis of criteria for plasticity of soils has done. It has been established that when solving the axisymmetric problem of soil resistance of the roadway to shear, many analytical criteria give the same result with the classical two-parameter Mohr – Coulomb criterion. A modified three-parameter Mohr – Coulomb criterion is proposed, which, unlike the original criterion, contains the third parameter of the material (d). Depending on the value of this parameter (d), the modified criterion can take the form of the original Mohr – Coulomb criterion (with $d = 0.5$) or the original Tresca criterion (with $d = 0$). For all other values of the parameter (d), varying in the range of $0 > d > 0.5$, the tangential stresses by the modified criterion are bigger, than the Mohr – Coulomb criterion, but smaller than the Tresca criterion. Therefore, the modified criterion covers a large range of tangential stresses, and its value can be determined from the results of tests of soils for triaxial compression. The analysis of such experimental data showed that the value of the third parameter (d) should be taken depending on the size of the axial deformation of the sample, taken as the limiting value. The value of the ultimate deformation is advisable to take in the range from 8 to 12%.

Keywords: Road · Soil · Granular material · Mohr - Coulomb criterion

1 Introduction

Verification of the roadbed and the layers of the pavement bases of granular materials for shear resistance is one of the traditional calculations of road structures according to the strength criteria. The essence of checking the resistance of soils and granular materials to shear consists in comparing the tangential stresses arising from the impact of the transport load with the limiting values of these stresses.

This calculation is based on the original Mohr – Coulomb plasticity condition. The limiting surface of this criterion, built in the space of principal stresses, is a pyramid, and the projection of this pyramid on the deviator plane is a hexagon with three compression angles and three extension angles.

The parameters of the material as grip and the angle of internal friction is determined on the basis of data processing triaxial tests. In the case of a triaxial test, is

© Springer Nature Switzerland AG 2020
V. Murgul and M. Pasetti (Eds.): EMMFT 2018, AISC 982, pp. 787–797, 2020.
https://doi.org/10.1007/978-3-030-19756-8_75

necessarily measured the sample's vertical deformation, assuming that the sample fails when it is deformed by $\varepsilon_1 = 15\%$ (Russian standard requirement) or $\varepsilon_1 = 20\%$ (US, UK and European Union standards).

From the analysis of these limit state it follows, that with appearance in a layer of pavement or roadbed, zone limit state with depth of 10 cm the sediment of surface will be 15–20 mm. Such deformations are big, they exceed the requirements for the depth of irregularities formed on the surface of the coating [1]. Therefore, there is a need to develop special methods for calculating road structures according to flatness criteria [1] with the ability to predict irreversible deformations [2] or to improve the method for calculating road construction by shear resistance [3, 4].

At the same time, two problems are solved. First, the possibility of modifying the Coulomb – Mohr criterion [5] or replacing it with another plasticity condition with higher tangential stresses [3, 4] is considered. Secondly, they develop methods for calculating ultimate loads [6, 7]. Such methods are based on:

– on the classical solution of a system of nonlinear differential equations, which includes equilibrium equations and the limit state equation that closes the entire system [6]. The Mohr – Coulomb plasticity criterion, which has various forms of writing, is usually used as the limit state equation [8, 9].
– on numerical methods of mechanics [7].

When solving the axisymmetric problem, it suffices to use only the strength criterion, in which tangential stresses are calculated through the principal stresses. For such a problem relevant work aimed at improving the methods for calculating the principal and tangential stresses [10, 11], as well as work aims at the experimental determination of parameters of soils and granular materials in granular medium mechanics formulas [12–14]. Modified formulas for calculating stresses are an alternative to traditional solutions, but they do not have the disadvantages of the latter.

Therefore, the development of a new plasticity condition and methods for calculating the principal stresses are urgent tasks. In this paper, the authors will consider the derivation of the equation of the limit equilibrium of the three-parameter Mohr – Coulomb criterion.

2   Materials and Methods

The Mohr – Coulomb plasticity condition is widely used in the calculations of soils and granular materials for shear resistance. This criterion can be written in the principal stresses, components of the stress tensor, or using invariants of the stress tensor and the deviator.

In Table 1 shows various forms of the original two-parameter Mohr – Coulomb criterion and the Tresca criterion limit state equation [8–10].

From data analysis Fig. 1 it follows that the direction of the line of action of the maximum principal voltage is given by the rotation of the axis of symmetry of the load. The rotation of the axis of symmetry of the load leads to the need to select the design section of the road pavement. This section is the axis of symmetry of the load. In this section, the directions of the main axes and the axes of the coordinate system are the same. Therefore, we have the equality $\sigma 1 = \sigma z$ and $\sigma 3 = \sigma x$. Thus, in order to solve the axisymmetric problem, the strength criterion should be written in the main stresses.

Table 1. Tresca and Mohr – Coulomb plasticity criteria.

Name of the criterion and source of borrowing	Mathematical expression of the criterion
1. Criterion Tresca [15]	$\frac{\sigma_1 - \sigma_3}{2} = c_u$
2. Mohr – Coulomb criterion in main stresses [8, 10]	$\frac{1}{\cos\varphi} \cdot \frac{\sigma_1 - \sigma_3}{2} - \text{tg}\varphi \cdot \frac{\sigma_1 + \sigma_3}{2} = c$ or $\sigma_1 \cdot (1 - \sin\varphi) - \sigma_3 \cdot (1 + \sin\varphi) = 2 \cdot c \cdot \cos\varphi$
3. Mohr – Coulomb criterion in invariant form [8, 9]	$\sqrt{J_2} = \left(\frac{I_1 \cdot \sin\varphi}{3} + c \cdot \cos\varphi \right) \cdot \left(\cos\Theta - \frac{\sin\Theta \cdot \sin\varphi}{\sqrt{3}} \right)$

Where σ_1 and σ_3 – the maximum and minimum principal stresses, Pa; c_u – undrained cohesion, Pa; φ – the angle of internal friction (friction angle), °; c – cohesion, Pa; J_2 – the second invariant of the stress deviator tensor, Pa^2; I_1 – the first invariant of the stress tensor, Pa; Θ – Lode angle, °.

The classic idea of the stress state of the pavement and the roadbed is shown in Fig. 1.

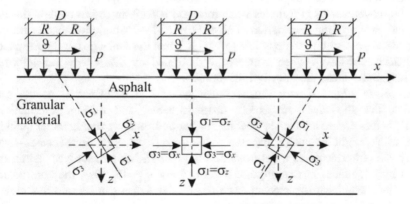

Fig. 1. The scheme of calculation of principal stresses at the base of the pavement of a granualar material.

The material parameters of the Tresca and Mohr – Coulomb criteria are determined from triaxial tests. The criterion for the end of the test is the destruction of the sample or its deformation to the limit value. The test results show that in the process of increasing the deviator $\sigma_d = \sigma_1 - \sigma_3$, deformations localize, causing the nucleation and subsequent development of slip areas [16, 17]. In Fig. 2 shows the stages of development of slip areas, recorded during trials of loam by triaxial compression [16].

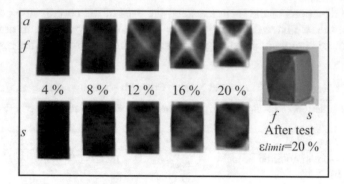

Fig. 2. Stages of development of slip areas under triaxial compression of rectangular samples loam with dimensions $4 \times 4 \times 8$ cm.

In the analyzed experiment, our colleagues [16, 17] made rectangular samples of clay for testing. ($W_{LL} = 0{,}62$; $I_p = 0{,}33$ and $\rho = 2{,}69$ g/cm^3). Samples were tested in a rubber sheath, over which a grid with 2 mm square cells was applied (see Fig. 3a). Before the test, to determine the initial location of the grid nodes on both sides of the sample, the surface of which is conventionally designated f and s (Figs. 2 and 3c), was filmed the video [17]. The same video survey was performed during the test with the vertical deformation of sample, corresponding to 4, 8, 12, 16, and 20% (Fig. 3b) [16, 17]. The movement of grid nodes was processed according to the method developed two years earlier in the dissertation of Higo [18], and the simulation results are shown in Fig. 3d. As a result, it was possible to establish that the first signs of the formation of slip areas are observed when samples are deformed by 8%, and when deformation $\varepsilon_1 = 12\%$, these areas are already fully formed.

Images of Fig. 2 clearly illustrate that with axial deformation of the sample $\varepsilon_1 = 8\%$, the slip areas are barely distinguishable, and during deformation by $\varepsilon_1 = 12\%$, these areas are clearly visible. Further deformation occurs rather quickly, as a rule, with a slight increase in the stress deviator $\sigma_d = \sigma_1 - \sigma_3$. In this case, a sample of clay soil either acquires a deformation of $\varepsilon_1 = 20\%$, or is destroyed before deformation by this value. In the range of deformations $\varepsilon_1 = 12$–20%, they are related to stresses by a nonlinear dependence, as a result of which it can be said that clay soils flow in this range of deformations.

Therefore, the failure of the sample should be taken from the condition of its deformation by a value of $\varepsilon_1 = 8$–12%. This conclusion implies that when calculating soils and granular materials by shear resistance, it is necessary to use another plasticity condition with higher tangential stresses instead of the Mohr – Coulomb criterion. Such a replacement will lead to the fact that the total thickness of the layers located above the calculated element of the road construction will be greater than when designing the road construction according to the Mohr – Coulomb criterion.

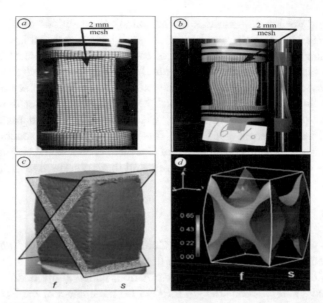

Fig. 3. Illustration explaining a procedure of the experiment described in the works [16, 17]: *a* – mesh photo before deformation; *b* – mesh photo at deformation of sample by 16%; *c* – experimental slip areas and localization of deformations; *d* – simulation of slip areas and localization of deformations

In paper [19], we considered the possibility of using, instead of the Mohr – Coulomb criterion, another analytical plasticity condition. The Lade [20–22], Drucker – Prager [23–25], and Matsuoka – Nakai criteria were considered as analytical criteria. In Fig. 4 shows the projections of the limit surfaces of these criteria on the deviatoric plane.

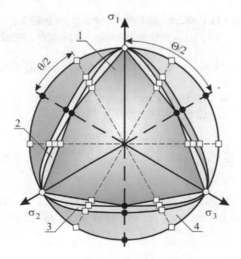

Fig. 4. Criteria in the deviator plan [19]: 1– Mohr – Coulomb [5, 8, 9]; 2 – Matsuoka–Nakai; 3 – Lade – Duncan [20–22]; 4 – Drucker – Prager [23–25].

Considering the limiting surfaces of the analytical conditions of plasticity on the deviatoric plane (see Fig. 4), we note that they intersect with Mohr hexagon in its compression angles. That is, in the angles of compression of the Mohr hexagon all the conditions of plasticity give the same result. To the compression angles of the Mohr hexagon correspond to the stress state characterized by the main stresses $\sigma_1 > \sigma_2 = \sigma_3$, which occurs along the axis of symmetry of the load distributed over the circular area. This cross-section taken by road industry experts as a calculation for assessing the stability of soils against shear (see Fig. 1). This implies that for this section under stress state $\sigma_1 > \sigma_2 = \sigma_3$ replacing the criterion of Coulomb - Mohr other analytical condition of plasticity is not practical, for the reason that all these criteria give the same result.

3 Results

In paper [19], we showed that the original Coulomb – Mohr condition can be relatively easily obtained from the O. Mohr strength criterion. To do this, it is sufficient to substitute the dependencies into the Mohr criterion, which allow one to calculate the limits of uniaxial compression strength through the parameters of the Coulomb's limit line, which is understood the grip and the angle of internal friction. The textbooks on the discipline "resistance of materials" give the criterion O. Mohr, recorded for stretching, and in our case, this criterion must be written down for compression. Therefore, considering the derivation of a modified plasticity condition, we write the original strength criterion for O. Mohr for stretching and for compression. In this case, the difference between the forms of recording this criterion for the compression and stretching is visually illustrated. So, respectively, under tension and compression, the limiting state according to criterion O. Mohr is described by equations in the form:

$$\sigma_1 - k_{Mten} \cdot \sigma_3 = R_{ten}; k_{Mten} = R_{ten}/R_{comp} \tag{1}$$

where σ_1 and σ_3 – the maximum and minimum principal stresses (in this case, the tensile stress), Pa; R_{ten} and R_{comp} – tensile and compressive strength limits, Pa.

$$\sigma_1 - k_{Mcomp} \cdot \sigma_3 = R_{comp}; k_{Mcomp} = R_{comp}/R_{ten} \tag{2}$$

Limit of strength in uniaxial compression and tension are determined through parameters limit the direct Coulomb - Mohr by formulas

$$R_{comp} = \frac{2 \cdot c \cdot cos\phi}{1 - sin\phi} = 2 \cdot c \cdot \sqrt{\frac{1 + sin\phi}{1 - sin\phi}}; R_{ten} = \frac{2 \cdot c \cdot cos\phi}{1 + sin\phi} = 2 \cdot c \cdot \sqrt{\frac{1 - sin\phi}{1 + sin\phi}} \tag{3}$$

where φ – the angle of internal friction (friction angle), °; c – cohesion, Pa.

Let us introduce in dependence (3) the third parameter d, so that

$$R_{comp} = 2 \cdot c \cdot \left(\frac{\cos\phi}{1 - \sin\phi}\right)^{2 \cdot d} = 2 \cdot c \cdot \left(\frac{1 + \sin\phi}{1 - \sin\phi}\right)^{d} \tag{4}$$

$$R_{ten} = 2 \cdot c \cdot \left(\frac{\cos\phi}{1 + \sin\phi}\right)^{2 \cdot d} = 2 \cdot c \cdot \left(\frac{1 - \sin\phi}{1 + \sin\phi}\right)^{d} \tag{5}$$

From the analysis of formulas (5) and (6), it follows that for $d = 0.5$, dependencies (4) and (5) take the form of traditional formulas (3), and for $d = 0$, expressions (4) and (5) are converted to view

$$R_{comp} = R_{ten} = 2 \cdot c \tag{6}$$

Thus, varying the parameter d from 0 to 0.5 allows one to obtain the connection of the limit of strength with the parameters of the Mohr – Coulomb limit line, which gives intermediate results between those calculated from dependences (3) and (6). Substituting the expressions (4) and (5) into the Mohr criterion for compressing (2), dependencies, we get

$$\frac{1}{2} \cdot \left(\sigma_1 \cdot \left(\frac{1 - \sin\phi}{1 + \sin\phi}\right)^{d} - \left(\frac{1 + \sin\phi}{1 - \sin\phi}\right)^{d} \cdot \sigma_3\right) = c \tag{7}$$

At $d = 0.5$, Eq. (7) takes the form of the Mohr – Coulomb criterion; at $d = 0$, dependence (7) turns into Tresca strength theory. Thus, a decrease in the value of the parameter d leads to an increase in the tangential stress determined by the left side of the criterion (7). At $d = 0.5$, the tangential stresses calculated by formula (7) coincide with the tangential stresses of the original Coulomb – Mohr criterion. When $d = 0$, Eq. (7) is converted to the form of the Tresca criterion, and the tangential stress reaches the value of the maximum tangential stress and is determined by half the difference between the maximum and minimum main stresses. The dependences for the calculation of tangential stresses for various values of the parameter d are given in Table 2.

Table 2. Formulas for calculating tangential stresses.

Value of parameter d	Formula
$d = 0,5$	$\tau = \frac{1}{2} \cdot \left(\sigma_1 \cdot \sqrt{\frac{1 - \sin\varphi}{1 + \sin\varphi}} - \sigma_3 \cdot \sqrt{\frac{1 + \sin\varphi}{1 - \sin\varphi}}\right)$
$0 < d < 0,5$	$\tau = \frac{1}{2} \cdot \left(\sigma_1 \cdot \left(\frac{1 - \sin\varphi}{1 + \sin\varphi}\right)^{d} - \left(\frac{1 + \sin\varphi}{1 - \sin\varphi}\right)^{d} \cdot \sigma_3\right)$
$d = 0$	$\tau = (\sigma_1 - \sigma_3)/2$

4 Discussion

When constructing the circles of limit stress for the criterion described by Eq. (7), it can be noted that they reach the maximum size at $d = 0.5$. Demonstrating this feature, the limiting equilibrium Eq. (7) is solved with respect to the magnitude of the maximum principal stress σ_1, implying that for given c, φ, d and σ_3, the calculated maximum principal stress is the limit for this soil. Solving Eq. (7) regarding σ_1, we get the formula for calculating its limit value:

$$\sigma_{1_{limit}} = 2 \cdot c \cdot \left(\frac{1+sin\phi}{1-sin\phi}\right)^d + \left(\frac{1+sin\phi}{1-sin\phi}\right)^{2 \cdot d} \cdot \sigma_3 \qquad (8)$$

The diameter of each circle of the limit stress is determined by the deviator of the limit stress, and the radius of the circle of the limit stress is calculated as half the diameter. Thus, the diameter and radius of the circles of the limit stress are calculated by the formulas:

$$D_{limit} = \sigma_{1_{limit}} - \sigma_3; R_{limit} = \frac{D_{limit}}{2} = \frac{\sigma_{1_{limit}} - \sigma_3}{2} \qquad (9)$$

To construct the circles of limit stresses, we assume that the cohesion and the angle of internal friction at a limited deformation of the sample of 15% are 30 kPa and 20°, respectively. We will vary the parameter d and accept $d = 0.5$ for circles of the first kind, $d = 0.3$ for circles of the second kind, and $d = 0$ for circles of the third kind. The circles of limit stress are constructed for two values of the minimum main stress $\sigma_3 = 50$ kPa and $\sigma_3 = 200$ kPa. Given such initial data, using formula (8), we calculate the values of the limiting value of the maximum principal stress $\sigma_{1_{limit}}$. For the circles of the first kind for the case $\sigma_3 = 50$ kPa and with the value of the third parameter $d = 0.5$, $d = 0.3$ and $d = 0$, we get that $\sigma_{1_{limit}} = 187.669$ kPa, $\sigma_{1_{limit}} = 150.987$ kPa, $\sigma_{1_{limit}} = 110$ kPa. For the circles of the limit stresses of the second kind, calculated for the case of $\sigma_3 = 200$ kPa, and at the same values of the parameter d, we obtain that $\sigma_{1_{limit}} = 493.61$ kPa (for $d = 0.5$), $\sigma_{1_{limit}} = 381.036$ kPa (for $d = 0.3$) and $\sigma_{1_{limit}} = 160$ kPa (with $d = 0$). Using formulas (9), we calculate the diameter and radius of each of the six circles of limit stress. The abscissa of each of the centers of the circles will be determined by half of the sum of the limit maximum main stress and the corresponding to it minimum main stress, that is, by the formula:

$$A = \frac{\sigma_{1limit} + \sigma_3}{2} \qquad (10)$$

Analyzing formula (10), we indicate that the value of A is equal to the value of the normal stress acting on the area, along which the maximum tangential stress acts, which in the graph is determined by the radius of the circle of limit stress. In Fig. 5 shows the circles of limiting stresses constructed according to the calculation data, which are designated by numbers 1–3 (circles of the first kind, corresponding to $\sigma_3 = 50$ kPa) and 4–6 (circles of the second kind for the case $\sigma_3 = 200$ kPa). To each of the three pairs of circles held the limit straight line.

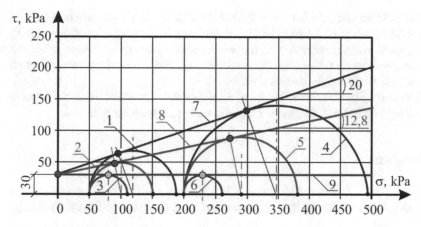

Fig. 5. Circles of limit stress: *1–3* at $\sigma_3 = 50$ kPa and d 0,5; 0,3 and 0; *4–6* at $\sigma_3 = 200$ kPa and d 0,5; 0,3 and 0.

From the analysis of the limit straight lines constructed in Fig. 5, it follows that they are characterized by the same cohesion, but different angles of internal friction. Therefore, the parameter d allows you to vary the radius, diameter and abscissa of the center of the circles of limit stresses. In experiments on the triaxial compression of soils, such variation occurs at a change in limiting deformation. The circle of ultimate stresses decreases with decreasing deformation taken as a limiting value. The construction of the Mohr – Coulomb limit straight lines to the circles of the limit stresses obtained from the analysis of experimental data leads to the fact that both the angle of internal friction and the cohesion change. From the data in Fig. 5 follows that only the angular coordinate of the limit straight lines changes, and the cohesion remains the same. This feature of the limit straight lines to the circles of the limit stresses constructed by the proposed criterion (7) and the identical formula (8) distinguishes them from their counterparts, built to the circles of the limit stresses of the original Mohr – Coulomb plasticity criterion. This feature of the proposed limiting balance equation allows using cohesion value and angle of internal friction which are regulated by various government standards and regulations.

The advantage of this difference is lies in the possibility of using a reliable and reliable database of cohesion and angle of internal friction of various dispersed soils, and on the other hand sets the task of experimentally determining the value of the parameter d. The method for determining this parameter will be discussed in a separate article.

5   Conclusions

The analysis and theoretical studies carried out allow asserting that a new three-parameter Mohr criterion, Coulomb, has been developed, in which the limiting state equation contains the third parameter d. Depending on the value of this parameter, the new criterion is capable of acquiring the form of the traditional two-parameter

Mohr-Coulomb criterion (case $d = 0.5$) and the classical Tresca criterion (case $d = 0$). For all other values of the third parameter, varying in the range $0 > d > 0.5$, the values of tangential stresses according to the new criterion have intermediate values between the maximum tangential stress (Tresca criterion) and tangential stresses according to the original Mohr-Coulomb criterion.

The new criterion allows the use of reliable databases about value cohesion and the value angle of internal friction for various soils and granular materials.

References

1. Gercog, V.N., Dolgikh, G.V., Kuzin, N.V.: Calculation criteria for road pavement evenness. Part 1. Substantiating the flatness standards of asphalt pavement. Mag. Civil Eng. 5(57), 45–57 (2015)
2. Aleksandrov, A.S., Semenova, T.V., Aleksandrova, N.P.: Analysis of permanent deformations in granular materials of road structures. Road Bridges 15, 263–276 (2016)
3. Kalinin, A.L.: Application of modified yield criteria for calculation of safe pressures on the subgrade soil. Mag. Civil Eng. 4(39), 35–45 (2014)
4. Chusov, V.V.: Prospects of application of empirical conditions of soil plasticity and determination of their parameters by triaxial tests. Vestnik Volgogradskogo gosudarstvennogo arhitekturno-stroitel'nogo universiteta. Seriya: Stroitel'stvo i arhitektura 42(61), 49–57 (2015)
5. Dudchenko, A.V., Kuznetsov, S.V.: The modified Mohr – Coulomb and Drucker – Prager models. influence of eccentricity on hysteresis loop and energy loss. Int. J. Comput. Civil Struct. Eng. 13(2), 35–44 (2017)
6. Hambleton, J.P., Drescher, A.: Modeling wheel-induced rutting in soils: indentation. J. Terrramech. 45(6), 201–211 (2008)
7. Du, Y., Gao, J., Jiang, L., Zhang, Y.: Numerical analysis of lug effects on tractive performance of off-road wheel by DEM. J. Braz. Soc. Mech. Sci. Eng. 39(6), 1977–1987 (2017)
8. Xu, X., Dai, Z.-H.: Numerical implementation of a modified Mohr–Coulomb model and its application in slope stability analysis. J. Modern Transp. 25(1), 40–51 (2017)
9. Labuz, J.F., Zang, A.: Mohr–Coulomb failure criterion. Rock Mech. Rock Eng. 45(6), 975–979 (2012)
10. Aleksandrov, A.S., Dolgih, G.V., Smirnov, A.V.: Improvement of calculation of stresses in the earth bed and layers of road clothes from granulated materials. Part 1. Analysis of decisions and a new method. In: IOP Conference Series: Materials Science and Engineering, vol. 463, pp. 1–11 (2018)
11. Badanin, A.N., Bugrov, A.K., Krotov, A.V.: The determination of the first critical load on particulate medium of sandy loam foundation. Mag. Civil Eng. 9(35), 29–34 (2012)
12. Steven, B.D.: The development and verification of a pavement response and performance model for unbound granular pavements. Ph.D. thesis, 291 p. University of Canterbury (2005)
13. Gonzalez, A.: An experimental study of the deformational and performance characteristics of foamed bitumen stabilised pavements. Ph.D. thesis, 392 p. University of Canterbury (2009)
14. Lunev, A., Sirotyuk, V.: Plate load test of base taken from coal ash and slag mixture in experimental tray and on experimental section of embankment. In: IOP Conference Series: Materials Science and Engineering, vol. 451, pp. 1–7 (2018)

15. Yu, M., Li, J., Ma, G.: Yield condition. In: Structural Plasticity. Advanced Topics in Science and Technology in China, pp. 32–63 (2003)
16. Higo, Y., et al.: A three-dimensional elasto-viscoplastic strain localization analysis of water-saturated clay. Geo-Research Institute, Osaka, Japan, vol. 86, pp. 3205–3240 (2006)
17. Oka, F.: Computational modeling of large deformations and failure of geomaterials. In: XVI ICSMGE, Osaka 2005, no. 1, pp. 47–94. Millpress (2005)
18. Higo, Y.: Instability and strain localization analysis of watersaturated clay by elasto-viscoplastic constitutive models. Ph.D. thesis, Kyoto University, Japan (2003)
19. Aleksandrov, A.S., Kalinin, A.L.: Improvement of shear strength design of a road structure Part 1 Deformations in the Mohr – Coulomb plasticity condition. Mag. Civil Eng. **59**(7), 4–17 (2015)
20. Ma, Z., Liao, H., Dang, F.: Unified elastoplastic finite difference and its application. Appl. Math. Mech. **34**(4), 457–474 (2013)
21. Ma, Z., Liao, H., Dang, F.: Effect of intermediate principal stress on strength of soft rock under complex stress states. J. Central South Univ. **21**(4), 1583–1593 (2014)
22. Veiskarami, M., Ghorbani, A., Alavipour, M.: Development of a constitutive model for rockfills and similar granular materials based on the disturbed state concept. Front. Struct. Civil Eng. **6**(4), 365–378 (2012)
23. Alejano, L.R., Bobet, A.: Drucker-prager criterion. Rock Mech. Rock Eng. **45**(6), 995–999 (2012)
24. Zhang, L., Liu, D., Song, Q., Liu, S.: An analytical expression of reliability solution for Druker-Prager criterion. Appl. Math. Mech. **29**(1), 121–128 (2008)
25. Zhu, J., Peng, K., Shao, J.F., Liu, H.: Improved slope safety analysis by new Druker-Prager type criterion. J. Central South Univ. **19**(4), 1132–1137 (2012)

Versatile Dynamics Simulator: Dedicated Particle Dynamics Software for Construction Materials Science

Vladimir Smirnov$^{(\boxtimes)}$ [iD] and Evgenij Korolev [iD]

Moscow State University of Civil Engineering, Yaroslavskoe shosse 26,
129337 Moscow, Russia
smirnov@nocnt.ru

Abstract. Design and discovery of materials guided by theory and computation is the current trend in materials science. The majority of modern building materials can be thought of as dispersions. Thus, particle systems should be considered as reasonable representations of both compositions and composites. Nowadays, there are numerous software packages available for modeling the particle dynamics. However, nearly all of such packages are targeted only to nano- and microscale spatial levels. For the R&D in building materials science, some specific functionality for macroscale modeling, along with simplified pairwise potentials, but complicated initial distributions and topology analysis methods have to be implemented. In the present article, we briefly represent the pathway from the problem of modeling the building materials as particle systems, through the general method of particle dynamics to the task of multiscale modeling, where we need unified modeling methods and single software implementation. Then, we introduce the Versatile Dynamics Simulator – dedicated software implementation of classical dynamics that was designed taking into account the demands of building materials science. Design goals and current state of the software are described.

Keywords: Building materials · Dispersions · Particle dynamics ·
Simulation software

1 Introduction

The term "modeling" is very general and may denote quite different techniques, which, in turn, can be classified with a lot of criteria. The primary criterion is the set of properties of interest, i.e. type of similarity.

Despite the practical orientation, in materials science of building materials (construction materials science), like in many other fields, modeling has been used all the time. In the beginning there was only "physical and chemical" modeling – production and testing the samples. The similarity was the composition and technology (between test sample and bulk building material). Later, this type of modeling – ad hoc approach – was extended by proper application of statistics and experimental design; estimation

© Springer Nature Switzerland AG 2020
V. Murgul and M. Pasetti (Eds.): EMMFT 2018, AISC 982, pp. 798–808, 2020.
https://doi.org/10.1007/978-3-030-19756-8_76

of parameters often performed for regression models derived from the knowledge about structural characteristics [1]. Still, this type of modeling can hardly be designated as mathematical one.

The starting point of mathematical modeling in construction materials science is the application of several semi-empiric relations that not only express the observed behavior of material, but also take into account some general suppositions about structure of composition (mix of components) and composite (building material). From that moment, there were constant attempts to move away from domination of empirics in construction material science. It was obvious that only by means of structural models the full power of mathematical modeling – arising from its universality and ability to examine the object with broad set of control parameters – can be unleashed. The role of mathematical models and their tight integration with numerical experiments is emphasized by the term "computational materials science" – interdisciplinary field spanning condensed matter physics, mechanics, chemistry, surface science [2] and – last but not least – numerical methods [3] and software engineering [4]. The net effect of modeling is the reduction of time needed for the transition from research to man-ufacturing [5]. The modern trend is multiscale modeling, which bridges materials properties from the atomic level to macroscopic length scales [6–8] and allows to derive macroscopic performance from microscopic foundations [9]. Nowadays the computational material science and multiscale modeling are adopted in the framework of Integrated Computational Materials Engineering (ICME) paradigm [4], where they are integrated with knowledge-guided decision making, data-driven reasoning [10] and machine learning [11, 12]. Thus, numerical modeling is widely used both in current materials science [13] and, in particular, in materials science of building materials. In latter case, promising applications are particle packing [14–17], mix design [18, 19], micromechanics [20, 21] and studying the effects of operating conditions to structure and properties [22]. In several contemporary research works, the full sequences of multiscale modeling of concretes (from nanoscale up to macroscopic scale, e.g. [23]) were presented.

Complexity of the models in construction materials science ranges from simple geometric representations to dynamic systems with rigid body collision detection and further to the multiphysics and multiphase chemistry. Complexity of the corresponding software tools reflects the complexity of underlying models. The proper choice of the software – be it ready-made and well known or original and brand new – is critical to efficiency of the simulation. In spite of this, intricate details of software setup are usually omitted [24] in many research papers. Such practice is common and well adopted in academic area. This is why there is a need in a number of papers dedicated to the details of specific software tools.

In the present work, we are focusing on one specific, albeit widely applicable modeling method and describe our software implementation that may be suitable for the simulations in construction materials science. The primary purpose of the work is to demonstrate distinctive characteristics of our implementation that makes is suitable for modeling from nanoscale up to macroscopic scale, and also to underline several fea-tures of the input language syntax which makes the setting up simulation quite comfortable.

2   Building Materials and Compositions as Particle Systems

No doubt, the majority of modern building materials at every single stage of their existence are dispersions. Because of this, particle systems can be considered as the most reasonable representations of both compositions and composites.

It is known that simple geometric models in form of particle systems (without mid- and long-range pairwise interactions), coupled with adequate probabilistic setup, can be used for the simulation of particle packing and optimization of mix design in construction materials science [18]. Complicated schemes for contact forces allows application of rigid-body particle systems in the process of complex modeling, e.g. in fracture mechanics [25].

The transition from rigid-body particle system to dynamical system with pairwise interactions seems obvious. Thus, the pure geometrical and probabilistic task of packing was evolved [17, 19] into transient time-domain problem of particle dynamics.

Modeling of particles' motion under internal and external forces is the particle dynamics method, and it is dated back to XIX century. The model [26–29] is the second law of the classic dynamics:

$$m_i\ddot{\mathbf{r}}_i - k_i(\dot{\mathbf{r}}_i - \mathbf{u}_i) = -\nabla U_i, i = \overline{1,N} \tag{1}$$

where mi is the mass of i-th particle, $r_i = (x_i; y_i; z_i)$ is the position of particle, k_i is the parameter that depends on form of the particle and viscosity of the dispersion medium, $\mathbf{u}_i$ is the velocity of the dispersion medium at point $\mathbf{r}_i$, U_i is the scalar potential at the point $\mathbf{r}_i$ and N is the number of particles.

The model (1) is usually rewritten as a system of ordinary differential equations:

$$\begin{cases} \dot{\mathbf{r}}_i = \mathbf{v}_i \\ \dot{\mathbf{v}}_i = g + \frac{1}{m_i}\left(\sum_{\substack{j=1 \\ j \neq i}}^{N} \mathbf{F}_{ij} + \mathbf{F}_{i,b} + \mathbf{F}_{i,e} \right) \end{cases} \tag{2}$$

where

$$\mathbf{F}_{ij} = -\mathbf{F}_{ji} = \frac{\mathbf{r}_{ij}}{r_{ij}}F_{ij}(d_{ij}) = \frac{\mathbf{r}_{ij}}{r_{ij}}F_{ij}(r_{ij} - R_i - R_j) \text{ and } \mathbf{F}_{i,b} = \frac{\mathbf{n}_{i,b}}{n_{i,b}}F_{i,b}(n_{i,b} - R_i) \tag{3}$$

are pairwise and boundary forces, m_i is the mass of the i-th particle, R_i are radii, $\mathbf{g}$ is the gravitational acceleration, $F_{i,b}$, $b = \overline{1,N_b}$ is the boundary force, N_b is the number of boundary planes, $\mathbf{n}_{i,b}$ is the normal of i-th boundary plane, and $F_{i,e} = 6\pi\eta R_i(\mathbf{u}_i - \mathbf{v}_i)$ is the force of viscous friction.

The model (2) is very general in nature. It can describe interacting particles from nanoscale level (at the scale where the quantum effects can be neglected) up to celestial scale. But the spatial scale still heavily affects [8] the choices of efficient integration methods, as well as selection of desirable procedures of pre- and postprocessing.

Nowadays, there are numerous software packages available for modeling the particle dynamics. However, almost any package is targeted to nano- and microscale spatial levels. This is reflected by common term "molecular dynamics".

3   Open Source Molecular Dynamics Packages

The modeling task is the solution of (2) with selected initial and boundary conditions. During the last decades, there was a considerable progress in development of efficient integration schemes, algorithms, and parallel implementations of particle dynamics. The most widely accepted particle dynamics packages are GROMACS [30] and LAMMPS [31, 32]. These open source software packages [33, 34] are in constant development since mid 1990s. Current online documentation [35] of LAMMPS is comprehensive, well written, and built with commonly used Sphinx system [36]. In our university we are using LAMMPS documentation in education process on constant basis.

Of cause, both GROMACS and LAMMPS can successfully be used for particle dynamics problems in construction materials science. But their usage, especially at upper spatial levels, is not always reasonable.

As was already mentioned, software design depends on spatial scale. The GROMACS and LAMMPS packages, as many others, are targeted primarily to nano- and, on rare occasions, to micro-scale spatial levels. The original LAMMPS papers [31, 32] focus on the details of integration scheme which is efficient for problems with lot of particles and short-range pairwise interactions. The offered integration algorithm is suitable for parallel implementation on high performance computing (HPC) hardware – both on shared and distributed memory systems.

In constructional materials science, we constantly encounter cases when we either have to set cutoff distance of pairwise potential to values comparable with the spatial size of the modeled problem (microscale), or to extend the central-force model (1) with elements taken from rigid-body dynamics (macroscale) [8]. Because of this, using the existing software leads to high time consumption for exploration of features and extending the functionality. Moreover, we often have to use complicated initial distributions and specific postprocessing methods suitable for registration of many topological parameters during the force-driven evolution of colloids and coarse-grained dispersions. At the same time, complex pairwise potentials and high amounts of particles are usually not needed for obtaining the informative conclusions.

4   Versatile Dynamics Simulator

Due to lack of the single, "light-weight" and portable software implementation suitable for our modeling needs, about twenty years ago we have started [37] our own implementation.

Early versions of the software were mentioned in several local publications (including [37]) which are not available online now. As of [26], numerous algorithms for modeling the macrostructure of building materials were implemented. Massive source code refactoring, extending and cleaning the syntax of input language were performed during 2018. Many simulation codes were integrated into single software implementation – Versatile Dynamics Simulator (VDS).

Design goals of VDS are:

- adequate representation of forces arising due to ionic shells, steric conditions and technological actions;
- integration scheme which is suitable for both large and small time steps; adaptive (self-tuning) time step;
- flexible representation of boundaries, including moving and oscillating ones;
- preprocessing methods which are suitable for complex initial states ("state" incorporates initial positions, radiuses and masses of particles);
- postprocessing methods for monitoring scalar parameters of spatial distribution, including topological and percolation parameters;
- reproducibility of entire transient phase trajectory – even in case of stochastic modeling;
- simple visualization facilities which are enough for quick decisions about the current computation setup, complemented with postprocessing methods which allow advanced visualization after the computation is finished;
- computation setup in form of plain text written in problem-oriented language with context-free (almost), "comfortable" syntax (e.g., no sensitivity to indentation, etc.) and low susceptibility to errors;
- ability to work as batch job;
- modular and extensible source code organization;
- high portability between compilers and platforms.

The description of VDS is better to be started from the last goals. Both the choice of software development tools and decisions about architecture of VDS were made long time ago under heavy influence of rather dissimilar sources [38, 39].

There are many claims in [38], which contradict common myths about programming practices. In particular, statement that ANSI C is portable, but C++ is not, being already true in 1990s, over time only increased its importance. Current C++ "standard" libraries, as well as numerous frameworks (written in "object-oriented" languages with extensive use of "high-level abstractions") can turn the long-term maintenance of the software into nightmare. With long time span of maintenance in mind, the only choice is old-fashioned ANSI C ("Unix's central technology" [39]) library calls complemented with several tens of platform-specific calls that are absolutely necessary for efficient implementation. The VDS is designed in such a way that both types of service calls are encapsulated into light-weight object oriented (despite the pure C language) middleware framework – LibV [40]. As of current, source code of VDS is within the source tree of LibV and this situation is not going to change in near future. VDS (with LibV) can be compiled to 32-bit static executable by ancient tools (WATCOM C 11, Microsoft Visual C 6.0, Borland C++ Builder 3) and run under Microsoft Windows NT 4.0 (there was no support for Microsoft Windows 9x from the beginning). The same source code can be compiled with modern Microsoft Visual C, GCC 8.x (including Cygwin) or Clang 7.x to 64-bit executable and run under Microsoft Windows 10, modern Linux and FreeBSD. And in latter case there is a notable (up to 40%) performance gain. It must also be stressed that in case of Microsoft Visual C we are talking only about C compiler, without huge and resource demanding "Visual Studio" development environment.

Despite the goal of visualization during computation, the VDS itself – like many other numerical packages – is free from graphical user interface (GUI). Without burden of GUI, there is no need of complicated frameworks that obsoletes over time. The importance of GUI is often overestimated. In academic world, all we need is to express the theoretical data (results of basic research and analytical deductions) in a form that is easy to comprehend as a whole. For example:

```
vec3     z {constant; parameters [0,0,0];}
vec3     gravity {constant; parameters [0,0,-9.8];}
float    pairwise1 {morse; parameters [2e-20,2e-6,1e-6];}
float    boundary1 {morse; parameters [-1e-21,15e-6,5e-6];}
domain   inner_sphere { minimal_distance 5e-6;
    sphere [0,0,0] 40e-6; }
boundary outer_sphere {sphere [0,0,0] 50e-6 boundary1;}
particle_system micro 128 {
   random_seed 2; radius dirac [1.4e-6,4e-6;9,1];
   density heaviside [1002,1005,1009;8,2,0];
   shells {percolation true; percolation_distance 3e-6;}
   environment {density 1000; viscosity 1e-6; velocity z;}
   pairwise pairwise1; boundary outer_sphere;
   domain inner_sphere; unary_accel gravity;
   monitor shot_periodic; monitor write_maxscript;
   monitor write_connmatrix;
}
goal micro {
   soft_particles true; sp_repulsion_factor 1.4;
   ms_global_scale 1e+6;
   time_limit 6; realtime_limit 600; RK_tolerance 1e-3;
   maximal_step 1e-1; minimal_step 1e-9;
   saves_per_second 20;
   view { width 1024; height 1024; rotz_dps 60;
     antialias true; variable_linewidth true; }
}
```

Listing above is written in somewhat clumsy form to save space. With proper indentation, syntax of setup will become very clean and easy to understand.

Functionality of built-in visualization facilities is illustrated by Fig. 1, a. This figure was in form of raster image, but VDS is able to decode OpenGL feedback buffer and can produce vector images.

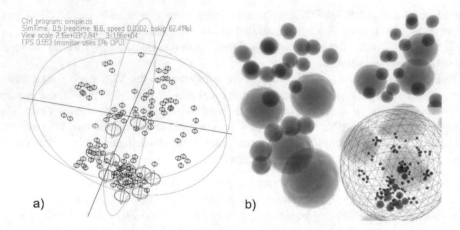

Ctrl program: simple.ds
SimTime 0.5 [realtime 16.6, speed 0.0302, bskip 62.41%]
View scale 2.19e+03^2.84^   3=1.96e+04
FPS 0.553 [monitor uses 0% CPU]

a)   b)

Fig. 1. Visualization: built-in (a) and performed with external tool (b).

Visualization can be further enhanced with external tools. Currently, VDS produces scene descriptions in the form that is acceptable by Autodesk 3DS MAX. Broad functionality of 3DS MAX opens unlimited possibilities for visualization at postprocessing stage (Fig. 1, b).

A number of scalar parameters can be monitored during computation: amount of aggregates and amount of particles in each aggregate; average and standard deviation of neighbor particles; average and standard deviation of distance from i-th particle to several closest neighbors; average and standard deviation of particles' kinetic and potential energy and so forth.

Values of monitored parameters are stored in plain text (comma separated values) file and can be processed later by any desirable tool. As an example, in Fig. 2, there is a kinetic dependence of size of largest aggregate. Both Figs. 1 and 2 were produced during computation with setup shown above.

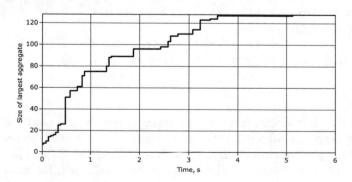

Fig. 2. Example of kinetic dependence produced during computation.

Many methods of initial distributions are available at automated preprocessing stage. In the setup mentioned above, there is a line:

```
radius dirac [1.4e-6,4e-6;9,1];
```

This line means: generate initial distributions of particles' radii as values of random variable with probability density

$$fR = 0.9 \cdot \delta R - 1.4\mu m + 0.1 \cdot \delta R - 4\mu m \qquad (4)$$

where δ – Dirac delta function.

Our software is capable of generating complex spatial distributions. Type (sum of delta functions, constant, piecewise constant and normal) and parameters of spatial probability densities can be specified for any coordinate. Spatial boundaries can be specified in form of constructive solid geometry (CSG) description:

```
union {
  cylinder [-101,0,0;1,0,0] [30,202];
  union {
    cylinder [0,-101,0;0,1,0] [30,202];
    union {
      cylinder [0,0,-101;0,0,1] [30,202];
      sphere [0,0,0] 60;
    }
  }
}
```

As of current, binary tree is used for inner representation of CSG. Because of this, number of operands for any Boolean operation, including union and intersection, should be 2. For any description similar to presented above, it is possible to generate distributions with different probability densities along any coordinate (Fig. 3).

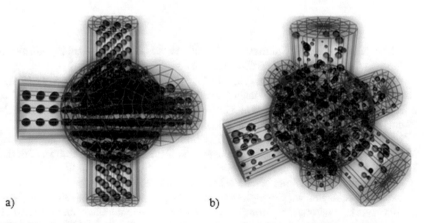

a)   b)

Fig. 3. Regular lattice (a) and distribution with constant density (b) are confined by CSG.

The solver of ordinary differential equation is the critical part of most modeling software. Several low-order integration methods are considered as suitable for particle dynamics: midpoint (second order Runge-Kutta), Verlet and similar. Our choice, though, is general-purpose embedded Runge-Kutta-Fehlberg "4–5" method. If compared with more common methods, this one is of higher computational cost. But our experience shows that automated error-based control of time step is the crucial advantage in almost any case.

5 Summary and Conclusions

Competition with packages similar to GROMACS or LAMMPS is pointless. In many fields and in most cases, ready-done well-developed software with reach set of features outperforms home-made "toy programs" in orders of magnitude. Nevertheless, it is reasonable to design and implement new algorithms for particle dynamics and to share the results with open source and academic communities. After all, it is the way how open source works.

We have developed the software which simplifies our research – and for us it is the definitive argument to use, maintain and develop this software. Presently, some critical features (e.g. proper parallel algorithms) are either missing or under development. Still, today we can state that our attempt to "design a small but capable… system with a clean… interface" [39] that was made twenty years ago is succeeded. The Versatile Dynamics Simulator can be used for modeling different dispersions in construction materials science – from colloids up to coarse grained systems.

The objectives of the further development are, among others, parallel algorithms, massive testing and verification and implementation of the advanced visualization features.

This work is supported by Ministry of Science and Higher Education of Russian Federation, project "Theoretical and experimental models of functional composites based on prime nanomaterials" 7.6250.2017/8.9.

References

1. Gladkikh, V., Korolev, E., Smirnov, V., Sukhachev, I.: Modeling the rutting kinetics of the sulfur-extended asphalt. Procedia Eng. **165**, 1417–1423 (2016)
2. Chen, L.-Q., Kalinin, S., Klimeck, G., Kumar, S., Neugebauer, J., Terasaki, I.: Design and discovery of materials guided by theory and computation. Comput. Mater. **1**, 15007 (2015)
3. Zhang, L., Ren, W., Samanta, A., Du, Q.: Recent developments in computational modelling of nucleation in phase transformations. Comput. Mater. **2**, 16003 (2016)
4. Matouš, K., Geers, M., Kouznetsova, V., Gillman, A.: A review of predictive nonlinear theories for multiscale modeling of heterogeneous materials. J. Comput. Phys. **330**(1), 192–220 (2017)
5. Boosting materials modelling. Nature Mater. **15**, 365 (2016)
6. Beaber, A., Gerberich, W.: Strength from modelling. Nature Mater. **9**, 698–699 (2010)
7. Evstigneev, A., Smirnov, V., Korolev, E.: Design of nanomodified intumescent polymer matrix coatings: theory, modeling experiments. MATEC Web Conf. **251**, 01033 (2018)

8. Smirnov, V., Evstigneev, A., Korolev, E.: Multiscale material design in construction. MATEC Web Conf. **106**, 03027 (2017)
9. Marzari, N.: The frontiers and the challenges. Nature Mater. **15**, 381–382 (2016)
10. Reddy, S., Gautham, B.P., Das, P., Yeddula, R., Vale, S., Malhotra, C.: An ontological framework for integrated computational materials engineering. In: Mason, P., et al. (eds.) 4th World Congress on Integrated Computational Materials Engineering. The Minerals, Metals & Materials Series, pp. 67–69. Springer, Cham (2017)
11. Zhang, Y., Ling, C.: A strategy to apply machine learning to small datasets in materials science. Comput. Mater. **4**, 25 (2018)
12. Homer, E.: High-throughput simulations for insight into grain boundary structure-property relationships and other complex microstructural phenomena. Comput. Mater. Sci. **161**, 244–254 (2019)
13. Janssens, K., Raabe, D., Kozeschnik, E., Miodownik, M., Nestler, B.: Computational Materials Engineering. Academic Press, Burlington (2007)
14. Bernal, J.D., Finney, J.L.: Random close-packed hard-sphere model. II. Geometry of random packing of hard spheres. Discuss. Faraday Soc. **43**, 62 (1967)
15. Scott, G.D., Kilgour, D.M.: The density of random close packing of spheres. J. Phys. D Appl. Phys. **2**, 863 (1969)
16. Jodrey, W.S., Tory, E.M.: Computer simulation of isotropic, homogeneous, dense random packing of equal spheres. Powder Technol. **30**(2), 111–118 (1981)
17. Bezrukov, A., Stoyan, D., Bargiel, M.: Spatial statistics for simulated packings of spheres. Image Anal. Ster. **20**, 203–206 (2001)
18. Xu, R., Yang, X.H., Yin, A.Y., Yang, S.F.: A three-dimensional aggregate generation and packing algorithm for modeling asphalt mixture with graded aggregates. J. Mech. **26**(2), 165–171 (2011)
19. Fu, G., Dekelbab, W.: 3-D random packing of polydisperse particles and concrete aggregate grading. Powder Technol. **133**(1–3), 147–155 (2003)
20. Nguyen, V.P., Stroeven, M., Johannes, L., Sluys, L.J.: Multiscale failure modeling of concrete: micromechanical modeling, discontinuous homogenization and parallel computations. Comput. Methods Appl. Mech. Eng. **201–204**, 139–156 (2012)
21. Palkovic, S.D., Kupwade-Patil, K., Yip, S., Büyüköztürk, O.: Random field finite element models with cohesive-frictional interactions of a hardened cement paste microstructure. J. Mech. Phys. Solids **119**, 349–368 (2018)
22. Su, H., Hu, J., Li, H.: Multi-scale performance simulation and effect analysis for hydraulic concrete submitted to leaching and frost. Eng. Comput. **34**(4), 821–842 (2018)
23. Eftekhari, M., Mohammadi, S., Khanmohammadi, M.: A hierarchical nano to macro multiscale analysis of monotonic behavior of concrete columns made of CNT-reinforced cement composite. Constr. Build. Mater. **175**, 134–143 (2018)
24. Korolev, E., Inozemtcev, A., Evstigneev, A.: Methodology of nanomodified binder examination: Experimental and numerical ab initio studies. Key Eng. Mater. **683**, 589–595 (2016)
25. Stroeven, P., Stroeven, M.: Assessment of packing characteristics by computer simulation. Cem. Concr. Res. **29**(8), 1201–1206 (1999)
26. Korolev, E.V., Smirnov, V.A.: Using particle systems to model the building materials. Adv. Mater. Res. **746**, 277–280 (2013)
27. Kiselev, D.G., Korolev, E.V., Smirnov, V.A.: Structure formation of sulfur-based composite: the model. Adv. Mater. Res. **1040**, 592–595 (2014)
28. Gladkikh, V.A., Korolev, E.V., Smirnov, V.A.: Modeling of the sulfur-bituminous concrete mix compaction. Adv. Mater. Res. **1040**, 525–528 (2014)

29. Korolev, E.V., Tarasov, R.V., Makarova, L.V., Smirnov, V.A.: Model research of bitumen composition with nanoscale structural units. Contemp. Eng. Sci. **8**(9–12), 393–399 (2015)
30. Berendsen, H.J.C., van der Spoel, D., van Drunen, R.: GROMACS – a message-passing parallel molecular-dynamics implementation. Comput. Phys. Commun. **91**, 43–56 (1995)
31. Plimpton, S.: Fast parallel algorithms for short-range molecular dynamics. J. Comput. Phys. **117**, 1–19 (1995)
32. Plimpton, S., Hendrickson, B.: A new parallel method for molecular dynamics simulation of macromolecular systems. J. Comput. Chem. **17**, 326–337 (1996)
33. Gromacs. http://www.gromacs.org. Accessed 31 Jan 2019
34. LAMMPS Molecular Dynamics Simulator. https://lammps.sandia.gov. Accessed 31 Jan 2019
35. LAMMPS Documentation. https://lammps.sandia.gov/doc/Manual.html. Accessed 31 Jan 2019
36. Overview — Sphinx. http://www.sphinx-doc.org. Accessed 31 Jan 2019
37. Korolev, E.V., Proshin, A.P., Smirnov, V.A.: Investigation of stability of aggregates in composites. News High. Educ. Inst. **4**, 40–45 (2002). (in Russian)
38. Grinzo, L.: Zen of Windows 95 Programming: Master the Art of Moving to Windows 95 and Creating High-Performance Windows Applications. Coriolis Group, Scottsdale (1995)
39. Raymond, E.S.: The Art of Unix Programming. Addison-Wesley, Boston (2003)
40. LibV Development Central. http://libv.org. Accessed 31 Jan 2019

Effect of Blank Curvature and Thinning on Shell Stresses at Superplastic Forming

Oleksandr Anishchenko[1]([envelope]) [iD], Volodymyr Kukhar[1] [iD],
Viktor Artiukh[2] [iD], Anatoliy Trebukhin[3] [iD], and Natalia Zotkina[4] [iD]

[1] Pryazovskyi State Technical University, Street Universytets'ka 7,
Mariupol 87555, Ukraine
alexander.anishchenko@gmail.com
[2] Peter the Great St. Petersburg Polytechnic University, Polytechnicheskaya 29,
195251 St. Petersburg, Russia
[3] Moscow State University of Civil Engineering, Yaroslavskoye shosse 26,
129337 Moscow, Russia
[4] Tyumen Industrial University, Volodarskogo Street 38,
625000 Tyumen, Russia

Abstract. The paper studies the relationship between principal stresses, curvature radii and shell thinning during superplastic forming of sphere-shaped domes from sheet blanks of several Al-Mg alloys and Sn-38% Pb alloy, which is the basis for refining the calculations of the power characteristics of the process. When using the Lame's superellipse to describe the curvature of the shells, it was established that the principal stresses, especially the tangential stress, depend on the principal curvature radii. It is shown that the intensity of stresses also depends on the thinning of the shells during superplastic forming. It is revealed that the higher the level of superplastic properties of the material of the blank, the less the principal stresses and effective stresses depend on the difference between the principal radii of curvature. It has been established that when calculating the power mode of superplastic forming of shells, it is unacceptable the assumption of uniform thinning of the blank during forming, since the calculation error can reach 130%.

Keywords: Superplastic forming · Blank curvature · Uniform thinning ·
Lame superellipse · Principal stresses · Al-Mg alloys · Babbitt

1 Introduction

The use of superplastic forming (SPF) materials is widely applicable in the production of shells, products such as boxes, hemispheres and bodies of rotation [1–3]. The advantage of SPF method over the known methods for the production of shells, including bodies of rotation, is the controlled saving of the material of the blanks, especially while forming dimensional articles [4–8]. Prediction of the form change of shells in SPF includes information about the type and analytical description of plastic flow curves (hardening) of materials that are previously obtained experimentally, also applying physical and geometric similarity methods [1, 9–14]. After processing the

© Springer Nature Switzerland AG 2020
V. Murgul and M. Pasetti (Eds.): EMMFT 2018, AISC 982, pp. 809–817, 2020.
https://doi.org/10.1007/978-3-030-19756-8_77

graphs, the approximation coefficients are found for the formula that describes the curves of high-speed hardening of the material [10, 11, 15–19]:

$$\sigma = k\zeta^m \tag{1}$$

where σ – intensity of flow stresses; k – material's properties index; ζ – intensity of strain rates; m – strain rate hardening exponent (is a sensitivity index to evaluate the dependence of flow stress σ on strain rate ζ).

In engineering analysis of the stress-strain state and the power parameters of SPF the Laplace equation is usually used [1, 18–21]:

$$\sigma_m/r_1 + \sigma_\theta/r_2 = p/S, \tag{2}$$

where σ_m, σ_θ – principal (meridian and tangential) stresses in the shell during forming; r_1, r_2 – radii of curvature of the shell in meridian and tangential directions; p – deformable medium pressure; S – current value of thickness of the shell along the contour.

When analyzing the SPF process (Fig. 1), forming zones of blanks with a uniform or predetermined wall thickness that are free from contact with the stamp surface are often represented as spherical surface areas with an uneven (sometimes even) wall thickness [1, 22–24]. Such approach simplifies the determination of r_1 and r_2, and, therefore, further calculations of stresses and forces. However, actual contours of the

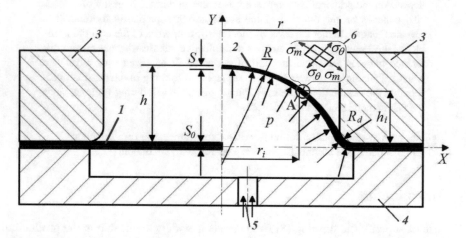

Fig. 1. Design scheme of SPF process: 1 – sheet blank before forming with initial thickness S_0; 2 – blank after forming with variable thickness S, height h and radius of curvature R; 3 – upper clamping die with radius r and working radius R_d; 4 – lower die; 5 – hole for gas supply and making a forming pressure p in die cavity; 6 – stress-strain state at an arbitrary point A (top view), located at a distance determined by an arbitrary radius r_i and height h_i.

shells differ from spherical ones [25–27]. This frequently requires to apply compensating elements in the system "deformable medium - blank" [28, 29] and introduces margins of error in the calculations, stipulating the need for their improvement.

Radii of curvature of free forming surfaces of the shells depend on the temperature, degree, strain rate, and superplastic properties of the material of the blanks. The principal radii of curvature r_1, r_2 are determined from various formulas that approximate experimental data. The distinction of the approach we are developing, is the assumption, previously substantiated in works [18, 19], that the most universal formula approximating the contour of a deformable blank at all stages of SPF is Lame's superellipse [18, 19, 30]:

$$(x/a)^p + (y/b)^q = 1, \tag{3}$$

where p and q – exponents; $x = r_i/r, y = h_i/r, a = r/r = 1, b = h/r$ (ri, hi, r, h – current and maximum values of the base radius and shell height respectively).

Depending on the values of p, q, a and b, formula (3) the entire set of contours in dimensionless coordinates that have sheet blanks at various stages of SPF are described.

Taking into consideration formula (2), the relative radii of curvature $r_m = r_1/r$ and $r_\theta = r_2/r$ are determined by the formulas [19]:

$$r_m = \frac{\left[1 + (bp/q)^2 a^{-2p/q} x^{2(p-2)} (a^p - x^p)^{(1-q)/q}\right]^{3/2}}{-(bp/q)a^{-p/q} x^{p-2}(a^p - x^p)^{(1-2q)/q}[(p-1)(a^p - x^p) + ((q-1)/q)px^p]}; \tag{4}$$

$$r_\theta = -ba^{-p/q}(a^p - x^p)^{1/q}\left[1 + (bp/q)^2 a^{-2p/q} x^{2(p-1)}(a^p - x^p)^{(2-2q)/q}\right]^{1/2}. \tag{5}$$

In papers [19, 31], it is shown that for SPF domes from blanks of variable thickness (BVT) and with different superplastic properties, the radii of curvature r_m, r_θ significantly differ from the spherical cap radius; they can be infinitely large and have extremums along the shell contour. Attention should be paid to the correlation between changes in the values of r_m, r_θ and thinning of the blank at SPF [23, 24, 31].

2   The Purpose of the Research

The objective of the research is to study the relationship between the principal stresses, the radii of curvature and the thinning of the shells at SPF of spherical domes to adjust the calculations of the power parameters of the technological process.

3 Materials and Methods of Research

To estimate the effect of the blank thickness on the stress state of the shell during the SPF, we used previously obtained experimental data and data from other authors [18–20, 25, 26].

Thinning, i.e. the relative distribution of thickness $z = S/S_0$ along the relative radius of the base x of shells with a relative height of $y = 1$ was approximated by the equations presented in Table 1 for different materials: AMG6 (analogs: A 95456, A 95556 according to UNS, USA) – Fe < 0.4%, Si < 0.4%, Mn = 0.5...0.8%, Ti = 0.02...0.1%, Al = 91.1...93.68%, Cu < 0.1%, Mg < 5.8...6.8%, Zn < 0.2%; AlMg5 DIN 1725 (analogs: ALMG5, ER 5356/AWS A5.10) – Si < 0.25%, Mn = 0.15%, Cr = 0.12%, Fe = 0.4%, Mg = 5.0%, Al = the rest.

On the basis of Eq. (2) and von Mises criterion, in works [19, 31] dependences convenient for analyzing the influence of the curvature of shell contours on the distribution of principal stresses σ_m, σ_θ and their intensities σ_e were obtained. For the relative values of the thickness (see Table 1) and the radii of curvature of the shells, these equations take the following form:

$$\sigma_m = pr_\theta/2_z, \quad \sigma_\theta = \left(2 - r_\theta/r_m\right)pr_\theta/2_z, \tag{6}$$

$$\sigma_e = \left(pr_\theta/2_z\right)\left[\sqrt{\left(r_\theta/r_m\right)^2 - 3\left(r_\theta/r_m\right) + 3}\right] \tag{7}$$

Assume that the shell contour is part of a sphere ($r_\theta/r_m = 1$), and the shell thinning is uniform ($z = 0.5$), formula (7) is simplified to the following form:

$$\sigma_o = pr_\theta/2_z, \text{ i.e. } \sigma_o = pr_\theta, \tag{8}$$

Figure 2a shows the distribution of the ratio of principal stresses σ_θ/σ_m along the radius of the shells base. The calculations did not take into account the presence of the interface radius of the deformable shell and the flange, i.e. in the interval x = (0.95... 1.0)r, the calculated data were unreliable and are not shown in the graphs.

For superplastic babbitt Sn-38%Pb with $m = 0.6$, the principal stresses are approximately equal (margin of error does not exceed 18%). With the deterioration of superplastic properties in alloys, the calculated values of σ_θ prevail over the values of σ_m due to a significant increase in the radius of curvature R [18, 19], especially for babbitt with $m = 0.25$.

In blanks of variable thickness the distribution $\sigma_\theta/\sigma_m = f(x)$ has the form of the parabola with a minimum in the range $x = 0.6...0.7$, i.e. in the places of interface of the central and peripheral zones of the blank with variable thickness.

Table 1. Formulas of approximation of the shell thickness at SPF.

Shell parameters	Approximating function	Shell parameters	Approximating function
alloy *AMG6*, BVT $m = 0.38$	$z_2 = -1.94x^4 + 5.03x^3 - 3.31x^2 + 0.38x + 0.25$	alloy *Sn-38%Pb*, $m = 0.25$	$z_4 = 0.88x^4 - 0.80x^3 + 0.44x^2 + 0.06x + 0.24$
alloy *Sn-38%Pb*, $m = 0.60$	$z_3 = 1.09x^4 - 0.93x^3 + 0.17x^2 + 0.05x + 0.43$	alloy *AlMg5* $m = 0.42$	$z_5 = 1.25x^4 - 1.82x^3 + 1.16x^2 + 0.00x + 0.23$

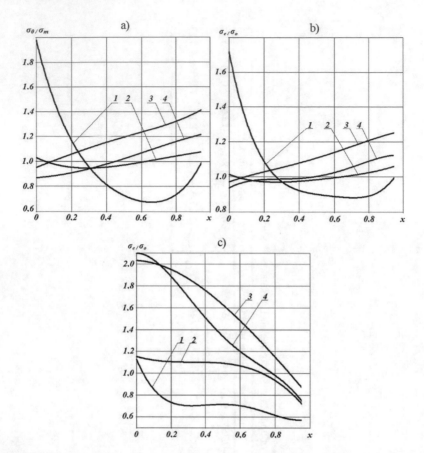

Fig. 2. Distribution of the ratios of the principal stresses and effective stresses (stresses intensities) along the relative radius of the base of the shells from: 1 – blanks of variable thickness; 2 – *Sn-38%Pb* alloy with $m = 0.60$; 3 – *Sn-38%Pb* alloy with $m = 0.25$; 4 – *AlMg5* alloy.

4 Discussion of Research Results

At the intersection points of curves 1–4 with ordinate $\sigma_\theta/\sigma_m = 1$, the contours of the shells coincide with the contour of the sphere. For babbitt shells with $m = 0.6$ and blanks of variable thickness, there are two such points in each one: $\sigma_\theta/\sigma_m = 0.09$ and 0.62, 0.27, and 0.95, respectively. Babbitt shells with $m = 0.25$ and *AlMg5* alloy have the ordinate $\sigma_\theta/\sigma_m = 1$ at $x = 0.09$ and 0.44, respectively.

The differences in σ_θ and σ_m values along the contour of the shells determine the stress intensity (effective stresses) values σ_e, which differ from the similar parameters of σ_o for spherical shells (Fig. 2b). With the observed geometric similarity of the corresponding graphs in Fig. 2a and b spread of σ_e/σ_o values for all shells is less than the range of σ_θ/σ_m values. With the margin of error of no more than 10%, it can be confirmed that it is acceptable to calculate the intensity of the stresses in the babbitt shells with $m = 0.60$ assuming that its contour is spherical and the equality $\sigma_\theta = \sigma_m$.

However, for babbitt shells with $m = 0.25$ and $AlMg5$ alloy, the deviations σ_θ/σ_m and σ_e/σ_o are more significant: $\sigma_\theta/\sigma_m = 0.95...1.41$ and $0.84...1.22$, $\sigma_e/\sigma_o = 0.97...$ 1.25 and $0.93...1.13$ respectively. In the shells from BVT, the minima of σ_e/σ_o and σ_θ/σ_m are reached in some places of the contour and make 0.88 and 0.66.

In some works [1, 22, 32–34], when determining the power parameters of SPF, the contour sphericity and uniformity of the shells thinning were simultaneously assumed. According to our calculations, such assumptions lead to the fact that even in babbitt shells with a high level of superplastic properties ($m = 0.60$), the deviation of σ_e/σ_o values from a unity reaches $+0.15...-0.26$, i.e. margin of error reaches 41% (Fig. 2c). For shells from blanks of variable thickness, the range of deviations is even greater: $+0.14...-0.43$ (margin of error – 57%). In the pole sections of domes made of $AlMg5$ alloy and babbitt with $m = 0.25$, values $\sigma_e/\sigma_o > 2$, and in the areas of blank pressing, $\sigma_e/\sigma_o = 0.76; 0.87$, i.e. margin of error of calculation reaches 130%.

Hence the graphs in Fig. 2c show unacceptability for the calculations of the assumption about the uniformity of blanks thinning at SPF.

5 Conclusions

The principal stresses, in particular the tangential stress, depend on the principal radii of curvature. The intensity of stresses also depends on the thinning of the shells at superplastic forming. The higher the level of superplastic properties of the blank material, the less the main stresses and their intensity depend on the difference between the principal radii of curvature. When calculating the power mode of SPF shells, the assumption of uniform thinning of the blank during forming is unacceptable, since margin of error of the calculations can reach 130%.

Acknowledgments. The reported study was funded by RFBR according to the research project №19-08-01252a "Development and verification of inelastic deformation models and thermal fatigue fracture criteria for monocrystalline alloys". The authors declare that there is no conflict of interest regarding the publication of this paper.

References

1. Giuliano, G.: Superplastic Forming of Advanced Metallic Materials: Method and Applications, 377 p. Woodhead Publishing Ltd., Oxford, Cambridge, Philadelphia, New Delhi (2011)
2. Zakhariev, I.Y., Aksenov, S.A.: Influence of a material rheological characteristics on the dome thickness during free bulging test. J. Chem. Technol. Metall. 52(5), 1002–1007 (2017)
3. Barnes, A.J.: Superplastic forming 40 years and still growing. J. Mater. Eng. Perform. 22 (10), 2935–2949 (2013). https://doi.org/10.1007/s11665-013-0727-4
4. Puzyr, R., Haikova, T., Majerník, J., Karkova, M., Kmec, J.: Experimental study of the process of radial rotation profiling of wheel rims resulting in formation and technological flattening of the corrugations. Manuf. Technol. 18(1), 106–111 (2018). https://doi.org/10.21062/ujep/61.2018/a/1213-2489/MT/18/1/106

5. Puzyr, R., Kukhar, V., Maslov, A., Shchipkovsky, Y.: The development of the method for the calculation of the shaping force in the production of vehicle wheel rims. Int. J. Eng. Technol. (UAE) **7**(4.3), 30–34 (2018). https://doi.org/10.14419/ijet.v7i4.3.20128

6. Anishchenko, A.S., Andryushchenko, A.P.: Rotary flaring of faceted flairs on pipe blanks. Sov. Eng. Res. **5**, 54–55 (1991)

7. Orlov, G.A., Kotov, V.V., Orlov, A.G.: Simulation of the behavior of pipes with variable wall thickness under internal pressure. Metallurgist **61**(1–2), 106–110 (2017). https://doi.org/10.1007/s11015-017-0461-5

8. Pereira, D.A., Batalha, M.H.F., Carunchio, A.F., Resende, H.B.: An analysis of superplastic forming to manufacture aluminum and titanium alloy components. Revista IPT: Tecnologia e Inovação **1**(3), 63–73 (2016)

9. Smyrnov, Y., Belevitin, V., Skliar, V., Orlov, G.: Physical and computer modeling of a new soft reduction process of continuously cast blooms. J. Chem. Technol. Metall. **50**(6), 589–594 (2015)

10. Kitaeva, D., Kodzhaspirov, G., Rudaev, Y.: On the dynamic superplasticity. Mater. Sci. Forum **879**, 960–965 (2017). https://doi.org/10.4028/www.scientific.net/MSF.879.960

11. Kitaeva, D.A., Rudaev, Y.I.: On the threshold stress in superplasticity. Tech. Phys. **59**(11), 1616–1619 (2014). https://doi.org/10.1134/S1063784214110127

12. Moroz, M., Korol, S., Chernenko, S., Boiko, Y., Vasylkovskyi, O.: Driven camshaft power mechanism of the vehicle diesel engine fuel pump. Int. J. Eng. Technol. (UAE) **7**(4.3), 135–139 (2018). https://doi.org/10.14419/ijet.v7i4.3.19723

13. Markov, O.E., Perig, A.V., Markova, M.A., Zlygoriev, V.N.: Development of a new process for forging plates using intensive plastic deformation. Int. J. Adv. Manuf. Technol. **83**(9–12), 2159–2174 (2016). https://doi.org/10.1007/s00170-015-8217-5

14. Kukhar, V., Artiukh, V., Prysiazhnyi, A., Pustovgar A.: Experimental research and method for calculation of "upsetting-with-buckling" load at the impression-free (Dieless) preforming of workpiece. E3S Web Conf. **33**, 02031 (2018). https://doi.org/10.1051/e3sconf/20183302031

15. Aksenov, S., Sorgente, D.: Characterization of stress-strain behavior of superplastic titanium alloy by free bulging tests with pressure jumps. Defect Diffus. Forum **385**, 443–448 (2018). https://doi.org/10.4028/www.scientific.net/DDF.385.443

16. Ganesh, P., Senthil Kumar, V.S.: Finite element simulation in superplastic forming of friction stir welded aluminium alloy 6061-T6. Int. J. Integr. Eng. **3**(1), 9–16 (2011)

17. Efremenko, V.G., Shimizu, K., Pastukhova, T.V., Chabak, Yu.G., Kusumoto, K., Efremenko, A.V.: Effect of bulk heat treatment and plasma surface hardening on the microstructure and erosion wear resistance of complex-alloyed cast irons with spheroidal vanadium carbides. J. Frict. Wear **38**(1), 58–64 (2017). https://doi.org/10.3103/S1068366617010056

18. Anishchenko, A., Kukhar, V., Artiukh, V., Arkhipova, O.: Application of G. Lame's and J. Gielis' formulas for description of shells superplastic forming. MATEC Web Conf. **238**, 06007 (2018). https://doi.org/10.1051/matecconf/201823906007

19. Anishchenko, O.S., Kukhar, V.V., Grushko, A.V., Vishtak, I.V., Prysiazhnyi, A.H., Balalayeva, E.Yu.: Analysis of the sheet shell's curvature with lame's superellipse method during superplastic forming. Mater. Sci. Forum **945**, 531–537 (2019). https://doi.org/10.4028/www.scientific.net/MSF.945.531

20. Anishchenko, A., Kukhar, V., Artiukh, V., Arkhipova, O.: Superplastic forming of shells from sheet blanks with thermally unstable coatings. MATEC Web Conf. **238**, 06006 (2018). https://doi.org/10.1051/matecconf/201823906006

21. Shats'kyi, I.P.: Closure of a longitudinal crack in a shallow cylindrical shell in bending. Mater. Sci. **41**(2), 186–191 (2005). https://doi.org/10.1007/s11003-005-0149-z

22. Jovane, F.: An approximate analysis of the superplastic forming of a thin circular diaphragm. Int. J. Mech. Sci. **10**(5), 405–427 (1968). https://doi.org/10.1016/0020-7403(68)90005-2
23. Kim, Y.H., Lee, J.M., Hong, S.S.: Optimal design of superplastic forming processes. J. Mater. Process. Technol. **112**(2–3), 167–173 (2001). https://doi.org/10.1016/s0924-0136 (00)00880-3
24. Abhijit, Dutta: Thickness-profiling of initial blank for superplastic forming of uniformly thick domes. Mater. Sci. Eng. A. **371**(1–2), 79–81 (2004). https://doi.org/10.1016/S0921-5093(03)00632-4
25. Lechten, J.-P., Patrat, J.-P., Baudelet, B.: Analyses theorique et experimentale du gonflement dans le domaine de superplasticite. Revue de Physique Appliquee **12**(1), 7–14 (1977). https://doi.org/10.1051/rphysap:019770012010700
26. Vitu, L., Boudeau, N., Malecot, P., Michel, G., Buteri, A.: Comparaison de trois modeles pour le post-traitment de mesures issues du test de gonflement libre de tubes. 22-ème Congrès Français de Mécanique, Lyon, 24 au 28 Août 2015, pp. 67–78 (2015). (in French)
27. Anishchenko, A.S., Feofanov, YuV, Bogun, A.B.: Hot expansion of precise ring forgings. Chem. Pet. Eng. **11**, 33–35 (1992)
28. Kukhar, V., Balalayeva, E., Nesterov, O.: Calculation method and simulation of work of the ring elastic compensator for sheet-forming. MATEC Web Conf. **129**, 01041 (2017). https://doi.org/10.1051/matecconf/201712901041
29. Balalayeva, E., Artiukh, V., Kukhar, V., Tuzenko, O., Glazko, V., Prysiazhnyi, A., Kankhva, V.: Researching of the stress-strain state of the open-type press frame using of elastic compensator of errors of "Press-Die" system. In: Advances in Intelligent Systems and Computing, vol. 692, pp. 220–235 (2018). https://doi.org/10.1007/978-3-319-70987-1_24
30. Sadowsky, A.J., Rotter, A.M.: Exploration of novel geometric imperfection forms in buckling failures of thin-walled metal silos under eccentric discharge. Int. J. Solids Struct. **50** (5), 781–794 (2012). https://doi.org/10.1016/j.ijsolstr.2012.11.017
31. Deshmukh, P.V.: Study of superplastic forming process using finite element analysis. University of Kentucky Master's theses. Paper 367, 82 p. (2003). http://uknowledge.uky.edu/gradschool_theses/367
32. Grebenisan, G., Bogdan, S.: Parameterized finite element analysis of a superplastic forming process, using Ansys®. MATEC Web Conf. **126**, 03001 (2017). https://doi.org/10.1051/matecconf/201712603001
33. Radyuk, A.G., Gorbatyuk, S.M., Gerasimova, A.A.: Use of electric-arc metallization to recondition the working surfaces of the narrow walls of thick-walled slab molds. Metallurgist **55**(5–6), 419–423 (2011). https://doi.org/10.1007/s11015-011-9446-y
34. Levandovskiy, A.N., Melnikov, B.E., Shamkin, A.A.: Modeling of porous material fracture. Mag. Civ. Eng. **1**, 3–22 (2017). https://doi.org/10.18720/MCE.69.1

Investigation of Degree of Internal Defects Closure in Ingots at Forging

Sergey Kargin[1,4(✉)] ⓘ, Viktor Artiukh[2,4] ⓘ, Vladlen Mazur[3,4] ⓘ,
Dmitriy Silka[3,4] ⓘ, and Natalia Meller[4,5] ⓘ

[1] Pryazovskyi State Technical University, Mariupol 87500, Ukraine
vbudarl1973@gmail.com
[2] Peter the Great St. Petersburg Polytechnic University, Polytechnicheskaya 29,
195251 St. Petersburg, Russia
[3] LLC «Saint-Petersburg Electrotechnical Company»,
196603 St. Petersburg, Russia
[4] Moscow State University of Civil Engineering, Yaroslavskoye shosse 26,
129337 Moscow, Russia
[5] Tyumen Industrial University, Volodarskogo Street 38,
625000 Tyumen, Russia

Abstract. Study results of mechanism of internal (axial) defect closure in ingot at forging are shown depending on: form of head (peen), form of ingot, ingot feed and degree of reduction. Intensity of closing of axial defect was chosen as a numerical characteristic of the investigation objective during the factor experiment. Dependence is obtained that describes connection between the degree of axial defect closure inside ingot and the main technological parameters of forging. It is shown that the biggest influence on the degree of axial defects closure in ingots at forging has the tool shape and the degree of reduction.

Keywords: Ingot · Internal (axial) defect · Reduction · Dependence · Forging · Variation interval · Efficiency

1 Introduction

At forging, it is required to ensure high quality of metal, eliminate various defects and provide raise of metal mechanical properties as well as to produce forgings with necessary shape and dimensions.

Elimination of internal defects is the main task for getting high quality forgings. Process of defects elimination happens in two stages: closure of defect prior to the moment when its two opposite walls are closed down and then welding of the closed down walls of the defect.

The studies [1–6] show that internal defects closure in metal at plastic deformation depends on different factors, mainly on: shape of ingot, degree of deformation, value of relative feed and shape of tool.

However, mechanism of defects closure at forging operations has not been fully investigated. For prediction of defects closure, it is necessary to possess data regarding

© Springer Nature Switzerland AG 2020
V. Murgul and M. Pasetti (Eds.): EMMFT 2018, AISC 982, pp. 818–824, 2020.
https://doi.org/10.1007/978-3-030-19756-8_78

alternation of defects dimensions at forging. Various process factors have different influence on type of deformation along section of ingot and also on internal defects closure [7–14].

Hence finding out influence of such factors on internal defects closure in ingots appears to be completely reasonable.

2 Purpose of the Work

Objective of this paper is to find out dependence of the degree of ingot internal defects closure on the basic technological parameters of forging.

For quantitative evaluation of relations between factors influencing type and value of deformations distribution and also internal defects closure, a statistic model was developed for process of forging of a cylindrical ingot into two, three and four-radial forgings with application of methods of experiment mathematical modeling [15].

3 Research Material

Intensity of closing of axial hole (defect) was chosen as a numerical characteristic of the investigation objective during the factor experiment.

This index satisfies all requirements set for optimization parameter because it is a numerical, single-valued and statistically determined value which measures efficiency of object of investigation quite satisfactorily. We determined the dependence of the value of intensity of axial hole (defect) closure ε_{hole} on four factors: relation of head (peen) radius and ingot radius R_{head}/R_{ingot} (x_1 factor), degree of reduction of ingot by convex head ε (x_2 factor), feed of ingot ψ (x_3 factor) and number of symmetrical radial beams along generatrix of ingot in its cross-section m (x_4 factor).

Values of the factors were varied at two levels. The levels and the intervals of factors variation are shown in Table 1.

Table 1. Levels and intervals of factors variation for establishing dependence of intensity of axial hole closing on technological factors.

Factors	R_{head}/R_{ingot}	ε	ψ	m
Designation of factors x_i	x_1	x_2	x_3	x_4
Bottom level, x_{ib}	0.6	0.07	0.5	2
Top level, x_{it}	1	0.21	1.5	4
Zero level, x_{iz}	0.8	0.14	1	3
Variation interval, Δx_i	0.2	0.07	0.5	1

Coded values of z_i factors are connected with natural values of x_i by relation:

$$z_i = (x_i - x_{iz})/\Delta z_i. \tag{1}$$

Coded values of factors at the top z_{it} and the bottom z_{ib} levels:

$$z_{it} = (x_{it} - x_{iz})/\Delta x_i = +1, z_{ib} = (x_{ib} - x_{iz})/\Delta x_i = -1. \tag{2}$$

Taking into account that variation intervals are quite wide and the specified factors levels are almost extreme, problem of compilation of a statistic model is reduced to a search of an interpolation formula of dependence of axial hole closure on technological factors.

Mathematical model of the process:

$$y = f(x_1, x_2, x_3, x_4), \tag{3}$$

is written below in a form of a linear equation of regression with due regard to all interactions:

$$
\begin{aligned}
y = a_0 + a_1x_1 + a_2x_2 + a_3x_3 + a_4x_4 + a_{12}x_1x_2 + a_{13}x_1x_3 + a_{14}x_1x_4 \\
+ a_{23}x_2x_3 + a_{24}x_2x_4 + a_{34}x_3x_4 + a_{123}x_1x_2x_3 + a_{124}x_1x_2x_4 + a_{134}x_1x_3x_4 \\
+ a_{234}x_2x_3x_4 + a_{1234}x_1x_2x_3x_4,
\end{aligned} \tag{4}
$$

where a_i are coefficients of the equation in natural scale determined as:

$$a_j = \sum_{i=1}^{N} y_i x_{ji} \Big/ N, j = 0, 1 \ldots k, \tag{5}$$

where $N = 2^k$ is number of tests in the experiment (k is number of factors).

In coded scale, the regression equation has the following view:

$$
\begin{aligned}
y = a_0' + a_1'z_1 + a_2'z_2 + a_3'z_3 + a_4'z_4 + a_{12}'z_1z_2 + a_{13}'z_1z_3 \\
+ a_{14}'z_1z_4 + a_{23}'z_2z_3 + a_{24}'z_2z_4 + a_{34}'z_3z_4 + a_{123}'z_1z_2z_3 \\
+ a_{124}'z_1z_2z_4 + a_{134}'z_1z_3z_4 + a_{234}'z_2z_3z_4 + a_{1234}'z_1z_2z_3z_4,
\end{aligned} \tag{6}
$$

where a_i' are equation coefficients in the coded scale determined as:

$$a_j' = \sum_{i=1}^{N} y_i z_{ji} \Big/ N, j = 0, 1 \ldots k. \tag{7}$$

Plan of factorial experiment is represented in the coded scale by a planning matrix 2^4 (refer to Table 2). Each experiment was accompanied with five measurements of intensity of axial hole closure (values $y_1 \ldots y_5$) while obtained average results (y) were used for determination of the regression dependence.

Table 2. Planning matrix of the two-level factorial experiment (24) for determination of dependence of axial hole closure intensity on technological factors.

No.	Factors																y
	z_0	z_1	z_2	z_3	z_4	z_{12}	z_{13}	z_{14}	z_{23}	z_{24}	z_{34}	z_{123}	z_{124}	z_{134}	z_{234}	z_{1234}	
1	+1	−1	−1	−1	−1	+1	+1	+1	+1	+1	+1	−1	−1	+1	−1	+1	0.0105
2	+1	−1	−1	−1	+1	+1	+1	−1	+1	−1	−1	−1	+1	+1	+1	−1	0.0275
3	+1	−1	−1	+1	−1	+1	−1	+1	−1	+1	−1	+1	−1	+1	+1	−1	0.0110
4	+1	−1	−1	+1	+1	+1	−1	−1	−1	−1	+1	+1	+1	−1	−1	+1	0.0330
5	+1	−1	+1	−1	−1	−1	+1	+1	−1	−1	+1	+1	+1	−1	+1	−1	0.1333
6	+1	−1	+1	−1	+1	−1	+1	−1	−1	+1	−1	+1	−1	+1	−1	+1	0.1995
7	+1	−1	+1	+1	−1	−1	−1	+1	+1	−1	−1	−1	+1	+1	−1	+1	0.2500
8	+1	−1	+1	+1	+1	−1	−1	−1	+1	+1	+1	−1	−1	−1	+1	−1	0.4011
9	+1	+1	−1	−1	−1	−1	−1	−1	+1	+1	+1	+1	+1	+1	−1	−1	0.0225
10	+1	+1	−1	−1	+1	−1	−1	+1	+1	−1	−1	+1	−1	−1	+1	+1	0.0600
11	+1	+1	−1	+1	−1	−1	+1	−1	−1	+1	−1	−1	+1	−1	+1	+1	0.023
12	+1	+1	−1	+1	+1	−1	+1	+1	−1	−1	+1	−1	−1	+1	−1	−1	0.0600
13	+1	+1	+1	−1	−1	+1	−1	−1	−1	−1	+1	−1	−1	+1	+1	+1	0.2500
14	+1	+1	+1	−1	+1	+1	−1	+1	−1	+1	−1	−1	+1	−1	−1	−1	0.3750

Transition from the coded values to natural ones was performed as:

$$x_1 = (R_{head}/R_{ingot} - 0.8)/0.2; x_2 = (\varepsilon - 0.14)/0.07;$$
$$x_3 = (\psi - 1)/0.5; x_4 = (m - 3)/1 \tag{8}$$

For automation of the value determination process of regression, dependence of axial hole closure intensity ε_{hole} on technological factors a software was developed (refer to Fig. 1) by means of integrated development environment Borland Delphi 7.

As a result, the dependence of intensity of axial hole closure ε_{hole} on: the relation R_{head}/R_{ingot}; the degree of ingot reduction by convex heads ε; the feed of ingot ψ; number of symmetrical radial beams along generatrix of ingot in its cross-section m:

$$\varepsilon_{hole} = 0.0251 - 0.095 \cdot R_{head}/R_{ingot} - 0.180 \cdot \varepsilon - 0.012 \cdot \psi - 0.013 \cdot m$$
$$+ 0.960 \cdot R_{head}/R_{ingot} \cdot \varepsilon - 0.018 \cdot R_{head}/R_{ingot} \cdot \psi + 0.029 \cdot R_{head}/R_{ingot} \cdot m$$
$$- 0.012 \cdot \varepsilon \cdot \psi + 0.034 \cdot \varepsilon \cdot m + 0.015 \cdot \psi \cdot m + 0.452 \cdot R_{head}/R_{ingot} \cdot \varepsilon \cdot \psi \tag{9}$$
$$- 0.001 \cdot R_{head}/R_{ingot} \cdot \varepsilon \cdot m - 0.055 \cdot R_{head}/R_{ingot} \cdot \psi \cdot m - 0.126 \cdot \varepsilon \cdot \psi \cdot m$$
$$+ 0.686 \cdot R_{head}/R_{ingot} \cdot \varepsilon \cdot \psi \cdot m.$$

To simplify the mathematical model, it has to be written in a form of a linear equation of regression taking into account interactions only of the first order:

$$y = a_0 + a_1 x_1 + a_{12} x_2 + a_3 x_3 + a_4 x_4 + a_{12} x_1 x_2 + a_{13} x_1 x_3$$
$$+ a_{14} x_1 x_4 + a_{23} x_2 x_3 + a_{24} x_2 x_4 + a_{34} x_3 x_4. \tag{10}$$

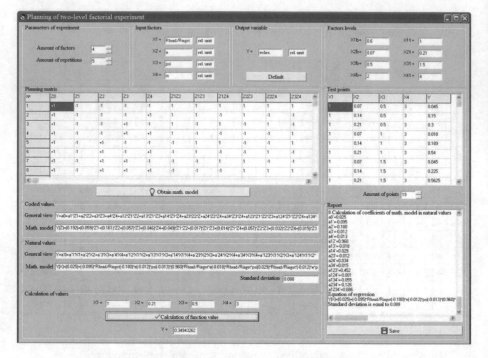

Fig. 1. Software interface of the two-level factorial experiment.

Then, the dependence of intensity of axial hole closure $\varepsilon_{hole} = f(R_{head}/R_{ingot}, \varepsilon, \psi, m)$ on the main factors of forging will be represented in the following view:

$$\varepsilon_{hole} = 0.582 - 0.570 \cdot R_{head}/R_{ingot} - 3.452 \cdot \varepsilon - 0.339 \cdot \psi - 0.104 \cdot m$$
$$+ 3.466 \cdot R_{head}/R_{ingot} \cdot \varepsilon - 0.169 \cdot R_{head}/R_{ingot} \cdot \psi + 0.070 \cdot R_{head}/R_{ingot} \cdot m \quad (11)$$
$$+ 1.617 \cdot \varepsilon \cdot \psi + 0.455 \cdot \varepsilon \cdot m + 0.031 \cdot \psi \cdot m.$$

For obtained regression Eqs. (8) and (10) standard deviation was determined which value should not exceed 0.15…0.2:

$$\sigma = \sqrt{\sum_{i=1}^{N} (\bar{y}_i - \tilde{y}_i)^2 \Big/ N_0}, \quad (12)$$

where $\bar{y}_i$ are specified values; $\tilde{y}_i$ are values obtained by means of regression Eqs. (9) and (11); N_0 is number of values.

In the process of modeling by means of the developed software, value of standard deviation was found to be equal to 0.088 (refer to Fig. 1) that proves adequacy of the obtained functional dependences (9) and (11). Thus, in order to simplify the calculations, it is possible to apply the obtained regression dependence (11) instead of the formula (9).

Analysis of the equation and obtained data shows that intensity of the axial hole closure depends on all four factors of the process [16–20]. The biggest influence being exerted by the shape of the profiled sides and the reduction degree.

4 Conclusions

1. The dependence was obtained that describes connection between the degree of axial defect closure inside ingot and the main technological parameters of forging.
2. It was determined that the biggest influence on the degree of axial defects closure in ingots at forging has the tool shape and the degree of reduction.

Acknowledgments. The reported study was funded by RFBR according to the research project №19-08-01252a 'Development and verification of inelastic deformation models and thermal fatigue fracture criteria for monocrystalline alloys'. The authors declare that there is no conflict of interest regarding the publication of this paper.

References

1. Sokolov, L.M., et al.: Technology of Forging: Textbook for Higher Technical Educational Establishments. Donbass State Engineering Academy (DSEA), Kramatorsk (2011). (in Ukrainian)
2. Markov, O.E., et al.: Implementation of energy-efficient technology of forging process of big ingots without shrinkage. Plast. Work. Metals **10**, 33–36 (2011). (in Russian)
3. Markov, O.E.: Efficient diagram of big ingots forging with application of forging drawing. Plast. Work. Metals **8**, 44–48 (2012). (in Russian)
4. Tyurin, V.A.: Innovation forging technologies with application of macro-shifts. Plast. Work. Metals **11**, 11–20 (2007). (in Russian)
5. Sokolov, L.M., et al.: Technological techniques ensuring working out of ingot axial zone at forging. Forg. Stamp. Prod. **2**, 25–27 (1985). (in Russian)
6. Kotelkin, A.V., et al.: Patent 1263413 of USSR, B 21 J 5/00. Method of modelling of ingot internal defects closing, Bul. 38, 15 October 1996. (in Russian)
7. Kukhar, V., Artiukh, V., Serduik, O., Balalayeva, E.: Form of gradient curve of temperature distribution of lengthwise the billet at differentiated heating before profiling by buckling. Procedia Eng. **165**, 1693–1704 (2016). https://doi.org/10.1016/j.proeng.2016.11.911
8. Kukhar, V., Artiukh, V., Butyrin, A., Prysiazhnyi, A.: Stress-strain state and plasticity reserve depletion on the lateral surface of workpiece at various contact conditions during upsetting. In: Murgul, V., Popovic, Z. (eds.) International Scientific Conference Energy Management of Municipal Transportation Facilities and Transport, EMMFT 2017. Advances in Intelligent Systems and Computing, vol. 692, pp. 201–211. Springer, Cham (2018). https://doi.org/10.1007/978-3-319-70987-1_22
9. Balalayeva, E., Artiukh, V., Kukhar, V., Tuzenko, O., Glazko, V., Prysiazhnyi, A., Kankhva, V.: Researching of the stress-strain state of the open-type press frame using of elastic compensator of errors of "Press-Die" system. In: Murgul, V., Popovic, Z. (eds.) International Scientific Conference Energy Management of Municipal Transportation Facilities and Transport, EMMFT 2017. Advances in Intelligent Systems and Computing, vol. 692, pp. 220–235. Springer, Cham (2018). https://doi.org/10.1007/978-3-319-70987-1_24

10. Kargin, S., Artyukh, V., Ignatovich, I., Dikareva, V.: Development and efficiency assessment of process lubrication for hot forging. In: Conference Series: Earth and Environmental Science, vol. 90, p. 012190 (2017). https://doi.org/10.1088/1755-1315/90/1/012190

11. Kukhar, V., Artiukh, V., Prysiazhnyi, A., Pustovgar, A.: Experimental research and method for calculation of 'upsetting-with-buckling' load at the impression-free (Dieless) preforming of workpiece. E3S Web Conf. **33**, 02031 (2018). https://doi.org/10.1051/e3sconf/20183302031

12. Artiukh, V., Kukhar, V., Balalayeva, E.: Refinement issue of displaced volume at upsetting of cylindrical workpiece by radial dies. MATEC Web Conf. **224**, 01036 (2018). https://doi.org/10.1051/matecconf/201822401036

13. Anishchenko, A., Kukhar, V., Artiukh, V., Arkhipova, O.: Superplastic forming of shells from sheet blanks with thermally unstable coatings. MATEC Web Conf. **239**, 06006 (2018). https://doi.org/10.1051/matecconf/201823906006

14. Anishchenko, A., Kukhar, V., Artiukh, V., Arkhipova, O.: Application of G. Lame's and J. Gielis' formulas for description of shells superplastic forming. MATEC Web Conf. **239**, 06007 (2018). https://doi.org/10.1051/matecconf/201823906007

15. Lunev, A.V.: Mathematical Modelling and Planning of an Experiment. Technical University Publishing House, St. Petersburg (2006). (in Russian)

16. Levandovskiy, A.N., Melnikov, B.E., Shamkin, A.A.: Modeling of porous material fracture. Mag. Civ. Eng. **1**, 3–22 (2017). https://doi.org/10.18720/MCE.69.1

17. Solomonov, K.N.: Application of CAD/CAM systems for computer simulation of metal forming processes. Mater. Sci. Forum **704–705**, 434–439 (2012)

18. Solomonov, K.: Development of software for simulation of forming forgings. Procedia Eng. **81**, 437–443 (2014)

19. Solomonov, K., Tishchuk, L., Fedorinin, N.: Simulation of forming a flat forging. J. Phys: Conf. Ser. **918**, 012038 (2017)

20. Gorbatyuk, S.M., Osadchii, V.A., Tuktarov, E.Z.: Calculation of the geometric parameters of rotary rolling by using the automated design system Autodesk Inventor. Metallurgist **55**(7–8), 543–546 (2011). https://doi.org/10.1007/s11015-011-9465-8

Experimental Evaluation of Bearing Capacity of Multifaceted Pipes Filled with Self-compacting Concrete Under Axial Compression

Igor Garanzha[1]([⊠]) [iD], Liliya Shchykina[2] [iD], and Farida Suyunova[1] [iD]

[1] Moscow State University of Civil Engineering,
Yaroslavskoe sh. 26, 129337 Moscow, Russia
garigo@mail.ru
[2] Donbass National Academy of Civil Engineering and Architecture,
Derzhavina Street 2, Makeevka 86123, Ukraine

Abstract. The paper describes the principle of mechanical tests aimed at determining the physical and mechanical properties of self-compacting concrete B20 and B30 (according to Spanish methodologies). The bearing capacity and the nature of the destruction of 6- and 12-sided composite (concrete-filled tubes) elements under axial compression has been experimentally studied. The order of destruction of materials and specific types of destruction of multifaceted composite elements, such as the development of plastic deformations of steel pipe and the severance of longitudinal welds, were established. The results of determining the bearing capacity of concrete-filled poles, indicating to its overestimation by the analytical and the finite element calculation to 30%, are compared.

Keywords: Multifaceted composite poles · Self-compacting concrete ·
Mechanical tests · Axial compression ·
Physico-mechanical characteristics of concrete ·
Nature of destruction · Bearing capacity

1 Introduction

The creation of reliable, durable and economical structures, as well as structures that meet the requirements of aesthetics and have a low land acquisition, is the main goal of the design of buildings and structures. Pipe concrete structures are not an exception in this direction, which will be used as load-bearing elements of urban infrastructure in the nearest future, such as supports for distribution electric networks (up to 110 kV), mobile communication towers, lighting poles, racks for the contact network of city electric transport, etc. [1–9]

One of the main prerequisites for the qualitative design of building structures, and composite in particular, is the degree of study of their actual work under load, features of the stress-strain state, the level of load-bearing capacity, the nature of the destruction. Nowadays both domestic and foreign scientists have conducted a fairly wide

© Springer Nature Switzerland AG 2020
V. Murgul and M. Pasetti (Eds.): EMMFT 2018, AISC 982, pp. 825–835, 2020.
https://doi.org/10.1007/978-3-030-19756-8_79

range of experimental researches in which they explored both life-size concrete structures and laboratory models with a geometric scaling relative to the full-scale structures [10–22].

In this work, experimental researches were carried out in two stages, aimed at determining the basic physical and mechanical characteristics, which were proposed by compositions of self-compacting concrete (SUB), followed by the study of the bearing capacity of composite structures based on steel multi-faceted pipes filled with SUB axial compression.

2 Experimental Determination of Physical and Mechanical Characteristics of Self-compacting Concrete

At the first stage, mechanical tests of cylindrical samples of self-compacting concrete (B20 and B30) under Central compression were carried out in order to experimentally determine the real values of their physical and mechanical characteristics for the proposed self – compacting concrete grade B20 and B30-prismatic strength of concrete R_b (MPa) and its modulus of elasticity E (MPa). The experimental formulation of the SUB was adopted for the research and it used components extracted in the province of Alicante (Spain), such as sand, cement, water and macadam, the properties of which are different from the regional characteristics of domestic counterparts.

Subsequently, the obtained parameters can be used in the calculation of structures using self-compacting concrete (particularly concrete pipes) both by analytical methods and by modeling in finite element analysis systems (ANSYS, ABAQUS, NASTRAN, Lira, etc.).

Mechanical tests of concrete samples are performed in accordance with the normative documents:

– for the determination of compressive strength → UNE-EN 12390-3:2001 (1) [23];
– for the determination of modulus of elasticity → UNE 83316:1996 (2) [24].

Cylindrical samples were taken as test samples made of self-compacting concrete (component dosages are given in Table 1) with diameter d = 150 mm and height H = 2d = 300 mm. The above-mentioned geometric characteristics of the samples are taken in such a way that the problem $2 \leq H/d \leq 4$ is solved.

Table 1. Experimental formulation of self-compacting concrete of strength classes B20 and B30.

Concrete class	Cement 32,5 (M400), kg	Macadam (fr. 5..10), kg	Sand, kg	Plasticizer viscocrete 5720		Water, liter	$\frac{W}{C}$	Concrete's fluidity, D_c, mm
				kg	%			
1	2	3	4	5	6	7	8	10
B20	200	600	900	3.8	1,9	100	0.5	540
B30	250	600	900	4.4	1,75	110	0.44	560

According to [23], determining the compressive strength of concrete, three samples under pressure are tested at a linearly increasing load at a rate of 0.2…1 MPa/s before their destruction. The final result of R_b is obtained by finding the average strength obtained from three samples.

There are 2 stages in the process of experimental determination of the elastic modulus of concrete according to [24], each of which includes 4 cycles of the same name:

- loading of the sample from the initial voltage $\sigma_1 = 5$ kPa/cm2 to 1/3 of its prismatic (cylindrical) strength σ_{pr};
- the exposure of the sample under load (at $1/3\sigma_{np}$) for 30 s;
- unloading to the stress value in the sample σ_1;
- holding the sample at a voltage σ_1 for 30 s.

The above set of cycles is repeated twice. After repeated holding of the sample at a voltage σ_0, the last loading cycle of the sample is performed, at the end of which with the help of a strain gauge device and sensors, the stress value σ_1 (at the beginning of the cycle) is respectively fixed and σ_2 (at the end of the cycle), as well as the corresponding strains ε_1 and ε_2.

Determining the modulus of elasticity of concrete in accordance with the normative document [24], the scheme of the test (the process of "load-unloading") is presented in the form of a graphical algorithm in Fig. 1. Using the formula (1), the elastic modulus of the concrete modulus E is determined.

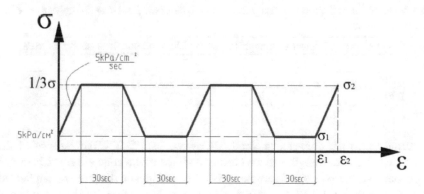

Fig. 1. Graphical algorithm for determining the concrete elasticity modulus.

$$E_c = \frac{\Delta\sigma}{\Delta\varepsilon} = \frac{\sigma_2 - \sigma_1}{\varepsilon_2 - \varepsilon_1} \tag{1}$$

σ_1 and σ_2 are the stresses of the concrete sample taken in accordance with the algorithm shown in Fig. 1;

ε_1 and ε_2 - the deformation obtained by readings from load cells of type LY1x (which are the products of the company "HBO", US) through an indicator of internal deformation of the material "R3" (Fig. 2). Due to the fact that the maximum accepted size of the fraction of a large filler is 10 mm, load cells with a base of 30 mm are used in the work to ensure the correct deposition.

Fig. 2. Experimental determination of the elastic modulus of self-compacting concrete.

The results of the experimental determination of the above-mentioned physical and mechanical characteristics are given in Table 2.

Table 2. Experimental physical and mechanical characteristics of self-compacting concrete under compression

Concrete class	Prismatic strength R_b, MPa	The modulus of elasticity E, MPa
1	2	3
B20	10.5	45411.7
B30	15.9	49238.1

According to the domestic methods described in [25–27], the essence of determining the prismatic strength and the elastic modulus of the concrete consists in testing by gradual loading (steps) of samples-prisms or samples-cylinders of standard sizes with axial compressive load. In the first case, before destruction, it is in determining the prismatic strength and in the second case it is up to the level of 30% of the destructive load N_u with the measurement of the load of the samples of their deformations, in determining the modulus of elasticity.

The prismatic strength and the elastic modulus of the concrete are calculated by the loads N_u and 0,3 N_u, longitudinal and transverse relative elastic-instantaneous deformations ε_1 and ε_2 determined in the course of tests. The experimental determination of the basic physical and the mechanical characteristics of the concrete occurs without unloading samples during testing, which is a fundamental difference between domestic and European methods.

3 Experimental Research of the Bearing Capacity of the Multi-faceted Concrete

At the second stage of work, the experimental researches of the bearing capacity of multi-faceted concrete elements are accomplished using self-compacting concrete in central compression. For research, the samples with height $H = 500$ mm and diameter of the circumscribed circle $D = 200$ mm are made with the number of faces $n = 6$ and 12 of steel grade C235 wall thickness $t = 3$ mm. As a filler, self-compacting concrete was used, the proposed compositions for B20 and B30. The concrete part of the samples was loaded with a linearly increasing load and the structure was brought to destruction. The number of the models −8. As a result, there were determined:

- breaking load N_{lim} (bearing capacity of elements);
- the nature of the destruction of samples (Fig. 3).

$n = 6$

$n = 12$

Fig. 3. The nature of concrete samples destruction with the number of faces n.

The analysis of the results of testing of pipe-concrete multifaceted showed the sequence of destruction of materials in the composite structure, in which the limit state occurs first for the steel pipe. Characteristic types of destruction were the development of plastic deformation and rupture of longitudinal welds.

Experimental results of the bearing capacity of multifaceted concrete elements (with the number of faces $n = 6$ and 12), as well as their comparison with the results obtained by the "universal" analytical method [10], and in the system of finite element analysis ANSYS are shown in Fig. 4 and Table 3, respectively.

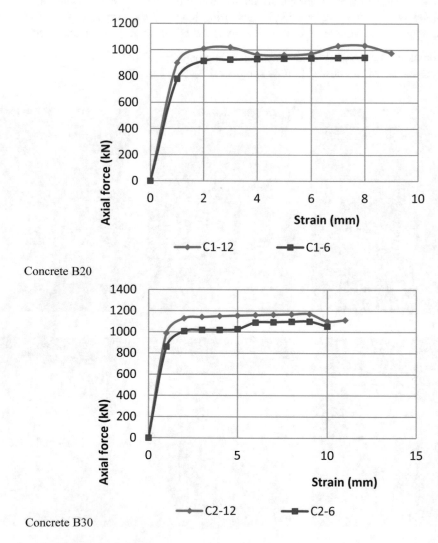

Fig. 4. Graphical representation of the experimental results that determine bearing capacity of multifaceted concrete-filled pipe.

Table 3. Comparison of analytical, experimental and numerical results that determine the bearing capacity of multi-faceted concrete samples

Sample code*	N_{exp}, kN	Type of destruction	N_{analyt}, kN	$\frac{N_{exp}}{N_{analyt}}$	$\frac{N_{exp}}{N_{ANSUS}}$	N (ANSYS), kN
1	2	3	4	5	6	7
C1-6-1	662	Plastics	788.7	0.840	0.703	942
C1-6-2	690	Weld seam	788.7	0.875	0.733	
C1-12-1	840	Plastics	910.9	0.922	0.813	1033.7
C1-12-2	858	Plastics	910.9	0.945	0.830	
C2-6-1	790	Plastics	844.4	0.936	0.718	1100.2
C2-6-2	804	Weld seam	844.4	0.952	0.731	
C2-12-1	836	Plastics	975.7	0.857	0.714	1170.6
C2-12-2	884	Weld seam	975.7	0.806	0.755	

* as the encoding of the form XX-Y-Z for the test samples, the following designations are used:

- XX → C1, C2 – concrete classes B20 and B30 respectively;
- Y = 6, 12 - the number of steel pipe faces;
- Z = 1, 2-sample sequence number.

When creating models of the objects under study in ANSYS, the following features of finite element modeling of this computational complex are used [13, 28–43]:

- for the possibility of taking into account the nonlinear properties of the materials that make up the composite structure, special volumetric finite elements Solid 65 for concrete and Solid 85 for steel are used;
- FE grid of the computational model is created as "ordered" based on the elements of hexagonal prismatic shape;
- the contact between steel and concrete under working together is considered (when concrete pipe design accommodates the volumetric stress state) through the use of special finite elements CONTA174 (contact element) and TARGET 170 (the element of penetration), as well as the introduction of the coefficient of friction μ between the materials.

The ANSYS finite elements used in the simulation are shown in Fig. 7. To simulate the physically nonlinear nature of the steel work, the law of its isotropic hardening was used, implemented in the form of bilinear and multilinear approximation of the steel deformation diagram (Ramberg-Osgood's curve), shown in (Figs. 5 and 6).

$$\phi = sin^{-1}\left(\frac{3}{1+2\frac{f_{cc}}{\sqrt{3}}}\right) \tag{2}$$

$$c = f_{cc} - 5\sqrt{3} \cdot \left(\frac{3 - sin\,\phi}{6\,cos\,\phi}\right) \tag{3}$$

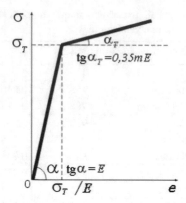

Fig. 5. Bilinear method of curves approximation of steel deformation

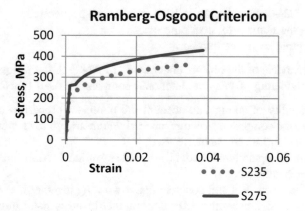

Fig. 6. Multilinear method of curves approximation of steel deformation

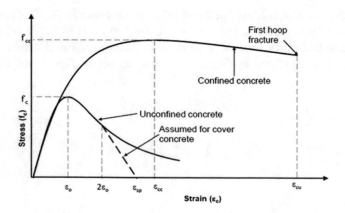

Fig. 7. Strain Mander's model for concrete

- f_{cc} – design resistance to compression of concrete in the pipe;
- f_c – design resistance to compression of concrete without compression by a metal shell;
- ε_0 – deformation corresponding to f_c;
- ε_{cc} – deformation corresponding to f_{cc} (Fig. 8).

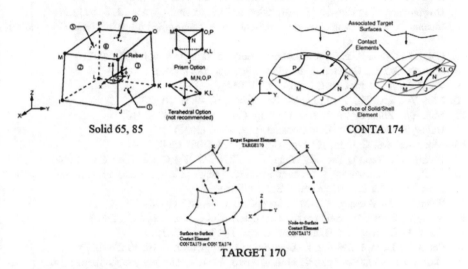

Solid 65, 85   **CONTA 174**

TARGET 170

Fig. 8. Finite elements for the simulation of tube-confined concrete in ANSYS.

4 Conclusions

1. According to the Spanish norms, the prismatic strength of R_b of self-compacting concrete of classes B20 and B30 at axial compression and its modulus of elasticity E by mechanical tests of samples are experimentally determined. Values for R_b below the same normative value by $\approx 5...7\%$, the modulus E was above the average value (30000Mpa for heavy concrete) approximately $50...60\%$.
2. According to the results of experimental researches of multi-faceted concrete samples the breaking load is determined, which should be taken as the bearing capacity of structures in central compression.
3. The nature of the destruction of structures under compression is determined, which confirms the prerequisites of numerical studies on the type and sequence of destruction of the components of multi-faceted concrete.
4. The analysis of the results showed that in comparison with the experimental results, the data, obtained by the analytical "universal" method, overestimate the bearing capacity of multifaceted concrete elements $\approx$ by $6...17\%$, which indicates a sufficient level of convergence, and the calculation data in ANSYS – by about $17...30\%$, which suggests the need for additional adjustment of the finite element computational model.

References

1. Gorokhov, E., Vasylev, V., Garanzha, I., Leshchenko, A.: Met. Const. J. **19**(2), 80 (2013)
2. Garanzha, I.: Air Lines J. **6**(3), 20 (2013)
3. Garanzha, I.: Mod. Steel Timb. Const J. **2**, 50 (2013)
4. Garanzha, I., Vatin, N.: Appl. Mech. Mater. J. **633–634**, 971 (2014)
5. Duvanova, I., Salmanov, I.: Construction of Unique Build. Const. J. **6**, 103 (2014)
6. Shaohuay, T.: A new application experience of concrete-filled tubes in China. Conc. Reinf. Conc. J. **3**, 24 (2001)
7. Cai, S.: Modern Steel Tube Confined Concrete Structures, vol. 358 (2003)
8. Storozhenko, L.: Concrete-filled pipe structures, 182 (2005)
9. Garanzha, I.: Metal Const. J. **20**(1), 53 (2014)
10. Min, Yu., Zha, X., Ye, J., Li, Y.: Eng. Const. J. **49**(3), 10 (2013)
11. Min, Yu., Zha, X., Ye, J., She, C.: Eng. Const. J. **32**, 1053 (2010)
12. Hajjar, J., Gourley, B.: Eng. Const. J. **122**, 1336 (1996)
13. Johansson, M., Gylltoft, K.: Eng. Const. J. **128**, 1081 (2002)
14. Han, L.-H., Hui, Q., Tao, Z., Wang, Z.-F.: Thin-Walled Const. J. **47**, 857 (2009)
15. Trull, V., Sanzharovsky, R.: Experimental researches of load capacity of steel concrete-filled tube. Izvest. Vuz. Build. Arch. J. **3**, 30 (1968)
16. Schneider, S., Alostaz, Y.: Cons. Steel Res. J. **45**, 352 (1998)
17. Xiao, Y., Tomii, M., Sakino, K.: Trans. Jpn. Conc. Inst. J. **8**, 542 (1986)
18. Chen, J., Weiliang, J.: Thin-Walled Const. J. **48**, 724 (2010)
19. Perea, T., Leon, R., Hajjar, J., Denavit, M.: Eng. Const. J. **10**, 1957 (2013)
20. Onishchenko, O., Pichugin, S., Onishchenko, V.: High efficiency technologies and complex structures in the constructions, 404 (2009)
21. Storozhenko, L., Pents, F., Korshun, S.: Concrete-filled structures of industrial buildings, 202 (2008)
22. Ermolenko, D.: Three-dimensional stress-strain state of concrete-filled tubes element, 316 (2012)
23. Ensayos de hormigón endurecido. Parte 3: Determinación de la resistencia a compresión de probetas, 19 (2001)
24. Ensayos de hormigón. Determinación del módulo de elasticidad en compression, 16 (1996)
25. Concretes. Methods of prismatic compressive strength, modulus of elasticity and Poison's ratio determination, 14 (2008)
26. Building materials. Concrete. Methods for determination of prismatic compressive strength, elastic modulus and Poisson's ratio, 20 (2010)
27. Leshchinsky, M., Skramtaev, B.: Tests of the concrete strength, 272 (1973)
28. Ellobody, E., Young, B.: Nonlinear analysis of concrete-filled steel SHS and RHS columns. Thin-Walled Const. J. **44**, 930 (2016)
29. Hu, H., Huang, C., Wu, M.: Const. Eng. J. **129**, 1329 (2013)
30. Tao, Z., Uy, B., Liao, F., Han, L.: Nonlinear analysis of concrete-filled square stainless steel stub columns under axial compression. Cons. Steel Res. J. **67**, 1732 (2011)
31. Dai, X., Lam, D.: Cons. Steel Res. J. **66**, 555 (2010)
32. Georgios, G., Lam, D.: Cons. Steel Res. J. **60**, 1068 (2004)
33. Portoles, J., Romero, M., Filippou, F., Bonet, J.: Eng. Const. J. **33**, 1593 (2013)
34. Fam, A., Qie, F.S., Rizkalla, S.: Eng. Const. J. **130**, 640 (2004)
35. Stolarsky, T., Nakasone, Y., Yoshimoto S.: Engineering Analysis with ANSYS Software, vol. 473. MPG Books Ltd., Bodmin (2006)
36. Basov, K.: ANSYS: user's guide, 640 (2005)

37. Kaplun, A., Morozov, E., Olfereva, M.: ANSYS in the hands of an engineer: a practical guide, 272 (2003)
38. Konyukhov, A.: Fundamentals of structural analysis in ANSYS, 102 (2001)
39. Chigarev, A., Kravchuk, A., Smalyuk, A.: ANSYS for engineers: a reference guide, 510 (2004)
40. Morozov, E.: ANSYS in the hands of an engineer. Mech. Destr. 456 (2010)
41. Basov, K.: ANSYS in examples and tasks 224 (2002)
42. Lyubimov, K.: Use of ANSYS system to the solution of problems of continuum mechanics, 277 (2006)
43. Lukyanova, A.: Simulation of a contact problem using ANSYS, 52 (2010)

Fire Simulation of Light Gauge Steel Frame Wall System with Foam Concrete Filling

Andrey Shukhardin[1] , Marina Gravit[2] , Ivan Dmitriev[2(✉)] ,
Gleb Nefedov[1] , and Tatiana Nazmeeva[3]

[1] Andrometa LLC, Engels St. 9/20, 249032 Obninsk, Kaluga Region, Russia
[2] Peter the Great St. Petersburg Polytechnic University,
Politechnicheskaya St. 29, 195251 St. Petersburg, Russia
marina.gravit@mail.ru, i.i.dmitriev@yandex.ru
[3] Steel Construction Development Association,
Ostozhenka St. 19/1, 119034 Moscow, Russia

Abstract. There is a fire resistance investigation of the light steel thin-walled wall system with foam concrete filling (low density of 200 kg/m^3) in a fixed formwork of glass-magnesite sheets. The wall fragment model in the PC SOFiSTiK presents dependence temperature gradient in the section from fire duration. According to the results of the calculation and modelling, the wall system has the fire resistance limit more than 120 min in standard fire conditions. The standard fire tests of natural wall samples confirmed the correctness of the design model. The samples did not reach a critical state. This result correlates with the simulation model. In the article, we designed the standard bearing wall solution for buildings of the second degree of fire resistance (REI 90). It is recommended to repeat the tests before reaching the limit state due to the theoretical validity of the actual fire resistance of REI 120.

Keywords: Fire design · Fire safety · Fire resistance · Light steel frame · Cold-formed section · Thin-walled structure

1 Introduction

The construction of both affordable and comfortable housing is one of the main urban problems of our time. In addition to the convenience and ease of installation, these structures must comply with sanitary, hygienic and heat protection requirements, ensure a high level of building energy efficiency. The frame construction with standardized template solutions for load-bearing and enclosing structures should be regarded as the way to solve this engineering problem. The low-rise buildings constructed on the basis of the LGSC frame can be considered as one of the most environmentally friendly and energy-efficient options for cottage construction [1].

The system application, which is based on the supporting LGSC (Light Gauge Steel Construction) frame with insulating filler, is currently gaining popularity due to the large number of the structure advantages, such as low metal consumption, factory assembly, ease of transportation. This construction method has long development

© Springer Nature Switzerland AG 2020
V. Murgul and M. Pasetti (Eds.): EMMFT 2018, AISC 982, pp. 836–844, 2020.
https://doi.org/10.1007/978-3-030-19756-8_80

history in Europe and the USA [2–5]; however, it is only in beginning stage in Russia [6, 7]. The regulatory document describing the rules of design LGSC has been recently published [8]. The term LGSC is understood as building construction from cold-formed galvanized profiles with a thickness less than 4 mm, which are produced by the cold forming method from rolled sheet metal. They have a high bearing capacity with low metal consumption due to the elaborated section shape [9, 10].

The regulatory fire resistance is one of the main problems in the metal frame design of buildings and structures [11–14]. The high thermal conductivity of steel and the small value of the reduced thickness can lead to a sharp increase in the temperature of the supporting structures during a fire and, as a consequence, the deterioration of the mechanical characteristics [15–20]. In the European regulatory document EN1993-1.2 [21] for cold-formed sections (class 4), it is recommended to take a critical heating temperature of 350 °C. In most cases, it can be accepted as unjustifiably low, because, for example, under certain boundary conditions, beams made of high-strength steel can retain their carrying capacity at temperatures up to 700 °C [22–24].

The standard walls and partitions solutions with a framework of thin-walled profiles mostly use non-combustible facing materials (plasterboard sheets, mineral wool plates). Their fire performance has been investigated in studies both numerical and experimental ways. Another variant is to use lightweight monolithic foam concrete as a filler, which in addition to the heat insulating material performs fireproofing functions for metal structures. Also, an additional barrier to the flame spread is a permanent glass-magnesite formwork, which has high fire resistance.

According to study [24], despite the low tensile strength, the foam concrete in the structure composition increases the system durability during compression and bending work. The aim of the article is to substantiate the constructive solution of a wall with a load-bearing LGSC framework with foam concrete filler for application in buildings of the second fire resistance degree with a fire resistance limit REI 90.

The following tasks are solved:

1. The simulation of the wall fragment section and prediction of the structure behaviour under the standard fire conditions.
2. The conduction of standard fire tests and confirmation of the design model.

2 Methods

We used the Hydra module of the SOFiSTiK software package (ver. 2018) to analyse the temperature distribution over the cross section of the wall structure. Data entry was carried out through a text editor Teddy.

The structure investigation was under the standard fire test according to GOST 30247.0-94 (ISO 834), which is described by Eq. (1).

$$T_t - T_0 = 345 \log(8t + 1) \tag{1}$$

Where:

- t - time, min
- T_t - temperature in the furnace at time t, °C
- T_0 - temperature in the furnace at t = 0, °C
- The initial temperature inside the room is T_0 = +20 °C

The study object is a bearing wall 190 mm thick with a framework of steel C-shaped profiles (150 × 45 × 1.6). The filler is foam concrete with a density of 200 kg/m³ in fixed formwork of glass-magnesite sheets of 10 mm. The test wall sample has dimensions 3060 × 3150 × 190 mm (Fig. 1).

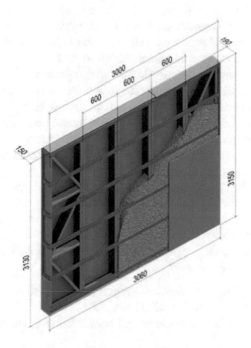

Fig. 1. Schematic of the wall test sample.

An external load (12.42 ton per m) is applied from above and then transmitted to four steel panel columns with a 600 mm step (load on the rack 9.5 ton).

For the simulation, we consider the central fragment of the wall, which includes the supporting frame column (Fig. 2).

We make a static calculation for a preliminary assessment of the structure fire resistance. It is based on determining the time to reach the critical temperature of the building structure. The critical temperature is understood as the heating temperature of the structure section, at which the limit state is expected to occur on the basis of R due to the loss of strength or stability of the structure.

We assume that the rod undergoes a central compression.

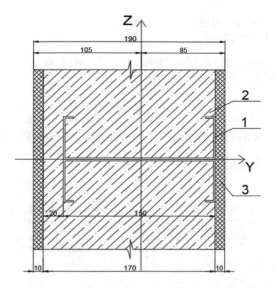

Fig. 2. Simulated fragment of the wall sample. 1. C-shaped LGSC profile; 2. Foam concrete; 3. Glass-magnesite sheet.

The coefficients taking into account changes in the regulatory resistance and the elastic modulus of steel are calculated by formulas (2) and (3).

$$\gamma_t = \frac{N}{F \cdot R_0} \tag{2}$$

$$\gamma_e = \frac{N \cdot l_0^2}{\pi^2 \cdot E_0 \cdot J_{min}} \tag{3}$$

Where N – Normative load, kg; F – The cross-sectional area, cm²; R_0 – the initial regulatory metal resistance, kg/cm². For steel C350 R_0 = 3500 kg/cm²; E_0 – the initial elastic modulus of the metal, kg/cm², E_0 = 2100000 kg/cm²; l_0 – estimated length of the rod, cm; J_{min} – the smallest moment of inertia of the rod section, cm⁴.

$$\gamma_t = \frac{9500}{(2 \cdot 4.06) \cdot 3500} = 0.33$$

$$\gamma_e = \frac{9500 \cdot (0.5 \cdot 315)^2}{3.14^2 \cdot 2100000 \cdot 19.48} = 0.58$$

According to [28], the critical temperature depends on the smallest value of the two calculated coefficients.

$$T_{kr} = 600 \,^\circ C$$

Standard fire tests were conducted according to Russian State Standards GOST 30247.0 and GOST 30247.1.

Thermal Characteristics of Materials. The thermal conductivity coefficient λ_t [W/(m · K)] varies according to the formula (4):

$$\lambda_t = A + Bt \tag{4}$$

Where: A – Initial thermal conductivity coefficient, W/(m · K); B – Coefficient of thermal conductivity change during heating, W/(m · K^2); t – Temperature of the material heating, K (Table 1).

Table 1. The coefficients of thermal characteristics of materials.

№	Steel	Foam concrete	Glass-mag. sheet
A	78	0.041	Nonlinear dependence according to [29]
B	−0.048	0.00019	
C	310	748	
D	0.48	0.63	

The coefficient of heat capacity c_t [J/(kg · K)] varies according to the formula (5):

$$c_t = C + Dt \tag{5}$$

Where: C – The initial coefficient of heat capacity, J/(kg · K); D –coefficient of heat capacity change during heating, J/(kg · K^2); t – Temperature of the material heating, K.

3   Results and Discussion

The calculated cross section was specified in the SOFiSTiK software package and then splitted into a grid of finite elements (Fig. 3).

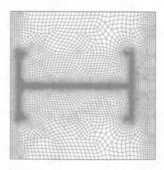

Fig. 3. Finite element mesh of the calculated section.

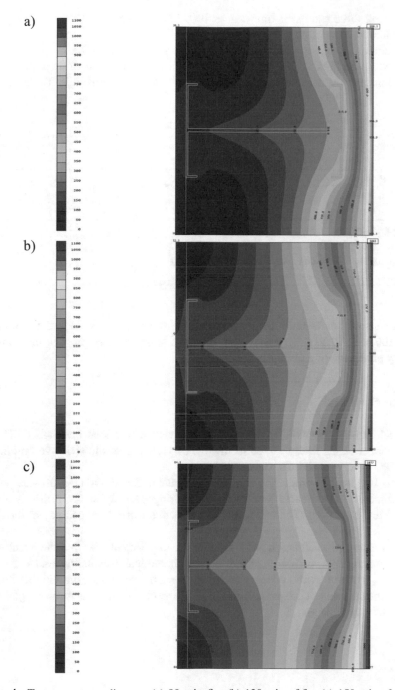

Fig. 4. Temperature gradient on (a) 90 min fire (b) 120 min of fire (c) 150 min of fire.

We took three estimated time values (90, 120 and 150 min) to reach the critical temperature and therefore fire resistance.

Simulation of the temperature distribution is shown in Figs. 4a–c.

The standard fire tests of samples were conducted on the basis of theoretical studies by the company Andrometa LLC. The test results are summarized in Table 2.

Table 2. Fire test results.

Observation result	Time of 1 sample, min	Time of 2 sample, min
Test beginning	0	0
Deflection 2 mm	17	15
Deflection 6 (5) mm	54	69
Formwork Collapse	70	83
Steam release from formwork joints	100	90
Test stopped by agreement with the Andrometa LLC	102	100

The actual deflection of the structure does not exceed the allowable: $\Delta = L/100 = 315/100 = 3.15$ cm. The limiting state of loss of bearing capacity, integrity and insulating ability is not reached.

4   Conclusions

The article substantiates the applicability of the wall with a load-bearing LGSC frame and a foam concrete filler in fixed formwork of glass-magnesite sheets in buildings of the second degree of fire resistance (REI 90).

There was made a simulation of a wall fragment in the SOFiSTiK software package and predicted the believable time of losing bearing capacity at fire impact under the standard fire conditions. The conducted fire tests confirmed the accuracy of the design model.

Comprehensive calculation and results of fire experiments showed the effectiveness of this construction. In accordance with the theoretical calculation and carried out simulation, it can be predicted that this design has a fire resistance limit higher than that stated in the sample test program. Additional fire tests of these structures are planned to confirm this hypothesis.

References

1. Orlova, A.V., Zhmarin, E.N., Paramonov, K.O.: Power efficiency of houses from light-gauge steel structures. Constr. Unique Build. Struct. **6**(11), 1–13 (2013)
2. Bolina, F., Christ, R., Metzler, A., Quinino, U., Tutikian, B.: Comparison of the fire resistance of two structural wall systems in light steel framing. DYNA (Colombia) **84**(201), 123–128 (2017). https://doi.org/10.15446/dyna.v84n201.57487
3. Veljkovic, M., Johansson, B.: Light steel framing for residential buildings. Thin-Walled Struct. **44**(12), 1272–1279 (2007). https://doi.org/10.1016/j.tws.2007.01.006
4. Vatin, N., Sinelnikov, A., Garifullin, M., Trubina, D.: Simulation of cold-formed steel beams in global and distortional buckling. Appl. Mech. Mater. **633–634**, 1037–1041 (2014). https://doi.org/10.4028/www.scientific.net/AMM.633-634.1037
5. Al Ali, M., Isaev, S.A., Vatin, N.I.: Development of Modified formulae for detection the welding stresses in the welded steel cross-sections. Mater. Phys. Mecha. **26**(1), 9–15 (2016)
6. Sovetnikov, D.O., Videnkov, N.V., Trubina, D.A.: Light gauge steel framing in construction of multi-storey buildings. Constr. Unique Build. Struct. **3**(30), 152–165 (2015)
7. Campian, C., Chira, N., Iuhos, V., Pop, M., Vatin, N.: Structural upgrading of steel columns for overhead power lines. Procedia Eng. **165**, 876–882 (2016). https://doi.org/10.1016/j.proeng.2016.11.787
8. Russian Set of Rules SP 260.1325800.2016 Cold-formed thin-walled steel profile and galvanized corrugated plate constructions. Design rules. Russia
9. Nazmeeva, T.V., Vatin, N.I.: Numerical investigations of notched c-profile compressed members with initial imperfections. Mag. Civil Eng. **62**(2), 92–101 (2016). https://doi.org/10.5862/MCE.62.9
10. Vatin, N.I., Nazmeeva, T., Guslinscky, R.: Problems of cold-bent notched c-shaped profile members. Adv. Mater. Res. **941–944**, 1871–1875 (2014). https://doi.org/10.4028/www.scientific.net/AMR.941-944.1871
11. Gravit, M., Dmitriev, I., Ishkov, A.: Quality control of fireproof coatings for reinforced concrete structures. IOP Conf. Ser: Earth Environ. Sci. **90**(1), 012226 (2017). https://doi.org/10.1088/1755-1315/90/1/012226
12. Gravit, M., Zybina, O., Vaititckii, A., Kopytova, A.: Problems of magnesium oxide wallboard usage in construction. In: Advances in Intelligent Systems and Computing, vol. 692, pp. 1093–1101 (2018). https://doi.org/10.1007/978-3-319-70987-1_118
13. Gravit, M., Lyulikov, V., Fatkullina, A.: Possibilities of modern software complexes in simulation fire protection of constructions structures with Sofistik. In: MATEC Web of Conferences, vol. 193, p. 03026 (2018). https://doi.org/10.1051/matecconf/201819303026
14. Gravit, M.V., Nedryshkin, O.V., Ogidan, O.T.: Transformable fire barriers in buildings and structures. Mag. Civil Eng. **77**(1), 38–46 (2018). https://doi.org/10.18720/MCE.77.4
15. Correia, A.J.P.M., Rodrigues, J.P.C.: Fire resistance of steel columns with restrained thermal elongation. Fire Saf. J. **50**, 1–11 (2012). https://doi.org/10.1016/j.firesaf.2011.12.010
16. Hamins, A., McGrattan, K., Prasad, K., Maranghides, A., McAllister, T.: Experiments and modeling of unprotected structural steel elements exposed to a fire. Fire Saf. Sci. (2005). https://doi.org/10.3801/iafss.fss.8-189
17. Heinisuo, M., Jokinen, T.: Tubular composite columns in a non-symmetrical fire. Mag. Civil Eng. **5**(49), 107–120 (2014). https://doi.org/10.5862/MCE.49.11
18. Schaumann, P., Tabeling, F., Weisheim, W.: Numerical simulation of the heating behaviour of steel profiles with intumescent coating adjacent to trapezoidal steel sheets in fire. J. Struct. Fire Eng. **7**(2), 158–167 (2016)

19. Schaumann, P., Bahr, O., Kodur, V.: Numerical studies on HSC-filled steel columns exposed to fire. In: Tubular Structures XI, pp. 411–416 (2017). https://doi.org/10.1201/9780203734964-50

20. Koh, S.K., Mensinger, M., Meyer, P., Schaumann, P.: Fire in hollow spaces: short circuit as ignition sources and the role of ventilation. In: 2nd International Fire Safety Symposium, IFireSS 2017, vol. 781 (2017)

21. EN1993-1.2. Design of steel structures. General rules. Structural fire design. European Committee for Standardization, Brussels, Belgium, June 2004

22. Kankanamge, N.D., Mahendran, M.: Mechanical properties of cold-formed steels at elevated temperatures. Thin-Walled Struct. **49**, 26–44 (2011). https://doi.org/10.1016/j.tws.2010.08.004

23. Schwarz, I., Slatinka, M., Jandera, M.: Structural fire behaviour of Z purlins. Appl. Struct. Fire Eng. 129947 (2015). https://doi.org/10.14311/asfe.2015.021

24. Arrais, F., Lopes, N., Vila Real, P.: Parametric study on the fire resistance of steel columns with cold-formed lipped channel sections. Appl. Struct. Fire Eng. 129947 (2017). https://doi.org/10.14311/asfe.2015.022

Road Construction, Foundation Soil

Strengthening High Embankments on Weak Soils of Geosynthetic Materials

Sergei Kudryavtsev$^{(\boxtimes)}$, Tatyana Valtseva, Viacheslav Kovshun,
Zhanna Kotenko, and Vladlen Stefanuk

Far Eastern State Transport University, Seryshev Str. 47,
680021 Khabarovsk, Russia
prn@festu.khv.ru

Abstract. A high embankment that approaches the bridge over the Amurskaya river branch was designed in a framework of an international project establishing a free economic zone on the Bolshoi Ussurusky Island. The embankment of 18 m high is made of fine river sand. The embankment base is composite and zoned in clay soils, plastic and sluffing shale. The thickness of week soils in the basement is from 2.5 m to 6.0 m. The considered project is of high criticality rating according to the Russian standards and regulations. The embankment target life is no less than 80 years. Designing earth structures in difficult engineering conditions is individual. The promising calculation methods, modern software and high-tech durable geosinthetic materials were used while developing rational and effective reinforcement structures on a weak basement. The materials used have long-standing parameters providing a guaranteed life span up to 120 years. Anchoring reinforcement meshes of integral geogrids of E'GRID brand provided long-term durability of the embankment and basement reinforcement. The theoretical researches and calculations were performed to assess deformation data of the structure and the weak basement. An overall assessment of general and local stability of the structure and the weak basement reinforced by the geogrid meshes was given. The numerical modeling method applying the program complexes «FEM models» and «PLAXIS» was used for calculation of the structure and the basement behavior. The scientific and expert approach to specific development of difficult geotechnical structures allowed creating technical solutions that provided necessary strength of the reinforcement. The structures designed and the measures taken guarantee operational reliability of the high embankment of sand on the weak inhomogeneous basement. The constructed reinforced complex will work as a whole structure up to 80 years.

1 Introduction

Designing and construction transport geotechnical projects on weak inhomogeneous basement is always connected with high risk including difficult climatic conditions of the Far East, regular submergences and increasing transportation loads. The Bolshoi Ussuriysky Island constitutes a part of a big river creekbed. It has a pronounced weak inhomogeneous basement composed of overwatered saturated loam (binder) soils.

© Springer Nature Switzerland AG 2020
V. Murgul and M. Pasetti (Eds.): EMMFT 2018, AISC 982, pp. 847–854, 2020.
https://doi.org/10.1007/978-3-030-19756-8_81

Physical and mechanical properties of the soils are potentially changeable due to amounts of precipitations, levels of floods and their frequency during climatic seasons.

Designing the structures of high responsibility requires not only professional construction works at high level but also scientific and theoretical research. Preliminary calculated designs of the embankment confirm that the most perspective solutions of construction projects in the given conditions are connected with rational usage of modern geosynthetic materials that has been technically and scientifically supported. The geosynthetic materials used provide a long-term sustainable work of the design that has been made of local construction materials including river sand. The properties of geosynthetic materials must correspond to the working conditions of the design. They also must meet the requirements of stability parameters and quality of reinforcement joints during the whole period of their exploitation – no less than 80 years.

Several variants of the reinforced embankment structure on a weak basement were considered in designing and research.

Advanced methods of numerical modeling that describes the behavior of high embankment on weak inhomogeneous base were considered in determining technical and performance criteria of the embankment's several variants.

The geomaterials which are supposed to be used as anchoring and reinforcement meshes in structural elements were researched on their properties.

The results of all the researches done and assessment of economic efficiency of the designed variants helped to determine the most rational design of the high bottomland embankment made of local river sand. The hi-tech anchoring and reinforcement meshes E'GRID provide a reliable reinforcement of inner and sloping parts of the embankment and its weak basement.

2   Calculation Methods of the Structures

Numerical modeling of road embankment variants on a weak base of the Bolshoi Ussuriysky Island and behavior of these structures was carried out with the programming complex «FEM models». The complex was developed by geotechnicians of industrial research company "Georeconsruction- basedesign", St. Petersburg, Russia.

An elastoplastic model with the Mohr-Coulomb yield criterion was used to describe behavior of the structures. The numerical calculations correspond well to the traditional engineering calculation methods. They provide fair precision in description of stresses and deformations in the structures.

Besides numerical modeling, the facilities of the programming complex PLAXIS 2D V8 were used to reveal the stability of the variants designed.

3   Research of Geogrids E'GRID and Their Properties

Assessment of main parameters of geogrids E'GRID was done in a laboratory of geotechnical materials at a federal enterprise "Rosdornii". A diagram of the physical and mechanical properties is shown in Fig. 1.

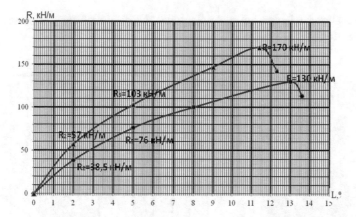

Fig. 1. Research results of physical and mechanical properties of geogrids E'GRID130R and E'GRID170R.

The geometrical parameters of the geogrids E'GRID is relevant to the Technical Specifications and tolerance errors. Besides, the geogrids E'GRID have necessary geometrical parameters for the reinforcement materials used in national and international construction of reinforced ground structures.

In further calculations the basic technical characteristics of E'GRID meshes as an anchoring and reinforcing material are taken only as long-term parameters typical only for this brand. All long-term parameters of the material were received in terms of large-scale continuous research (more than 400 days) in 4 climatc chambers carried out by the recognized English laboratory «NewGrids Limited».

The functional connections are represented in logarithmic formulae:

$$Log(LN) = -0.2027 - 0.0321 \, Log(T) \tag{1}$$

$$LN = UCLS/QCLS \tag{2}$$

The components of the functional connections 1 and 2 for climatic conditions at 20 °C are given in Table 1.

Table 1. The components of the functional connections 1 and 2.

Design life	75 years	100 years	120 years
	$10^{5.817}$ h	$10^{5.943}$ h	$10^{6.022}$ h
UCLS/QCLS	0.4079	0.4041	0.4018

Figure 2 shows a laboratory-scale plant and the results of creep-rupture stress tests of the geogrid:

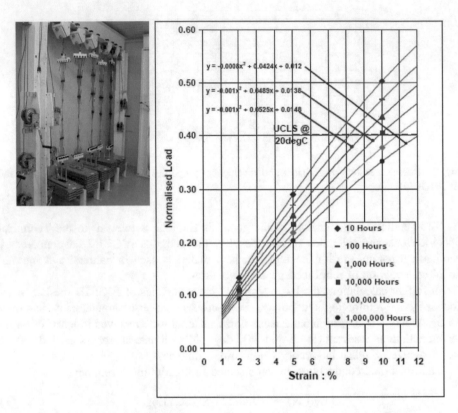

Fig. 2. Laboratory-scale plant and the results of creep-rupture stress tests of the geogrid.

All intermediate parameters of the creep-rupture stress tests of the geogrid E'GRID were used in calculations and numerical modelling of reinforced structures according to the temperature and the given life span.

4   Calculation and Alternate Designs of Embankment Structure

When designing the embankment several alternate designs of a high embankment made of sand and reinforced by geogrids were considered (Fig. 3).

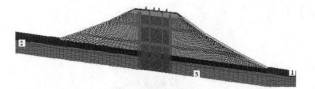

Fig. 3. Structural design of original embankment made of sand: 1 - plastic silt sandy loam; 2 - yielding silt sandy loam; 3 - medium-grained sand

Types of geogrids and their placement are shown in Fig. 4.

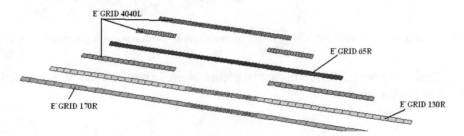

Fig. 4. Types of geogrids E'GRID used for reinforcement.

A rational structure design strengthened with reinforcing meshes of geogrids E'GRID for the embankment on a weak inhomogeneous base is shown in Fig. 5:

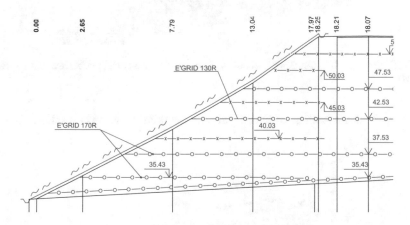

Fig. 5. Reinforcement scheme of structure design.

Figure 6 shows isolines of vertical deformations (m) of the structure reinforced with geogrids E'GRID.

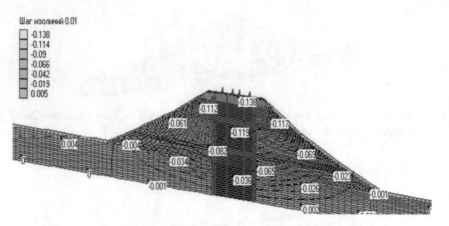

Fig. 6. Isolines of vertical deformations (m) of the structure reinforced with geogrids E'GRID.

Figure 7 shows isolines of horizontal deformations (m) of the structure reinforced with geogrids E'GRID.

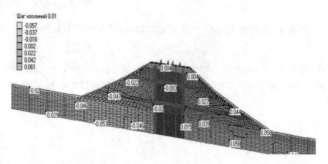

Fig. 7. Isolines of horizontal deformations (m) of the structure reinforced with geogrids E'GRID.

Figure 8 shows isolines of horizontal stresses (kPa) in geogrid meshes.

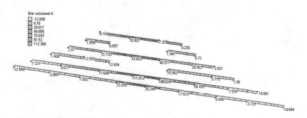

Fig. 8. Isolines of horizontal stresses (kPa) in geogrid meshes.

The results of numeric modeling of the projected structure made of a high embankment on a weak base strengthened with reinforcing geogrids E'GRID were taken into account for assigning the parameters of stress-deformed state in the structure and the reinfocement elements.

5 Determining Height Limit Values of the Reinforced Embankment

The results of finding stability received by PLAXIS 2D version V8 for three basic cross profiles were used to determine approximate minimal height of embankments reinforced by geomaterial for given conditions.

According to the relationships shown in Fig. 9, reinforcement of sand embankments is necessary only above 7,0 m high. However, due to the possibility of soil strength decrease because of the reasons mentioned it is recommended to set the embankment height limit of 5,0 m if it is not completely reinforced, while the base reinforcements should be done for the embankments higher than 3,5 m.

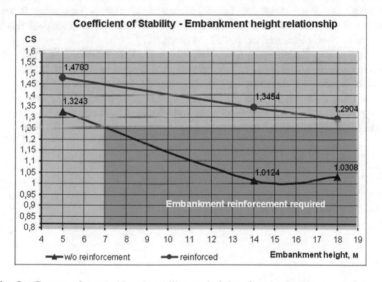

Fig. 9. Curves of general/local stability vs height of embankment on weak base.

6 Conclusion

Total deformation of the reinforced structures has decreased due to the increasing rate of elastic deformations by 30–50% as well as decreasing of shearing deformations. This provides functional reliability of a high embankment.

The results of the numerical modeling done and determining local and general stability confirm efficiency the design meaningful measures on decreasing deformation of embankments made of sandy soils on a weak base.

To prevent a treat of significant permanent deformations and to provide a functional stability of the structure and its weak base, it is reasonable to strengthen the embankments higher than 3,5 m according to the calculations and the accepted design solutions.

References

1. Valtseva, T.Y., Kudryavtcev, S.A., Berestyanyy, Y.B., Goncharova, E.D., Mikhailin, R.G.: Motorway structures reinforced with geosynthetic materials in polar regions of Russia, pp. 502–506 (2014)
2. Kudriavtcev, S., Berestianyi, I., Goncharova, E.: Engineering and construction of geotechnical structures with geotechnical materials in coastal arctic zone of Russia, pp. 562–566 (2013)
3. Zhusupbekov, A., Kudryavtsev, S.A., Berestyanyy, U.B., Arshinskaya, L.A., Valtseva, T.U.: Developing design variants while strengthening roadbed with geomaterials and scrap tires on weak soils. In: Proceedings of the International Workshop on Scrap Tire Derived Geomaterials - Opportunities and Challenges, IW-TDGM 2007, International Workshop on Scrap Tire Derived Geomaterials, pp. 171–178 (2008)
4. Berestyanyy, U.B., Kudryavtsev, S.A., Valtseva, T.U., Goncharova, E.D., Mikhailin, R.G.: Geotecnical materials in designs of highways in cold regions of Far East. In: Proceedings of the 14th International Conference on Cold Regions Engineering, pp. 546–550 (2009)
5. Valtseva, T.U., Kudryavtsev, S.A., Berestyanyy, U.B., Barsukova, N.V.: Practice of use of positive properties of geosynthetic materials on building objects in severe climatic conditions of the Far East of Russia, pp. 423–427 (2006)

Experimental Studies of the Gravity-Type Foundation Windage in a Wind Tunnel

Olga Poddaeva⬤, Pavel Churin⬤, and Julia Gribach$^{(\boxtimes)}$⬤

Moscow State University of Civil Engineering, Yaroslavskoe Shosse 26,
129337 Moscow, Russia
js-995@mail.ru

Abstract. The gravity-type foundation (GTF) is a floating platform, which during the operation period is held on the seabed due to its own weight and bonds with the ground. These structures are actively used on the sea shelf that contains a significant amount of natural resources. The main advantage of such structures is the possibility of towing them over long distances and placing them in a working position at the place of operation at sea without the use of expensive lifting equipment and vehicles, as well as the possibility of reuse in a new place. Building constructions of oil and gas plants located on the GTF are built in specialized ports. The article is devoted to an experimental study of the windage and wind load parameters of gravity-type foundation in a boundary layer wind tunnel, taking into account the screening of the wind flow from the coastline and port facilities located on it. This information is primarily necessary for specialists in hydraulic engineering to select the optimal scenario for mooring a floating structure during the construction of building structures on it. In the text of the article, the methodology of experimental modeling used is described, the layout design for testing is described, and their results are presented. The results are presented in the form of dimensionless aerodynamic coefficients in tabular and graphical form.

Keywords: Gravity-type foundation · Windage · Wind load · Mooring · Aerodynamics · Experimental studies

1 Introduction

With the development of science and technology, oil and gas fields on the sea shelf are being actively developed. One of the main problems of this direction is the construction of structures of offshore platforms: it is important to observe a number of conditions for creating a reliable and sustainable object. First, it is important to properly design the base [1].

Depending on the depth and purpose of the wells, stationary offshore bases are divided into embankments, bulk or frozen islands, pile and large block foundations, gravity and frame type platforms. In connection with the rapid pace of development of marine spaces, the depth at which the construction of structures is carried out is increasing. The most stable and reliable bases are gravity-type foundations that can be installed to a depth of 150 m. The main advantage of these structures is their large

© Springer Nature Switzerland AG 2020
V. Murgul and M. Pasetti (Eds.): EMMFT 2018, AISC 982, pp. 855–863, 2020.
https://doi.org/10.1007/978-3-030-19756-8_82

mass, making it possible to install them on the seabed without additional fastening in the form of piles [1, 2]. Own buoyancy and the presence of a ballasting system make it possible to tow gravity-type foundations over long distances, install them in a working position at the place of operation at sea without the use of expensive lifting, and transport vehicles. In addition, an advantage is the possibility of re-using the platform in a new place, increased vibration resistance and fire resistance, high resistance to marine corrosion, slight deformation under the influence of loads, and higher protection from pollution of the sea.

At the same time, one of the major drawbacks of this type of structures is the complexity of the scientific and technical support during their design. Since the main building structures located on the GTF are built in specialized port facilities, the main issue of scientific and technical support is a competent assessment of the GTF mooring. One of the main characteristics studied during the mooring simulation is wind impact, which in turn directly depends on the windage structure [1–3].

The purpose of this work is an experimental study of the windage and wind load parameters on a floating structure [4, 5]. Only in this way can one obtain the wind load characteristics with regard to the shielding surfaces [6, 7].

The study was conducted on the example of the projected GTF in the Arctic zone of the Russian Federation. The proposed plan includes the construction of a separate building-conductive object, including the terminal based on the gravity-type founda-tions and the appropriate port infrastructure. It is planned to build a coastal stationary terminal for the extraction, storage and shipment of liquefied natural gas and stable gas condensate on gravity-type foundations [1, 4, 8, 9].

2   Materials and Methods

The study was carried out in the Unique scientific installation of the Large Research Gradient Wind Tunnel of NRU MGSU (see Fig. 1) according to the following algo-rithm [10, 11]:

1. a model of the port complex and the gravity-type foundation under study is installed in the wind tunnel, with an integrated force-torque sensor for performing weight measurements.
2. "zero" (in the absence of flow) readings of force-torque sensor are read, which is necessary to take into account the magnitude of the initial displacement due to loading the sensor with the weight of the model's structure.
3. the flow velocity in the wind tunnel is set to 10.0 m/s. Force-torque sensor readings are recorded into two files: a) load readings with a frequency of 1000 Hz, and b) averaged load readings for the entire duration of the recording program.
4. the angle of attack varies from 0° to 360° in 45° steps. For each angle, step No. 3 is repeated.

Fig. 1. Unique scientific installation Large research gradient wind tunnel NRU MGSU

When conducting experimental studies of wind exposure, the following similarity criteria were met:

- geometric similarity of the layout and the object under study;
- compliance of the wind flow parameters with the field conditions of the development site (the studies were conducted in a gradient flow, subject to the condition of self-similarity according to Reynolds).

3 Results

At the first stage of work, a climatic analysis of the terrain was carried out. The climate of the region under consideration is characterized by relatively mild winters with frequent storms and cool, wet summers with fogs and frequent but light precipitations [11].

The maximum wind speed for the observation period is 40 m/s in the SW direction, the maximum gust is 42 m/s.

The wind region is IV (SP 20.13330.2011), the standard value of wind pressure is 0.48 kPa (48 kg/m^2).

For experimental studies, a model of the object under investigation was developed and manufactured. Taking into account the size of the working part of the aerodynamic tube, the scale of the layout of 1: 300 was selected from the maximum possible condition for blocking the flow. The investigated model was installed on an automatized rotary table located in the working area of the aerodynamic pipe (Figs. 2 and 3).

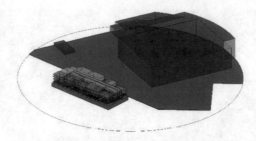

Fig. 2. Three-dimensional model of the object under study

Fig. 3. Model of the object under study in the working area of the wind tunnel

The models are made of steel studs, plexiglass and hard plastic according to the source data provided. At the design stage, a simplified three-dimensional model of the object under study was created taking into account the surrounding buildings. The finished parts were glued with chemically pure dichloroethane and epoxy glue.

At the stage of manufacturing the model, a 6-component force-moment sensor was used in the structure under study, which is used to measure the aerodynamic forces and moments. The sensor is attached to a rigid metal plate at the base of the layout.

The process modeling is based on the observance of the equality of dimensionless similarity criteria in accordance with the requirements of the similarity theory.

At a distance of 0.5 m from the edge of the model installed in the working areas, the Pitot-Prandtl tube is installed, which is a device for determining the total and static pressures in the flow during the experiments. The full pressure channel is connected to a differential digital pressure gauge, which serves to determine the air flow rate. This provided an additional test of the wind tunnel and increased reliability of the experiment.

The main measurement cycle was performed at a flow velocity of $V\infty = 10.0$ m/s, which corresponds to Re = 0.38 * 106. This value of Re is in the zone of self-similarity.

In other words, the dimensionless aerodynamic coefficients of the scale model obtained in a wind tunnel should be identical to the corresponding values in natural conditions (of course, while observing the geometric similarity of the model).

Tables 2 and 3 summarize all the experimental data on the aerodynamic coefficients Cx, Cy, Cz, CMx, CMy, and CMz when the angle of attack β varies from 0° to 360° in 45° steps (see Fig. 4), for two different levels of water according to the Baltic system of heights (BSH) - 324 and 244 cm.

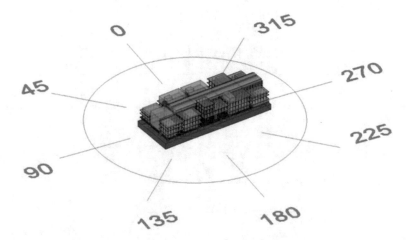

Fig. 4. Considered angles of attack of the wind flow

Figures 5 and 6 show graphs of the dependence of the aerodynamic coefficients on the angle of attack. The reliability of the results of experimental modeling is confirmed by the use of equipment that has calibration certificates issued by an authorized organization. The calculation of the aerodynamic coefficients and loads on the real object was carried out at the GTF windage shown in Table 1.

Table 1. Model and real object windage

Angle of attack	Model, [m^2]	Real object, [m^2]
0	0.177002	15930.18
45	0.211020	18991.8
90	0.075404	6786.36
135	0.204380	18394.2
180	0.177002	15930.18
225	0.211020	18991.8
270	0.075404	6786.36
315	0.204380	18394.2

Table 2. A summary of the experimental data on the aerodynamic coefficients Cx, Cy, Cz, CMx, CMy and CMz when the angle of attack varies from 0° to 315°, annual sea level = −324 cm according to the BSH.

Angle of attack	Cx	Cy	Cz	Cmx	Cmy	Cmz
0	0.73	0.7	−0.04	−0.4	−0.01	−0.03
45	−0.7	1.06	0.04	−0.78	−0.53	0.01
90	−1.41	0.63	0.06	−0.55	−0.6	0.01
135	−1.78	−0.36	0.14	0.56	−0.35	−0.01
180	−0.47	−1.27	0.26	1.95	0.01	−0.01
225	1.58	−1.4	0.33	1.79	0.24	0.01
270	1.67	−0.19	0.1	0.03	0.84	−0.01
315	1.22	0.77	0.14	−0.77	0.8	−0.01

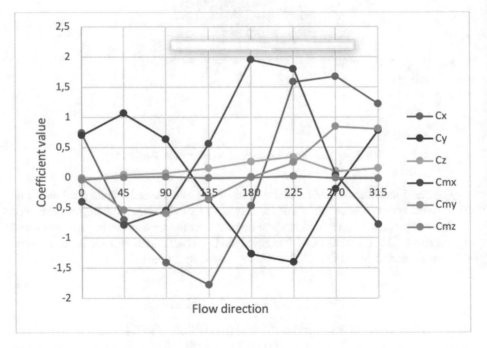

Fig. 5. The graph of the dependence of the aerodynamic coefficients Cx, Cy, Cz, CMx, CMy, and CMz when changing the angle of attack from 0° to 360° annual sea level = −324 cm by BSH

Table 3. A summary of the experimental data on the aerodynamic coefficients Cx, Cy, Cz, CMx, CMy, and CMz with a change in the angle of attack from 0° to 360°, annual sea level = −324 cm by BSH.

Angle of attack	Cx	Cy	Cz	Cmx	Cmy	Cmz
0	0.75	0.78	0.02	−0.46	0.24	−0.44
45	−0.94	0.98	0.11	−0.75	−0.96	0.01
90	−1.87	0.51	0.12	−0.53	−1.24	0.10
135	−2.06	−0.51	0.19	0.57	−0.60	−0.18
180	−0.72	−1.37	0.32	1.97	0.08	−0.07
225	1.52	−1.46	0.36	1.87	0.51	0.25
270	1.54	−0.26	0.15	0.04	1.70	−0.13
315	1.09	0.64	0.22	−0.87	1.80	−0.25

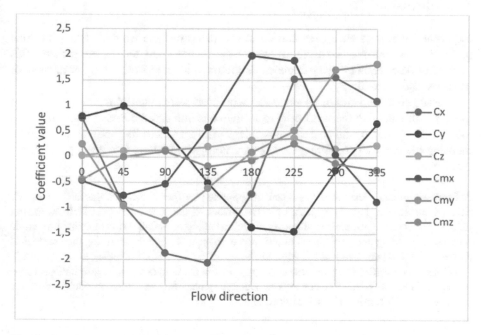

Fig. 6. The graph of the dependence of the aerodynamic coefficients Cx, Cy, Cz, CMx, CMy, and CMz when changing the angle of attack from 0° to 360°, annual sea level = −324 cm by BSH

4 Discussion

In the process of conducting experimental studies, force-moment loads in a dimensionless form (aerodynamic coefficients Cx, Cy, Cz, CMx, CMy, and CMz) were determined. The maximum values of the aerodynamic coefficients Cx = 1.8 and Cy = 1.46, characterizing the GTF windage at level of 244 cm in BSH, were obtained at angles of attack of 315° and 225°, respectively. The maximum values of the aerodynamic coefficients Cx = 1.78 and Cy = 1.4, which characterize the windage of the GTF at level of −324 cm in BSH, were obtained at attack angles of 135° and 225°, respectively.

The obtained data are extremely important when choosing the optimal scenario of mooring the studied gravity-type foundation with the aim of the subsequent construction of structures on it. In addition to the wind impact assessment, the wave impact and coastal sea currents will also have a significant impact. The wave impact assessment is usually performed in specialized wave pools, the influence of sea currents is estimated either analytically, on the basis of long-term field measurements in the port water area, or on the basis of the results of numerical modeling in certified software systems.

5   Conclusion

Experimental studies have been carried out to determine the windage and wind load parameters on the projected gravity-type base in the wind tunnel. At present, this method of determining these parameters is the most optimal from the point of view of accuracy and reliability.

Certain characteristics of windage and wind load were obtained taking into account the effect of shielding the wind flow from the coastline and the massive port facilities located on it, which will make it possible to further select the optimal scenario of mooring the gravity base during the construction period taking into account all safety requirements.

Acknowledgments. The work was financial supported by the Ministry of Education and Science of the Russian Federation within the framework of the state #7.6075.2017/BCh, Project «Investigation of the phenomena of aerodynamic instability of building structures in aero-elastic statement, including the development of an innovative methodology for analyzing meteorological data to refine the parameters of the wind load».

All tests were carried out using research equipment of the unique scientific installation «Large Research Gradient Wind Tunnel» of The Head Regional Shared Research Facilities of the Moscow State University of Civil Engineering.

References

1. Olsen, T.O.: Concrete gravity platforms. In: Encyclopedia of Maritime and Offshore Engineering (2018)
2. Yeganeh, A., Prendergast, L.J., Kenneth, G.: Performance testing of a novel gravity base foundation for offshore wind. In: Civil Engineering Research Association of Ireland (2016)
3. Alerechi, L.W., Oluwole, O.O., Odunfa, K.M.: Finite element analysis of concrete gravity based platform subjected to sudden crash (impact) load using ANSYS. Covenant J. Eng. Technol. (2018)

4. Gaydarov, N.A., Zakharov, Y.N., Ivanov, K.S., Semenov, K.K., Lebedev, V.V.: Numerical and experimental studies of soil scour caused by currents near foundations of gravity-type platforms. In: International Conference on Civil Engineering, Energy and Environment (2014)
5. Bredmose, H., Skourup, J., Hansen, E.A., Christensen, E.D., Pedersen, L.M., Mitzlaff, A.: Numerical reproduction of extreme wave loads on a gravity wind turbine foundation. In: 25th International Conference on Offshore Mechanics and Arctic Engineering (2006)
6. Sigbjörnsson, R., Smith, E.K.: Wave induced vibrations of gravity platforms: a stochastic theory. Appl. Math. Model. **3**(4), 155–165 (1980)
7. DiBiagio, E., Myrvoll, F., Hansen, S.B.: Instrumentation of gravity platforms for performance observations. In: Proceedings of the International Conference on the Behaviour of Offshore Structures (1976)
8. Yushkov, A., Nudner, I.: Computational investigation of turbulent flow impact on non-cohesive soil erosion near foundations of gravity type oil platforms. In: Mathematical and Information Technologies (2016)
9. Gaydarov, N.A., Ivanov, K.S., Lebedev, V.V.: Modeling cohesionless and cohesive soils erosion near oil platforms of gravity type. In: Stability and Control Processes (2015)
10. Churin, P.S., Poddaeva, O.I., Fedosova, A.N., Pomelov, V.Y.: Analysis of domestic and foreign regulatory and scientific and technical documents in the field of wind influence on buildings and structures that are part of hazardous production facilities. In: Materials Science and Engineering (2018)
11. Churin, P.S., Poddaeva, O.I., Pomelov, V.Y.: Experimental studies of wind impact on coke chambers. In: MATEC Web of Conferences (2018)

Using of Ash and Slag from Power Plants for Embankments Construction

Miloš Marjanović(✉) ⓘ, Veljko Pujević ⓘ, and Sanja Jocković ⓘ

Faculty of Civil Engineering, University in Belgrade,
Bulevar kralja Aleksandra 73, 11000 Belgrade, Serbia
mimarjanovic@grf.bg.ac.rs

Abstract. This paper presents the results of laboratory research of physical and mechanical properties of fly ash and mixtures of fly ash and slag from landfills of thermal power plants "Nikola Tesla" and "Kostolac" in terms of their use for road embankments construction. Following mechanical material properties were tested: unified compressive strength (UCS), shear strength, deformability, CBR, and resistance to frost. Results of the study have shown that waste materials from Serbian thermal power plants can be successfully used as the building and reconstruction material for road subgrade. This can bring multiple economic benefits, such as: saving of natural materials, lower energy consumption, lower pollution of environment, and less waste landfill areas.

1 Introduction

Because the construction of transport infrastructure requires significant amounts of natural building materials (sand, gravel, crushed stone etc.), replacement of these materials with ash or slag is of great environmental, economical and practical significance. Any reduction of waste material deposits and their reuse brings multiple benefits such as: saving of natural resources, preventing ecosystem change, reduction of pollution of soil, underground and surface waters and air. Also, by using replacement materials, energy is saved for exploitation and crushing standard stone material.

During 2014–2015, Institute for testing materials IMS and Faculty of Civil Engineering in Belgrade have conducted the broad study [1, 2] of use of ash and slag from thermal power plants of Electric Power Industry of Serbia (JP EPS). The aim of this study was to perform the optimum number of laboratory tests of ash and slag from thermal power plants Nikola Tesla A/B (Obrenovac) and Kostolac A/B (Kostolac), in order to evaluate the possibility of use of tested materials for embankments construction, according to current technical requirements. The fly ash from silos and ash/slag mixture from power plant landfills were tested within this study.

2 Laboratory Tests and Analysis of Results

Within laboratory testing program, in general, the following material properties were evaluated: chemical properties, physical properties, mechanical properties with and without addition of hydraulic binder, development of mechanical properties over time.

© Springer Nature Switzerland AG 2020
V. Murgul and M. Pasetti (Eds.): EMMFT 2018, AISC 982, pp. 864–871, 2020.
https://doi.org/10.1007/978-3-030-19756-8_83

Results of mechanical properties tests that are relevant for assessment of quality of embankment construction were chosen and presented. The following laboratory tests for the determination of the physical and mechanical properties were used: specific gravity, grain size distribution, moisture-density relationship (Proctor compaction test), unconfined compressive strength (UCS), direct shear, CBR and odometer tests [3]. All tests were performed on two or three specimens, according to SRPS standards.

Three types of waste materials from Serbian thermal power plants were tested:

- Ash-slag mixture from landfills in thermal power plants "Kostolac A" and "Kostolac B" (designated TEKO AB)
- Fly ash from silos of thermal power plant "Nikola Tesla A" (TENT A)
- Fly ash from silos of thermal power plant "Nikola Tesla A" (TENT B)

According to EN 14227-4, all tested ashes are aluminum-silicate ashes (class V), with pozzolanic properties, and without self-cementing properties. These types of ashes are mostly used with the addition of binder (cement or lime) [4]. The influence of addition of hydrated lime and cement was evaluated within this study. Laboratory testing was performed on four combinations of materials and binders (Table 1). Minimum percentage of used cement was determined in order to maintain uniformity of the mixture with minimum cement consumption, while the minimum amount of lime was chosen from the condition that the pH value of mixture shall be 12.4, securing the optimum conditions for the hydration process.

Table 1. Used % of binders

Material	Binder	% of binder
TEKO AB	Cement	2%
TENT A		4%
TENT B	Lime	5%
		7%

Materials without binders were tested first, in order to establish initial physical and mechanical properties (etalons). These properties were later used to compare with the properties of treated material. In order to evaluate and compare the stabilization results with different amount of binders, it was important to prepare specimens with the same initial conditions. Testing specimens were prepared by compaction at optimum moisture content from Standard Proctor compaction test. Appropriate dry amount of material and binders (based on % of material dry weight) were mixed with ash thoroughly to produce a homogeneous mixture, and then water was added to reach the optimum moisture content. After mixing, mixture was compacted without delay. Specimens with the addition of cement were kept hermetically closed in plastic wrap prior to testing, while the specimens with lime were not hermetically closed.

Development of mechanical properties of stabilized material over time was tested on specimens that were 1.7 and 28 days old. Unconfined compressive strength (UCS) tests were also performed on test specimens after 15 min, 1 h, 2 h and 4 h after compacting, in order to obtain a more detailed information on the strength increment over time. In order to test the resistance of the stabilized material to frost, UCS tests were performed on samples 28 days old, after 15 freezing and melting cycles.

3   Test Results

3.1   Ash-Slag Mixture

Test results for ash-slag mixture TEKO AB are presented as representative.

Unconfined Compressive Strength (UCS). Results of UCS testing for ash-slug mixture TEKO AB with addition of lime and cement are presented in Fig. 1. Expected increase of strength was obtained. With addition of cement, the strength increase was 2.5–8 times, depending on % of cement. With addition of lime this effect is higher - UCS was increased 10–12 times.

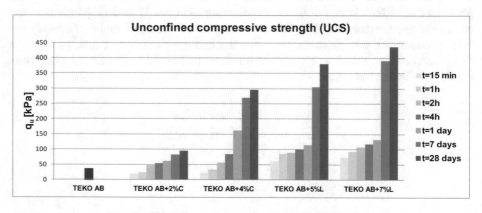

Fig. 1. Unconfined compressive strength (TEKO AB)

Effective Shear Strength Parameters. Effective shear strength parameters (c', φ') from direct shear test are displayed in Fig. 2. With addition of binders, internal friction angle was increased about 10–25%. By addition of cement, cohesion was increased up to 230%, and about 40% with addition of lime.

California Bearing Ratio (CBR). Results of CBR tests are displayed in Fig. 3. Etalon CBR value for tested mixture TEKO AB is 24%. With addition of binders, the increase of CBR over time is clearly observed. The same trend was observed with an increase of % of binder. The total increase is higher with the addition of lime (CBR value up to 150–200%).

Parameters of Compressibility. Compressibility (oedometric) modulus (M_v) for stress level of 100–200 kPa is displayed in Fig. 4. The increase in stiffness with time and with % of binder is obvious - the compressibility modulus is increased 1.5–3.5 times, depending on the amount of binder.

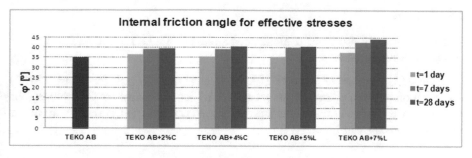

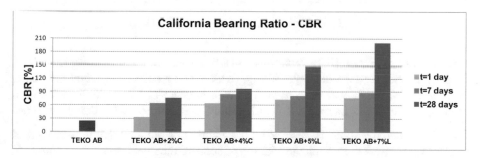

Fig. 2. Effective shear strength parameters (TEKO AB)

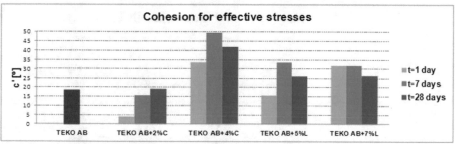

Fig. 3. California bearing ratio (TEKO AB)

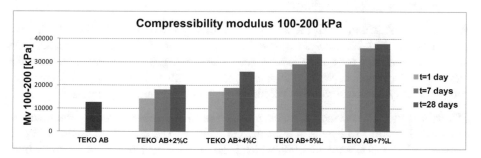

Fig. 4. Compressibility modulus (TEKO AB)

3.2   Fly Ash

Test results for fly ash from thermal power plant "Nikola Tesla B" (TENT B) are presented as representative.

Unconfined Compressive Strength (UCS). Results of UCS tests for fly ash TENT B with addition of lime and cement are presented in Fig. 5. This ash, without binders, has very low UCS. With addition of binders, UCS was increased 4–11 times (with cement) and 12–18 times (with lime).

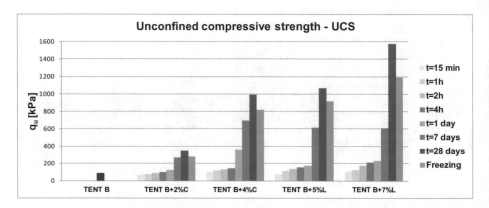

Fig. 5. Unconfined compressive strength (TENT B)

Effective Shear Strength Parameters. Effective shear strength parameters (c′, φ′) from direct shear test are displayed in Fig. 6. Obtained results show slight increase of internal friction angle over time, as well as with increase of binder %. By addition of cement, cohesion was increased 2–4 times, and 2.5–3.5 times with addition of lime.

California Bearing Ratio (CBR). With the addition of binder, there is clear increase of CBR values over time. With the addition of cement, the increase of 10–11 times was obtained (CBR = 115–134%), depending on the % of binder. Similar effects were found with the addition of lime, with the increase of 13–16 times (CBR = 161–197%). The obtained results are displayed in Fig. 7.

Parameters of Compressibility. Compressibility (oedometric) modulus (M_v) for stress level of 100–200 kPa is displayed in Fig. 8. With the addition of cement, the increase of stiffness over time is obtained. The same effect was obtained with the increase of % of cement. With addition of lime, these effects are lower. The increase of compressibility modulus with the addition of cement is around 3–5 times.

Effects of Frost. Resistance of stabilized fly ash to frost was tested by the determination of UCS after 15 cycles of freezing and melting. The results are given in Fig. 5. Resistance index is around 75–86%, which is acceptable, because of the generally high total values of UCS.

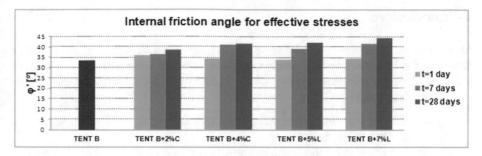

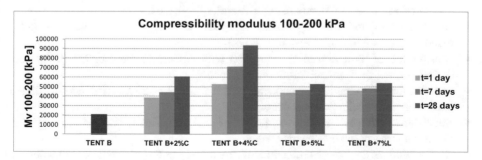

Fig. 6. Effective shear strength parameters (TENT B)

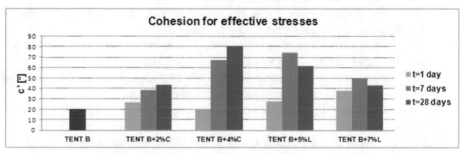

Fig. 7. California bearing ratio (TENT B)

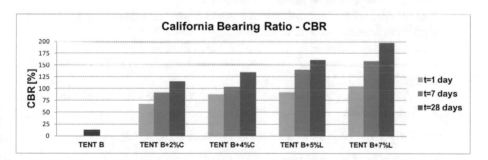

Fig. 8. Compressibility modulus (TENT B)

4 Conclusions

In order to evaluate the applicability of fly ash and ash-slag mixtures in road construction, the laboratory tests of stabilized waste materials were performed. Based on the analysis of the laboratory results, which in line with [5–7], the following conclusions have been made:

- Fly ash and slag from thermal power plants have acceptable mechanical properties, without addition of binders:
  - effective shear strength parameters are in the ranges of $\varphi' = 34$–$35°$ and $c' = 18$–20 kPa. Regarding the embankments stability, these parameters satisfy the design requirements for slopes of the embankment;
  - according to parameters of compressibility (compressibility modulus M_v), ash and slag can be categorized as soils of low compressibility;
  - CBR values are varying, based on the ash origin, and are in the ranges of 12–24%, which are acceptable values for embankments.
- With addition of binders, the increase of all mechanical properties was observed for all tested materials:
  - the increase of effective friction angle was observed over time and with the increase of binder amount ($\varphi'_{28\ days} = 39$–$44°$). The increase of effective cohesion is even higher ($c'_{28\ days} = 40$–80 kPa). These strength parameters allow the steeper slopes of the embankments;
  - with addition of binder CBR values were increased and there is general trend of significant increase over time ($CBR_{28\ days} = 75$–200%);
  - frost resistance index is in range of 75–86%. Despite the fact that the required index for some materials is 80%, obtained results can be considered as acceptable
  - parameters of compressibility are increasing over time and with the increase of amount of binder.

The following use of ash and slag is recommended [8, 9, 10]:

- stabilization of soft subsoil layer,
- construction of embankments using soil stabilized with ash (with or without cement or lime),
- construction of embankments using ash and with the ash core,
- construction of transition layer using ash with addition of cement and lime (with mandatory proof of frost resistance index),
- construction of formation layer using ash with binders (cement or lime), for new designed track or in rehabilitation/reconstruction,
- construction of formation layer using soil stabilized with ash, for new designed tracks, or in rehabilitation/reconstruction,
- stabilization using ash in the construction of the hydraulically bonded bearing layer under concrete or asphalt bearing layer,
- construction of a bridge-embankment transition area - cement stabilization is defined in railway standards, and utilization of ash should also be investigated for this stabilization.

Acknowledgements. Authors acknowledge the JP EPS (Electric Power Industry of Serbia) as a customer of the study "Use of the fly ash and slag from JP EPS thermal power plants in railways", who financially supported this very important research topic. The results of this study have justified the mass use of ash from Serbian thermal power plants in the construction of transport infrastructure subgrade.

References

1. Study: Application and placement of ash generated in Electric Power Industry of Serbia power plants Institute for testing materials, Belgrade, Serbia (2011). in Serbian
2. Study: Use of the fly ash and slag from JP EPS thermal power plants in railways. Institute for testing materials, Faculty of Civil Engineering, University in Belgrade, Belgrade, Serbia (2015). in Serbian
3. American Coal Ash Association: Fly ash facts for highway engineers. Technical report No. FHWA-IF-03-019 (FHWA), (2003)
4. Baykal, G., Edinçliler, A., Saygili, A.: Highway embankment construction using fly ash in cold regions. Resour. Conserv. Recycl. **42**(3), 209–222 (2004)
5. Vukićević, M., Popović, Z., Despotović, J., Lazarević, L.: Fly ash and slag utilization for the Serbian railway substructure. Transport **33**(2), 389–396 (2016)
6. Vilotijević, M., Vukićević, M., Lazarević, L., Popović, Z.: Sustainable railway infrastructure and specific environmental issues in the Republic of Serbia. Tech. Gazette **25**(4), 516–523 (2018)
7. Popović, Z., Lazarević, L., Vukićević, M., Vilotijević, M., Mirković, N.: The modal shift to sustainable railway transport in Serbia. In: MATEC Web of Conferences, vol. 106, p. 05001 (2017)

Mathematical Modeling of Water-Saturated Soil Basements

Tatyana Maltseva$^{(\boxtimes)}$ ⓘ

Industrial University of Tyumen, Volodarskogo Str. 38, 625000 Tyumen, Russia
maltv@utmn.ru

Abstract. Based on the proposed mathematical models, the stress and strain state of earth foundations consisting of saturated soils was calculated under different loads applied on the original ground and under certain combinations of loads (interinfluence of foundations). The paper gives the results of calculations of stressed and deformation states of loaded saturated bases through time. Calculation formulas for the final ground settlements are obtained that meet the stabilized state of the bases, skipping the description of the consolidation process, which responds to the requests of designers. All calculations show the unloading effect of pore water on the reduction of stresses and strains that arise in the soil skeleton. The calculation results were compared with experimental data.

Keywords: Water-saturated soil basements · Stresses and strains · Stabilized state of the bases · Unloading effect of pore water · Soil skeleton · Sediment

1 Introduction

The enlarging of urban infrastructure and civil-industrial development of new territories are carried out on saturated clay soils or moss and muck lands, which have almost no drain. For today one of the relevant issues is the study of stress-strain state of weak bases and their carrying capacity [1–3]. The description of the process of saturated soil consolidation by means of filtration consolidation models causes vanishing of excess pore pressure after a finite period of time. Therefore, the saturated soil is considered afterwards as single-phase soil demanding the models of deformable solid body mechanics for description of its stress and strain state. Field experiments showed [4] that total stresses almost do not change through time, so they can be described in terms of elasticity theory. However, the displacements obtained according to the elasticity theory are several times different from the results of the field experiment. The article simulates the contribution of pore water to the stressed and strain state of the soil skeleton provided with viscoelastic properties. The results of the research can be used in the design and calculation of sediment of engineering facilities erected on weak grounds, which will consequently help making safe and economical decisions, developing new technologies aimed at enhancing the carrying capacity of pore water [5, 6].

© Springer Nature Switzerland AG 2020
V. Murgul and M. Pasetti (Eds.): EMMFT 2018, AISC 982, pp. 872–884, 2020.
https://doi.org/10.1007/978-3-030-19756-8_84

The subject of the study is saturated soil, behavior of which under load is described in terms of the theory of linear hereditary viscoelasticity without considering the aging of the material. According to the field experiment, there are residual excess pore pressures after the end of the soil consolidation process [7]. The models of the theory of elasticity and plasticity do not describe this pore pressure, because the soil is considered as single-phase one. According to linear filtration models (Terzagi K., Gersevanov N. M., Bio M., Florin V. A., Zaretsky Yu. K., Ter-Martirosyan Z.G. and others) residual excess pore pressures become zero.

According to the viscoelastic version of the kinematic model (one-dimensional case) proposed by L. Maltsev, the experimental pore pressure curve (Fig. 1) is accurately described with a broken line.

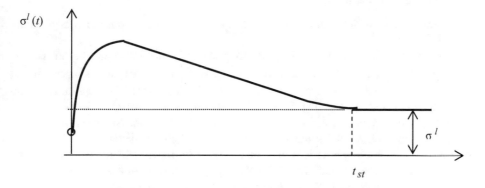

Fig. 1. The qualitative nature of the pore pressure curve.

Here t_{st} is the initial moment of stabilization. For example, when testing a 1-meter-high saturated clay sample, the consolidation process ends at moment $t_{st} = 95$ dey. The horizontal line on the graph corresponds to the residual excess pore pressure. The dashed line continues the horizontal line up to the initial time value. The kinematic model exists in two versions: elastic and viscoelastic. The elastic version of the kinematic model corresponds with the replacing of the actual graph with a horizontal line that is compliant with the residual excess pore pressure $\sigma^i(t > t_{st})$, up to the initial point in time. The elastic version of the model has no time parameter. In the viscoelastic version of the kinematic model, the solution is divided into two stages:

1. the solution of the corresponding problem is found through the elastic problem statement;
2. the reassignment system is introduced (Volterr principle) and the approximate transition from the image to the original is performed for a fixed point of spatial coordinates using the polyline method, according to which the system of linear algebraic equations is composed.

Its order coincides with the number of links of the polyline. The division of the problem into two stages allows at the first stage in some cases to obtain an analytical solution of the problem due to absence of time parameter.

In nonlinear filtration models (of Florin V. A., Kosterin A. V., and others) that take into account the initial pore pressure gradient, two zones are introduced: active and passive, the last of which has no water filtration. Finding the boundary separating these two zones and moving in time complicates the solution of the problem, but still does not take into account the contribution of residual excess pore pressure to the strain state of the saturated base. The article summarizes the kinematic model for the spatial case and, in contrast to the filtration models, describes the residual excess pore pressure and its effect on the soil skeleton deformation during stabilization.

2   Problem

Basic premises of the model:

1. Saturated soils with the presence of hydraulically continuous water inside pores are considered.
2. The relative strains of the solid (mineral particles, subscript s) and liquid (porous water, subscript l) phases are small $\left|\varepsilon_i^s\right| \leq 0.01; \left|\varepsilon_i^l\right| \leq 0.01$, in one geometrical point, the model has two material points: liquid and solid. The continuity of the body is ensured by applying the Cauchy relations to each of the phases.
3. Six hypotheses of the theory of elasticity are valid for the skeleton of the soil. Tangential stresses on the model occur only in the soil skeleton.
4. For pore water, a hypothesis is introduced about the linear dependency between the partial derivative of the residual excess pore pressure and the relative linear deformation along each of the coordinate axes, and the mention is made about the linear dependency between the relative strain and the change of the relative porosity along the coordinate axis. Instead of equality of pressures on the horizontal and lateral facets, the model provides pore water with the characteristic according to which those pressures are different, similar to water in the soil skeleton. The hypothesis of homogeneity is preserved, orthotropy is introduced instead of isotropy. For example, in the case of semi-decomposed peat mechanical constants of pore water are different along and across the remainder of the stem. The difference persists when choosing directions along and across the layer, therefore properties of water placed into soil pores differ from properties of common water. The linear hereditary theory of viscoelasticity is applied not only to the physical equations for the soil skeleton, but also to the equations of the state of pore water. In the viscoelastic version of the model, the Boltzmann law is used for skeleton of soil, and an analogue of the Boltzmann law is introduced for the pore water.
5. The interaction of the two phases is described by the kinematic hypothesis, according to which one phase releases part of its volume for the other phase, therefore the relative linear strains of the soil skeleton and pore water along each of the coordinate axes are opposite in sign and directly proportional.

2.1   Problem Definition

The system of equations of the kinematic model includes: equilibrium Eq. (1) taking into account volume forces K, Hooke's physical law with mechanical characteristics modified due to the presence of water in the pores (2), physical equations of state for pore water (3), Cauchy Eq. (4), kinematic equations of interaction of phases (5). Direct tension stresses in the soil skeleton and compressing stresses in pore water are considered positive, so the expression $\sigma_{ij}^s - \sigma_{ij}^l \delta_{ij}$ corresponds to the sum of the stresses. Along each of coordinate axis, the increment of the total normal stress is balanced by the increment of the tangential stress arising in the soil skeleton. Summation with respect to repeated index is used. The system of equations is:

$$\left(\sigma_{ij}^s - \sigma_{ij}^l \delta_{ij}\right)_{,j} + K_i = 0, \qquad \delta_{ij} = \begin{cases} 0, i \neq j \\ 1, i = j \end{cases} \tag{1}$$

$$\sigma_{ii}^s = (2G + b_i)\varepsilon_{ii}^s + \lambda\theta, \quad \theta = \varepsilon_{ij}^s \delta_{ij}; \quad b_i = \frac{E_i^l}{\aleph_i^2} \quad \sigma_{ij}^s = G\varepsilon_{ij}^s, i \neq j \tag{2}$$

$$P_{ij}^l = E_i^l \varepsilon_{ij}^l \delta_{ij}, \quad P_{ij}^l = h_i \sigma_{ij,j}^l \delta_{ij} \tag{3}$$

$$\varepsilon_{ij}^s = \frac{1}{2}\left(u_{i,j}^s + u_{j,i}^s\right), \quad \varepsilon_{ij}^l = u_{i,j}^l \delta_{ij} \tag{4}$$

$$\varepsilon_{ii}^s = -\aleph_j \varepsilon_{ij}^l \delta_{ij} \tag{5}$$

where u_i^s, u_i^l - displacement of solid and liquid phases; $G = \frac{E^s}{2(1+v)}$, $\lambda = \frac{vE^s}{(1+v)(1-2v)}$, E^s - stress-strain modulus, v - Poisson ratio, G - shear modulus, λ - Lame constant introduced for soil skeleton, non-dimensional parameters $\aleph_i$, dimensional parameters h_i (m), E_i^l (MPa) describe the possible orthotropy of pore water. All the parameters of the model are determined according to the experiment with a two-phase sample, to tray tests or to field tests. By transforming the equilibrium Eq. (1), we obtain a system of differential equations, written through the displacement of particles of the soil skeleton (index s is omitted):

$$-\left((G + \lambda)\frac{\partial\theta}{\partial x_i} + G\Delta u_i + b_j\frac{\partial^2 u_i}{\partial x_i^2}\delta_{ij} + c_j\frac{\partial u_i}{\partial x_i}\delta_{ij}\right) = K_i, \; c_i = \frac{E_i^l}{h_i\aleph_i} \tag{6}$$

which differs from the known Lame equations by two additional terms in each equation with parameters b_i, c_i.

Let us introduce three differential vector-operators: (a) Lame operator $A = (G + \lambda)graddiv + G\Delta$; (b) operator $B = \left(b_i\frac{\partial^2}{\partial x_i^2}\right)_{i=1}^3$, which reflects the change of the three diagonal elements in the mechanical constant tensor of the fourth rank; (c) operator $C = \left(c_i\frac{\partial}{\partial x_i}\right)_{i=1}^3$, corresponding to the equations of state for pore water (3).

For a finite simply connected spatial area bounded by a piecewise smooth surface and placed on a part of the surface S_2 (water-permeable), excess residual pore pressure becomes zero, the static boundary condition is written only for the soil skeleton, so the mixed boundary conditions are:

$$u|_{S_1} = 0, \; t^{(v)}|_{S_2} = q(x_1, x_2) \tag{7}$$

$$t^{(v)} = \sum_{i,k=1}^{3} \left(\sigma_{ik} + b_i \frac{\partial u_l}{\partial x_i} \right) \cos(v, x_k) e_i$$

$$= \sum_{i,k=1, k \neq i}^{3} ((2G + b_i)\varepsilon_i + \lambda \varepsilon_{ik} + \sigma_{ik}) \cos(v, x_k) e_i \tag{8}$$

$t^{(v)}$ – the stress operator in the soil skeleton, which differs from the analogous elasticity theory operator by changing the mechanical constants due to b_i terms. The kinematic boundary condition is the main and the static condition is natural.

According to the viscoelastic version of the one-dimensional kinematic model according to the Boltzmann law, for physical equations the following equations are valid:

$$\varepsilon^s(t) = \int_0^t \Pi_s(t - \tau) d\sigma^s(\tau), \; \varepsilon^l(t) = \int_0^t \Pi_l(t - \tau) dP^l(\tau) \tag{9}$$

where $\Pi_s(t)$, $\Pi_l(t)$ - creep functions.

Introduction to the system of equations (1)–(5) of the corresponding integral relations leads to a system of integrodifferential equations.

In the Laplace-Carson images based on the Borel theorem, the resultant of two functions is replaced by the product of their images.

$$[\varepsilon^s(t)]^* = [\Pi_s(t)]^* [\sigma^s(t)]^*, \; [\varepsilon^l(t)]^* = [\Pi_l(t)]^* [P^l(t)]^*, \tag{10}$$

In the simplest case, the parameters v, $\aleph$, h will be assumed constant. The redefinition rule (Volterr principle) is used to go from solving a problem in terms of elastic statement to a solution in terms of viscoelastic statement:

$$E^s \to \frac{1}{[\Pi_s(t)]^*}, \; E^l \to \frac{1}{[\Pi_l(t)]^*}, \; G \to G^*, \; \lambda \to \lambda^* \tag{11}$$

$$\varepsilon^s \to (\varepsilon^s)^*, \; \sigma^s \to (\sigma^s)^*, \; \sigma^l \to (\sigma^l)^*, \; \varepsilon^l \to (\varepsilon^l)^*, \; P^l \to (P^l)^* \tag{12}$$

then the equations of the viscoelastic problem in the images coincide with Eqs. (1)–(5) and accordingly with (6).

The division of the solution into two stages for a system of fixed spatial points allows abandoning the solution of the system of integral-differential equations.

3 Results and Discussion

Let us consider the class of problems on loading a viscoelastic two-phase semi-plane with different types of loads and give a comparison with laboratory and field experiments.

The differences in the statements of the Flaman problem in the theory of elasticity and in the paper lies in the fact that in the theory of elasticity the Flamann problem is considered for an no bounded semi-plane. At the same time. when a normal stress is decomposed into the case of two phases (a saturated base), a bounded simply connected area is introduced.

In the polar coordinate system (θ, r), let us introduce a semi-cylinder of small radius ρ and replace the line load F with radial stresses σ_r distributed over its surface (Fig. 2).

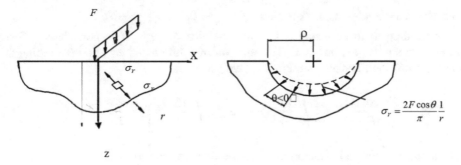

Fig. 2. Replacing of the line load F with radial stress.

Normal total stresses

$$\sigma_r = \sigma_r^s - \sigma_r^l, \sigma_r = -\frac{2F\cos\theta}{\pi}\frac{1}{r}, \tau_{r\theta} = \sigma_\theta = 0 \tag{13}$$

satisfy the equilibrium equations and the equation of strain compatibility in the polar coordinate system under zero boundary conditions (with the exception of the point of application of the load F). The Flamann problem under additional conditions $\sigma_\theta = \tau_{r\theta} = 0$ is the one-dimensional stress state problem, which allowed the use of a one-dimensional version of the kinematic model when decomposing radial stress

$$\sigma_r^s - \sigma_r^l = -\frac{2F\cos\theta}{\pi}\frac{1}{r}, \varepsilon_r^s = \frac{1}{E}\sigma_r^s, \varepsilon_r^l = \frac{h}{E^l}\frac{\partial\sigma_r^l}{\partial r}, E = E^s + \frac{E^l}{\aleph^2} \tag{14}$$

$$\varepsilon_r^s = \frac{\partial u_r^s}{\partial r}, \varepsilon_r^l = \frac{\partial u_r^l}{\partial r}, \varepsilon_r^s = -\aleph\varepsilon_r^l, \sigma_\theta = \sigma_\theta^s - \sigma_\theta^l = 0, \tau_{\theta r} = 0 \tag{15}$$

On the daylight surface of the semi-plane ($r = \rho$), the excess residual pore pressure is zero, while in the depth of the massif it is nonzero, so water moves along the radius from the area of increased pressure to the area of reduced pressure opposite to the movement of the skeleton particles of the soil u_r^s. Displacements of pore water particles in a direction which is orthogonal to the radius u_θ^l are caused by lateral deformations of the soil skeleton. The system of equations (14) has been reduced to solving a linear equation of the first order with respect to the strain of the soil skeleton:

$$\frac{\partial u_r^s}{\partial r} + a^2 u_r^s = -\frac{2F}{\pi E}\frac{\cos\theta}{r} - a^2 C(\theta),\ a^2 = \frac{E^l}{Eh\aleph}\left(\frac{1}{M}\right), \tag{16}$$

To find the integration constant $D(\theta)$, which appeared as a result of solving this equation, and $C(\theta)$, which was a consequence of the physical equation

$$\varepsilon_r^l = \frac{h}{E^l}\frac{\partial \sigma_r^l}{\partial r}, \tag{17}$$

two boundary conditions are specified: $\sigma_r^l\big|_{r=\rho} = 0;\quad u_r^s\big|_{r=R} = 0.$

According to these conditions, a limited area $\rho \le r \le R$ is considered, where strains and stresses are non-zero. The solution of a differential equation describing the expansion of the Flamann solution into two phases is:

$$u_r^s = \frac{2F\cos\theta}{\pi E}\left[e^{-a^2 R}\int_\rho^R \frac{e^{a^2 r}}{r}dr - e^{-a^2 r}\int_\rho^r \frac{e^{a^2 r}}{r}dr\right],\ -\frac{\pi}{2} \le \theta \le \frac{\pi}{2}, \tag{18}$$

The stresses in the soil skeleton are determined according to Hooke's law

$$\sigma_r^s = -\frac{2F\cos\theta}{\pi}\left[\frac{1}{r} - a^2 e^{-a^2 r}\int_\rho^r \frac{e^{a^2 r}}{r}dr\right],\ -\frac{\pi}{2} \le \theta \le \frac{\pi}{2}, \tag{19}$$

The first term corresponds to Flaman's solution for a single-phase body, the second term in brackets describes the unloading effect of the liquid phase.

Based on the formula (11) the values u_r^l (m), σ_r^l (MPa) are obtained.

From the graphs (Fig. 3) it follows that with increasing depth, the liquid phase gradually assumes the total pressure, the stresses in the soil skeleton decrease faster than the stresses found by the solution of Flamann. At low total stresses, the stresses in the skeleton are almost zero; therefore, the liquid phase exhibits its properties more strongly. The graph of vertical displacements for the points of the day surface that corresponds to a single-phase body, shows a substantially slow decrease as compared to the graph of the strain function, constructed taking into account the influence of pore water (Fig. 4). The effect of the liquid phase on the solid phase is manifested by the "raising" of particles of the soil skeleton over the daylight surface near the action of force F, which is followed by a rapid decrease to zero.

Tangential movements in the soil skeleton are:

$$u_\theta^s = r \sin \theta \int_\rho^r \frac{1}{r^2} b(r) dr, \, b(r) = \frac{2F}{\pi E} \left[e^{-a^2 R} \int_\rho^R \frac{e^{a^2 r}}{r} dr - e^{-a^2 r} \int_\rho^r \frac{e^{a^2 r}}{r} dr \right] \quad (20)$$

Based on the fundamental solution of (18), (19), together with Trefilina E. R. we obtained the solution of the problem of a uniformly distributed load's effect on a two-phase viscoelastic semi-plane.

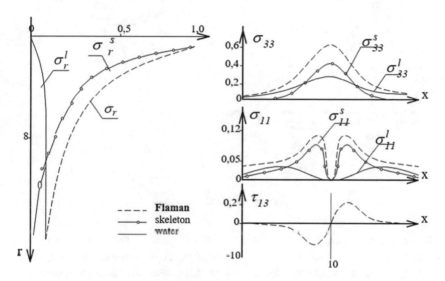

Fig. 3. Radial, vertical, horizontal normal and tangential shearing stresses.

The problem has a transition from the solution in the images to the original for a fixed point of coordinates using the approximate method of broken lines. The stress-strain state of the soil massif from the action of nearby objects and the base of the road was calculated in the elastic version.

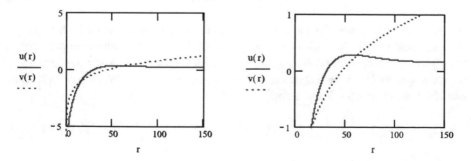

Fig. 4. Vertical movement of the surface points.

According to Flamann's solution (- - -), according to kinematic model (—).

As the example, the graphs for vertical stresses in the liquid and solid phases are given. According to the Flamann's solution in the case of a uniformly distributed load of $2b$ width with the parameters $a^2 = 0.02$ (1/m), $\rho = 0.02$ m, $q/\pi = 1$ MH/m, $b = 10$ m (Fig. 5) in sections (a) $z_1 = 0.5b$, (b) $z_2 = b$, (c) $z_3 = 1.5b$. It follows that with increasing depth, the pore pressure increases.

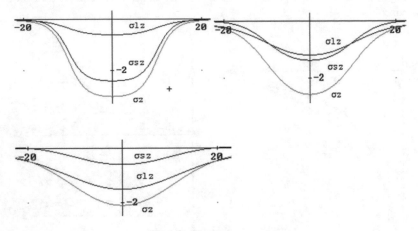

Fig. 5. Vertical stresses.

In the cross section $z = 1.5b$, two points of coordinates were fixed ($z = 1.5b$, $x = 0$), ($z = 1.5b$, $x = b$), for which the change in pore pressure through time is shown. In the elastic solution for $\sigma_{33}^l(z_3 = 1.5b, x = 0)$, in accordance with the Volterr principle, the resignation $a^2 \to (a^2(p))^*$ was applied, the solution of the viscoelastic problem in the images was obtained:

$$(\sigma_{33}^l)^* = \frac{2q}{\pi}(a^2)^* \int_{-b}^{b} \frac{(1,5b)^3}{r^3} e^{-(a^2)^* r} \left[\ln\frac{r}{\rho} + \int_{\rho}^{r} \frac{e^{(a^2)*\xi} - 1}{\xi} d\xi \right] d\eta \qquad (21)$$

$$r = \sqrt{\eta^2 + (1,5b)^2} \qquad (22)$$

Based on the processing of the experiment conducted by V. A. Demin with a sample of saturated peat having dimensions $d = 0.257$ m and $h = 0.292$ m when the pressure on the piston base is $\sigma_0 = 0.01$ MPa and a water lock height is 0.3 m, the dependency of universal parameter of the kinematic model from time was obtained $a^2(t)$ with using Heaviside function $h(t)$. It has a form of a broken line:

$$a^2(t) = a(0)\left(1 - \sum_{i=0}^{5} (\beta_i - \beta_{i+1})(t - T_i)h(t - T_i)\right), \qquad a(0) = 0,581 \qquad (23)$$

$$T0 = \beta0 = \beta6 = 0, T1 = 0, 1, T2 = 1, T3 = 5, T4 = 25, T5 = 60, \beta1 = 4, 068,$$
$$\beta2 = -0, 352, \beta3 = -0,066, \beta4 = -0, 014, \beta5 = -7, 262.10 - 3, t = \bar{t}/\text{day}.$$

In the images according to Laplace-Carson we have

$$\left(a^2(p)\right)^* = p \int_0^\infty e^{-pt} a^2(t) dt = a(0) \left(1 - \sum_{i=0}^5 (\beta_i - \beta_{i+1}) \frac{1}{p} e^{-pT_i}\right),$$

$$0 \leq p \leq \infty \tag{24}$$

Figure 6 shows the actual graph of this function.

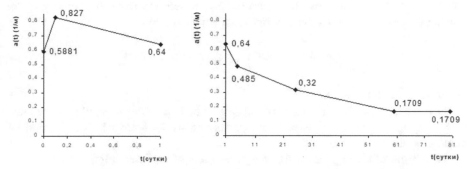

Fig. 6. Generalized parameter of the kinematic model $a^2(t)$.

In accordance with the method of broken lines, the desired original $\sigma_{33}^l(t)$ is represented as a broken line:

$$\bar{\sigma}_{33}^l(t) = \sigma(0)\left(1 - \sum_{i=0}^5 (\gamma_i - \gamma_{i+1})(t - T_i)h(t - T_i)\right), \quad \gamma_0 = \gamma_6 = T_0 = 0, \tag{25}$$

where the parameters $\sigma(0)$, γ_i are searchable, the cross-points T_i are given as for the function $a_2(t)$ or this spline in Laplace-Carson images has the following written form

$$\left(\bar{\sigma}_{33}^l\right)^* = \sigma(0)\left(1 - \sum_{i=0}^5 (\gamma_i - \gamma_{i+1}) \frac{1}{p} e^{-pT_i}\right) \tag{26}$$

To determine the unknown parameters, a system of linear algebraic equations is compiled by coincidence on the system of points p_j of the broken line $(j = 1, \ldots, 5)$ in images (26) with the known right-hand side (21, 22)

$$\bar{\sigma}_{33}^{l*}(p = p_j) = \sigma_{33}^{l*}(p = p_j), p_j = \frac{\ln T_j - \ln T_{j-1}}{T_j - T_{j-1}}, \quad T_0 = 10^{-5} \tag{27}$$

When $p = \infty$ the terms of the sum in formula (26) become zero, so $\sigma(0) = \sigma_{33}^{l*}(p = \infty) = 1.206$. After some transformations, a system of linear algebraic equations was obtained, the order of which coincides with the number of links in the broken line:

$$\begin{cases} \sum_{i=1}^{5} \gamma_i \left(e^{-p_1 T_{i-1}} - e^{-p_1 T_i} \right) = p_1 \left[\frac{\sigma_{33}^{l*}(p_1)}{\sigma(0)} - 1 \right] \\ \ldots\ldots\ldots\ldots \\ \sum_{i=1}^{5} \gamma_i (T_i - T_{i-1}) = \frac{\sigma_{33}^{l*}(0)}{\sigma(0)} - 1 \end{cases} \tag{28}$$

The last equation of the system is different from the previous ones, since it corresponds to the point $p = 0$. The matrix of the system is poorly conditioned, therefore the authors proposed a method to improve its conditionality, which was reduced to clarifying the points of collocation from solving a transcendental equation

$$\frac{1}{m} \left(e^{-p_j T_{i-1}} - e^{-p_j T_i} \right) = \left(e^{-p_j T_i} - e^{-p_j T_{i+1}} \right), \tag{29}$$

where m is a natural number.

m was chosen so that the matrix of the system had an almost triangular appearance. According to the numerical experiment performed by Parfenova T. V., the optimal value $m = 6$.

As a result of solving the SLAE, the parameters of the broken line were obtained: $\gamma 1 = -0.806$, $\gamma 2 = 0.122$, $\gamma 3 = -0.017$, $\gamma 4 = -0.008666$, $\gamma 5 = -0.026$.

Figure 7 shows the graphs of pore pressures changing over time at points: (a) $x = 0$ $z_1 = 0.5b$ (—), $z_2 = b$ (···), $z_3 = 1.5b$ (- - -); (b) $x = b$ $z_1 = 0.5b$ (—), $z_2 = b$ (···), $z_3 = 1.5b$ (- - -).

A comparison of the solution in the kinematic model with the solution of Korotkin V. G. is carried out according to the model of filtration consolidation. Comparison with this solution showed a qualitative similarity for the time-intermediate state of the consolidation process. The parameters of the kinematic model were determined as a result of testing a large-sized sample of saturated clay, then a comparison was made of the theoretical forecast for the new model with the results of testing this clay in a tray, carried out by Nabokov A. V., and with calculations obtained from other models. The theory successfully conforms to the calculations and experiment.

Comparison of pore pressure with an experiment conducted on a sample of water-saturated peat by Amaryan L.S. showed that the largest discrepancies are observed at the maximum points. The maximum of them is 18%, when the function reaches the asymptotic value (residual pore pressure) the discrepancy is no more than 5%.

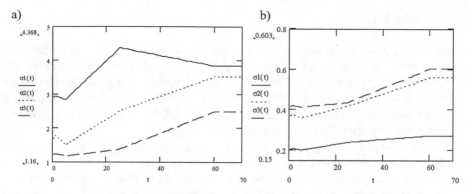

Fig. 7. The change of pore pressure over time.

Theoretical processing of the Zekhniev's field experiment was carried out according to the data from the lower pore pressure sensor (the furthest from the surface). The theoretical forecast was made for the other two sensors (Fig. 8). For the middle sensor №1 the maximum discrepancy was 10%, and the discrepancy in residual pore pressure was 6.7%. For the upper sensor №3, the maximum discrepancy was 23%, and for residual pore pressure – 7%.

Thus, the theory satisfactorily conforms to the results of the field experiment.

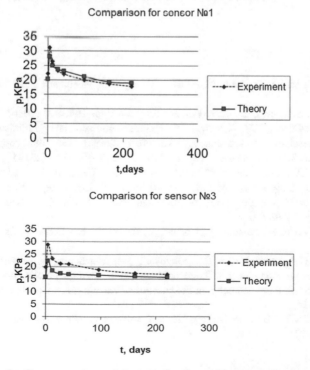

Fig. 8. The comparison of theoretical values with experimental ones.

4   Conclusion

- After the consolidation process is completed, modeling of the stress and strain state of saturated soil taking into account residual excess pore pressure reduces to systems of two or three linear differential equations, which are a generalization of the well-known Lame equations differing from them by additional terms describing pore water. The process of soil consolidation is described using a system of integral-differential equations, which is solved in two stages.
- The mechanical characteristic of saturated soil (parameter of one-dimensional model) is determined as a non-monotonic function of time.
- In the Flamann problem, the stressed and strain state is decomposed into two phases. Comparing the vertical strains of daylight surface points for single-phase and two-phase bodies, it follows that the effect of pore water on the soil skeleton is manifested in rapid decrease of these bodies.
- The calculation of the stress-strain ground base subjected to the influence of closely located objects (6 m and more) showed that with the removal of objects from each other, the normal stresses in the soil skeleton that are found due to the new model, decay 40% faster than the stresses found due to Flamann's solution. This led to a decrease in sediment skeleton of the soil by 26% compared with the same solution.
- Calculations of the model for a stabilized state, taking into account residual excess pore pressure, conform quite well (with maximum difference of 7%) to known results, and calculations of the model for consolidating saturated soil conform to known results satisfactorily (with maximum difference up to 24%).

References

1. Usmanov, R., Mrdak, I., Vatin, N., Murgul, V.: Applied Mechanics and Materials, pp. 633–634 (2014)
2. Maltseva, T., Saltanova, T., Chernykh, A.: Procedia Eng. **165**, 839 (2016)
3. Bay, V.F., Mal'tseva, T.V., Krayev, A.N.: Metodika rascheta slabogo glinistogo osnovaniya, usilennogo peschanoy armirovannoy po konturu podushkoy s krivolineynoy podoshvoy (in Russian). Nauchno-tekhnicheskiy vestnik Povolzh'ya, **5**, 108–111 (2014)
4. Vorontsov, V.V., Chikishev, V.M., Ogorodnova, Y.V., Lipikhin, A.S.: Eksperimental'nyye issledovaniya raboty slabogo glinistogo osnovaniya, armirovannogo geotekstil'nym materialom pod deystviyem polosovoy nagruzki (in Russian). Nauchno-tekhnicheskiy vestnik Povolzh'ya **3**, 88–93 (2014)
5. Korsun, V., Vatin, N., Franchi, A., Korsun, A., Crespi, P., Mashtaler, S.: Procedia Eng. **117**, 970 (2015)
6. Novikov, Y.U.A., Nabokov, A.V.: Issledovaniye napryazhenno-deformirovannogo sostoyaniya vodonasyshchennogo glinistogo osnovaniya, usilennogo peschanymi armirovannymi po konturu svayami pod lentochnym fundamentom. Vestnik grazhdanskikh inzhenerov **3**(44), 133–136 (2014)
7. Mal'tseva, T.V.: Matematicheskaya teoriya vodonasyshchennogo grunta. Vektor buk, Tyumen' (2012)

Asphalt Concrete Using Polymer Waste from the Factories of Siberia

Galina Vasilovskaya$^{(\boxtimes)}$ (iD), Maria Berseneva$^{(\boxtimes)}$ (iD),
and Eugene Yanaev$^{(\boxtimes)}$ (iD)

Siberian Federal University, Svobodny avenue, 79,
660041 Krasnoyarsk city, Krasnoyarsk Region, Russia
vasgv2ln@mail.ru, mari-leonm@yandex.ru,
EYanaev@sfu-kras.ru

Abstract. The article presents the results of studies of asphalt concrete with the use of polymer waste from Siberian factories. There were examined: rubber crumb on general purpose rubber obtained by shredding old automobile tires and butyl rubber crumb from a tire repair plant in Krasnoyarsk, devulcanized coagulum from Krasnoyarsk synthetic rubber plant, atactic polypropylene from chemical plant in Tomsk. Using infrared spectroscopy, the nature of the interaction of polymer waste with bitumen was studied. It is defined that only butyl rubber crumb chemically interacts with bitumen. The remaining waste with bitumen forms a heterogeneous two-phase mixture. The basic properties of polymer waste as well as mineral aggregates for asphalt concrete and bitumen are determined. Waste products from the Krasnoyarsk Chemical Metallurgical Plant was used as a mineral powder in asphalt concrete. Chemical and x-ray diffraction analyses were used to explore the composition of this powder. Polymer bitumen binders were prepared with polymeric wastes, and their main physical and mechanical parameters were determined. Based on the studies conducted, the optimal compositions of polymer-bitumen binders were selected. Using local mineral aggregates and optimal compositions of polymer-bitumen binders, asphalt-polymer concrete mixtures were calculated and prepared, which were tested for basic physical and mechanical parameters. It was determined that all polymer additives improve the characteristics of asphalt concrete, but the best performance is asphalt concrete with butyl rubber crumb. Experimental work on the developed composition of asphalt concrete was conducted. The resulting compounds are recommended for "Krasnoyarskavtodor" for production implementation on the roads of the Krasnoyarsk region.

Keywords: Bitumen · Asphalt concrete · Polymeric waste ·
Physical and mechanical properties · Areas of Siberia ·
Polymer-bitumen binder · IR spectra · Diffractogram · Mineral powder ·
Crushed stone · Sand

1 Introduction

In connection with the design, construction and operation of road asphalt concrete pavements in areas of Siberia characterized by harsh climatic conditions, the problem of increasing their reliability has become of great importance. The peculiarity of the

© Springer Nature Switzerland AG 2020
V. Murgul and M. Pasetti (Eds.): EMMFT 2018, AISC 982, pp. 885–899, 2020.
https://doi.org/10.1007/978-3-030-19756-8_85

materials operating conditions in these areas is that they are characterized by very low winter temperatures and rather high summer temperatures with a long period of time with temperatures below 0 °C.

Such harsh weather conditions impose high requirements to the materials and structures of road surfaces. Namely, they should have the deformability required for these regions in the frost and heat resistance at increased summer temperatures. Asphalt materials on bitumen binder do not meet the requirements imposed on them, since the temperature range of the working capacity of bitumen is almost entirely in positive temperatures. Asphalt concrete deformability problems are considered in the works of Boguslavsky, Kandhal, Zawadzki [1–9].

Therefore, there is a need to improve the characteristics of bitumen. Comprehensive studies of the properties of bitumen were delivered by V.A. Zolotarev, S.V. Popadak and others. The problem of improving the quality of bitumen is currently being solved in two ways: improving the technology of bitumen production and combining bitumen with various additives that increase their physicomechanical properties. The most promising is a method for improving bitumen with polymer additives. In this way, one can either significantly improve the operational characteristics of bitumen, or obtain a new material with completely different physicomechanical and chemical properties. The problems of modifying the properties of bitumen are dealt with by V.A. Zolotarev, J. Pilat, I. Gawel, L.M. Gokhman.

To improve the quality of bitumen, it is necessary to change its structure and mechanical characteristics in the direction of approximating them to those of high polymers that are distinguished by their ability to maintain high deformability at sufficiently low negative temperatures and, at the same time, do not soften or lose strength when heated. Currently, bitumen is modified with rubber, latex, polyethylene, rubber regenerate, etc.

However, polymers are expensive and scarce materials. Therefore, a very promising is the use of waste polymers in the composition of road asphalt. The use of polymer wastes to modify a bitumen binder makes it possible to increase its performance, solve an important environmental problem associated with waste management, and reduce bitumen consumption, which is especially important nowadays in the face of rising prices and a shortage of crude oil. There has also been used mineral waste of industrial production. These issues are addressed by researchers.

The purpose of this work was to study the possibility of using the waste of polymers from Siberian factories in the compositions of polymer-bitumen binder for road asphalt concrete.

2   Materials and Methods

In the work, the following were used as waste polymers:

1. Rubber crumb on general purpose rubbers obtained by grinding old tires with solitary cord inclusions from the Krasnoyarsk tire repair plant. The granulometric composition of rubber crumb is given in Table 1.

Table 1. Granulometric composition of rubber crumb.

Sieve sizes, mm	Fractional tailings, %	Full tails, %	Passed through sieve, %
2.5	–	–	100
1.25	56.0	56.0	44.0
0.63	17.0	73.0	27.0
0.315	16.0	89.0	11.0
0.14	8.9	97.9	2.1
0.071	2.0	99.9	0.1
0	0.1	100.0	0

2. Butyl rubber crumb, obtained by grinding the plugs from the old medical vesicles. The granulometric composition of butyl rubber crumb is given in Table 2.

Table 2. Granulometric composition of butyl rubber crumb

Sieve sizes, mm	Fractional tailings, %	Full tails, %	Passed through sieve, %
1.25	27.75	27.75	72.25
0.63	51.25	79.0	21.00
0.315	19.75	98.75	1.25
0.14	1.25	100	–
0.071	0	–	–
0	0	–	–

3. Devulcanized coagulum is a waste of rubber production of the Krasnoyarsk plant of synthetic rubber. In appearance, these are pieces of rubber-like mixture.
4. Atactic polypropylene is a by-product in the production of polypropylene at the Tomsk Chemical Plant. Atactic polypropylene is produced in the form of plates powdered with talcum powder. Atactic polypropylene is a rubbery substance with a molecular weight of 20,000–40,000 amu, has an amorphous structure. Properties of atactic polypropylene are presented in Table 3.

Table 3. Properties of atactic polypropylene

Name of indicators	Characteristics
1. Form	Amorphous light gray substance
2. Volatile matter content, %	3
3. Ash content, %	3
4. Content of isotactic fraction, %	40
5. Immersion depth of needle at 25 °C, degree of penetration	60
6. Melting point, °C	130
7. Viscosity at 180 °C, SDRs	10000

The choice of these wastes for the modification of the bitumen binder is explained by the fact that the rubber crumb has an organic affinity with the components of the bitumen and with their physical and mechanical interaction, a new material is obtained that differs favorably from the bitumen. Light oil fraction of bitumen with the introduction of the rubber vaporize and enter it due to the partial swelling and dissolution. dissolution. Thus, rubber devulcanization occurs. This has a stabilizing effect on bitumen, protects it from premature aging. A large amount of waste is generated in the production of synthetic rubbers. The use of rubber production waste in the form of coagulum is also an effective direction in obtaining polymer-bitumen binders for asphalt concrete, since its combination with bitumen does not cause difficulties. Atactic polypropylene belongs to the least reactive class of polyolefins, that is, it is less susceptible to various factors, including atmospheric. Also, atactic polypropylene is most widely used from waste polyolefins. As a rule, polyolefins have a low brittleness temperature. These types of polymers are the least scarce and relatively cheap.

5. Bitumen of BND 60/90 grade from the Achinsk refinery was used as a binder in asphalt concrete. The properties of bitumen are given in Table 4.

Table 4. BND 60/90 bitumen properties

Name of indicators	Characteristics
1. Softening temperature by the method of "Ring and ball",°C	50
2. Immersion depth of needle at 250 °C, degree of penetration	80
3. Extensibility, cm, at 25 °C	75
at 0 °C	7.5
4. Fraas brittleness temperature, °C	−15

6. As a filler in asphalt concrete was used limestone flour, which is a waste product of the Krasnoyarsk Chemical and Metallurgical Plant (KMZ). The chemical composition of limestone flour is given in Table 5.

Table 5. Chemical composition of limestone flour

Chemical composition, %				
$CaCO_3$	CaO	LiO_2	Free alkaline	Total
98.16	1.4	0.24	0.2	100

As can be seen from the table, the main mineral of limestone flour is calcite mineral, which is part of carbonate rocks, which should ensure good adhesion of bitumen with mineral powder due to the processes of chemisorption at the bitumen - stone phase boundary. X-ray analysis of limestone flour was also performed.

Diffractogram of limestone flour is shown in Fig. 1.

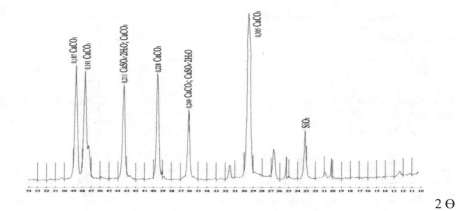

2Θ

Fig. 1. Diffractogram of limestone flour

Analysis of the diffractogram showed that calcite is also the main mineral of limestone flour, which confirms the results of chemical analysis.

The grain composition of limestone flour was determined. The test results compared to the requirements of GOST R 52129-2003 are given in Table 6. The table shows the arithmetic mean values obtained by testing three samples.

Table 6. Grain composition of limestone flour

Sieve sizes, mm	Fractional tailings, g	Fractional tailings, %	Full tails, %	Passed through sieve, %	GOST requirements
1.25	0	0	0	100	Not less than 100
0.63	0	0	0	100	–
0.315	0.01	0.01	0.01	99.99	Not less than 90
0.16	5.81	5.81	5.82	94.18	–
0.071	14.78	14.78	20.60	79.40	70 to 80
Bottom	79.40	79.40	100.00	0	–
Total	100	100			

As can be seen from the table, according to the grain composition, limestone flour meets the requirements of GOST R 52129-2003. Table 7 shows physical and mechanical properties of limestone flour compared to the requirements of GOST R 52129-2003 for mineral powder MP−1.

Table 7. Physical and mechanical properties of limestone flour.

Name of indicator	Dimension	GOST requirements
1. Porosity, %	33.21	Not more than 35
2. Swelling of samples from a mixture of powder with bitumen, %	0.92	Not more than 2.5
3. Moisture% by weight	0.24	Not more than 1.0
4. Bitumen intensity, g	65	–
5. Total specific effective activity of natural radionuclides, Aeff, Bq/kg	27	Not more than 740

As can be seen from Tables 5, 6 and 7, KMZ powder meets the requirements of GOST R 52129-2003 for mineral powders MP-1 and can be used in the manufacture of asphalt concrete for the construction of roads within the territory of settlements and prospective development areas.

7. Crushed stone and crushed sand from the Berezovsky open pit in the city of Krasnoyarsk were used as coarse and fine aggregates.

Determination of the physicomechanical properties of waste polymers, bitumen, mineral aggregates, polymer bitumen binder and asphalt concrete was carried out according to the methods of the corresponding State Standards.

To study the interaction of bitumen with polymer waste, IR spectroscopy was used. The IR spectra were recorded on a Specord 75 IR spectrophotometer in the range of 400–4000 cm^{-1}, where vibrations of bonds of various groups of bitumens are manifested. The recording conditions of the spectra are selected in such a way that the hardware distortions are minimized. Samples were recorded in a thin layer between KBr plates. Solid samples were prepared in a potassium bromide matrix with a fixed hanging of the substance. In the figures of the IR spectra, the horizontal wavelengths are plotted, and vertically, the intensity of the absorption by the substance of the infrared part of the spectrum.

3   Results

First, the effect of selected polymeric wastes on the properties of bitumen was studied. In order to study the nature of the processes occurring during the interaction of bitumen with polymeric wastes, studies were carried out using IR spectroscopy of raw material samples: bitumen and polymeric wastes, as well as composite binders obtained from them.

Figure 2 shows the infrared spectrum of BND 60/90 bitumen. This spectrum is represented mainly by the absorption bands of the vibrations of CH2 - and CH3 - aliphatic groups in the range of 2880–2960 cm^{-1} and 1375–1460 cm^{-1}. An analysis of their intensities, as well as the presence of a doublet of the bands 720 and 745 cm $^{-1}$, indicates that paraffin hydrocarbons with an odd number of hydrocarbon atoms in the chains are the main constituent of bitumen. Along with paraffin chains of normal structure, naphthenic-aromatic structural fragments are present in significant amounts in the composition of the bitumen under study. This is evidenced by the presence of absorption bands of 3060, 1600, 1515, 810, and 870 cm^{-1} in the spectrum of bitumen.

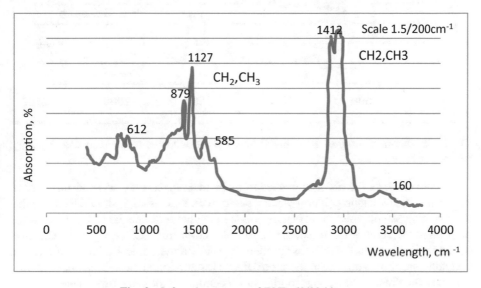

Fig. 2. Infrared spectrum of BND 60/90 bitumen

Oxygen-containing products are represented in the structure of bitumen by functional groups:

- OH (absorption band 3380 cm^{-1});
- carbonyl (complex absorption band 1695 cm^{-1}), apparently included in the composition of ketones;
- ethereal (absorption band 1030 1300 cm^{-1}).

It should be noted that the amount of oxygen-containing products in bitumen, as shown by the results of spectroscopic studies, is very small, 1.7% to 2.0%.

Figure 3 shows the IR - spectra of the original bitumen (spectrum "a") and mixtures of bitumen with butyl rubber powder (spectrum "b"). Comparing the spectra shows their significant difference in the range of 1500–400 cm^{-1}. The "b" spectrum contains absorption bands at 1015, 450, and 1225 cm^{-1} and a doublet of 1370–1380 cm^{-1} bands, which are absent in the spectrum of the original bitumen. The observed bands are present in the spectrum of butyl rubber, which proves the introduction of this additive in bitumen. The analysis of the strip contours indicates the presence of a chemical bond between the fragments of butyl rubber chips and functional groups of bitumen.

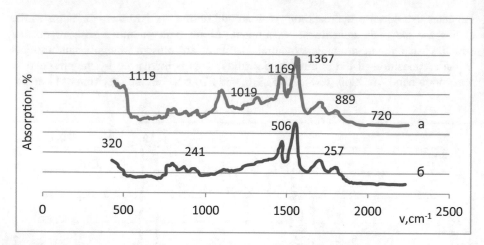

Fig. 3. IR spectra of original BND 60/90 bitumen and mixtures of bitumen with 7% butyl rubber powder

Figure 4 shows the absorption spectra of bitumen (spectrum "a") and mixtures of bitumen with 7% atactic polypropylene (spectrum b). First of all, it should be noted that the resulting sample mixture is heterogeneous in volume. Evidence of the latter is the poor reproducibility of the IR spectra of samples taken from different points. This may be due to the fact that the resulting product is mainly a mechanical mixture of bitumen and additives. To some extent, the proof of this assumption is the significant scattering (background difference) when recording the spectrum. At the same time, changes in the ratio of the intensity of the bands in the region of 720–870 cm^{-1}, as well as 1380 and 1465 cm^{-1}, indicate the presence of CH3-groups chemically bound with bitumen in the product. However, the question of the nature of the connection is unequivocally difficult to solve, since in our opinion the main mass is a mechanical mixture.

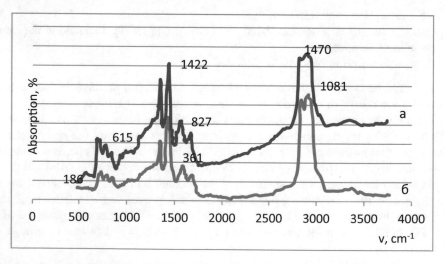

Fig. 4. Absorption IR spectra of BND 60/90 bitumen and mixtures of bitumen with 7% atactic polypropylene.

Figure 5 shows the IR spectra of crumb rubber, Fig. 6 shows a mixture of bitumen with 8% rubber crumb, and Fig. 7 shows the spectra of devulcanized coagulum and Fig. 8 shows the spectra of bitumen with 8% of coagulum. Analyzing the obtained spectra, we can conclude that this waste does not chemically interact with bitumen, since the spectra of bitumen with these wastes do not contain new absorption bands characteristic of the waste, but only bands characteristic of bitumen are observed. The bands that appear are so insignificant that they cannot correspond to the formation of new chemical compounds. And the fact that the absorption bands corresponding to polymers are not observed on the IR spectra of the mixtures is explained by the fact that the concentration of polymers in the bitumen is insignificant and the bulk is bitumen. Thus, rubber crumb bitumen compositions and devulcanized coagulum are physical mixtures and are heterogeneous two-phase systems.

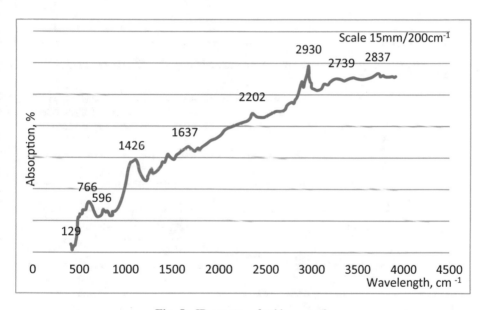

Fig. 5. IR spectra of rubber crumb.

Compositions of polymer bitumen binders were prepared on selected polymeric wastes. For this, bitumen was injected with a different amount of waste. The minimum dosage of waste was 2% of the mass of bitumen, and the maximum dosage was determined by their technological compatibility with bitumen and amounted to 15 to 20%.

The rubber crumb was pre-screened through a 5 × 5 mm sieve, and the devulcanized coagulum was crushed into a crust of 2 to 3 mm. The temperature of mixing bitumen with crumb rubber and coagulum was 180 to 200 °C, mixing time was 2 to 3 h. The preparation of compositions with atactic polypropylene and butyl rubber chips

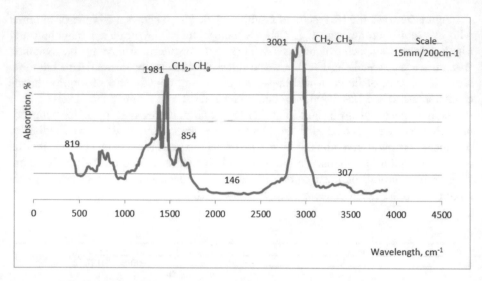

Fig. 6. IR spectra of BND 60/90 bitumen mixture with 8% rubber crumb.

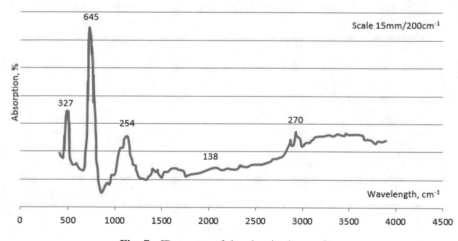

Fig. 7. IR spectra of devulcanized coagulum.

was carried out as follows. The required amount of polymer was introduced into the evaporated and heated to a temperature of 160 to 180 °C bitumen with continuous stirring. Addition of the subsequent portion was carried out after melting the previous one. The total mixing time of bitumen with atactic polypropylene and butyl rubber crumb was 1.5 to 2 h.

The study of physicomechanical properties of the polymer-bitumen binder consisted of determining its deformability at a negative temperature, which was

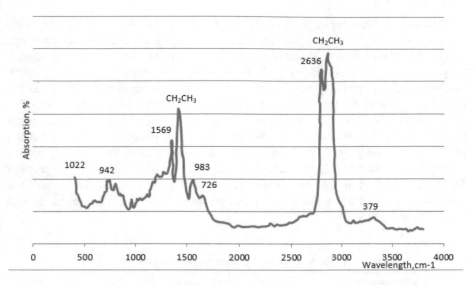

Fig. 8. IR - spectra of BND 60/90 bitumen mixture with 8% devulcanized coagulum.

characterized by the temperature of brittleness, water absorption and standard characteristics: softening temperature, penetration and tensile properties. The properties of the polymeric bitumen compositions are given below in Table 8.

Table 8. Properties of polymer bitumen binders

Amount (%) and type of additive	Softening temperature, °C	Penetration, °P		Extensibility, cm		Water absorption under vacuum % wt.	Fraas brittleness temperature, °C
		at 25 °C	at 0 °C	at 25 °C	at 0 °C		
2 APP	55.8	8.5	2.0	6.3	0.2	0.21	−21.3
5 APP	69.3	9.0	3.2	7.2	0.4	0.53	−25.0
10 APP	74.6	22.3	13.0	8.8	0.5	1.85	−26.0
15 APP	81.5	45.0	18.0	–	0.5	1.53	–
20 APP	87.4	32.8	26.0	–	–	2.24	–
5 RC	48.5	11.0	6.5	5.0	0.3	0.5	−19.0
10 RC	50.7	16.0	8.1	6.0	0.4	0.3	−21.0
15 RC	67.0	22.0	7.4	8.0	0.3	0.04	−20.0
5 C	50.8	7.0	4.0	3.5	0.2	0.9	−18.0
10 C	66.3	9.0	5.0	4.8	0.3	0.4	−19.0
15 C	75.4	10.0	6.0	5.4	0.4	0.3	−20.0
5 BRC	60.5	13.8	11.3	12.3	2.8	0.2	−25.0
10 BRC	71.2	16.7	7.8	14.2	4.6	0.3	−28.0
15 BRC	78.4	26.9	6.5	15.6	6.9	0.4	−32.0

where: APP – atactic polypropylene; RC - rubber crumb; C - devulcanized coagulum; BRC - butyl rubber crumb.

Studies have made it possible to conclude that the introduction of all polymer waste leads to an increase in the softening temperature and a decrease in the brittleness temperature of the bitumen. Reduced water absorption occurs before certain dosages of polymeric waste. Compositions with high dosage of waste have a large water absorption. This is probably due to the poor distribution of large amounts of waste in the mixture and in the preparation of heterogeneous compositions. The best performance was obtained with compounds with butyl rubber crumb and atactic polypropylene. On the basis of the studies carried out, taking into account the best performance of the polymer-bitumen binder with the lowest cost, the following optimal formulations were selected, on which hot fine-grained asphalt-concrete was prepared for the upper layers of road pavements. The amount of polymer-bitumen binder was selected by an experienced method. The compositions of polymer bitumen binders are given below in Table 9.

Table 9. Polymer bitumen binder compounds

Compound #	1	2	3	4	5
Quantity and type of polymer additive	4 RC	5 APP	5 C	5 BRC	–
Quantity of polymer-bitumen binder, %	7.5	7.5	7.0	7.0	6.0

To compare the properties, the asphalt concrete composition was prepared on "pure" bitumen without polymer additives (composition No. 5).

The selection of the mineral part of the asphalt concrete was carried out using the limiting curves for dense mixtures with a slip coefficient of 0.7–0.9. When selecting the compositions, we proceeded from the need to obtain dense mixtures with a minimum content of mineral powder and bitumen. The composition of the mineral part of the asphalt concrete after selection turned out to be as follows (% wt.): Crushed stone −45.3; sand −44.8; mineral powder −9.9. The laboratory process for preparing asphalt polymer concrete consisted of preparing bitumen and preparing a polymer bitumen binder, sieving and heating mineral materials, dosing ingredients and mixing them. The preparation of bitumen consisted in its melting, dehydration by heating at 100–110 °C until the evolution of foam ceased and heating to a temperature of 150–160 °C. The preparation of the polymer-bitumen binder was carried out by mixing bitumen heated to operating temperature with polymer waste.

The preparation of the mineral components consisted in the heating of rubble and sand up to 180 °C. Mineral powder is not heated. The mixing temperature of the polymer-bitumen binder with mineral materials was 160–170 °C. Determination of the properties of asphalt polymer concrete was carried out according to the methods of GOST 12801-2013. The properties of selected compounds of asphalt polymer concrete in comparison with the requirements of GOST 9128-2013 for the II-nd road-climatic zone, type A, I-th grade and with the properties of asphalt concrete on "pure" bitumen are given below in Table 10.

Table 10. Properties of asphalt concrete

Name of indicators	Composition numbers					
	1	2	3	4	5	GOST's norm
1. Strength at compression, MPa: at 20 °C	4.3	3.0	3.55	3.65	2.75	Not less than 2.5
at 50 °C	1.15	1.1	1.17	1.25	0.95	Not less than 1.1
at 0 °C	10.5	9.63	10.87	8.75	11.5	Not more than 11.0
2. Water resistance coefficient	0.89	0.85	0.85	0.97	0.79	Not less than 0.9
3. The coefficient of water resistance with prolonged water saturation	0.85	0.79	0.82	0.87	0.72	Not less than 0.85
4. Water absorption, % of volume	0.76	0.66	0.71	0.57	1.01	–

4 Discussions

From Table 10 above it can be seen that in terms of strength at 20 °C, all asphalt-polymer concrete compositions meet the requirements of GOST and have the best performance compared to asphalt concrete on "pure" bitumen. The strength at 50 °C, which characterizes its heat resistance and the ability to form rut in summertime, is greater with a compound with butyl rubber crumb. Also, the strength at 0 °C, which characterizes its deformability at negative temperatures, is greater for compounds with butyl rubber crumb. The coefficient of water resistance does not meet the requirements of GOST compounds with crumb rubber, atactic polypropylene and coagulum. Thus, on the basis of the conducted research, it is possible to choose asphalt polymer concrete based on butyl rubber crumb, which by all indicators meets the requirements of GOST and far exceeds these indicators for ordinary asphalt concrete.

Experimental work was carried out on the developed composition of asphalt concrete. In laboratory conditions, asphalt concrete was prepared, which was placed in holes in the asphalt concrete pavement cleared from dirt and dust in the SFU. In the course of the work, it was established that asphalt-concrete mixtures fill gaps in pavements with a dense and uniform layer and adhere well to old asphalt concrete. Observations of the experimental site for the past period showed no visual changes in the coatings. The resulting composition of asphalt concrete was recommended to Krasnoyarskavtodor for the production introduction on the roads of the Krasnoyarsk region.

5 Conclusions

1. A study was conducted on polymeric waste from Siberian factories. We studied rubber crumb on general-purpose rubber, obtained by grinding old automobile tires, butyl rubber rubber-new crumb of a Krasnoyarsk tire repair plant, devulcanized

coagulum of the Krasnoyarsk synthetic rubber plant, and atactic polypropylene of the Tomsk chemical plant.

2. Using the method of IR - spectroscopy studied the nature of the interaction of bitumen with waste. It has been established that only butyl rubber crumb chemically interacts with bitumen. With the remaining waste, bitumen forms heterogeneous two-phase systems.

3. With the use of selected polymeric wastes, polymeric binder binders for asphalt concrete were prepared. After analyzing the properties of the binders, optimal compositions were selected.

4. Using local mineral aggregates and selected compositions of polymer-bitumen binders, asphalt concrete compositions were prepared and prepared for the upper layers of the pavement. The calculation of the composites was carried out according to the limiting curves for dense mixtures with a slip coefficient of 0.7 to 0.9.

5. As a mineral powder in asphalt concrete, waste from the production of the Krasnoyarsk Chemical-Metallurgical Plant was used. The composition of this waste was studied using chemical and radiographic analyzes.

6. From the prepared asphalt mix, samples were molded and tested for basic physical and mechanical characteristics. Analysis of the obtained data showed that all polymeric wastes improve the properties of asphalt concrete on "pure" bitumen and exceed the requirements of GOST.

7. The best indicators for all properties showed asphalt concrete with butyl rubber crumb. This composition of asphalt concrete was prepared under laboratory conditions and laid in the pavement on the territory of the Siberian Federal University. Observations of the test site showed that the new layer from the test composition adheres well to the old coating and no cracks appeared on it after a year of operation.

8. The developed composition of asphalt-polymer concrete was recommended to Krasnoyarskavtodor for introduction on the roads of the Krasnoyarsk region.

References

1. Rahman, A.A., Huang, H., Ding, H., Xin, C., Lu, Y.: Fatigue performance of interface bonding between asphalt pavement layers using four-point shear test set-up. Int. J. Fatigue **121**, 181–190 (2019)
2. Jaczewski, M., Judycki, J., Jaskula, P.: Asphalt concrete subjected to long-time loading at low temperatures - deviations from the time-temperature superposition principle. Constr. Build. Mater. **202**(3), 426–439 (2019)
3. Ameri, M., Mansourkhaki, A., Daryaee, D.: Evaluation of fatigue behavior of asphalt binders containing reclaimed asphalt binder using simplified viscoelastic continuum damage approach. Constr. Build. Mater. **202**(4), 374–386 (2019)
4. Li, J., Liu, J., Zhang, W., Liu, G., Dai, L.: Investigation of thermal asphalt mastic and mixture to repair potholes. Constr. Build. Mater. **201**(8), 286–294 (2019)
5. Wan, Y., Jia, J.: Nonlinear dynamics of asphalt–screed interaction during compaction: application to improving paving density. Constr. Build. Mater. **202**, 363–373 (2019)

6. Franesqui, M., Yepes, J., García-González, C.: Improvement of moisture damage resistance and permanent deformation performance of asphalt mixtures with marginal porous volcanic aggregates using crumb rubber modified bitumen. Constr. Build. Mater. **201**, 328–339 (2019)
7. Slebi-Acevedo, C., Lastra-González, P., Pascual-Muñoz, P., Castro-Fresno, D.: Mechanical performance of fibers in hot mix asphalt: a review (review). Constr. Build. Mater. **200**, 756–769 (2019)
8. Mohammadafzali, M., Ali, H., Sholar, G.A., Rilko, W.A.: Effects of rejuvenation and aging on binder homogeneity of recycled asphalt mixtures. J. Transp. Eng. Part B: Pavements. **145** (1), 04018066 (2019). Baqersad, M.aEmail
9. Tarbay, E.W., Azam, A.M., El-Badawy, S.M.: Waste materials and by-products as mineral fillers in asphalt mixtures. Innovative Infrastruct. Solutions **4**(1), 5 (2019)

Studying the Effect of Cellulose Containing Stabilizing Additives on the Bitumen Properties in SMA

Dmitrii Iastremskii[✉] and Tatiana Abaidullina

Tyumen Industrial University, Volodarskogo 38, 625000 Tyumen, Russia
yaster.dmitry@yandex.ru

Abstract. One of the methods for developing asphalt concrete with high performance and technical characteristics is the introduction of semifunctional stabilizing additives to the composition of stone mastic asphalt. Determining the optimal gauging is made by empirical identification. This is due to the lack of a proper regulatory and technical basis and functional dependencies linking the technical characteristics and the material composition of the additives with the expected effect. The results of studying the effect and mechanism of the complex cellulose containing stabilizing additive on the properties of a bituminous binder are presented. It is shown that it is necessary to set an optimal gauging of the additive in order to obtain a continuously reinforced binder with a given set of properties, since its introduction changes the basic properties of bitumen, such as ductility, viscosity and thermal resistance. It has been proven that the degree of change in properties depends on the nature of the additive, its material composition and geometrical parameters of cellulose fiber. The structural features of continuously reinforced bitumen are determined using the method of fluorescence microscopy. It has been determined that the ability of additives to prevent the delamination of the dispersed system is due to the selective adsorption of the active components of bituminous binder on the fibers. The effect at the microstructure level is defined by the content of the stabilizing additive in the composition of asphalt concrete, and also by the dispersion level of the fibers and the interaction intensity with the binder.

Keywords: Viscosity · Penetration · Stabilizing additive · Bitumen binder

1 Introduction

One of the ways to increase the service life of asphalt concrete layers of road structures is by using the stone mastic asphalt (SMA). Polymer-dispersed reinforcement of such asphalts is provided by the introduction of polymer-containing, cellulose or other fibrous stabilizing additives (SA) [1–3]. The microstructure of asphalt concrete is formed by combining bituminous binders, polymers, fillers and fibers. The properties of the composite are determined by processes occurring in contact between the liquid and solid phases. Therefore, they depend on the degree of filling, dispersion and surface activity of the fillers, the binder concentration and other factors. The surface activity of the fiber in relation to the binder and the roughness of fiber's surface are important

© Springer Nature Switzerland AG 2020
V. Murgul and M. Pasetti (Eds.): EMMFT 2018, AISC 982, pp. 900–907, 2020.
https://doi.org/10.1007/978-3-030-19756-8_86

when using various fibrous components in a stabilizing additive [4]. It is rather difficult to single out an additive, the modifying ability of which substantially prevails over the properties of other additives, from a number of known stabilizing additives due to the methodological peculiarities of studying bituminous composites. The relevance of the study lies in the development of the structural technological concept of using the stabilizing additives, the study of their influence on the physical and mechanical properties of the bituminous binder.

The main function of SA for SMA is to ensure the stability of the structure by increasing the thickness of the bituminous film, ensuring the presence of free bitumen in the intergranular space of the stone mastic asphalt mixes (SMAM) during its short-term storage and transportation to the place of laying. It also helps preventing leakage of organic binding to the surface coatings due to the formation of gelatinous mastic [1–7]. It is necessary to objectively evaluate the effect of a stabilizing additive on the physical and mechanical characteristics of the bituminous binder in order to identify the effect mechanism of cellulose fibers on the bitumen matrix structure of the asphalt concrete.

2 Materials and Methods

The results of experimental studies on the structure and physical and mechanical properties of bitumen with stabilizing additives are exposed in the paper. The studies were performed using BND 90/130 bitumen of the Omsk Oil Refinery OAO "Gazpromneft", the main technical characteristics of which are presented in Table 1.

Table 1. Characteristics of viscous oil bitumen mark BND 90/130

Ser. no.	Indicator	Standard according to GOST 22245	Actual values
1	Needle penetration depth at 25 °C	91–130	121
2	Needle penetration depth at 0 °C	no less than 28	40
3	Softening temperature by ring and ball, °C	no less than 43	45.1
4	Extensibility at 25 °C, cm	no less than 65	98
5	Extensibility at 0 °C, cm	no less than 4.0	4.5
6	Brittleness temperature, °C	no less than −17	−28
7	Flash point, °C	no less than 230	272
8	Change of softening temperature after warming up, °C	no less than 5	5
9	Penetration index	In an amount of −1.0 to +1.0	−0.4

The complex cellulose-containing stabilizing additive (CSA) [8] was used for the structuring of bitumen. It consists of cellulose fibers with the tape structure of filaments 1.1–1.4 mm long. This additive was selected due to the fact that stone mastic asphalts made with it have high operational and technical characteristics and are resistant to rutting under prolonged loading. The technical characteristics of the CSA are presented in Table 2.

Table 2. Characteristics of stabilizing additives

Ser. no.	Additive	Characteristics	Value
1	CSA	Pour density, kg/m^3	500–560
		Average fiber length, mm	1.4
		Average fiber diameter, mm	0.029
		Moisture, %	4.6
		Thermal stability, %	4.6
		Fiber content with length in amount of 0.1 to 2.0	90

The gauging of additives was varied in the range from 0.3 to 0.5% in order to study the CSA effect on the properties of bitumen, bitumen without additives was adopted as the control composition.

Before the CSA was introduced into bitumen, it was crushed in a paddle mill to a fibrous state, imitating the mechanical effect of the mixer blades on an asphalt mixing plant. After that, the prepared fiber was introduced into bitumen heated to 150–160 °C and mixed in a laboratory mixer until homogeneous state for 10 min.

The set of standard methods [9–13] has been used in order to conduct research on the effect of additives on the physical and mechanical properties of bitumen. However, standard empirical tests do not represent the effectiveness of stabilizing additives, do not simulate the technological features of preparing the stabilization-modified binder, and do not allow studying the structural features of the composite. In order to study the interaction processes between the bitumen and components of a stabilizing additive, and to determine their interaction and impact on the structure formation, bitumen compositions with the considered stabilizing additive were studied using fluorescence microscopy. The operating principle of the "BiOptic S-400" fluorescent microscope is as follows: under the influence of exciting light, fluorochromes of polymeric substances begin to give off light, which is yellow-green luminescence in the range from 500 to 650 nm [14–17].

In order to prepare the samples for the research, bituminous binder samples with a structure-forming additive were boiled in distilled water for 60 min, then cooled to room temperature and dried for 24 h. Then the required sample volume was taken and a fluorescent microscopic analysis was performed with digital microphotograph recording.

3 Results

Bituminous binder with a complex cellulose-containing stabilizing additive is a slow-moving, viscous-plastic mixture with a black sheen. After the final distribution of the cellulose fiber at the termination of mixing, a portion of the fibers was deposited at the bottom of the tin container (determined visually), in which the mixing was performed.

Rheological properties indicators of the bitumen with CSA are given in Table 3. Graphic correlations between penetration, softening temperature, tensile properties of bitumen and the amount of additive added in SMA are presented in Figs. 2, 3 and 4.

Table 3. Properties indicators of bitumen with additives

Ser. no.	Indicator	The content of the additive in bitumen, %	Unit	CSA
1	Needle penetration depth at 25 °C	0	mm^{-1}	121
		0.3	mm^{-1}	93
		0.4	mm^{-1}	89
		0.5	mm^{-1}	83
2	Softening temperature by ring and ball	0	°C	45.1
		0.3	°C	47
		0.4	°C	49
		0.5	°C	53
3	Bitumen extensibility at 25 °C	0	cm	98.0
		0.3	cm	77.8
		0.4	cm	74.2
		0.5	cm	72.6

The correlation between the indicators of the needle penetration depth in bitumen with CSA at 25 °C and the content of the additive in it is illustrated in Fig. 1.

The average bitumen penetration value for the control sample is 120 mm. With the introduction of 0.3% additive, the needle penetration depth of the bituminous binder with the CSA fibers is reduced by 28 mm. Thus, it can be assumed that CSA fibers absorb oils and resins more intensively, increasing the viscosity of bitumen. A subsequent increase in cellulose fiber concentration confirms this assumption. At the maximum accepted gauging of SA fiber 0.5%, the needle penetration depth for bituminous binder with CSA fibers is reduced by 38 mm.

The dependence of bitumen softening temperature values on the amount of SA is presented in Fig. 2.

According to the ring and ball method, the softening temperature of bitumen is 45 °C for the control sample. After introducing 0.3% of CSA, the softening temperature increases by 4 °C. When the gauging exceeds 0.3%, the heat resistance growth rate continues. Thus, the required softening temperature for a bituminous binder with a 0.5% CSA is 52.5 °C. However, viscosity seriously changes at dosages exceeding 0.3% according to the results of the study, which can lead to changing the preparation modes of

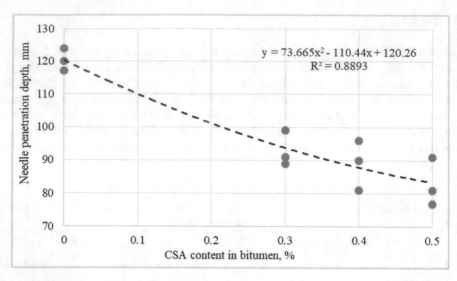

Fig. 1. The correlation between the bitumen penetration and the content of additives

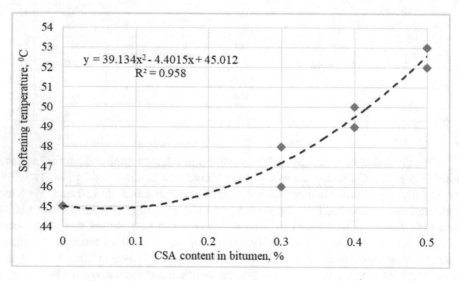

Fig. 2. The dependence of the bitumen softening temperature from the additive content

the mixture. Figure 3 shows the dependence of the bitumen extensibility at 25 °C on the additive content.

It is found that increasing the gauging of the additive leads to the decrease of bitumen extensibility at 25 °C. The bituminous binder extensibility is reduced by 20.2 cm with a 0.3% content of CSA cellulose fiber. The subsequent increase in the content of the additive leads to the continuous increase of the bitumen viscosity. At 0.4%, the extensibility of the bituminous binder with the CSA fibers decreases by

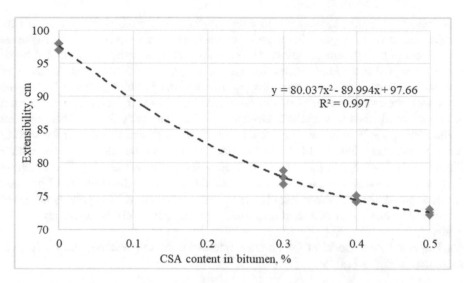

Fig. 3. The dependence of the bitumen extensibility on the additive content

another 3.6 cm and is 74.2 cm. With a further increase in the content of the additive up to 0.5%, the bitumen binder with the CSA fibers has a ductility of 72.6 cm.

According to the results of conducted research, the use of CSA has a positive effect in terms of improving heat resistance with a slight change of other characteristics. Increasing heat resistance will ensure the stability of the asphalt concrete at elevated temperature conditions, but the extensibility begins to decline rapidly after increasing gauging above 0.3%, which does not meet the conditions of the laying and compacting technological processes for asphalt concrete. Thus, CSA arguably improves the properties of bitumen, extends the range of plasticity and increases its durability.

The structure of the bituminous binder and the interaction effects on the bitumen-fiber interface are reflected in the presented photomicrographs (Fig. 4).

Fig. 4. Electronic microphotography of a bitumen composite with CSA; a – x10 multiplicity; b – x20 multiplicity.

The distribution process of cellulose fiber from stabilizing additives in bitumen can be represented as follows. Under the action of high temperature and due to the penetration of bitumen components into the granules, they swell and disintegrate with mechanical stirring. The swelling process means impregnating fibers with the maltenes component of the bituminous binder. The cellulose ability to swell is determined by its composition and structure. Cellulose is a natural polymer and is a fibrous substance of fibrillar capillary-porous structure. During mixing, the bitumen structural elements react with the macromolecules of the fibers and enter into an adsorption interaction with them inside the fibers on the free surfaces of the elementary fibrils. When bitumen is introduced, swelling is caused by the distribution of oils and resins over the surface and into the micro-bundles of the fibers. Each fiber in the bundle is covered with a bitumen film, as a result, the volume of micro-bundles increases [14]. The resulting fluorescent photomicrographs show that the cellulose fiber in the structure of bitumen forms a spatial network, adsorbing bitumen films on its surface. This process is caused by intermolecular interactions and diffusion. The thickness of the adhesive layer is 1–3 nm.

The conditional model of the bitumen interaction on the cellulose fiber surface is presented in Fig. 5 [18].

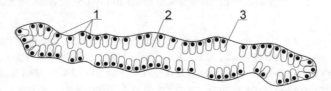

Fig. 5. Interaction model of bitumen on the cellulose fiber surface. I - bitumen structure formation centers; II - cellulose fiber; III - bitumen film.

The interactions on the interface surface are one of the most important factors affecting the strength of the crushed stone mastic asphalt. Non-additive properties appear as a result of the formation of an extended film structure (synergistic effect), such as increase in the strength and thermal resistance of the composite.

4   Conclusions

1. It has been found that the reinforcing effect is determined by the fiber saturation degree of the bitumen binder, as well as by the reinforcement dispersion level, which is particularly due to the diameter of the fibers used, and by the homogeneity degree of the bitumen composite.
2. The necessity of optimizing the composition of the bitumen binder is substantiated according to the condition of ensuring the asphalt concrete stability based on it. The composition has to be optimized for shear and climatic effects, so the optimal gauging of the complex cellulose-containing stabilizing additive is 0.4% of the bitumen mass.
3. It has been found that there is an increase in the thermal resistance of the bitumen binder with an increase in the concentration of CSA. Also, the smaller is the

diameter of fibers, the greater is the influence, so the softening temperature of bitumen increases in direct proportion to the increase in the concentration of CSA fibers by 10 °C.

4. A pattern proving that the introduction of SA significantly (by 24–27%) reduces the bitumen ductility has been determined. The pattern is caused by processes occurring on the contact surface of the fibers and depends on the content and dispersion of the fibers.

5. Cellulose fiber in the structure of bitumen forms a spatial network, adsorbing bitumen films on its surface due to intermolecular interactions and diffusion. The formation of such system increases the viscosity of bitumen and the number of elastic bonds building up its resistance to the effects of temperature deformations.

References

1. Kostin, V.I.: Shhebenochno-mastichnyj asfal'tobeton dlja dorozhnyh pokrytij (Stone-Mastic Asphalt for the Road Surface), 67 p. NNGASU, N. Novgorod (2009)
2. Yady'kina, V.V., Tobolenko, S.S., Trautvajn, A.I.: Stabiliziruyushhaya dobavka dlya shhebenochno-mastichnogo asfal'tobetona na osnove otxodov cellyulozno-bumazhnoj promy'shlennosti. Izvestiya vy'zov. Stroitel'stvo, no. 2, pp. 31–36 (2015)
3. Yady'kina, V.V., Gridchin, A.M., Trautvajn, A.I.: Vliyanie stabiliziruyushhix dobavok iz otxodov cellyulozno-bumazhnoj promy'shlennosti na svojstva shhebenochno-mastichnogo asfal'tobetona. Vestnik BGTU im. V.G. Shuxova, no. 6. pp. 7–11 (2013)
4. Kiryukhin, G.N.: Coatings of crushed stone-mastic asphalt concrete. Elite, Moscow (2009). GN Kiryukhin, EA Smirnov
5. Yastremsky, D.A., Abaidullina, T.N.: Collected Materials of the International Scientific Practical Conference, vol. 3, pp. 268–273. RIO FGBOU V Tyumen Industrial University, Tyumen (2016)
6. Yastremsky, D.A., Abaidullina, T.N., Pakhomov, I.A.: Collected Materials XV Scientific and Practical Conference of Young Scientists, Graduate Students, Applicants and Undergraduates, pp. 213–218. TyumGASU(2015)
7. Yastremsky, D.A., Abaidullina, T.N., Chepur, P.V., Gladkikh, V.A.: Bull. Moscow Automob. Highway State Tech. Univ. (MADI) 2(49), 63–70, (2017)
8. Yastremsky, D.A., Telesov, A.N., Telesov, A.A., Telesov, P.A.: Patent RU 2632839, MPK C08L95/00, C04B 26/26
9. Solomentsev, A.B., Baranov, I.A.: Constr. Reconstr. 4, 53–58 (2010)
10. Solomentsev, A.B., Baranov, I.A.: Constr. Reconstr. 4, 56–62 (2011)
11. Yastremsky, D.A., Chepur, P.V., Abaidullina, T.N.: Fundam. Res. 9, 96–101 (2016)
12. Yastremsky, D.A., Chepur, P.V., Abaidullina, T.N.: AIP Conference Proceedings, vol. 1800, Article number 020002 (2017). https://doi.org/10.1063/1.4973018
13. Yastremskij, D.A., Abajdullina, T.N., Zimakova, G.A.: Issledovanie vliyaniya stabiliziruyushhix dobavok na svojstva bituma v ShhMA. Vestnik Moskovskogo avtomobil'nodorozhnogo gosudarstvennogo texnicheskogo universiteta (MADI) 4, 63–70 (2018)

Damage Accumulation in Asphalt Concrete Under Compression

Natalya Aleksandrova[ID], Vasiliy Chusov[(✉)][ID], and Yuriy Stolbov[ID]

Siberian State Automobile and Highway Academy,
Mira 5, 644080 Omsk, Russia
chysow@gmail.com

Abstract. The analysis of methods for determining the damage of solids from external loads is performed, and the possibility of their application to asphalt concrete is estimated. As a result, it was found that the principles of deformation and energy equivalence are the most acceptable methods for determining damage. The strength criteria and conditions for plasticity of asphalt concrete are given, obtained by the authors by entering damage into the equations of the limiting state of solids. The method of testing asphalt concrete by compression is described, and the rules for processing the results in determining damage are given. As a measure of the stress state, a new relative characteristic has been introduced k_σ, determined by the ratio of the stress difference and the limit of proportionality to the difference of strength and proportionality limits. It is proved that if this characteristic is equal to one, the damage reaches the highest effective value, which is less than one. A further increase in damage occurs when the load on the sample decreases. Thus, an effective range of damage variation has been established, within which accumulated damage does not lead to the exhaustion of the strength resource of the asphalt concrete sample. Mathematical models describing the relationship of damage with a new measure of the stress state k_σ are obtained.

Keywords: Damage material · Asphalt pavement · Limit of proportionality · Strength limit

1 Introduction

The growth of motorization and the increase in the traffic flow of heavy and very heavy trucks causes the relevance of work aimed at developing methods for calculating asphalt concrete pavements for resistance to fatigue failure. Traditionally, this calculation is performed by comparing the maximum tensile stress from bending with the strength of asphalt concrete. The strength of asphalt concrete on tensile bending is determined taking into account the fatigue function of the number of loads, including constant coefficients, the value of which depends on a number of factors. Regulatory documents of the Russian Federation to such factors include:

1. The residual porosity, according to which asphalt concrete is classified into high-density, dense, porous and highly porous.

© Springer Nature Switzerland AG 2020
V. Murgul and M. Pasetti (Eds.): EMMFT 2018, AISC 982, pp. 908–918, 2020.
https://doi.org/10.1007/978-3-030-19756-8_87

2. The viscosity of bitumen used for the preparation of asphalt mixture, in accordance with which viscous bitumens are classified into brands.
3. Road-climatic zone characterized by climatic parameters that affect the aging of bitumen and the occurrence of defects in asphalt concrete that affect long-term strength.
4. Crushed stone content in cold asphalt mixes.

Despite the large amount of research carried out by experts of the road sector of the Russian Federation, the regulatory method of calculation has significant drawbacks. First, the criterion of the greatest normal tensile stresses is applied, which incorrectly evaluates the strength of materials. The disadvantages of the maximum normal stress criterion are well known and are described in the literature on strength criteria and plasticity conditions. For the case of a complex stress state, this criterion is incorrect for any material [1–5]. Secondly, the applied fatigue function does not depend on the magnitude of the stress. This function is continuous and decreasing, but it postulates a decrease in strength from an arbitrarily large number of loads. In fact, asphalt is able to withstand only a certain number of loads. This number is called the limit, it depends on the magnitude of the tensile stresses during bending or deformations caused by these stresses. This argument is confirmed by the works in which the models for calculating the limit number of loads, depending on the listed by us and other factors, are given [6–9].

    In our work, we propose to go a different way, given the fatigue of asphalt concrete by introducing measures of the theory of accumulation of damage. Such measures is continuity C and damage D. Both measures are related to each other, so they can be calculated.

    The advantage of this approach is that the fault can be determined from the ratio of parameters damaged and undamaged body for example elastic modulus. From this it follows that knowing the amount of damage accumulated to a certain period of exploitation of the asphalt concrete pavement and the initial elastic modulus of the asphalt concrete, it is possible to calculate the elastic modulus of the damaged asphalt concrete. This modulus of elasticity of asphalt concrete, accounting the accumulated damage, can be used in the calculations of pavements for all strength criteria. Such a module is applicable when calculating the total modulus of elasticity on the surface of the asphalt concrete layer. It can be used to calculate the average modulus of elasticity of the upper layer of two-layer systems, in which the resistance of asphalt concrete layers to bending tensile and the resistance of the material of the lower layer to shear are checked. In any of these three calculations, an increase in damage leads to a decrease in the elastic modulus of asphalt concrete and the need to increase the thickness of the pavement or to replace the layer material with a material with higher mechanical properties (elastic modulus, bending tensile strength, cohesion, internal friction angle). Thus, the application area scope of damage functions in the calculations of road structures is much larger compared to the fatigue function used in the standard calculation method for determining the fatigue strength of asphalt concrete in bending tensile.

2   Materials and Methods

Measures of continuity and damage of the material are offered L.M. Kachanov and Yu. N. Rabotnov [10–12]. These measures are related to each other, so that they can be determined through one another. The relationship of continuity and damage is described by the formulas

$$D + C = 1; \quad D = 1 - C; \quad C = 1 - D \tag{1}$$

Given the dependencies (1) in the calculations, it is enough to enter only one of these two measures, for example, damage, which is determined by the formula [13]:

$$D = \frac{A - A_{eff}}{A}; \quad \sigma_{Dij} = \frac{\sigma_{ij}}{1 - D}, \tag{2}$$

where A – is the overall cross-section area, m^2; A_{eff} – is the effective cross-section area, m^2; σ_{ij} – components of the stress tensor, Pa; σ_{Dij} – components of the effective stress tensor, Pa.

The first formula from expressions (2) reveals the physical essence of a scalar D. The effective cross-sectional area A_{eff} is the area not affected by defects (pores, microcracks, etc.). The cross-sectional area A is understood to mean the entire area consisting of the effective area and the total damage area. If $A = A_{eff}$, then the cross section is integer, there are no defects on it, and D = 0. If $A_{eff} = 0$, then the cross section is completely occupied by defects, and the body is destroyed, which is characterized by $D = 1$. Thus, damage is the ratio of the total area occupied by defects, to the cross-sectional area of the whole (intact) body. The state of the body is described by a scalar value D, varying in the range of $0 \leq D \leq 1$.

The second of the dependencies of formulas (2) postulates the principle of equivalence of stresses in a damaged and continuous body. According to this principle, any component of the stress tensor of a damaged body σ_{Dij} can be determined through the same component of the stress tensor of a solid and a scalar variable of damage D. Using the components of the stress tensor of a damaged body, it is possible to calculate any invariant characteristic of it. Using these characteristics, it is possible to convert the strength criterion of a continuous medium into the strength criterion of a damaged medium. In [14, 15] we have proposed criteria for the calculation of asphalt concrete coatings on the resistance to tensile bending and shear resistance. Such criteria are given in the Table 1, which presents a similar criterion for a solid body for clarity.

Application of the criteria Table 1 requires the determination of damage. A large number of works are known in which the geometrical approach was used in experimental studies. The essence of this approach is to measure the total area occupied by defects within the element selected from the cross section of the sample. The difficulty lies in determining the area of each individual defect, which is caused by the small size of the damage. Especially difficult to determine the width of microcracks. Therefore, inaccuracies arose, and in some cases the scalar value D exceeded unity, which does not correspond to the restrictions imposed by formulas (2). Therefore, other approaches were developed, which received the general name of the principles of equivalence of

Table 1. Strength criteria for solid and damaged materials for the calculation of asphalt concrete pavements for resistance on tensile bending and shear [14, 15].

Purpose of the criterion	Name of the criterion	Mathematical type of criterion
1. Calculation of the resistance to tensile from bending	1.1. Mohr Criterion for solid medium	$\sigma_1 - k_{Mohr} \cdot \sigma_3 = R_{tb}$; $k_{Mohr} = R_{tb}/R_c$
	1.2. Mohr Criterion for damaged medium	$\frac{\sigma_1 - k_{Mohr} \cdot \sigma_3}{1-D} = R_{tb}$
	1.3. Pisarenko-Lebedev Criterion for solid medium	$k_{Mohr} \cdot \sigma_i + (1 - k_{Mohr}) \cdot \sigma_1 = R_{tb}$
	1.4. Pisarenko-Lebedev Criterion for damaged medium	$\frac{k_{Mohr} \cdot \sigma_i + (1 - k_{Mohr}) \cdot \sigma_1}{1-D} = R_{tb}$
2. Calculation of shear resistance in asphalt concrete	2.1. Mohr–Coulomb Criterion for solid medium	$\frac{\sigma_1 - \sigma_3}{2 \cdot cos\phi} - tg\phi \cdot \frac{\sigma_1 + \sigma_3}{2} = c$
	2.2. Mohr–Coulomb Criterion for damaged medium	$\frac{\sigma_1 - \sigma_3}{2cos\phi \cdot (1-D)} - tg\phi \cdot \frac{\sigma_1 + \sigma_3}{2 \cdot (1-D)} = c$

where σ_1 and σ_3 – the maximum and minimum principal stresses, Pa; k_{Mohr} – material parameter of Mohr; R_{tb} – tensile strength from bending, Pa; R_c – compressive strength, Pa; σ_i – the intensity of the normal stress (the intensity of Mises), Pa; c – cohesion, Pa; φ – the angle of internal friction, °.

physical states of damaged and solid bodies. Dependencies obtained on the basis of the principles of deformation and energy equivalence have gained the greatest popularity. Formulas for determining damage according to these and other equivalence principles are given in Table 2, borrowed from paper [10].

Analyzing the work of specialists in the field of damage evolution of various materials, we note that formulas obtained by applying the principles of equivalence of deformations [16, 17] and the energy equivalence of damaged and continuous media [18] have become widespread. Apparently, this is due to the ease of determining the elastic modulus of the material in the laboratory.

From the analysis of the strength criteria from Table 1 follows that for their use in calculations of asphalt concrete pavements it is necessary to determine the elastic module of damaged and intact asphalt concrete during bending (for the Mohr and Pisarenko-Lebedev criteria) and for compression (for the Mohr-Coulomb criterion). It should be borne in mind that various structural elements of the road with asphalt concrete pavement are calculated on the impact of various loads. The coating within the carriageway is designed on the impact of a dynamic short-term multiple-applied load. Asphalt pavements located on sections of roads within parking lots, public transport stops, intersections with railway tracks in the same level, near traffic lights, are counted on the impact of prolonged single and repeated loads.

Of course, to cover such a large area of calculations of asphalt concrete coatings, it is necessary to do a lot of experiments on various techniques. Therefore, in this paper,

Table 2. Formulas of equivalence principles for calculating scalar value D [11].

Name of the equivalence principle	Mathematical expression	Recommendations on the choice of the principle depending on the type of damage				
		Brittle	Ductile	Creep	Low cycle fatigue	High cycle fatigue
Damage area	$D = \frac{A_D}{A}$	Try to see	Good	Good	Try to see	Try to see
Density of damaged and solid bodies	$D = \left(1 - \frac{\rho_D}{\rho}\right)^{\frac{2}{3}}$	Do not try	Good	Try to see	Try to see	Do not try
Deformations (elastic modulus)	$D = 1 - \frac{E_D}{E}$	Good	Very good	Very good	Very good	Do not try
Ultrasonic wave speeds	$D = 1 - \frac{\vartheta_D^2}{\vartheta^2}$	Very good	Good	Good	Try to see	Try to see
Strain energy	$D = 1 - \sqrt{\frac{E_D}{E}}$	No recommendation information				
Micro-hardness	$D = 1 - \frac{H_D}{H}$	Good	Very good	Good	Very good	Try to see

where ρ_D and ρ – density of the damaged and solid, g/cm³; E_D and E – elastic modulus of damaged and solid medium, Pa; ϑ_D and ϑ – speed of propagation of ultrasonic waves in a damaged and solid medium, m/s; H_D and H – microhardness of the damaged and solid body.

the definition of damage is considered, through the elastic modulus of asphalt concrete upon application of a single load. When determining the elastic modulus or modulus of elastoplastic deformation, one should take into account that the modulus can be secant, tangent [19], and also piecewise linear and nonlinear. The rules for determining such modules are shown in Fig. 1.

The secant modulus is determined by the arctangent of the angle (β) of the slope of the straight line drawn through the origin and the point, selected on the graphical dependence of the deformation on stress, to the axis of deformations. That is, $E = \text{arctg}\beta$. The determination procedure is given in Fig. 1, a. To determine the tangent module, to the selected on the graphical dependence strain from stress the point, a tangent is constructed, and the module determines the arctangent of the angle (β) of the slope of this tangent to the deformation axis.

The rules for constructing a tangent are explained on Fig. 1, b. To determine the piecewise linear modulus, it is necessary to consider the increments of deformations $\Delta\varepsilon$ and stresses $\Delta\sigma$ from two successively applied loads. The module is determined by the ratio of the increment of stress to the increment of deformation, that is $E = \Delta\sigma/\Delta E$. The determination procedure is given in Fig. 1, c. The results of determining a piecewise linear module can be interpreted as a nonlinear module, what is shown in Fig. 1, d. In each Fig. 1, a; 1, b; 1, c and 1, d given characteristic limits of stress, the limit of proportionality p_{pr} and limit strength R_c. The proportionality limit characterizes the largest stress value at which the linear relationship between stress and strain is maintained. This means that the first straight section of any of the graphs is elastic

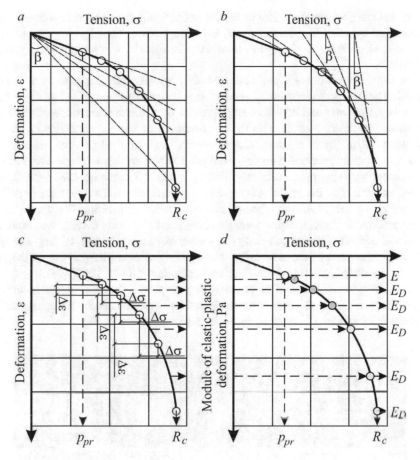

Fig. 1. Compression charts: *a, b* and *c* – schemes for determining the secant, tangent, and piecewise linear module; *d* – nonlinear module.

deformation. Since Hooke's law postulates the complete recovery of the elastic deformation after removal of the load, it can be said that the stresses varying in the range of $0 < \sigma \le p_{pr}$, do not cause damage. Therefore, the value of stress $\sigma = p_{pr}$ is the largest value at which the modulus of an undamaged (solid) body E can be determined. At stresses $p_{pr} < \sigma \le R_c$ the modulus of the elastoplastic deformation of the damaged body ED is determined. From this it follows that the level of the stress state at which damage occurs is expediently characterized by the ratio:

$$k_\sigma = \frac{\sigma - p_{pr}}{R_c - p_{pr}} \tag{3}$$

Dependence (3) is convenient because when $k_\sigma = 0$, the damage is also zero $D = 0$, and when $k_\sigma = 1$, the damage reaches the value corresponding to the exhaustion of material strength, but it is less than 1, that is, when $k_\sigma = 1$ $D < 1$. This is explained by the fact that after the stress reaches its tensile strength, the deformation continues, but is accompanied by a fall in load. However, knowledge about variation of damage within the stress state level $0 \le k_\sigma \le R_c$ is quite sufficient for strength and plasticity calculations.

Hot, fine-grained, dense asphalt-concrete mix type B, mark I is used as the starting material. The components used to prepare the asphalt mix: bitumen, crushed stone, sand, mineral powder and additives were previously tested to determine whether their controlled parameters comply with the requirements of the RF state standards. After confirming the quality of bitumen, crushed stone, sand, mineral powder and additives, an asphalt-concrete mix and samples were prepared from them. These samples were tested according to the procedures specified by the GOST standard 12801-98, in order to determine whether the asphalt-concrete mix controlled parameters comply with the requirements of the GOST standard 9128-2013. After confirming the quality of the asphalt concrete mixture, samples were prepared from it to study damage accumulation. Cylindrical asphalt concrete samples with a diameter and height of 10 cm were made for testing. The production of samples was carried out according to the method regulated by the state standard of the Russian Federation (GOST 12801-98).

The damage test was carried out with the help of a measuring and computing complex, which included a tensile machine and software installed on a computer. The test included 3 stages. Illustrations of each stage are shown in Fig. 2.

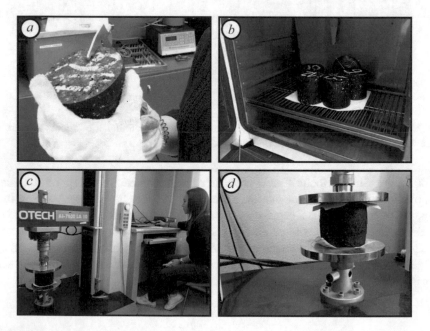

Fig. 2. Illustrations of asphalt concrete test stages: a – measurement of the geometric dimensions of the sample, which are entered into the test program as source data b – thermostating of samples to the required temperature (10 °C, 25 °C and 50 °C); c – testing and working of the operator; d – deformation of the sample with the formation of defects.

3 Results

Samples manufactured for testing are divided into three groups. The first group of samples tested at a temperature of 10 °C, the second – 25 °C and the third – 50 °C. The number of samples in each group was the same and amounted to 12 units. In Fig. 3 shows the graphic dependences of deformations on pressures when testing samples 17 and 28 at asphalt concrete temperatures of 25 °C and 50 °C, respectively.

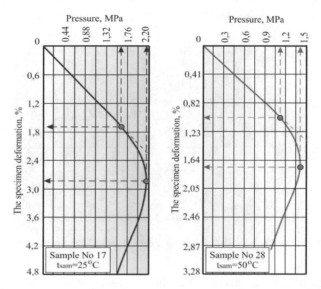

Fig. 3. Dependence of the vertical deformation of samples on pressure, tested at a temperature of asphalt concrete 25 °C and 50 °C.

The test results of each group of samples processed by mathematical statistics. The test results of asphalt samples, depending on their temperature during testing, are grouped into three data sampling. For each sample, a check was made for the presence of gross errors and the determination of statistical characteristics: the mean value, the rms value, the coefficient of variation, the interval of variation of a random variable, the determined characteristic, and the calculated values of the determined characteristics limiting the interval of variation.

Table 3 shows the limits of proportionality and strength of asphalt concrete.

Table 3. The actual and required grain composition of the mixture and the content of bitumen.

Test temperature, °C		10	25	50
Expected value	Proportionality limit, MPa	3,52	1,53	0,98
	Strength limit, MPa	5,27	2,15	1,32

At stresses and strains corresponding to the proportionality limit, elastic module of undamaged asphalt concrete are determined. For the remaining load stages, from the applying of which the stresses exceed the proportionality limit, the module of elasto-plastic deformation of the damaged asphalt concrete are determined. Using the principle of equivalence of deformations, asphalt concrete damage is calculated. Samples of damage of asphalt concrete when testing samples at different temperatures were processed by statistical methods. As a result, mean values and intervals of variation of the actual damage value are determined. The dependence of the average damage values on the level of the stress state is given in Fig. 4.

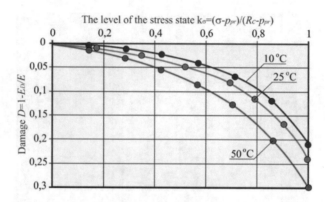

Fig. 4. Dependence of damage from the level stress state and temperature of asphalt concrete.

The experimental data presented in Fig. 4 were processed by the methods of mathematical analysis. The purpose of processing was to mathematical modeling of the damage from the level of the stress state. Since the lower limit of variations in the level of stress state and damage is zero, then not all functional dependencies are suitable for approximation or linearization of experimental data. As a result, mathematical modeling is performed by third-degree polynomials.

As a result, it was established that the dependence of damage on the temperature of hot, dense, fine-grained asphalt type B, mark I on the level of the stress state is described by dependence.

$$D = a \cdot \left(\frac{\sigma - p_{pr}}{R_c - p_{pr}}\right)^3 + b \cdot \left(\frac{\sigma - p_{pr}}{R_c - p_{pr}}\right)^2 + c \cdot \frac{\sigma - p_{pr}}{R_c - p_{pr}}, \tag{4}$$

where a, b and c – temperature dependent material parameters.

Model parameters (4) $a_,$, b and c are approximated depending on the temperature of the asphalt concrete. In the temperature range from 10 to 50 °C, to approximate these coefficients, it suffices to use second-degree polynomials.

4 Discussion

As a result of the research, it was established that damage can be used in the strength criteria as a parameter that increases the magnitude of stresses or invariant characteristics of the tensor and stress deviator.

Determination of asphalt concrete damage is advisable to carry out on the basis of applying the principle of deformation equivalence, calculating difference of one and the ratio of the modules of damaged and intact asphalt concrete.

However, the task of further research should be the development of a theoretical and experimental method for calculating and determining damage when exposed to repeated loads. Such studies will allow to design asphalt concrete pavements not only in places of parking and car stops, but on the carriageway of roads.

References

1. Wagnoner, M.P., Buttlar, W.G., Paulino, G.H.: Disk-shaped compact tension test for asphalt concrete fracture. Exp. Mech. **45**(3), 270–277 (2005)
2. Aleksandrov, A., Aleksandrova, N., Chusov, V.: Ways of application of the provisions of mechanics of bodies with cracks to the calculation of asphalt concrete on strength and plasticity. In: MATEC Web of Conferences, vol. 239, pp. 1–8 (2018)
3. Babich, D.V.: A statistical strength criterion for brittle materials. Strength Mater. **43**(5), 573–582 (2011)
4. Tsybul'ko, A.E., Romanenko, E.A., Kravchenko, E.V.: A new strength criterion for isotropic materials, taking account of their elastic or deformational properties. Russ. Eng. Res. **28**(11), 1047–1050 (2008)
5. Huang, J., Zhao, M., Du, X., et al.: An elasto-plastic damage model for rocks based on a new nonlinear strength criterion. Rock Mech. Rock Eng. **51**(5), 1413–1429 (2018)
6. Das, A.: Structural design of asphalt pavements: principles and practices in various design guidelines. Transp. Dev. Econ. **1**(1), 25–32 (2015)
7. Obando-Ante, J., Palmeira, E.M.: A laboratory study on the performance of geosynthetic reinforced asphalt overlays. Int. J. Geosynth. Ground Eng. **1**, 5 (2015)
8. Wang, X.R., Yin, B.Y., Luo, W.B.: Fatigue damage analysis of an asphalt mixture based on pseudostiffness. Strength Mater. **50**(5), 764–771 (2018)
9. Aleksandrova, N.P., Chusov, V.V., Stolbov, Y.V.: Criteria of strength and plasticity of asphalt concrete with the account of effect of accumulation of the damage while influence of re-load. In: IOP Conference Series: Materials Science and Engineering, vol. 463, pp. 1–10 (2018)
10. Lemaitre, J.: A Course on Damage Mechanics, 2nd edn. Springer, Heidelberg (1996)
11. Murakami, S.: Continuum Damage Mechanics A Continuum Mechanics Approach to the Analysis of Damage and Fracture. Springer, Netherlands (2012)
12. Altenbach, H., Maugin, G.A., Erofeev, V.: Mechanics of Generalized Continua. Springer, Heidelberg (2013)
13. Ambroziak, A., Klosowski, P.: Survey of modern trends in analysis of continuum damage mechanics. Task Q. **4**, 437–454 (2006)
14. Aleksandrova, N.P., Chysow, V.V.: The usage of integral equations hereditary theories for calculating changes of measures of the theory of damage when exposed to repeated loads. Mag. Civ. Eng. **2**(62), 69–82 (2016)

15. Aleksandrova, N.P., Aleksandrov, A.S., Chusov, V.V.: Application of principles of theory of damage accumulation to calculation of asphalt-concrete coatings. In: IOP Conference Series: Materials Science and Engineering, vol. 262, pp. 1–7 (2017)
16. Granda Marroquín, L.E., et al.: Cumulative damage evaluation under fatigue loading. Appl. Mech. Mater. **13–14**, 141–150 (2008)
17. Tsiloufas, S.P., Plaut, R.L.: Ductile fracture characterization for medium carbon steel using continuum damage mechanics. Mater. Sci. Appl. **3**, 745–755 (2012)
18. Shen, J., et al.: Material damage evaluation with measured microdefects and multiresolution numerical analysis. Int. J. Damage Mech. **23**(4), 537–566 (2014)
19. Augustin, P., Oneţ, T.: Elastic deformation of concrete. Determination of secant modulus of elasticity in compression. Acta Technica Napocensis: Civ. Eng. Archit. **55**(2), 190–204 (2012)

Author Index

© Springer Nature Switzerland AG 2020
V. Murgul and M. Pasetti (Eds.): EMMFT 2018, AISC 982, pp. 919–922, 2020.
https://doi.org/10.1007/978-3-030-19756-8